Computing in Civil Engineering

VOLUME 1

Proceedings of the
Second Congress held in conjunction with
A/E/C Systems '95

Sponsored by the Committee on Coordination Outside ASCE of the
Technical Council on Computer Practices of the
American Society of Civil Engineers

Atlanta, GA
June 5-8, 1995

Edited by J.P. Mohsen

D0746524

Published by the
American Society of Civil Engineers
345 East 47th Street
New York, New York 10017-2398

ABSTRACT

These proceedings consist of papers presented at the Second Congress on *Computing in Civil Engineering* held in Atlanta, Georgia, June 5-8, 1995. The proceedings cover major areas of concern: 1) databases; 2) expert systems and artificial intelligence; 3) computing in construction; 4) computers in civil engineering education; 5) computing in transportation; 6) geographic information systems; 7) computing in structures; 8) environmental and hydraulic engineering; and 9) computing in geotechnical engineering. Within these broad topics, the book contains papers on subjects such as: 1) Evolutionary systems and genetic algorithms; 2) neural networks in construction and transportation; 3) pen computing; 4) GIS for geotechnical and earthquake engineering; 5) graphical user interfaces; and 6) object-oriented programming. Additionally, the book contains papers addressing current uses of computers in civil engineering undergraduate education and the impact of multimedia information of civil engineering education.

Library of Congress Cataloging-in-Publication Data

Computing in civil engineering: proceedings of the Second Congress held in conjunction with A/E/C Systems '95 / spon sored by the Committee on Coordination Outside ASCE of the Technical Council on Computer Practices of the American Society of Civil Engineers; edited by J. P. Mohsen.
 p. cm.
 Includes indexes.
 ISBN 0-7844-0088-1
 1. Civil engineering—Data processing—Congresses. I. Mohsen, J. P. II. American Society of Civil Engineers. Technical Council on Computer Practices. Committee on Coordination outside ASCE. III. Congress on Computing in Civil Engineering (2nd: 1995: Atlanta, Ga.)
 TA345.C6476 1995   95-15012
 624'.0285—dc20   CIP

FOREWORD

This volume contains papers submitted for presentation at the Second Congress on Computing in Civil Engineering, II-CCCE, held in Atlanta, Georgia, June 5–8, 1995. The Second Congress was sponsored by the ASCE Technical Council on Computer Practices (TCCP), and was organized by a subcommittee of TCCP's Committee on Coordination outside of ASCE. The organizing subcommittee included:

J. Crozier Brown, Conference Chair
J. P. Mohsen, Agenda Chair
Glen Fromm, Publicity Chair
W. Tracey Lenocker, Vendor Liaison
Morton B. Lipetz
Louis F. Cohn
Leroy Emkin

II-CCCE was a continuation of a series of annual conferences on computing in civil engineering that began in 1978. The Congress is the primary ASCE-wide forum for computer issues in civil engineering. The Second Congress consisted of several multiple session symposia on a variety of topics as listed below.

Symposium on Databases
H. Craig Howard, Symposium Leader
Symposium on Artificial Intelligence and Expert Systems
Tomasz Arciszewski, Symposium Leader
Symposium on Computing in Structures
Celal N. Kostem, Symposium Leader
Symposium on Environmental/Hydrology
Kenneth P. Brannan, Symposium Leader
Symposium on Computing in Construction
Iris Tommelein, Symposium Leader
Symposium on Computers in Civil Engineering Education
Teresa M. Adams, Symposium Leader
Symposium on Geographic Information Systems
Teresa M. Adams, Symposium Leader
Symposium on Computing in Transportation
John R. Stone, Symposium Leader
Symposium on Computing in Geotechnical Engineering
Mohamad Hussein, Symposium Leader
Symposium on Computing in Pavement Engineering
Javed Alam, Symposium Leader

Each of the papers in this volume has been accepted for publication by the proceedings editor. All papers are eligible for discussion in the *Journal of Computing in Civil Engineering* and are eligible for ASCE awards.

Much of the effort in organizing the received abstracts, accepted papers, and preparation of the Congress Agenda was done through the able assistance of the staff of the Civil Engineering Department at the University of Louisville. Particular appreciation is hereby expressed to Ms. Leanne Whitney, Ms. Debbie Jones, and Ms. Gail Graves. Without their diligence, these proceedings would not have been possible.

J. P. Mohsen
Editor

CONTENTS

SESSION 01M2: PRODUCT MODELING FOR INTEGRATION AND COLLABORATION I

SESSION 02M2: CASE-BASED REASONING I

SESSION 03M2: INFORMATION DELIVERY SYSTEMS IN CONSTRUCTION

SESSION 06M2: STRUCTURAL MODELING, ANALYSIS, AND DESIGN I

SESSION 10M2: AUTOMATION (OR NOT) IN HIGHWAY ENGINEERING

SESSION 11M2: GRAPHICAL USER INTERFACES

SESSION 01M3: PRODUCT MODELING FOR INTEGRATION AND COLLABORATION II

SESSION 02M3: CASE-BASED REASONING II

SESSION 03M3: INFORMATION MANAGEMENT IN CONSTRUCTION AND TRANSPORTATION I

SESSION 04M3: THE INTEGRAL PROJECT: OVERVIEW, DESIGN CONCEPTS AND APPLICATIONS

SESSION 06M3: STRUCTURAL MODELING, ANALYSIS, AND DESIGN II

SESSION 08M3: CURRENT USE OF COMPUTERS IN CE UNDERGRADUATE EDUCATION

SESSION 10M3: OPTIMIZATION AND COMPUTER MODELING IN TRANSPORTATION

SESSION 01M4: MODELING OF FORM, FUNCTION, AND BEHAVIOR

SESSION 02M4: CASE-BASED DESIGN EDUCATION

SESSION 03M4: INFORMATION MANAGEMENT IN CONSTRUCTION AND TRANSPORTATION 2

SESSION 04M4: THE INTEGRAL PROJECT: SOFTWARE DESIGN AND IMPLEMENTATION

SESSION 06M4: FINITE ELEMENT MODELLING

SESSION 08M4: INTERDISCIPLINARY EDUCATION EFFORTS

SESSION 09M4: GIS FOR INFRASTRUCTURE AND TRANSPORTATION ENGINEERING

SESSION 01T1: INTEGRATED OBJECT-ORIENTED SYSTEMS

SESSION 02T1: MACHINE LEARNING I: GENERAL ISSUES

SESSION 03T1: SIMULATION APPLICATIONS IN CONSTRUCTION I

SESSION 05T1: PAVEMENT ANALYSIS, DESIGN, AND EVALUATION

*Manuscript not available at time of printing.

xii

SESSION 03T4: FUZZY LOGIC AND EXPERT SYSTEMS IN CONSTRUCTION

SESSION 04T4: WATER SUPPLY AND POLLUTION CONTROL

SESSION 08T4: IMPACT OF MULTIMEDIA INFORMATION ON CE EDUCATION AND LEARNING EXPERIENCE

SESSION 09T4: GIS FOR GEOTECHNICAL AND EARTHQUAKE ENGINEERING

SESSION 01W1: MULTIMEDIA/DOCUMENT MANAGEMENT I

SESSION 02W1: COLLABORATIVE/CONCURRENT ENGINEERING I

SESSION 03W1: PLANNING AND SCHEDULING TECHNIQUES

SESSION 04W1: DECISION SUPPORT AND ENVIRONMENTAL MONITORING

SESSION 06W1: SEISMIC DESIGN OF STRUCTURES

SESSION 02W2: COLLABORATIVE/CONCURRENT ENGINEERING II

SESSION 03W2: PEN COMPUTING AND BAR CODES

SESSION 04W2: DATA MANAGEMENT AND DECISION SUPPORT SYSTEMS IN ENVIRONMENTAL ENGINEERING

SESSION 10W2: UNCERTAINTY AND RISK IN PROJECT MANAGEMENT: EMERGING SOLUTIONS TO REAL WORLD PROBLEMS

The Value of a Unified Approach to Product and Process Modeling

Hossam El-Bibany[1], A. M. ASCE

Abstract

This paper supports the concept of a unified computer representational model coupled with a unified computational methodology for building systems that integrate product and process knowledge. It presents a parametrics methodology and argues that the unified approach could be used for both design and construction, separately or in an integrated structure that adds value to the current design and construction processes.

Introduction

Product and/or process models have always been considered the foundation for building computer systems. From simple databases to more complicated AI-based systems, the structure and function of the systems are built on various models of the underlying knowledge.

This has proven successful when a computer system is conceived to solve a problem based on a well defined body of knowledge. As a structural design system, for example, uses a mathematical computational model as its core, a knowledge-based system uses a well defined body of knowledge for its reasoning process. In each of these cases, the problem to be solved is well defined beforehand.

But what if the problem and its underlying body of knowledge is not available? And what if more bodies of knowledge need to be incorporated in the system at a later stage? Can we build a system, may be in the form of a shell, that is capable of handling these requirements? And what are the elements of the professional environment that would dictate such requirements?

This paper argues that the virtual concurrent Architectural/Engineering/Construction (AEC) environment needs such approach. The need stems from a value-added analysis and not from an automation goal. The paper starts by describing the values needed by the nature of project organization in the AEC industry. It then describes the formal unified parametrics modeling with its added value to the overall process. The

---

[1] Assistant Professor, Department of Architectural Engineering, PennState University, University Park, PA 16802.

companion paper [El-Bibany 95] describes the standards of a system architecture that is built on a unified formal parametrics representation.

The Concurrent Environment: What is the Value?

Elements of a concurrent AEC environment that create the required added value include:

• The project organization is different from one project to another.

• With one specialty an organization may need to use different bodies of knowledge to perform the same task on different projects. For example, a structural design organization may need to use different codes and design methodologies based on project location and nature. Similarly a construction organization may need to use different construction methods based on project location or resource availability.

• When working concurrently, project participants make decisions and assumptions that often affect the responsibility of others.

• The various project participants don't always join the group at the same time. The fact that this may force one organization to redo its work shows that the various organizations are interdependent in their knowledge environment.

The previous points, when analyzed based on basic organizational theory (e.g., [Thompson 67] and [Galbraith 77]), prove to have tremendous effect of the concurrent process, its efficiency, and the quality of the resulting product.

So the value is not in the automation of the process. It is mainly in the creation of a new tool that can contribute to the total quality of the process.

A Unified Formal Methodology: Why?

With the requirements set in the previous section, the question that arises is: can one build a system that can be used for such different areas, from structural design to construction management for example? And why is this question important?

Notice that if this is not the case then everytime two systems are built based on different representational and computational models, the task of integrating between them is either reduced to simple data integration (input-output interface) or a separate computer system with special coordination and knowledge integration strategies needs to be built. Furthermore, if the latter solution is chosen, the knowledge integration and coordination strategies need to be changed from one project to another based on the points discussed in the previous section.

The need then arises to think if such unified model (representational and computational) would be computationally sound and complete. System soundness depends on the formality of the unified model. A model with formal representation and sound computational foundation, like formal logic may provide an answer. For a formal methodology, the model has to keep its soundness with the introduction with new bodies of knowledge.

Parametrics as a Unified Formal Modeling and Computational Methodology

Research in various fields have proved the brittleness of traditional knowledge-based methodologies in open-knowledge applications [Lenat and Guha 90]. The brittleness results from two main factors: implicit knowledge representation into predicate names, and the failure to model the real world with detailed levels of knowledge abstraction. A third factor arises in case the knowledge-base will be used, maintained and updated by people other than its developer: the use of hierarchies and inheritance for domain knowledge representation. To augment the knowledge base with a new domain knowledge item, the system's maintainer needs to understand how representation and inference are based on the hierarchical representation [El-Bibany 94].

Parametrics as a formal representation when coupled with a general constraint-management methodology provides an alternative [El-Bibany and Paulson 94]. The constraint-management methodology has to include global analyses of constraint sets in order to avoid local constraint satisfaction problems.

The methodology is based on the view of the world in terms of entities and their relationships. There are two things that need to be captured: the *knowledge* that is encapsulated within the entities and the *relationships* among the entities [El-Bibany 92]. Both could be represented using parametrics.

Once the problem is formulated in terms of constraints, a constraint-management system analysis the constraint set, develops a plan of action for the solution of the set, solves the set for the required parameters and keeps track of the solution network in a directed graph structure that can be used to propagate further value changes.

Notice that if the constraint-management system is powerful enough to solve an arbitrary set of parametric constraints then the overall representation and computational model provides the following advantages:

- It allows to add to or delete from the constraint set without any change in methodology. So if a constraint set representing structural design requirements is added to another representing the construction plan and schedule with a supplement of constraints relating relationships between some attributes of the structural member and the required resources for construction, then using the same computational methodology, the overall set may be solved.

- It allows to divide the constraint set into smaller subsets (interrelated or not) for handling the computational complexity and distributing the computational process.

- Through the management of various constraint sets and their relationships, it allows the representation of domain strategic logic.

- It simplifies the creation of a unified representational models for domain entities with standards of internal structure and function representation and external entity relations. The unified entity model may be then used to represent any type of entity, physical (crane, beam, room) or conceptual (activity, crew productivity) over the overall life cycle of a building.

The parametrics methodology will then need to be imbedded in a collaborative computer system architecture that forms the structure of shell that simplifies the human interaction in various organizational settings. The shell can be used by various organizations to set with their internal knowledge and to integrate their knowledge with other organizations. A brief description of the system architecture is given in [El-Bibany 95]. For a detailed description, the reader is referred to [El-Bibany 92].

References

El-Bibany, H., 1995, "A Unified Architecture for Concurrent Engineering," *Proceedings of the Second ASCE Congress on Computing in Civil Engineering*, June, 1995, Atlanta, Georgia.

El-Bibany, H., 1994, "Knowledge Issues in Modeling," *Proceedings of the First ASCE Congress on Computing in Civil Engineering*, June, 1994, Washington, D.C.

El-Bibany, H., and Paulson, B.C., 1994, "Collaborative Knowledge-integration Systems: A Tool for AEC Design, Management and Coordination," *Microcomputers in Civil Engineering*, Special Issue on Innovative Research, Vol. 9, No. 1, February, pp. 29-40.

El-Bibany, H., 1992, *Architecture for Human-Computer Design, Management and Coordination in a Collaborative AEC Environment,* Ph.D. Thesis, Civil Engineering Department, Stanford University, Stanford, CA, June.

Galbraith, J. R., 1977, *Organization Design*, Addison-Wesley, Reading, MA.

Lenat, D. B., and Guha, R. V., 1990, *Building Large Knowledge-Based Systems, Representation and Inference in the Cyc Project*, Addison-Wesley Publishing Company, Inc., Reading, MA.

Thompson, J. D., 1967, *Organizations in Action: Social Science Bases in Administrative Theory*, McGraw-Hill, New York.

Models of Construction Process Information

Thomas Froese, Member, ASCE[1]

Abstract

Computer-integrated construction (CIC) requires data standards, or common information models through which computer systems can exchanged project information. High-level "core" models are required as unifying references for the more detailed, application-specific models used for the actual information exchange. A variety of core models have been developed in the area of *construction project and process information*. This paper introduces several such models from a variety of CIC projects and presents some key points of comparison. The overall objective is the eventual emergence of generally accepted standards in this area.

Introduction

Computer-integrated construction (CIC) will enable the widespread communication of project information among all project participants and all project life cycle stages using computing technologies. A cornerstone of the emerging CIC technologies is data standards, or common information models through which heterogeneous computer systems can exchanged project information. In order to accommodate the breadth of application areas within architecture, engineering, and construction (AEC), researchers are developing layered approaches with detailed, discipline-specific models built upon higher-level, more generic and broadly applicable models. The high-level models are used both to provide a consistent approach among the detailed models and to directly support information exchange between different discipline areas.

We have been investigating high-level or "core" models of construction processes to support the eventual emergence of standard models in this area (Froese 1992, 1993, 1994a-d). To a large degree, the challenges in this area lay not in modeling the information per se, but in finding solutions that are responsive to the wide variety of existing and emerging models for different applications—that are able to translate information among these different areas, for example. Over the past few years, a number of significant CIC projects have been carried out that incorporate some type of core AEC project or construction process model. While the lack of a uniform approach among these projects indicates that standard models have not yet arisen, this is to be expected as meaningful standards evolve slowly. A more encouraging observation is

---

[1] Assistant Professor, Department of Civil Engineering, University of British Columbia, Vancouver, B.C., Canada, V6T 1Z4, tfroese@civil.ubc.ca

that the quantity and quality of construction process models is increasing, indicating a greater awareness of the necessity and requirements of such models. This paper introduces the core construction process models developed in several recent CIC projects and presents some key points of comparison. The objective is to continue the dialog of comparing approaches and moving towards eventual common solutions.

Core Construction Process Models

This section introduces several conceptual models of construction processes. In many cases, the models shown are only the central process-related portions of larger models and care should be taken in interpreting their content without a more complete description of the research projects.

Information Reference Model for AEC: The Information Reference Model for AEC (IRMA) was developed at a workshop in 1992 as an exercise in combining several independent AEC project models into a single model of construction information (Luiten et al. 1993). While this model was intended more as a reference and comparison tool than an end product, it has served as a useful vehicle for further conceptual development. The model identifies the central objects in AEC projects. These are shown in Figure 1 which, like the other figures in the paper, is in an informal EXPRESS-G format (heavy links represent sub-type relationships, other links represent the associations described by their labels). The relationships among the objects is shown in figure 2. In 1993, the model was the topic of *IRMA-tica'93*, an electronic mail conference (Froese 1993). While there has been no collaboratively developed revision to the original model, some of the improvements discussed during the conference were incorporated into the AEC Process View Model shown in Figure 3 (Froese 1994b).

Building Project Model: The Building Project Model (BPM) was developed by Luiten as part of his recent Ph.D. thesis on Computer Aided Design for Construction (Luiten 1994). The goal was to provide a conceptual model that integrates product, activity and resource information. The model focuses on two main abstractions: the relationship between products and activities (as illustrated in the portion of the model shown in Figure 4), and concretisation, or the progression of an object from initial requirements, to a proposed object, and finally to an actual realized object.

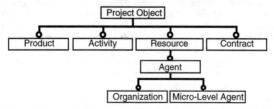

Figure 1. Subclass ("type-of") hierarchy of the Information Reference Model for AEC.

Figure 2. Relationships in the Information Reference Model for AEC.

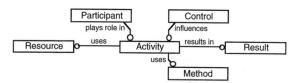

Figure 3. An AEC Process View Model.

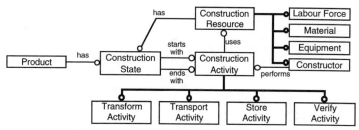

Figure 4. A Portion of the Building Project Model.

ICON: The Information/Integration for Construction (ICON) project (Aouad et al. 1994) is aimed at investigating the feasibility of establishing a framework for integrating information systems in the construction industry. It has applied thorough information technology techniques such as the Information Engineering Method, object-oriented analysis and design, and computer aided software/systems engineering (CASE). Figure 5 shows a portion of ICON's Construction Planning Object Model. Other construction-process-related models developed within ICON include a tendering object model, a procurement systems model, and numerous activity models that depict the specific tasks involved in performing an AEC project.

ATLAS Large Scale Engineering Project type Model: The ATLAS project aims at the development, demonstration, evaluation and dissemination of architectures, methodologies and tools for Computer Integrated Large Scale Engineering (LSE) (Tolman, Bakkeren and Böhms 1994). The ATLAS LSE Project type Model (PtM) is an abstract model of LSE project information. If forms the middle of three layers since it draws upon more general STEP resource models and can be specialized for different engineering sectors such as buildings or process plants (it can also be used directly for exchanging information *between* different sectors). The model is also intended to be specialized for different *views*, i.e., for use by different project participants such as architects, project managers, etc. A portion of an initial release of the ATLAS LSE PtM is shown in Figure 6.

Generic Reference Model for Life Cycle Facility Management: A Generic Reference Model (GRM) for life cycle facility management has been proposed (Reschke and Teijgler 1994) based on work done by EPISTLE and its members (notably processBase, PISTEP, and SPI-NL), on the IDEF0 function modeling standard, and on the modeling techniques of Shell ICT (West 1993). The model is a generic conceptual model of engineering project information that is applicable across many application domains and life cycle phases, and is intended to provide a contribution to the discussion around the interoperability of Application Protocols within STEP. Figure 7 shows a portion of this model.

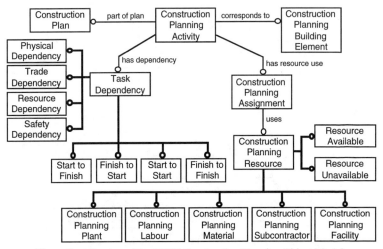

Figure 5. A Portion of the ICON Construction Planning Object Model.

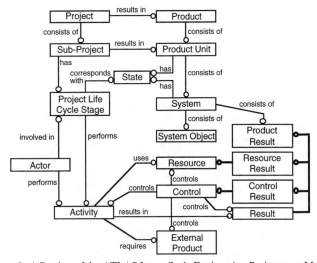

Figure 6. A Portion of the ATLAS Large Scale Engineering Project type Model.

STEP: The largest and perhaps most significant product modeling standardization effort is the International Standards Organization (ISO) draft standard 10303, Standard for The Exchange of Product Model Data, or STEP (see NIPDE). Numerous organizations world-wide are developing STEP to provide a computer-interpretable unambiguous method for exchanging product data to and from any system. While there

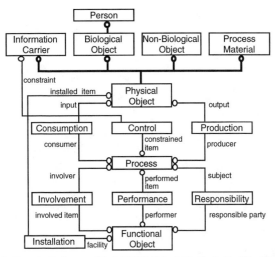

Figure 7. A Generic Reference Model (GRM) for Life Cycle Facility Management.

are no components of STEP that provide conceptual core models of project processes to date, the AEC building and construction group within STEP is currently preparing a draft building construction core model (ISO 1994). This model or some similar effort that results in an official STEP standard may offer the best opportunity for a widely adopted model of construction processes. The objective of this paper is to contribute discussion to just such efforts.

Core Modeling Issues

This section discusses some of the central similarities and differences among the models introduced above. We first compare different types of models, then discuss the central entities included in each model, next compare central relationships among the entities, and finally comment briefly on the modeling mechanisms used in the models.

Types of Models: There are many different types of information models and the intended role of any specific model is not always clear. Generally, the models described above are both *"core models"* (rather than *"application models"*) and *"conceptual models"* (rather than *"aspect models"* or *"classification models"*, for example).

Core models are intended to be high-level models that provide a unifying reference for more detailed *application models* that will be constructed on top of them. Unlike application models, core models are generally *not* intended to be instantiated for representing actual data (though they can be for exchanging information between different application areas). While some of the models described above are intended to directly support actual implementations (such as the ATLAS and ICON models), even these can be described as model segments that play "core" roles for other, more specialized sections of the models.

Conceptual models provide formal definitions of the basic entities and relationships required to represent information about some domain: construction projects in this case.

Aspect models are more detailed since they provide all of the specific attribute definitions required to fully represent the actual domain information. Those models described above that are intended to support specific implementations tend to provide the attribute definitions, while those intended to act only as references for other modeling projects do not.

Classification models adopt some simple conceptual model for representing entities and use this to develop extensive categorization or classification breakdowns of domains for the purpose of providing classification systems or enumerating all of the specific elements of the domains (examples within the AEC domain include the MASTERFORMAT classification and number system developed by the Construction Specifications Institute (CSI) and Construction Specifications Canada (CSC), the ISO Classification of Information in the Construction Industry Proposal (ISO 1993) that prepossess several categorization dimensions, or the Integrated Building Process Model (Sanvido 1990) which uses IDEF0 models to itemize the steps involved in constructing a project). Some of the models described in this paper provide some degree of classification breakdowns of certain entities—several of the models, for example, categorize different types of resources (labor, material, equipment, etc.)—while others provide no classification beyond what is required to distinguish between the central conceptual entities. Generally, a conceptual model should break down an entity into sub-types only when the various sub-types are distinguishable by different relationships and attributes within the model rather than being different in name only.

Central Elements of the Models: All the models adopt the *"activity,"* or some similar term, as the central element for representing construction process information. This is in contrast, for example, to representing construction process information as attributes of the *product* elements. The other two concepts that the models universally include as central entities are *"products"* (or some representation of the actual artifact being constructed) and *"resources"* (things employed in the construction process). Beyond these three major conceptual elements, the different models each offer various other elements such as representations of project participants and their roles, controls (constraints on other objects), costs, time, and quality.

Central Relationships in the Models: While the major elements are quite consistent across the various models, the basic relationships among these elements are not. Of particular interest is way in which products and processes are related. One approach (used in the ICON model, for example) is to adopt a simple relationship between products and processes such that processes *"correspond to"* certain products. Most of the models, however, strive for a richer relationship between the two.

Another approach is to draw upon the IDEF0 technique of modeling processes (this is most observable in the GRM model, though the documentation of both the BPM and ATLAS models discuss the IDEF0 approach). IDEF0 is perhaps the most common methodology used in the data modeling community for activity modeling, or identifying the processes involved in some domain. IDEF0, shown in Figure 8, provides a simple model consisting of processes as the central concept with inputs, outputs, controls, and mechanisms (or resources) as flows among processes. Given this approach, the relationships between products and processes is that products act as both inputs to and outputs from processes. In use, however, the distinction between whether objects used in an process are inputs, mechanisms, or even controls is not always well defined. The GRM seems to provide clear definitions for these

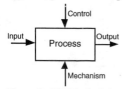

Figure 8. IDEF0 Activity Modeling Notation

aspects of the model by representing inputs as physical objects that are *consumed* by the process, outputs as physical objects that are *produced* by the process, mechanisms as functional objects (i.e., functional definitions of objects independent of the specific physical forms used to meet the functional requirements) that *perform* the process (such as a person or a tool), and controls as physical information carrier objects that *constrain* the process. However, even these definition may not be as clear as they first seem. If we attempt to use these definitions to classify the various objects that are used to carry out a process, for example, we may find objects that are consumed, but not depleted (such as electricity); objects that are partially consumed (such as forms with a fixed number of reuses or operating costs of equipment); some objects that are made unavailable for other uses (such as crews) and others that are used but still available elsewhere (such as information or management resources); and objects that may be termed inputs since they are prerequisites for the process but are not *used* by the process (such as preceding components that provide structural support). Likewise, processes may result in outputs that do not fit the *production* relationship, such as processes that modify previously existing objects, or even simply transport or store objects.

Still another approach, adopted by the BPM and the IRMA models, is to focus on the definition that "activities change the state of objects." Thus activities have relationships to object that represent the starting and ending states of project objects. This approach requires that various states of product or resource objects always be defined, and again, it is unclear how non-transformation processes such as "storage" should be represented.

Finally, some of the models, such as the ATLAS LSE PtM and the AEC Process View Model, choose instead to explicitly represent the more generic concept of a "result". The relationship, then, is that processes result in results, and results can be subtyped to provide more specific meanings. In ATLAS, the subtypes are objects that are produced by the process (states are also used, but at a higher aggregation level than the level of activities), in the AEC Process View Model, the exact nature of the product-process relationship is left for specific applications models to determine by refining the results object.

While these relationships, particularly the product-process relationship, may seem like a small detail, they do stand as some of the major inconsistencies among the models. The are of interest, too, because they appear to be central to hopes of providing a high degree of integration between construction process information and all other information from the project life cycle. Yet, a link between product information and process information is *virtually non-existent* in any current common AEC software. The solution to the best way to relate the concepts, then, will likely need to wait until various approaches have been utilized and tested in practical CIC systems.

Modeling Mechanisms: It is important to note that information standards must not only define the basic conceptual entities and relationships that make up a domain such as construction projects, but they must also formalize many more generally applicable characteristics of the data structures used to represent and communicate information—the *modeling mechanisms*. Examples of modeling mechanisms include *composition* (breaking objects into their sub-parts), *concretisation* (representing the problem-solving life cycle from requirement to planned and realized object), *versionning*, multiple *representations* of objects, etc. Some of the models described above include extensive descriptions of proposed modeling mechanisms while others rely on even more generic modeling kernels to define these mechanisms. A specific discussion of required modeling mechanisms is beyond the scope of this paper.

Conclusion

This paper has introduced several versions of core conceptual models of construction processes from a variety of CIC research projects. Some central similarities and differences among the models where discussed, most notably dealing with the approach to representing and relating project products and processes. The overall objective of this comparison—and indeed of many of the modeling efforts themselves—is to contribute towards an eventual emergence of generally accepted standards in this area that will enable the widespread exchange of information among all project participants and project life cycle stages.

Acknowledgments: We gratefully acknowledge financial support for this work by the Natural Sciences and Engineering Research Council of Canada.

References

Aouad, G., et al. (1994). *Integration of Construction Information, Final Report.* University of Salford.

Froese, T. (1992). *Integrated Computer-Aided Project Management Through Standard Object-Oriented Models,* Ph.D. Thesis, Dept. of Civil Eng. Stanford University.

Froese, T. (editor) (1993). *Transcripts of IRMA-tica'93, An International E-Mail Conference on the IRMA model: an Information Reference Model for AEC,* Available for anonymous ftp or gopher file transfer from irma.hs.jhu.edu.*

Froese, T. and Paulson, B. (1994a). "OPIS, An Object-Model-Based Project information System," *Microcomputers in Civil Engineering,* Vol. 9, Feb, pp. 13-28.

Froese, T., (1994b). "Developments to the IRMA Model of AEC Projects," *Computing in Civil Engineering: Proc. of the First Congress,* ASCE, Vol. 1, pp. 778-785.

Froese, T. and Yu, Q. (1994c). "StartPlan: Producing Schedule Templates Using IRMA", *Computing in Civil Engineering: Proc. of the First Congress,* ASCE, Vol. 1, pp. 63-70.

Froese, T. (1994d). "Information Standards In The AEC Industry," *Canadian Civil Engineer,* Vol. 11, No. 6 (Sept.), pp. 3-5.

ISO (1993). *Classification of Information in the Construction Industry.* ISO TC59/SC13, Document N35E, 1993.

ISO (1994). *Building Construction Core Model, Project Proposal.* ISO TC184/SC4/WG3 Document N341, October 1994.

Luiten, G., Froese, T., Björk, B.-C., Cooper, G., Junge, R., Karstila, K., and Oxman, R. (1993). "An Information Reference Model for Architecture, Engineering, and Construction." *Proc. of the First Int. Conf. on the Management of Information Technology for Construction,* World Scientific, Singapore, 391-406.

Luiten, G. (1994). *Computer Aided Design for Construction in the Building Industry.* Ph.D. Thesis, Fac. of Civil Eng., Delft University of Technology, The Netherlands.

National Initiative for Product Data Exchange (NIPDE), A World Wide Web Home Page at URL: http://www.eeel.nist.gov/nipde/Intro.html.

Reschke, R. and Teijgler. H. (1994). *Generic Reference Model for Life Cycle Facility Management (a proposal).* ISO TC184/SC4/WG3 Document N351, Rev.1

Sanvido, V. (1990). *An Integrated Building Process Model,* Computer Integrated Construction Research Program, Dept. of Architectural Engineering, Pennsylvania State University, USA, Technical Report No. 1.

Tolman, F., Bakkeren, W. and Böhms, M. (1994). *Atlas LSE Project type Model.* ESPRIT Project 7280—ATLAS/WP1/Task 1500 Document D106-Ic, April 1994.

West, M. (1993). *Developing High Quality Data Models.* Shell Int. Petroleum , The Hague, Report IC 91-077 T4, Dec. 1993.

* References by Froese may be available through the World Wide Web at URL: http://www.civil.ubc.ca/www/usr/froese/home.html

General Product Model and Domain Specific Product Model in the A/E/C Industry

Kenji Ito, Member AIJ and JSAI[1]

Abstract

Many researchers have been attempting to realize a generic product modeling in order to achieve the integrated information systems for the A/E/C industry. However, it is difficult to describe the generic or common product model which can support the generation, sharing and maintaining project data during the various phases of the A/E/C project life cycle from planning to design, construction and management of the facility. Then, there is not enough generic product model which describe the actual building project, and no one proved the usefulness of generic product modeling in the real business world. On the other side, the clients' requests are becoming more diverse and extensive than ever, it is necessary to concentrate company whole efforts and to fortify cooperation among different divisions in the A/E/C industry, quickly. Therefore, domain specific product model has been implemented in order to realize the information sharing among the domain specific or organization specific applications in the A/E/C industry. This domain specific product modeling approach can not solve the basic problems to share the whole project information by the all project participants, but can solve the current problems which is increasing efficiency of particular domain tasks.

The scope of this research is to compare the methodology and usefulness of generic product model and domain specific product model in the A/E/C industry in order to propose the current problems of product modeling research.

This paper describes the advantage and disadvantage of generic product model and domain specific product model from both of R&D viewpoint and information sharing viewpoint in the A/E/C industry.

[1] Research Engineer and System Engineer, Information Systems Dept., Information Systems Div., Shimizu Corporation. No. 1-2-3, Shibaura, Minato-Ku, Tokyo 105-07, Japan.

13

As the conclusion of this paper, discussion of the future direction of product modeling and process modeling in the industry for the information sharing and data exchange will be introduced.

1. Introduction

Many computer applications, which support the production of facility, have been developed for the A/E/C industry. However most of these applications can only be utilized within narrow application domains or narrow business area. Efforts have been attempted to integrate these application software by linking systems and providing data transfer interface so that data communication among applications of different domains can be realized by non-flexible ways. Then a product model that can properly describe a facility and is accessible by multiple participants of different disciplines is a very important ingredient for integration for A/E/C industry.

Many researchers have been attempting to develop a product model or a project model using object-oriented methodology and some of these models support providing the integrated information system environment. The author has been proposing the object-oriented project model called PMAPM (Object-Oriented Project Model for A/E/C Process with Multiple-Views) [Ito 90, 91, 94] which supports multiple participants of A/E/C process. This object-oriented project model is composed by both of general product and process model. However, there are few results to apply a general product or project model to the actual A/E/C project in the industry.

2. General Product Model

The constructed facility has many faces or aspects according to the process of project. In general, for each project, there is one facility and the object representing the facility has only one physical value but many functional values depending on the various views of the information by the participants of different disciplines. It means that each participant has own viewpoint, own data requirement, own data format and own computer-based applications, and every participants are expecting more information by less data input about the constructed facility. Therefore, developing the general or common product model which supports the various participants of project are needed and this product model should be a kernel of the integrated system environment in future. This general product model will support the all participants viewpoint, data requirement and data format. Then any domain specific product model or any application specific product model is defined as a subset of general product model and there are flexible dynamic link among general product model and sub model. According to this definition, the author has been proposing PMAPM which composes the several kind of methods, functions, demons and interface modules in order to support the combination of separated models such as the process model and the general product model. The author showed the usefulness and power of PMAPM as the prototype environment of integrated systems. However, it is very difficult to apply the whole concept of PMAPM to the real project in Shimizu. Because there are many existing computer-based applications in the industry and most of these applications have no object-oriented data model or interface to the object-oriented product model. Then the author has been trying to apply sub part of PMAPM's concept to the industry environment as a domain specific product model. [Ito 95a, 95b]

3. Domain Specific Product Model

There are two types of domain specific product models as follows:

• Domain specfic product model which depend on the business work flow, such as product model for design, product model for construction and so on.

• Domain specfic product model which depend on the project information flow, such as product model for schedule, product model for cost, product model for resorces and so on.

These domain specific product models are very useful for domain related applications and it is easier to apply the existing applications or actual project than general product model. Moreover, the research on domain specific product model is a good theme for both of research center and industry. Because most of these research are short-term resaerch than general product model research and it is easy to forcaste and to prove the research results not only as a prototype system but also as a actual system in the industry. However, according to the progress of each research and development of the domain specific product model, we will find the several type of product model and several type of data description in each domain. Because, each domain specific product model adopts the best data structure or data description for each domain and process dependent or domain dependent viewpoint of each participants will be included in the product model structure or inside of domain applications. It means that we have to start the new research project that is a creation of the integrated general product model or the integrated cross domain product model from many type of domain specific product models. For example, in Shimizu, there are some research on domain specific product or process model such as [Ito 95a, 95b], [Yamazaki 94] and so on. Basic concept of the product model in these research are similar but current structure or data description of model are differrent. Therefore, we should try to integrate the result of each domain specific product model as a cross domain product model near future. Furthermore, these approach will influence to the construction of the integrated system environment in the industry. On the other side, this issue is very important to the standard of data exchage between the industry. It is similar to consider the standard model for data exchange by top-down style or bottom-up style.

4. Conclusion and Discussion

This paper described an advantage and disadvantage of the general product model and domain specific product model from both of R&D and information sharing viewpoint with the example in Shimizu. The product model is one of the key issues of realizing integrated systems environment in A/E/C industry. Thus the future direction of research and development about the general product model and domain specific model should be discussed with many researcher and engineer. Then the author hopes that this paper provides the chance to discuss the proper evaluation of general product model research and domain specific research. Furthermore, realizing the information exchange and coordination research between the general product modeling and domain specific modeling.

References

[Ito 90]   K. Ito, K. Law and R. Levitt, "PMAPM: An Object Oriented Project Model for A/E/C Process with Multiple Views," *CIFE Technical Report*, No. 34, Stanford University, July 1990.

[Ito 91]   K. Ito, "Design and Construction Integration Using Object Oriented Project Model with Multiple Views," *Proceedings on Construction Congress II*, ASCE, pp. 336-341, Boston, U.S.A., April 1991.

[Ito 94]   K. Ito, "Integrated System Environment Using An Object-Oriented Project Model with Multiple Views *Proceedings of The First ASCE Congress on Computing in Civil Engineering*, ASCE,, pp. 1204-121, Washington D.C., U.S.A., June 1994.

[Ito 95a]   K. Ito, "An Object Model for Integrated Construction Planning and Safety Prevention Database," *Proceedings of The Second ASCE Congress on Computing in Civil Engineering*, ASCE, Atlanta, U.S.A., June 1995. (will be appeared)

[Ito 95b]   K. Ito, "Object-Oriented Scheduling Model for Integrated Schedule Information from Design to Construction Management," *Proceedings of The Second ASCE Congress on Computing in Civil Engineering*, ASCE, Atlanta, U.S.A., June 1995. (will be appeared)

[Yamazaki 94] Y. Yamazaki, "An Object-Oriented Process Modeling to Support Cooperative Planning in Industrialized Building Construction Project," *Proceedings of the First ASCE Congress on Computing in Civil Engineering*, ASCE, pp. 998-1005, Washington D.C., U.S.A., June 1994.

Functional Units, Design Units, Technologies:
The Components of Case-Based Design in the SEED System

Ulrich Flemming[1], Robert Coyne[2], James Snyder[3]

Abstract

SEED is a collection of modules to support the early phases in building design. It provides uniform case-based design capabilities across modules and problem decompositions within a module and automates indexing and retrieval. These mechanisms are based on a triad of shared concepts: functional units, design units, and technologies.

1   Introduction

SEED aims at developing the prototype of a software environment to support the early phases in building design. Emphasis is placed on providing support for the rapid generation of internal design representations, including design alternatives, that can be evaluated in terms of multiple criteria. Case-based design, that is, the retrieval of past solutions in new problems contexts and their adaptation to this context, is a fundamental capability to be provided under this emphasis. In (Flemming et al. 1994), we presented the requirements for cased-based design within the SEED context and outlined a conceptual approach toward meeting these requirements. The same paper also indicates how we place our work in relation to ongoing work in case-based building design at other institutions.

The present paper elaborates on the representation, storage and retrieval of cases. Section 2 introduces briefly the context for case-based design in SEED. At the heart of case-based design in SEED is a triad of concepts we refer to as *design units, functional units,* and *technologies.* These concepts are introduced in section 3. Sections 4 and 5 describe how an object-oriented representation of functional units, design units, and technologies together with the associated reasoning processes can support all operations necessary for case-based design uniformly across the modules of SEED.

2   Case-Based Design in the SEED Context

SEED is a collection of task-specific *modules* (see Flemming et al. for a more detailed description). Each SEED module attempts to relieve designers especially from more routine design generation and evaluation tasks. A prerequisite for this is that the requirements currently at work are explicitly represented in the module. A SEED module therefore contains three generic components: a *problem specification component* that allows designers to specify and modify dynamically and interactively design requirements; a *generation compo-*

---

1. Professor, Department of Architecture and Engineering Design Research Center (EDRC), Carnegie Mellon University, Pittsburgh, PA 15213. The Engineering Design Research Center is an NSF-supported Engineering Research Center.

2. Research Scientist, EDRC, and Adjunct Assistant Professor, Department of Architecture, Carnegie Mellon University, Pittsburgh, PA 15213.

3. Research Assistant, Department of Architecture and Engineering Design Research Center (EDRC), Carnegie Mellon University, Pittsburgh, PA 15213.

nent that supports the rapid generation of design representation that satisfy the specified requirements; and an *evaluation component* able to evaluate design representations against these requirements.

A *database* supports work in the individual modules in multiple ways, for example, by storing temporary design versions and design variants within a specific project or by storing cases that can be re-used across projects. Figure 1 depicts the resulting generic architecture of a SEED module.

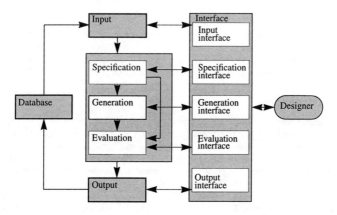

Figure 1: Generic architecture of a SEED module

Three modules with this generic architecture are currently under development: (1) SEED-Pro, a module that support the generation of an architectural program in terms of the basic functional units a planned building must contain; (2) SEED-Layout, which supports the generation of schematic layouts of the functional units in a program; and (3) SEED-Config, which supports the generation of a 3-dimensional configuration of spatial and physical building components based on schematic layouts

The case-based design capabilities in SEED must satisfy the following requirements:

• The reuse of solutions is ubiquitous in building design and extends across all phases and many tasks, from programming to the design of construction details. The retrieval and adaptation of past solutions are therefore capabilities to be offered *across modules*.

• The problem addressed in a module may be decomposed into a hierarchy of (sub)problems. For example, the layout of a building may be decomposed into the layouts of individual floors, of zones on a floor, of rooms within a zone, and of equipment or furniture in a room. Standard solutions may exist for problems at any level in such a hierarchy. As a result, retrieval capabilities for past solutions should be available in SEED also *across hierarchical problem decomposition levels within a module*.

- *A solution to any subproblem in a problem hierarchy should also be retrievable by itself.* For example, the layout of public restrooms developed for an office building may be perfectly reusable in an educational facility, whereas the overall organization of the two buildings may be very different.

- The indexing and retrieval mechanisms should be *uniform across modules and problem decompositions within a module.* This makes not only the development effort more effective because it can be shared between modules, but allows modules to share cases.

- Case-based design in SEED starts from the premise that *its case memory accumulates as a side-effect of a firm's normal design activities.* Any indexing and other types of processing needed to prepare a case for storage in the case library must occur in the background and remain hidden from designers to the largest possible degree.[1]

- Upon retrieval, a *solution must be immediately available for interactive or automatic editing and adaptation.*

3   Functional Units, Design Units, Technologies

Every SEED module uses an internal representation that is particularly suited to the tasks addressed in the module. But the internal representations are based on a shared conceptualization of problem specifications and solutions. This shared conceptualization provides also a sound basis for a uniform indexing and retrieval mechanism to support case-based design across modules and problem decompositions within a module.

The shared conceptualization relies on two central constructs, design units and functional units. *Design units* are the basic spatial or physical entities that make up the representation of a design; examples are a room and a wall. Design units are the primary focus of attention during form generation, which concentrates on specifying their shape, location and non-geometric attributes.

A design unit is generally multi-functional, that is, has to accomplish more than one purpose. A *functional unit* collects all of the requirements that a design unit has to satisfy in a single construct. For example, a wall may have to provide load-bearing capacities, sound insulation, thermal resistance, visual privacy and light reflectance. All of these requirements are collected in SEED in a single location, the functional unit associated with that wall. Functional requirements are normally expressed in the form of constraints and criteria; examples are the lateral and vertical load a wall has to be able to resist, its minimum thermal resistance, sound absorption etc.

A functional unit defines a design problem, namely that of shaping, placing and otherwise determining the properties of a design unit that satisfies the respective requirements in a given design context. A functional unit, together with a design *context*, constitutes a problem specification in SEED.

A functional unit can contain *constituent functional units* as illustrated in Figure 2. It shows a functional unit 'firestation' containing the constituent units 'administrative wing',

---

1. This will not prevent designers from also adding personalized indices to a case, a feature not discussed in the present paper

'apparatus room' and 'dormitory wing', where the administrative wing contains a unit 'kitchen', 'dining room' etc. SEED-Layout interprets constituent relations strictly as spatial containment relations: the constituents of a functional unit will always be allocated within the design unit associated with that functional unit. SEED-Config interprets constituent relations more loosely as part-of relations that may include spatial containment as a special case. In each case, the hierarchical decomposition of a functional unit into constituent units leads to a parallel decomposition of the resulting design problem into subproblems through several abstraction levels, where at each level, the problem consists of shaping and placing design units that satisfy the requirements of the associated functional units. The solution of a design problem at some level in this hierarchy establishes the context for problem specifications at the next lower level.

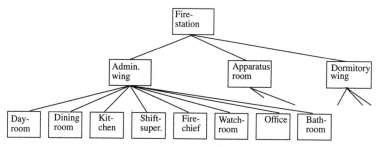

Figure 2: Example of a functional unit hierarchy in SEED-Layout

We have found it useful to add a third construct, *technology*, to the concepts shared between modules in SEED. In general, a technology is a collection of computational mechanisms to create design units that satisfy the requirements of an associated functional unit in a given context. An example of a technology available in SEED-Config is a set of production rules that generate the representation of external walls (design units) between specific rooms and the outside (context), where the wall consists of appropriately spaced metal studs, insulation, gypsum board on the inside and brick veneer etc. on the outside and satisfies a collection of load-bearing, performance etc. requirements (functional unit). SEED-Layout offers at the present time a single technology to allocate spatial functional units, which generates only arrangements of rectangles.

The triad design unit, functional unit and technology is sufficient to indicate how the requirements for case-based design specified in the preceding section can be met, in principle, in SEED. A case consists generally of

- a solution in the form of a collection of design units that may be hierarchically decomposed

- an associated index in the form of a context and a functional unit that also may be hierarchically decomposed

- the technologies that were used to create the solution in the context, given the functional requirements collected in the functional unit.

- (possibly) an explicit evaluation of the solution against the requirements, which may be used, for example, to break ties between alternative cases during retrieval or to exclude solutions from consideration.

Since modules share the underlying constructs of design unit, functional unit and technology, a conceptually uniform indexing and retrieval mechanism across modules becomes possible, in principle at least (subject to translations in both directions between a module and the case library). The same mechanism is applicable across problem decompositions within a module because the representations of problem specifications and of solutions are formally identical across levels. This makes it possible to index subproblems and retrieve them independently using a uniform mechanism.

Conversely, the explicit problem specification that exists in a module can be used to compute automatically an index when a designer feels that a solution that has been generated with the help of the module is worth entering into the case library and tells the module to do this. The computation of the index itself and the translation of the solution into the schema required by the case library can then happen in the background. This may be complemented by the designer adding comments to the case or embellishing it in other ways.

Finally, a solution is retrieved in a form that makes it immediately amenable to manual or automatic adaptation using any one of the generation modes available in the given module, while the technology retrieved together with the case offers a set of generative mechanisms that are particularly promising because they were used originally in the generation of the solution.

The following sections indicate how the general approach outlined in the present section is being implemented in the first SEED prototype. They use illustrations mainly from SEED-Layout, which is currently the best understood and most fully implemented module.

4   Indexing and Retrieval

A generic matching and retrieval mechanism as outlined in the preceding section becomes conceivable, in principle, for the first SEED prototype because it will be implemented based on an object-oriented approach in which all major entities handled by a module are objects with attributes that can be compared with the attributes of other objects. We use the term *object* here in a conceptual, generic sense (independent of a specific programming language or implementation environment) to denote an entity characterized by attributes with values. An object may belong to a type or class hierarchy through which it inherits attributes and values from other objects. For example, a spatial functional unit 'administrative wing' may inherit attributes from a super unit 'massing_element'.

We call an object *structured* if it has parts or constituents. An example are the constituents of a functional unit that define a problem hierarchy as described above. We use the term *structure* in the following to refer solely to such hierarchical part-of or constituent structures. These constituent hierarchies have to be distinguished from the inheritance or is-a hierarchies to which individual objects may belong. Given this type of structured object representation, comparisons of a current problem specification (which we call a *target index*) with a case index can proceed, in principle, object-by-object and attribute-by-attribute. Details are given in (Flemming, 1994).

The degree to which the structure of the target index and that of a retrieved case index match determines largely the degree to which a module can actively support the refinement or adaptation of a retrieved case because this indicates also how closely the retrieved solution solves the current problem. We are particularly interested in cases whose index *refines* the target index; that is, it transitively preserves the hierarchical dependencies between the constituents of the target index, but adds constituents at certain levels, thus expanding the hierarchy in a way that is compatible with the target index. Figure 3 shows as an example a problem spec-ification that refines the one shown in figure 2. It groups the rooms in the administrative wing into two zones and adds a hall and a lobby to the administrative zone. An associated solution is shown in Figure 4. These refinements may reflect decisions a designer made while planning the fire station and illustrate the typical modifications an initial architectural program under-goes as the design progresses. *Refinement may be a desired side effect of the retrieval of a case* because it saves time not only in terms of finding an initial solution, but also in terms of time spent on expanding and refining the problem specification itself.

As a consequence, the comparisons made during search through the case library should not discriminate against a case when its index refines the current problem structure; on the contrary, they may favor it for that very reason. That is, a case with a compatible index may be preferred over another case if it has more structure, everything else being equal. The latter restriction is important because a designer dealing with a one-bedroom apartment, for example, may not be interested in a 3-bedroom apartment, let alone an apartment building. Problem attributes like size restrictions, budget, etc. will prevent cases with such over-refine-ments from being retrieved.

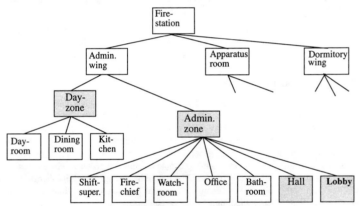

Figure 3: A problem structure that refines the structure shown in Figure 2

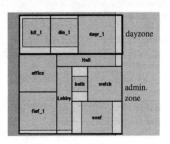

Figure 4: Layout corresponding to the problem shown in Figure 3

5   Case Adaptation

When the designer asks a module to find good candidate solutions in the case library to solve the problem currently specified, the module tries to find solutions that allocate at least all the functional units currently specified. But in the case of refinement, such solutions contain design units associated with functional units that are not yet specified. After retrieval, the new design units become part of the current solution and the associated functional units (which are preserved in the index and retrieved with the solution) become part of the current problem specification, where they are available for inspection and editing. That is, *case adaptation may include adaptation of the problem specification*, a situation encountered also in other domains.

A SEED module supports the interactive editing and expansion of problem specifications. The operations underlying these capabilities are immediately available after a case has been retrieved to incorporate refinements in the retrieved problem structure into the current one. That is, they form base from which automatic or interactive adaptation mechanisms can be constructed.

SEED-Layout, for example, offers designers operations to

*   add an object to a list attribute

*   delete an object in a list attribute

*   aggregate units in a *constituents* list into a new constituent

*   disaggregate units in a *constituents* list

*   move a functional unit to a different *constituents* list

*   edit other types of attributes of an object.

An object present in the current specification, but not matched in the case index, requires no adaptation. But the designer should be informed because this object may also be missing from the solution in some form (a functional unit not allocated, a constraint not satisfied). An object present in the current specification and matched in the case index to an object in an equivalent location in the constituent hierarchy also requires no adaptation, except when the values of certain corresponding attributes differ. In this case, the editing capabilities of the module can be invoked to change the attributes. A functional unit present in the case index, but not matched in the current problem specification, that *refines* its structure can be inserted with the above commands. In each case, the designer's agreement should be sought before execution of the adaptation.

A more difficult situation occurs when an object present in the current specification matches an object in the case index, but not in an equivalent location in the constituent hierarchy. The evaluation component of a SEED module allows designers to evaluate a solution with respect to the requirements of an associated problem specification. The solution retrieved from a case can be evaluated in this way before, during, or after adaptation of the current problem specification. This will bring out all existing discrepancies, and the generation commands available in a module are immediately executable to adapt the solution so that these discrepancies are eliminated; alternatively, the problem specification can be modified to achieve the same effect.

In order to support the allocation of design units, SEED-Layout offers designers a range of commands, some of which are listed below:

- <u>add</u> a design unit in a location specified by the designer and associate with it a functional unit not yet allocated

- <u>remove</u> a design unit

- <u>edit</u> the dimensional attributes of a design unit

- <u>change</u> the functional unit associated with a design unit

- <u>generate</u> the next or all alternatives to allocate selected functional units in the current solution.

These commands can again be used as building blocks to implement interactive or automated adaptation procedures. For example, the remove operation can be used to eliminate functional units that do not exist in the problem specification or that are incorrectly allocated, or the edit operations can be used to adapt the dimensional attributes of a design unit to satisfy certain constraints.

At any time, a design unit may have no functional unit associated with it, in which case all constraints on its shape and location are relaxed. This suggests an interesting extension of the retrieval procedure outlined in the preceding section. If we allow functional units to be comparable - on an optional basis - not only with subclasses, but also with superclasses, we may be able to retrieve layouts based on size and dimensional constraints alone; for example, we may retrieve a narrow, linear scheme for a narrow site from a project dealing with building functions that differ from the ones currently under consideration.

Acknowledgments. The development of SEED is sponsored by Battelle Pacific Northwest Laboratories, the US Army Corps of Engineers Construction Engineering Research Laboratory (USACERL), the National Institute of Standards and Technology (NIST), the Engineering Design Research Center (EDRC) at Carnegie Mellon University, the Australian Research Council and the University of Adelaide.

References

Flemming, U., R. Coyne, R. Woodbury (1993) "SEED: A Software Environment to Support the Early Phases in Building Design" in *ARECDAO93* (Proc. IV Int. Conf. on Computer Aided Design in Architecture and Civil Engineering, Barcelona, Spain) 111-122

Flemming, U. (1994) "Case-Based Design in the SEED System" in *Knowledge-Based Computer-Aided Architectural Design*, G. Carrara and Y. E. Kalay, ed.s, New York: Elsevier, 69-91

Flemming, U, Coyne, R., Snyder, J. (1994) "Case-Based Design in the SEED System" in *Proc. 1st Congress on Computing in Civil Engineering*, Washington, DC, June

Using Case-Based Reasoning for Design Media Management

Mary Lou Maher [1]

Abstract

The development of case-based reasoning systems for structural engineering design projects requires the identification and representation of project data that can assist in a new design situation. The development of a representation system that can be used during and after the design project raises issues concerning design media management. Since project data is inherently multimedia; including text reports, graphical images, photographs, CAD drawings, decision variables, etc., case-based reasoning can be viewed as an approach to design media management. An approach to incorporating multimedia into data management techniques for the use and reuse of project data requires a data model that accommodates engineering project data. Two models are presented here: the encapsulated data model an the layered data model.

1. Introduction

Case-based reasoning is a problem solving paradigm using artificial intelligence techniques that uses previous cases as the basis for new problem solving episodes. Case-based reasoning involves the development of a case memory representation of previously solved problems that are retrieved for the solution of similar new problems. The representation of design cases is based on knowledge representation paradigms such as object-oriented models, rule-based models, and dependency graphs. An important aspect of case-based reasoning is the indexing and retrieval of cases based on a description of a new problem. Thus case-based reasoning is similar to querying a database manager where the user does not know which specific case should be retrieved. Applications of and extensions to case-based reasoning for design projects has been explored and reported in [Maher and Zhang, 1993; Maher and Balachandran, 1994].

Collecting design information for the development of a case-based reasoning system requires an understanding of the types of data that are used to describe design projects. Design information is inherently multimedia, where information is distributed among drawings, text-based reports, databases, and more recently, video, virtual reality representations, and animation files. Currently, multimedia design data

---

[1] Associate Professor, Department of Architectural and Design Science, University of Sydney, NSW 2006 Australia, mary@archsci.arch.su.edu.au

is not comprehensively managed. For example, CAD drawings are stored in separate files and in separate directories. Some drawings are stored on disk, others on tape. The retrieval of a CAD drawing of a current project is done by the people involved in editing the files. The retrieval of a CAD drawing of a past project involves identifying the way the project was archived, locating the device on which the drawings were stored, and restoring the file or viewing microfiche, or in some cases looking at the printed version of the CAD drawing. As CAD drawings become linked with CAD models, video, animation, and non-graphic databases there is a need for multimedia management information systems.

This paper looks at how current developments in case-based reasoning in design and multimedia databases can be drawn together to form a comprehensive approach to the use and reuse of design project information. The next section gives a brief introduction to design case representation and retrieval. The following section provides an approach to modelling multimedia design data. The final section is a discussion of the issues that need to be addressed before such systems become practical.

2. Design Case Representation and Retrieval

One of the characteristics of design is that designers rely extensively on past experience in order to create new designs. The experience may be their own, that is, specific design problems they have encountered before, or the experience may be the documented experience of others. The use and reuse of this experience assumes that the experience can be recalled at the appropriate time. Case-based reasoning techniques provide paradigms for the use of experiential knowledge in memory to aid in the solution of problems and the performance of tasks. The representation of design cases varies from sets of attribute-value pairs, to collections of indexed drawings, to multimedia hypertext. The major issues to be addressed representing design cases include:

What information is stored in a case?
How is the information represented?
How is the information indexed?

Another characteristic of design is the phenomena of exploration in the early stages of design configuration. A designer starts with an ill-structured and partially defined specification, and through a process of exploration, the specifications can change and become more detailed. The view of design as problem solving suggests that design can be modelled as a search process. However, as search is typically treated as a goal directed process, there is a need to qualify the model of design as search to allow exploration. In exploration, the goal(s) change as the space is searched. As a case-based reasoning system, exploration implies that the designer has only a partial specification of the relevant case to be retrieved and that the specification may change in response to the case(s) retrieved. The major issues to be addressed in design case retrieval include:

How is a design specification represented?
What are the strategies for design case retrieval?
How can the specifications change over time?

The development of a case-based design system for the structural design of buildings, CASECAD (Maher and Balachandran, 1994), provides the basis for discussing these issues. In developing CASECAD a set of projects from Acer Wargon Chapman Associates were used as the source of project data. The information was presented as a set of construction drawings complemented by interviews with the project engineers. The development of the case representation was preceded by the author's experience in developing knowledge-based design systems for structural design that made use of generalised descriptions of structural systems and prototypes (Maher, 1988; Gero, 1990). As a result, the cases in CASECAD are a combination of features as attribute-value pairs and CAD drawings for illustration.

The information stored in a design case includes the function, behaviour, and structure features of each case. This information provides a more comprehensive view of the case than would have been included if only the design description as a physical form was included. The use of function, behaviour, and structure as a guide for collecting information ensured that information regarding the purpose of the design case, the expected or actual behaviour or performance criteria, and the structural form was included. Although this information was not readily available from the drawings that are used to document the project, the information was determined to be necessary for identifying when the case is relevant for a new design situation.

The information in a case is represented as a hierarchy of objects with pointers to drawings for illustration. Each object included a set of features, text description, and one or more graphic files organised according to function, behaviour or structure. The object-based representation allowed the case-based reasoner to do pattern matching for case retrieval. The cases are indexed only by the features and not by the drawings. It is assumed that the specifications of a new design problem are given as desired features rather than as a sketch or CAD drawing.

The development of a case base and the implementation of retrieval strategies has lead to a consideration of data models for representing design project data. The representation of design cases needs to include the various types of information associated with a design project. The retrieval of design cases needs to be flexible and lead to the retrieval of one or more design cases. The consideration of database management techniques provides a formal basis for data storage and retrieval and the concept of a query language. The query language could then become the basis for building a reasoning system that assists the designer in exploring the casebase. This also leads to a more comprehensive view of case-based reasoning in engineering and the development of databases that support the life cycle of the project as well as the reuse of project data.

3. Models for Multimedia Design Databases

Multimedia information systems is a relatively new field with many research areas not yet addressed. The speed with which the technology is advancing makes it difficult to identify principles that are not technology-specific. The focus in this paper is on semantic data models for organising multimedia design databases, the implementation of these models in current database management systems, and the implications of the models on the data queries.

The major consideration in the development of multimedia data models is the nature of multimedia information and its difference to conventional information systems. In conventional information systems the models include network, hierarchical, relational, and object-oriented data models where the data is stored as

ASCII characters and the models provide a structure for organising the information. In multimedia information systems the format for storing the data varies from one media to another, for example, postscript is a standard representation used to store information in a format independent of the software that generated it, gif is a standard format for storing digital graphical images, jpeg compression is used to store video files in less space then the digital format recorded, etc; and that some of the media is time-based, such as video and sound. The various formats used in multimedia information representations differ from conventional database management representations because they are not easily inspected by the people that use them. A person can see and understand the information in a relational table but cannot understand a postscript file without the use of another program to display the information. The formats used to store multimedia documents can be considered *subsymbolic*, and the information in a conventional database can be considered *symbolic*. A subsymbolic representation requires a computer program to interpret the information and display it for a person to understand.

Two models are presented that build on the notion of subsymbolic and symbolic representations of information. One model uses different layers of data to distinguish between the symbolic and subsymbolic data; this is referred to as the *layered model*. The second model encapsulates the subsymbolic data within a symbolic object, referred to as the *encapsulated model*.

3.1 The layered model

Figure 1 illustrates a layered model for representing the subsymbolic and symbolic data at different conceptual levels in which the mapping from one level to the next is done when a query is made to the database. This model allows flexibility in mapping from the data in the symbolic level to many different subsymbolic files, depending on the semantics of the symbolic data. This model can be implemented in a conventional database management system where the symbolic data is stored as tables or objects. The subsymbolic data is referenced by file name, creating a link between the symbolic and subsymbolic data.

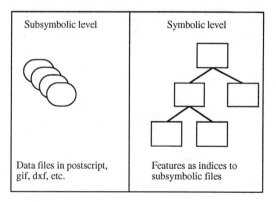

Figure 1: The layered model separating symbolic and subsymbolic data

CASECAD used a layered model for case representation. The casebase contains a set of structural design projects. Information in case memory is represented in two modalities: symbolic and graphical. The graphically represented examples are indexed symbolically so that they can be retrieved given a problem specification in symbolic form. The Case-Base Manager (CBM) provides facilities for creating, browsing, modifying, displaying and saving information associated with design cases. A mouse-based editing facility allows any text or graphics to be modified either through a text editor or directly through the CAD system. Access to the files and interfaces of two CAD programs is available: AutoCAD and XFIG. AutoCAD is a general purpose CAD system where the graphical descriptions of design cases are created and stored as DXF files. XFIG is a general purpose 2D drawing program. The graphical descriptions of abstract behaviours of subcases are created and stored as FIG files.

CASE NAME: 370Pitt
text-description:
"370 Pitt street is a $20 m office building completed in 1989. Its 17 storeys rest partly over tunnels forming a link in Sydney's underground railway system. Retail outlets and car parking facilities are located on ground and upper ground levels, while each typical floor provides approximately 1000 m2 of office space. There is no basement due to the presence of the underground city circle rail loop immediately below the building. Lateral stability is provided by the eccentric core and the western facade panels."
FUNCTION
    purpose : "To resist the primary loads of gravity and lateral force action"
    objective : "To maximise the use of precast elements for the construction"
    support-building-type : office
    support-grid-geometry : rectangular
    resist-wind-load : 1.5 kPa
    resist-gravity-load : 6 kPa
BEHAVIOUR
    total-net-area-of-usable-space : 15000 m2
    construction-cost : 20 millions
    total-number-of-occupied-floors : 17
    xfig-file : "370-pitt/general.fig"
STRUCTURE
    overall-length : 38 m
    overall-width : 29 m
    total-building-height : 70 m
    material: precast-concrete, insitu-concrete
    location-of-core: eccentric
    length-of-core: 11 m
    width-of-core: 11 m
    number-of-stories: 17
    floor-to-floor-height: 4 m
    floor-system: precast-panels-supported-by-beam-shells
    acad-file: "370-pitt/general.dwg"

Figure 2. Example of a case where graphic data is accessed by filename

An example of the symbolic description of a case in CASECAD is shown in Figure 2. The case describes the structural system of a building at 370 Pit Street. The attributes (shown in *italics*) and their values comprise the symbolic data that can be stored and managed in a conventional database management system. The behaviour of the structural system is illustrated as a bending moment diagram in an xfig file, referenced as the value of an attribute *xfig-file*. The layout of the structural system is illustrated in an AutoCAD drawing file, referenced as a value of an attribute *acad-file*.

3.2 The encapsulated model

Figure 3 illustrates the encapsulated model, a model using an object-oriented approach where each object contains symbolic and subsymbolic information. In this model the symbolic data encapsulates the subsymbolic data. The symbolic data can include information about the content of the subsymbolic data. For example, if the subsymbolic data is an image of a site, the symbolic data can give information about the location, size, and use of the site. This provides a direct mapping from a group of symbolic data to one image or subsymbolic representation. This model can be implemented as large objects in Postgres.

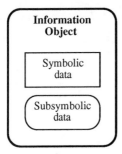

Figure 3: The encapsulated model for multimedia objects

The case base of CASECAD can be re implemented using the encapsulated model. The information stored in the object shown in Figure 2 is reorganised into four objects: the building object, the CAD object, the XFIG object, and the text object. The attribute-value pairs, comprising the symbolic data related to the semantics of the building, are stored in an object called *building*. The CAD object is defined to represent CAD drawings and contains symbolic data about the specific CAD drawing and an attribute for storing the CAD file as a large object. Similarly, the behaviour drawing is stored in an object representing XFIG drawings, containing both symbolic information about the drawing and an attribute for storing the .fig drawing as a large object. Finally, the text description of the building is stored as a postscript file in an object created to represent symbolic data about postscript files.

Object definition examples in POSTGRES are shown in Figure 4. The create command shows how the object classes are defined. The append command shows how the data is added to create object instances. The object BUILDING contains symbolic data about the building, including the function, behaviour, and structure properties shown in Figure 2. The CADdrwg object contains symbolic information about CAD drawings generally including properties related to the name of the building, the content of the drawings, the name of the computer program that generated the drawing, etc. One of the properties of the CADdrwg object is a reference to a large object in which the drawing data is stored. Each object has methods that provide procedures for creating, changing, and managing both the symbolic and subsymbolic data.

```
create BUILDING                    append BUILDING
(name = char16,                    (name = "370 Pitt St",
purpose = text,                    purpose = "To resist the primary loads ... action",
objective = text,                  objective = "To maximise the use precast elem...",
spp_bldg_type = text,              spp_bldg_type = "office",
...                                ...
net_space = float4,                net-space = 15000,
...                                ...
overall_length = float4,           overall_length = 38)

methods = addbuilding)

create CADdrwg                     append CADdrwg
(name = char16,                    (name = "370 Pitt St",
lcontent = text,                   content = "3D model of 370 Pitt St",
CAD_program = text,                CAD_program = "AutoCAD",
...                                ...
CADdrwg = large_object,            CADdrwg = "370-Pitt/general.dwg")

methods = add_drwg,
display_drwg, edit_drwg)

CLASS XFIG
create XFIG                        append XFIG
(name = char16,                    (name = "370 Pitt St",
lcontent = text,                   content = "structural system bending moment diag",
CAD_program = text,                Xfig_program = "xfig",
...                                ...
XFIGfile = large_object,           xfig_file = "370-Pitt/general.fig")

methods = add_drwg,
display_drwg, edit_drwg)

CLASS TEXT_DESC
create POSTCRIPT_OB                append POSTCRIPT_OB
(name = char16,                    (name = "370 Pitt St",
lcontent = text,                   content = "Building description for 370 Pitt St",
Postcript_ed = text,               Postcript_ed = "Ghostview",
...                                ...
PS_file = large_object,            PS_file = "370-Pitt/general.text")

methods = display_file,
add_file)
```

Figure 4. The building data using the encapsulated model

The use of the encapsulated model to represent the multimedia data allows the definition of specialised media objects that include properties related to the management of the specific type of media. These specialised objects provide a means of adding, modifying, and retrieving multimedia data in a way that is compatible with conventional database management systems.

4. Discussion

Two models for organising multimedia design data are presented: the layered model and the encapsulated model. Both models can be implemented in conventional datbase management systems, however, some issues are not addressed by conventional data management. Identifying the organisation of the data and the facilities for querying the database requires a language that maps onto the data model. The most common language used in conventional database management systems is SQL, developed for the relational data model. Such languages provide flexibility in allowing the data from different tables to be combined or projected onto another set of attributes. This provides a flexibility in the retrieval so that the data presented to the user can be determined by the user, not predetermined by the database administrator. The information in a multimedia document cannot necessarily be combined and projected onto a different set of attributes as is possible when the data is stored in a homogeneous form as it is in a relational database. By identifying semantic models for organising multimedia design data that recognise the difference between the symbolic data stored in conventional database management systems and the subsymbolic data stored in files of digital data, we can begin to address these issues.

Acknowledgements. This work is supported by the Australian Research Council. The author acknowledges the support and assistance of the Key Centre of Design Computing. Dong Mei Zhang is acknowledged for her conception and implementaiton of CADSYN and M. Balachandran for his conception and implementation of CASECAD.

References

Gero, J. S. (1990). Design Prototypes: A Knowledge Representation Schema For Design, *AI Magazine*, **11**(4), 26-36.

Maher, M. L. (1988). Engineering design synthesis: a domain independent representation, *Artificial Intelligence for Engineering Design, Analysis and Manufacturing* **1**(3): 207-213.

Maher, M.L. and Balachandran, M. (1994) A Multimedia Approach to Case-Based Structural Design, *ASCE Journal of Computing in Civil Engineering*, **8**(3): 359-376.

Maher, M.L. and Zhang, D.M. (1993). CADSYN: A Case-Based Design Process Model, *(AI EDAM) Special Issue on Case-Based Design Systems*, Academic Press **7**(2): 97-110.

FHWA Information Technology Infrastructure

Edward G. Willey

This document explains where FHWA is today in implementing information technology; where we are heading; and how our direction relates to the current situation and trends in the computer and telecommunications industries. There is also a brief review of the major information systems being developed or enhanced by the agency. As far as possible, we have kept the discussion non-technical and where special terms are used we have attempted to define them.

The major areas discussed in this paper are:

- the overall information "architecture" of the agency (where data is collected, stored and processed, and why);
- major applications projects;
- communicating outside FHWA; and
- data communications and the FHWA wide-area network.

INFORMATION ARCHITECTURE

The FHWA uses a "Classic" model for most of its information processing. The corporate repository for critical nationwide information is the **Transportation Computer Center (TCC)** mainframe for a number of excellent reasons including high security; a single point of access (the AASHTO Value Added Network {AASHTO VAN}) for outside (State DOT) users; 24 hr/day availability; and offsite backup and disaster recovery. Internal working documents such as correspondence, budget, and policy documents are largely kept on **local area networks**, where they can be shared and backed up.

Chief, Information Systems Division, Federal Highway Administration, 400 7th Street, SW, WDC 20590

The FHWA currently is exploiting two types of **"client-server"** processing. The first type uses the mainframe as a data processor and the personal computer (PC) for formatting and presenting the processed data. A more complex form of client-server processing is being implemented for the Comprehensive Transportation Information and Planning System (CTIPS), where the data to be accessed will be queried, collected and consolidated from States and FHWA through AASHTO VAN.

MAJOR INFORMATION SYSTEMS PROJECTS

Projects in progress for major FHWA programs include:

- Intelligent Vehicle Highway System (IVHS)
- Vehicle Travel Information System (VTRIS)
- Highway Performance Monitoring System (HPMS)
- Office of Motor Carriers Management Information System
- Office of Motor Carriers Intrastate Processing

Projects for internal and administrative systems include:

- Training Management System
- Electronic Signatures
- Graphical User Interface for the Fiscal Management Information System (FMIS)
- FMIS: Demo Projects
- FMIS: Surface Transportation Program (STP) Enhancements
- FHWA Executive Information System (EXIS)
- Reports from the Departmental Accounting and Financial Information System Batch Control File
- Cash Payment Redesign
- Electronic Data Sharing (EDS) through AASHTO VAN
- Comprehensive Transportation Information and Planning System (CTIPS)

COMMUNICATING OUTSIDE FHWA: E-MAIL AND OTHER CONNECTIONS

<u>E-Mail:</u> The DOT Interdepartmental Network (IDN) mail system continues to expand. We currently see no insurmountable barriers to communicating with anyone else in the USDOT. Every agency is not completely "wired" as FHWA is; but we anticipate E-mail exchange with some offices in all other modes. The E-mail system provides gateways to the government-wide X.400 and the Internet mail systems.

<u>Other</u> <u>Outside</u> <u>Communications:</u>   We are looking to extend
the States' reach through TCC using **AASHTO VAN** as a
transport mechanism.   The first projects are to give
States the capability for Electronic Federal-Aid payments
and access to the Electronic Policy Reference System
(EPRS) which runs on a Novell local area network (LAN) in
Headquarters and the FHWA Electronic Bulletin Board
(FEBBS).

<u>FEBBS</u> is for internal and external users to access
various FHWA information. After a period of declining
calls due to internal FHWA traffic shifting to the wide-
area network (WAN), usage has picked up again. NHTSA and
FTA also post some information on the system. FEBBS is
accessible via Internet, directly and through a DOT
Internet data server.

WAN AND DATA COMMUNICATIONS

We currently have connected all FHWA Headquarters and
field offices' local area networks to the FTS 2000 network
to form an FHWA WAN. This allows the exchange of E-mail
and files of information among FHWA employees as well as
access to TCC for mainframe processing.

A Broker for Delivering and Accessing Environmental Regulations

Mark A. Krofchik[1], James H. Garrett, Jr.[2], M. ASCE,
and Steven J. Fenves[3], Hon. M. ASCE

Abstract

As increased attention is being paid to the effects that our decisions make on the environment, environmental regulations are continually being created, issued, and revised. Users of these regulations, such as product designers, often do not know which regulations are applicable, when these applicable regulations have been updated, or which newly written regulations apply to their design context.

The general approach that we are taking to meet these needs is to develop a *broker* for locating, accessing, and using environmental regulations. The primary idea of the broker is to provide a a computer–based mechanism by which distributed sources of information (in terms of both geographical and subject distribution) can be provided to a distributed collection of users without the users having to know where that information resides.

A variety of enabling technologies now exist that make implementation of the above described broker possible: the World Wide Web, a distributed hypermedia environment that operates on files stored on computers accessible from the Internet; various search facilities that search over the Internet; billing facilities for supporting electronic commerce; representation and evaluation facilities for formal models of regulations that are accessible over the Internet; and various machine learning facilities. A prototype broker has been developed on the World Wide Web and incorporates a software tool called Inquery, which provides text searching capabilities over provisions in environmental regulations.

1 Introduction

1.1 The Problem

With ever increasing global attention being paid to environmental issues, regulatory agencies throughout the world are continually creating, issuing, and revising environmental regulations. For the purposes of this paper, we use the term environmental regulation to encompass all types of formal documents, such as regulations, codes, standards, and specifications, that impose environmental requirements on private and public projects and organizations. Users of these regulations, ranging from corporate environmental policy makers to product designers, are drowning in a "sea of environmental regulations." Such users often do not know which existing regulations

---

[1] Research Assistant, Dept. of Civil and Env. Engrg., Carnegie Mellon University
[2] Associate Prof., Dept. of Civil and Env. Engrg., Carnegie Mellon University
[3] Sun Company Univ. Prof., Dept. of Civil and Env. Engrg., Carnegie Mellon University

are applicable, when applicable regulations have been revised, or which new regulations apply. It is unrealistic to assume that a user is able to easily identify all specific regulations that are applicable to his or her problem, what specific provisions within these regulations address aspects of the problem, and what procedures are to be followed to ascertain that the applicable provisions are met.

To deal with this growing body of applicable regulations, many large corporations have personnel dedicated to being aware of all regulations that affect their company. These groups may write internal corporate documents which essentially outline or summarize these regulations. The internal documents are then made available to designers within the corporation who are responsible for the compliance of the specific products that they are developing. However, while this procedure relieves designers of the burden of being aware of the ever changing state of regulations, the burden is shifted to one of being aware of the still large body of internal corporate documents describing relevant, interpreted regulatory requirements. Simply reading through each and every document may still take a designer a significant amount of time. Hence, even with this internal system for compliance within a corporation, it is still unrealistic for a designer to locate and be aware of every internal document that applies to his or her project.

Designers in small corporations have, in general, fewer resources at their disposal than do their counterparts in large corporations. Hence, these designers likely have more responsibilities when designing a new product. One such responsibility may be to assure product compliance with applicable environmental regulations. Currently, these designers must manually and directly search and access the applicable regulations.

1.2 Motivation

Distribution of environmental regulations represented in electronic form is becoming increasingly prevalent. Some regulatory agencies are making their regulations available through various CD–ROM and On–line services. By doing so, all users of these services have, in principle, immediate access to the most up–to–date versions of the text of these documents. Some of these systems only provide search over hypertext links, while others offer the capability of full–text search over the entire text of regulations. However, neither of these capabilities is very effective for locating the *subset of applicable provisions* within the regulations.

Users of environmental regulations need considerable assistance in: (1) identifying applicable regulations, and provisions within those regulations, for a given design context; and (2) determining the compliance of a product design with the applicable provisions of regulations.

A number of enabling technologies has now become available which may play a role in providing this assistance in accessing and using environmental regulations. Among these are: (1) the World Wide Web, a distributed hypermedia environment that operates on files stored on computers accessible from the Internet; (2) various search facilities that search over the Internet; (3) billing facilities for supporting electronic commerce; (4) representation and evaluation facilities for formal models of regulations; and (5) various machine learning techniques for acquiring elements of knowledge, such as useful classifications of text passages.

1.3 Organization

This paper presents a proposed solution to the problem and describes a prototype system that we have developed. The concept and functionality of the proposed solution are first discussed. Then several key enabling technologies are described. Following these descriptions, a working prototype incorporating these enabling technologies is illustrated. The final section of the paper discusses future plans for this research.

2 Proposed Solution: A Broker

The general approach that we are taking to meet the needs of regulation users is to develop a *broker* for locating, accessing, and using environmental regulations. The primary idea of the broker is to provide a a computer–based mechanism by which distributed sources of information (in terms of both geographical and subject distribution of environmental regulations) can be provided to a distributed collection of users without the users having to know where that information resides.

2.1 The Working Environment of the Broker

To better understand how the proposed broker should function, we first describe the environment in which the broker is intended to operate. The two primary organizations involved in this environment are *regulatory agencies* and *regulation users*. Regulatory agencies, geographically and organizationally dispersed, issue regulations that contain a variety of specific provisions (limitations, definitions, recommendations, etc.) that are applicable for various design contexts. For many of these regulatory agencies, the only source of revenue comes from the sale of these regulations, while for governmental organizations, these regulations are issued for free or at nominal costs to users. These regulatory agencies usually issue their regulations in paper form but, more recently, many have turned to some form of electronic media (CD–ROM or On–Line services) to distribute their regulations. Regulation users, such as product design companies whose products must conform to applicable regulations, governmental bodies charged with enforcing these regulations, or consulting companies that interpret these regulations and offer advice to other organizations, are geographically and organizationally dispersed and are concerned about different issues. Currently regulation users must maintain individual "knowledge bases" about the existence of various regulations and how they apply and it is their responsibility to identify all applicable regulations and ensure compliance.

2.2 The Broker Concept

Regulation users need assistance in: (1) identifying applicable regulations, and provisions within those regulations, for a given design context; and (2) determining the compliance of a product design with the applicable provisions of a regulation. The basic concept of the broker is to provide distributed access to environmental regulations in electronic form over a wide–area network, support a fine–grained search down to the provision level of these regulations for a given design context, and provide a mechanism by which regulatory agencies can acquire revenue from users accessing their regulations. This concept is schematically illustrated in Figure 1.

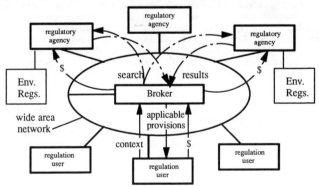

Figure 1 – Broker Network

Both regulatory agencies and regulation users reside on this network and are brought to each other's attention via the broker. Persons in regulatory agencies use broker modeling tools to model their regulations into a hypertext/computable form and serve that model to those who need and pay for it. The broker maintains the locations of these modeled regulations and serves as an interface between the creators and users of these regulations. Regulation users describe their design problem context to the broker and request it to generate a list of applicable regulations and provisions. When this transaction takes place, revenue from the user of the broker to the particular regulatory agencies is automatically transferred over the same network.

2.3 Necessary Functionalities of the Broker

We expect the broker to provide three major groups of functionalities: (1) regulatory agency support; (2) regulation user support; and (3) broker management.

Regulatory Agency Support. The first set of functionalities is support for *regulatory agencies* in representing their regulations into a form (or forms) that can support: (1) identification of applicable regulations and provisions for a specific design context; and (2) assistance in evaluating provision compliance. To this effect, the regulations need to be represented in: (1) hypertext form, where each provision is associated with a collection of classifiers; and (2) computable form, which can be evaluated to determine regulation compliance of a collection of design attributes. Hence, regulatory agency support must include assistance in: (1) representing the entire regulation document in hypertext form; (2) creating a collection of classifier trees for classifying the provisions of the regulation; (3) mapping local classifiers, unique to the particular regulation, to generic classifiers; (4) creating a glossary; and (5) representing provisions in computable form.

Regulation User Support. The second set of functionalities is support for *regulation users* in using the regulations served by the broker. This set of functionalities includes: (1) assistance in identifying the applicable regulations and, more specifically, their provisions, registered with the broker that are relevant to their problem context; (2) assisting a regulation user in evaluating the computable form of provisions for a collection of problem attributes (e.g., a product design description); (3) providing a way for regulation users to pay the regulatory agencies for the use of their regulation, if necessary; and (4) providing a way for regulation users to send input, comments, and/or criticisms about particular regulations to the regulatory agencies. This last functionality will allow regulatory agencies to receive input from the people who read and use their regulations. With this two–way communication between regulation users and regulatory agencies, possible revisions to improve regulations can be initiated by the regulation users.

Broker Support. The third set of functionalities is associated with managing the broker and its databases. This set of functionalities deals both with assisting regulatory agencies in registering their regulations with the broker and also with overall broker maintenance. The broker support needs to include: (1) assistance in registering regulations with the broker; (2) verifying that new regulations are properly represented to be compatible with the broker; (3) verifying that the new regulations are accessible by the broker; (4) developing and maintaining generic classification systems for use in searching for relevant provisions; (5) assuring that regulations are secure and tamper–resistant; (6) announcing new or updated regulations to the user community; and (7) managing the charging of regulation users and the payment of regulatory agencies.

3 Enabling Technologies

Within the last few years, a number of enabling technologies has become available that make the implementation of the broker concept both feasible and practical. The World Wide Web is a distributed hypermedia environment that operates on files stored on computers accessible from the Internet (Vetter 94). Search facilities now exist, such as Inquery, that provide the capability of full–text searching, classifier–based searching, or a combination of the two (Callan 92). Internet–based billing facilities, such as Netbill, now exist that support secure electronic commerce in information goods over the World Wide Web (Cox 94). Each of these enabling technologies is described in the following three subsections.

3.1 World Wide Web and Web Clients

The World Wide Web is an environment which uses hypermedia techniques to access and display information from the Internet. The World Wide Web is a client/server system, in which *clients* request data from *servers* that actually locate and manage the data and supply it to the clients. Currently, several World Wide Web clients exist which request and display information from World Wide Web servers. Developed by the National Center for Supercomputing Applications (NCSA), Mosaic is one such World Wide Web client. Currently, one of the fastest growing class of tools on the Internet, it is likely that World Wide Web clients like Mosaic will become the universal way of accessing global information (Smith 94). Because of its popularity, client/server system, world–wide access, and hypermedia capabilities, the World Wide Web was chosen as the implementation environment for the prototype broker.

3.2 Search Facilities

Currently, there are several information retrieval systems available that perform text searching of documents. We are using Inquery, developed at the University of Massachusetts. Inquery is a probabilistic Information Retrieval (IR) system that retrieves, from a collection of documents, those documents that are relevant to a keyword query. In addition to this full–text searching capability, Inquery allows for field searching, where the search is only conducted on text found within fields associated with each document in the collection (Callan 92). As related to the broker concept, fields are used to add classifiers to the provisions of environmental regulations. Then, by creating a query, a regulation user can effectively search for applicable provisions over the full–text, over classifiers within fields, or over a combination of both.

Another information retrieval approach that we are exploring is the Minicode Generator, developed by the National Research Center of Canada. The Minicode Generator is a computer–based facility for providing automated assistance in using the National Building Code of Canada (NBCC) (Vanier 93). Given a small set of characteristics and classifiers of the building to be checked, the Minicode Generator identifies the subset of provisions that are applicable by eliminating all provisions that are *not* related to the given query. With the broker concept, the Minicode paradigm can be used to eliminate those environmental regulations and provisions that are found *not* to match a given query.

3.3 Network Billing Facilities

Currently being developed at Carnegie Mellon University, Netbill is a business model, set of protocols, and software implementation for commerce in information goods and other network delivered services. Netbill is expected to provide a means for customers to purchase information goods from information vendors over the Internet. A Netbill server will maintain accounts for both customers and merchants. These accounts will be linked to conventional financial institutions. A Netbill transaction transfers the information goods from merchant to user, and debits the customer's account and credits the merchant's account for the value of the goods. Users may be charged on a per item basis, or by a subscription allowing unlimited access (Sirbu 95). The incorporation of the functionalities of Netbill into the broker concept would provide a secure means for financial transactions to occur between regulation users and regulatory agencies over the World Wide Web. It would also provide better service to regulation users since the users would only pay for what they use.

3.4 Other Enabling Technologies

Other enabling technologies are currently being explored to deliver additional functionalities in the broker. Two of these include the NewsWeeder Project and the "standards processing system."

NewsWeeder, currently under development at Carnegie Mellon University, is a machine learning facility which filters all Internet news messages that are irrelevant to a user's customized profile of interest (Lang 94). Irrelevant words are first screened out of news messages, leaving small subsets of words, represented as word vectors, to represent the messages. The word vectors are then used by the system to determine a document's importance to a particular user's interests. Related to the broker concept, this automatic classification of documents may eventually be extended to directly provide automated classification of provisions of environmental regulations.

The "standards processing system" being developed by Kiliccote and Garrett provides a software approach for computer–aided support in using regulations and design systems (Garrett and Kiliccote 95). The system is composed of the following major components that interact with one another using Internet protocol: (1) an *evaluation module*, which is resident within the design system and manages the evaluation of a design with respect to applicable design standards; (2) a *data server*, which acts as a front–end between the database of the design system and the standards processing servers; (3) a *standards processing client*, which displays the results of evaluation to the designer and supports access to the standards processing servers over the Internet; and (4) a *standard processing server*, which evaluates a given design to check whether it satisfies the requirements of a specific design standard. The design standard with which a specific standard processor server evaluates a given design must be in a formal, computable form. This work, which begins to provide a facility for determining conformance of a design with applicable design standards, can be used to provide a conformance evaluation facility in the broker.

4   Prototype Broker

To begin to investigate the potential of the enabling technologies described in Section 3, we have developed a prototype broker incorporating the capabilities of the World Wide Web and Inquery. Figure 2 illustrates the entry point onto the World Wide Web (referred to as the Broker Homepage) of the prototype broker. Three options, or hyperlinks, allow a user to register new environmental regulations, find applicable environmental regulations, or to learn about the broker.

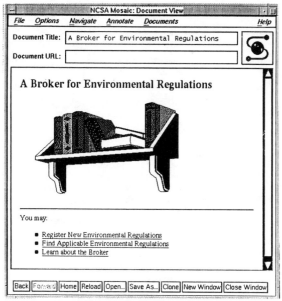

Figure 2 – Broker Homepage

To find applicable regulations registered with the broker, a regulation user first clicks the second hyperlink in Figure 2 and page with a subject–search field appears. For illustrative purposes, suppose he or she wants to locate regulations that deal with landfills. Upon entering the query for "landfills," a list of all registered landfill regulations appears, as shown in Figure 3. At this point the user can traverse any of the regulations located via hypertext links from this level down to the individual provisions. Alternatively, the user may choose to search for individual provisions of these applicable regulations that pertain to a certain subject or a keyword query. As shown in Figure 3, the user may click the hyperlink to perform an Inquery search over the provisions of the relevant regulations. Choosing this option, the user is prompted for an Inquery query, which is a keyword query over (1) the full–text of the provisions, (2) the classifiers associated with the individual provisions, or (3) a combination of the two. For example, a user may enter a query to locate all provisions containing the term "final cover" in the text. The result of entering this full–text query is illustrated in Figure 4. The provisions containing the words "final cover" are listed with a hyperlink to the text of each specific provision.

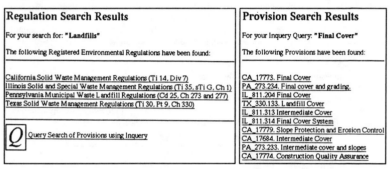

Figure 3 – Regulation Search Results   Figure 4 – Provision Search Results

Williams and Holtz also developed "a standard document server" using the distributed hypertext facilities supported by the World Wide Web (Williams 93). However, their server only supports classifier–based searching, and does not integrate the more recent text searching facilities that have become available.

5 Future Plans

We plan to continue to develop the broker prototype by: (1) incorporating Netbill to provide secure Internet commerce among regulatory agencies and regulation users; (2) providing support tools for regulatory agencies to author their regulations and provisions into a hypertext/computable form; and (3) testing the broker on selected user groups to gain feedback concerning desirable functionalities of the system. Additionally, we intend to explore the feasibility of implementing an actual broker on the World Wide Web. We are investigating how this broker concept can become a useful and widely accepted means of locating, accessing, and using applicable environmental regulations and provisions. For example, we aim to present the current broker prototype to regulatory agencies and discuss with them the possibility of distributing their regulations over such a system.

6 Conclusions

Systems that distribute environmental regulations in electronic form are becoming more and more widespread. A major reason for the increasing popularity of existing CD–ROM and On–line services is that they attempt to provide easy access to applicable sections of these regulations through hypertext and text–searching capabilities. While this new assistance to regulation users is a step in the right direction, it still lacks the ability to locate and provide access to all provisions relevant to a user's problem. Accessing applicable provisions of environmental regulations and determining compliance of a design with these provisions are two major capabilities that are not available in present systems.

We propose a broker to deliver existing hypertext and text–searching capabilities while incorporating additional capabilities such as supporting automated billing and conformance evaluation. The broker resides on a wide–area network between regulatory agencies and regulation users. It serves applicable regulations and provisions to regulation users and, in turn, transfers revenue from the regulation users to the regulatory agencies. Several enabling technologies have recently been developed which will provide the capabilities to implement such a broker. These technologies include the World Wide Web, various text searching facilities, network billing facilities, regulation representation and evaluation facilities, and machine learning facilities.

To demonstrate our broker concept, we have begun to develop a prototype broker using the World Wide Web as the implementation environment to provide hypertext and world–wide access, and using Inquery to provide full–text searching, classifier–based searching, or a combination of the two.

7 Acknowledgments

This work has been supported through the Environmental Research Program Grant, awarded by the IBM Corporation, and by the National Science Foundation (Grant No. DDM–8957493). We gratefully acknowledge a grant from the Bureau of National Affairs of their CD–based collection of federal and state environmental regulations for use in conducting this experiment. We also gratefully acknowledge the help that we have received from the developers of Inquery in integrating Inquery into the World Wide Web environment.

8 References

Callan, J. P., Croft, W. B., and Harding, S. M. (1992). "The INQUERY Retrieval System." *Proc. of the Third National Conference on Database and Expert Systems Applications*, Valencia, Spain, 78–83.

Cox, B. T. H. (1994). "Maintaining Privacy in Electronic Transactions." Master's Thesis, Department of Information Networking, Carnegie Mellon University, August.

Garrett, Jr., J. H., and Kiliccote, H. (1995). "Computer–Aided Support for Modeling, Accessing, and Evaluating Design Standards." *Proc. of the 1995 NSF Design and Manufacturing Grantees Conference*, LaJolla, CA, January 5–7, 29–30.

Lang K. (1994). "NewsWeeder: An Adaptive Multi–User Text Filter." Research Summary, Department of Computer Science, Carnegie Mellon University, August 31.

Sirbu, M., and Tygar, J. D. (1995). "Netbill: An Internet Commerce System Optimized for Network Delivered Services." *to be published in the Proc. of IEEE Computer Communications Conference*, March.

Smith, R. J., and Gibbs, M. (1994). "Navigating the Internet." *Sams Publishing*, Indianapolis, IN.

Vanier, D. (1993). "Minicode Generator: A Methodology to Extract Generic Building Codes." *Proc. of CAAD Futures 1993*. Flemming and Van Wyk, eds. North–Holland Publishers.

Vetter, R. J., Spell, C., and Ward, C. (1994). "Mosaic and the World Wide Web." *Computing Practices*, October, 49–57.

Williams, F. (1993). "A Standard Document Server." Master's Thesis, Department of Civil and Environmental Engineering, Carleton University.

Integrating CADD and KBS for Structural Steel

Gregory P. Pasley[1], Student Member, ASCE
W. M. Kim Roddis[2], Member, ASCE

Abstract

SteelTeam is a computer-based tool under development that integrates knowledge-based systems with a computer-aided design and drafting environment. SteelTeam acts as a communication tool by providing a medium for the electronic transfer of data in the steel building industry. In addition, SteelTeam acts as an intelligent assistant aiding the parties involved with the design, detailing, fabrication, and erection of steel buildings. SteelTeam provides expert feedback representing concerns of the parties involved in all phases of the steel building design cycle. This feedback is in the form of knowledge-based systems that help the parties involved avoid downstream problems by addressing various concerns at an early stage in the design.

Introduction

Communication is an important part of engineering a quality product. It is difficult to balance the constraints and concerns of the different parties involved in engineering design and manufacturing without quality communication. Because of the high interdependency among the activities of design, detailing, fabrication, and erection of steel buildings, the ability to communicate and coordinate is crucial to the economy and success of

---

[1]Graduate research assistant, University of Kansas, Dept. of Civil Engineering, Lawrence, KS 66045. E-mail: gpasley@kuhub.cc.ukans.edu

[2]Associate Professor, University of Kansas, Dept. of Civil Engineering, Lawrence, KS 66045. E-mail: kroddis@kuhub.cc.ukans.edu

the overall product. Most successful designs require the investigation of several design alternatives. The design teams must balance the constraints, concerns, and desires of the other team members to obtain the optimum design. This is not possible without frequent and quality communication between the team members. Problems often arise when different members in the design process are not aware of what suits the other members as the best solution to the problem under consideration (Ricker 1992; Thornton 1992). Unfortunately, there is normally little interaction between the parties involved in the various phases of the steel building design. The preliminary design, and often the entire design, is complete before the fabricator or erector are brought under contract. This situation can result in inadequate incorporation of fabrication and erection concerns during the design phase.

The computer-based tool currently under development, SteelTeam, integrates knowledge-based systems (KBS) with a computer-aided design and drafting (CADD) environment to aid and simulate the communication process in the steel building industry. SteelTeam functions as a communication tool by providing an electronic means to transfer data between the different parties involved in the design, detailing, fabrication, and erection of steel buildings. The electronic transfer of data between the designer and the detailer results in increases in productivity, speed, and accuracy ("Electronic" 1991). Merely providing the drawing information in an electronic format results in time savings and a reduction in human error, leading to reduced project costs. By providing an electronic link between the designer and fabricator, the relationship between these two parties becomes more of a partnership.

SteelTeam also acts as an intelligent assistant to aid in the design decision making process. The knowledge bases in SteelTeam address design, fabrication, constructibility, section availability, and completeness concerns. SteelTeam uses these knowledge bases to represent the expert advice of members of the design teams who may not normally be present during the early stages of design, or whose expertise may otherwise be unavailable. This provision of knowledge at an early phase in the design cycle simulates the upstream communication between the design team members which helps avoid downstream problems in the design process.

There are several successful research applications that integrate CADD and KBS. An early example is the LSC Advisor (Dym et al. 1988) with a KBS operating on top of a geometric database contained in a CADD system. Examples of systems combining graphics and expert systems include BERT and Evaluator, both using rule-based shells in combination with CADD packages (Jain et al. 1988). CIFECAD is a high-level user interface tool developed in AutoCAD that is used for data capture during building design (Ito et al. 1990). Systems using an object-oriented approach include PMAPM (Ito et al. 1990), using an object-oriented information model to facilitate information sharing in the design, construction, and management of a facility, as well as IntelCAD (Calvert et al. 1991), which integrates an object-oriented inference engine with AutoCAD (AutoCAD 1992). Finally, the Interdisciplinary Communication Medium, ICM, uses KBS and graphical modelling techniques to aid communication in collaborative conceptual design. SteelTeam combines KBS, CADD, an object-oriented product model, and collaborative problem solving to create a tool that facilitates communication in the steel building industry.

Platform Description

SteelTeam is developed to run inside of AutoCAD (AutoCAD 1992) on a PC-compatible computer. AutoCAD is chosen because of its widespread use and modest platform requirements. AutoCAD is also chosen because of the existence of several built-in programming tools. One of these tools is the availability of the programming language AutoLISP. AutoLISP is a subset of the programming language LISP. LISP has been used with great success in artificial intelligence programming. AutoLISP is being used to develop the KBS in the SteelTeam system.

AutoCAD also allows the easy construction of a graphical user interface through the use of programmed dialogue boxes. AutoCAD supports the use of an object-based data model. This allows data to be attached to graphic entities that are viewed on the AutoCAD drawing. There are attribute value slots which are filled each time an instance of an object is created (Abelson and Sussman 1985). This information can then be accessed by both the users of the system and the KBS present within SteelTeam.

System Architecture

Figure 1 shows the architecture of the SteelTeam system showing the two primary parts: 1) the Design Object Model Construction Module and 2) the knowledge bases and shell. The first portion focuses on the development of the object-based data structure that allows for the definition and storage of the design documents and any information that is stored with them. The second portion involves the definition and use of the knowledge bases.

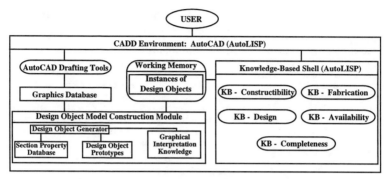

Figure 1. SteelTeam System Architecture

The user accesses the SteelTeam system from within the AutoCAD drafting environment, having access to all the tools normally available within AutoCAD in addition to special commands that aid in the construction of the SteelTeam design document. To enter drawing information into SteelTeam, the user issues a command from a pull-down menu or at the command line which activates the portion of the program labeled the Design Object Model Construction Module. This module prompts the user for information regarding the member that is being created and converts this information into instances of the design object. The instances of the design objects are stored as extended entity data (AutoCAD 1992) of the graphical entity used to represent the member in the drawing document. This information is stored in the working memory of the system. The representation of the building elements follows the object-oriented convention of encapsulating data structures along with the procedures needed to manipulate the data within objects. The information regarding the section size of the structural member is linked to a data base (AISC 1990) that contains the section properties for rolled steel sections. This information is also attached to the member data for later use. Once the information is in the working memory of the system, it can be accessed by the KBS inside of SteelTeam or by the user through a graphical user interface.

The user can query the knowledge-based shell once a portion of a building has been defined in SteelTeam. The knowledge-based shell allows access to the various knowledge-bases that represent design, fabrication, constructibility, section availability, and completeness concerns. The knowledge-based shell is in AutoLISP and uses a backwards chaining inference engine. The rules are written in a pseudo-code language that is read by the inference engine. An example of a rule that determines whether special fabrication of a member is required is shown as an and/or diagram in Figure 2. The KBS access information from the working memory of the system in order to make decisions based on heuristics contained in the knowledge bases. The SteelTeam system takes a passive role notifying design team members of the existence of possible conflicts or concerns along with the information leading to these selections. SteelTeam does not have the ability to alter design documents or implement its suggestions, leaving the responsibility for changes with the designer.

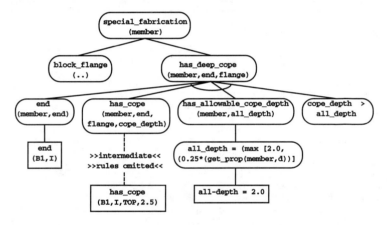

Figure 2. And/Or Diagram of Special Fabrication Rule

System Use

The user begins each drawing by defining data objects representing structural members and their connections. For example, the "Create Beam" command is selected from a pull-down menu or entered at the command line within AutoCAD. The user then provides information regarding the beam entity through a series of dialogue boxes. The information is stored with the graphic entity representing the member on the drawing. The object-based data structure with instance information given for a typical beam is shown in Table 1. Note that information regarding end reactions can either be provided by the

user, or a method can be invoked to determine the end
reactions. There is similar flexibility when entering
other attribute values for the document.

Object	Attribute	Sample Instance
Beam		
	External Name	"B1"
	Shape Section	"W14x99"
	Elevation	0.0
	End Connection Types	Shear
	End Supporting Reactions	BY METHOD
	-- etc. --	

Table 1. Sample Beam Object Data Structure

Once the drawing is entered, information regarding
the primary structural members and connections is stored
with the SteelTeam drawing document. The end document
not only contains the geometric information found on
conventional design documents, but also additional information regarding reactions, connectivity, and other
user-defined attributes for each of the members. The
SteelTeam design document must contain all of the information that is communicated between the various parties
in the design, detailing, fabrication, and erection of
steel buildings.

The presence of knowledge in the form of KBS allows
SteelTeam to act as a decision support tool. SteelTeam
helps the engineer understand the effects that early
decisions have on the final design characteristics, such
as constructibility and ease of fabrication. The user
queries the knowledge-based shell through a series of
dialogue boxes by identifying which issues the user would
like investigated. For example, a user investigating
constructibility concerns would receive advice that
column splices should be located within five feet of the
floor elevation to eliminate falsework. SteelTeam uses
its knowledge bases to simulate upstream communication.
This provision of knowledge at an early stage in the
design allows the designer to make changes that result in
a reduced project cost.

The user also has the option of exploring all of the

concerns present in the various KBS. SteelTeam notifies
the user when a potential problem is found on the design
document.	SteelTeam identifies the member in question,
the cause for the concern, and also allows the user to
query the system for the rules that caused this observation to be made. This allows the users of the system to
check the decisions of SteelTeam, and allows users with
less experience to gain insight into the reasoning process.

The final SteelTeam system is used by the designer
to develop the design documents. The designer then uses
the system to check the design under the scrutiny of the
other parties that may not be present in the current
project organization, but whose knowledge is captured in
the KBS inside of SteelTeam. The document is then passed
along to the detailer or fabricator who can access information about each member in addition to the geometric
description of the members. If changes must be made, the
detailer or fabricator can check the changes against the
concerns of the other parties in the design using
SteelTeam. If there are specific questions, this information can be attached to the document for transmittal to
the designer. Once the final design reaches the constructor, the information contained on the design document not only shows the structural layout, but also
contains additional information about the members that
allows the final design to follow the intentions of the
original design.

Preliminary Results

Several benefits can already be seen from the
SteelTeam system. The data exchange mechanism allows the
transmittal of a more complete set of information. this
increased access to better information allows more informed decisions to be made by all parties involved in
the steel building design. SteelTeam performs a valuable
completeness check on the information that is being
transmitted between parties. The expert knowledge that
is available to all parties in the design process allows
engineers to look ahead in the design process to determine the effects that early decisions have on the final
project. SteelTeam provides a two-way communication
network for the member of the building design team,
allowing early and continued exchange of information
between the designer, detailer, fabricator, and erector.

Acknowledgements

This research is supported by the National Science
Foundation under NSF Grant MSS-9221977. Additional

support is provided by the Center for Excellence in Computer-Aided Systems Engineering at the University of Kansas.

References

Abelson, H., and Sussman, G. J. (1985). "Structure and Interpretation of Computer Programs." The MIT Press, Cambridge, MA.

AISC Database; version 1.08. (1990). American Institute of Steel Construction, Inc., Chicago, IL.

AutoCAD User's Guide; release 12. (1992). Autodesk, Inc., Sausalito, CA.

Calvert, T., Dickinson, J., Dill, J., Havens, W., Jones, J., and Bartram, L. (1991). "An Intelligent Basis for Design." Proc., Pacific Rim Conf. on Communications, Computers, and Signal Processing, IEEE, New York, NY, 371-375.

Dym, C. L., Henchey, R. P., Delis, E. A., and Gonick, S. (1988). "A Knowledge-Based System for Automated Architectural Code Checking." Computer-Aided Design, 20(3), 137-145.

"Electronic Transfer Brings Together Engineer and Fabricator." (1991). Modern Steel Design, 31(1), 53-57.

Fruchter, R., Clayton, M. J., Krawinkler, H., Kunz, J., and Teicholz, P. (1993). "Interdisciplinary Communication of Design Critique in the Conceptual Design Stage." Proc., 5th Int. Conf., Computing in Civil and Bldg. Engrg., ASCE, New York, NY, 377-384.

Ito, K., Law, K. H., and Levitt, R. E. (1990). "PMAPM: An Object Oriented Project Model for A/E/C Process with Multiple Views." Tech. Rep. No. 34, CIFE, Stanford Univ., Stanford, CA.

Jain, D., and Maher, M. L. (1988). "Combining Expert Systems and Cad Techniques." Microcomputers in Civil Engrg., 3(4), 321-331.

Ricker, D. T. (1992). "Value Engineering and Steel Economy." Modern Steel Construction, 32(2), 22-26.

Thornton, W. A. (1992). "Designing for cost efficient fabrication." Modern Steel Construction, 32(2), 12-20.

Integrating Structural Design With Analysis Results

by

Margaret T. Neggers, Member, ASCE [1]

Abstract

Two major phases of the structural engineering process are analysis and design. Since each of these tasks involves repetitive procedural processing, they have been automated by engineers and developers. However, analysis and design are integrally tied and benefit from an iterative process. Linking data between these two tasks is imperative to the process. Once analysis results can be integrated with design procedures, and vice versa, engineers will gain efficiency in the automated structural engineering process.

Problem Statement

Currently, integrating analysis and design can be accomplished by either purchasing existing structural modeling programs that have integrated design and analysis, or investing in in-house development to bridge between an analysis program and a chosen design program.

The analysis process is fairly well defined. Any generic finite element analysis program may suit the analysis needs of the engineering office. On the other hand, the design process is fairly volatile. There are major influences on the design needs of a particular office. Code requirements vary by material and by vicinity. On top of that, code requirements are constantly evolving based on current research findings. Since some code requirements are open to interpretation, each engineer within an office may have distinct views on how a design should be performed.

---

[1] Senior Associate, The Premisys Corporation
P.O. Box 10042, Chicago, IL, 60610; Phone (312) 828-0034

The problem from the commercial development point of view is that the process of creating input and processing output for design programs is generally embedded in the analysis program. Any modifications to the design program require a re-compilation and re-release of the program. By removing the hard-coded design processing from the analysis program, design programs may be maintained separately. The software developer only needs to modify and re-release the design program files. The structural engineer may also integrate their own design programs with the structural modeling program.

The implications of the following method of integrating design with analysis results is that third party developers, super-users within a design firm, or the authors of a design program may provide integration capabilities directly with analysis results. There is no need to wait for the next release of the program, or to request an enhancement to the program to take advantage of the pre and post processing design capabilities for a new design program.

An Implementation

An approach to integrating design programs with analysis results is presented using IBM's Architecture & Engineering Series (A&ES) Structural Modeling System and Workbench. The Structural Modeling Program is a generic pre- and post-analysis processor. Design pre- and post- processing is accomplished using the Workbench program. Preprocessing tasks include overriding default design input parameters and mapping analysis load cases to design load cases. Postprocessing tasks include updating section sizes for analysis, grouping members by governing members, calculating material quantities, listing members, generating schedules, and graphically depicting member sizes on the drawings.

Any design program that integrates with the Workbench will benefit from all of its pre- and post-processing capabilities. The integration is possible through a set of design format files, which define the design program input and output parameters. A utility, called the Design Program Installation program is provided to install design programs into the Workbench. Integration is accomplished outside the Structural Modeling Program and the Workbench. Since all the design format files are ASCII based, the integrator can use any text editor to create the files and integrate their design programs with the analysis results available in the Workbench.

The steps to integrating a design program with the Workbench are:

❑ Evaluate the input and output parameters for the design program.
❑ Create an option table format file.
❑ Use the form generator program to create data forms for the option table.

❏ Create the input and output format files.
❏ Create the list and design record format files.
❏ Use the provided program to compile the previous files into a format readable by the Workbench.

Evaluate the input and output parameters for the design program: The first step to integrating a design program within the Workbench is to outline all input and output parameters surrounding the design program. Input parameters from the Workbench to the design program come from the geometric model of the structure, the loads placed on the members, and the analysis results. Output parameters from the design program to the Workbench include designed sizes and efficiencies of the design. This is simply a planning stage and will facilitate the installation steps.

Create an option table format file: The option table is an organizational tool used by the Workbench to outline design program input. The Workbench provides capabilities to manipulate the option tables. Option tables may be added, deleted, changed, and listed. Option tables are assigned to members, which store the option index number as an attribute.

Below is a listing of a sample model option and girder tables. Each option table added to the model represents one of the installed design programs. Since girders 1 and 2 are assigned to option table 1, they will be designed as a concrete beam, using the criteria set in option table 1. Girder 3 will be designed with the criteria set in option table 2, as a concrete beam. Since option 3 is a stlbeam option, girders 4 and 5 will be designed as stlbeams.

Option Index Number	Design Program
1	conbeam
2	conbeam
3	stlbeam

Girder Index Number	Option Assignment
1	1
2	1
3	2
4	3
5	3

The option table criterion is categorized into Geometry, Properties, Loads, Analysis Results, and Design Parameters. The geometry definition includes member lengths, support lengths, and section configurations. The property definition includes information about the material of the member. The loads and analysis results include information about loads along the length of the member and end forces information from the analysis program.

Design parameters include special options that represent optional capabilities offered by the design program. For example, the steel beam design program can either design for the optimal weight of a section or check an existing size.

PROGRAM	conbeam
DESCRIPTION	"Concrete Beam Design"
ELEMENT TYPE	girder
LOAD CASES	Dead Live Windx Windy Temp1 Temp2 Quakex Quakey
UNITS	inch pound Rad Sec Fah

The first few lines of the option file contain general information about the design program, including the program name and its description. This information is used to register the design program with the Workbench. The element type indicates what types of members may be designed. Load cases outline design load cases that the design program handles. The units line indicates what type of units the input and output file will have.

Variable	Type	Length	Unit Type	#	Default Value	After Action
conc_flag	text	10	none	1	property	property: 2 3 4 specified: -2 -3 -4
fprimec	real	0	stress	2	4000.0	none
wc	real	0	unit_wgt	3	145.0	none
fct	real	0	stress	4	0.0	none
stl_flag	text	10	none	5	property	property: 6 7 specified: -6 -7
fy	real	0	stress	6	60000.0	none
Es	real	0	stress	7	29000000.0	none

The Option Format for Material

The option format file defines the fields in the option table. Flags, such as conc_flag are used to indicate the source of the concrete material properties. Flags such as these allow the users to override the values obtained from the property definitions in the structural model. The unit type indicates how to handle unit conversion for the field when creating the input file or reporting its value. The after action is used to manipulate the form input. In the above listing, if the value in the conc_flag field is "property", then fields 2, 3, and 4 are disabled. If its value is "specified", fields 2, 3, and 4 will be enabled.

Use the form generator program to create data forms for the option table: Data forms are used within the Workbench program to report and alter the design input parameters.

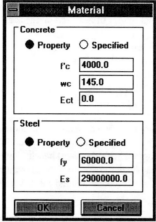

The Option Form

There is a one-to-one relationship between the option table and the data fields in the option form. The field number field in the option table represents the input field number in the option form. The user will see this form when they manipulate the material fields in option tables that have the type "conbeam".

Create the input and output format files: The input format file outlines the format of the design program input file. The Workbench will use this format file as a template when it creates an input file for a member.

```
\*header *project *user *charge *firm *office
\*pstart beam *memnumber *baynumber *flnumber
\*control aci
\*trace 3
\*beam *beam_type *nomwidth *nomdepth *t_slab *clear_webs *isolated_bf
*isolated_tf
*bforces
\*interld
*bloads
\*span *length *left_end *right_end *l_sup_wid *r_sup_wid
\*exposure *exposure
\*torsion *torsion
\*deflection *deflect *sf1 *sf2 *xi
\*material *fprimec *fy *wc *Es *fct
\*cover *cover_t *cover_b *cover_s
\*barconfig *stirrup *bar_low_c *bar_high_c *bar_low_t *bar_high_t
\*barover *As_max *eff_d *dprime
\*doubleopt *dub_des_opt *Aprimes *alpha *As
\*pend
\*end
```
The Input Format File

All words that are preceded by an '*' are keywords. Each of these keywords will be replaced when the input file is created. As an example, the keyword "*project" will retrieve the project name from the structural modeling environment. The conbeam design program shown in the example also works with a keyword based input file. The \ indicates to ignore the following character and represent it exactly as is in the input file. For member number 45, bay number 5, floor number 3, the line from the input format file,

*pstart beam *memnumber *baynumber *flnumber

will become

*pstart beam 45 5 3

in the input file.

Keywords are provided for geometric, property, load, and analysis information. Design parameters that are not normally stored with the structural model are stored in the option table for the design program. When the Workbench finds a keyword that it does not recognize, it looks in the option table provided by the design program.

The output file format outlines member design information from the design program output file. Each design program must create a member design summary either at the end of its output or in a separate summary file. The Workbench reads the output file until it finds the word PROGRAM followed by the >> and the program name.

\# Concrete beam program output format definition
\#

PROGRAM	>> conbeam >>
DESIGNATION	>> *designation >>
T/B STEEL BARS	>> *numtop *topsize *numbot *botsize >>
STIRRUPS	>> *numstir *stirsize *stirspace >>
TORSION	>> *al >>
EFFICIENCY	>> *eff >>
GOVERNING	>> *asreq *combo >>

The Output Format File

The keywords, again indicated by asterisks, represent internal variables that will store information about the member's design. In the above example, the word that appears in the location indicated by the keyword "*designation" will be placed in the design record variable "designation".

Create the design and list record format files: The design record format file describes the data structure to represent the designed member. The design record contains all variables found in the output file format. This information is used to perform post-processing activities, such as listing, drawing, and updating sizes of members.

Variable Name	Type	Length	Unit Type
designation	text	8	none
numtop	integer	0	none
topsize	integer	0	none
numbot	integer	0	none
botsize	integer	0	none
numstir	integer	0	none
stirsize	integer	0	none
stirspace	real	0	length
al	real	0	area
eff	real	0	none
asreq	real	0	area
combo	integer	0	none

The Design Record

The list record format file is a template for member listings. The fields below indicate where to place and how to convert the variable contents.

Variable Name	Type	Unit Type	Row,Field	Format
drindex	integer	none	1,1	%4d
memNumber	integer	none	1,5	%4d
designation	text	none	1,10	%-8.8s
numtop	integer	none	1,20	%2d
topsize	integer	none	1,23	#%2d's
numbot	integer	none	1,30	%2d
botsize	integer	none	1,33	#%2d's
numstir	integer	none	1,40	%2d
stirsize	integer	none	1,43	#%2d's
stirspace	real	length	1,49	@%5.2f
al	real	area	1,56	%6.2f
eff	real	none	1,64	%5.2f
combo	integer	none	1,70	%-9.9s

The List Format

In the list format shown, the contents of the botsize variable (size of bottom rebar) will be placed in row 1 starting at field 33 as '# 5's'.

Use the Design Program Installation Program to compile all the files into a binary format: The Design Program Installation Program "compiles" the ASCII files into a binary file, checking for incompatibilities or errors. The Workbench uses this binary file to perform all pre- and post- processing activities on the members.

Summary

The Achilles' heel in the structural engineering automation process has always been in the area of design. The problem affects both practicing structural engineers and engineering software developers. The structural engineer wants to integrate new design programs with their analysis results. The software developer tries to keep their design programs up to date with code provisions from several jurisdictions and interpretations. The generic and flexible approach to integrating analysis results with design programs provides a mechanism to integrate a design program quickly, and serves both parties well.

Computer-Aided Structural Modeling

Chris A. Merrill[1], M. ASCE and Michael E. Pace[2], M. ASCE

Abstract

The Computer-Aided Structural Modeling (CASM) computer program is designed to aid the structural engineer in the preliminary design and evaluation of structural building systems through the use of 3-dimensional (3-D) interactive graphics. CASM allows the structural engineer to quickly evaluate various framing alternatives in order to make more informed decisions in the initial structural evaluation process. CASM's power and ease of use lies in its 3-D representation of a building which incorporates the generation of loads, structural layout, and preliminary design of the structural framing members. CASM uses U.S. Army Corps of Engineers (Corps) design criteria and is limited to low rise buildings. CASM requires Microsoft Windows 3.1 or later for modeling, structural layout, and load generation, and Microsoft Excel 3.0 or later for the preliminary design of structural members.

Background

The need for a structural modeling system to rapidly model a building, generate loads, layout and size structural members, and report the quantities in a form easily used by estimators has long been recognized as a critical need within the Corps. This modeling need is magnified due to the design level review process used within the Corps. A critical juncture in this review process is the

---

[1]Civil Engineer, Information Technology Laboratory, U.S. Army Engineer Waterways Experiment Station, 3909 Halls Ferry Rd, Vicksburg, MS 39180-6199

[2]Civil Engineer, Information Technology Laboratory, U.S. Army Engineer Waterways Experiment Station, 3909 Halls Ferry Rd, Vicksburg, MS 39180-6199

preliminary or 35% concept design level. At this point, three viable building alternative designs are required. This is intended to insure the selection of the most economical structure that satisfies all design constraints. Designing three building alternatives under time and budget limitations to a level where meaningful costs can be determined and compared is difficult to accomplish. Therefore, in 1987 the development of a simple, computerized structural engineering modeling system that engineers could use to try out multiple design ideas without doing extensive, time consuming calculations was undertaken. The development effort was completed in 1994 with the CASM, version 5.00 software release. Funds for the development of CASM were supplied by the Directorate of Military Programs through the Research, Development, Test, and Evaluation program at the Waterways Experiment Station. The work was administered by the Information Technology Laboratory and the Building Systems Task Group of the Computer-Aided Structural Engineering (CASE) Project provided guidance during the development. Development of CASM was performed by Wickersheimer Engineers, Inc.

Overview of Capabilities

CASM is designed to allow the structural engineer to quickly evaluate various framing alternatives in order to make more informed decisions in the initial structural evaluation process. CASM provides a 3-D representation of the building volume; automatic lateral and gravity load generation for the 3-D model; system framing and structural design capabilities; and the capability to export data to commercial structural analysis and design programs. These capabilities are elaborated upon in the following discussion. More detailed coverage of the capabilities of CASM can be found in Wickersheimer, et al. (1994a,b).

Basic Geometry Modeling

CASM has powerful 3-D modeling capabilities geared towards low rise building systems. CASM models the geometry of a building using 3-D shapes such as cubes, cylinders, prisms, spheres, and domes. The outer shell of the building is modeled using these basic shapes. Complex building geometries can be modeled by manipulating the basic volumes using commands like stacking, joining, duplicating, and intersecting. The resulting 3-D model provides dimensional data such as roof slopes, eave heights, ridge lengths, and bay sizes for the analysis and design functions of the program.

Previously existing DXF files (e.g. from Autocad) can be used to aid in the preparation of the model. CASM has the ability to read in DXF files and use them as reference files. The reference file is used as a template in combination with the basic building volumes to construct the model. The user may snap to points on the reference file to extrude or alter the basic building volumes. By snapping to points on the reference file, it is possible to build the model without ever having to enter a dimension.

CASM's modeling process does not use a standard right hand xy coordinate system. CASM models the building exterior and interior using dimensions that the structural engineer would have readily available. That is, dimensions such as north-south and east-west dimensions, roof slopes, wall heights, member spacings, and member offsets. Menus are provided during the various modeling operations that display the dimensions of the shape or member under consideration. The menus are dynamically updated so that changes in the model are reflected in the menu and vice versa. In Figure 1 a building is shown in the initial stages of the modeling process where the shell or exterior of the building is being formed.

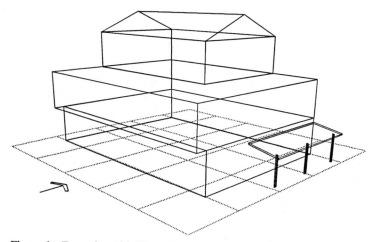

Figure 1. Formation of 3-D model.

Once the shell of the building is formed, the interior can be filled in with structural elements. The interior can be divided into bays along user defined grid lines. The bays can consist of structural elements such as decking, columns, beams, girders, joists, trusses, and walls.

Basic Design Criteria

The user can enter information directly into or retrieve information from a user definable design criteria database. The database can be tailored for particular areas and buildings so that required design criteria will be readily available. The design criteria includes information about the project, regional design information, and site specific design information.

The project information includes items such as the project name, location, type of construction, occupancy, and design code being used to calculate snow, wind, and seismic loadings. The structural provisions contained in CASM concerning load computations follow Corps criteria.

Regional information pertains to information such as ground snow loads, basic wind speeds, temperature data, annual rainfall, seismic zone, and frost depth. Site specific information deals with such items as the function of the specific building, exposure to the climate, whether the roof is heated, classification of the type of soil, and soil bearing pressures.

Load Capabilities

Typically, structural analysis programs require the engineer to input load magnitudes prepared in advance. CASM has powerful 3-D load generation capabilities. All dead loads, occupancy loads, minimum roof live loads, snow load, wind loads, and seismic loads can be computed using the user specified design criteria. Any dimensional data needed for the computation of the loads is pulled from the 3-D building model.

The user can select and construct dead and live loads from several user definable menus of building materials. These loads can then be applied to any desired area of the building volume.

Snow, wind, and seismic loads are automatically calculated in 3-D using the user specified design criteria. Wind loads may also be calculated for components and cladding and open roof structures. Figure 2 shows a section cut of the snow loading for a building.

Structural Layout and Preliminary Design

The structural layout and design capabilities of CASM provide great flexibility to the structural engineer in the creation of framing options within floor, roof, and wall planes. The engineer can easily and rapidly experiment with various framing schemes by modeling structural elements such as beams, girders, joists, girts, columns, walls, shear walls, and trusses. Figure 3 shows a floor framing and dead load diagram.

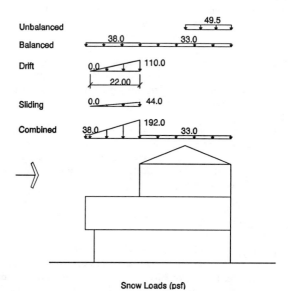

Snow Loads (psf)

Figure 2. Section cut showing generated snow loads.

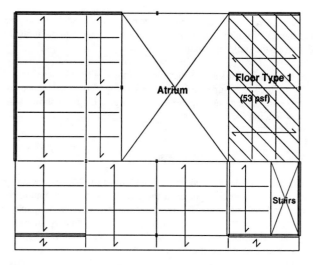

Figure 3. Structural framing and dead load diagram.

Once a particular structural system has been modeled, the user can apply loads to the building geometry from a list of user defined load cases. Loads applied to the model can be distributed to individual members or components based on user selected methods. These include distribution by tributary area, continuous beam models, patterned loading for mufti-span beams, and rigid or flexible diaphragm assumptions for lateral loads.

Structural components such as shear walls with openings, continuous beams, 2-D frames with various bracing options, and trusses can be selected and analyzed. The shears and moments can be computed and used in the subsequent design of the member using Microsoft Excel spreadsheets. At the present time, 17 design spreadsheets are available to design:

a. bar joists.

b. composite beams.

c. concrete beams, columns, decks, pan joists, planks, and isolated footings.

d. steel light gage C channels, beams, columns, form decks, joist girders, roof decks, trusses, girts, and wall studs.

An example of the loading and analysis of a beam is shown in Figure 4 and 5.

Although CASM uses a 3-D model, only elements on 2-D planes are analyzed. For preliminary design, this is an appropriate assumption for low-rise buildings. Depending on the complexity of the building and loadings, the output from CASM could also be considered a final design. If complex building geometry or loadings render the analysis assumptions inadequate for final design, the user can have CASM write an output file for the commercial package STAAD III.

Design Report Documentation

CASM provides an extensive range of output for documenting the preliminary design process. To compare the economic feasibility of several framing alternatives, CASM provides a quantity takeoff of the materials based on building material (steel, concrete) and structural element (beams, joists, walls, etc.). CASM also provides a printout of the design criteria used for the building; lists of materials detailing the composition of floors, roofs, and walls; and load and design calculations. Graphics consisting of section cuts; shear and moment diagrams; framing layouts; and loading diagrams can be printed for inclusion in the design document.

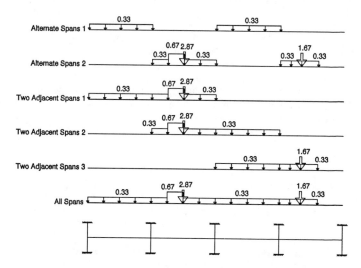

Figure 4. Example of patterned loading for continuous beam.

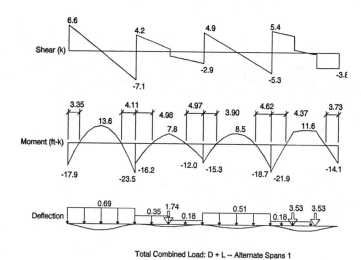

Total Combined Load: D + L -- Alternate Spans 1

Figure 5. Analysis results of continuous member.

Summary

CASM has been designed with the philosophy of creating a system that is intuitive to the structural engineer and allows him the opportunity to perform "What if?" scenarios to foster creative design and maximize economic gain.

CASM has been distributed throughout the Corps of Engineers, other governmental agencies, the private sector, and the academic community with favorable comments. In the past two years, CASM has been used over 100 times by 9 Corps offices on 12 different projects for the Army, Navy, and Air Force. The usage of CASM should continue to grow as more engineers become familiar with the capabilities of CASM.

References

Wickersheimer, D., McDermott, G., Taylor, K., and Roth, C. (1994a). "Tutorial Guide: Computer-Aided Structural Modeling (CASM), Version 5.00," Instructional Report ITL-94-1, U.S. Army Engineer Waterways Experiment Station, Vicksburg, MS.

Wickersheimer, D., McDermott, G., Taylor, K., and Roth, C. (1994b). "User's Guide: Computer-Aided Structural Modeling (CASM), Version 5.00," Instructional Report ITL-94-2, U.S. Army Engineer Waterways Experiment Station, Vicksburg, MS.

CADD IN HIGHWAY DESIGN:
WHAT HAPPENED TO JUST PUSH A BUTTON?

JOHN C. VANCOR, P.E.[1]
ASCE Member

As the use of CADD has become commonplace in the engineering profession, highway design projects have been viewed as ideal CADD applications. The intensive geometric computations, application of numerical standards, and the repetitive nature of the associated drafting and detailing seemed a perfect match with the powerful automated design capabilities of the available software.

Years ago when the CADD revolution came to highway design, claims by vendors, reinforced by users' optimism, led to unrealistic expectations. It was perceived by many, including the author, that we were on the threshold of being able to "just push a button" to design and draft a highway project.

In that scenario, improved aerial survey methods would create an accurate digital ground model of existing conditions. The design software would apply design standards to general design parameters set by the design engineer. These general parameters would include such factors as design speed, maximum degree of curvature, and minimum length of vertical curves.

A three-dimensional digital terrain model would be created for the proposed design incorporating predetermined typical sections. These computer-generated models could be used to develop accurate estimates of quantities of pavement, gravel, earthwork, clearing, and landscaping materials.

[1]Senior Engineer, Fay, Spofford & Thorndike, Inc., 288 South River Road, Bedford, NH 03110

With this tool, recordkeeping would be revolutionized. Check prints and bundles of record plans would be replaced by electronic files. The digital terrain model would become a standard deliverable, part of a three-dimensional database, recording construction projects. The database could be available for use during construction and for future planning and engineering studies.

CADD's revolutionary impact on highway design has been dazzling. State DOTs and private firms have seen dramatic benefits in productivity and quality. Impressive as these advances are, the use of CADD often falls short of the lofty expectations of a few years ago.

This paper does not propose a solution which will achieve those early expectations, but rather is intended to promote discussion. Several related issues will be examined from the perspective of a highway design engineer employed by a consulting engineering firm committed to the use of CADD.

DESIGN STANDARDS

Critical to the level of automation in a design process is the coding of design standards into the software. However, differing design standards among state DOTs present a challenge to software developers. Either users must have flexibility to customize to a particular set of standards, or the amount of design automation must be limited.

Even if design standards are coded into the software, the type of work being performed on many projects negates much of the advantage. Due to environmental and funding constraints, many projects involve overlay and widening existing facilities which were constructed years ago to different design standards. Many projects are located in urban or environmentally-sensitive areas where the need to minimize impacts forces the use of non-standard geometrics, superelevation transitions, and slope tie-ins.

Often non-standard treatments are needed to address environmental issues. Examples include inclusion of noise barriers, special slope conditions at wetlands, and inclusion of Best Management Practices for erosion and sedimentation control.

DESIGN PROCESS

Differences between the optimal computer-aided design process and the actual design process which many projects follow, lead to inefficiencies in the total CADD project.

The optimal computer-aided design process would likely consist of the following steps:

* The existing ground model is generated and checked from aerial and/or electronic ground survey.

* The proposed digital terrain model is developed applying design standards and user set design parameters.

* The resultant design is carefully checked to identify areas where special treatments are necessary due to geometric constraints or the need to minimize impacts.

* The design is modified to incorporate revisions and preliminary drainage design is performed. Public hearing plans are developed.

* Comments from the hearing process are incorporated and final drainage design including design of roadside ditches, performed.

* Two-dimensional drawings of plans, profiles, and cross-sections are developed from the digital terrain model and finished drafting performed. Quantities are taken off from the computer model, and final submissions are made.

It is apparent that this process cannot be reduced to a simple "push a button" process, but rather involves interaction with the design engineers, reviewing authorities, and the public.

In practice, many projects do not follow the relatively straightforward process described above. With input from many sources, and changing conditions, design revisions are incorporated throughout the final design process and often into the construction phase, as well.

After the point in the design process where information is called from the digital terrain model to develop plans, profiles, and cross-sections, any design revision would have to be caught up with both these two-dimensional drawings and the terrain model, as well. This results in an inefficient duplication of work. The alternative is not to update the digital terrain model. This would then preclude the future use of the model as a record database. This points to a conflict between the goal of using CADD to streamline design and gain efficiency; and the desire to maintain an electronic record model.

SUBMISSION STANDARDS

For submission of final drawings in electronic form to be successful, state DOTs must have strict standards in place.

The effort to develop strict submission standards may conflict with the goal of many designers to gain a competitive edge by in-house customization of design procedures.

Standards such as rigid naming conventions may inadvertently limit customization options within the design software, or limit the ability to link drawings with other software products such as spreadsheets or word processing packages.

At present, there are many different software products performing various highway design tasks. These products have differing capabilities, hardware requirements and costs. The use of different software products, often running on different platforms, presents problems related to the submission of drawing files too. If a state DOT tries to develop standards which apply not just for that DOT's preferred system, but other major systems as well, they may be forced to settle for a lowest common denominator set of standards which could ultimately weaken the advantages of the state's preferred design system.

THE HUMAN SIDE

Within the industry, the fear is often expressed that powerful automated design software could empower technicians to replace experienced design engineers. Design could be performed by underqualified personnel who do not fully understand the finished product.

The reality appears to be the opposite. The retention of experienced personnel is critical to the success of a CADD project. The tasks which CADD systems perform best are routine and repetitive. Tasks ill-suited for automation include identifying and solving the design issues which occur on all projects where unusual conditions exist. Thus, the tasks least compatible with automation are the tasks where experienced design engineers and detailers are needed most.

Implementation of CADD has in many ways made the job of managers more difficult. With the design and drafting work being performed by a smaller team of personnel, scheduling has become less flexible. With a limited number of proficient operators and workstations, gone is the ability to make up for a sliding schedule by assigning more and more workers to the plans and cross-sections.

QUANTITY TAKE OFF

On highway projects, not only is the design developed, but a set of bid documents must be prepared and accurate estimates of quantities taken off. CADD has tremendous potential to estimate quantities but its use involves trade-offs in efficiency.

Many systems develop rough cut and fill quantities with extreme efficiency. While these quantities are adequate for preliminary design purposes, a significant effort is necessary to develop the proposed digital terrain model to a level where final design accuracy is obtained. Depending on the typical sections used, accounting for details such as slope rounding and angle points in the bottom of the pavement structural section may be time consuming.

Several systems offer the potential of storing attribute data for use in automated take offs. The use of attribute data involves trade-offs. If attributes are stored to facilitate quantity take offs, savings on the estimate must be weighed against the effort to develop the attribute database.

The number of trees above a certain diameter to be removed during construction could be considered as an example. The diameter could be stored as an attribute of the tree object, and a sorting routine could determine the number of trees within the project area. The effort of creating and maintaining the attribute database should be weighed against the effort to manually take off the quantity.

Sometimes the most direct computer method for quantity take off is not consistent with the standard method. For example, it may be simpler to get a single prismoidal volume between the proposed ground surface and the existing ground surface to calculate earthwork quantities. The standard method is to use average areas on cross-sections.

Although it may seem sensible to review and revise take off methods to make the best use of automated computer technology, it must be remembered that the methods in use are accepted as the best methods for checking and tracking quantities during construction. Considering that the cost of design is a relatively small portion of the total project cost, it is unlikely that changes in methods of measurement and payment will be enacted quickly to facilitate savings in design.

ELECTRONIC RECORDS

An often-stated goal is to require submission of the digital terrain model for use during construction and future planning and engineering studies.

As discussed above, there is a significant extra cost involved in developing the digital terrain model to match the final design.

It is clear that such a model could be an asset during construction. The model could be useful for tracking quantities and developing detouring schemes. The potential for use following construction is not as clear.

In order to reflect as-built conditions, as-built survey would need to be incorporated into the model. As a database of terrain models develops, the interface between models would need to be edited as well. Responsibility for the accuracy of such a database could become an issue. Considerations related to ethics and liability are involved in this responsibility.

Responsibility related to a two-dimensional plan is generally understood; whether the plan is a stamped mylar or a CADD drawing. If a change is made after a plan is stamped, it would be relatively easy to check against a record copy. However, no easy check could be made against a three-dimensional digital terrain model of a design. With each round of editing performed by others, responsibility for accuracy will become harder to determine.

SUMMARY

The impact CADD has had in the field of highway design over the last decade has been incredible. Each month, trade journals highlight projects that demonstrate gains in productivity and quality through the use of CADD.

The author believes the failure to reach the "just bush a button" scenario should not be considered a negative indication, but rather stands to highlight conflicts in the goals of a few years ago; the role of engineering expertise, and intuition; and the important role many government agencies and the public have in the design process.

Application of Photolog Laser Videodisc Technology in Highway Safety Research

Jun Wang,[1] Charles C. Liu,[2] M. ASCE and Jeffrey F. Paniati[3]

Abstract

Roadway safety research requires a variety of roadway inventory data that, coupled with accident and traffic information, can be used to examine the safety effects of different highway designs and to evaluate the effectiveness of various highway safety treatments. Existing databases often lack detailed information about the characteristics of the roadway. Collection of these data is usually difficult since it is a time consuming and expensive process. This paper demonstrates how applications can be developed to collect the necessary roadway data using an existing Photolog Laser Videodisc (PLV) system. The computerized procedure reported on here can be used to develop inventory files which can be integrated with accident databases for the conduct of highway safety research.

The utility of such integration is illustrated through a real world study examining the safety impact of rural multi-lane highway designs. The reported technique should benefit highway safety researchers, as well as transportation engineers, who require supplemental data collection to support their analyses. This effort is the first of a series of FHWA-funded efforts to develop data collection tools for use with the PLV. In today's information age, it is imperative that automated techniques be devised to integrate the various available information sources (e.g., photolog, GIS, CAD drawings, etc.) in an efficient and effective manner, to facilitate a systematic and coherent decision making process.

---

[1] Transportation Engineer, LENDIS Corporation, 6300 Georgetown Pike, McLean, VA 22101

[2] President, LENDIS Corporation, 6840 Melrose Drive, McLean, VA 22101

[3] Program Manager, Federal Highway Administration, 6300 Georgetown Pike, McLean, VA 22101

Introduction

The conduct of high quality roadway safety research requires access to computerized databases that include not only accident information but also information concerning roadway geometrics, traffic volumes, roadside conditions and intersection configurations. One of the major goals of highway safety research is to establish the relationship between roadway accidents and various highway geometric design elements. To this end, a variety of roadway inventory data along with the accident and traffic volumes information at both the accident and non-accident locations are needed. However, information on many important roadway features that are believed to be associated with causation of roadway accidents are not conventionally collected and built into the databases. Safety analysts and highway engineers are often challenged to collect and integrate these data in an efficient and economical manner. The use of Photolog Laser Videodisc (PLV) technology provides a promising way to meet this challenge.

Many transportation agencies maintain photologs (or videologs) of the roadways under their jurisdiction. A photolog is a series of sequential images taken form a moving vehicle of approximately driver's eye level to provide a permanent visual record of the roadway network at a given time. Most States use an automated vehicle to film the State-maintained highway system in both directions in 35 mm color film. A photograph is taken every 0.01 mile (16.09 m) with the camera for optimum coverage of the highway and roadside development. These photolog images are used by engineers to view a roadway of interest without having to physically travel to that location. The utility of photologs, however, has been limited by the sequential nature of the film. Locating an image is a cumbersome and time consuming process. As such, agencies have began to use laser videodiscs to store the photolog images. Videodisc-based photologs allow the images to be randomly accessed in seconds under the control of a microcomputer.

This paper describes the development and application effort of using PLV system to effectively collect supplemental roadway inventory data from videolog images and to incorporate these data into the Highway Safety Information System (HSIS), a multi-state highway safety database, for the conduct of highway safety research. Specifically, the development of a Longitudinal Roadway Data Collection (LRDC) computer program under the PLV system is reported. Its application is also illustrated through a real world study examining the safety impact of rural multi-lane highway designs using data from an HSIS State.

HSIS Database

The HSIS is a multi-state highway safety database developed and maintained by the Federal Highway Administration (FHWA) and Highway Safety Research Center (HSRC) of the University of North Carolina (Council and Paniati, 1990). The system consists of accident, roadway inventory, and traffic volume data for a selected group of

States. Currently, the HSIS contains the data form 1985 through 1993 for Illinois, Maine, Michigan, Minnesota and Utah. Data from California, North Carolina and Washington are being added to the system. The system uses raw data already collected by the States and converts the data to a common computer (Statistical Analysis System) format. The data files in the system are categorized as accident file, vehicle file, occupant file, roadlog file, and traffic file. For some States, supplemental intersection, interchange, and guardrail inventory files are also available. All police reported accidents are included in the accident file, and for each accident, a variety of details are recorded including date and location of accident, road and environmental conditions, accident type, and number and severity of injuries. The roadway inventory file contains information on major geometric or cross-section variables (e.g., number of lanes, lane width, pavement type, shoulder width or type, etc). The traffic file contains the average annual daily traffic volume (AADT), percentage of commercial vehicles, etc. Using a location-based linking system (i.e., route number and milepost), these files can be linked to obtain the number, rate, severity and type of accidents that have occurred on specific highway sections over a given period of time. Safety analysts and highway engineers are using the HSIS to access these data for a variety of highway safety studies, ranging from basic problem identification to modeling efforts that attempt to estimate the impact of design and/or operational decisions.

It should be noted that while the HSIS contains a wealth of information on both accidents and roadways, data on unconventional items (e.g., minor intersections, access points, roadside conditions, etc.) are not included in the existing HSIS roadway files. In the past safety analyses either had to be conducted without access to important roadway data or expensive supplement field data collection had to be undertaken. This paper presents a new alternative: the use of PLV-based data collection programs.

Photolog Laser Videodisc (PLV) System

Currently, the HSIS Laboratory is equipped with PLV systems for two of the HSIS States, Minnesota and Michigan. The Minnesota system was selected for this effort because of the availability of a flexible software development tool called SuperHIWAY. SuperHIWAY is a proprietary software system originally developed for the operation of the Connecticut PLV system (Burns, 1990). It was subsequently adapted to the Minnesota PLV system by the FHWA.

The personal computer (PC) which controls the PLV station operates under the control of the SuperHIWAY computer program. This program can accept user input through either a computer keyboard or a graphics tablet. As such, it is user-friendly and does not require users to have extensive experience using computers. However, the software also offers advanced features and capabilities that experienced programmers can take advantage of, to accomplish very sophisticated tasks of their own design. There are more than 80 different commands that SuperHIWAY Version 1.5 will accept and process. These commands form the SuperHIWAY language. The record of a series of

SuperHIWAY commands that, as a group, perform some large desired function inside the computer is called a "Program File" (or "PF" for short). For example, a large program file called MIGO.PF (Minnesota General Observation Tool) was developed for displaying Minnesota highway images on a PLV station. The LRDC (Longitudinal Roadway Data Collection) program discussed in the following sections is also a large PF that can be called within the MIGO.PF to perform data collection functions.

The photolog images on the laser videodisc can be displayed on a video monitor by using the PLV viewing station. A PLV viewing station consists of a laser video monitor, a laser videodisc player, a GraphOver 9500 video image and graphics processor, a graphics tablet with a "puck", and a personal computer. Figure 1 shows the components of the PLV viewing station.

Development of the LRDC Program

As mentioned earlier, there is a need to develop a data collection function within the PLV system so that the highway safety researchers can easily collect roadway inventory data that do not exist in the HSIS database. Also, the collected data should be directly integrated into the HSIS database for a safety analysis. After carefully examining data collection capability of the PLV system and present HSIS research needs, it was determined that the prototype data collection program should focus on collection of longitudinal roadway information (e.g., roadside objects, roadside hazard rating, intersection/driveway access points, roadway median data, and guardrail information). The objective of this program development effort was to develop and calibrate a computer program allowing users to collect longitudinal roadway data from the PLV images in an interactive manner. By executing the program in the SuperHIWAY software environment, the user will view the videodisc images, select certain roadway sections based on interested data items, and record appropriate information into a separate data file. As such, the developed data collection program was named LRDC program.

During the course of the program development, several principles were established to ensure the program would be truly geared to the needs of safety researchers and feasibly designed with current PLV system ability. These program design principles are as following:

- The program should focus on collection of simple data item information that can be easily identified and distinguished by data collectors from the PLV images (e.g., type of objects, roadside hazard rating, type of intersections).

- The major purpose of the output data file is to merge/link the collected data into HSIS data sets to perform a safety analysis. Therefore, any tasks that can be

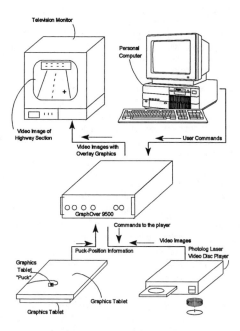

Figure 1. Components of the PLV Station

accomplished at the later data merge and manipulation stage should not be included in the LRDC program (e.g., generate statistics, calculate section length, etc). In addition, the output data file should be in ASCII format so that it can be easily converted into the SAS format and merged with HSIS data sets.

- As with most highway related databases, there are two types of data contained in the HSIS. Much of the roadway data is recorded by segments. That is, data is tied to a homogeneous section (within a homogeneous section all variables remain constant) identified by a route number and beginning and ending milepost. Conversely, the accident data (and some roadway inventory data) are recorded as point data. That is, each record is identified by a single milepost. The two different data types present a special merging/linking problem because it is not always a simple point-to-point or section-to-section match. In order to be compatible with the HSIS database, the LRDC output data file should be either "point" based data (each record represents a point location with a milepost field) or "section" based data (each record represents a roadway segment with a beginning milepost and an ending milepost). For each data record, a roadway ID (i.e., route system and/or route number) and milepost (beginning and ending milepost if section data) must be automatically recorded along with other data fields.

- For some studies, a fixed length section format is required for the data collection. This data collection function is specially designed to collect data at a pre-set fixed distance interval. In this case, the videodisc photolog plays back images for a pre-selected distance interval (e.g., every 0.1 mile or 0.16 km), after which the system pauses and waits for data input. Information such as the roadside hazard rating can be obtained by this type of data collection procedure.

- It is anticipated that the user will apply the program in more than one session for a particular project. Therefore, the researcher or data collector should be able to "configure" the program once, and store the configuration choices in a data file for subsequent uses by the program. The configuration procedure would allow for identifying data items and data values, setting data format (whether the output data file is "point" or "section" data structure), assigning input keys and other characteristics such as name of the output data file, whether a fixed length intervals are imposed, etc.

Figure 2 illustrates the frame work of the LRDC program. It should be noted that the Figure 2 only shows the data collection function that LRDC program drives. The significant part of the data collection has to "navigate" the visual database and get the right picture on the screen in front of the researcher. This is done by the MIGO program which provides overall operation of the Minnesota's visual database. The LRDC is called within MIGO by activating a "sidecar" application (the term "sidecar" as the side-car on a two-person motorcycle, where the driver rides the motorcycle and a passenger rides in the side-car). In this concept, MIGO is the driver and motorcycle that gets the passenger to the location where data can be taken by the "sidecar" program (LRDC). The current version of LRDC employs computer keyboard input but it could be easily adapted to work in a "point and click" mode using the "puck."

Using a simple SAS program, the data files generated by LRDC can be easily transferred into SAS data sets. Because of the compatibility of the data structures between the PLV data file and the HSIS database, the collected PLV data can be readily liked with HSIS roadway file via a common linking system (i.e., route system, route number and milepost).

Application of the LRDC Program

The LRDC program was successfully applied in an HSIS multi-lane cross section design study. This effort attempts to examine the safety effects of various roadway cross

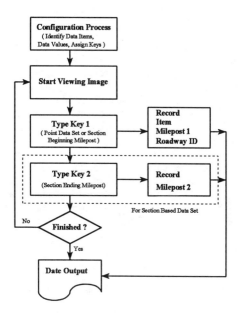

Figure 2. Flow Chart of LRDC Program

sectional designs for rural, non-freeway, multi-lane highways. In this study, two major design factors, roadside condition and intersection/driveway access, were believed to affect safety; but they are not available from the existing HSIS database. This information was collected from PLV images using the LRDC program. For the "roadside condition" variable, a roadside hazard rating scheme developed by Zegeer et al. was used for data collection and analysis. This roadside hazard rating is a subjective measure of the hazard associated with the roadside environment (Zegeer et al., 1987). The rating values indicate the accident damage likely to be sustained by errant vehicles on a scale from one (low likelihood of an off-roadway collision or overturn) to seven (high likelihood of an accident resulting in a fatality or severe injury). Fixed-length (in 0.1 mile or 0.16 km intervals) "section" data on average roadside hazard ratings and "point" data on intersection, driveway, and interchanges were collected from the PLV images. In this sample application of the LRDC program, more than 475 miles (760 km) of rural, multi-lane highways were covered. After the data were collected, two SAS data files (one roadside hazard rating file and one intersection/driveway file) for Minnesota's rural, multi-lane, non-freeway roadways were assembled and they became the permanent HSIS supplemental files.

Summary and Conclusions

This paper presents the development and application of a procedure for collecting supplemental roadway inventory data from an existing PLV system and integrating the data into a multi-state accident database for the conduct of highway safety research. This computerized procedure can efficiently and effectively collect additional roadway inventory data from a PLV system, in a linkable format, to supplement existing files. The reported application development initiates a series of efforts to develop other useful PLV applications. These include more sophisticated PLV measurement tools, as well as advanced image recognition and analysis approaches.

The concept and technique reported in this paper can certainly apply to any of similar systems that do not necessarily using the same software (i.e., SuperHIWAY). The technique described in this paper should benefit highway safety researchers, as well as transportation professionals in general, who typically depend on more than one information source to pursue their analyses.

Acknowledgments

The authors wish to thank Mr. David D. Burns for his advice and assistance in developing the PLV applications described herein.

References

Burns, D. D., *SuperHIWAY: User Reference Manual*, Version 1.5. Needham, MA, June 1990.

Council, F. M. and Paniati, J. F., The Highway Safety Information System, *Public Roads*. Vol. 54, No. 3. Federal Highway Administration, Washington, D. C., December 1990.

Zegeer et al. *Safety Cost-Effectiveness of Incremental Changes in Cross-Section Design -- Informational Guide*, Report No. FHWA-RD-87-094, Federal Highway Administration, Washington, D. C. December 1987.

A LOW TECH SOLUTION
FOR A HIGH TECH IVHS PROBLEM
by Thomas Harknett, P.E.

Abstract

The introduction of electronic toll collection (ETC) on existing toll facilities is an important current application of one element of IVHS. With the initiation of ETC, there is a need for another IVHS element—quick, effective communication to assist the motorist in the toll plaza, which becomes more complicated and confusing as we move into the modern electronic world. However, as will be demonstrated, sometimes a low-tech solution to the communication problem is more effective than a high-tech one. The following paper discusses one such instance where a low tech solution was selected; of course, we did need a high tech approach to prove it!

Background

With the retrofitting of ETC more commonplace nationally, the issue of necessary signage changes generally has been neglected or given short shrift. The New York State Thruway Authority has installed a read-only system at four cash toll barriers in southern and western New York State. Eventually the Authority will install a read/write ETC so that fully automated collection will be possible for all users whether paying simple cash tolls or paying on a mileage basis. At most cash barriers, passenger car tolls of less than $1.00 are charged and these passenger cars have the option to use either the exact change or full service lanes. With the arrival of the IVHS technology electronic toll collection, three payment methods are available requiring a very different approach to signage guiding drivers to the correct lanes.

Prior to ETC, a series of signs provided advance warning of the toll plaza ahead as well as advance information on the amount and method of paying tolls. The signs, depicted in Figure 1, for the NYSTA's two-way Spring Valley toll barrier, informed drivers that exact-change coin machines for the 40-cent toll were located on the left side of the toll plaza. The signs followed highway standards: green with white letters, all according to the Manual of Uniform Traffic Control Devices (MUTCD). Additional signage regarding speed limits and interstate route

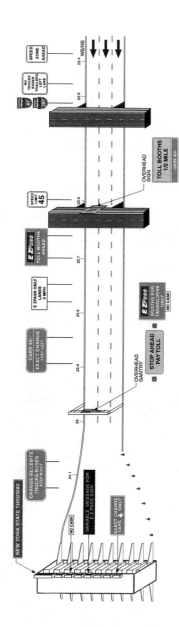

Existing Signage
Spring Valley Toll Plaza WB
figure 1

emblems was also present.

The inclusion of ETC added a third choice for drivers, complicating their decision and requiring revised and new messages. During the ETC installation phase, supplementary signage was added to alert users that electronic payment could be made. Lanes accepting electronic payment were dubbed E-ZPass by a seven-agency regional group representing virtually all of the toll agencies in New York, New Jersey and Pennsylvania. Further constraining the signage possibilities was a pre-agreed purple logo, the color of which conflicted with the MUTCD. Because the NYSTA varies the number of lanes by peak period, direction of travel and general level of usage, a flexible system had to be implemented to switch from one operational setup to another. To alleviate the concerns regarding speeding and those related to lane reversal, the ETC lanes were located together in the middle of the toll plaza. Exact-change lanes remained on the left. It had been expected that once ETC was operational and the driver familiarization period over, drivers would learn the new system and fall into predictable behavior. Many months after installation of E-ZPass, drivers remained confused and occasionally made erratic maneuvers in the toll plaza. Some unsure drivers would enter the center lanes of the plaza, move to the left after they determined they were in the wrong lane, only to find out that exact change was needed and they were still in the wrong lane. The final course correction required a multiple-lane change toward the right, finally arriving at the desired booth.

The Signage Improvement Study

In order to reduce the frequency of undesirable erratic maneuvers, the NYSTA quickly began a study to improve signage at the toll plaza. This involved reviewing existing signage and standards and preparing a recommendation for the new operation. Since it was obvious that the existing signage was not working for the driver, it was decided that it was essential that the study should clearly convey the concept from the driver's perspective. Thus, good illustrations of the proposed signage concept were essential to assess its merits and provide top management with a clear understanding of the system.

Two variations of signage improvements existed. The first consisted of modifying signs on the toll plaza canopy itself with "high-tech" variable message signs on the top plus additional signage in advance of the toll plaza. The second variation involved revamping the entire set of approach signs and rethinking the canopy signage. After a short time, the first and lesser alternatives, were discarded. It was considered ineffective. Efforts then focused on the second alternative.

The General Concept The concept that emerged from the study of second alternative efforts involved the use of color coded signs with logos for each payment option. This led to choosing the color, logo design and sign position for the approach to and at the plaza. Because the composite of these concepts is difficult to convey to top management, several graphic presentations were provided. First, exhibits with the help of a Macintosh computer and a Bubble Jet Printer were prepared juxtaposing different colors so that color contrast could be studied. The colors studied involved all colors in the MUTCD regardless of the stated purpose (these colors therefore represented those that could be ordered from sign vendors). Three distinct colors were selected that gave drivers a good contrast. Sign logos were studied in a similar manner with variations in horizontal and vertical dimensions, as well as shape and character recognition. These ideas were also created on the Macintosh and printed out on the Bubble Jet. A black-and-white version of the final logos selected is shown below in Figure 2. The signs are all the same size, vary in color and have distinctive logos that can be spotted at a distance.

Frequency and Type of Signage Next efforts focused on the frequency and type of advance signage and treatment in the plaza area. The overall concept for the approach signage included each payment type sign three times: twice along the road and once overhead at the entrance to the toll plaza. The approach signs are not required to be variable as they designate the toll plaza area (left, center or right) for each type of toll payment, rather than individual lanes.

Plaza treatment consisted of high visibility pavement markings and canopy signage. Plaza pavement markings were proposed to be enhanced by establishing wide painted areas to channelize traffic and assist in organizing queuing. This involved reviewing existing signage and standards and preparing a recommendation would also assist the driver to make an earlier decision, further reducing erratic maneuvers near the toll booths.

Two alternatives of canopy signage were considered: electronic or drum-type variable message system that rotated fixed signs. The electronic signs could be flipped disc or the newer fiber optic variety but these had limitations regarding the uniformity of the signage "look." A drum-type system was selected instead since it allows use of fixed message signs which could present sign color, layout and logo shapes consistent with advance signage. This similarity is more effective in obtaining a uniform driver response since it does not require interpretations and translation of images that would have resulted had the electronic variable message alternative been selected. The drum-type system, also, was an available and tested technology. A composite of the entire concept was then prepared (see Figure 3).

Video Imaging Presentation

Additional graphic presentations were prepared to better capture how the driver would see the proposed plan. This involved using video imaging to present the entire concept in still and moving video format. Video imaging is a technique that electronically superimposes an image onto a photographic medium. Still versions of the plaza approach at critical locations were developed from photographs that were scanned into the computer; then the new signs were imaged

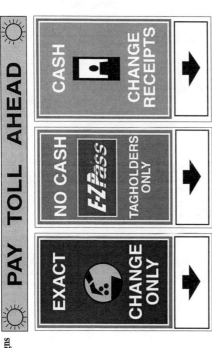

Plaza Approach Signs

Canopy Signs

**Conceptual Design of Accepted Logos
Spring Valley Toll Plaza WB**
figure 2

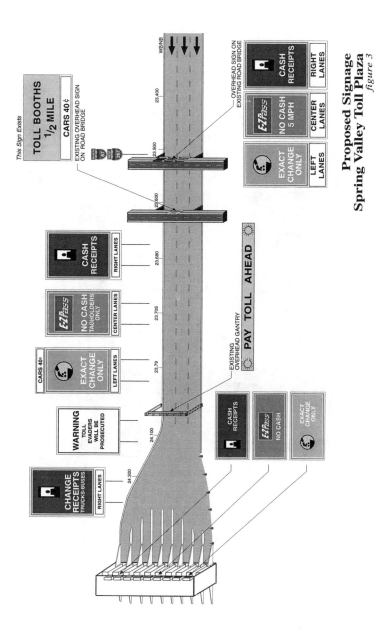

Proposed Signage
Spring Valley Toll Plaza
figure 3

onto the electronic file. The result was a picture, in color, showing the concept under study. Two of these pictures with and without the new signs are shown in Figure 4.

For the video-imaging tape graphics, with the use of two Pentium personal computers, individual frames of a video tape of the plaza's approach were altered to render the new concept. Since the eye changes focus 30 times a second, a smooth motion picture of the signage concept would need to be rendered 1,800 times for each minute of film needed. Thank goodness for the computer age! The signage concept video was then run at different speeds to approximate driving up to and through the toll plaza at 60, 50 and 40 miles per hour. After top management viewed the video tape, they immediately understood the concept and its effectiveness. As a result of the above, management agreed to a course correction that adopted the simple, low-tech IVHS solution.

Applicability

The presentation of a high-tech, video-imaged product has numerous applications, each time allowing decision makers to immediately see what an improvement would look like without building it. This technique could apply to the study of highway signage on freeways and expressways, especially near complex interchanges where sign clutter needs to be minimized. In the NYSTA's case, the high-tech, video-imaging technology led to a comprehensive low tech solution that considered all the variables. Sometimes high-tech is not the solution. In this case, the implementation of a low-tech solution is better for the driver and for the operating agency.

Recommended Signage

Existing Signage

**Spring Valley Toll Plaza
With and Without New Signs**
figure 4

WINDOWS OF OPPORTUNITY FOR EXISTING TOLL PLAZA
ETC (Electronic Toll Collection) RETROFITS

Richard J. Gobeille, P.E.

Abstract

Over the last several years ETC (Electronic Toll Collection) has been heralded as a great solution for reducing congestion and controlling rising toll collection costs. Reality shows that this is not always the case and may, sometimes, increase both the congestion at existing toll plazas and the cost to collect tolls. What causes the increased congestion and costs, and computer models to predict these impacts are the topics of this paper.

Introduction

The author has been directly involved in many aspects of researching and testing ETC since 1989. Beyond studying technology, the institutional, operational and cost issues of implementing ETC for several toll agencies have been reviewed. The results have been surprising. ETC is not a guarantee of reduced congestion or costs for the toll agencies and their patrons.

The number of lanes dedicated to ETC use, the average capacity of a toll collection lane, and the percentage of toll paying vehicles that use ETC impact the potential for ETC to reduce congestion. The required number of ETC lanes should match the hourly number of vehicles paying tolls electronically. This, however, cannot always be achieved. Mismatches between toll lane types and approaching patron payment preferences can introduce throughput constraints not related to capacity. For example, the number of ETC patrons can exceed the capacity of the ETC lanes, causing congestion and delays for ETC patrons. The converse is also true where the number of non-ETC patrons can exceed the capacity of non-dedicated lanes.

Many toll roads currently operate highly automated toll collection systems that do not include ETC. Typically, there would be a mix of manual and ACM

91

(Automatic Coin Machine) equipped lanes. Systems with modest tolls of $0.50 or lower could have 75% of their lanes operating in the unmanned ACM mode. The ACM lanes typically operate with a cost between $0.01 and $0.03 per toll transaction. The ACM transactions are the transactions that ETC would replace. Therefore, to save toll collection costs, the ETC transaction must cost less than the ACM transaction. Obviously, this is a low per transaction cost to achieve.

The same is not true for a totally manned toll plaza where the ETC transaction would compete with the manual transaction. Costs for manual collection range from $0.15 to $0.25 or more per transaction; this cost range is not unreasonable for an ETC transaction to achieve. On a cost per transaction basis, ETC would likely be effective for this type of facility.

Not withstanding the above, ETC does offer great potential for reducing toll plaza congestion and collection costs. But , the potential also exists in several scenarios for the implementation of ETC to have a negative effect on the toll collection process; therefore, a careful study for each ETC scenario should be undertaken to estimate its true value to the toll agency.

Toll Plaza Throughput

When implementing ETC in existing toll plazas it is essential to determine the effect of ETC on plaza throughput. An obvious analysis is that the capacity of a toll plaza will be increased with the addition of dedicated ETC lanes. An increase in capacity, however, does not necessarily result in an increase in throughput. Unlike a token, the concept of dedicated lanes precludes the ETC patron from using any other lane. This introduces a capacity constraint based on the patrons' method of payment. Because of this, the throughput of a toll facility is then dependent on the approaching market distribution of patrons between ETC, token, exact change and cash. At any given time the throughput capacity can be considerably less than the theoretical capacity of the toll plaza.

Throughput vs. Capacity

A three-lane plaza currently operating at capacity during peak hour will be used to exhibit this problem. Assume three automatic coin machine (ACM) lanes can process 900 vehicles per hour (plaza capacity 2,700). One of the lanes is then converted to a dedicated ETC lane at 1200 vehicles per hour (plaza capacity now 3,000). In this analysis the approaching vehicles will be equal to the theoretical capacity of the toll plaza (3,000). On day one, 10% of the approaching vehicles has an ETC tag (300 vehicles). The remaining 90% (2,700) wish to use the automatic lanes. The throughput capacity is then 300 (approaching market) ETC vehicles and 1,800 (capacity of the ETC lanes) vehicles through the automatic lanes. The plaza can process a total of 2,100 vehicles per hour with the defined approaching market distribution. This is a 22% **reduction** in the plaza throughput after a 11% increase in the theoretical plaza capacity. This results in increased congestion.

Conversely, on day 100, 67% of the approaching vehicles have a tag (2,000 vehicles). The throughput capacity in this case is 1,200 ETC vehicles and 1,000 vehicles through automatic lanes. The plaza can process 2,200 vehicles per hour. a 19% **reduction** in the plaza throughput capacity. These example show the importance of matching lane configurations to approaching market penetrations to prevent congestion.

Throughput Model

A model was developed to analyze the impacts of lane operations, market penetrations of ETC, and dedicated ETC toll collection lanes. Several base assumptions central to the model follow. Some of which relate to the original existing conditions at the toll plaza being reviewed. All comparisons are made to the capacity of the toll plaza prior to the introduction of ETC. The future approaching traffic volume will be equal to the new theoretical capacity of the toll plaza.

The model considers four types of existing lanes: first, staffed lanes; second, Automatic Coin Machine (ACM) lanes; third, token only lanes; and fourth, user defined lanes. For each lane type the engineer must develop the vehicle processing rate for each lane type in the plaza being studied. Vehicle processing rates depend on the toll rate being charged and patron characteristics unique to each plaza. For example, a token only lane with no gate can process 900 vehicles per hour (VPH). The addition of a gate on the token only lane reduces this to 500 VPH. Developed rates for each of the lanes are input by the user into the model.

Policy issues enter the model at this point. The operator of the toll facility must establish which lanes will be replaced by the new dedicated ETC lanes. Besides selecting lanes, the operational parameters of the lane must be established to estimate the processing rate for the ETC lanes. Depending on the use of gates and posted speed limits, the processing rate for ETC lanes can range from 600 to 1500 VPH. Once all of the parameters have been established, the model is ready to be run.

Figure 1, on the following page, shows the results of a model run for the three lane toll plaza discussed earlier. It may be seen from the graph that the range of market shares for ETC which result in an increased throughput capacity for the facility is very limited. Only ETC market shares from 30 to 50 percent will increase throughput for this example.

Figure   2 represents the full output for a more typical toll facility. The Sample toll plaza has a total of seven toll lanes. It normally operates with one staffed manual lane (450 VPH), one staffed lane selling tokens (300 VPH), three automatic coin machine lanes (800 VPH each) and two token only lanes (900 VPH each) for a total theoretical capacity of 4,950 vehicles per hour.

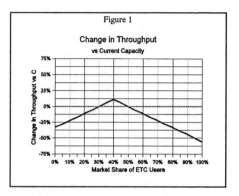

The Sample Toll Authority decided that it would operate its dedicated lanes at 20 MPH and estimated that 1200 (VPH) would be processed though each ETC lane. They also decided to eliminate tokens. The new plaza configurations and lane capacities are listed in Table 1 below for one, two and three dedicated ETC lanes.

Table 1 ASCE Sample Toll Plaza Lane Configurations					
Dedicated ETC Lanes	Staffed Lanes	Automatic Lanes	Token Only Lanes	Staffed Sales Lanes	Total Lanes
0 (0)	1 (450)	3 (2,400)	2 (1,800)	1 (300)	7 (4,950)
1 (1,200)	2 (900)	4 (3,000)	0 (0)	0 (0)	7 (5,300)
2 (2,400)	2 (900)	3 (2,400)	0 (0)	0 (0)	7 (5,700)
3 (3,600)	2 (900)	2 (1,600)	0 (0)	0 (0)	7 (6,100)

The results of the model runs show increased throughput capacity when the approaching market share of ETC ranges from 15 to 30 percent for one dedicated lane, 30 to 55 percent for two dedicated lanes and 40 to 80 percent for three dedicated lanes. From the graph it may be seen that the operation of the toll plaza should change to match the approaching ETC market throughout the day. At commuter peak hours the market share of ETC will likely be high and warrant three dedicated lanes. Saturday beach traffic would have a much lower market share of ETC and best operate with one dedicated lane. This example shows the value of the model in predicting a "window of opportunity" for increasing toll plaza throughput.

FIGURE 2
ETC TOLL PLAZA THROUGHPUT MODEL

TOLL PLAZA PARAMETERS

Plaza Name:
ASCE Sample
Plaza Direction:
Northbound
Prepared By:
R. Gobeille
Prepared Date:
01-Feb-95

	VPH	Lane Type	Existing Plaza Lanes	Number of Dedicated Lanes		
				1	2	3
	450	Manual Lane	1	2	2	2
	800	Automatic Lane	3	4	3	2
	900	Token Only Lane	2	0	0	0
	300	Staffed Sales Lane	1	0	0	0
		Total non-ETC Lanes	7	6	5	4
		Total VPH w/o ETC	4950	4100	3300	2500
	1200	ETC Lanes		1	2	3
		ETC VPH	0	1200	2400	3600
		TOTAL ALL LANES	7	7	7	7
		TOTAL VPH	4950	5300	5700	6100
		Lane Count		ok	ok	ok

TOTAL THROUGHPUT

Number of Dedicated ETC Lanes	ETC MARKET SHARE										
	0%	10%	20%	30%	40%	50%	60%	70%	80%	90%	100%
1	4100	4630	5160	4910	4380	3850	3320	2790	2260	1730	1200
2	3300	3870	4440	5010	5580	5250	4680	4110	3540	2970	2400
3	2500	3110	3720	4330	4940	5550	6040	5430	4820	4210	3600

CHANGE IN THROUGHPUT VS CURRENT CAPACITY

Number of Dedicated ETC Lanes	ETC MARKET SHARE										
	0%	10%	20%	30%	40%	50%	60%	70%	80%	90%	100%
1	-17%	-6%	4%	-1%	-12%	-22%	-33%	-44%	-54%	-65%	-76%
2	-33%	-22%	-10%	1%	13%	6%	-5%	-17%	-28%	-40%	-52%
3	-49%	-37%	-25%	-13%	-0%	12%	22%	10%	-3%	-15%	-27%

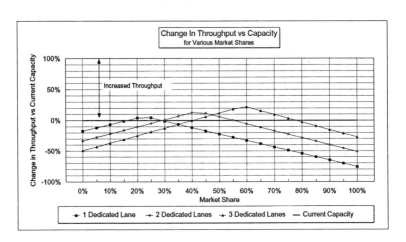

Toll Plaza Operating Costs

To analyze the cost of ETC compared to an existing system, a cost per toll transaction should be developed. In this way, the changes in patron payment patterns will be accurately reflected in the cost for toll collection with ETC. For example, if a toll agency is planning to replace tokens with ETC, then the current costs associated with tokens must be considered when determining the actual costs of the new toll collection with ETC.

Development of Operating Cost

The first step in developing operating costs is to determine the capital investment required to develop the system. Several categories for capital cost are considered in the ETC evaluation. Lane costs include material and labor costs for providing all of the in-lane equipment. Plaza costs include material and labor costs for a computer system to collect ETC and Video Enforcement Systems (VES) data, a communications link to headquarters, and advance signing for the various combinations of toll collection lanes. Systems integration costs cover the cost of consultants to design and implement the ETC and VES systems. Tag purchases are considered a capital cost item and frequently over 100,000 tags must be purchased. The introduction of a new toll payment method requires some up-front marketing that must be done to make the public aware of the ETC system. The capital costs will be depreciated and since ETC is computer based, a five year useful life is used in this analysis.

Four additional categories for operating costs are considered. The system, once installed, will require an ongoing maintenance program. An annual cost of 20% of the capital cost will be used for all equipment. Maintenance on the software will be based on 15% of the system integration cost.

Enforcement costs will vary depending on the enforcement level selected for each scenario. A minimal staff of ten people with an associated burden of 50% and an allowance for miscellaneous expenses such as mailings and Department of Motor Vehicle lookups for license plate identification is included in the enforcement costs.

An administrative staff of four will be required to oversee the operation of ETC. It will consist of one administrator, one auditor, two staff members, and a burden of 50% on their salaries. Transaction costs are intended to recover the costs of obtaining payments from the patron. This may be a cost to a bank, a clearing house, a franchise, a credit card company or an internal cost for accounting for patron funds.

A tag store is a location at which the patron may obtain tags and add value to their accounts. The number of stores will vary based upon the geographical area covered by the system.

Operating Cost Model

An operating cost model has been developed to quickly analyze what impact the introduction of ETC will have on the operating costs of a toll system. Inputs to the model include the number of toll plazas with ETC lanes, the number of toll lanes which are to be ETC equipped, the number of tag stores, estimates of annual ETC transactions and the number of tags required for purchase.

These parameters are then entered into the model that contains a range for costs for each item. The ranges are carried through the analysis and presented in the results. The field is currently under development and many costs are changing as the market matures. For example, some credit card companies are offering attractive transaction costs if a toll authority has an exclusive contract for debiting customer accounts from their unique cards.

Table 2 shows the results of a model for a system that will install 100 lanes of ETC equipment with an estimated sixty million annual ETC transactions. The table shows a range of costs from seven to twelve cents per ETC toll transaction. For the new system to be cost effective to the toll agency, the toll transactions that are being replaced by ETC should cost more per transaction than the seven to twelve cents calculated.

If this facility had all manual toll collection, it would be more cost effective to collect tolls with ETC. Staffed lanes typically cost fifteen cents or more per transaction. On the other hand, if this facility had automatic coin machines, ETC would increase the cost of toll collection. Automatic coin machines typically cost one to two cents per transaction.

Summary

In summary, ETC offers many opportunities for decreasing congestion and costs to the toll collection community. There also are opportunities for ETC to increase congestion and costs when retrofitting existing toll plazas. The two computer models discussed offer reality checks so that the "windows of opportunity" for the beneficial implementation of ETC can be quickly identified before major investments of time and money are made. The models developed have been successfully used to identify facilities that do not warrant consideration of ETC and those which would gain from such implementation.

TABLE 2

ETC SYSTEM COST MODEL

3	Total Number of Plazas
50	Total Number of Lanes
2	Total Number of Tag Stores
2	Enforcement Level: Minimum (1) to Maximum (4)
30,000	Total Estimated Number of N/B Daily Patron Trips
67,000	Total Estimated Number of N/B Daily Transactions
5	Economic Life For Capital Depreciation (ETC)
10	Economic Life For Capital Depreciation (Non-ETC)

2	Tags vs Daily N/B Trip Ratio
67%	Percentage of Tags in Circulation
2%	Violation Rate
10%	Fine Rate
48,910,000	Estimated Annual Transactions (Round Trip)
60,000	Total Estimated Tags Required

	Low Cost	High Cost	% of Low Cost	% of High Cost
CAPITAL COST				
ETC Lane Cost	$1,500,000	$2,187,500	20.899%	19.440%
Non-ETC Lane Cost	$1,687,500	$2,625,000	23.511%	23.328%
ETC Plaza Cost	$600,000	$1,500,000	8.359%	13.330%
Non-ETC Plaza Cost	$270,000	$360,000	3.762%	3.199%
ETC Enforcement	$300,000	$400,000	4.180%	3.555%
ETC System Integration	$1,000,000	$1,500,000	13.932%	13.330%
ETC Tags	$1,320,000	$1,680,000	18.391%	14.930%
ETC Marketing	$500,000	$1,000,000	6.966%	8.887%
Sub-Total Capital Cost	$7,177,500	$11,252,500		
(Non-refundable Tag Deposit)	($600,000)	($600,000)	-8.359%	-5.332%
ETC OPERATING COST				
Depreciation of ETC Capital Costs	$1,044,000	$1,653,500	30.566%	27.440%
Maintenance	$277,500	$330,000	8.124%	5.476%
Enforcement	$1,075,000	$1,805,000	31.473%	29.955%
Transaction Cost	$564,100	$1,572,300	16.515%	26.093%
Tag Store	$455,000	$665,000	13.321%	11.036%
Sub-Total Operating Cost	$3,415,600	$6,025,800		
Total Cost Per Transaction	$0.070	$0.123		

Creating Windows Based Engineering Software Applications Using Microsoft Visual Basic

A. B. Savery, Member, ASCE[1]

Abstract

Microsoft Visual Basic is a computer applications development tool that can be used to automate calculations. This program makes it possible for users in small design offices to quickly create very attractive, user friendly, customized software, with excellent quality graphical interfaces.
Visual Basic uses the Basic language with a greatly enhanced visual operating system in the Windows environment. The program provides graphic boxes and drop down menu buttons that can be used to input data, and store often used data. Other graphical tools allow the placement of code that operates on the data and can deliver results to output boxes, or separate forms. Visual Basic may be used for repetitive engineering calculations to give quick and user friendly solutions. In addition, applications developed with Visual Basic can be exported to as many other computers as desired by creating an executable file of the program instructions.
Visual Basic also has graphics tools that can be used to visually enhance the application. Attractive pictures and descriptions may be created using Windows compatible graphics products, and then pasted into a picture box in the application being developed. Windows version 3.0 or newer is required. Three examples of developed programs will be demonstrated with an explanation of how they were written.

---

[1]Structural Engineer, Eastman Chemical Company, Building 54D, Kingport, TN   37662

Introduction

The advent of the Windows operating system has made computing much more user friendly and allowed for large productivity increases. New software now must be mouse driven and graphically augmented in order to be competitive. Windows has become a standard for software development. Until recently the graphical interfaces were time consuming to create and were mainly created by companies that specialized in the development of software using programs such as C++ or Pascal. For this reason small offices could not afford the time necessary to develop their own customized software for Windows. If it was not offered on the market, old software tools like Quick Basic might be used to write specialized software, or hand calculations would be continued to be used. This situation has changed now with the advent of Visual Basic. I have written three computer programs in Visual Basic that automate my work and greatly increase my productivity and that of other engineers in our office. These computer programs automate the design of concrete beams, steel beams, and the calculation of building code seismic loads. I have only just scratched the surface of potential applications in my own work.

Plant Engineering

Through the use of lap top computers the potential to greatly increase productivity where rapid design solutions are needed in the field is now available. With the proper software an engineer now can take a lap top computer into the field and design steel framing on the spot and provide crafts with designs immediately. An important aspect of such software though is the importance that it be kept as simple as possible for working in an adverse environment in the field. For this reason complex finite element programs are not the best solution for this problem. For many designs involving the design of steel framing all that is required is software for simple beam design considering unbraced length and stress and deflection checks. The first program example below makes it possible to perform such on the spot plant designs.

Automation of Simple Steel Beam Design

The first software example is for the design of simple steel framing using Allowable Strength Design. This program automates many of the ASD formulas in the AISC steel manual and contains the properties for many of the common wide flange shapes. This computer program was

written using Visual Basic and automates the formulas for
calculating allowable stress for unbraced beams, the
requirements for compact sections, allowable stresses for
columns, and the formulas for combined stress. Examples
of output are presented in Figures 1, 2, and 3. Figure
1 shows the format of the first form in this computer
program with the required input data necessary to
calculate the allowable stress for an unbraced beam in
accordance with the AISC formulas. Once the allowable
stress is known for a simple beam with distributed load
it is a simple next step to calculate the allowable load,
which is given as output on the form. Figure 2 shows the
form used to analyze uniformly loaded beams for
additional information such as deflection and reactions.
This form allows the designer the option to calculate
loads from allowable stress or stress from loads. It
also allows the recalculation of allowable stress for
different values of unbraced length. Figure 3 shows the
column design form with the typical input required to
design a simple column. The outputs are the stresses and
stress ratio. The user can also branch to subroutine
forms to calculate stress interaction ratios when bending
is combined with axial stress.

Visual Basic Programming Procedures

Visual Basic is object based with operations
performed using input boxes attached to other objects
called forms. Input boxes are used to input data that
is then operated on using basic code attached to command
box objects. Each object has many properties that can be
set using the properties window. Some of the properties
that may be set are the color and shape of the object,
the name of the object used as a variable when writing
the code, or a caption showing the title of the object.
Objects have a property of visible or invisible and can
be made to suddenly appear when user specified conditions
are satisfied. When a program is started the code is
executed whenever a double mouse click occurs on a
command button. Examples of some of my programs shown in
the figures contain many of the different types of
objects available. The Button type objects shown are the
command click objects, and the Basic code is attached to
these. Figure 4 shows an example of such code. This
Basic code is for the calculation of Cb used to calculate
allowable stress for unbraced beams. This code is
attached to the Cb button shown on the form in Figure 1.
The code was displayed with a double click of the mouse
button while in the programming mode. In the run mode a
double click executes the code. Answers are delivered to
other boxes.

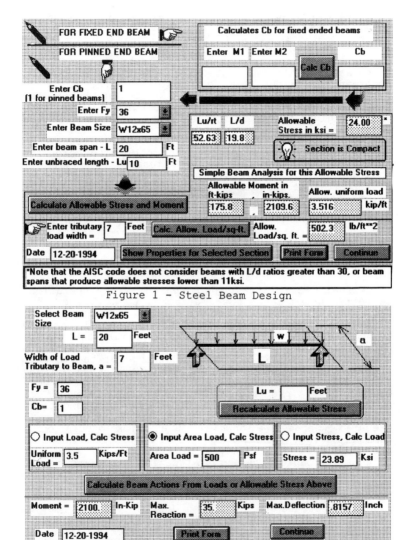

Figure 1 - Steel Beam Design

Figure 2 - Steel Beam Design for Uniform Load

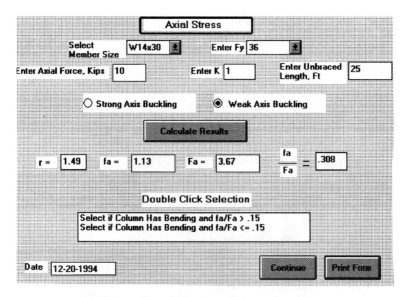

Figure 3 - Column and Beam Design

```
Sub Command1_Click ()
Mom1 = Val(M1.Text)
Mom2 = Val(M2.Text)
Cbx = 1.75 + 1.05 * (Mom1 / Mom2) + .3 * ((Mom1/Mom2) ^2)

If Cbx > 2.3 Then Cbx = 2.3
Cbf.Text = Format$(Cbx, "#.#")
End Sub
```

Figure 4. Example of Visual Basic Code

Automation of Concrete Beam Design

These computer programs were written with the idea
in mind that we would automate the calculations that are
most often performed in our office. Since we often
design and check existing, and new concrete beams with
bottom reinforcement, this is the application that was
developed. Figure 5 shows the concrete beam analysis
form. This form contains drop down menu buttons for
selecting input that is limited in choice.

Regular input boxes are used for parameters that are more variable such as the beam dimensions. The output includes information that allows a quick hand check of the results and a nice graphical image to clarify the input requirements.

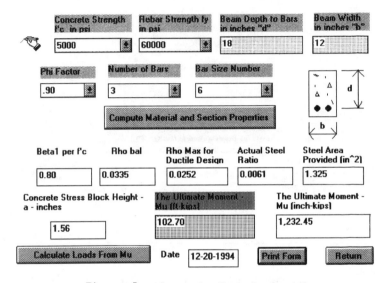

Figure 5 - Concrete Beam Analysis

Automation of the SBC Code Seismic Calculations

The 1994 version of the Standard Building Code (SBC) in section 1607.4 contains equations for the calculation of base shear, and floor load. Some of these equations are as follows: $V = C_s W$ where $C_s = (1.2A_v S)/(R\ T^{2/3})$. I decided to automate these in order to make it possible to quickly generate floor loads for buildings for my work. The purpose was to take some of the drudgery from this work and to make it possible to perform parametric studies of different effects that cause changes to the floor load calculations. The use of Visual Basic made it possible to create a program that is very user friendly, and increased my productivity when making these calculations. Figures 6, 7, and 8 show the forms used in the program.

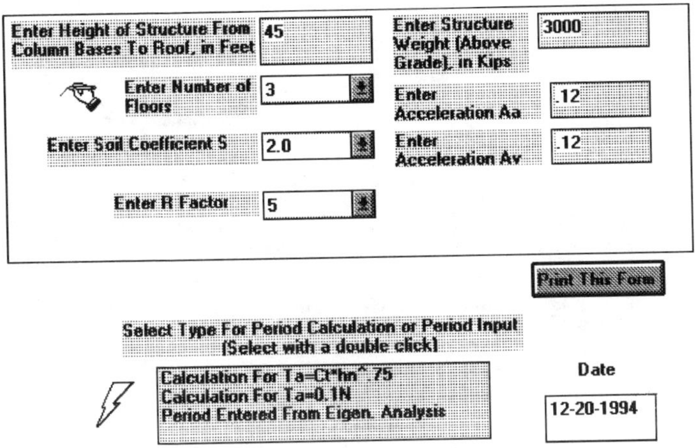

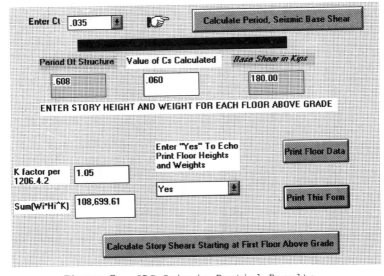

Figure 6 - SBC Code Seismic Load Input

Figure 7 - SBC Seismic Partial Results

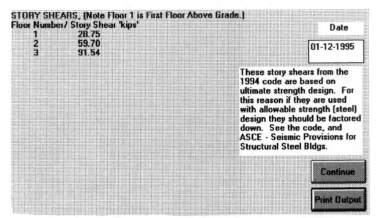

STORY SHEARS, (Note Floor 1 is First Floor Above Grade.)
Floor Number/ Story Shear 'kips'
    1   28.75
    2   59.70
    3   91.54

Date

01-12-1995

These story shears from the 1994 code are based on ultimate strength design. For this reason if they are used with allowable strength (steel) design they should be factored down. See the code, and ASCE - Seismic Provisions for Structural Steel Bldgs.

Continue

Print Output

Figure 8 - SBC Seismic Floor Load Results

Summary

I have provided several examples of completed Visual Basic computer programs that have been used to automate my work. Also a description of how such programs are developed was provided. From this information I have attempted to show that such programs are attractive, user friendly and can increase productivity considerably. Visual Basic makes it possible for engineers who do not normally have time to program complicated interfaces to automate some of their applications and make them look very advanced. Further advantages are the ability to share the software in your office and to make it possible to do design work in the field using a lap top computer. Eastman Chemical Company and the author do not warranty the computer code, procedures, or opinions expressed in this article.

References

1. Microsoft Visual Basic, "Programmers Guide", 1993 Microsoft Corporation.

2. American Institute of Steel Construction, Inc., "Manual of Allowable Stress Design", 9th Edition.

3. Standard Building Code, 1994 Edition, Southern Building Code Congress International.

Visualization for a Shoreline Evolution Model

David A. Leenknecht[1], Member, and Wayne W. Tanner[2]

Abstract

Coastal Engineering practice often includes the application of numerical models for studying longshore sand transport and resultant shoreline evolution. The U.S. Army Corps of Engineers (CE) frequently employs the GENESIS model which is a one-line finite-difference model of such processes. This type of model has diverse and sizeable data requirements, and produces substantial amounts of information requiring analysis. A recent trend in computer-based applications is the implementation of graphical user interfaces (GUIs). This paper summarizes the development of a GUI to aid the setup, execution, and analysis of simulations performed with the GENESIS model. It will discuss the background of the model and framework in which it is couched, some design goals, and considerations about the choices made in the GUI-model implementation. A summary is provided of the model data requirements, as well as editing and visualization capabilities of the GUI.

Background of ACES and GENESIS

The GENESIS model was developed in response to a CE need for improved technologies for predicting shoreline evolution as a function of longshore sediment transport. Details of the model development and theory are documented

---

1 Research Hydraulic Engineer, U.S. Army Engineer Waterways Experiment Station, Coastal Engineering Research Center, 3909 Halls Ferry Road, Vicksburg, MS 39180.

2 Computer Scientist, U.S. Army Engineer Waterways Experiment Station, Coastal Engineering Research Center, 3909 Halls Ferry Road, Vicksburg, MS 39180.

elsewhere (Hansen and Kraus 1989). The model has been implemented on DOS-based microcomputers (Gravens 1992) and supercomputers (Cialone et al. 1994), with corresponding capabilities for temporal and spatial resolution and performance. Implementations were based upon script files and numerous separate programs and files for preparing model input and analyzing output at discrete times. These activities required an extensive knowledge not only of actual model procedures and data, but also considerable knowledge of auxiliary procedures, programs, and data file types, formats, and contents preceding and succeeding the actual simulation. Contemporary trends in computer interface techniques and increased processing and graphic capabilities, particularly in the UNIX and RISC-based engineering workstation market, encouraged a re-casting of modeling approaches to enhance the engineering and minimize data handling chores.

Efforts in casting several CE coastal engineering technologies in predecessor interface environments, specifically the Automated Coastal Engineering System (ACES) on PC-DOS based platforms (Leenknecht et al. 1992, 1995) resulted in substantial cost savings and efficiency in utilizing such technologies. A decision was made to cast several models including GENESIS in a successor product (ACES 2.0) which has, as a principal focus, the goal of simplifying and making intuitive the application of more complex CE coastal models on powerful deskside environments. Early emphasis has focused on nearshore wave transformation models (Ebersole et al. 1986 and Jensen 1983), as well as other coastal process-oriented technologies which are currently incorporated in the ACES 2.0 system. As part of the effort, direct access to large databases of hindcast waves and winds is included, with the early implementations using CE Wave Information Study data such as those created by Hubertz et al. (1991).

Goals and Considerations

The GUI developed within ACES 2.0 for the GENESIS model has the following goals: 1) provide a more powerful and intuitive approach to model setup, 2) provide a software environment which centralizes model-related activities, 3) provide more powerful graphic capabilities for model setup and analysis of simulation results, 4) maximize standardization of software, procedures, and data developed for sharing between several modeling technologies embodied and envisioned within the ACES 2.0 environment, and 5) capitalize on the graphics, floating-point performance, networking, and multi-tasking capabilities of RISC-based workstations.

The implementation uses two separate codes: the GENESIS model (written in FORTRAN), and the GUI (written in C, X Windows, Motif, and a proprietary graphics package). Most implementations within the ACES 2.0 system utilize this strategy of using separate codes for model and interface for a number of reasons: 1) development and extension of the technologies can proceed independently, 2) the model is commonly originally developed and supported in FORTRAN, 3) GUI software is widely implemented in C, 4) event-driven processes elementary to GUI environments are much harder to implement in FORTRAN, and 5) model execution can be distributed to more powerful workstations (or supercomputers) on the network using client-server approaches or simpler public domain network queuing environments. The choice of UNIX/C/X/Motif has allowed the GUI and model to be used in a client-server arrangement directly on host workstations, and remotely on X-terminals, PCs, or other workstations on networks, maximizing opportunity for use with a heterogeneous hardware environment. This could only be accomplished by adhering to common network-based non-proprietary standards.

An important aspect of this implementation of GENESIS is the use of standard data files developed for all models and GUIs within the ACES 2.0 system. Generic file types have been developed which can embody large temporal and spatial databases of hydrodynamic, geographic, and geophysical information, and also serve as file vehicles for simulation inputs and results. The files are written using Network Common Data Form (NetCDF), a standard data protocol (Rew et al. 1993). Files based upon NetCDF are binary yet transportable among dissimilar computers, direct access with mechanisms for handling large amounts of data with single procedure calls, and self-descriptive with multidimensional structures of data of many types with descriptive meta-data. They are easily accessible from FORTRAN and C codes using public domain libraries.

Data Summary, GUI, and Visualization Capabilities

The GENESIS model requires data which may be classified as: "spatial domain," wave time series, longshore sediment transport constants, lateral boundary conditions, and other simulation configuration data. Given descriptions of shoreline location, structure location and features, basic sediment attributes, boundary condition assumptions, and offshore or nearshore wave data; the model responds with an estimate of the shoreline evolution as a function of longshore sediment transport, driven by wave forcing. Traditionally, the model is applied deterministicly, being sensitive not only to wave magnitudes and directions, but also the sequence and timing of events. Simulation lengths usually span a period of several years in prototype time. An illustration of the data flow between the GUI and model is shown in Figure 1.

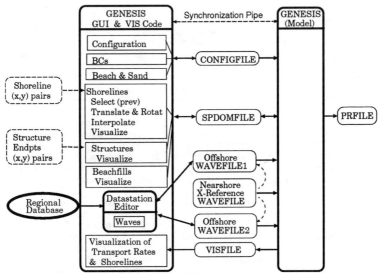

Figure 1. GENESIS GUI and Model Data Flow

Descriptions of the "spatial domain" include an initial (and optionally a reference) shoreline, discretized in a local coordinate system (Y shoreline coordinates as a function of uniformly spaced X coordinates along a baseline oriented roughly parallel to the shoreline). The GUI establishes shoreline information in a variety of fashions. It can read ASCII files containing ordered (but potentially irregularly spaced) shoreline coordinates digitized in local, State Plane, or UTM projections. From this information, the GUI can perform a least-squares fit to estimate a local coordinate system orientation. It can perform coordinate rotations and translations, and subsequent interpolations to create a uniform grid (in X) of the shoreline suitable for use in the rectilinear finite-difference model. Estimated or specified rotation and translation parameters can also be used to place other spatial domain data (such as a reference or comparison shoreline) as well as structure coordinates in the same local coordinate system. The GUI will identify problem coordinates (such as multiple values of Y for a given X)

resulting from poor discretizations or distortions from rotations and translations. Alternatively, shoreline descriptions can be edited in tables within the GUI to facilitate idealized shoreline descriptions, point removal, or editing. The interface can also extract shorelines from previous simulation files. Graphic comparisons of interpolated shorelines to the original data within the GUI provides an intuitive inspection of the quality of the interpolated shoreline ultimately applied in the model.

Additional spatial domain data collected and displayed by the GUI includes structures pertinent to shoreline protection, such as groins, detached breakwaters, and seawalls. Structures may be specified by a variety of means. The model treats structures as linear segments with some restrictions concerning type, orientation, and location. The GUI performs many of the idealizations required by the model. Similar to shorelines, structure endpoint coordinates may be read from digitized ASCII files, specified in local, State Plane, or UTM systems, and rapidly converted to local model coordinates with the same transformations available from shoreline preparations. Tabular entry of structure locations (by coordinates or internal model "cell" indexing) is also provided. The GUI allows graphical insertion, deletion, and editing of structures using a mouse. Zooming, continuous display, tracking, and reporting of cell and mouse coordinates, indices, and structure features (such as length) facilitate structure editing in this mode. Seawall specification is enhanced by a procedure which allows placement by graphically windowing on a region of shoreline and specifying an offset (in Y) of the seawall location. The model internally deals exclusively with cell indices, and the GUI facilitates rapid and intuitive determination and visualization of the local model coordinates and indices for structures in the model domain. Beach fills are specified on form subpanels.

Wave time series constitute the largest data sets used by the model. The GUI can directly access large wave databases (typically regional 20-year hindcasts at 3-hour intervals), and it permits editing and filtering of the wave data by a variety of criteria. Frequently used filters include time slicing, elimination or flagging of offshore travelling wave events, and wave events that fail minimum sediment transport threshold criteria. The wave editor can prepare two separate wave components, has similar capabilities for winds and water levels, and is used extensively within other GUIs in the ACES 2.0 package for models with related data requirements. Waves are managed and converted in any of six propagation direction conventions which include oceanographic, meteorologic, and four model-dependant conventions.

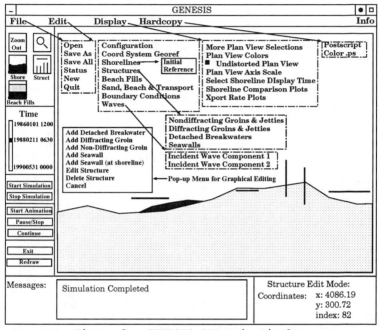

Figure 2.   GENESIS GUI Main Display

The model requires additional non-graphical data. These include choices for several lateral and seaward boundary conditions; sand, beach and longshore transport attributes; and basic information regarding file names, sampling intervals for printing or animation of results, time steps for the difference scheme, and general data for model configuration and execution. Such data are provided using forms within the GUI. A view of the main display for the interface is at Figure 2. Data for the various menu items are specified on subpanel forms. The plan view displays shorelines structures, beach fills, and grid features within the graphics area of the main display. Intuitive point and click operations common to GUI environments are used to select and control all features of the interface.

Simulation results include shoreline positions at sampling periods and longshore transport rates on an annual basis. The GUI can initiate and cancel the model execution, and display the results. Animations of the shoreline evolution are presented in the main graphics area, and can be viewed as a sequence of color filled polygons of sea/land position, or an overlaid trace of the simulated envelope of shoreline locations. Structures and beach fills are displayed. Many attributes such as zoomed regions, cell boundary and center locations, distortion and scaling options, and color selections for all features can be selected. Shoreline positions can be examined at specific times and the evolution can be stepped forward or backwards in time. This feature is useful for examining model behavior during large storm wave conditions, near stability limits, transport reversals, and shoreline evolution near structures.

Additional graphic displays are provided by the interface for comparison of pertinent simulation results. Initial, final, maximum (seaward), minimum (landward), and reference shoreline locations determined by the simulation may be compared. Also provided are comparisons of gross, net, left-, and right-directed longshore transport rates (computed on a per annum basis) for selected periods of the simulation.

Conclusions & Summary

The environment developed for the GENESIS shoreline evolution model provides an easy and intuitive approach for preparing, managing, and analyzing simulations. It consolidates a large number of data preparation activities, and provides powerful graphic and animation features providing an intuitive and engineering-oriented environment for examining and editing information required and produced by a simulation. It utilizes standardized procedures and data files which are part of a comprehensive system of technologies being built for CE coastal engineering application on UNIX-based engineering workstations.

Acknowledgements

This work was conducted at the Coastal Engineering Research Center, U.S. Army Engineer Waterways Experiment Station, under the Coastal Research Program. The Automated Coastal Engineering System work unit supported the effort. The authors would like to thank M.B. Gravens, and A. Sherlock for their contributions to the implementation, and E.F. Thompson, and J.M. Smith for review of this document. Permission was granted by the Chief of Engineers to publish this information.

Appendix I. References

Cialone, M.A., Mark, D.J., Chou, L.C., Leenknecht, D.A., Davis, J.E., Lillycrop, L.S., Jensen, R.E., Thompson E.F., Gravens, M.B., Rosati, J.D., Wise, R.A., Kraus, N.C., Larson, P.M., and Smith, J.M. (1994). "Coastal Modeling System (CMS) User's Manual," *Instructional Report CERC-91-1*, Coastal Engineering Research Center, U.S. Army Engineer Waterways Experiment Station, Vicksburg, MS.

Ebersole, B.A., Cialone, M.A., and Prater, M.D, (1986). "Regional Coastal Processes Numerical Modeling System, Report 1, RCPWAVE - A Linear Wave Propagation Model for Engineering Use," *Technical Report CERC-86-4*, Coastal Engineering Research Center, U.S. Army Engineer Waterways Experiment Station, Vicksburg, MS.

Gravens, M.B. (1992). "A User's Guide to the Shoreline Modeling System," *Instructional Report CERC-92-1*, Coastal Engineering Research Center, U.S. Army Engineer Waterways Experiment Station, Vicksburg, MS.

Hansen, H., and Kraus, N.C. (1989). "GENESIS: Generalized Model for Simulating Shoreline Change," *Technical Report CERC-89-19*, Coastal Engineering Research Center, U.S. Army Engineer Waterways Experiment Station, Vicksburg, MS.

Hubertz J.M., Driver, D.B., and Reinhard, R.D. (1991). "Hindcast Wave Information for the Great Lakes," *WIS Report 24*, Coastal Engineering Research Center, U.S. Army Engineer Waterways Experiment Station, Vicksburg, MS.

Jensen, R.E. (1983). "Methodology for the Calculation of A Shallow-Water Wave Climate," *WIS Report 8*, Coastal Engineering Research Center, U.S. Army Engineer Waterways Experiment Station, Vicksburg, MS.

Leenknecht, D.A., Szwalski, A.S., and Sherlock, A. (1992). "Automated Coastal Engineering System, Technical Reference, Version 1.07," Coastal Engineering Research Center, U.S. Army Waterways Experiment Station, Vicksburg, MS.

Leenknecht, D.A., Sherlock, A., and Szuwalski, A.S. (1995). "Automated Tools for Coastal Engineering," *Journal of Coastal Research*, Vol. 11, No. 2, (in publication).

Rew, R., Davis, G., and Emmerson, S. (1993). *NetCDF User's Guide, An Interface for Data Access*, Unidata Program Center, University Corporation for Atmospheric Research

Windows Software for Settlement of Shallow Foundations on Granular Soils

Don J. DeGroot[1], Michael D. Raymond[2], Associate Members, ASCE,
and Alan J. Lutenegger[3], Member, ASCE

Abstract

This paper describes a Windows v3.1 based application developed as a design tool for estimating the settlement of shallow foundations on granular soils. Background information on how Windows programs are developed and the challenges faced in developing one specifically for foundation engineering design are described. The methodology used by the program to allow for the calculation of settlement based on user input information on foundation geometry and loading, subsurface stratigraphy and soil property information from any of six different in situ test methods is presented. This includes a listing of the number of different settlement calculation methods available in the program for each in situ test method. Specific features of the program that allow the user to easily conduct comparisons among results from different settlement methods and how it allows for data modification based on engineering judgement and/or local practice are also described.

Introduction

The design of shallow foundations is typically governed by two criteria: stability and settlement. However, for shallow foundations founded on granular soils, settlement is usually the critical design criterion since these soils generally have good bearing capacity characteristics. In such cases, engineers must be able to accurately estimate

---

[1]Assistant Professor, Department of Civil and Environmental Engineering,
University of Massachusetts, Amherst, MA 01003-5205
[2]Assistant Civil Engineer, Foster Wheeler Environmental Corporation,
470 Atlantic Ave., Boston, MA 02210-2208
[3]Associate Professor, Department of Civil and Environmental Engineering,
University of Massachusetts, Amherst, MA 01003-5205

footing settlements in order to produce efficient designs that will meet allowable settlement requirements of a proposed structure. The estimated settlement depends on a number of parameters, including: footing geometry, footing location, footing stiffness, contact pressure, soil profile & layering, water table depth, compressibility, and soil type.

Due to the difficulty of obtaining undisturbed specimens of granular soil for laboratory testing, soil properties are typically estimated using results of in situ tests (e.g., Standard Penetration Test, Cone Penetrometer Test, etc.). As a result, most common settlement methods are based on empirical relationships developed between in situ test results and/or soil properties and foundation performance. Results from some of these in situ tests are used to correlate with the compressibility of the soil layers. Unfortunately, no single method of predicting the settlement of shallow foundations on granular soil has been shown to consistently provide more accurate results than others and therefore several methods are often used and the results compared. This is due to the difficulty in estimating granular soil properties, uncertainty in the empirical correlations used for each method, and the natural variability of soils

While conventional personal computer software can deal effectively with many engineering problems there is still a need for software designed to handle the specific calculations involved in estimating settlement (Christian 1994). This is primarily a result of the limited market for such programs. There are a number of public domain or commercially available programs for estimating the settlement of shallow foundations on granular soils (e.g., Knowles 1990; Urzua 1991; Goudreault and Fellenius 1992), however, none of these are Windows based programs. The appeal of taking advantage of many of the user friendly features of the Windows operating system provided the motivation for developing the program described herein. Undoubtedly, others are also developing similar programs, (e.g., The Geotechnical Engineering Group at Texas A&M) but each program will differ in philosophy and functionality.

This paper describes a software package for estimating the settlement of shallow foundations on granular soils for individual site specific foundations. The program is titled "Settlement of Shallow Foundations on Granular Soils" or SSFGS and operates in a Windows environment. The advantages and disadvantages of programming in Windows and using Windows applications are briefly reviewed. This is followed by a description of the basic philosophy that was followed to develop the SSFGS program and also a presentation of its key functional aspects.

Development of Windows Based Programs

There are several advantages that Windows programs have over traditional Disk Operating System (DOS) programs for both the user and the programmer. With a Windows program, the user interacts with a graphical user interface (GUI) which is similar regardless of the program being executed, thus enabling the user to use new programs with little training. The GUI contains objects such as drop down menus, power bars, icons, and dialog boxes that speed up communication between the user and the computer. The Windows environment also allows a user to conduct true multitasking (i.e., run several programs at once). Programs can also be developed that enable a user to copy data to the system's clipboard memory area which can then be easily pasted from the clipboard into another program that is running concurrently. Applications developed for Windows ensure compatibility with a large number of input and output devices. This means that whatever printer driver and display driver are being used for running Windows can also be used for running the Windows application. This significantly eases the task for the programmer of a Windows application because it does not require the development of numerous display and output drivers to cover the multitude of different monitors and printers currently available. This is not the case for DOS based applications which must come with all of the necessary drivers to allow the program to successfully run on a variety of different hardware configurations.

In a Windows environment, the programmer also has the advantage of using memory beyond the 640K DOS barrier. Windows allows applications to access all installed memory and any available virtual memory (i.e., memory swapped with the hard disk drive). Additionally, Windows handles all memory management to ensure optimal system performance which frees the programmer from dealing with this task.

All of these aspects enable programmers to develop Windows applications that enhance the speed at which users can perform tasks with a specific application including initially learning how to use it. However, developing Windows applications is significantly different and more difficult than programming DOS based applications. This is in large part due to the need for learning many new programming concepts the most significant of them being event driven programming. Even when creating the most basic of applications, the programmer has to consider many important issues such as drawing graphics in resizeable windows and user interaction with windows, buttons, icons, menus, etc. all in a multitasking environment. Furthermore, the program needs to manipulate the nearly 600 functions in the Windows Application Programming Interface. This makes the development process much more challenging, but when successful, produces an end product that is typically much easier to use compared to DOS based programs.

Selection of a Programming Language

The first step in developing a Windows application is selecting a programming language that allows one to create and maintain dependable applications faster and more easily. Borland Turbo C++ v4.0 was selected as the programming language for developing the SSFGS program. While programming in Windows is typically more difficult, as noted above, there are many aspects of using Borland C++ that simplify the programming task. For example, Borland C++ v4.0 allows for visual programming, provides a Windows Integrated Development Environment, and 16- and 32-bit programming. It includes ObjectWindows v2.0, Object Windows Library, debugging tools, and other object-oriented programming technology.

SSFGS Program Philosophy

The underlying philosophy in developing the SSFGS program is that estimating settlement of shallow foundations on granular soils is almost entirely dependent on information from in situ tests. Accordingly, the program was developed under a structure that is governed by the types of in situ tests that are commonly used in geotechnical engineering practice. These include: Standard Penetration Test (SPT), Cone Penetration Test (CPT), Pressuremeter Test (PMT), Drive Cone Test (DCT), Plate Load Test (PLT) and the Dilatometer Test (DMT). During a given program session the user is requested to select one of these six in situ test methods and all subsequent input information and calculations to follow are specific to the particular in situ test selected. However, at any time the user may switch in situ test method, input the new data, and then continue with the session using the basic project data such as footing dimensions, loading, etc. (i.e., this information remains in active memory).

Another important issue that governed development of the program was the desire to make it be a tool that can be used to assist engineers while conducting settlement calculations. It is not intended to be a full or even partial substitute for engineering judgement. In other words, the SSFGS program can be looked upon as a user friendly number cruncher but nothing more. All important decisions such as which settlement method(s) to select for a given in situ test, which correction methods to use or not use are all controlled by the user. The program was intentionally written so as not to perform tasks of this nature that require engineering judgement.

With this philosophy in mind, the SSFGS program was developed with the following major objectives:

1. Allow calculation of settlement based on data from six in situ test methods.
2. Provide flexibility for the engineer to use engineering judgement.
3. Allow the engineer to select amongst a number of different published settlement methods in order to permit a comparison.
4. Provide a convenient design tool that allows the engineer to easily answer "what if" questions (e.g., quickly conduct parametric studies).
5. Provide an easy to use graphical interface for the engineer.

In the SSFGS program, these objectives are met for a given design problem by facilitating user input/identification of the following key aspects of the design process:

1. Foundation geometry and loading.
2. Subsurface stratigraphy.
3. Input of measured properties from the site investigation.
4. Selection of the appropriate settlement calculation method(s).

SSFGS Program Overview

The SSFGS program was created as a Windows application to provide an easy to use Graphical User Interface (GUI). The interface contains objects such as drop down menus, power bars, icons, and dialog boxes that speed up communication between the user and the computer. The program has a main Window that includes menu functions and a button bar that are activated by either a mouse or a specific sequence of keyboard commands. The main button bar contains specific menu commands and are provided in the SSFGS Window to allow more rapid execution of the more commonly used commands. There are three basic steps used to run the program: (1) providing input data for a specific design problem, (2) selecting the appropriate settlement prediction method(s), and (3) managing data output from the program.

The program also includes a help system that is similar to all other Windows help systems. However, items specific to the SSFGS program have been incorporated in the help system. All of the different methods for computing settlement for each in situ test type are described in the help system including a complete reference for the method, how the method works, assumptions, and any other relevant information.

SSFGS Data Input

User identified input data required by the SSFGS program for a given design problem consists of:

1. Foundation geometry.
2. Foundation loading.
3. Site conditions (e.g., depth to water table, layer unit weight data, etc.).
4. In situ test type (e.g., SPT, CPT, etc.).
5. In situ test data (e.g., N from SPT; q_c from CPT; etc.).

Once this information is input the user has the following options to complete a settlement calculation:

1. Selection of relevant correction methods for the in situ test data.
2. Selection of one or more calculation methods.
3. Input of relevant factors associated with a given calculation method.
4. Print out of input data, intermediate calculations, and final results.

The data input process is greatly simplified with the use of dialog boxes in the GUI. The dialog boxes allow the user to input the specific data in any order and allows data to be changed at any time during a design session. For example, if the calculated settlement based on the input data is too large, the user can easily reopen the foundation geometry dialog box, change the footing dimensions, and recalculate the settlement. Specific dialog boxes that correspond to a number of the buttons on the main window button bar include: (1) Site Conditions Data, (2) Footing Data, (3) Subsurface Data, (5) Select or Change In Situ Test Type, (6) Select Settlement Method(s), and (7) View Intermediate Calculations. Two other methods can be used for entering data including: (1) importing a text file, and (2) pasting a data set from the Windows clipboard.

SSFGS Data Output

Optional output information from the SSFGS program for a given design session consists of the following items:

1. Listing of the input data.
2. Listing of the intermediate calculations for each method selected.
3. Listing of the estimated settlement for each method selected.

The output process is greatly simplified by the SSFGS program which handles most of the required tasks. The user only needs to select the print command button on the main window button bar to generate a hard copy of the desired information.

Settlement Calculation Methods

Table 1 lists the number of settlement calculation methods, by in situ test method, that are available in the program. Lutenegger and DeGroot (1995) provide details on all of the different settlement calculation methods that are available in the program and also give example calculations using data from the FHWA footing load tests conducted at Texas A&M University (Briaud and Gibbens 1994).

In Situ Test Method	Calculation Methods
Standard Penetration Test (SPT)	21
Cone Penetrometer Test (CPT)	7
Pressuremeter Test (PMT)	4
Drive Cone Test (DCT)	1
Plate Load Test (PLT)	5
Dilatometer Test (DMT)	2

Table 1. Settlement Calculation Methods Available in the SSFGS Program.

Figure 1 shows the Subsurface Data dialog box for the SPT test and gives an example of the type of features built into the program that explicitly require the use of engineering judgement. It is well known that the SPT blow count (N) values are affected by many factors and it is up to the engineer to decide whether to correct the measured N values and, if so, which correction method to use. For example, several methods are commonly used in practice to correct the SPT blow count values for the effect of overburden pressure. Depending on which correction is used, different results can be obtained for the same design problem. The SSFGS program was developed with the option of correcting the overburden pressure by selecting one of three popular methods: Teng (1962); Peck, et al. (1974); and Liao and Whitman (1986). It was, however, also considered important to provide the engineer with the option of using a given settlement calculation method as it was originally presented. Accordingly, the engineer can instruct the program to perform a calculation as originally presented or if based on their experience they prefer to use a specific overburden correction method they can also make this selection. These options highlight the advantage of having the program developed in a Windows environment wherein the user can easily switch among dialog boxes and change data to determine the effect on predicted settlement.

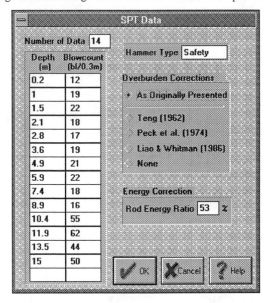

Figure 1. The SPT Subsurface Data Dialog Box

Summary

The SSFGS Windows application is a very flexible and easy to use program allowing for data input by dialog boxes, import files, or by pasting from the Windows clipboard. It provides simple methods for data output such as viewing on screen, printing directly to the printer, or by pasting output data into another Windows program. The settlement of shallow foundations on granular soils can be calculated using in situ data from six different in situ test methods. Options are available for users to select different correction methods relevant for a particular in situ test method and/or calculation method. The use of dialog boxes allows a user to quickly conduct parametric studies by easily varying footing dimensions, calculation method, correction factors, etc. In this regard, the program provides an engineer with a valuable design tool but still requires the use of engineering judgement to come up with a final settlement estimation.

Acknowledgements

The SSFGS program was developed as part of a project on settlement of shallow foundations on granular soils that was sponsored by the Massachusetts Highway Department (MHD). The authors thank, Helmet Ernst, Nabil Hourani, and Laura Krusinski of MHD for their input. The opinions and findings contained herein are those of the authors and do not necessarily reflect that of MHD.

References

Briaud, J.-L. and Gibbens, R.M. (1994). *Test and Prediction Results for Five Spread Footings on Sand.* Geotechnical Special Publication No. 41, ASCE, New York.

Christian, J.T. (1994). "Software for Settlement Analysis." *Vertical and Horizontal Deformations of Foundations and Embankments*, Geotechnical Special Publication No. 40, ASCE, A.T. Yeung and G.Y. Felio, Eds, Vol. 2, 1718-1729.

Goudreault, P.A. and Fellenius, B.H. (1993). *UNISETTLE, Version 1.0 - Background and User Manual*, Unisoft, Ottawa, Canada.

Knowles, V.R. (1990). *CSANDSET - Settlement of Shallow Footings on Sand (I0030).* U.S.A.E. Waterways Experiment Station., Vicksburg, MS.

Liao, S.S. and Whitman, R.V. (1986). "Overburden Correction Factors for SPT in Sand." *Journal of the Geotechnical Engineering Division*, ASCE, 112 (3), 373-377.

Lutenegger A.J. and DeGroot, D.J. (1995). *Settlement of Shallow Foundations on Granular Soils.* Univ. of Mass. Transportation Center Research Report, Amherst, MA.

Peck, R.B., Hanson, W.E., and Thornburn, T.H. (1974). *Foundation Engineering.* John Wiley & Sons, New York.

Teng, W. (1962). *Foundation Design.* Prentice Hall, New Jersey.

Urzua, A. (1991). *SAF-I - User's Manual.* Prototype Engineering, Winchester, MA.

Developing a Field Inspection Reporting System for the Construction Industry

Anthony D. Songer[1] and Eddy M. Rojas[2]

ABSTRACT

Construction site availability of advanced portable computing, multimedia, and wireless communications allows, even encourages fundamental changes in many jobsite processes. One process amenable to such change is field inspection. Using emerging technologies and information systems, the Field Inspection Reporting System (FIRS) project suggests new processes or "reengineers" the traditional AEC field inspection process. This paper addresses work to date on the FIRS system, an ongoing component of the Construction Information Technology 21 project at the University of Colorado, Boulder.

INTRODUCTION

The Construction Information Technology 21 project at the University of Colorado's Construction Engineering and Management program addresses site level implementation of emerging technologies and information systems. One construction site process which demonstrates a large propensity for change under a reengineering effort is that of field inspection. On any construction project, inspection data must be collected and processed by a variety of parties in order to control the process and protect their particular interests. By nature, field inspection processes are time consuming. Data are not well organized, flexible, or easily accessible. Additionally, inspection information is often not provided in a useful format or timely basis.

Integrating emerging and information technologies provides the mechanism to fundamentally change existing field inspection processes. FIRS reengineers this process adding value in a variety of ways. These include:

---

[1] Assistant Professor, Department of Civil, Environmental, and Architectural Engineering, University of Colorado at Boulder, Boulder, CO 80309-0428. songer@bechtel.colorado.edu.
[2] Graduate Student, Department of Civil, Environmental, and Architectural Engineering, University of Colorado at Boulder, Boulder, CO 80309-0428. rojase@bechtel.colorado.edu.

- Reduced the cost of the inspection process.
- Reduced field inspection time.
- Reduced cycle time of corrective actions.
- Support Total Quality Management techniques.
- Facilitated access to inspection information.
- Reduced the paper work involved in the inspection process.

This paper describes the underlying process framework and technical platform of FIRS.

FRAMEWORK

A detailed system analysis of field inspection results in essential models of the processes and data involved in the system. These essential models address the question of "what" the system provides. They do not address "how" the system performs.

Inspection processes were modeled by building data flow diagrams (DFDs). These diagrams (see Figure 1) illustrate data flow (arrows), data storage (end-open boxes), system agents (boxes) and the processes that respond to and change data (circles). Figure 1 shows the top level DFD of the inspection system. Five major processes are identified in the diagram: Capture Project Information, Capture Item Information, Distribute Information, Update Item Information, and Generate Reports. A system analysis of each major process provides detailed DFDs.

Data are modeled using Entity-Relationship diagram (ERD). An ERD is a modeling tool that depicts the associations among different categories of data within an information system.

The result of the system analysis phases of the project is DFD's and an ERD which provide deep understanding of the system requirements. For additional information concerning the development of DFD's and ERD's readers are referred to [Whitten et al. 1994].

TECHNICAL PLATFORM

The system requirements specified in the analysis phase are the foundation for the development of the target technical platform. Components of the platform discussed include system design, input devices, interface, data architecture, and reports.

System Design: FIRS is based in a client/server architecture that supports cooperative processing. Inspection information is not processed at only one site, but cooperatively in several sites that are connected by modems. (See Figure 2)

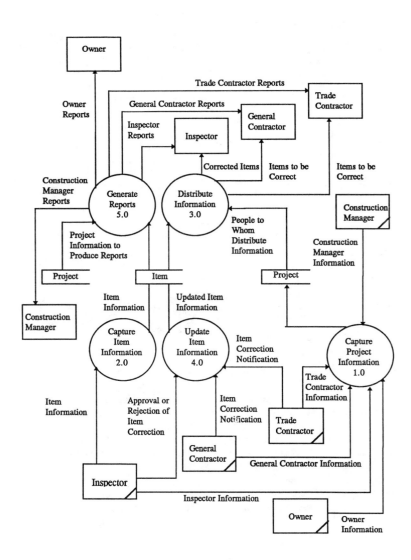

Figure 1: Overview Data Flow Diagram

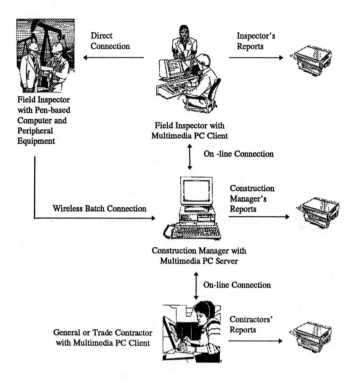

Figure 2: Field Inspection Reporting System Layout

The server provides storage for the database and performs all the statistical calculations necessary to produce reports. It is administrated by the Construction Manager. General and trade contractors as well as inspectors have on-line access to the system through PC multimedia clients. System clients provide the interface, input, edition, and printing capabilities. All the data is collected by inspectors using portable computers and peripheral equipment.

Input Devices: There are several alternatives for data collection devices. The basic system consists of a portable computer with a sound card for voice recording. The system can growth with a voice recognition capable computer, a digital camera, and/or a digital video input.

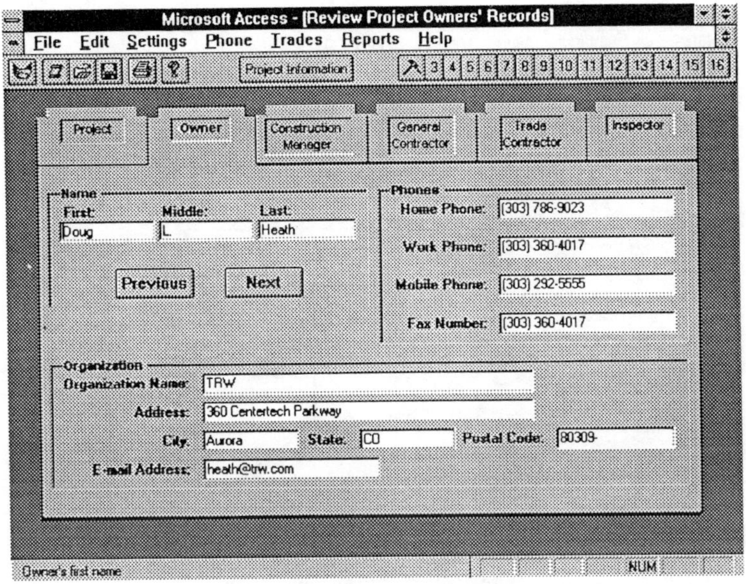

Figure 3: Project Information Screen

As an illustration, the Dauphin DTR-2 is a pen-based computer with voice recognition. It has a 80486SLC2 microprocessor that runs at 50 Mhz., 4 Mb. of RAM expandable to 16 Mb., a hard disk drive of 128 Mb., a battery life of 2 hours, and 2 PCMCIA slots.

User Interface: The user interface uses a task-centered design process. Under this approach, a process is structured around specific tasks that the user wants to accomplish with the system. Figure 3 illustrates the screen related with the task of finding general project information.

FIRS is a construction engineering management system developed for architects, engineers, construction managers, and technical personnel. Therefore, FIRS is designed with a technical user in mind. These users need to access information as direct and as soon as possible. The navigation capabilities of the system accomplish both objectives. Toolbar shortcuts and menu selection provide direct data access. Speed is provided by using a folder interface style that allows the user to access several pieces of information with only one click of the mouse. For example, Figure 3 shows the project information screen. This screen contains information about the project, the owner, the construction manager, the general

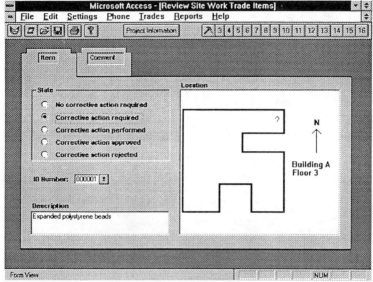

Figure 4:   Item Information Screen

contractor, the trade contractors, and the inspectors.   All the information is accessible just by selecting the desired page in the folder by clicking on its tabs with the mouse.

Data Architecture: FIRS mirrors the construction industry organization of work. The item, which is the basic unit of information, is classified by trade.   The trades correspond to the Construction Specification Institute's master format.   Each item has attributes, such as number, states, description, location, and comment   (See Figure 4).   Comment is one attribute represented by three media formats: audio, picture, and video.   (See Figure 5)

The item number is an identifier that is used to differentiate among all the items in an particular trade. The description is a brief explanation of what the item is.   The location is a graphical representation of the item's location in the building's plans. The states are different conditions in which an item can be in a specific point of time.   They are:

• No corrective action required.
• Corrected action required.

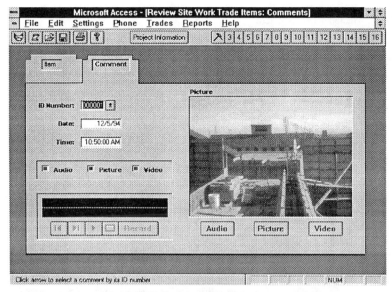

Figure 5: Comment Information Screen

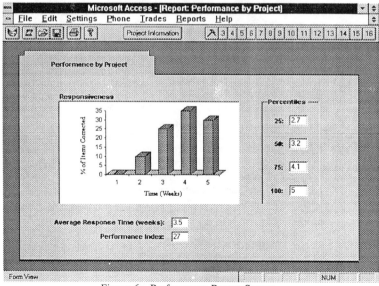

Figure 6: Performance Report Screen

- Corrective action performed.
- Corrective action approved.
- Corrected action rejected.

Each state can be associated with one or more comments. The comment is a multimedia resource to describe the state of an item. Figure 5 shows an example of the comment screen. It includes a picture of the Chemical Stores and Lab. Addition Project at the Engineering Center at the University of Colorado at Boulder. The screen allows the navigation among audio, picture, and video media as shown.

Comments are input into the system by the inspectors using portable computers and by trade contractors using PC Clients. Audio is recorded with a PCMCIA sound card in portable computers, or with a conventional sound card in a PC Client. Pictures are taken with a digital camera and storage as a file in portable computers. Video is recorded with a conventional video camera connected with a PCMCIA digital video card in portable computers.

Reports: Inspection reports are an essential component of the system. They present information that supports ongoing decision-making processes as well as total quality management strategies. Some of these strategies are related with performance measurements of general contractors, trade contractors, and inspectors that can be used in pre-qualification processes. Figure 6 illustrates an example of a performance report.. It includes parameters such as responsiveness and performance index. As an illustration, responsiveness is a measure of how a trade contractor respond to an inspector's request. It measures the time between the corrective-action-required state and the corrective-action-performed state of an item. The results are plotted in a bar chart.

CONCLUSIONS
The FIRS project studies the construction inspection process from a generic point of view. The reengineering effort focuses on the general inspection process itself and not in the particular tasks that are performed as part of any specific inspection. These particular tasks constitute the control mechanism of a construction process, any change in these tasks must be a consequence of reengineering the specific construction process and not the inspection process. The FIRS project is currently in a prototype stage. A simulation to compare FIRS with a current inspection process is under way. A pilot implementation and an evaluation of the system are scheduled for January of 1996.

REFERENCE
Whitten, J., Bentley, L., and Barlow, V. (1994) "System Analysis and Design Methods." Third Edition, Richard D. Irwin, Inc., USA.

The Hydrologic Modeling System (HEC-HMS): Design and Development Issues

William Charley, Art Pabst and John Peters[1]

Abstract

The Hydrologic Engineering Center's Hydrologic Modeling System (HEC-HMS) is a software package for precipitation-runoff simulation. Software development and architecture issues associated with development of HEC-HMS are described. The software's object-oriented structure and the role of its graphical user interface are presented.

Introduction

The Hydrologic Engineering Center (HEC) of the US Army Corps of Engineers develops software for application in the disciplines of hydrologic and hydraulic engineering, and water resource planning. A project named NexGen is underway to develop "next-generation" software to replace current, widely-used software such as HEC-1 (for precipitation-runoff simulation) and HEC-2 (for steady-flow water surface profile computations).

The HEC-HMS, which will replace HEC-1, is intended for precipitation-runoff simulation using observed or hypothetical (design) precipitation. The user can select from a variety of technical options for each of the major computational elements: precipitation specification, loss estimation, excess-to-runoff transformation, and hydrologic routing. The program can be used to simulate runoff from complex subdivided watersheds, and can utilize distributed rainfall, which is now available from a new generation of weather radars. A basin "schematic" capability enables the user to configure the hydrologic elements of a watershed (such as subbasins, routing reaches, reservoirs, diversions) graphically, and to access editors and simulation results from schematic components. Requirements for the HEC-HMS include the following:

---

[1]Respectively, Hydraulic Engineer; Chief, Technical Assistance Division; and Senior Hydraulic Engineer; Hydrologic Engineering Center, 609 Second St., Davis, CA 95616-4687

o   state-of-the-art engineering algorithms
o   comprehensive Graphical User Interface (GUI)
o   substantial use of graphics
o   operational in native X-window and Microsoft Windows environments
o   substantial use and manipulation of time series data
o   inter-program data exchange
o   an interface to Geographic Information Systems
o   use of existing computational algorithms written in FORTRAN
o   extensible and easy to maintain
o   distribution of the software unrestricted, with no run-time license fees

Languages, Toolkits and Libraries

Historically, HEC software was developed in FORTRAN, primarily by engineers. Development environments followed the typical progression of mainframe to mini-computer to personal computer and workstations. In preparation for NexGen software development, the "world" of windowing environments, event programming, GUI's, the C language, and finally, object oriented programming with C++ were explored. Prototype applications were developed. Initial skepticism with object-oriented programming turned into significant support for this technology; HEC-HMS is being developed with object oriented techniques.

To facilitate GUI development and porting to the requisite platforms, a commercial multi-platform toolkit was acquired. Although adoption of such a toolkit adds significantly to an already steep learning curve, the effort to develop and maintain separate platform-specific versions of source code is avoided. Some graphics are being developed with relatively low-level calls to routines in the multi-platform toolkit. A commercial graphics library was acquired to facilitate development of time series graphics. Versions of the graphics library are available for the various target platforms.

A specialized management system designed for efficient handling of time series data has been under progressive development since 1979. The Data Storage System (HEC-DSS) makes use of a library of routines that have capability to read and write variable-length, named records in a direct access file. Storage and retrieval of time series data is accomplished with blocks of data of pre-specified size based on the interval of the data. To facilitate use of HEC-DSS in object-oriented applications, a set of time series manager classes were developed that utilize the HEC-DSS library.

Existing HEC software contain a base of well documented and tested FORTRAN algorithms for performing hydrologic computations. The algorithms will be useful in the new software and have been incorporated into a library labeled *libHydro*. Thus, development of HEC-HMS draws on mixed languages (C++, C and FORTRAN), and utilizes a variety of libraries.

HEC-HMS Architecture

Figure 1 illustrates the internal architecture of HEC-HMS. Although linked into a single executable, there is a clear separation between the GUI and the simulation engine, where all computations are managed and performed. This permits independent development of the GUI and the engine, and facilitates the utilization of an alternative GUI in the future, should this become desirable. The GUI has access to objects within the engine through public interfaces. There are no references to the GUI from the engine, except for calls to a generic error message dialog box. The user interacts with the GUI through the windowing system, whether that is on an X device connected to a UNIX workstation, or Microsoft Windows on a PC.

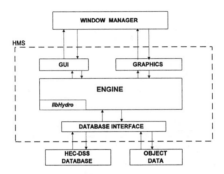

Figure 1 Software Architecture

For similar reasons, clear separations are maintained between the simulation engine and the graphics module, and the engine and the database interface module. The engine uses objects that interface to the database, and objects that perform graphics, but these objects could easily be modified to access a different database system or graphics package.

Currently, time series and similar data are stored using HEC-DSS, while persistent object data, such as parameters and coefficients, are stored in ASCII text files. The HEC-DSS provides a convenient and efficient way of entering, storing, retrieving and displaying series type data. ASCII text files provide a convenient means for testing the simulation engine independent of the GUI. It is anticipated that the text files may be replaced by a database in the future.

The engine is comprised of three major components: the project manager, the precipitation analysis model, and the basin runoff model. The project manager handles the control of the simulation time window, the utilization of precipitation and basin runoff models, file names, and various other management tasks. The precipitation analysis model computes subbasin average precipitation from either historical gaged data or from design storms that are frequency-based or that utilize Standard Project Storm criteria. The basin runoff model uses this precipitation to compute subbasin discharge hydrographs, which can be routed through river reaches or reservoirs, and diverted or combined with other hydrographs.

Figure 2 illustrates use of objects in HEC-HMS. The BasinManager object creates, manages and destroys the various HydrologicElement objects that comprise the simulated watershed. When a compute is requested by the user, the BasinManager finds the hydrologic object which is acting as the outlet (all links, except for diversions, eventually point to this object), then sends it a message to compute. Because the outlet object requires hydrographs from objects above it for its computation, it requests objects upstream of itself to compute, which in turn request hydrographs from their upstream links. Thus, the request is propagated to all hydrologic objects that constitute the watershed.

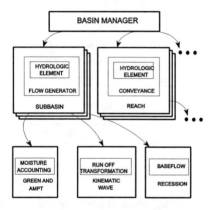

Figure 2 HEC-HMS Objects

Hydrologic classes inherit from a fundamental base class titled HydrologicElement. Data members of this class contain information pertinent to all element types, such as name, description, type, etc. The classes Conveyance, FlowGenerator, Node, and SinkBase each inherit from the HydrologicElement base class to provide certain types of connectivity functions. For example, the Conveyance class allows a link to one upstream object (from which to retrieve an inflow hydrograph), and one downstream object. The FlowGenerator class does not allow links to an upstream object; it can only link to downstream objects. Member functions of these classes provide a variety of capabilities, such as establishing and deleting links to other objects, and determining the cumulative basin area. The primary hydrologic classes, which inherit from these intermediate classes, are Reach and Reservoir (from Conveyance), Source and Subbasin (from FlowGenerator), Diversion and Junction (from Node), and Sink (from SinkBase). The data members of a derived class include data associated with its base class as well as data defined directly within the derived class.

Member functions of these classes provide the capability of setting and accessing data parameters, computing discharge hydrographs, and performing other desired object behaviors.

The primary hydrologic objects generally use "process" objects to implement the different computational methods. For example, Subbasin objects instantiate objects from a moisture accounting class, a runoff transform class, and a baseflow class. The object instantiated depends on the hydrologic method used. A GreenAmpt object or a InitialConstant object might be instantiated for moisture accounting, depending on the method selected by the user. A Snyder object, or Clark object, or Kinematic Wave object might be instantiated for the runoff transform. A process object has as data members the unique parameter data required for the particular method, and a member function to compute (e.g., compute excess given the precipitation). In many cases the "compute" member functions call routines from libHydro to perform the actual computation.

Each process class inherits from a base class for the process group that contains capabilities common to all the method classes within that group. For example, the moisture accounting base class defines time series objects to retrieve the subbasin precipitation and store the computed excess precipitation that are needed by all derived classes. Likewise the runoff transform base class defines time series objects that obtain the excess precipitation and store the computed hydrograph for all types of derived transform classes.

The TimeSeries class inherits from the DataManager class, which accesses the HEC-DSS database software. DataManager contains several "static" member functions, one of which points to buffers of data retrieved and stored. When an object retrieves (or stores) a set of data that is already in a memory buffer, no actual file access is required.

As previously indicated, persistent storage of data parameters and coefficients is presently achieved with ASCII text files. As an example of objects interacting and working with each other, when the user requests to save data parameters in persistent storage, the BasinManager sends a message to each hydrologic object to save its own data. After saving its data, each hydrologic object in turn sends a message to its processor objects to save their data. To re-establish a "model" from a persistent data file, a data loading object recreates hydrologic objects via the BasinManager, and sets their data parameters and coefficients through their public interfaces. In this procedure, as far as the engine is concerned, the objects are created and parameters set just as if this action were being done by the user through the GUI.

Graphical User Interface

The GUI is the window through which the user interacts with HEC-HMS. It enables specification of information to be retrieved or stored (e.g., data files), specification of application-specific information (both data and task instructions), and viewing of results. The GUI enables the user to easily and effectively perform the various types of analysis for which the program is capable.

With the GUI, the user can define, change, control, and view a model's configuration, inputs and results. The multiple windows shown in Figure 3 illustrates some of the screens that comprise the GUI for an example watershed. The screens are, in a

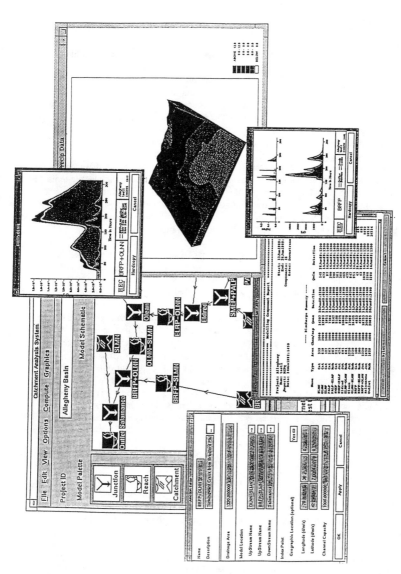

Figure 3 Graphical User Interface

counter clockwise order from the upper left: 1) the schematic configuration of hydrologic elements that make up the example Allegheny Basin model; 2) a data editor for entering or changing information for a junction object; 3) a tabular report of critical flows for each hydrologic element; 4) a graphic of subbasin precipitation and flow results; 5) a perspective view of a precipitation field over the entire basin; and 6) a graphic of the flows entering and leaving a junction. Each of these windows make up a portion of the complete GUI. The user may open any number of graphical and tabular windows to display reports and model information for any hydrologic element of interest.

The user-controlled schematic representation of the components of a hydrologic system is a key element of the GUI. The schematic employs icons to represent subbasins, routing reaches, reservoirs, diversions, etc., and their topological connections. In creating a new watershed model, the user selects an element type (e.g., subbasin) from a popup menu and drags it to a desired location on the schematic background. Other elements are established in a similar fashion, as are connecting links between elements. An element can then be selected, and access to a data editor for that element is provided as an option from a popup menu. After a simulation has been executed, an element can again be selected, and a popup menu provides access to a display of simulation results for that element.

Navigation through the schematic is facilitated with a view finder window that shows a miniature view of the entire schematic, and a frame around that portion of the schematic presently visible. The user can move the frame to bring other portions of the schematic into view. Capability is also provided to "collapse" portions of the schematic so that unwanted detail can be hidden from view.

The schematic capability is required in a number of NexGen software products besides HEC-HMS, such as those for simulating reservoir systems and river hydraulics, and for analyzing flood damage. Hence the approach for developing the capability was to develop a Schematic Model Library (SML) which consists of C++ classes for general application. The SML links with libraries from the multi-platform toolkit so that the schematic capability is portable across platforms.

Where several GUI windows display the same data value or information related to a data value, it is necessary to provide a mechanism to assure that the contents of all windows remain current. If, for example, the drainage area in the data edit window were changed, then the tabular report, which shows each element's area, would need to be updated. All model results that are affected by the drainage area value would no longer be current. To provide a mechanism to systematically handle the update of windows, "observer" objects are used. An observer object allows a window to register its interest in being notified when specific model data is changed. Thus, the object that generates the report could use an observer object to notify a hydrologic element object of its interest in knowing of a change in drainage area. The report object would then be able to update the particular data value, or regenerate the complete report.

While the user is able to interact with the model through the GUI to accomplish her or his modeling needs, it is frequently desirable to repeat a sequence of model interactions over and over again with different model parameters. Under this scenario the GUI as the means of carrying out repeated operations can become the user's greatest frustration. A similar need for an alternate to the GUI model control capability occurs when a model must be operated unattended, or under the control of another higher level modeling process that sees the hydrologic model as only a contributing component. In each of these cases the ability to drive the model from a script of instructions which include both control and data values is needed. In its full implementation the model design will permit the user to define any number of macro scripts that can be invoked to simplify often repeated model GUI sequences. An example might be a tool bar allowing the selection of a macro to trigger the routine generation of six hardcopy plots and four reports to summarize model results.

The GUI design requires careful consideration of many issues such as, the mental image a user will have of the problem solving steps, logical navigation through those steps, the organization of related data into specific GUI screens, the aesthetic layout of the information on each screen, look and feel consistent with the parent windowing system, and adequate handling of error conditions.

Closing Comment

The current architecture and development plans for HEC-HMS reflect the set of requirements listed in the Introduction and experience to date in model development. While the learning curves have been steep and initial development has progressed much more slowly than was originally anticipated, we find that we are now able to extend existing modeling capabilities and continue model development in a reasonably efficient and straight forward manner. We also anticipate that software maintenance will be facilitated, and that future adoption of alternative graphics, GUI or database features will not require a major rewrite of the computational engine or other components.

Acknowledgements

Anthony Slocum has contributed significantly to the design of HEC-HMS and is the developer of the Schematic Model Library. Paul Ely is the developer of libHydro.

Application of Integrated Product and Process Development in Facility Delivery

Philip Lawrence[1], Associate Member, ASCE, Beth Brucker[1], Michael Case[1], Rajaram Ganeshan[2], Michael Golish[1], Eric Griffith[1], and Jeffrey Heckel[2]

Abstract

The following is a description of an application of Integrated Product and Process Development (IPPD) principles to the design and construction of Army facilities. The Collaborative Engineering Team at the Construction Engineering Research Laboratories of the US Army Corps of Engineers has developed a hardware and software test bed to demonstrate how IPPD techniques can be applied to the design and construction of facilities. The test bed uses a virtual teaming architecture centered upon an object-oriented representation provided by the Agent Collaboration Environment (ACE) and Design++. The information exchange between these two systems is accomplished using an approach called the Virtual Workspace Language (VWL) which incorporates the Knowledge Interchange Format (KIF) and the Knowledge Query and Manipulation Language (KQML). Common frame libraries for the electronic modeling of product and process information has been developed to provide a shared ontology with which all participants can create objects and communicate information. Agents which use this shared representation to exchange information have been developed for project management, architecture, design, construction, and operation and maintenance of a standard US Army fire station facility. Also, legacy software such as Computer Aided Design (CAD) tools have been integrated into the system.

1. Introduction

The current practice in facility delivery is a disjointed process whereby islands of information are created and data is often lost or misinterpreted when translated between participants [Fischer, 1994]. One of the most common approaches to facility delivery is the design-award-build process [Haviland, 1987]. In accordance with this process, the owner contracts architects and engineers to create a design for a facility, once the design is fully developed and documented, proposals for construction are solicited. The owner then awards the construction contract to one or more builders and retains the previously-involved parties to administer this contract. The delivery of a facility therefore involves a collaborative effort between participants. Lack of coordination and integration between these participants can lead to inefficiencies in the form of quality and/or productivity loss, and eventually to design flaws, the resolution of which often involves litigation [Loss, 1991]. Studies indicate that computer supported collaboration can increase coordination as well as decrease overall number of errors in collaborative work.

The Artificial Intelligence community has spawned many techniques such as frame, rule, object oriented value, and agent systems for the computer-based modeling and

---

[1]Principle Investigator, Collaborative Tech. Division, USACERL, Champaign, IL.
[2]Researcher, Collaborative Technologies Division, USACERL, Champaign, IL.

utilization of product information. These techniques present a powerful aide in the computer support of collaborative work. While utilization of these techniques to model facility product information can alleviate much of the integration problems mentioned above, doing so fails to consider the process of facility construction. In recent years, researchers in the area of manufacturing have determined that productivity as well as product quality could be increased if these modeling techniques were used for concurrent product and process development [Kambhampati, 1990]. To automate the concurrent product and process development via computer, it is necessary to electronically model the process along with the product. The methodology which integrates these product modeling techniques and process engineering has come to be known as Integrated Product and Process Development (IPPD).

In manufacturing, IPPD has been used to concurrently address product and process plan concerns. It is the hypothesis of researchers at the U.S. Army Construction Engineering Research Laboratories (USACERL) that process plans in manufacturing are significantly analogous to facility construction plans; therefore, applying IPPD to facility delivery will provide similar benefits to those seen in manufacturing. The Collaborative Technologies Division of USACERL is investigating the application of an IPPD approach to the design, construction and maintenance of facilities[3]. An automation test bed has been developed that will serve as an example of the application of IPPD in facility design and construction.

This paper describes an integrated approach to product modeling and process engineering utilizing IPPD and collaborating technologies which has been applied to the facility delivery process. Section 2 gives a background and description of the collaboration technology used in the test bed. Section 3 describes the test bed. In section 4, the preliminary results of the test bed are discussed. The paper ends in section 5 with a discussion of the project results and future work.

2. Background and Enabling Technologies

The test bed uses a virtual teaming architecture centered upon an object-oriented representation provided by ACE and Design++. A common representation has been developed to provide a shared set of concepts (ontology) which facilitates the creation of objects and communication of information. This representation contains both product and process information. Agents which use this shared representation have been developed. Also, legacy tools such as Computer Aided Design (CAD) have been integrated into the system. Combining these capabilities, it is possible to create, manipulate, utilize and communicate information about a facility design. The remainder of section 2 describes the components of the test bed demonstration which make this possible.

2.1 Agent-based Systems

The term "agent" used loosely can mean any piece of software (software agent) or person (user agent) that can possibly effect change to electronic data and/or can interact with other agents [Kautz, 1994]. Software agents are software systems that operate within an electronic environment and can sense the state of and effect

---

[3]This paper describes research being performed on computer supported collaborative engineering. Former research focused on computer automation of concurrent engineering. It should be noted that the IPPD approach is closely related to, but is not concurrent engineering. The process of concurrent engineering is subsumed by the topic of collaborative engineering and is one of many processes which can be implemented in the re-engineering of a process such as facility delivery. Re-engineering of the facility delivery process would involve the restructuring of an entire industry and is therefore beyond the scope of the discussion in this paper.

changes to this environment. A software agent can possess rules that are used to diagnose a problem, check for errors or otherwise automate a task. Agents differ from expert systems in that typically only one simple task is performed by a agent whereas expert systems tend to have large, complex, interacting rulebases. In this respect, agents serve as modular and easily-maintainable facilitators of work.

Software environments which allow the creation of agents are called "agent-based systems". Agent-based systems are well suited for use as integration platforms. Complex systems can be gradually implemented using many small agents. The resulting system is easy to maintain since agents may be changed internally or replaced without affecting the remaining agents in the system.

2.2 Product and Process Modeling

In order for agents to perform work, an electronic model of the data (a "product model") on which to operate must exist. In facility delivery, it is essential that this model be expressive enough to represent the complexities of facility components and their interactions. Object-oriented approaches to product modeling provide a powerful mechanism to model complex data, and are therefore often used in domains such as facility delivery. Frame systems extend the object-oriented paradigm to allow users to create objects (instances of frames) "on the fly", without having to recompile an application. Relations (e.g.: instance-1 is "Connected-to" instance-2) and constraints (e.g.: the height of instance-1 should be less than the height of instance-2) are often used with frame systems to represent interactions between instances.

Normally, product modeling is used to represent the physical artifacts (e.g.: doors, walls, windows, pumps and pipes) that exist in a domain. However, tasks such as those which exist in a construction plan can also be represented by frame systems. Using product modeling in this manner is called "process modeling". The test bed incorporates an agent-based system (the Agent Collaboration Environment) and a design environment (Design++) to provide integrated product and process modeling. These two systems are described in the following paragraphs.

Agent Collaboration Environment: The Agent Collaboration Environment (ACE) is an agent-based software environment developed at USACERL which was specifically designed to support the delivery and sustainment of facilities. ACE supports the *Discourse Model* of collaboration [Case, 1995]. The Discourse Model uses knowledge level agents [Genesereth, 1988], that manipulate artifacts in a workspace. An agent represents the user process by manipulating or reacting to artifacts present in their workspace. The primary role of an agent in ACE is as an assistant that uses heuristic rules and a powerful checklist facility to automate routine tasks, thus enhancing productivity and ensuring repeatable work process quality.

Although most agents act under the user's direction, they can also run in the background and act in an advisory capacity. Experienced users can store their knowledge in agents for use by others. The true strength of ACE, however is tool integration. For example, ACE offers the possibility of blurring the distinction between data in CAD drawings, analysis programs, and contract specifications. This capability is described further in section 2.4. ACE provides a central database of frames, relations, and constraints that reduces redundant data input and the associated risk of human error and improves document consistency.

Design++: Design++ is a commercially available object oriented value based frame system. It is a design environment that assists researchers and developers in capturing corporate knowledge as end-user application. Design++ supports product models, libraries, tight CAD integration, GUI tools, and value propagation. Product models are based on part-of relations. Users can define relations between data. The libraries define objects, slots, values, inheritance, rule propagation as a when-modified demon, and facets. With tight CAD integration, CAD objects reflect changes in the graphical representation of the product information as attributes are manipulated through rule propagation. Finally, a suite of applications tools assist in defining GUI interactions and rule propagation behavior.

2.3 Shared Ontology

In order to share information between workers using design environments and agent-based systems, it is necessary to provide either a common representation (a shared ontology) or a translation from one representation to another. For heterogeneous closely-coupled collaboration (participants working closely together), the shared ontology method of information sharing is appropriate. In this approach, workers must agree upon a set of concepts to be contained in the ontology. Using the shared ontology as a guideline, identical libraries of frames can be generated in disparate design environments and agents systems which incorporate frame systems. This methodology was utilized by the test bed.

2.4 Visualization

When large amounts of data exist (which is usually the case in the facility delivery process), it is essential that this information is able to be displayed in a graphical manner [Sriram, 1992]. Therefore, an important element of the systems used in the test bed is their ability to graphically represent objects that are defined in libraries of frames. The ability to communicate with CAD systems is referred to as "CAD interface". ACE and Design++, although the implementation is dissimilar, use a similar methodology for the graphical display of the electronic data. The CAD interface in these systems provides a two-way link from the product model representation to a graphical representation in CAD. Groups of lines and arcs can be displayed in a CAD drawing to represent individual instances of data. The CAD interface provides a mechanism for changes made in the CAD drawing to be automatically reflected in the associated objects. Likewise, if a user edits an object directly, the rendering can be re-drawn to reflect these changes.

2.5 Communication

Using the Discourse Model, the individual workspaces used by each user of ACE are joined together into *Virtual Workspace (VWS)* that supports asynchronous and distributed collaboration. Although users are working in ACE on their own computers, there appears to be one large workspace. A set of workspaces that are joined together in a virtual workspace is called a *group*. Workspaces communicate using an electronic mail system called *VWS-Mail*. Users can specify an "interest set" which serves as a filter of mail messages allowing the user to receive only mail relevant to their task. Design++ was extended to include the VWS metaphor allowing agents developed in Design++ to collaborate with other agents. The kernel of the VWS architecture was implemented within the Design++ development environment.

The information exchange between applications in the two systems is accomplished using an approach called the *Virtual Workspace Language (VWL)* [Case, 1995]. The VWL is a protocol for communication between programs having disparate representations. The VWL incorporates the Knowledge Interchange Format (KIF) and the Knowledge Query and Manipulation Language (KQML) [Finin, 1994]. KQML performatives describe messages concerning information or knowledge being communicated from one program to another.

3. Test Bed

The intent of the test bed is to illustrate how IPPD principles and collaboration technologies can be used to facilitate the facility delivery process. The overall goal is to build a collaborative environment for professionals in the Architecture,

Engineering and Construction field. To create such an environment, the test bed utilizes the following: (1) workspaces that allows participants to create and manipulate product and process information, (2) a common communication protocol with which participants can share information and be part of a collaborating group, (3) a capability for participants to use software agents, (4) an open architecture that allows participants to link existing analysis tools into the collaborative group environment, and (5) a linkage between industry standard CAD graphics programs and a database of design information.

A scenario was chosen which would serve as a basis for the development of the test bed. The scenario involved the interaction of participants representative of those in the facility delivery process. The test bed demonstrates how interaction between project managers, architects, construction managers, and operation & maintenance personnel might take place under an IPPD implementation. With the scenario as a guideline, a shared ontology was created, frame libraries were developed in ACE and Design++, and agents to aide each of the participants were developed. The test bed was then demonstrated and results analyzed. The remainder of section 3 describes the domain selection, scenario creation, and software agents respectively.

3.1 Domain Selection

The scope identified for the test bed was the concept design phase of facility delivery. The facility design phases are divided into three distinct parts, Schematic Design (0- 10%), Design Development (11-35%), and Construction Documents (36%-100%) [Haviland, 1987]. Within the Corps of Engineers, the Schematic Design and Design Development phases are combined as the Concept Design phase. The Concept Design phase was chosen for the test bed because of the potential impact of decisions made at this time on the life cycle cost of a facility. Decisions made in the Concept Design phase can greatly affect life cycle cost of the facility and subsequent change made to the decisions made during this phase can be extremely costly.

The facility type selected for the demonstration was a Fire Station from the US Army Facility Standardization Program. The Fire Station's manageable size and complexity made it a desirable example building type. The Fire Station is divided into three areas; administration, dormitory and apparatus. Each area includes several sub requirements, resulting in a compact, complex facility type. These requirements make the firestation an interesting example of the necessary interactions between facility delivery personnel.

3.2 Scenario Creation

The test bed scenario revolves around the concept design phase (up to 35% design) of the design of a standard U.S. Army Fire Station. To develop a realistic test bed scenario using the fire station, drawings, specifications, and design review comments of two fire stations were studied. Along with ten other facilities types, the fire station design review comments were analyzed using a categorization scheme developed for another study [Brucker, 1995]. The categorization scheme classifies each comment by its location in the documents, problem type, building system area, and related disciplines. Examples of errors spotted during the review include designs which provide interior access to mechanical rooms (a violation of Army regulations) and include exhaust systems in the apparatus are which are innapropriate for the fire truck engines type. Examples such as these were identified as having occurred due to a lack of coordination and were used to develop the scenario.

The scenario begins with the a project manager creating a project group and using the project management agent to create building requirements for a new fire station facility. These requirements were broadcast via VWS-Mail to all members of the project group. The architect, upon receiving the requirements, used the architect agent to create a preliminary design which was broadcast to the group. The O&M

and construction personnel received the design and used their agents to help spot design mistakes and to develop construction plans respectively. The demonstration ended with the construction planning agent providing a visualization of the construction plan.

3.3 Software Agents

A shared ontology which included descriptions of architectural zones, building components, construction time and cost components, and functional requirements was created. Libraries of frames mimicking the shared ontology were developed in the two environments ACE and Design++. Using the common library representation, heterogeneous agents could then be created to communicate about objects. The following paragraphs describe the agents created for the test bed.

Project Management Agent: In this test bed scenario the project manager is responsible for setting up the work group, establishing a project schedule, and monitoring project milestones. The Project Manager provides initial functional requirements for the building using required government documents. The project manager uses project management software (Microsoft Project) to plan the project. After resource leveling and scheduling activities and milestones, the project management agent broadcasts the work assignments, through ACE, to the project participants and requests acceptance of the work assignments. Participants acknowledge the assignment on their local workstations.

Architect Agent: The agent that assists the architect in the demonstration is the _Standard Facility Designer (SFD)_. SFD has knowledge about facilities in the Facility Standardization Program of the Corps of Engineers. SFD is an agent developed in the Design++ development environment with the VWS extensions. When the functional requirement for a fire station has been identified and communicated to the SFD through the VWS, SFD can quickly produce a preliminary design. The factors influencing the design are the number of station companies, commander location and site considerations. The preliminary design includes representations of room and architectural zone information, walls, doors, windows, roof structure, and exterior enclosure. After SFD produces a preliminary design, the architect modifies the design as required per professional judgment. When the architect is ready to share the preliminary design with other interested participants, she publishes the information through the VWS kernel.

Operation and Maintenance Agent: The Operation and Maintenance (O&M) agent was created to demonstrate how error checking agents can automate verifying the correctness of existing design as per specifications in documents such as Corps of Engineers guide specifications and/or Engineering Technical Letters (ETL). The O&M agent was written in ACE and contains rules which detect errors in the design of mechanical rooms. Once errors are detected, the agent will notify the user of the error and display an ETL (in the form of a Windows help document) to the user and bring them to the exact point in the ETL where reference is made to the detected error. This agent has the capability of utilizing the CAD interface to help the user locate faulty elements of a design. When an object such as a door violates one of the rules in the O&M agent, the agent can change the color of the offending object in a CAD drawing. In this manner, the user working in CAD can be notified of the exact location of errors in the design drawing.

Construction Planning Agent: The construction planning agent uses a model-based planning approach to generate a preliminary construction plan including both time and cost components based on the design generated by the architect [Ganeshan, 1995]. The agent uses knowledge-bases to generate activities (and associated quantities of work) required to install component types and select construction methods for each activity. The resources required for each activity in the plan are obtained from the MCACES unit price databases. These databases also provide productivity and unit cost (labor, material and equipment) information which is used to compute activity durations and costs. The construction planning agent provides an interface with traditional project management software (Microsoft Project) which is used to compute start and finish times using the Critical Path Method and view and manipulate the schedule. A schedule visualization capability allows the user to step through any portion of the construction plan. This capability is provided using the CAD interface and the interface with Microsoft Project. The agent provides feedback from construction time and cost perspective. It also helps planners and contractors to understand design intent, order long lead time components, check interferences, check schedules, and can help coordinate different trades. The agent also includes a capability to select alternatives for building systems that have not been determined in the architect's design. The alternatives are selected from a database of pre-defined assemblies used in previous projects. This capability allows the agent to propose alternatives that may have been overlooked by the designers.

4. Results

A demonstration of the test bed was presented at USACERL in which members of the Collaborative Technologies Division acted as facility delivery personnel and used the software agents to facilitate their work. The test bed demonstrated how collaboration technology and IPPD principles can be applied to the facility delivery process. Participants representitive of those encountered during the design and construction of a facility were able to create, modify, share and visualize electronic information and to be aided by agents. Software agents were written in ACE and Design++ which demonstrated powerful capabilities to facilitate work. This method of collaboration allows early feedback during the facility delivery process.

The VWS proved to be an effective vehicle for the sharing of information. By declaring interest sets, participants were able to filter the incoming VWL statements, thereby reducing the imported information to a minimally-necessary set.

Although the shared ontology approach enabled the sharing of information, this approach proved to be inefficient, in that, libraries of frames mimicking this ontology had to be created for each software environment. The frame libraries were prone to inconsistencies.

The use of CAD interfaces makes it unnecessary to send CAD drawings from one participant to another. Design information shared between participants can be re-displayed using drawing agents. This will provide a decrease in information loss between participants using this technology.

5. Discussion and Future Work

Implementation of IPPD in the facility delivery process shows potential to significantly improve the quality of decision making, documentation and the facility itself. Through agent-assisted collaboration, the extended group can work as a coordinated team. Software agents can assist personnel by helping them make more informed decisions as well as by automating repetative tasks. However, further research is necessary to fully investigate the use of collaborative technologies and to validate the expected improvements in efficiency and design quality.

A scenario for a new test bed involving a more complex building type is currently under development. The scope of the demonstration will be extended to include agents aiding in roof design and material selection, as well as energy analysis. Along with the increased scope, the following modifications to the test bed scenario will be considered:

- *Conflict:* Although the software systems used in the test bed provide a mechanism to detect, negotiate and resolve conflicts among participants, this capability was not utilized. Future work will include an in-depth examination of how conflicts can be resolved among the agents and users.

- *Object server:* To address the inefficiencies of the shared ontology approach, future research will test the feasibility of an "object server". The object server will be responsible for providing frame descriptions of unknown objects that a participant receives via the VWS, thereby extending the ontology in a user's workspace. It is quite possible that these representation schemes will become standard within a discipline and/or domain and that an object server method of sharing representations will effectively facilitate the sharing of objects.

- *Assemblies:* It is often desirable to copy portions of an existing design into new work to reduce duplication of effort. Future research will investigate how an IPPD approach can facilitate storing and sharing of groups of objects or "assemblies".

References

[Brucker, 1995] Brucker, B. A., Golish, L. M., Ts.Ang, A., "An Insight Towards Measuring the Design Process, Phase I: Content Analysis of Army Design Reviews," Environmental Design Research Association conference proceedings, 1995.

[Building Systems Design, 1992] Building Systems Design, 1992 "MCACES Gold Version 5.2 User Manual," BSD, Inc., Atlanta, GA.

[Case, 1995] Case, M. P. and Lu, S. "The Discourse Model for Collaborative Engineering Design," to appear in the Int. Journal for Computer Aided Design, 1995.

[Finin, 1994] Finin, T., Weber, J., Genesereth, M., "Specification of the KQML Agent-Communication Language," draft request for comments, University of Maryland, College Park, MD, February, 1994.

[Fischer, 1994] Fischer, M., Froese, T., Phan, D. H. D., "How do Data Models and Integration Add Value to a Project," Computing in Civil Engineering, Proceedings of the First Congress in Conjunction with A/E/C Systems '94, American Society of Civil Engineers, 1994, pp.992-997.

[Ganeshan, 1995] Ganeshan, R., Stumpf, A., Chin, S., Liu, L., Harrison, W., "Integrating Object-Oriented CAD and Rule-Based Technologies for Construction Planning," Technical Report, U. S. Army Construction Engineering Research Laboratories, in preparation, 1995.

[Genesereth, 1988] Genesereth, M. R., and Nilsson, N.J., "Logical Foundations of AI," Palo Alto, CA, Morgan Kaufmann, 1988, pp. 314.

[Haviland, 1987] Haviland, D., "The Project," Architect's Handbook of Professional Practice, The American Institute of Architects, Volume 2.1, 1987.

[Kambhampati, 1990] Kambhampati, S., and Cutkosky, M. R., "An Approach Toward Incremental and Interactive Planning for Concurrent Product and Process Design," Proceedings of the ASME Winter Annual Meeting, Issues in Design/Manufacture Integration, DE-Vol. 29, November 25-30, 1990, pp. 1-8.

[Kautz, 1994] Kautz, H., Selman, B., Coen, M., Ketchpel S., "An experiment in the Design of Software Agents," Proceedings of the Twelfth National Conference on Artificial Intelligence, AAAI-94, Seattle, WA, 1994.

[Loss, 1991] Loss, F., Earle, K. W., "Performance Failures in Buildings & Civil Works," Architecture and Engineering Performance Information Center, School of Architecture, University of Maryland, College Park, MD, 1991.

[Sriram, 1992] Sriram, D., "Papers on Computer-Aided Collaborative Engineering," Intelligent Engineering Systems Laboratory, Dept. of Civil and Environmental Engineering, 253, M.I.T., Cambridge, MA, 1992.

Control Concepts for Integrated Product and Process Models

Chang-Ho Lee[1] and Richard Sause[2], Member, ASCE

Abstract

Design product and process models are used to organize and represent information and activities involved in design. They are used in the development of computer integrated design systems. An entity-based approach is one way to integrate product and process models. The paper presents certain control concepts for entity-based integrated product and process models. The control concepts describe how product entities and process entities are created during the course of the design process. The concepts are illustrated using a structural frame design example, and a formal notation for the control concepts is suggested.

Introduction

Computer integrated systems for structural design support engineers by representing design information and helping to manage the design process. Formal models of the design product and process serve as conceptual tools in the development of computer integrated systems (Sause et al. 1992). A product model describes information created during the design process, and a process model describes the associated activities. Product and process models can be integrated into a single model. An entity-based approach has been proposed for this integrated model (Hong and Sause 1994). A model of this type is defined by a "generalization" model which includes product entity categories and process entity categories. When the generalization model is applied to a specific design project, an "instance" model is created, which includes specific product entities and specific process entities. The paper shows how an instance model is developed from a generalization model, and develops certain control concepts that are needed. A formal notation for representing the control concepts needed for the entity-based model is suggested.

Example of Entity-based Integrated Model

A structural frame design example is shown in Figure 1 for a frame designated as "frame A". The design constraints of the frame are shown in Figure 1a. The frame

---

[1]Graduate Student, Dept. of Civil and Env. Eng., Lehigh University, Bethlehem, PA 18015
[2]Assistant Professor, Dept. of Civil and Env. Eng., Lehigh University, Bethlehem, PA 18015

should carry the applied loads, and the clear space should be maintained. Two types of frames, a rigid frame and a braced frame, are considered in Figure 1b. Each frame alternative includes members such as girder G1, column C1, and brace Br1. Alternatives for each member are considered. For example, WF girder 1 and joist girder 1 are developed for girder G1. One frame alternative as well as one alternative for each member should be selected and developed to complete the design of frame A. The selected frame alternative is shown in Figure 1c. The information and activities involved in the design of frame A can be represented using an entity-based integrated product and process model (Hong and Sause 1994). In the modified E-R diagram developed for these entity-based models, the representation shown in Figure 2 is used for product and process entity categories and product and process entities (i.e., instances).

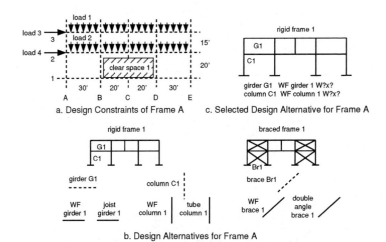

a. Design Constraints of Frame A   c. Selected Design Alternative for Frame A

b. Design Alternatives for Frame A

Figure 1. Example of Design of Frame A

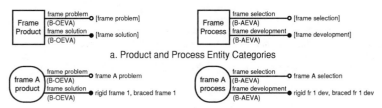

a. Product and Process Entity Categories

b. Product and Process Entities

Figure 2. Entity Representation

Figure 3 shows an entity-based integrated product and process generalization model for frame design. A text-based version of the entity representation is used in Figure 3. For example, a text-based version of the frame product entity category in Figure 2a is shown in Figure 3a. The product entity categories are called object entities, and the process entity categories are called activity entities. The integrated model includes only a few of the entity categories needed for frame design, and the attributes of some activity entity categories are not shown due to space limitations. The product entity categories include: (1) problem entity categories that describe design problems and (2) solution entity categories that describe proposed solutions to design problems (Hong and Sause 1994). Thus there are frame problem and solution entity categories in Figure 3a. Similarly the process entity categories include: (1) selection entity categories that describe the generation, comparison, and selection of a number of design alternatives for a design problem and (2) development entity categories that describe the development of a single design altenative for a design problem. Thus there are frame selection and development

```
OBJECT_ENTITY [frame product]
frame problem   (B-OEVA)   SV   [frame problem]
frame solution  (B-OEVA)   MV   [frame solution]

OBJECT_ENTITY [frame problem]
grid lines      (B-OEVA)   MV   [grid line]
loads           (B-OEVA)   MV   [load]
clear spaces    (B-OEVA)   MV   [clear space]

OBJECT_ENTITY [frame solution]
SUPERCATEGORY OF [rigid frame] AND [braced frame]
girders         (B-OEVA)   MV   [girder]
columns         (B-OEVA)   MV   [column]
joints          (B-OEVA)   MV   [joint]

OBJECT_ENTITY [rigid frame]
SUBCATEGORY OF [frame solution]
girders         (B-OEVA)   MV   [girder]
columns         (B-OEVA)   MV   [column]
joints          (B-OEVA)   MV   [joint]

OBJECT_ENTITY [braced frame]
SUBCATEGORY OF [frame solution]
girders         (B-OEVA)   MV   [girder]
columns         (B-OEVA)   MV   [column]
joints          (B-OEVA)   MV   [joint]
braces          (B-OEVA)   MV   [brace]
```

a. Product Entity Categories

```
ACTIVITY_ENTITY [frame process]
frame selection    (B-AEVA)   SV   [frame selection]
frame development  (B-AEVA)   MV   [frame development]

ACTIVITY_ENTITY [frame selection]
formulation              (B-AEVA)   MV   [fr prob formulation]
alternative identification  (B-AEVA)   MV   [fr alt identification]
alternative development     (B-AEVA)   MV   [fr development]
comparison                  (B-AVA)    MV   [fr alt comparison]
elimination or selection    (B-AVA)    MV   [fr alt elim or selection]
selected alt completion     (B-AEVA)   SV   [fr development]

ACTIVITY_ENTITY [frame problem formulation]

ACTIVITY_ENTITY [frame alternative identification]

ACTIVITY_ENTITY [frame development]
SUPERCATEGORY OF [rigid frame dev] AND [braced frame dev]

ACTIVITY_ENTITY [rigid frame development]
SUBCATEGORY OF [frame development]

ACTIVITY_ENTITY [braced frame development]
SUBCATEGORY OF [frame development]
```

b. Process Entity Categories

Figure 3. Integrated Product and Process Generalization Model for Frame Design

```
OBJECT_ENTITY frame A product
frame problem   (B-OEVA)   SV   frame A problem
frame solution  (B-OEVA)   MV   rigid frame 1, braced frame 1

OBJECT_ENTITY frame A problem
grid lines      (B-OEVA)   MV   grid line A, B, C, D, E, 1, 2, 3
loads           (B-OEVA)   MV   load 1, 2, 3, 4
clear spaces    (B-OEVA)   MV   clear space 1

OBJECT_ENTITY rigid frame 1
girders   (B-OEVA)   MV   girder G1, ...
columns   (B-OEVA)   MV   column C1, ...
joints    (B-OEVA)   MV   joint 1, ...

OBJECT_ENTITY braced frame 1
girders   (B-OEVA)   MV   girder G1, ...
columns   (B-OEVA)   MV   column C1, ...
joints    (B-OEVA)   MV   joint 1, ...
braces    (B-OEVA)   MV   brace Br1, ...
```

a. Product Entities

```
ACTIVITY_ENTITY frame A process
frame selection    (B-AEVA)   SV   frame A selection
frame development  (B-AEVA)   MV   rigid fr 1 dev, braced fr 1 dev

ACTIVITY_ENTITY frame A selection
formulation              (B-AEVA)   MV   fr A  prob formulation 1
alternative identification  (B-AEVA)   MV   fr A alt identification 1
alternative development     (B-AEVA)   MV   rigid fr 1 dev, brcd fr 1 dev
comparison                  (B-AVA)    MV   fr alt comparison 1
elimination or selection    (B-AVA)    MV   fr alt elim or selection 1
selected alt completion     (B-AEVA)   SV   rigid frame 1 dev

ACTIVITY_ENTITY frame A problem formulation 1

ACTIVITY_ENTITY frame A alternative identification 1

ACTIVITY_ENTITY rigid frame 1 development

ACTIVITY_ENTITY braced frame 1 development
```

b. Process Entities

Figure 4. Integrated Product and Process Instance Model for Design of Frame A

entity categories in Figure 3b. Figure 4 shows an integrated product and process instance model for the design of frame A. It is obtained by applying the integrated product and process generalization model in Figure 3 to the design of frame A. The instance model also includes only a few of the entities needed for the design of frame A due to space limitations.

Evolution of Instance Model

The integrated product and process instance model in Figure 4 represents the information which is created and the activities which are carried out during the design process. However, the model does not show how the information is created and the activities are carried out *over time*. Figures 5 and 6 show two possible examples of how the information is created and the activities are carried out during design of frame A. Each figure represents an evolution of the instance model for the design of frame A because the entities included in the model and the state of the entities in the model change during the design process. For example, in Figure 5a1 the frame solution attribute of the frame A product entity does not yet have a value, but in Figure 5a2 this attribute has two values, rigid frame 1 and braced frame 1. Figures 5 and 6 show only a small part of the evolution of an integrated product and process instance model for the design of frame A. The attributes of many of the process entities are not shown due to space limitations.

Example 1
(1) State 1 (Figures 5a1 and 5b1): The product entities, frame A product and frame A problem, are created when the process entities, frame A process, frame A selection, and frame A problem formulation 1, are initiated and completed. The attributes of the frame A problem formulation 1 process entity are not shown due to space limitations.
(2) State 2 (Figures 5a2 and 5b2): The product entities, rigid frame 1 and braced frame 1, are created when the frame A alternative identification 1 process entity is initiated and completed.
(3) State 3 (Figures 5a3 and 5b3): The layouts of rigid frame 1 and braced frame 1 are created and stored in the corresponding product entities when the process entities, rigid frame 1 development and braced frame 1 development, are initiated and completed.

Example 2
(1) State 1 (Figures 6a1 and 6b1): Same as in Example 1.
(2) State 2 (Figures 6a2 and 6b2): The rigid frame 1 product entity is created when the frame A alternative identification 1 process entity is initiated and completed.
(3) State 3 (Figures 6a3 and 6b3): The layout of rigid frame 1 is created and stored in the rigid frame 1 product entity when the rigid frame 1 development process entity is initiated and completed.
(4) State 4 (Figures 6a4 and 6b4): The braced frame 1 product entity is created when the frame A alternative identification 2 process entity is initiated and completed.
(5) State 5 (Figures 6a5 and 6b5): The layout of braced frame 1 is created and stored in the braced frame 1 product entity when the braced frame 1 development process entity is initiated and completed.

The two examples differ in the time that product entities are created. In Example 1, both the rigid frame 1 and braced frame 1 product entities are created when the frame A alternative identification 1 process entity is initiated and completed. In Example 2, the rigid frame 1 product entity is created when the frame A alternative identification 1 process entity is initiated and completed, but the braced frame 1 product entity is created when the frame A alternative identification 2 process entity is initiated and completed. The difference in the time that product entities are created is due to differences between the examples in the sequence that process entities are initiated. In example 1, the frame A problem formulation 1 process entity is initiated in Figure 5b1. It is followed by the frame A alternative identification 1 process entity, the rigid frame 1 development process entity, and the braced frame 1 development process entity in Figures 5b2 and 5b3. That is, the formulation, alternative identification, and alternative development attribute values are initiated in sequence (with two values for the alternative development attribute). In Example 2, the alternative identification and alternative development attribute values are initiated iteratively (with two iterations) in Figures 6b2 through 6b5.

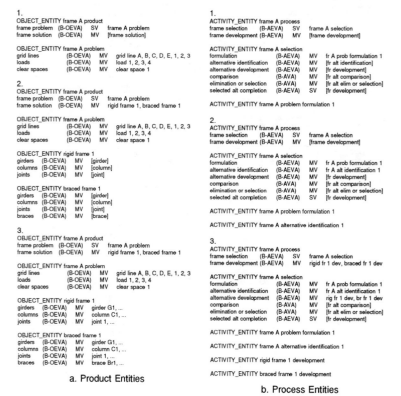

a. Product Entities

b. Process Entities

Figure 5. Example 1 of Evolution of Integrated Instance Model

1.

OBJECT_ENTITY frame A product
frame problem (B-OEVA) SV frame A problem
frame solution (B-OEVA) MV [frame solution]

OBJECT_ENTITY frame A problem
grid lines (B-OEVA) MV grid line A, B, C, D, E, 1, 2, 3
loads (B-OEVA) MV load 1, 2, 3, 4
clear spaces (B-OEVA) MV clear space 1

2.

OBJECT_ENTITY frame A product
frame problem (B-OEVA) SV frame A problem
frame solution (B-OEVA) MV rigid frame 1

OBJECT_ENTITY frame A problem
grid lines (B-OEVA) MV grid line A, B, C, D, E, 1, 2, 3
loads (B-OEVA) MV load 1, 2, 3, 4
clear spaces (B-OEVA) MV clear space 1

OBJECT_ENTITY rigid frame 1
girders (B-OEVA) MV [girder]
columns (B-OEVA) MV [column]
joints (B-OEVA) MV [joint]

3.

OBJECT_ENTITY frame A product
frame problem (B-OEVA) SV frame A problem
frame solution (B-OEVA) MV rigid frame 1

OBJECT_ENTITY frame A problem
grid lines (B-OEVA) MV grid line A, B, C, D, E, 1, 2, 3
loads (B-OEVA) MV load 1, 2, 3, 4
clear spaces (B-OEVA) MV clear space 1

OBJECT_ENTITY rigid frame 1
girders (B-OEVA) MV girder G1, ...
columns (B-OEVA) MV column C1, ...
joints (B-OEVA) MV joint 1, ...

4.

OBJECT_ENTITY frame A product
frame problem (B-OEVA) SV frame A problem
frame solution (B-OEVA) MV rigid frame 1, braced frame 1

OBJECT_ENTITY frame A problem
grid lines (B-OEVA) MV grid line A, B, C, D, E, 1, 2, 3
loads (B-OEVA) MV load 1, 2, 3, 4
clear spaces (B-OEVA) MV clear space 1

OBJECT_ENTITY rigid frame 1
girders (B-OEVA) MV girder G1, ...
columns (B-OEVA) MV column C1, ...
joints (B-OEVA) MV joint 1, ...

OBJECT_ENTITY braced frame 1
girders (B-OEVA) MV [girder]
columns (B-OEVA) MV [column]
joints (B-OEVA) MV [joint]
braces (B-OEVA) MV [brace]

5.

OBJECT_ENTITY frame A product
frame problem (B-OEVA) SV frame A problem
frame solution (B-OEVA) MV rigid frame 1, braced frame 1

OBJECT_ENTITY frame A problem
grid lines (B-OEVA) MV grid line A, B, C, D, E, 1, 2, 3
loads (B-OEVA) MV load 1, 2, 3, 4
clear spaces (B-OEVA) MV clear space 1

OBJECT_ENTITY rigid frame 1
girders (B-OEVA) MV girder G1, ...
columns (B-OEVA) MV column C1, ...
joints (B-OEVA) MV joint 1, ...

OBJECT_ENTITY braced frame 1
girders (B-OEVA) MV girder G1, ...
columns (B-OEVA) MV column C1, ...
joints (B-OEVA) MV joint 1, ...
braces (B-OEVA) MV brace Br1, ...

a. Product Entities

1.

ACTIVITY_ENTITY frame A process
frame selection (B-AEVA) SV frame A selection
frame development (B-AEVA) MV [frame development]

ACTIVITY_ENTITY frame A selection
formulation (B-AEVA) MV fr A prob formulation 1
alternative identification (B-AEVA) MV [fr alt identification]
alternative development (B-AEVA) MV [fr development]
comparison (B-AVA) MV [fr alt comparison]
elimination or selection (B-AVA) MV [fr alt elim or selection]
selected alt completion (B-AEVA) SV [fr development]

ACTIVITY_ENTITY frame A problem formulation 1

2.

ACTIVITY_ENTITY frame A process
frame selection (B-AEVA) SV frame A selection
frame development (B-AEVA) MV [frame development]

ACTIVITY_ENTITY frame A selection
formulation (B-AEVA) MV fr A prob formulation 1
alternative identification (B-AEVA) MV fr A alt identification 1
alternative development (B-AEVA) MV [fr development]
comparison (B-AVA) MV [fr alt comparison]
elimination or selection (B-AVA) MV [fr alt elim or selection]
selected alt completion (B-AEVA) MV [fr development]

ACTIVITY_ENTITY frame A problem formulation 1
ACTIVITY_ENTITY frame A alternative identification 1

3.

ACTIVITY_ENTITY frame A process
frame selection (B-AEVA) SV frame A selection
frame development (B-AEVA) MV rigid fr 1 dev

ACTIVITY_ENTITY frame A selection
formulation (B-AEVA) MV fr A prob formulation 1
alternative identification (B-AEVA) MV fr A alt identification 1
alternative development (B-AEVA) MV rig fr 1 dev
comparison (B-AVA) MV [fr alt comparison]
elimination or selection (B-AVA) MV [fr alt elim or selection]
selected alt completion (B-AEVA) SV [fr development]

ACTIVITY_ENTITY frame A problem formulation 1
ACTIVITY_ENTITY frame A alternative identification 1
ACTIVITY_ENTITY rigid frame 1 development

4.

ACTIVITY_ENTITY frame A process
frame selection (B-AEVA) SV frame A selection
frame development (B-AEVA) MV rigid fr 1 dev

ACTIVITY_ENTITY frame A selection
formulation (B-AEVA) MV fr A prob formulation 1
alternative identification (B-AEVA) MV fr A alt identification 1, 2
alternative development (B-AEVA) MV rigid fr 1 dev
comparison (B-AVA) MV [fr alt comparison]
elimination or selection (B-AVA) MV [fr alt elim or selection]
selected alt completion (B-AEVA) SV [fr development]

ACTIVITY_ENTITY frame A problem formulation 1
ACTIVITY_ENTITY frame A alternative identification 1
ACTIVITY_ENTITY rigid frame 1 development
ACTIVITY_ENTITY frame A alternative identification 2

5.

ACTIVITY_ENTITY frame A process
frame selection (B-AEVA) SV frame A selection
frame development (B-AEVA) MV rigid fr 1 dev, braced fr 1 dev

ACTIVITY_ENTITY frame A selection
formulation (B-AEVA) MV fr A prob formulation 1
alternative identification (B-AEVA) MV fr A alt identification 1, 2
alternative development (B-AEVA) MV rig fr 1 dev, brcd fr 1 dev
comparison (B-AVA) MV [fr alt comparison]
elimination or selection (B-AVA) MV [fr alt elim or selection]
selected alt completion (B-AEVA) SV [fr development]

ACTIVITY_ENTITY frame A problem formulation 1
ACTIVITY_ENTITY frame A alternative identification 1
ACTIVITY_ENTITY rigid frame 1 development
ACTIVITY_ENTITY frame A alternative identification 2
ACTIVITY_ENTITY braced frame 1 development

b. Process Entities

Figure 6. Example 2 of Evolution of Integrated Instance Model

Flexible Design Process

Figure 7 shows the flexibility that is possible in the sequence of initiation of the attributes of the frame selection process entity category (from Figure 3b). The possible sequences are obtained from the study of the evolution of instance models such as Figures 5 and 6. The formulation attribute value (i.e. frame problem formulation) is initiated first. It is followed by the alternative identification attribute value. Another formulation attribute value can be initiated (to reformulate the problem if needed) or the alternative development attribute value can follow the completion of an alternative identification attribute value. Then other formulation, alternative identification, and alternative development attribute values can be initiated or the comparison attribute value can be initiated. After the elimination or selection attribute value is initiated, other formulation, alternative identification, and alternative development attribute values can be initiated. Finally the selected alternative completion attribute value is initiated.

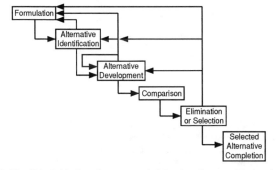

Figure 7. Flexible Initiation Sequence of Selection Process Entity Attributes

Representation of Flexible Design Process

A context-free grammar shown in Figure 8 is used to represent the flexible sequence in Figure 7. A context-free grammar is a widely accepted notation for specifying the syntactic structures of a programming language (Wood 1987). It is used in the paper for specifying the initiation sequence of process entity attributes. A context-free grammar is described by listing its productions. The grammar in Figure 8 includes five productions. Each production consists of the left side of the production, an arrow and the right side. The right side is separated by the symbol l. It indicates alternative right sides. A lowercase word such as "formulation" is the name of an attribute. An uppercase letter such as A and B leads to another production. S, the left side of the first production, is the start symbol of the grammar.

Figure 9 shows how one initiation sequence of the selection process entity attribute values can be derived from the grammar in Figure 8. The derivation starts from the first production with one of its alternative right sides. The uppercase letter B leads to another production, and the letter is replaced by one of the right sides of the production for B. The letter D is similarly replaced. Finally one possible initiation sequence of attribute values is obtained (the last line in Figure 9). The formulation attribute value is initiated first. It is followed by the alternative identification and alternative development attribute values iteratively. Then the comparison, elimination or selection, and selected alternative completion attribute values are initiated.

```
S -> formulation alt_id A alt_dev B comparison elim_or_sel C selected_alt_com
  | formulation alt_id alt_dev B comparison elim_or_sel C selected_alt_com
  | formulation alt_id A alt_dev comparison elim_or_sel C selected_alt_com
  | formulation alt_id A alt_dev B comparison elim_or_sel selected_alt_com
  | formulation alt_id alt_dev comparison elim_or_sel C selected_alt_com
  | formulation alt_id A alt_dev comparison elim_or_sel selected_alt_com
  | formulation alt_id alt_dev B comparison elim_or_sel selected_alt_com
  | formulation alt_id alt_dev comparison elim_or_sel selected_alt_com
A -> formulation alt_id A | formulation alt_id
B -> D B | D
D -> A alt_dev | alt_dev | alt_id A alt_dev | alt_id alt_dev
  | formulation alt_id A alt_dev | formulation alt_id alt_dev
C -> D B comparison elim_or_sel C | D comparison elim_or_sel C
  | D B comparison elim_or_sel | D comparison elim_or_sel
```

Figure 8. Grammar for Design Process

```
S -> formulation alt_id alt_dev B comparison elim_or_sel selected_alt_com
S -> formulation alt_id alt_dev D comparison elim_or_sel selected_alt_com
S -> formulation alt_id alt_dev alt_id alt_dev comparison elim_or_sel selected_alt_com
```

Figure 9. Example of Derivation in Grammar

Conclusions

The paper has presented control concepts for integrated product and process models which focus on the sequences in which process entity attribute values (i.e. other more detailed activities or actions) are initiated over time during the design process. The control concepts provide flexibility to the development of integrated product and process instance models as the design process proceeds. The control concepts are discussed for only a few of the entities needed for the frame design process, and more detailed descriptions of control concepts are needed to cover the complete frame design process.

Acknowledgements

This research has been supported by the Engineering Research Center for Advanced Technology for Large Structural Systems (ATLSS) at Lehigh University under NSF Grant ECD-8943455. Opinions, findings, and conclusions are those of the authors and do not necessarily reflects the views of NSF.

References

Hong, N. K. and Sause, R. (1994), "Toward Integrated Models for Structural Design", *First Congress on Computing in Civil Engineering*, ASCE

Sause, R., Martini, K., and Powell, G. H. (1992), "Object-Oriented Approaches for Integrated Engineering Design Systems", *Journal of Computing in Civil Engineering*, ASCE, 6(3), pp. 248-265

Wood, D. (1987), *Theory of Computation*, John Wiley & Sons, Inc.

Case Combination and Adaptation of Building Spaces

Ian F.C. Smith [1] D. Kurmann [2] and Gerhard Schmitt [2]

Abstract

Case based design involves processes of retrieval and adaptation. When more than one case contributes to the design, an additional process, case combination, is employed. We focus upon adaptation and combination of cases. Recent work involving building design has developed methods for adapting exact cases through a geometrical parameterization which occurs at run time according to the characteristics of the new context. Three methods have been developed for geometrical modification of cases, dimensional adaptation, topological adaptation and case combination. This paper reviews methods developed for run-time parameterization of spatial constraints and presents a new system, IDIOM - Interactive Design using Intelligent Objects and Models. In IDIOM, case adaptation and combination are applied incrementally on building spaces in order to interactively compose floor layouts. While the IDIOM approach more closely resembles the use of cases by some designers, reuse of parts of buildings sacrifices the holistic integrity of the original building. Integrity is restored through user interaction and through activation of design knowledge as building parts are assembled.

1 Introduction

Designers employ previous designs in order to cope with complexity. It is often difficult to consider all aspects of a design, including necessary tradeoffs, in an adequate manner. Previous designs provide ready made solutions, albeit for different contexts. Throughout our paper it is assumed that the designer *knows* which design solutions are appropriate; the focus of this contribution is case adaptation.

---

[1] Federal Institute of Technology (EPFL), LIA/DI, 1015 Lausanne, Switzerland
[2] Federal Institute of Technology (ETH), CAAD, 8093 Zurich, Switzerland

Using previous design solutions transforms synthesis tasks, involving unreliable abductive processes, into an analysis of the differences between the original design context and the new context. Since analysis involves deduction, it is inherently more reliable and easier than abduction for humans (and machines) to carry out, particularly when a closed-world assumption is unrealistic. Close-world assumptions are almost always unrealistic for complex design tasks.

Precompiled or generalized design cases can be used rapidly and effectively for many routine design tasks. However, the task of generalizing previous design solutions carries the assumptions that all new design contexts are known *a priori* and that designers will be able to manipulate all parameters of complex designs. Therefore, when generalizations are simple, design support is weak since results are predictable and when generalizations are complex, design support is also weak because designers are unable to adapt them effectively when they have many degrees of freedom. Moreover, it is difficult to ensure that the parameterization of complex design solutions is appropriate for all situations.

Much work in case-based design was motivated by the possibility of reusing design solutions. This work includes contributions towards mechanisms and hydraulic devices [19], human-computer interfaces [1], structural building frames [14], assembly sequences [15], structural connections [20], mechanical components [2], software [13] and architectural aspects of buildings [3, 6, 8]. Several shortcomings have been identified in this work. Firstly, it is often assumed that once a designer has identified which case is relevant to the current problem, the necessary adaptation of the old case is obvious and needs little computer support. When research has addressed adaptation, it has been limited to symbolic considerations [9, 10, 18]. Indeed, most case-based reasoning research has focused on discrete variables and therefore, there is little relevance to the spatial aspects treated in this work.

Recent work in Switzerland involving building design has developed methods for adapting *exact* cases through a geometrical parameterization which occurs *at run time* according to the characteristics of the new context [11]. Dimensional adaptation is supported in order to maintain the original topology of the case. Topological adaptation has been attempted for special cases of buildings using techniques such as shape grammars [12]. However, this has proven difficult to generalize for a range of buildings. In addition, run-time parameterization of topological adaptation knowledge for several abstractions has not been possible. This has lead to an investigation of combining two cases [5].

This paper reviews methods developed for run-time parameterization of spatial constraints and presents a system called IDIOM which applies these techniques to intelligent objects in an interactive manner. The use of dynamic constraint satisfaction methods with constraints derived from assumptions is examined and finally, the work is compared with related studies.

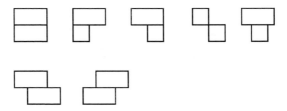

Figure 1: Examples of primitive topologies for rectangular spaces

2 Intelligent Objects - IDIOM

Development of IDIOM has drawn upon on algorithms elaborated during the
CADRE project [11]. In CADRE, cases are created automatically from CAD
drawing files and this results in a spatial topology and a set of base parameters.
Dimensional constraints on these parameters are then generated from an evaluation of the topology. For example, Figure 1 defines some primitive topologies for
two architectural spaces. Constraints are formulated in order to define conditions
for maintenance of topology, see Fig. 2. Fig. 2(a) shows the constraints related
to a primitive topology. Figure 2(b) gives an example of the constraints for two
spaces with a door between them.

Additional constraints are added to this set in order to take into account
of the environment around the building and structural engineering parameters.
Structural engineering constraints ensure that dimensional adaptation can be performed without violating criteria for strength and serviceability. Typically, these
constraints are the same for a range of topologies of similar types of structures.
All of these constraints form the *base-parameterization* which is stored with the
case.

An advantage of this approach is that, provided design criteria can be expressed as algebraic constraints, the process is independent of the discipline which
generated the requirement. Thus, horizontal integration of disciplines such as
structural engineering and architecture is possible through obtaining a parameterization which is simultaneously valid for both disciplines.

While CADRE is able to support dimensional adaptation efficiently, attempts
to carry out topological adaptation have not been as successful. This is partly
because it is difficult to represent and manipulate topological knowledge in a
generic manner. Thus far, success has been limited to certain classes of buildings
[12].

Subsequent research into case-based building design is motivated by two factors. The first factor arose through discussions with building designers. These
discussions resulted in the observation that although they frequently reuse designs, they rarely wish to adapt whole building cases. Often, the cases which are
most useful are spaces and collections of spaces [17].

The second factor is that most design domains cannot be modelled completely.

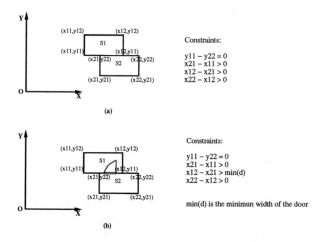

Constraints (a):

$y11 - y22 = 0$
$x21 - x11 > 0$
$x12 - x21 > 0$
$x22 - x12 > 0$

Constraints (b):

$y11 - y22 = 0$
$x21 - x11 > 0$
$x12 - x21 > \min(d)$
$x22 - x12 > 0$

$\min(d)$ is the minimun width of the door

Figure 2: Examples of parameterization with constraints on spaces.

Building designers know this intuitively since their designs are often influenced by complex considerations of social, political and economic factors. As a result, it is often the source of frustration when a design system performs automatic design, often providing just one design solution. A much better role for computer systems is to provide support for defining *allowable spaces of acceptable designs*. When exploring these spaces designers are able to introduce *their interpretation* of what is not modelled through adding constraints or through identifying design solutions for further development.

These two factors lead to the definition of an **intelligent object** that is used in this paper: an intelligent object is a *part of a real case* which can be interpreted for each new design task using models of their function, behavior and structure. Models provide explicit representations of physical principles, thereby avoiding the brittleness associated with traditional rule based systems.

For a design system containing intelligent objects to be successful, interactive design features are of great importance. Designers need tools for :

- identifying design spaces

- navigating within design spaces

- changing design-space shape according to problem specific interpretations

- creating and modifying links between models and design objects

Such needs are fulfilled in IDIOM through the use of advanced interfaces. One of the interfaces which is having an impact on the development of IDIOM is the software prototype 'Sculptor'. Sculptor creates a design environment that provides direct and intuitive access to a three-dimensional design model through interactive modelling of intelligent objects in a virtual design space [21]. Different kinds of knowledge are attributed to objects, thus improving interaction with other objects. Objects contain knowledge of themselves and their environment. Such aspects offer new possibilities for participatory design in which all building partners can examine new objects and make decisions. Built-in knowledge used for the direct interaction with an architectural model includes:

- Gravity. An object falls down if it is not supported by another objects or the ground.

- Collision detection provides a very intuitive way to examine building designs. Modification of objects may happen only under valid conditions, i.e., when solid objects are not intersecting

Constraints and relations between objects are specified interactively in three dimensions. This functionality of Sculptor allows the creation and manipulation of 3D scenes and models. In addition, Sculptor demonstrates other new techniques of man-machine interaction. The prototype supports specification of objects, models with attributes such as form, geometry, color, material, texture etc. It permits different points of view and functions for walk and fly through in 3D space. Using Sculptor, objects can be grouped together hierarchically. Objects, groups and virtual worlds can be changed by scaling, resizing, rotating, reshaping and moving them in space. All manipulations and scenes are rendered in real-time [16].

We have developed a system called Interactive Design using Intelligent Objects and Models (IDIOM) in an attempt to build upon successful aspects of CADRE and Sculptor. A dictionary definition for the word "Idiom" is

A phrase which means something different from the meaning of the separate words

This definition provides a useful analogy for a central goal of the IDIOM system, which is to provide support for incremental composition of designs while allowing specifications of the design space to include holistic considerations of groups of objects. These extra specifications originate for two sources; they are activated when certain groups of objects are present in the design and they are introduced by the designer incrementally as the design is composed.

Objects in IDIOM currently consist of spaces which are labeled according to their function. They contain windows, doors and furniture. The base parameterization of a case includes parametric relationships which define acceptable positions of doors, windows and furniture in order for the room to fulfill functional requirements. Since objects are parts of real cases, values for all parameters are included.

Designers proceed by selecting a case and placing it in the design environment. If the environment already contains a case, a neighborhood relationship with another object is declared according to primitive topologies such as those illustrated in Fig.1. The declaration of a new neighborhood relationship initiates a new parameterization. As objects are added to the design, the parameterization continues to grow. Additional constraints are activated when certain objects are grouped together in order to establish more global design criteria. For example, constraints related to circulation are not activated until at least three spaces are already part of the design. Also, users may interpret the case at any stage through adding and removing constraints. These holistic attributes provide the relationship between the attributes of the IDIOM system and its dictionary definition.

3 Limitations and relation to other work

The dimensional parameterization described is limited to rectangular spaces and elements. Complex non-linear constraints slow the system down to the point where interactive design becomes difficult. For use in real time, constraints should be formulated to be as simple as possible. The current implementation of IDIOM allows for linear and simple non-linear constraints.

Since this work is focused on run-time adaptation of values of continuous variables which represent spatial design attributes, there is very little relation to work of other groups in this area. Over the last few years, several German groups have been developing a system for building objects within the scope of a project called FABEL, run by GMD, St. Augustin, Germany [4]. Their concern has been storage and retrieval of cases and since in our project, we have concentrated on object combination and adaptation, we see this work as complementary. Other work includes the SEED project [7] where large numbers of cases are stored and indexed for retrieval using functional units. Although a case editor is a available for adaptation, no other computational support is reported. Our approach is different due to our capabilities to adapt complex objects through run-time parameterizations within an intelligent user interface. An extension to CADSYN [22] employs constraint satisfaction techniques for verification and repair of adapted designs. CADSYN ensures only local consistency between constraints, thereby limiting its effectiveness to simple constraint networks where risks of divergence, looping and empty solution spaces are low. Our experience with geometric design has revealed that relevant constraint networks are highly interdependent, and therefore local consistency approaches are unreliable.

4 Conclusions

Incremental parameterization of intelligent objects results in effective support for geometric design. The IDIOM system provides a useful framework for such parameterization. Interactive parameterization of design cases is supported, thereby allowing users to introduce their interpretation of those factors which cannot be modelled explicitly by computers. Such a feature is essential for supporting the activities of professionals.

5 Acknowledgments

The work described in this paper is funded by the Swiss National Science Foundation (NFP23 and SPP-IF). The authors would like to thank Boi Faltings for contributing to the concept of intelligent objects as well as Christian Frei, Kefeng Hua, Claudio Lottaz and Ruth Stalker for their ideas, collaboration and implementation of the IDIOM system.

References

[1] J. Barber et al, "ASKJEF: Integration of case-based and multi-media technologies for interface design support" *Artificial Intelligence in Design '92*, Kluwer, Dordrecht, NL, 1992

[2] "DEJAVU : A case-based reasoning designer's assistant shell" *Artificial Intelligence in Design '92*, Kluwer, Dordrecht, NL, 1992

[3] S. Bakhtari and B. Bartsch-Spörl, "Our perspective on using CBR in design problem solving" 1st European Workshop on Case-Based Reasoning, Kaiserslaulten, 1993

[4] S. Bakhtari, K. Börner, B. Bartsch-Spörl, C-H. Coulon, D. Janetzko, M. Knauff, L. Hovestadt and C. Schlieder, "EWCBR93 : Contributions of FABEL" Fabel Report No. 17, GMD, Sankt Augustin, 1993

[5] B. Dave, G. Schmitt, B. Faltings and I. Smith "Case-based design in architecture" *Artificial Intelligence in Design '94* 1994

[6] E.A. Domeshek and J.L. Kolodner "A case-based design aid for architecture" *Artificial Intelligence in Design '92*, Kluwer, Dordrecht, NL, 1992

[7] U. Fleming, "Case-based design in the SEED System", 1st Computing Congress, American Society of Civil Engineers, Washington, 1994

[8] A.K. Goel, J.L. Kolodner, M. Pearce and R. Billington "Towards a case-based tool for aiding conceptual design problem solving" DARPA 1991 Case-Based Reasoning Workshop, Morgan Kaufmann, San Mateo, CA, 1991

[9] K. Hammond, G. Collins, G. Fisher, A. Goel, T. Hinrichs, and A. Kass, Panel on Case Adaptation, DARPA 1989 Case-Based Reasoning Workshop, Morgan Kaufmann, San Mateo, CA, 1989

[10] T.R. Hinrichs and J. Kolodner "The roles of adaptation in case-based design" Ninth National Conference on Artificial Intelligence, AAAI Press, 1991

[11] Hua, K. "Case-based design of geometric structures" Thesis No 1270, Swiss Federal Institute of Technology, Lausanne, 1994.

[12] K. Hua, I. Smith, B. Faltings, S. Shih and G. Schmitt, "Adaptation of Spatial Design Cases" *Artificial Intelligence in Design '92*, Kluwer, Dordrecht, NL, 1992

[13] N.A.M. Maiden, "Case-based reasoning in complex design tasks" 1st European Workshop on Case-Based Reasoning, Kaiserslauten, 1993

[14] M.L. Maher and D.M. Zhang, "CADSYN: Using case and decomposition knowledge for design synthesis" *Artificial Intelligence in Design '91*, Butterworth Heinemann, Oxford, 1991

[15] P. Pu and M. Reschberger "Assembly sequence planning using case-based reasoning techniques" *Artificial Intelligence in Design '91*, Butterworth Heinemann, Oxford, 1991

[16] G. Schmitt, F. Wenz, D. Kurmann and E. van der Mark, 'Die kurze physische Praesenz der Architektur', Ars Electronica 94, Intelligente Ambiente - Vol. 1, Linz, Austria, 143-149.

[17] G. Schmitt, "Design reasoning with cases and intelligent objects" IABSE Colloquium, Beijing, International Association of Bridge and Structural Engineering, 1993 pp 77-87

[18] K. Sycara "Using case-based adaptation for plan adaptation and repair" DARPA 1989 Case-Based Reasoning Workshop, Morgan Kaufmann, San Mateo, CA, 1988.

[19] K. Sycara and D. Navinchandra "Influences: A thematic abstraction for creative use of multiple cases" DARPA 1991 Case-Based Reasoning Workshop, Morgan Kaufmann, San Mateo, CA, 1991

[20] J. Wang and H.C. Howard, "A design-dependent approach to integrated structural design" *Artificial Intelligence in Design '91*, Butterworth Heinemann, Oxford, 1991

[21] F. Wenz, D. Kurmann, Interaktive Installation 'ImPuls' in the art and media exhibition 'Kuenstliche Spiele' (artificial games), Medialab, Munich, in: G. Hartwagner, S. Iglhaut, F. Rtzer (Eds.), Boer, Munich, Germany, 1993, p. 346-349

[22] D.M. Zhang and M.L. Maher, "Using case-based reasoning for the synthesis of structural systems" *IABSE Colloquium Beijing*, International Association of Bridge and Structural Engineering, pp 143-152 1993.

Case-Based Scheduling Using Product Models

Ren-Jye Dzeng[1] and Iris D. Tommelein[2], AM ASCE

Abstract

Architects and contractors who repeatedly design and build the same kind of facilities may reuse previously developed facility designs and construction schedules. This research develops an automatic planning system, named CasePlan, that reuses old schedules to develop new schedules based on the similarity between old designs and new designs. Project designs are represented as product models that are instantiated from generic product models by specifying values for the attributes in the models. The models along with the schedules are stored as cases that can be retrieved and reused to schedule similar projects at a later stage.

Similarity assessment is probably the largest bottleneck to be overcome in order to persuade field practitioners to use CasePlan. This paper focuses on how CasePlan assesses similarity between cases and how it searches for similar cases. Examples of boiler erection projects will be used to illustrate the concepts.

Introduction

Contractors plan new projects based on their individual experience and practice. We are developing a computer-based case-based planner, named CasePlan, that automates the planning process based on product specifications and the past project data that contractors have collected over the years. CasePlan retrieves best cases from its case library to choose construction alternatives for a new project. "Best" is determined by the similarity metrics CasePlan users define. This paper emphasizes the similarity metrics and case retrieval in CasePlan.

We have collected schedules from industry, including construction projects of boilers for coal-fired power plants and US Kits-of-Parts post-offices around the Michigan area. We specified generic models for boilers and post-offices from which new projects can be instantiated. For brevity, simplified boiler examples will be used for the discussion. At the end, we lay out our plan for evaluating and improving CasePlan's similarity assessment.

---

[1] PhD. Cand., Civil and Envir. Engrg. Dept., Univ. of Michigan, Ann Arbor, MI 48109-2125
[2] Asst. Prof., Civil and Envir. Engrg. Dept., Univ. of Michigan, Ann Arbor, MI 48109-2125

Literature Review

Many computer-based construction planners have been developed using artificial intelligence programming techniques, Builder (Cherneff et al. 91), Ghost (Navinchandra et al. 88), SIPEC (Kartam and Levitt 90), to name a few. These systems formulate planning knowledge by means of rules and class hierarchies for a particular type of project and use it to construct new schedules based on design descriptions. They are **constructive planners** in that they always generate a plan from scratch for each given project. These planners do not consider individual project's characteristics such as contractor's practice and site constraints. They also have no means of recognizing that they may have planned similar projects previously or that plans can be reused. In contrast, our research recognizes that one can develop cases from previous projects and schedules, and reuse this knowledge by means of *case-based reasoning* (CBR). To the authors' knowledge, no construction planners adopt this approach.

A brief but comprehensive review on CBR systems is provided by Kolodner (1993). The systems are implemented for a variety of domains, including common-sense planning, design, diagnosis, etc. They adopt three steps in solving problems: (1) retrieving the most useful case(s), (2) reusing the retrieved case(s) to solve the new problem, and (3) storing the new problem and solution as a case.

The emphasis here is on case retrieval. The reviewed systems use some indices to reduce the search space for cases, then apply some matching and ranking processes to predict the most useful case(s). Examples of indices are goals, salient attributes of the problems, and factors that cause failure in solving problems. Some systems reuse a single case for the entire problem-solving process, and some reuse multiple cases. Some search for the most useful cases, and some reuse the first case they find satisfactory. Some require a perfect match between a reused case and the new problem to solve a problem, and some allow a partial match. Nevertheless, the matching and ranking process, if used, requires three steps: (1) determine correspondence of matching attributes, (2) assign importance of attributes in assessing similarity, and (3) determine degree of similarity between corresponding attribute values based on some similarity measurement.

Three types of retrieval strategies are available: numeric, heuristic, and a combination of the two. Most numeric strategies are based on a simple algorithm called nearest-neighbor matching, which will be discussed in detail later. Heuristic strategies can be based on heuristic rules to identify good-enough match or preference specifications to filter the case library for desired cases.

In summary, CBR provides a way to use knowledge specific to individual projects, which is essentially lacking from existing construction planners. Most CBR systems use similar concepts and approaches, but vary in their implementation and combination of these techniques to suit their intended domains. With no exception, CasePlan is similar to those CBR systems in many ways, but is also unique in that it tackles the planning problems in construction and uses product models as the base for CBR.

CasePlan Overview

CasePlan generates a schedule based on project descriptions and a library of cases (Dzeng and Tommelein 93, Tommelein and Dzeng 94). Instead of predefining constructive planning knowledge like existing construction planners, it uses product models for representing, retrieving, and adapting knowledge of cases that describe previously built projects and their associated schedules to generate a new schedule. CasePlan uses class hierarchies to organize its knowledge as shown in Figure 1. A *Case* comprises a *Project* and *Schedules*. A *Project* comprises a *Product*, a construction *Site*, and *Project-specifications*. In turn, a *Product* comprises different types of components, such as *Stoker* and *Drums*. Each component may comprise one or several subcomponents. Classes have descriptive attributes whose values are either supplied by the user or inferred by CasePlan. An existing project and its schedule are stored as a *Case*, where values of the attributes are known.

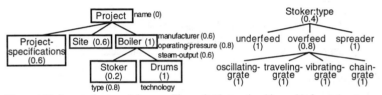

Figure 1 Boiler product model   Figure 2 Abstraction hierarchy for stoker types

Generic *Products* like boilers are predefined in CasePlan. A new project is specified by instantiating the appropriate generic *Product* and specifying the *Site* and *Project-specifications*. To generate a schedule for a new project, CasePlan compares the attribute values of the new *Project* with each existing *Project* in its case library, retrieves the best-match cases, and uses the planning knowledge embedded in the cases. Best match is defined by similarity functions that compute scores representing the degree of match between the new project and a case.

Planning knowledge refers to the information required to generate a schedule. It includes **construction technologies** used for components (represented as activity subnetworks), **construction methods** used for activities (describing the required resources, productivity, etc.), and the dependency relationships among these construction alternatives, product designs, and project constraints.

Examples

Our study in the construction projects of boilers and post-offices shows that contractors did reuse existing plans if the plans were available to them. Similar product designs have similar construction plans. The plans differ mostly in the scheduling information such as starting times and durations of activities.

A simplified boiler model with attributes is shown in Figure 1, and it will be used for the discussion. A boiler may comprise a stoker and drums. Table 1 shows three simplified projects. Project-1 is the new project to be planned; Case-A and -B are two cases available for reuse. Column Attributes lists attributes (e.g., *name*, and *manufacturer*) for objects shown in column Objects (ignore the numbers in parentheses for this moment). Values of the attributes for Project-1, Case-A and -B are listed in columns Project-1, Case-A, and Case-B, respectively. For example, for the attribute *type* of stokers, Project-1 has the value *chain-grate*, Case-A *spreader*, and Case-B *vibrating-grate*.

Objects	Attributes	Project-1	Case-A	C1	C2	C3	Case-B	C4	C5	C6
Project (0.6)	name (0)	Detroit Power Station	Detroit Power Station		0	0	Pittsburgh Power Station		0	0
Boiler (1)	manufacturer (0.6)	Zurn Energy	ABB-CE	0.6	0.36		Zurn Energy	1	0.60	
	operating-pressure (0.8)	17	22.5	0	0	0.30	11	1	0.80	0.94
	steam-output (0.6)	1,000	2,100	0.4	0.24		650	0.8	0.48	
Stoker (0.2)	type (0.8)	chain-grate	spreader	0.4	0.32	0.32	vibrating-grate	0.8	0.64	0.64
Drums (1)	technology		T(drums A)				T(drums B)			
						0.20				0.59

Table 1 Three simplified boiler projects

Case Retrieval

Given a new project to plan, CasePlan retrieves the cases that are likely to be the most useful ones to its plan generation process. It first determines its **case search space**, a set of cases where CasePlan searches for good ones, based on the index of cases. Then it matches the project against each of the cases in the search space to determine which ones should be reused.

Indexing of Cases

CasePlan indexes its cases by their product models. For example, cases for boilers are grouped separately from cases for post-offices. Such an indexing scheme has three purposes:

(1) To reduce search space. The indexing allows CasePlan to quickly identify the set of cases that have the same type of products as the new project. These cases are likely to be more useful than those with different products. For example, given a boiler project, CasePlan will only search the cases of boiler projects.

(2) To ensure appropriateness of the matching process. Matching between only the same type of products ensures that the matching is performed on the basis of the same set of components and attributes.

(3) To categorize planning knowledge. CasePlan is a knowledge-based system where the storage of knowledge is distributed in cases (Dzeng and Tommelein 94). With the index, the planning knowledge can be categorized based on products. For example, construction technologies applied to drums in the past projects can easily be presented.

Matching Between Project and Cases

CasePlan performs the matching at the component and project levels. At the component level, it finds the best-match case for each component based on component characteristics. The case is used to determine the component technology for the component. A **component technology**, represented as an activity network, describes a construction process for a component. All CasePlan activities have names comprising a verb phrase (e.g., *Connect-tubes-for*, *Lift*) and a noun referring to a component (e.g., *drums*). Activities in a component technology only differ in verbs and not in their nouns (e.g., *Connect-tubes-for-drums*, *Lift-drums*). For example, if the drums of Case-A are more similar to the drums of Project-1 than Case-B, the technology *T(drums A)*, shown in column Case-A in Table 1, will be used for the drums of Project-1.

At the project level, CasePlan finds the best matching case based on product characteristics (e.g., the boiler's *manufacturer* and *operating-pressure*) and construction environment (i.e., *Site* constraints and *Project-specifications*). The case is used to determine the product technology for the project. A **product technology**, represented as an activity network, describes the construction sequence of the product's components (e.g., Install-drums, Install-stokers).

CasePlan assumes matching correspondence of the components and attributes with the same names because the project and the matching case always have the same set of components and attributes. For example, *operating-pressure* in a boiler of a project corresponds to the *operating-pressure* in a boiler of a case. Of course, this assumption relies on user's ability to consistently specify a product, i.e., to use components and attributes with the same names in the same fashion.

Usually CasePlan can only find partial matching cases for a new project. To determine the degree of each match, users need to assign an importance value to each attribute and determine the metrics for measuring the similarity of matching values.

Importance Values of Components and Attributes

An importance value assigned to a component/attribute designates the amount of attention that CasePlan should pay to the component/attribute when it compares a case to a project. In our example, the importance value for each component and attribute is shown as a number in parentheses in both Figure 1 and Table 1. Higher values receive more attention during the matching. A component/attribute that has the highest value of 1 is extremely important. CasePlan will not continue evaluating the match of a case if the corresponding value is missing or the similarity value, as defined later, is 0. For example, CasePlan will not compare a case that has no drums with a project that has drums because the drums have an importance value of 1.

Conversely, a component/attribute that has the lowest value of 0 will be ignored during the matching. An unimportant component/attribute does not increase the degree of the match even when the corresponding values in a case and a project match each other perfectly. For example, projects with the same *name* (0) (e.g., Detroit Power Station) are not necessarily similar in their designs and construction processes. Conversely, boilers made by the same *manufacturer* (0.6) (e.g., Zurn Energy) are more likely to have similar designs and construction processes. As another example, the *type* of a *stoker* significantly affects its assembly and thus is assigned an importance value of 0.8. However, a *stoker* itself does not affect the sequence for constructing other boiler components. Thus, it is assigned an importance value of 0.2.

An attribute may be assigned different importance values for different tasks CasePlan performs. For example, the weight for a boiler component is very important when the task is to choose a construction method (e.g., lifting equipment). However, it is less important when the task is to choose a component technology.

Similarity Metrics for Attribute Values

In CasePlan, an attribute value can only be one of the following types, namely objects, numbers, or keywords (e.g., *chain-grate* or *underfeed* for the *type* of stoker). It can also be a **reason specification**, which is a Lisp expression that evaluates to a value. CasePlan only performs the matching on the evaluated values of reason specifications instead of directly on reason specifications. Matching between two objects is determined by the matching of their attribute values. Thus, CasePlan only needs to deal with two types of matching: keyword vs. keyword and number vs. number.

Common practice is to use an abstraction hierarchy to measure the similarity between two non-numeric values such as keywords. All legitimate keywords are defined in an abstraction hierarchy. The most specific common abstraction (MSCA) of two entities is computed. The more specific the MSCA, the better the match.

Figure 2 shows an abstraction hierarchy for different stoker *types*. One way to numerically compute the similarity based on the MSCA is to count the distance of the path starting from one node, passing through the MSCA, and ending at the other node. For example, the similarity value of *chain-grate* and *vibrating-grate* is 2, which is the distance from *chain-grate* to *overfeed*, and from *overfeed* to *vibrating-grate*.

Such a simple scheme performs poorly for CasePlan because the abstraction hierarchy is specified by human schedulers and is usually not perfectly balanced for desired similarity assessment. Nodes will tend to be physically farther from each other in the more densely populated parts of the hierarchy than in the more sparsely populated parts. A remedy to this problem is to assign a specificity value to each node as shown in parentheses in Figure 2. The similarity value of two nodes is determined by the specificity value of their MSCA. For example, the similarity value is 0.8 (specificity value of *overfeed*) for *chain-grate* and *vibrating-grate*, and 0.4 (specificity value of *Stoker:type*) for *chain-grate* and *spreader*. Thus, *chain-grate* is more similar to *vibrating-grate* than to the *spreader* type of stoker.

For the match between numeric values, one can compute their difference based on a qualitative scale or quantitative scale. For example, consider the boiler *operating-pressure*. As shown in columns Project-1, Case-A, and Case-B in Table 1, Project-1 has a value 17 Mega-Pascal (MPa) (2,500 psi), Case-A 22.5 MPa (3,300 psi), and Case-B 11 MPa (1,600 psi). If only the quantitative difference is measured, Case-A appears more similar to Project-1 than Case-B because it has a smaller quantitative difference from Project-1 (22.5 - 17 = 5.5 vs. 17 - 11 = 6). However, this conclusion about similarity is incorrect. Case-B and Project-1 are both designed to operate below the critical pressure of water (22.1 MPa or 3,206 psi) and have a similar configuration design. Case-A is designed to operate at a supercritical pressure and has a quite different configuration.

The previous example shows a situation where a qualitative comparison is more relevant than a quantitative one. CasePlan allows users to specify qualitative ranges for quantitative values in the attributes of generic product models. Each specification is a list of qualifier and value range lists. A specification example for *operating-pressure* is:

```
((sub-critical (nil 22.1)) (super-critical (22.2 nil)))
```

This means that a pressure under 22.1 MPa inclusive will be classified as *sub-critical*, and one above 22.2 MPa as *super-critical*. If no abstraction hierarchy is defined for *sub-critical* and *super-critical*, the similarity value of the boiler *operating-pressure* for Case-A is 0, and for Case-B is 1, as shown in columns C1 and C4 of Table 1.

For attributes where quantitative measurement works better, a range for possible difference needs to be specified to ensure the consistency on the similarity scale. A difference range is a list of value range and similarity scale lists. For example, the difference range for the attribute *steam-output* (in metric-ton/hr) of a boiler may be specified as follows:

```
(((0 100) 1)        ((101 500) 0.8)     ((501 1000) 0.6)
 ((1001 2000) 0.4)   ((2001 3000) 0.2))  ((3000 nil) 0))
```

Thus, boilers with a steam output difference less than 100 ton/hr (220,000 lb/hr) are considered no different; ones with a difference between 101 ton/hr and 500 ton/hr (222,000 and 1,100,000 lb/hr) are considered quite similar with a similarity value of 0.8, and so on. Those with a difference greater than 3,000 ton/hr (6,600,000 lb/hr) are considered no match at all. Thus, the similarity value of *steam-output* for Case-A is 0.4, and for Case-B is 0.8, as shown in columns C1 and C4 of Table 1.

Retrieval Based on Similarity

CasePlan's retrieval strategy is similar to the **nearest-neighbor matching**, which ranks the cases based on a similarity function that computes the sum of similarity value for the match of each attribute multiplied by the attribute importance value. CasePlan's similarity function is a normalized one, stated as follows.

$$\sum_{i=1}^{n} \frac{w_i \times \text{similarity} \ (a_i^p, \ a_i^c)}{\sum_{i=1}^{n} w_i}, \text{ where } w_i \text{ is the importance value of attribute i; n is}$$

the total number of attributes evaluated; similarity $(a_i^p, \ a_i^c)$ is the degree of similarity between the values of a project and a case for the attribute i.

For example, to determine which case is better for choosing a product technology for Project-1, similarity values are computed at the project level. For brevity, the matching for *Project-specifications* and *Site* is ignored. In Table 1, column C3 shows the similarity values between Case-A and Project-1 for *Project* (0), *Boiler* (0.30), and *Stoker* (0.32). At the bottom of the column is the final similarity value 0.20, which is obtained from $(0.6 \times 0 + 1 \times 0.30 + 0.2 \times 0.32)/(0.6 + 1 + 0.2)$. The value for Case-B is 0.59, shown in column C6. Thus, the product technology of Case-B will be chosen.

The values in column C3 and C6 are computed based on those in C1, C2 and C4, C5. C1 and C4 shows the similarity value for each attribute. C2 and C5 shows the multiplication of each attribute's similarity value and importance value. For example, the similarity value between the boilers of Project-1 and Case-A is 0.30, which is obtained from $(0.36 + 0.24)/(0.6 + 0.8 + 0.6)$. The similarity value for the attribute *manufacturer* in Case-A is 0.36, which is obtained from (0.6×0.6).

Determination and Evaluation of Similarity Assessment

Similarity assessment is probably the most difficult part to overcome to persuade people in industry to use CasePlan because it is the main information that is lacking from their existing documentation. Experts may have a feeling about which attributes are relatively important and which are not, and we have to rely on their experience.

Similarity functions can be poor for four reasons: (1) importance values for attributes are incorrect, (2) the degree of similarity between values are measured incorrectly, (3) some attributes that are important to similarity assessment are not included in the model, and (4) the matched attributes do not correspond to each other. Except for the fourth reason, which relies on users' consistency of using attributes, survey and experiments need to be conducted to find the satisfactory results.

Questionnaires are to be sent out to field experts to solicit opinions on the importance values of attributes and the similarity between values. A series of experiments will be conducted to determine the appropriateness of our similarity assessment and whether more complicated algorithms should be applied.

Conclusions

CBR provides a way to closely relate the construction planning problem to individual projects and contractors' practice. CasePlan may produce different schedules if it is used by different contractors and supplied with different cases. This approach is essentially different from the one taken by existing construction planners. The quality of the schedules produced by CasePlan strongly relies on the similarity assessment and the quality and quantity of its case library. We will be continuing to evaluate CasePlan's similarity assessment and investigate ways to improve it if necessary.

Acknowledgments

This research is funded by grant MSS-9215935 from the National Science Foundation (NSF), whose support we gratefully acknowledge. Any opinions, findings, conclusions, or recommendations expressed in this paper are those of the authors and do not necessarily reflect the views of NSF.

References

Cherneff, J., Logcher, R., Sriram, D. (1991). "Integrating CAD with Construction-Schedule Generation." *J. Computing in Civil Engrg.*, ASCE, 5 (1), 64-84.

Dzeng, R.J., Tommelein, I.D. (1993). "Using Product Models to Plan Construction." *Proc. 5th Intl. Conf. Comp. Civil and Bldg. Engrg.*, ASCE, 1778-1785, NY.

Dzeng, R.J., Tommelein, I.D. (1994). "Case Storage of Planning Knowledge for Power Plant Construction." *Proc. 1st Comp. Congr.*, ASCE, 293-300, NY.

Kartam, N.A., Levitt, R.E. (1990). "Intelligent Planning of Construction Projects." *J. Computing in Civil Engrg.*, ASCE, 4 (2), 155-176.

Kolodner, J. (1993). *Case-Based Reasoning.* Morgan Kaufmann, San Mateo, CA.

Navinchandra, D., Sriram, D., Logcher, R. (1988). "GHOST: project network generator." *J. of Computing in Civil Engrg.*, ASCE, 2(3), 239-254.

Tommelein, I.D., Dzeng, R.J. (1994). "Automated Case-Based Scheduling for Power Plant Boiler Erection: Use of Annotated Schedules." *11th ISARC*, 179-186, Elsevier Science.

A CASE-BASED MODEL FOR
BUILDING ENVELOPE DESIGN ASSISTANCE

K. Gowri* Member, S. Iliescu** and P. Fazio*** Member

ABSTRACT

Engineering design cases contain a wealth of implicit knowledge and information which are specific to particular design contexts. Attempts to generalize this information using knowledge-based techniques have been thwarted due to inherent limitations in knowledge extraction and representation. The present work proposes to integrate case-based reasoning in a knowledge-based environment to improve the design assistance offered to designers in selecting the materials and construction techniques for building envelope components. This approach can benefit by representing codes and standards using traditional knowledge representation paradigms and experiential knowledge using case-based reasoning techniques. The case-base provides the initial design solution which is adapted to meet the requirements of a new design context. This design model has been implemented in a software prototype and is being evaluated for further development.

INTRODUCTION

In North American climatic conditions, the building envelope plays an important role in protecting the indoor environment from the rapidly varying outdoor weather conditions. Traditionally the selection of materials and construction techniques for building envelope components were based on cost, aesthetics and past experience. But the need for durable and energy efficient envelopes has prompted designers to evaluate the thermal, moisture and air barrier characteristics of envelope assemblies. The evolution of energy efficiency standards and better understanding of building science concepts have offered several methods of evaluating building envelope performance. The present research aims at developing a systematic approach to building envelope design in which performance evaluation and code compliance checking will be integral to the process. In an effort to implement such a process, a knowledge-based system known as BEADS (Building Envelope Analysis and Design System) was developed (1). Field testing and attempts to refine BEADS knowledge base have shown that there are inherent limitations to generalizing and representing the building envelope design knowledge.

----

* Assistant Professor, ** Graduate Student ,*** Professor and Director, Centre for Building Studies, Concordia University, Montreal, Canada H3G 1M8

Designers often have preferences on materials and construction techniques backed by an argument based on previous design context which is analogous to a current design situation (2). But it is not always assured that design contexts will exactly be the same, thus requiring adaptation of the previous design. While adapting an earlier design detail, designers need to analyze the consequences of adaptation. Recent developments in case-based reasoning provide the means of computerizing the design process using case retrieval and adaptation in a more systematic and efficient manner. The present research focuses on integrating the case-based and knowledge-based approaches so that a candidate design case retrieved may be adapted using the appropriate knowledge.

CASE-BASED REASONING (CBR)

In general, a case is a contextualized piece of knowledge embedding a domain specific lesson about how to achieve a goal (3,4). A case-based reasoning system can only be as good as its memory of cases and adaptation strategies. The case memory not only includes the library of cases, but also a set of access procedures defined to store and/or retrieve information from that repository of cases. In the building envelope design domain, a case could include not only the full description of building envelope elements, but also some detailed knowledge about what strategies are to be applied to achieve energy efficiency, or guidelines derived from good design practice. We have to distinguish between cases that augment the systems knowledge and are worth to be remembered (that is, stored in the library of cases) and cases that duplicate existing information, without making a significant difference to what is already known. Therefore, a design instance to be kept in the case library, usually called prototypical or paradigm case must represent a specifics piece of knowledge tied to a context, recording that knowledge at the operational level, rather than at conceptual level. In addition, a paradigm case should record an experience that is different from expected, either by how it achieves a goal or by how it avoids a failure. It is important to note that the value of a case is directly linked to the appropriate circumstances in which the case was defined. Alternately, the case should be indexed by its functionality context that will govern the condition under which the case may be retrieved. For example, the building envelope descriptions for a high rise building should be indexed such that its retrieval for the design of a low-rise building will be, at best, unlikely if not impossible.

Starting with case retrieval as the primary process and ending with memory update, CBR has the following four major steps (5):

Step 1: Access and retrieval - Before accessing the case library, a target case must be assembled from the input data. In order to retrieve the most appropriate cases, similarity metrics and goal hierarchy must be provided. The similarity metrics are based on features of the target case which must be matched with the individual instances in the case library. The goal hierarchy defines the order of matching the features and weighting of the matched features for ranking the retrieved cases. One important implementation issue here is the efficiency of the search process which is dependent on the size of case library.

Step 2:   Mapping (Case transformation) - The retrieved case should be transformed to match the input requirements as much as possible by swapping the unmatched features using adaptation rules. For example, a curtain wall description retrieved from a case base may be adapted for insulation thickness but not for the structural or exterior cladding layer. The reliability and quality of adaptation rules will govern the success of the reasoning process. At this juncture, one can recognize the integration of the case-based and knowledge-based approaches in which the source case is provided by the case-based reasoner and case transformation is pursued using knowledge-based techniques.

Step 3:   Evaluation and repair - A design description retrieved and adapted must be verified for performance. Hence performance evaluation techniques such as simplified mathematical simulations or qualitative assessment methods must be used to determine the validity of the design solution. If there is an apparent non-compliance to performance, then that feature must be directed as having highest priority in the goal hierarchy when retrieving and ranking the cases. It may also be treated as the feature requiring attention during adaptation.   The evaluation and repair strategies provide the feedback to steps 1 and 2, if a retrieved case fails to meet the design context requirements.

Step 4:   Learning and updating - A case-based reasoner has to acquire the knowledge through the reasoning process and update the case memory incrementally. If the design solution generated is identified as a unique instance, or an improved one, then this case will be worth remembering and accordingly this should be saved either as a new case or modify an already existing case description in the library. The most important part of the case library update is choosing the ways to "index" the new case so that it will be retrieved when it can be most useful. If the learning mechanism identifies a pattern to generalize a feature, then an empirical rule may be added to the retrieval or adaptation processes.

It can be observed that the CBR approach is very useful in domains where there is a lot of information available, but not systematically organized into hierarchies of concepts and rules. In addition, CBR is valuable when the domain knowledge is incomplete, sparse and dynamically changing with design contexts. It is important to recognize that solving problems exclusively using cases could introduce reasoning bias as well as the potential to fail in novel situations which are very much different from those stored in the case library. The building envelope design process can be modeled such that individual envelope components are retrieved from a library of generic cases and later adapted using a knowledge base to meet the design context requirements.

INDEXING BUILDING ENVELOPE CASES

In developing a case-based model for building envelope design, we assume that a reasonable number of design cases are available and our first task then is to develop an appropriate mechanism for indexing and retrieval. The building envelope may be viewed as a conceptual object 'ENVELOPE' consisting of several components such as walls, roof, windows and doors. In addition to its components, the ENVELOPE object has a feature vector representing the geometry, size, occupancy, structural system, and energy efficiency requirements. These features are not necessarily linked and their relative importance in a design context can not be generalized. But these can be resolved based on a designers preference and accordingly assigning weighting factors during matching and retrieval. In general, the problem to solve is the following: given a collection of n unknown objects,

$$\text{ENVELOPE}_i = \{ \text{WALL}_i, \text{ROOF}_i, \text{GLAZING}_i, \textit{feature vector}_i \}$$

where i represents a design case, find a suitable indexing strategy for developing the case repository.

Earlier research in developing BEADS demonstrated that problem decomposition can be effectively used to represent an envelope by a combination of its components (6). Hence, rather than storing the ENVELOPE cases as complex entities described by large feature vectors, we assembled three different libraries of components, a first one for WALLs, a second one for GLAZINGs and a third one for ROOFs. Case decomposition allows to enforce economy of representation, since the same component may be linked to several ENVELOPE instances. To establish the set of indexes, we developed a functional classification of wall types, roof types and glazing types based on generic concepts as shown in Figure 1. The identifiers used for each WALL and ROOF type are mostly symbolic, and conforms to the current terminology used in the industry for the various layers and construction types. These identifiers serve as the primary index representing the envelope type. The ENVELOPE object has five indexes consisting of the building type, structure type, building height in number of stories, floor area and roof shape, which will be used as the secondary indexes for retrieval. Each envelope component is then represented by the generic functional layers, and instances of building materials are specified by their functional attributes and physical properties.

Building envelope cases are represented using the following two types of specifications: (i) content grid and (ii) context grid. The content grid captures the various layers present in an envelope element and the context grid keeps track of the relationships between cases based on the secondary indexes. While the content grid is linked to the composition of envelope elements, the context grid is linked to the indexing mechanism that designates under which circumstances the case should be retrieved. This organizational structure is derived from the concept of Thematic Organizational Units (TOP) aimed at linking cases that share common goals, plans, and goal/plan interactions (7).

CASE RETRIEVAL AND FILTERING

Similar to database search, retrieval of cases from a case library can be seen as a massive search problem, but the search process in CBR could settle for a partial best match rather than a full match. Partial match algorithms need to be directed in such a way that matching is only attempted on cases with potential for adaptation. Algorithms for searching a case library are associated with the

KNOWLEDGE BASE HIERARCHY

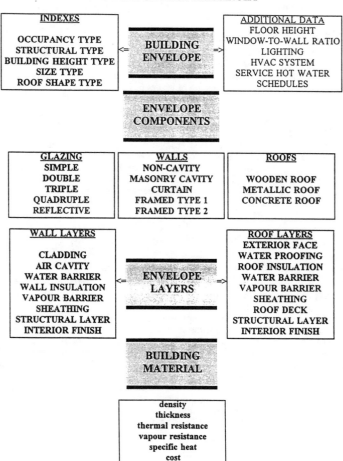

Figure 1: Functional Classification of
Envelope Components

organizational structure of the library. The building envelope case library has a combination of flat and network models to take advantage of hierarchical representation and linked lists of secondary indexes. The search process consists of initial breadth-first search to delineate all potential cases, followed by a neural network filtering to focus on a smaller subset of cases. A description of the search and filtering process of WALL library is presented below.

Each WALL object is defined by a collection of layers, each layer having a set of properties such as thermal resistance, vapour resistance, density and thickness. Typically, the sum of individual layer's material properties will yield the overall properties of a wall composition. Each individual overall wall property has an impact on wall performance. However, it is very difficult, if not impossible, to link the walls composition in terms of layers to its performance, since different walls assembled from different layers can have the same overall properties. Hence, it may be useful to retain the relative values in terms of percentage of the overall WALL performance. This approach will permit us to create an almost binary representation for each WALL instance by specifying threshold limits. For example, consider a vapour barrier layer within a wall assembly. Assuming a threshold limit of 0.1, the ratio of vapour resistance offered by this vapour barrier material to the overall vapour resistance of the wall assembly will be very close to 1, while other parameters such as thermal resistance may be 0. In general, a given WALL instance having m layers, each one described along p physical parameters, into a $m \times p$ matrix of numbers in the $[0, 1]$ interval:

$$\text{WALL}_i = \begin{matrix} C_{11} & C_{12} & C_{13} & \dots\dots & C_{1p} \\ C_{21} & C_{22} & C_{23} & \dots\dots & C_{2p} \\ \dots\dots\dots\dots\dots\dots\dots\dots\dots\dots\dots\dots \\ C_{m1} & C_{m2} & C_{m3} & \dots\dots & C_{mp} \end{matrix}$$

where C_{ij} is the scaled value of property j in layer i. That representation of WALL instances is very similar to pattern recognition techniques used for digitized images. If we consider the C_{ij} co-efficients as a variation in gray scales with '0' representing white and '1' representing black, then it would be possible to extract a "physical signature" for the wall using neural network (NN) classifiers. The physical properties selected to describe each layer include the density, specific heat, thickness, thermal resistance, and vapour resistance. The normalized values of these properties can be classified to assign a 3 bit class code. During the training phase, all cases in the library will be passed to the NN filter, so that the NN will "learn" how to classify a case based on its "physical signature". Subsequently, during the test phase, the preliminary description of the new target case will be submitted to the NN filter to obtain a preliminary classification of that case with which the scope of search can be limited to a smaller subset of instances.

In a typical envelope design situation, the constraints would include the overall thermal performance, moisture performance, overall thickness of components, material preferences for cladding, roof shape and structural system. Based on this incomplete description, a "target" instance of WALL can be assembled and searched in the WALL case library. While all cases that do not fit the indexes of the "target" instance are eliminated from the beginning, the remaining alternatives are further filtered using the physical signature assigned by the NN classifier.

Once a subset of suitable cases is found, the cases must be ranked by order of preference before proceeding to adaptation. A linear evaluation function based on weights associated with feature is used for ranking. These weights are determined as a combination of user preferences and ease of adaptation of the features under consideration.

CASE ADAPTATION

The unmatched features of a best matched case needs to be replaced using a transformation technique that will swap materials of similar functions and performance. This adaptation takes into consideration the class abstraction of the source case, to properly identify the layers to be replaced. The mapping order follows the relative layer of importance. For example, the structural layer will be mapped before the interior finish. After each layer mapping, we check that the "nearest neighbor" of the replaced layer does not introduce any compatibility problems. In addition, the replacement of any specific layer will be compared with the generic cases of construction to ensure admissibility. During this transformation, the retrieved case will be re-instantiated and verified for individual performance requirements. In building envelope design, there are three major categories of design requirements: thermal resistance, vapour resistance and material compatibilities. Thermal and vapour resistance requirements need to be verified using simplified mathematical simulation techniques whereas material compatibility requirements need to be verified using heuristic rules.

SOFTWARE ARCHITECTURE AND IMPLEMENTATION

The case-based approach to building envelope design has been implemented in a software prototype system named CBDEC (Case-Based Design of Envelope Components). This software system consists of five modules: (i) Constructor, (ii) Matcher, (iii) Modifier, (iv) Checker and (v) Learner. The Constructor assembles the target case by appending the user requirements with additional information available in the knowledge base. The knowledge base consists of ASHRAE 90.1 performance requirements, design weather data and building code requirements. The Matcher browses the case library to extract a subset of design cases which meet the matching criteria, and ranks the partially matched cases based on the evaluation function. A neural network filter and a breadth-first search engine have been developed for search and retrieval. The function of Modifier is to adapt the "best match" case to meet all the design context requirements. The adaptation is guided by a knowledge base to verfiy material substitutions and swapping of layers to conform to the current design practice. The Checker verifies through simulation and using a knowledge base of design information, that the modified case conforms to the energy efficiency, moisture performance and material compatibility requirements. A simple static condensation check of wall assemblies and compliance check of ASHRAE System Performance requirements are carried out. If this adaptation fails, then the next highest ranked case will be considered by the Modifier and the process continues. Finally, the case Learner saves a new completed design information by identifying the uniqueness of the design context. This software implementation is being carried out in the c++ environment taking advantage of object oriented programming and graphic utilities commercially available for the personal computers.

SUMMARY

The application of case-based reasoning to building envelope design promises to overcome many of the limitations in generalizing the design knowledge. The design model integrating the case-based and knowledge-based approaches would greatly enhance a designers ability to recognize and accept design solutions generated by computers. Though these designs are adapted by computational techniques, the initial case description comes from practioners. The software protoype and knowledge base is being evaluated for functionality and completeness. Further research is also pursued at the Centre for Building Studies in integrating the envelope design module with other sub-systems such as structure, HVAC and lighting.

REFERENCES

1. Fazio, P., Bedard, C. and Gowri, K., "Knowledge-Based System Approach to Building Envelope Design", Computer Aided Design, Vol. 21, No. 8, 1989.

2. Mackinder, M., "The Selection and Specification of Building Materials and Components", Institute of Advanced Architectural Studies Research Paper No. 17, The University of York, York, U.K., 1980.

3. Riesbeck, C.K. and Schank, R.C., "Inside Case-Based Reasoning", Lawrence Erlbaum Associates Publishers, Inc., Hillsdale, NJ, 1989.

4. Kolodner, J.L., "Case-Based Reasoning", Morgan Kaufmann Publishers, Inc., SanMateo, CA, 1993.

5. Hammond, K., "Case-Based Planning", Perspectives in Artifical Intelligence, Academic Press, 1989.

6. Fazio, P., Bedard, C. and Gowri, K., "Constraints for Generating Building Envelope Design Alternatives", in Evaluating and Predicting Design Performance, ed. by Y.E.Kalay, John Wiley & Sons, Inc., New York, NY, 1992.

7. Schank, R.C., "Dynamic Memory: A Theory of Learning in Computers and People", Cambridge University Press, New York, NY, 1982.

Developing Construction Database Models

Kenneth H. Dunne, M.ASCE[1]

ABSTRACT

This paper addresses the continuing development of a three-dimensional matrix model of construction resource data collection and retrieval for scheduling analysis and negotiations of construction claims.

It continues the ideas presented in *Contemporaneous Development of Field Data: Transportation Construction Claim Defense* (Dunne and Navarrete, 1994) and presents the mathematical model being developed for use on the new $30 MM twin MacArthur Causeway Bridges under construction for the Florida Department of Transportation in Miami, Florida.

INTRODUCTION

Focus is upon developing a specific computer database of the necessary, normal and routine information that is easily recognized by field engineers and inspectors, but may be neglected in preparing the daily reports of construction.

Each engineer and inspector is brought into the domain of computer analysis when prompted on the resource screen for the exact information normally seen on the hard-copy paper daily report which is filled out each day of Contract work.

Quality control of contemporaneous field-generated data and information, and it's retrieval within the model, as objective evidence for computer analysis will

[1] Resident Engineer, Frederic R. Harris, Inc., 15485 Eagle Nest Lane, Suite 220, Miami Lakes, Florida 33014

ultimately enhance the position of Transportation System Owners in reducing risk and liability brought about by transit construction claims.

Taking the collection of field data and storing it in a database for later analysis of schedule progress will improve efforts of transportation clients in preparing to defend against unwarranted construction claims on transit and bridge projects.

The achieved result anticipated at the end of the MacArthur Project in 1996 is the finalized development of a computer database model, and criteria for the quality control of field data and information collected and recorded contemporaneously in the jobsite field office. This database can then be effectively utilized to analyze and evaluate the merit of construction claims for additional costs and time.

DEVELOPING THE MODEL

Substantiating manpower, equipment, and material resources on the jobsite in a contemporaneous manner is necessary preparation for negotiating fairly and reasonably with the Contractor, concerning the direct and indirect costs of impacts sustained by the Contractor due to changes, additional work, or delays, and assessing the actual progress of construction.

Assignment of **available** resources is generally quite different from **planned** resources used by the Contractor in establishing activity durations of the Critical Path.

Duration is actually the time it takes assigned resources to travel through the activity and progress the work.

Recording the available **and applied resources** to each significant activity group will result in establishing a permanent quality record of historical performance on an up-to-date, or contemporaneous basis. This can then be extrapolated into a time-performance ratio, which is the basis for projecting actual activity completions.

Two case studies are used to illustrate model development.

Underwater bridge foundation construction data is presented first.

Followed by extensive wood formwork cycling for eleven concrete bridge deck placements, each 67' wide by 235 - 320' long.

Case Study I - Underwater Drilled Shaft Construction

Forty-two (42) shafts, 84" in diameter are drilled with rock augers through predominantly limestone and sandstone formations beneath Biscayne Bay in Miami, to depths of -(90-100) feet below sea level for each of the twin bridges.

A common foundation with the Metro-Dade Transit Agency's future Metromover Extension to Miami Beach utilizes the capacity of the two center drilled shafts to support the transit system superstructure.

The open, cylindrical shafts are then concreted by the tremie method in open water from water-borne cranes mounted on 45'x90' marine barges. Each shaft requires on average, pumping 190 cubic yards of 5500 psi concrete after the rebar cage is put into place within a four-hour time limit.

The drilling sub-contractor supplies two leased 150T capacity cranes, rotary drilling rig and land-based 50T capacity crane with drill rig, 1 superintendent, 1 foreman and 7 craft. Pick-up trucks, generators, boat, welding machines and miscellaneous tools are also equipment resources supplied by the sub-contractor.

The General Contractor has on-site two 250T cranes, one 350T ringer crane with 300' boom, nine barges, two tug boats and provides surveying for the drilled shaft locations in open water and final elevations.

The General Contractor has on average between 55-60 craft on-site each day. As-needed, labor resources are diverted to the sub-contractor during problems encountered, or lack of adequate manpower supplied by the sub-contractor on any given day.

Labor is shared during concreting operations, and crane equipment and tug time are backcharged to the sub-contractor when needed. Predominant labor for drilling is by the sub-contractor.

WRITING THE MODEL EQUATION

The model is based on calculating Tr which is the <u>time resource performance</u> for each selected CPM schedule activity under consideration.

Developing the **drilling** model equation for this continuing activity can be represented as follows:

$$Tr = \frac{(MpSUB + MpGC)}{Plan\ Mp} \times \frac{(EqSUB + EqGC)}{Plan\ Eq} \times lineal\ feet\ excavated/day = PERFORMANCE$$

Developing the **concreting** model equation in a similar manner,

Tr = $\frac{(MpGC+MpSUB)}{Plan\ Mp}$ x $\frac{(EqGC+EqSUB)}{Plan\ Eq}$ x *(Mt Conc)* x *L.F. Shaft* = *PERFORMANCE*

where Tr is time-resource performance, Mp is manhours(MH), Eq is Operating Equipment hours(OEH), and materials as unit quantity.

TrTOTAL = TrDRILL + TrCONCRETE = Foundation Construction Performance

The <u>unit ratio basis</u> is important in this case due to the Payment Item in the Contractor's Contract requiring payment on a lineal foot <u>of completed shaft only.</u>

It is also essential in developing the model to consider every activity as a <u>unit of available and assigned resources on the jobsite.</u>

SIMULTANEOUS ACTIVITIES WITH COMMON RESOURCES

Many activities occur simultaneously on each contract day, but the number of these activities completed is a direct function of availability of common manpower and equipment resources.

It is this <u>availability</u> of a crane, or front-end loader, or proper materials to support the assigned craft, to <u>progress the planned durations of every activity shown in the schedule, plus hundreds of problem-oriented and general support functions, that deprive a planned productivity schedule from being achieved.</u>

The Daily Reports of Construction (DRC) document all applied manpower and equipment on the jobsite each day, and record all material and equipment deliveries when they occur on any given day.

Tug boats trucked from Pittsburgh to Miami must be assembled on the jobsite before they can be assigned to any activity, just as heavy crane equipment must be assembled and mobilized onto barges for working on the water. This **assembly activity** is documented on the DRC as any other, and resource data entered into the database on those Contract Days affected.

On MacArthur Bridge, it took two weeks to assemble one tug using 9 craft and mechanics, and three weeks to assemble, mobilize and weld two barges together to support the 350T ringer crane to be used throughout the life of the project using 11 craft and two foremen full-time plus support equipment resources such as welding machines and other cranes to assemble the ringer crane.

REWORK ACCOUNTABILITY

All Contract Time must be delineated between production time, and time required to bring non-complying work into compliance.

Therefore, the model's equation is revised to account for resources expended fixing non-complying work. This one major component, known as *REWORK* can completely change the value of Tr if not properly documented.

Manpower and equipment resources assigned to REWORK are documented just the same as documenting contract work, however, it is negatively attributed to the **resources speed of travel through the activity, through the expression**

$$Tr = (1.0\text{-}TrREWORK) \; x \; (TrDRILL + TrCONC) \; = \; PERFORMANCE$$

Equipment breakdowns (EQBK), in hours, are also essential in establishing the **productivity ratio and performance** of the Contractors' work progress.

$$Tr = (1.0\text{-}TrEQBK) \; x \; (1.0\text{-}TrREWORK) \; x \; (TrDRILL + TrCONC) \; = \; PERFORMANCE$$

The ability to analyze and graphically interpret the thousands of data and notes collected and written on the daily reports of construction, would require thousands of manhours to input the data, by inexperienced personnel at a much later date, and at very high cost to the Owner.

By measuring the time it takes for a resource to travel through an activity on a contemporaneous basis, the empirical equation

$$Tr = \frac{Mp(a)xEq(a)xMt(a)}{\text{Duration of Activity (a)}}$$

expresses the interactive relationships between manpower and equipment. Materials are less interactive, excepting formwork and rebar.

Reinforcing steel, although common throughout, are detailed and shop-fabricated at different lengths, bends and weights, therefore making it virtually impossible to substitute required rebar not-on-site, for available material.

As Tr increases, delays are occurring. As Tr decreases, productivity is increasing.

Mp directly effects productivity. The number of activities scheduled for the day utilizing the same or similar manpower resources under evaluation, and

Overtime. It is documented in accordance with two primary databases: Certified Payrolls and Daily Reports of Construction.

ACCOUNTING FOR MANPOWER

The equation expressed above is a measure of resource flow through each activity.

Resources applied to many and simultaneous activities or tasks by the contractor will greatly alter the planned duration of the activity and prolong the project schedule if the activity is on the Critical Path.

Contemporaneously inputting directly to the resource screen the daily totals of number of craft, hours worked, activity worked and equipment utilized, will appear in the matrix. (Dunne and Navarrete, 1994).

For example, as 1 foreman and 7 craft are entered into the matrix for Contract Calendar Day 1 under MANPOWER, and so on for each calendar day of contract work, and the activities completed that day are noted in the matrix under ACTIVITY, and each is tied to TIME, expressed as each calendar day.

The resulting three-dimensional matrix generated over the life of the project, will then be able to sort by manpower, activity and time to develop the historical and contemporaneous model of the contractor's daily performance based on the contractor's applied resources on the jobsite.

The resulting empirical data writes the equation of the line representing sustained performance and ability to meet planned productivity and duration.

As Tr is measured for each activity in the contractor's schedule, the model identifies the manpower and equipment resources allocated to that specific activity, and a productivity ratio is established. This ratio is compared to the Planned Durations in the contractor's CPM schedule and percentage completion reported each month.

Series of **successive activities,** established as the Critical Path by the Contractor's scheduling software, are selected as the primary data acquisition boxes.

Significant concurrent activities are expressed in **shaded matrix boxes.**

The prevailing theory of this paper, as expressed by multiple Tr calculations, is that as the matrix grows upwards and longer as time increases, the activity base expands transversely, creating an interconnected, stable and inverted structure.

Component sides of boxes are the <u>CPM Scheduling interconnecting ties and interdependencies submitted by the Contractor at the beginning of the project.</u>

If all critical path boxes do not share a common side of <u>any 6 sides</u> when the <u>resources are applied</u>, then the Contractor has not resourced that activity, and the critical path will be delayed.

The coordinate axes x,y *(activity, resource)* represents the resources applied to each activity created in the database, resourced each Contract day along the z *(time) axis*.

Therefore, <u>every x,y face of the box must intersect along the z axis to be planar, and maintain the **continuity of critical activities on a resource basis**</u> to achieve a resource productivity ratio of 1.0.

If the activity is not resourced due to changed conditions, and resources are shifted to a concurrent activity, then the Contractor has mitigated the potential damage to the schedule, and it becomes the Owner's responsibility.

It is desirable to build the matrix from the bottom, up.

Obviously, as the mobilization begins and resources are recorded, the *natural process of construction will be established, by the Contractor.* From the foundation and long-lead items required for actual construction progress, the <u>as-built schedule will generated each day from this database.</u>

Correlation between the scheduling software, and the Harris database is <u>project-specific</u>, developed by each Harris Project Team with the Owner's engineering and construction management staff for concurrence and approval by Executive Management.

Direct correlation and database access are essential to the full implementation of the system.

The development of this CPM support database does not re-create the many positive aspects of CPM scheduling software, but enhances its performance through <u>contemporaneous data collection and retrieval using the theory of resource assignment and availability to confirm the schedule progress and Critical Path expressed by the General contractor in the Project Schedule.</u>

Case Study II - Bridge Deck Formwork Cycling

Before concrete placement can begin, the Contractor must form-out the spaces between the seven pre-stressed, post-tensioned, 145'-long concrete girders comprising the cross-sectional area of the composite deck structure.

Six spaces, 6'-0" wide x 145'per span x 2 spans per placement, amounting to over 1700 s.f. of formwork for each of the 15 spans over water must be in-place for 7 days after concrete placement by specification, before they can be stripped-out for re-use on succeeding spans.

1 foreman and 4 laborers are assigned and available for this task, along with an under-deck "stripping buggy", supported from the completed deck structure above. The crew stands on the lower platform and works overhead removing the whalers and plywood deck forms after concrete deck curing and compressive strength test results from cylinder breaks.

The construction economics of buying additional 500 sheets of 3/4" CDX Plywood and 1,000 L.F. of 2x6 double-back whalers with 1/2" threaded rods up through the top concrete beam flange to support the formwork on 6'-0" centers vs. additional manpower to strip-out the existing formwork at a faster pace and re-use it, must be evaluated against time lost for the slowest rate of productivity and associated **time delay costs.**

Speed of travel of the labor and equipment resources through the activity is the focal point of the Contractor's decision.

CONCLUSIONS

Tracking this speed of travel through an activity on the model developed in Tr for multiple activities provides a contemporaneous database for estimating project construction activity durations, and evaluating effects of time delays caused by changes and conflicts in plans and specifications.

The ability to visually integrate this information into an inverted, stable pyramid to demonstrate the necessity for stability of the model, can present Executive Management with overall views, and selected data resources of interest.

REFERENCES

1.   *Contemporaneous Development of Field Data: Transportation Construction Claim Defense*, ASCE 1st Congress on Computing in Civil Engineering, Washington, D.C., 1994.

CADD-Based Construction Information Management for Corps of Engineers Projects

Sangyoon Chin[1], Liang Y. Liu[2]
Annette L. Stumpf[3], Rajaram Ganeshan[3], Donald K. Hicks[4]

Abstract

A large amount of information is created throughout the life of a facility construction project, from design, construction, testing and startup to operation and maintenance. Effective operations, maintenance, and rennovation of the facility are dependent on high quality information collected during the facility life-cycle. It has been a challenge to efficiently capture and manage the project as-built information. This paper presents an on-going research effort to capture and manage construction information electronically throughout the life-cycle of a project. A conceptual data model and a prototype system, named CADD-based Construction Information Management System (CADCIMS) have been developed based on existing research on multimedia project control and documentation system (MULTROL) and CADD-based information retrieval system (CADCON). CADCIMS provides various ways to store and access the construction information based on CADD system. It allows the construction management team to take advantage of the CADD files received from the design team to build an information management framework for managing construction tasks. Furthermore CADCIMS uses the information framework to collect/update accurate as-built project information for operation/maintenance.

Introduction

Information in various formats is created throughout the life-cycle of a construction project, from design, construction, to facility operation and maintenance (O&M). Information is created and used during the project life-cycle by many participants at different times. Like most construction engineers, resident engineers in corps of

---

[1] Research Assistant, US Army Construction Engineering Research Laboratories (USA-CERL); Dept. of Civil Eng. University of Illinois at Urbana-Champaign, Urbana, IL , s-chin@cecer.army.mil

[2] Assistant Professor, Dept. of Civil Eng. University of Illinois at Urbana-Champaign, Urbana, IL 61801

[3] Principal Investigator, Design and Construction Team, Facility Management Division, USA-CERL

[4] Team Leader, Design and Construction Team, USA-CERL, P.O. Box 9005, Champaign, IL 61826

engineers projects need to keep track of design/construction changes and as-built information in order to control and monitor construction progress. Facility operators/maintainers need up-to-date as-built information and drawings which reflect the design/construction changes in the construction phase. Increasingly, facility operators/maintainers are using computer-based facility management systems. Collection and management of the various types of information in an electronic format during design and construction will enable construction engineers and facility operators/maintainers to access information when they need it. In order to provide a smooth information transfer from the construction phase to the facility O&M phase, it is necessary to develop tools to help resident engineers manage design and construction information electronically in a consistent manner.

This paper presents the development of a conceptual data model and an application, based on the data model and named CADCIMS, for processing design and construction information. CADCIMS is a prototype CADD-based construction information management system for design and construction information integration. CADCIMS is developed for the use in the Microsoft Windows™ environment by using CADD and a relational database management system. CADCIMS allows users to manage construction information in various forms in a more intuitive and consistent way. Project information can be stored and retrieved based on CADD objects, activities, and change orders. CADCIMS provides an integrated environment to collect, store, and retrieve project information from design to construction, and down to O&M conveniently and consistently.

Electronic Design files and Construction Information Management

As computer technology has advanced, many sophisticated CADD systems have been developed. More and more architects/engineers(A/E's) are choosing CADD as the method to develop design solutions, but A/E's and also constructors haven't taken full advantage of the capabilities of CADD systems (Stumpf et. al. 1994). Most drawings are done in 2D, although more and more A/E firms are experimenting with 3D design. Mostly due to liability and copyright issues, A/E's are hesitant to give the contractors CADD-generated drawing files. Therefore, most contractors plan and estimate costs and schedules based on paper-based design information. This paradigm of paper-based information exchange has its limitations and problems. Mistakes and inconsistencies are inevitable during the translation process from paper documents. Now that more and more A/E's are using CADD systems, it seems natural and beneficial to get the design files (CADD and specifications) electronically to start construction planning and control. There are other social or legal issues to be solved; however, in design-build or partnering projects, information integration has proven to help planning and control during construction. This information integration provides a good framework for capturing as-built information for the O&M phase, after the construction is completed.

Existing Computer-based Construction Management Tools

Primavera Project Planner ™, Expedition ™, and Timberline Collection of Estimating Software ™ are probably among the most widely used software for construction. Primavera Project Planner ™ is a project management software. It focuses on CPM scheduling, resource allocation and leveling, cost control and report generating. Primavera Expedition™ is a construction contract management software. It keeps track of most of administrative information among owner, contractors, and subcontractors during construction. The information includes submittals, changes/modifications, bid packages and contracts, purchase orders, daily logs, etc. Timberline Precision Collection of Estimating Software ™ is a cost estimating software integrated with AutoCAD™ and Primavera Project Planner™.

In addition to these commonly used computerized systems for project and control, several USA-CERL research prototype systems such as MULTROL, CADCON, and Schedule Generator were developed to enhance or complement the features provided by the commercial products. MULTROL was developed to enhance project scheduling systems with a multimedia capability, CADCON was developed to integrate the CADD objects with the as-built information, and Schedule Generator was developed to automatically generate construction activities and scheduling from design information.

This paper introduces a new conceptual data model, named CADCIMS, which integrates the construction information management framework with the design information. This framework for construction information management is used to integrate these commercial products and research prototypes for managing design and construction information. The following sections outline these prototype systems and describe the data model developed to integrate them.

A Multimedia Project Control and Documentation System, MULTROL

A prototype multimedia project control and documentation system (MULTROL) has been developed by the University of Illinois and US Army CERL. MULTROL provides documentation and management of the project information in various formats such as text, video, sound, and images (Liu et. al. 1994). Since MULTROL is intended to be used by a schedule engineer, the captured as-built information is associated with activities. Access to the as-built information starts from the selection of the activity. Users may input the daily activity description, cautions and problems, or lessons-learned in a text format. Video, sound, and images allow construction engineers to capture construction progress or problems more vividly. This system provides schedule engineers with fast and easy access to the as-built information so that they can detect potential delays and cost overruns.

A CADD-based Information Retrieval System, CADCON

CADCON is a prototype construction information retrieval system based on 3D CADD objects. This application was developed to see how CADD systems can be used effectively by resident engineers during construction (Stumpf et. al. 1993). A CADD system, AutoCAD R.12 for Windows, was integrated with a relational database management system (RDBMS), Microsoft ACCESS. A conceptual data model was developed to make the database more efficient. Linkage between the CADD system and the RDBMS makes it possible to retrieve the related construction information on activity, cost, as-built pictures, and specification documents by selecting a 3D object in the CADD system.

Schedule Generator

Schedule Generator (Ganeshan et. al. 1995) is a prototype application developed in a collaborative decision-making process environment (Golish 1994). The Schedule Generator uses a model-based planning approach to generate a preliminary construction plan including both time and cost components based on the design generated by the architect. The plan generation process has the following steps(see Ganeshan, et. al., 1995 for details): (1) defining construction zones and groups of components, (2) generating activities, (3) identifying construction methods for each activity, (4) assigning resources for each activity from MCACES (MCACES 1992) databases, and, (5) sequencing activities. Knowledge-bases are used to assist in various steps in this process. This system enables designers or constructors to evaluate what-if scenarios of design in terms of construction time and cost, and to verify the constructibility through the schedule visualization.

Integration of CADCON, MULTROL, and Schedule Generator

One of the shortcomings in CADCON is that it does not support data abstractions so that the detail level of activity and component must be predefined before construction begins. Therefore a lot of data input for component, activity, cost and their relationships was required. To overcome the problem of massive data input, the integration of Schedule Generator and CADCON was conducted to import component, activity, cost, construction method and their relationship directly without any effort to transform design information.

Another problem was that pictures, drawings, and specs are not stored in the database. They are stored in separate files. That is, only file names are stored in CADCON and MULTROL, because the formats for video, sound, images and documents are not supported by most general RDBMS. RDBMS has limited support for data types like video, sound, and images. This can cause data consistency problem since files can be casually deleted or moved to another directory, and the file-based storage is not very effective in a multi-user environment (Loomis 1995). The solution is to embed the multimedia information into the database table by using object linking and embedding (OLE) developed by Microsoft. Both MULTROL and CADCON allow users to access multimedia information while MULTROL is activity-based approach and CADCON is component-based. Therefore it was considered more effective to integrate MULTROL and CADCON so that users can access construction information from both view points.

CADD-based Construction Information Management System (CADCIMS)

As described in the previous section, CADCON, MULTROL and the Schedule Generator were integrated in order to eliminate the massive data input (components, activities, and cost information) and to directly integrate with the design information in CADD. A new data model was created along with a prototype system called CADCIMS. CADCIMS represents a new way of information integration which allows design in CADD to be integrated with construction information framework. CADCIMS model was developed to eliminate any potential information redundancy due to the integration. The purpose of the CADCIMS model is to build an information structure to capture, store, retrieve, manage the as-built information including multimedia data types, and to provide facility O&M with accurate construction information. The CADCIMS model is described by using an Entity-Relationship diagram (Korth and Silberschatz 1991) shown in Figure 1. The conceptual model was developed to organize the construction information such as component, cost, activity, changes. etc. Names on the box represent information types and names on the diamond represent relationships while lines and arrows represent the type of the relationship.

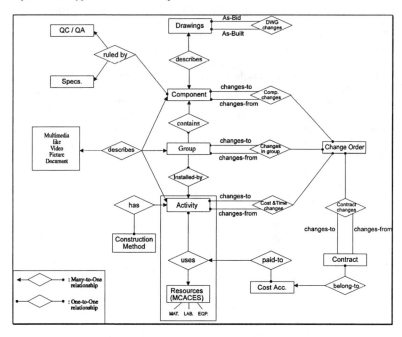

Figure 1. Integrated Conceptual Model for CADCIMS

Even though Schedule Generator supports the multiple data abstractions in component and activity, only the objects in the bottom level is considered. Instead, the group information and its relationships with components and activities are adopted to the conceptual model since Schedule Generator has the feature to group components according to a work area (zone) in order to associate with activities. It is sometimes better to deal with the group of components than with each single component. This will reduce time and effort and improve the accuracy of the database.

The activity information has a one-to-many relationship with the resource information because many resources are involved in one activity. An activity is performed in a certain way, which is the construction method in this model and influences the resources for the activity. Resource information such as labor, equipment and material was adopted from the MCACES database. MCACES is used as a cost estimating system by the US Army Corps of Engineers (MCACES 1992).

Change orders are considered as an information type to reflect changes during construction. Change orders usually involve changes in cost only or both cost and time (Hester et. al. 1991). Also changes in component information can occur. Therefore change relationships were set up for component, component group, activity and contract types, and these relationships are associated with the change order type. The component information and the component changes relationship will keep track of the history of changes, and the same mechanism can be applied to component group, activity, and contract information. The group information and changes in group relationship will be very useful when a change order contains changes in many components. To reflect changes during construction will help construction engineers monitor the construction progress more effectively by providing more accurate information , and will help capture information for facility O&M.

Multimedia information including video, sound, picture, document, and many other types, are supported by OLE in CADCIMS. Multimedia information will be embedded into the database table so that users do not have to manage both the database and files. This will improve the database consistency. Relationships are set up with component, component group, and activity so that multimedia information can be accessed from activity, component or component group. Supporting multiple relationships with activity, component, and component group enables users to access appropriate information from various viewpoints, and provides better semantics between multimedia information and activity, component or component group. Documents are also considered as one of the multimedia information so that manual, product information and warranty information for a specific component (e.g. HVAC equipment) can be stored in the database for use during facility O&M.

Specification and QC/QA (quality control/quality assurance) information are considered in the document level only. Drawings are considered as a separate type but changes between as-bid and as-built drawings are kept track of by the DWG changes relationship in Figure 1.

Based on CADCIMS conceptual model described in Figure 1, a database was built using MS ACCESS™ V. 2.0, a relational database management system in the MS Windows ™ environment. The database is integrated with a CADD system,

AutoCAD™R.12 for Windows, through DDE (Dynamic Data Exchange) and ODBC (Open DataBase Connectivity). GCLISP™, AutoLISP™, and Visual Basic™ have been used to build the interfaces among the Schedule Generator, the CADD system, and the database. Even though the general RDBMS had some difficulties in supporting multimedia data types, the introduction of object linking and embedding (OLE) developed by Microsoft has resolved this problem to some extent. Almost any kind of data including document, images, video, sound, etc. can be embedded as objects in the database. As a result, data for drawing and multimedia types in CADCIMS conceptual model can be stored in the relational database. Storing pictures, drawings, and specification documents in a database has more advantages than storing them in files. Current database technology provides better security and consistency, and can work better in a multi-user environment because the database management systems provide transaction management such as concurrency control. Figure 2 below shows a typical information management using CADCIMS.

Future Study

We envision the future version of CADCIMS will be based on an object-oriented database model instead of the relational database model. Our future research includes other types of construction information, which was not considered in current version of CADCIMS. By applying object-oriented concept modeling like object-oriented programming (OOP) and OODBMS, we expect CADCIMS will be more efficient and robust.

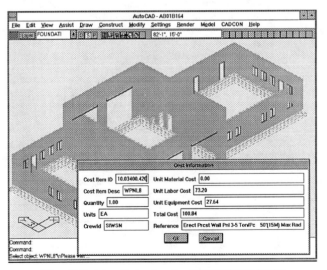

Figure 2. Prototype System based on CADCIMS Data Model

Conclusions

Based on existing research such as MULTROL, CADCON, and Schedule Generator, a conceptual data model and prototype CADD-based construction information management system (CADCIMS) was developed to capture and manage the as-built information during construction and to provide smooth information transfer from design to construction and to facility operation and maintenance. CADCIMS allows users to access construction information from both 3D CADD objects and the database. This system provides various ways to access the construction information. Users can access the desired information from different views, such as building component, activity, component group, drawing, contract and change orders. CADCIMS allows resident engineers to monitor and control construction more effectively because changes are directly reflected and also the impact of a change can be evaluated more precisely. Database consistency can be improved by embedding the multimedia information into the database. Also up-to-date construction information can be transferred to facility operation and maintenance (O&M) with minimal effort so that the information can provide the corporate knowledge of the facility for O&M, renovation, demolition, and future projects. CADCIMS demonstrates how design and construction information can be integrated, and that life-cycle information management from design, construction, and operation/maintenance may not be too far away.

References

CSI/CSC UniFormat (1992), The Construction Specification Institute, Alexandria, VA.

Ganeshan R., Stumpf, A., Chin, S., Liu, L.Y. and Harrison, B. (1995), *Integrating Object-Oriented CAD and Rule-Based Technologies for Construction Planning*, Draft US. Army CERL Technical Report.

Golish, M. (1993), *Architect's Associate: Applying Concurrent Engineering to Facility Design*, RDA Magazine, October 25.

Hester, W. T., Kuprenas, J. A. and Chang, T.C. (1991), *Construction Changes and Change Orders: Their Magnitude and Impact*, Construction Industry Institute, Source Document 66, October.

Korth, F. K. and Silberschatz, A. (1991), *Database System Concepts*, McGraw-Hill, Inc., 2nd Edition.

Liu L. Y., Stumpf, A. L. and Kim S. S. (1994), "Applying Multimedia Technology to Project Control," *Proceedings of the First Congress on Computing in Civil Engineering*, ASCE, June.

Loomis M. E. S. (1995), *Object Databases: The Essentials*, Addison-Wesley Pub. Co.

MCACES GOLD Ver. 5.2 User's Manual (1992), Building Systems Design Inc. Atlanta, GA.

Primavera Expedition Reference (1992), Primavera Systems, Inc.

Stumpf A., Ganeshan, R., Chin, S., Ahn, K. and Liu, L.(1993), *CADD Requirements to Support Corps of Engineers Construction Field Activities*, USACERL Technical Report, September 30.

Stumpf A. L., Liu L. Y., Chin, S. and Ahn, K. (1994), "Using CADD Applications to Support Construction Activities," *Proceedings of the First Congress on Computing in Civil Engineering*, ASCE, June.

USE OF CAD THREE DIMENSIONAL MODELLING AS A CONSTRUCTIBILITY REVIEW TOOL

John A. Kuprenas[1], Member, ASCE, Joseph T. Um[2], and Craig E. Booth[3]

ABSTRACT

This paper introduces a methodology to use a computer aided drafting (CAD) three dimensional model of a project to identify design conflicts as part of a preconstruction constructibility review. Early identification of these conflicts will allow engineers to make corrections to the design prior to the start of construction, thereby saving change order cost markups. A project case study demonstrates the implementation of this methodology and describes the types and results of the three dimensional conflict reviews. Ideas for future expansion of this work are explored.

INTRODUCTION

Design errors and omissions present a significant problem in the construction industry in that they are often manifested through conflicts between piping, equipment, and structural members. Even with excellent management, a high number of these conflicts result in construction change orders and doom a project to exceed the project budget and schedule. This paper introduces a methodology to use a CAD three dimensional model of a project to identify design conflicts as part of a preconstruction constructibility review. Early identification of conflicts allows engineers to redesign conflict areas prior to construction, hence, eliminating field change corrections and the premium cost associated with them.

A constructibility review is a comprehensive program used to enhance the overall objectives of a project. Accomplished through a systematic approach, a constructibility review optimizes construction related aspects of a project during all phases of a project ("Constructibility" 1991). Over the last decade, the work

[1]Const. Mgr., Vanir Const. Mgmt., 3435 Wilshire Blvd., Los Angeles, CA 90010 Lecturer, Dept. of Civ. Engrg., California State Univ., Long Beach, CA 90840
[2]Proj. Arch., Ellerbe Becket, Inc., 2501 Colorado Ave., Santa Monica, CA 90404
[3]Proj. Arch., Ellerbe Becket, Inc., 2501 Colorado Ave., Santa Monica, CA 90404

included in a constructibility review has extended beyond a simple review of plans and specifications.   In a thorough constructibility review, the overall project plan, project planning and design, the construction-driven schedule, project costs or estimates, and construction methods are optimized during the entire life of the project ("Constructibility" 1991) .   The financial merits of a constructibility review are well documented.   Savings of six to ten percent of the construction cost and ten to twenty times the cost of the review effort have been reported ("Constructibility" 1986, "Integrating" 1982).   Benefits of constructibility reviews also extend into schedule performance through shortened construction durations, fewer schedule delays, and reduced risk premiums that the contractors include in their bids ("How to" 1987).

Three-dimensional modelling has been done for centuries using actual physical models.   With the development of computerized drafting programs, three dimensional modelling thorough computer aided drafting is presently available to even the smallest design firm.   AutoCAD is one software package that is currently one of the more popular CAD software tools.   This use of three dimensional CAD in construction has been shown to benefit construction managers and designers.   A three dimensional model benefits planning and estimates, improves productivity, assists in construction scheduling, and can even be used in claim analysis (Mahoney and Tatum 1994, Morad and Beliveau 1991).   In addition, most CAD computer software is compatible with interference check software programs which automatically check in three dimensions for conflicts between drawing elements.

This work outlines a methodology to combine the above mentioned benefits of both a constructibility review and three dimensional CAD modelling.   Within this methodology, the inference check and planning strengths of the three dimensional CAD modelling are used to identify potential construction conflicts as part of the optimization of construction included in a constructibility review.   The methodology presented in this paper specifically is used to review design layouts and to identify design conflicts.   The methodology is implemented in the final design stages of a project.

CONSTRUCTIBILITY REVIEW METHODOLOGY

The project used to demonstrate this methodology is a central utility plant serving essential utilities to a six hundred bed rehabilitation hospital.   The area served by the plant consists of over 1.3 million square feet of hospital campus (built-out area by the year 2003) located on a fifty acre site.   The central utility plant is a 35,000 square feet, two-story building which will provide the hospital with chilled water, domestic hot water, steam, domestic cold water, heating hot water, soft cold water, emergency power, compressed air, electric service, oxygen, and an energy management control system.

The client for this project was particularly concerned about excessive changes orders and project schedule delays.   The plant construction schedule had no float since the new plant was scheduled to feed a new one hundred fifty bed patient building also under construction. Hence, the project owner was willing to spend money prior to construction to attempt to alleviate any potential delays. Based on the expected savings experienced for a typical constructibility review, the client under took a full constructibility review. One experimental and non-traditional element of the full review was the creation of a three dimensional constructibility review model and interference report.

Ideally, this three dimensional constructibility review would be conducted at the fifty to ninety percent construction document level of the project. Major systems layout would be complete by this stage of the design; yet there would still be time to review results of the review and make corrections prior to the job bidding. The three dimensional AutoCAD constructibility review from the following case study, however, was not completed prior to the start of the construction and comments from the review were not addressed by the architect or engineer prior to the bid. Although not of any immediate advantage to the owner, this undertaking is still of great value. Researchers can examine the resolution of all the conflicts identified by the three dimensional constructibility review. It can be determined whether any particular conflict ever became an issue, a contractor requests for information, a change order, or perhaps even a construction claim. By combining the analysis for every individual conflict, the overall efficiency of the methodology can be evaluated, and researchers can establish a dollar and time value for the review.

Model Construction
For the three dimensional constructibility review report, the facility was divided into six (6) areas - the boiler room, the chiller room, the cooling tower yard, the electrical room, the emergency generator room, and the equipment room. A series of separate reports were generated for each area and for major subsystems in each area. Subsystems in each of the six areas may include: equipment, heating and ventilation duct work, plumbing, roof drains, mechanical piping, and structure. Not every subsystem was included in every area, since a particular subsystem may not be required within a particular area of the utility plant. Elements included in the model which were not studied as part of the three dimensional constructibility review include: decking, architectural walls (exterior and interior) and floor slabs, and equipment pads.

Review Types
The **composite review** tests for the interferences between major subsystems. A composite report was performed for each area of the plant. Interferences between members of individual subsystems are not included in this report. For this report, subsystems were identified as "active" or "passive". Active subsystems had each element of the subsystem tested against each element of all other active and passive subsystems. Passive subsystem were tested for interference with active subsystems but not with other passive subsystems. For example, in the equipment room mechanical piping (active) was tested against all elements of equipment (passive) and

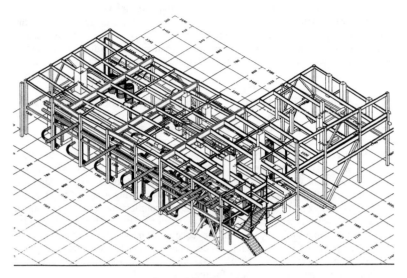

Composite Graphic

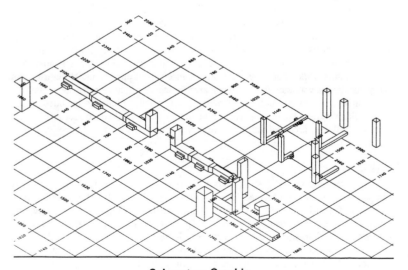

Subsystem Graphic

FIG. 1. Three Dimensional Constructibility Review Graphics

structure (passive) but equipment (passive) was not tested against elements of the structure (passive).

The second review, the **subsystem review**, tests each element of the subsystem against each other elements of the same subsystem. Separate reports are included for key subsystems. These reports show the results of testing each element of the subsystem against each other element of the same subsystem. Analysis of these reports shows that most interferences at this level are artifacts of the techniques used to create the model and not true interferences. For example, many pipe tees are reported as interferences. Some more substantial interferences were, however, discovered in this review.

CONSTRUCTIBILITY REVIEW RESULTS

Each three dimensional constructibility composite or subsystem review report includes both a graphic and printout. The **graphic** includes an isometric image of the area of subsystem or subsystems tested. A composite review graphic from the chiller room is shown in figure 1. The subsystems shown in this review graphic are equipment, ductwork, and mechanical (including chilled water supply and return, domestic cold water and steam). A single subsystem review graphic from the chiller room showing only the ductwork is also shown in figure 1. In both graphics, a grid has been included showing the x,y coordinates of the area in inches. The **printout** describes the test setup, summarizes the test results, and supplies detailed information regarding each conflict discovered by the interference check software. A sample conflict for the chiller room graphic review is shown in figure 2. The conflict printout identifies the subsystems involved in conflict and supplies the location of the conflict in x,y,z coordinates in inches. The layer names of the conflicting elements generally includes an abbreviation for the specific system to which the element belongs and begins which a single letter indicating the type of system. The conflict printout in figure 3 shows an interference between a heating and ventilation duct and a structural member.

```
---------------------------------------------------------------
Interference 7:

    Object #1:

        Filename ...  C:\ACAD\DWG\CH-ST.jsm
        Commodity ..  STRUCTURE
        Label ......     Entity Type : POLYLINE     Handle : NULL    Layer : S-RFLW

    Object #2:

        Filename ...  C:\ACAD\DWG\CH-HV.jsm
        Commodity ..  DUCT-WORK
        Label ......     Entity Type : POLYLINE     Handle : NULL    Layer : M-HV-FL

    Point: 1540.055286, 2057.242997, 323.152185
    Type: Hard
---------------------------------------------------------------
```

FIG. 2. Constructibility Review Conflict

The results of the composite review are shown in table 1. The review results are summarized by room. The chiller room had the by far the majority of conflicts. This room is inherently complex since it contains both the gas and electric chillers, extensive quantities of piping, and chilled water and steam. The equipment room also had a high number of conflicts. Like the chiller room, the equipment room has extensive piping. Table 2 shows the results of the subsystem review. The table shows the number of conflicts broken down by room and by subsystem. Again, the chiller room had the most conflicts. The subsystem with the most conflicts was "mechanical 1" which consists of chilled water supply and return, domestic cold water, and steam. Note that this subsystem review is independent of the composite review. Any conflicts found in the composite review are not re-identified in this subsystem review. The equipment room had the second most conflicts in the subsystem review. The area of particular concern would be the subsystem defined

TABLE 1. Results of Composite Review

Room (1)	No. of conflicts identified (2)	Active systems (3)	Passive systems (4)
Boiler Room	28	Ductwork Mechanical 1 Mechanical 3 Plumbing 1	Equipment Plumbing 2 Structure
Chiller Room	122	Ductwork Mechanical 1	Equipment Plumbing 1 Plumbing 2 Structure
Cooling Tower Yard	4	Mechanical 1	Equipment Plumbing 1 Structure
Electrical Room	5	Ductwork	Equipment Plumbing 2 Structure
Emergency Generator Rm.	15	Equipment Ductwork	Plumbing 2 Structure
Equipment Room	83	Ductwork Mechanical 1 Mechanical 3 Plumbing 1	Equipment Plumbing 2 Structure
Total (all rooms)	257	NA	NA

TABLE 2. Results of Subsystem Review

Room (1)	Subsystem self tested (2)	No. of conflicts identified (3)
Boiler Room	Ductwork	5
	Mechanical 1	1
	Mechanical 3	5
	Plumbing 1	24
Chiller Room	Ductwork	0
	Mechanical 1	120
Cooling Tower Yard	Mechanical 1	0
Electrical Room	Ductwork	4
Emergency Generator Room	Equipment	8
	Ductwork	0
Equipment Room	Ductwork	2
	Mechanical 3	4
	Plumbing 1	75
Total (all rooms)	NA	248

as "plumbing 1" which consists of domestic cold water supply and return and domestic hot water supply and return.

CONCLUSIONS

The constructibility review has proven its place in the construction industry. This work has outlined a methodology for the use of AutoCAD three dimensional model as a element of complete constructibility review. Within this new methodology two reviews are performed. A composite review shows interferences between major subsystems. A subsystem review tests each element of a subsystem against other elements within the same subsystem. This three dimensional constructibility review would ideally be conducted at the fifty to ninety percent construction document level of the project prior to the job bidding.

Review results can be expected to vary. In the worst case, this tool can function in simple manner as a visual tool to examine overall view of the structure under review. In the best cases, hundreds of potential construction conflicts could be identified, and the potential to save tens of thousands of dollars in construction changes exists. Future work from this research will be to expand the case study.

This expansion would trace the course of individual conflicts. Future work would also link three dimensional model with a knowledge based system, and construction scheduling. This would allow an analysis to include projected conflict impacts. It would also function as a change order analysis tool during construction.

APPENDIX. REFERENCES

"Constructibility: a primer." (1986).   *Publication 3-1,* Construction Industry Institute, Univ. of Texas at Austin, Austin, Tex.

"Constructibility and constructibility programs: white paper." (1991).   The Construction Management Committee of the ASCE Construction Division, *J. Const. Engrg. and Mgmt.*, ASCE, 117(1)

"How to implement a constructibility program." (1987).   *Publication 3-2,* Construction Industry Institute, Univ. of Texas at Austin, Austin, Tex.

"Integrating construction resources and technology in engineering." (1982). *Report B-1,* Business Roundtable, Aug., New York, N. Y.

Mahoney, J. J. and Tatum C. B. (1994). "Construction site applications of CAD." *J. Const. Engrg. and Mgmt.*, ASCE, 120(3), 617-631.

Morad, A. A. and Beliveau, Y. J. (1991). "Knowledge-Based planning system." *J. Const. Engrg. and Mgmt.*, ASCE, 117(1), 1-12.

THE INTEGRAL PROJECT: Overview

Richard M. Shane[1], Dave McIntosh[2], Edith A. Zagona[3], and Terry J. Fulp[4]

ABSTRACT

The INTEGRAL project is developing portable, reliable, and easy-to-use computer network software that will allow improved coordination between multipurpose reservoir operations and thermal plant schedules. The software is designed for geographically dispersed decision makers working for electric utilities and government agencies responsible for coordinated reservoir and power system operations. The project places special emphasis on developing an operational planning tool that will be sensitive to both power and nonpower requirements, including environmental commitments and objectives.

The project was initiated in 1992 by the Tennessee Valley Authority and the Electric Power Research Institute. The Bureau of Reclamation, in cooperation with the Western Area Power Administration, has since joined as a third cosponsor. The University of Colorado Center for Advanced Decision Support for Water and Environmental Systems located in Boulder Colorado is the contractor responsible for developing the general INTEGRAL software package.

With INTEGRAL software, all critical decision making locations in an organization will be able to work with a common set of models and historical, current, and forecast data so they may rapidly collaborate to compare operating alternatives and set reservoir releases and thermal plant schedules in the most efficient manner possible.

----

1. Technical Specialist, Reservoir Operations, Planning, and Development, Tennessee Valley Authority, Knoxville TN 37902
2. Project Manager, Electric Power Research Institute, Palo Alto, CA 94303
3. Research Associate, University of Colorado Center for Advanced Decision support for Water and Environmental Systems, Boulder, CO 80309-0421
4. Operations Research Analyst, Bureau of Reclamation, Boulder City, NV 89006

The software is comprised of two major components; a comprehensive, network-based, decision support code organized around a relational database management system; and a power and reservoir system modeling framework based on a modular, quick-change, "object-oriented" programming philosophy.

The decision support code, called TERRA, is designed to help users track power system status, anticipate problems, develop plans to avoid problems and coordinate planning efforts over a corporate-wide network connecting work stations located in decision making centers. To do this, it will perform the following specific functions:

<u>Manage current and historical system status data</u> -- System status information such as pool levels, turbine discharges, spills, hydrogeneration, water temperatures, etc. will automatically be added to data sets containing status information for the current operating year. Summaries of historical data will also be included.

<u>Maintain a system constraint data base</u> -- A data base containing current power and reservoir system constraints, commitments, and guidelines will be maintained. The data base will be updated as changes are made during daily scheduling activities or as new operating policies are adopted.

<u>Track system compliance</u> -- System status data will be continually checked for compliance with constraints, commitments, and guidelines.

<u>Manage model and analysis tool interfaces</u> -- The decision support code will provide customized data management interface code for each model called from within the decision support program. The code translates output files associated with a model into a form readable by the relational data base. If this output is to be used by another model or by an analysis program, it will be translated back from the data base into a file format readable by the second program. In this fashion, the data management system can facilitate communication between a variety of models and/or analysis programs.

Special care has been taken to allow tailoring of the decision support code to the needs of any utility. To this end, the system has the following attributes:

1. Standard ARC/INFO Geographic Information System coverages for display of facility locations and map features,
2. A standard wide-area communication platform (TCP/IP),
3. A GALAXY interface that allows access from either UNIX or WINDOWS computing environments, and
4. A widely accepted relational data base structure.

The power and reservoir system modeling framework called PRSYM represents a new approach to model building for river, reservoir, and power systems. The modeling framework concept produces modeling capability that is highly adaptable with little programmer support. No longer will a programmer be needed to select new modeling techniques, select a new modeling time step, apply the model to a new system, include new components in a model of an existing system, and change output analysis methods. Minimal programming will be needed to customize PRSYM to a new application or add improved modeling techniques to an existing application. PRSYM represents a major advance in modeling flexibility, adaptability, and ease of use. With PRSYM, users will be able to:

1. Visually construct a model of their reservoir configuration using "icon programming" with icons representing reservoir objects, stream reach objects, diversions, etc.,
2. Select appropriate engineering functions, standardized by the industry, to reflect object characteristics needed for schedule planning e.g. reservoir and stream routing methods,
3. Replace outdated functions with improved versions developed by the industry,
4. Develop and include functions that are unique to their system,
5. Experiment with operating policies, and
6. Use data display and analysis objects to customize data summary presentations.

It is expected that this PRSYM flexibility will result in significant software development and maintenance cost savings for the industry. Plans are being made to form a users group to manage code maintenance and object library distribution after project termination.

Both the Tennessee Valley Authority and the Bureau of Reclamation are moving toward full implementation of the INTEGRAL products. Design concepts and application details are covered more fully in the seven companion papers associated with the two INTEGRAL conference sessions.

INTEGRAL Project: TERRA Decision Support System

René F. Reitsma[1], Peter Ostrowski Jr.[2], Stephen C. Wehrend[1]

Abstract

Managing large power and water resource systems requires the coordination of multiple interests, resources, operational constraints and staff. In this paper, we explore some information flow requirements in a complex water resources management organization, at the Tennessee Valley Authority (TVA). From these requirements, we formulate a number of requirements and characteristics for a decision support system (DSS) for supporting organizational decision making.

Introduction: TVA Information Management Requirements

Table 1 represents parts of the organizational infrastructure at TVA.

Table 1: Organizations and their tasks within TVA.

The Power Control Center (PCC) is responsible for monitoring, scheduling and forecasting the demand and supply of electricity for TVA. PCC provides data and hydro power plant operating constraints to other groups within TVA for purposes of modeling, analysis and scheduling.
Reservoir Operations, Planning and Development (ROPD) is responsible for monitoring, scheduling and forecasting operations of the TVA reservoir system. ROPD produces reservoir release schedules which set the daily release constraints within which PCC can schedule hourly operations.
Other Water Management groups are responsible for monitoring water quality and ecological conditions on the Tennessee River and its tributaries, and the development and implementation of water quality improvement plans and programs.
The Engineering Laboratory (EL) provides engineering and research expertise and supports monitoring, modeling, operations, scheduling and forecasting for other groups within TVA.
The Operations Duty Specialist logs current fossil and nuclear power plant operations and provides various scheduled and forced maintenance data.

1. Center for Advanced Decision Support Water and Environmental Systems (CAD-SWES), Dept. of Civil, Environmental and Architectural Engineering, University of Colorado at Boulder.
2. Tennessee Valley Authority, Engineering Laboratory; Norris, TN.

Managing a complex water resources system such as the Tennessee/Cumberland watershed with an organization structured along sectors implies challenges to the coordination and flow of information through the organization. First of all, information needs to be provided to the right places in a timely fashion. The data from which information is derived, however, is collected over various intervals in time and space, often not coinciding with the time and spatial dimensions required at the time of decision making. As such, data needs to be systematically stored and archived, but in such a way that it can be easily manipulated as part of aggregation and disaggregation procedures. Both information and its constituent data, moreover, need to be shared. An information flow which supports decision making in both routine and crisis situations depends not only on proper generation and propagation of information, but also on the sharing of parts of this information by different departments and work groups. Some of this information needs to be generated on an ad hoc basis, for instance as a reply to a request from somewhere else in the organization, whereas other information, for instance the current state of all reservoirs in the system, can be automatically generated on either a regular or trigger basis.

Although the flow of information through an organization often requires combination and re-combination of various base data across the various sectors and layers of the organization, responsibilities for the generation of that base data often rest with specific people or parts of the organization. Automation and re-organization of information flow sometimes require changes in the distribution of base data generation, but especially in older organizations such as TVA, data generation and maintenance authority generally coincide with the organizational structure. As such, the information flow infrastructure needs to support the sector-specific generation of data and follow the organizational structure. In addition, it needs to support the use of data which goes across organizational structure and sectors.

TERRA Architecture

The above information management requirements imply a number of architectural characteristics. First, data collection must address multiple sites where data is generated and must be flexible enough to occur at any time or any regular schedule. In addition, ad hoc data loading facilities must be provided. Furthermore, since data entering the system are of diverse nature and are generated at many locations, strict logging and error reporting is necessary. Note that from a functional point of view it is not relevant whether or not the data physically resides at a central location. Modern database server systems allow developers to create a variety of centralized or distributed database structures, each of which acts logically as a single, organized database. All applications such as data storage and retrieval, data viewing, data aggregation and disaggregation or the initialization of a model with current system data, interact with this central data repository. This so-called data-centered architecture served as the basis for the TVA Environment and River Resource Aid (TERRA) decision support system. Details on the system's functionality were presented in Reitsma, Ostrowski and Wehrend (1994).

TERRA Design and Development

TERRA was designed, developed and implemented (fielded) over a period of approximately 24 months. The initial design activities employed a design technique known as problem-centered design. This technique was originally developed as a pure program design technique (Lewis, Rieman and Bell, 1991) but was adopted to the field of environmental DSS design. In problem-centered design, the design of a software system is approached as an abstraction problem. From elaborate real-world problem scenarios, system designers abstract the general classes of system functions, their dependencies and interrelationships. These are then tested and validated with additional scenarios until no more significant modifications are necessary.

Table 2 conceptualizes the complete design, development and implementation of the TERRA DSS as a structured process containing seven phases. The eighth phase, maintenance and further development, is an ongoing activity. For a more detailed description of this type of development process and each of its phases, refer to Reitsma et al. (forthcoming).

Table 2: TERRA development process.

Phase	Description
1. Needs Analysis	Inventory of users' needs (functions and data).
2.High-Level Design	Inventory of system functions, design of a system architecture and initial selection of software tools, development platform, programming languages, etc.
3. Detailed Functional Design	Detailed description of all system components, including control flow and user-interfaces.
4. Detailed Software Design	Design of the database data model and of software programs, data structures and algorithms.
5. Development	Population of the database and development of system source codes.
6. Initial Fielding	Pilot testing at TVA site.
7. Testing, Training and Fine Tuning	Further testing and revisions of the TERRA software and user training.
8. Maintenance and Further Development	Ongoing system maintenance and system software, hardware and data modifications.

Figure 1 maps the development phases of Table 2 onto relative expenditures in terms of human resources and time. The widths and order of the bars represent the periods of time spent on the first seven project phases. Bar heights represent the cost of each phase per unit of project time and are thus a measure for the rate at which project funds were spent during the various phases of the project. Rates have been indexed with cost rates for the development phase set to 100. Also, note that costs have been computed on the basis of total staff months without differentiating between the various billing rates for different staff. As a result, the rates for some phases are underestimated while those for others are overestimated. Multiplying bar heights with widths provides total cost per project phase. An analysis of the resulting cost distribution is currently being conducted.

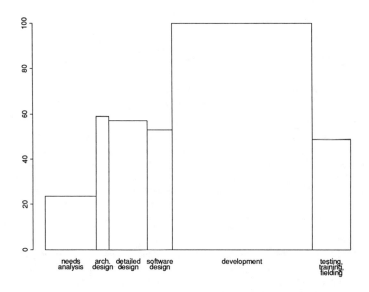

Figure 1. TERRA development phases versus cost rates.

References

Lewis, C., Rieman, J. and Bell, B. (1991) Problem-Centered Design for Expressiveness and Facility in a Graphical Programming System; Human-Computer Interaction; vol. 6; pp.319-355.

Reitsma, R., Ostrowski, P. and Wehrend, S. (1994) Geographically Distributed Decision Support; The Tennessee Valley Authority (TVA) TERRA System; in. Fontane, D.G., Tuvel, H.N. (eds.) Water Policy and Management, Solving the Problems; ASCE, New York, NY; pp. 311-314.

The INTEGRAL Project:
The PRSYM Reservoir Scheduling and Planning Tool

Edith A. Zagona[1], Richard M. Shane[2], H. Morgan Goranflo[3],
and Dieter Waffel[4]

Abstract

A generic river basin modeling tool for scheduling and planning requires a high degree of software flexibility to model any river basin, manage data input and output efficiently enough for near real-time operations and provide a selection of solution algorithms, all through a user-friendly interface. The Power and Reservoir System Model (PRSYM) provides an extensible, maintainable software framework which is not just a model, but a modeling environment to meet all the scheduling and planning modeling needs of managers and operators of utilities and river basins.

Introduction

PRSYM is a generalized river basin modeling environment which provides an electric utility or river basin manager with a tool for scheduling, forecasting and planning reservoir operations. PRSYM integrates the multiple purposes of reservoir systems, such as flood control, navigation, recreation, water supply, and water quality, with power system economics, and solves the system either descriptively with pure simulation or rule-driven simulation, or prescriptively with a goal programming optimization.

PRSYM is under development at CADSWES, the University of Colorado, under a joint sponsorship of the Electric Power Research Institute (EPRI), the Tennessee Valley Authority (TVA) and the U.S. Bureau of Reclamation (USBR).

PRSYM is the resident river basin simulation and optimization model for the Tennessee Environmental Resources and River Aid (TERRA), a decision support system which is also a joint EPRI/TVA/CADSWES endeavor. The model is "resident" by virtue of its immediate accessibility to all TERRA users and its direct link to the TERRA database. PRSYM will be the common model used by the River System

---

1. Research Associate, Center for Advanced Decision Support for Water and Environmental Systems (CADSWES), University of Colorado, Boulder, CO.
2. Technical Specialist, River System Operations Tennessee Valley Authority, Knoxville TN.
3. Technical Specialist, River System Operations Tennessee Valley Authority, Knoxville TN.
4 Manager, Market Management, Electrical System Operations, Tennessee Valley Authority, Chattanooga, TN.

Operations and the Electric System Operations groups in TVA to schedule operations for TVA's reservoirs and hydroplants.

The USBR/WAPA joined the 1994 and 1995 development as a second utility, ensuring the generic applicability of the modeling tool. The USBR/WAPA anticipates using PRSYM to replace both the existing long-term policy and planning model and the existing mid-term operations model of the Colorado River, and to implement several daily operations models for both the Upper and Lower Colorado Regions.

PRSYM is capable of functioning as a general scheduling or planning tool by virtue of four software characteristics: 1) an object-oriented software and modeling approach; 2) a data-centered design; 3) a multi-controller architecture and 4) a graphical user interface (GUI) and utilities which create not just a model, but a modeling environment.

Object-Oriented

The object-oriented nature of PRSYM is in both the software approach and the modeling approach. The software is object-oriented in that it is written in C++ and uses general object-oriented design concepts. The modeling approach is object-oriented in that the features of the river basin are represented by objects which can be manipulated on the workspace and which contain the code for modeling those features. PRSYM provides objects for storage reservoirs, power reservoirs, river reaches, confluences, canals, diversions, pumped storage facilities, and other features.

Each river basin feature object type (called an "engineering object" or "PrsymObj") contains *methods* and *slots*. Methods are the code for the engineering algorithms which are used to model the physical processes of the system. The user may select from a number of methods for modeling particular processes. For example, the power reservoir offers the user several methods for modeling hydropower generation including the total hydroplant method, the unit method, and several specialized methods previously developed by the sponsors.

Slots are variables which contain the data for the system. The slots hold input data, output data and parameter data needed by the various methods. Timeseries slots keep track of timeseries data, table slots hold tabular data, and scalar slots hold single-valued data. Each slot has a unit type assigned to it by the engineering code. Computations are carried out in a set of default units, but users may enter and view data in a unit of their choice. Slots are associated with the methods which use them and only the slots for selected methods are seen by the user in the GUI.

The user double clicks the mouse on an object on the workspace to open the object revealing a list of slots, the user's units for each slot and the current value. Also, indicators show if each slot is an input and if it is referenced in currently loaded rules or constraints. Double clicking on a slot brings up the Open Slot dialog. For time series slots, this is a series of numbers associated with points in time. From this dialog the user may designate any or all of the values as inputs or outputs; set user units; display format, precision, minimum and maximum limits on the slot; and set convergence stopping criteria for iterations. Table slots show multiple columns with column head-

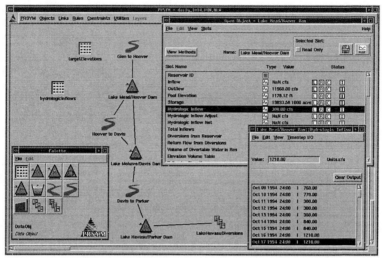

Figure 1. PRSYM Workspace with model of Lower Colorado River. Also shown are the Object Palette and the Open Object and Open Slot dialogs

ings and the same options for units and value displays.

Data objects are a special type of object on which users can create slots. Time series slots on data objects can be linked to timeseries slots on engineering objects to serve as input or catch the output values. These slots can be designated "read only" and not overwritten if they are not designated input. For example, the floodguide elevations for a reservoir could be stored in a data object in read-only mode, and the slot linked to the corresponding reservoir's poolElevation slot. The user can designate some of the values as inputs. These will be propagated to the object and become inputs for the simulation. The other values will be preserved for future use. Data objects may hold table slots which are defined by the user and hold data which is used in rules formulated by the user in the rule editor. Finally, data objects may hold expression slots, a subclass of timeseries slots. These are algebraic expressions containing other slot names as variables. The expressions are formed by the user in the expression editor. The expressions are evaluated at the end of the run for all timesteps. Special functions help the user form expressions, for example, to sum up the power generation from all the power reservoirs.

The user constructs the model through the GUI by selecting objects from a palette, dragging the objects with the mouse onto the workspace, naming the objects, and linking them together. Objects are linked together to form the topology of the river basin using the graphical link editor. Specifically, a slot on one object is linked to a slot on another object. During the simulation, the solution process involves propagation of the information among objects via the links. For example, the user may link the outflow of Reservoir1 to the inflow slot of Reservoir2. During the simulation, when Reservoir1's mass balance is computed and the outflow slot is set by the calculation, the value is propagated across the link and is set as the inflow to Reservoir2.

An object-oriented approach offers a number of advantages. The first is the convenient separation of engineering code from the rest of the software. Engineering objects inherit much of their behavior, such as time management, from the parent simulation object class, the SimObj. Hence, only engineering methods are on the engineering objects themselves, providing a nice separation of the water resource code from the other software. This allows water resource engineers to develop the engineering algorithms without having to deal with, or work around the managing software.

Secondly, the code is easily extensible. New objects and methods are easy to add if new features or processes must be modeled. Also, existing programs can be called, taking advantage of previous programming investments by the sponsors. It is also possible to add new controllers, or solution algorithms. The advantages of the architecture are described in more detail in the accompanying paper by Betancourt, et al.

Data-Centered

PRSYM is a data-centered modeling environment in the sense that a model represents a particular river basin by virtue of the data that the user enters while building and running the model. This data includes the names of the features of the basin and how they are linked together to form the topology; the methods selected for modeling the physical processes; the data which describe the features such as the storage-elevation curves and the power generation curves; the operating policies which are expressed through the rules or the constraints; and the data inputs which drive the solution, such as hydrologic inflows. Data can be entered into slots by entering the values directly, by importing ASCII files, or by executing data management interface routines.

A special feature of PRSYM is the Data Management Interface (DMI) which provides a means of using external programs to automatically load data into PRSYM. These routines are written by the user or the user's organization in any programming languages. They are invoked through the PRSYM GUI. This is an essential feature for implementing PRSYM as a near real-time operations model. For example, TVA plans to update their operational model at least once a day with the latest existing conditions (pool elevations, recent releases, etc.), predicted hydrologic inflows, scheduled hydroplant operations, and special operations. The data will come from databases and from other on-line programs. The updating of the model must happen quickly and automatically for the model to be useful in a real-time operations mode. The DMI may also be used to get historical data from files or databases to run long-term planning models. The DMI is discussed in greater detail in the accompanying paper by Betancourt, et al.

PRSYM may serve as a planning model of the Colorado River using a monthly timestep for eighty years, calculating hydropower generation according to the USBR's special power routines, using historical hydrologic inflows, and driven by the many operating rules known as the "Law of the River." Or, PRSYM could be a hydroplant scheduling model for the TVA system of rivers and reservoirs, employing a six-hour timestep for a two-week horizon, modeling the powerplants by the unit method and keeping track of what units are available on an hourly basis. Evaporation, bank storage, salinity and sedimentation of reservoirs are calculated for the Colorado River, but not for the TVA. The TVA routes flows through riverine reaches, keeps track of wedge storage in reservoirs, and calculates the thermal replacement value of the hydropower produced. These are not needed for the Colorado planning model. The differences between these implementations of PRSYM are strictly a matter of user input data.

Multiple Controllers

Several solution methods are available in PRSYM and the architecture can support any number of additional "controllers." The current implementation includes pure simulation, simulation plus water quality, rulebased simulation, and linear goal programming optimization. The controller is selected on the run control panel where the user also specifies the begin and end dates of the run, and the timestep size (see Figure 2)

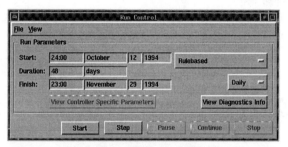

Figure 2. PRSYM Run Control Panel

Pure simulation solves an exactly specified problem. For example, if the user specifies inflows and reservoir releases for a basin, the model calculates storages, pool elevations and power generation. Each object has a number of methods known as "dispatch methods." These methods list the possible combinations of known and unknown slots which constitute a well-specified problem. For example, the dispatch method called "solveMB_givenOutflowHW" on a PowerReservoir object lists outflow and poolElevation as known slots, and inflow, storage and energy as slots which must be unknown to solve the mass balance equation.

The simulation begins with values being set on objects. Initially, the values are input values provided by the user. Each time a value is set on an object, that object examines its list of known slot values, called the "current set," and compares it to the list for each dispatch method. If one of the methods can be executed, the object notifies the controller that it can dispatch, and the controller puts it on the dispatch queue. The objects on the queue dispatch in turn. When an object dispatches, the dispatch method is executed and the object solves the algorithms which model its physical processes. For example, a PowerReservoir object calculates mass balance, hydro-power generation, turbine release, spill, total outflow and other associated values. River reaches route flow either upstream or downstream, calculating inflow or outflow.

The dispatch methods may also call the default user-selectable methods for modeling certain processes. For example on a Reach object, if the inflow is known, the solve_Outflow dispatch method calls the default method for routing. The user may have selected simple time-lagged routing, kinematic wave, Muskingum, impulse response, or others. The method selected by the user is registered as the default method and is executed by the dispatch method.

In the process of dispatching, an object calculates the values of all the output slots. When a value is set on a slot which is linked to another object, the value is immediately propagated across that link and set on the slot on the other object. If an outflow slot is set, it will be propagated to the connecting inflow slot on the downstream object and that object responds by checking its dispatch lists in light of the new known value. In this way, information traverses the system, and objects solve themselves, not needing to know anything about other objects in the system. If the user supplies too much data, PRSYM produces an overdetermination error and the simulation is halted. Underdetermined models are not currently flagged, but are evident if some objects do not dispatch for some timesteps.

Basin Feature:	Method Category	Methods
Power Reservoir	powerCalculation	*plantPower* *unitGeneratorPower* *CRSSPeakBasePower* *CRSSPeakPower* *LCRPower*
	tailwaterCalculation	*CRSSTailwaterBackwater* *CRSSTailwater* *TVAmeltonHillTW* *TWValueOnly* *TWbaseValluePlusLookupTable* *hooverTailwater* *tailwaterComper*
Reach	routing	*impulseResponse* *kinematic* *macCormack* *muskingumCunge* *timeLag* *variableTimeLag*

Figure 3. Selected objects and their method categories and methods available in the current PRSYM implementation.

In some cases the solution must iterate between objects. For example, turbine release is dependent on tailwater elevation which may, in turn, depend on the downstream pool elevation. The downstream pool elevation depends on the inflow from the upstream reservoir. In this case, the algorithms use an initial guess for tailwater elevation, compute turbine release and outflow, then the downstream reservoir dispatches and calculates its elevation. That elevation is propagated back up to the upstream reservoir's tailwater, and the iteration continues until the slots either converge or reach a maximum iteration limit. In the case of repeated dispatching, all slots are known, so an object uses the previous dispatch method.

The simulation plus water quality dispatcher is identical to the pure simulation except that after the simulation is completed, the water quality methods are executed on each object. Currently the water quality methods calculate temperature and salinity on Reaches and Reservoirs.

Figure 3 lists selected objects in the current version of PRSYM, the physical processes (method categories) that are modeled for each, and the user-selectable methods that are available for those categories.

Rulebased simulation uses the basic simulation model, but rather than having a completely specified problem, the model is driven by rules which set values. Each rule has a unique priority. User inputs implicitly have a priority of "0," the highest priority. At the beginning of the run, the user-input values are set and if any objects can dispatch, they go on the queue. When the queue is empty, the rule processor is called. The rule processor checks its rules and fires the highest priority rule which is applicable based on the current state of the system. When a rule fires it sets values in

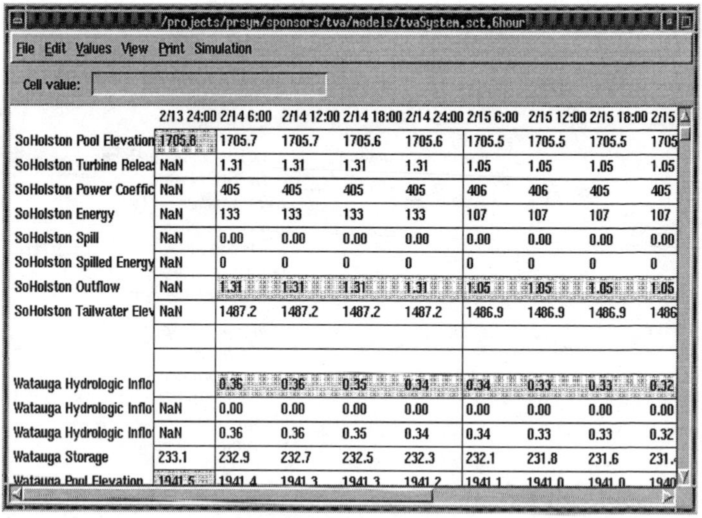

	2/13 24:00	2/14 6:00	2/14 12:00	2/14 18:00	2/14 24:00	2/15 6:00	2/15 12:00	2/15 18:00	2/15
SoHolston Pool Elevation	1705.8	1705.7	1705.7	1705.6	1705.6	1705.5	1705.5	1705.5	1705
SoHolston Turbine Releas	NaN	1.31	1.31	1.31	1.31	1.05	1.05	1.05	1.05
SoHolston Power Coeffic	NaN	405	405	405	405	406	406	405	405
SoHolston Energy	NaN	133	133	133	133	107	107	107	107
SoHolston Spill	NaN	0.00	0.00	0.00	0.00	0.00	0.00	0.00	0.00
SoHolston Spilled Energy	NaN	0	0	0	0	0	0	0	0
SoHolston Outflow	NaN	1.31	1.31	1.31	1.31	1.05	1.05	1.05	1.05
SoHolston Tailwater Elev	NaN	1487.2	1487.2	1487.2	1487.2	1486.9	1486.9	1486.9	1486
Watauga Hydrologic Inflo		0.36	0.36	0.35	0.34	0.34	0.33	0.33	0.32
Watauga Hydrologic Inflo	NaN	0.00	0.00	0.00	0.00	0.00	0.00	0.00	0.00
Watauga Hydrologic Inflo	NaN	0.36	0.36	0.35	0.34	0.34	0.33	0.33	0.32
Watauga Storage	233.1	232.9	232.7	232.5	232.3	232.1	231.8	231.6	231.
Watauga Pool Elevation	1941.5	1941.4	1941.3	1941.3	1941.2	1941.1	1941.0	1941.0	1940

Figure 4. Simulation Control Table

the model which may, in turn, cause objects to dispatch. In general, the rule processor is called on whenever the dispatch queue is empty. (The rule processor is described in more detail in the accompanying paper by Wehrend and Reitsma.) The rulebased simulation differs from pure simulation in that there could be legitimate conflicts in setting values. A reservoir release could be set by two conflicting rules. Logic dictates that the higher priority rule should set the value. However, when rules set values and the objects dispatch, a chain reaction of sorts results in the effect of that rule propagating over many objects. The effect of that rule, however, should not overwrite a value set by a higher priority rule. Thus, it is necessary to keep track of the priority of the last rule fired in setting all simulation values. The PRSYM rulebased simulator adds a priority value to each slot. When a value is set on an object for this controller, the slots are added to the current set in order of priority, beginning with the highest. Objects may dispatch several times using different methods as the rules change the values and their priority levels.

Optimization is the third main controller currently under development in PRSYM. The user enters prioritized constraints and objectives through a constraint editor. The solution is a linear goal programming formulation which interprets each priority level in turn as an objective, finds the optimal solution, then does not further violate that objective in subsequent lower priority objectives. The optimization problem, with initial conditions, mass balance and continuity constraints, upper and lower bounds, and linearization of nonlinear processes is all generated automatically by PRSYM. This controller is described in more detail in the accompanying paper by Eschenbach, et al.

In addition to the three main controllers, PRSYM also offers a multiple run management meta-controller. With this utility, many runs can be made, with the data automatically updated, and only the specified desired results saved. Special drivers communicate these results with spreadsheets or relational databases.

Modeling Environment

The PRSYM modeling environment is designed to give the user the tools needed to build and run simulations to meet the needs of scheduling and planning activities. This is achieved through graphical user interface tools, data management tools, plotting features and the simulation control table (SCT).

The SCT is a spreadsheet-like display of the data from a PRSYM model. The user constructs and configures an SCT, and can construct as many as needed to display various combinations of data. These can be iconified and brought up as different views on the model are needed. The SCT is interactive in that the user can specify inputs and run the simulation from this interface. Figure 4 shows part of an SCT for a TVA river basin model. The SCT gives operators a quick overview of the whole system, and was designed with many special features like control keys for operators who need to run simulations quickly.

Additional features for analyzing results and performing trade-off analysis are projected to be developed in the coming year.

Applications of Simulation-Based Reliability Assessment in Design

Pavel Marek (F.ASCE)[1], Milan Guštar[2], and Thalia Anagnos (M.ASCE)[3]

Abstract

A reliability assessment technique has been developed that uses Monte Carlo simulation and the power of modern personal computers to rapidly analyze complex problems that would be impossible to solve analytically. Variability of loads, material properties and cross-sectional dimensions are represented by bounded histograms. The reliability of structures represented by carrying capacity and serviceability of structural elements is expressed in terms of probability of failure. Safety measures such as Factors of Safety, the Reliability Index β, as well as the Design Point, the Load and Resistance Factors in Partial Factors Design are altogether bypassed. This computer-based fully-probabilistic procedure has the potential for application in designer's everyday work, and as an alternative in specifications for structural design.

Introduction

The structural reliability assessment concept of a limit state surface separating a multidimensional domain of random variables into safe and fail domains has been generally accepted and is used in the development of design specifications. In last four decades deterministic concepts based on "Safety Factors" and "Allowable Stresses" have been replaced in national and international codes by the Limit States Method allowing qualitative improvements in expressions of the loading, load-effect combinations, resistance of structural elements, components and systems, and of reliability, i.e., safety, serviceability, durability and economy. Two main trends in the interpretation of the Limit States Method in the specifications for structural design can be observed. First, a semi-probabilistic approach based on partial factors has been developed and applied in specifications such as AISC (1993) and the European Standard (1993). Second, advances in computer technology and reliability theory

[1] Visiting Professor, DrSc, Department of Civil Engineering, San Jose State University, San Jose, CA 95192-0083

[2] Consulting Eng., E.E., M.S., ARTech, Pisecka 10, 130 00 Praha 3, Czech Republic

[3] Professor, PhD, Department of Civil Engineering, San Jose State University, San Jose, CA 95192-0083

have made it possible to consider a qualitatively different fully probabilistic approach. This is reviewed in the handbook by Sundararajan et al (1994).

When considering multi-random-variable input resulting from statistical and probabilistic evaluation of data, the analysis of reliability based on analytical mathematical solutions can become extremely difficult or even impossible. The potential of the partial factor concept is very limited due to the substance of the analysis of input data and their interaction. Similarly, the fully probabilistic approach based solely on closed-form solutions is limited as well. Alternative approaches, which have become very efficient due to modern computer technology, use methods based on sampling and simulations. These methods can be based on Monte Carlo Simulation, or on special procedures such as Latin Hypercube Response Approximation. Today, tens of thousands steps of complicated calculations are conducted by a PC 586 in seconds.

In the last seven years, the authors have given special attention to the probability-based design of structures using Monte Carlo simulation. The proposed approach briefly explained next is based on **parameter generated histograms** while the common applications of the simulation technique are based mainly on **moment generated distributions**. For detailed review of the proposed method and procedures, including computer programs and examples, see Marek et al (1995).

Load Effects and their Combinations

A method for the evaluation of load effect combinations was proposed in the first research phase, see Marek and Guštar (1989), and Marek and Venuti (1990). An application of the method to a single-component variable load effect (e.g. axial force N in a structural member) is shown in Figure 1. Acceptable probabilities of exceedance could be defined for purposes of determining design load effects. Using the simulation-based computer program, ResCom™, a combination of practically any number of variable load effects from different sources can be analyzed. Analysis of load combinations with the ResCom™ program is limited to statistically independent loadings.

A similar computer program LoadCom™ allows direct comparison of the load effect combinations obtained by the simulation procedure with the multi-step and multi-component approach applied in ASCE (1993) and CSA (1989). This program is applicable to the analysis of the simultaneous effect of six different basic loads (dead load, long- and short-lasting live loads, snow, wind and earthquake). A sample printout is shown in Figure 2.

Currently research is focused on multi-component load effects. A simulation based method is proposed and two computer programs, M-Star™ and AntHill™, have been developed (Marek and Guštar, 1993). One application is the determination of

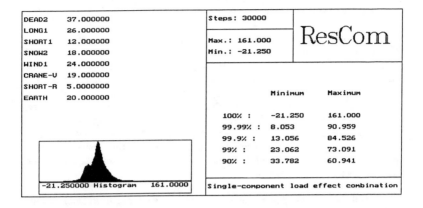

Figure 1. ResCom™ printout showing the bounded histogram that results from the combination of eight load effects. The magnitude of the load effect combination is determined for selected probability of exceedance.

Load	Nominal	Minimum	Maximum	Steps: 20000	LoadCom™
				Units: kN	
Dead	27.000000	24.300000	37.800000	Country: USA	
LongLast	66.000000	0.0000000	105.60000		
ShortLast	23.000000	0.0000000	36.800000		Minimum Maximum
Snow	15.000000	0.0000000	24.000000		
Wind	31.000000	-40.300000	40.300000	LRFD Common	-16.000000 149.50000
EarthQuak	28.000000	-28.000000	28.000000	ASD	-4.0000000 117.00000

	Minimum	Maximum
100%	-44.000000	272.50000
99.99%	-10.488235	179.41176
99.9%	-1.8000000	164.51764
99.0%	19.300000	140.93529

Example: Load Effect Combination, Single-Component Variable

Figure 2. Comparison of the load effect combination calculated according to ASCE (1993) with the probability based method using the LoadCom™ computer program

the extreme magnitude of a principal stress in a structural member, resulting from a combination of bending moment, normal force and shear force due to several independent variable loads. Such analysis of load effect combinations is not covered in current specifications. The application of the multi-step and multi-conditional approach according to, for example, ASCE (1993) would be extremely difficult or even impossible in the case of three and more component variables. The following example outlines the analysis of a three component variable (for details see textbook Marek et al, 1995).

Example: Determine the magnitudes of the principal stresses, σ_{1max}, corresponding to the probabilities $p[\sigma_1 \le \sigma_{1max}]$= **0.992, or 0.008.** Load effects are expressed by stresses due to the combination of several independent variable loadings: two different dead loads, two long-lasting loads, wind load, and earthquake are considered. The stresses are expressed by following equations:

$$\sigma_x = 40*D2 + 80*LL1 + 70*W + 60*EQ \qquad [N/mm^2] \qquad /1/$$
$$\sigma_y = -79*D3 + 50*LL2 + 50*W + 60*EQ \qquad [N/mm^2] \qquad /2/$$
$$\tau_{xy} = -20*D2 - 5*LL1 - 15*W - 40*EQ \qquad [N/mm^2] \qquad /3/$$

where D2, D3, LL1, LL2, W, and EQ are coefficients expressing the variability of individual stresses (for histograms representing the variable coefficients see Marek and Venuti, 1990). The investigated principal stress σ_1 is expressed by the equation:

$$\sigma_1 = (\sigma_x+\sigma_y)/2+((\sigma_x-\sigma_y)^2/4+\tau_{xy}^2)^{0.5} \qquad /4/$$

As can be seen from this equation, the principal stress is a three-component variable. The components are σ_x, σ_y, and τ_{xy}. In this example each component is a function of several independent variable loadings as defined in Equations /1/ to /3/. Since principal stress is a multi-component variable, the common multi-step and multi-component procedures reviewed in Wen (1990) are not convenient for application in design practice. However, the simulation based computing procedure can be applied. Using the computer programs **M-Star™** or **AntHill™**, the principal stress σ_{1max} = 5 N/mm^2 corresponding to the probability $p[\sigma_1 \le \sigma_{1max}]$= 0.992, and σ_{1max} = 175 N/mm^2 corresponding to the probability $p[\sigma_1 \le \sigma_{1max}]$= **0.008** respectively are obtained. For details see Marek et al (1995). This procedure may be applied also to four- and more-component variables. For example the principal stress may be evaluated at a point of a structural member if it is exposed to biaxial bending, biaxial shear force, axial force and to torque.

Resistance

The design resistance of a structural component can be evaluated considering variable mechanical properties of the material and variable geometrical properties of

the cross-section. Examples of the evaluation of resistance are found in Marek et al
(1990) and (1995).

Safety

Once the probabilistic distributions of the load effects and the resistance have
been developed, the next step is to assess the safety (reliability) of the structural
element. The safety function, SF, can be expressed by the equation

$$SF = (R - Q)$$ /5/

where R is the resistance , and Q is the load effect. The program AntHill™ generates
a set of dots corresponding to the interaction of individual components, R and Q.
This population of dots is created using the Monte Carlo method. With the assistance
of a personal computer, tens of thousands of dots ("ants") representing the possible
interactions of R and Q are expressed as a two dimensional histogram ("anthill", see
Figure 3) in a matter of seconds.

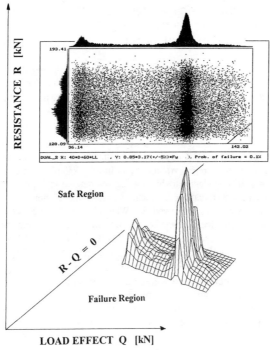

Figure 3. Safety assessment of a structural component by the AntHill™ computer
program. The line R-Q=0 separates the "anthill" into two parts: the failure region (to
the right), and the safe region (to the left).

The substance of the safety check based on the simulation technique is shown in Figure 3. The load effect Q of several time-dependent and mutually uncorrelated loads (e.g. considering bending moment at a particular cross-section of a structural component exposed to wind, live load and snow) is expressed by a bending moment Q_M. The resistance R is expressed by a moment R_M (considering variable yield stress, variable cross-sectional geometry, and other variables). In order to obtain the probability of failure, one needs to draw a line R-Q=0. This line separates the set of dots into two parts. The number of dots to the right of the line **R-Q=0** divided by the total number of dots is equal to the probability of failure. This calculation is performed by the AntHill™ program.

Simulation Based Reliability Assessment vs. Partial Factors Design

The proposed simulation based method was used for assessing the reliability of

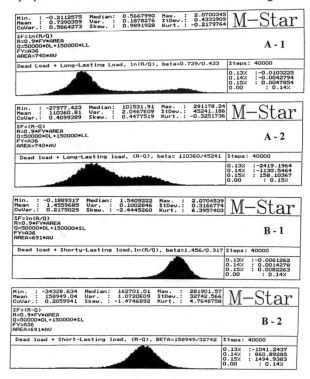

Figure 4. Analysis of the reliability index β for two load cases and two definitions of the reliability function.

the factor of safety in Allowable Stress Design, and the Reliability Index β applied in specifications, see Marek and Anagnos (1993). In the following example the reliability index β is evaluated using the simulation concept and M-Star computer program, see Marek et al (1993[a]).

In all four analyzed cases (Case A-1,2 and B-1,2 in Figure 4) a steel bar is exposed to tension. In all four cases (a) the total factored load effects are the same, (b) the cross-sectional areas, and the yield stresses of the material as well as their variations are the same, and (c) the probabilities of failure are the same.

The character of the live load differs (short-lasting load from occupancy is assumed in Cases B-1,2 and long-lasting load from storage in Cases A-1,2). Two different reliability functions are applied: the reliability function RF=ln(R/Q) in Cases A-1 and B-1, and the reliability function RF=(R-Q) in Cases A-2 and B-2. As obtained by the simulation technique, applying the method proposed earlier (Marek et al, 1993), **four very different values of the factor** β= 4.85, 4.59, 2.44, and 1.71 are obtained. It can be concluded that the shape of the reliability function RF and the reliability index β carry no information about the probability of failure. This suggests that the substance of design concepts based on the reliability index β should be reconsidered.

Summary and Conclusions

Dramatic improvement of the personal computers and of the simulation based procedures related to structural reliability assessment lead to qualitative new design concepts with the potential for application in designer's work. The computing potential of personal computers on every designer's desk permits the consideration of a reliability assessment format that bypasses altogether the common reliability measures such as Factor of Safety and the Index of Reliability β and corresponding deterministic and semi-deterministic concepts such as Allowable Stress Design and Load and Resistance Factor Design.

An alternative assessment procedure has been developed and proposed. This computer-based probabilistic alternative uses Monte Carlo simulation and parameter generated histograms in the assessment of the carrying capacity and the serviceability of structural elements. The reliability of structural components is expressed in terms of probability of failure. The proposed concept has been experimentally used in education of civil engineering students at San Jose State University with positive response of graduate and undergraduate students as well as of practicing engineers. At present, this concept serves as tool for better understanding of the individual variables and their interaction in the safety, durability and serviceability check of structures. The application of the simulation based reliability assessment concept in future specifications for structural design can be recommended for consideration.

Acknowledgments

The authors wish to acknowledge the support provided by San Jose State University, by Grant Agency of Czech Republic, by the American Institute for Steel Construction, Chicago, by National Science Foundation, Washington, and by MARTEC and ArTech Prague.

References

AISC (1993) Load and Resistance Factor Design Specification *for Structural Steel Buildings*, American Institute of Steel Construction, Chicago.

ASCE (1993) Minimum Design Loads for Buildings and Other Structures. *ASCE Standard 7-93*.

CSA (1989) *Limit States Design of Steel Structures*, CAN/CSA-S16.1-M89, *Canadian* Standards Association, Canada.

European Standard (1993) "Basis of Design and Actions on Structures", ENV 1991-1, CEN, Brussels, edited draft ,October 1993.

Marek, P., and Guštar, M. (1989) Assessment of the Combinations of the Response of Structures to the Loading (in Czech)., *Pozemni stavby No.2*, *SNTL*, Prague.

Marek, P., and Venuti, W.J. (1990) On Combinations of Load Effects. *J.. Construct. Steel Research 16, pp. 193-203*, London..

Marek, P., Venuti, W.J., and Guštar, M. (1990) Combinations of Design Tensile Yield Stresses in Build-up Sections (in French), *Constr. Metallique* No. 4,

Marek, P., Guštar, M., and Tikalsky, P.J. (1993) Monte Carlo Simulation = Tool for Better Understanding of LRFD, *J. Str. Div. ASCE*, No.5.

Marek, P., and Guštar, M. (1993) AntHill Method for Analysis of Multi-Component Variables, *Research Report, Department of Civil Engineering*, San Jose State University, California.

Marek, P., and Anagnos, T. (1993) Factor of Safety vs. Index of Reliability β vs. Probability of Failure in Structural Steel Design (1993), *Research Report, Dept. of Civ. Eng., San Jose State Univ.*, California, (submitted to AISC).

Marek, P., Guštar, M., and Anagnos, T. (1995[a]) Alternative Probability Based Reliability Assessment of Structures using Monte Carlo Simulation as a Tool. *CERRA - ACASP7* , Paris France , July 10-13, 1995

Marek, P., Guštar, M., and Anagnos, T. (1995). Simulation Based Reliability Assessment. (In Press). *CRC Press Inc., Boca Raton, Florida.*

Marek, P. and Guštar, M. (1988-94). ResCom™, M-Star™, AntHill™, DamAc™, and LoadCom™. HAB Int., PO Box 1932, Davis, CA 95617.

Sundararajan, C.(Raj), et al (1994) Probabilistic Structural Mechanics Handbook., *Chapman & Hall*, New York, NY..

Wen, Y.K. (1990) Structural Load Modeling and Combination for Performance and Safety Evaluation. *Elsevier.*

Design of R.C. Columns Subjected to Biaxial Bending Using the Direct Model Inversion (DMI) Method

A. Zaghw[1]

Abstract

The design of Reinforced Concrete columns subjected to biaxial bending is traditionally performed by trial and error. First concrete dimensions and reinforcement steel are assumed based on experience. If the section is not capable of resisting the applied forces a new trial section is used and the process is repeated until the designer finds a section with adequate capacity. This method can be time consuming, further it does not guarantee an economical design.

This paper explains a method of designing R.C. columns subjected to biaxial moment using artificial neural networks. The proposed method is shown to outperform traditional design methods in terms of speed, and economy.

1. Artificial Neural Networks

Artificial neural networks is a method of machine learning which is biologically inspired. Artificial neural networks are composed of processing units connected by links with variable weights. Each processing unit sums up all the weighted input it receives from other units. The weighted sum is then substituted in an activation function to calculate the output of the processing unit. The sigmoidal logistic function is used as an activation function in this study.

The processing units or nodes in the artificial neural network are arranged in layers. Each layer is usually connected only to the layers which lie before and after it. A four layer network composed of an input layer, an output layer, and two hidden layers is used in the course of this study.

Before using the neural network, it should be trained either by showing it some examples of input vectors and their desired output vectors "supervised learning", or by showing the neural network input vectors only and asking it to clusters similar vectors together "unsupervised learning". During training, the neural network weights are modified such that the network error is minimized. The most common algorithm which is used to train neural networks in supervised learning is the back-propagation algorithm [1]. Training is performed in this study by using a variation of the back propagation algorithm that employs the conjugate gradient method rather than the steepest descent to minimize the errors [2].

---

[1] Assistant Professor, Structural Engineering Department, Cairo University, Egypt.

2. Direct Model Inversion Method (DMI)

The purpose of model inversion is to find an inverse of a given system. For example, in the design of R.C. columns, if the column dimensions, reinforcement steel area, and neutral axis position and orientation are known, the maximum normal and moments around the two major axis that the section can resist becomes known. The inverse of this system, which is knowing the maximum normal and moments the column section should resist, what are the required concrete dimensions, reinforcement steel area, and neutral axis location? Clearly the inverse in this case is of more interest than the original system.

The inversion of the system is performed in this study using the direct model inversion method (DMI) [3]. This method trains a neural network in supervised learning to perform the inverse mapping of a given system. The advantage of using this method of inversion is that it does not require the creation of a forward model of the system before training as in the indirect model inversion case.

The system in this study is represented by a computer program, which takes as input the column dimensions, reinforcement steel area, and neutral axis position and orientation and outputs the maximum normal and moments around the two major axis that the section can resist. Training of the neural network is performed by sampling the space of possible inputs at random and using the system to produce the corresponding outputs. Each output of the system becomes the input of the neural network during training. In other words, input-output pairs to the system are used, respectively, as output-input pairs to the neural network. Figure 1 summarizes the DMI method. The common-input variables are those variables which are inputs to both the system and the neural network such as the material properties.

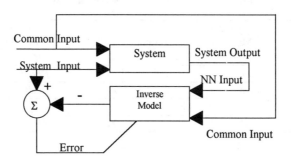

Figure 1. Direct Model Inversion

3. Description of the Neural Network

A feed forward neural network with 6 input units, 5 output units, and 2 hidden layers each consisting of 7 hidden units, is used in this study. Table 1 describes the input and output units of the neural network.

Table 1. Input and Output Units of the neural Network

Input Units	Output Units
f_c (Concrete Strength) [*]	d (depth of Concrete Section)
f_y (Yield Strength of Steel) [*]	A_{stb} (Top and Bottom Steel Area)
b (Width of Concrete Section) [*]	A_{srl} (Right and Left Steel Area)
M_X (Moment along the X axis)	θ (The counter clockwise angle of the NA with the horizontal)
M_Y (Moment along the Y axis)	Y_{NA} (The Vertical distance of the NA above the Concrete Centroid)
N (Normal Force)	

Table 2 describes the domain considered in this study for both the input and output units.

Table 2. Maximum and Minimum values for the Input and Output Nodes

	Input Nodes			Output Nodes	
Unit	Minimum	Maximum	Unit	Minimum	Maximum
f_c	1.4 ksi	10 ksi	d	6 in	60 in
f_c	28 ksi	100 ksi	A_{stb}	$f_c b d 10^{-5}$	$11 f_c b d 10^{-5}$
b	6 in	60 in	A_{srl}	$f_c b d 10^{-5}$	$11 f_c b d 10^{-5}$
M_X	NA	NA	θ	$-\pi/2$	0
M_Y	NA	NA	Y_{NA}	-80 in	20 in
N	NA	NA			

A program was written to calculate the capacity of the column from the concrete dimensions, reinforcement steel area, neutral axis position and orientation, and material properties. The program takes thus as input the output variables of the neural network, in

---

[*] This variable is an input to both the system and the neural network (common-input variable)

addition to the common input variables $(f_c, f_y,$ and $b)$, and computes the maximum normal, N, and moments around the two major axis, M_x and M_y the column section can resist. The computation of $(N, M_x$ and $M_y)$, was based on the following assumptions:
- The strain distribution is linear, i.e., plain sections remain plain after deformation.
- The strain in the topmost concrete fiber is -.003.

The stress distribution in the concrete is found by dividing the cross-section into a uniform grid. The strain in each grid is assumed uniform and equal to the strain in its center. The stress in each grid is also uniform and is obtained from the concrete stress-strain relation. The stresses and forces in the reinforcement steel is found similarly. Details of this method is found in [4].

The above program is used to obtain the data needed for both training and testing the neural network. The training data consists of 5,500 design cases, while the testing data consists of 2,750 design cases. The following section discusses the training results.

4. Discussion of the Results

The neural network was trained and tested on two different data sets, on a 486-66MHz personal computer. Figure 2, 3, and 4, show the capacity of 2,500 test columns as designed by the neural network versus the requested capacity. It is clear that in most of the cases the design proposed by the neural network accurately matches the requested capacity. The following example shows how the neural network can be used to design a reinforced concrete column subjected to biaxial bending.

Example
Find the concrete dimensions and reinforcement steel area for a column subjected to the following forces: $N = 1880$ kips, $M_x = 1530$ ft.kips., and $M_y = 440$ ft.kips. The material strength are $f_c = 3000$ psi and $f_y = 70,000$ psi.

Solution:
Assume $b = 20$in. Using the neural network the column section is:
$d = 44$ in. $A_{stb} = 11.61$ in^2 $A_{srl} = 11.35$ in^2 $\theta = -.322 \pi$ $Y_{NA} = -17$ in
This section has the following capacity:
$N = 1893$ kips, $M_x = 1600.2$ ft.kips., and $M_y = 476.7$ ft.kips which is very close to the requested capacity. Also, this method of design is much faster than the traditional trial and error procedure.

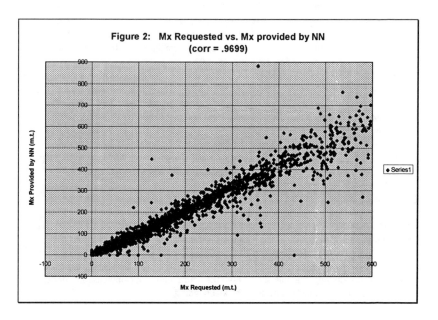

Figure 2: Mx Requested vs. Mx provided by NN (corr = .9699)

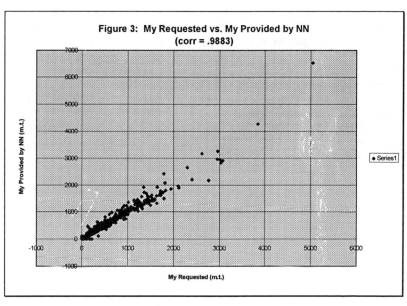

Figure 3: My Requested vs. My Provided by NN (corr = .9883)

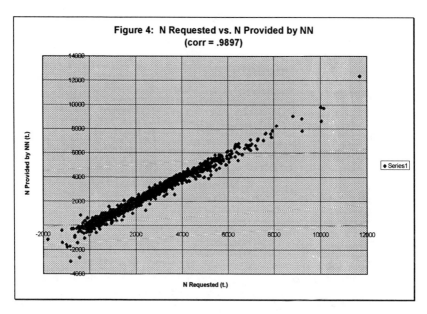

Figure 4: N Requested vs. N Provided by NN
(corr = .9897)

5. Conclusion

The use of neural networks can speed up the design process of components traditionally designed on trial and error basis such as reinforced concrete columns subjected to biaxial bending. The design is performed by creating an inverse model to an analysis program "system" as explained previously. The capacity of the designed component will be as close as possible to the requested capacity, which results in a more economical design than traditional trial and error approach.

6. References

[1] Rumelhart, D. E., Hinton, G.E., and Williams, R. J., "Learning Internal Representation by Error Propagation," Parallel Distributed Processing, Vol.1 pp.318-362, MIT Press, Cambridge, MA, 1986.

[2] Zaghw, A. H., and Dong, W., "An Automated Approach for Selecting the Learning Rate and Momentum in Back-Propagation Networks", Proceedings of the IEEE International Conference on Neural Networks, June 28-July 2, 1994.

[3] Jorgensen, C. C., "Distributed Memory Approaches for Robotic Neural Controllers", RIACS Technical Report 90.29. Research Institute for Advanced Computer Science, NASA-Ames, Moffett Field CA 94035, 1990.

[4] Harik, I. E. and H., Gesund, "Reinforced Concrete Columns in Biaxial Bending", in Concrete Framed Structutres Stability and Strength, pp.111-132, Elsevier Applied Science Publishers, 1986.

Status of Electronic Data Interchange for Steel Structures

D.W. McConnell, P.E.[1]
J.A. Bohinsky, P.E.[2]

Abstract

With electronic data interchange (EDI), a three dimensional data representation of a steel structure replaces a large portion of the information traditionally presented on engineering drawings. The "promise" of EDI raises the possibility of eventually eliminating the necessity of drawings for transfer of information between engineering and fabrication. Steel fabricators are now reading structural data directly into their computerized detailing software, to design connections, create shop and fabrication drawings, and produce CNC data for automated fabrication equipment. Other value added data, electronically transmitted between engineering, fabrication, and construction, is fueling interest in EDI. Much remains to be done in meeting the expectations for productivity and added value that EDI generates in sales presentations. This paper addresses the current state of the art and provides a "reality check" on practical use of EDI for steel structures.

Introduction

As EDI for steel structures has moved from concept to prototypical use on projects, interest has focused primarily on three areas: Sending an electronic data file description of a steel structure to the fabricator's computerized detailing system; Receiving "As Fabricated" data back from the detailing system to be read into the Plant Design System for interference checking; and Receiving "up to the minute" status data from the electronic tracking system for steel in fabrication, shipping, and construction. Progress of varying degrees is beginning to show between participants willing to join in the experiment.

[1]CAD/CAE Structural Engineer, Brown & Root, Inc.,
P.O. Box 4574, Houston, TX 77210-4574
[2]Civil/Structural Engineering Department Manager, Brown & Root, Inc.,
P.O. Box 4574, Houston, TX 77210-4574

The data format currently being transmitted is the Intergraph Steel Detailing Neutral File (SDNF). This neutral file is the result of efforts by Intergraph's Structural Engineers Special Interest Group (SESIG) to provide a common data exchange format between the engineering structural modeling software and steel detailing software used by fabricators. This file contains geometry, member and material data defining a structure to be fabricated. The data is sufficient to generate a model of most linear members. An arc element addition has been proposed by the Intergraph SESIG to cover curved members.

Current detailing software understands connections associated with this data are to be detailed at beam to beam, beam to column, or column splices. Miscellaneous connections not fitting these categories are currently handled interactively by the detailer using drawings issued by the engineer. As such, the file does not yet convey 100% of the information transmitted by traditional engineering drawings. Neither has a two way reading or writing of data been realized between the major commercial engineering design/graphics software and the steel detailing software.

Presently, data exchange based on the Intergraph SDNF is being pursued in actual practice with some degree of success. In the United States, this activity has been led by software developers (such as Intergraph, Design Data, SteelCad, CDS) AEC companies, (such as Brown & Root, Fluor Daniel, DuPont, Bechtel, etc.) and fabricators and detailers.

Credit should be given to the efforts of Intergraph's SESIG in pushing forward the concept of a standard data exchange for steel. The American Institute of Steel Construction (AISC) is beginning to look at activities in the exchange of data in the steel industry. They are also following the progress made in the European Community's development (STEP, Standard for the Exchange of Product Model Data, ISO 10303) of standard application protocols (AP). The STEP standard could possibly provide the AP to overcome the shortcomings of the Intergraph neutral file.

In Actual Practice

At Brown & Root, we have completed a pilot program to initiate use of EDI for steel by sending a test file from an actual structure to several fabricators. We were able to gain information on the current abilities of fabricators in using the neutral file in designing connection details and creating drawings. Results were received from the following participants in our program: AFCO, Cives, Kline, CTIW, Havens, Paxton & Vierling, and Hirschfeld Steel Fabricators. The two detailing systems we encountered in our pilot program were Design Data SDS/2 and SteelCad (other software also exists). Our first actual project use of EDI with steel structures will occur early in 1995.

Several other engineering firms and fabricators have been using the SDNF file in conjunction with IFC drawings. The fabricators are attempting to make use of these data files for the initial "bulk" input of structural information, primarily, the beams, columns, and bracing. Best use of the EDI process occurs when fabricator/detailers and engineering design companies have a close working relationship. Because EDI is still in an early stage of development, each party must be willing to share some of the overhead costs of moving the technology ahead.

In practice, drawings remain as the controlling contract document. In all cases, fabricators report that engineers maintain that the drawings provide the correct information regardless of the data file. This creates additional work for the fabricator in reconciling the information in two forms. The detailer reads in the neutral file data and regenerates the model with his detailing software. He then cross checks the detailing model against the drawings to see if discrepancies exists. Members and miscellaneous elements not represented by the data file must be identified and modeled. Corrections and additions are done interactively until the detailer is satisfied that his model corresponds to the drawings.

If the data file is accurate and the detailer is efficient at cross checking, the rapid input allowed by the data file can still yield time savings. Facing such a verification task, many fabricators elect not to use the SDNF and use the drawings only. Some detailers have expressed resistance at the idea of receiving a data file that does not have the same level of confidence from the engineer as do the IFC drawings. Use of EDI in some cases is driven by the desire to establish the capability from a marketing standpoint. Several fabricators say they generally benefit from the electronic data interface at the initial receipt of the data. Benefits dissapear when there are revisions to this original data.

The system in place does not handle revisions effectively between the parties. A new file must be read in for the complete structure containing modifications. The revision tracking process is clear cut when using drawings with revision marks, "clouds", issue dates, and signatures. Most fabricators using the EDI process to read the initial IFC data will revert to the tried and true drawings exchange process once revisions begin to flow. This is also caused by having entered part of the detail model interactively and not wanting to have to redo this manual activity.

One fabricator described their prototype revision handling system. This system made use of comparisons between earlier mill order reports generated from the neutral file by the detailing software. Elements showing changes were given new piece marks and the prior piece mark was dropped. This system works for minor revisions. Revisions with configuration changes generally require a new data file and re-creation of the model. Having a piece mark identification system the same in the engineering model and the detailing model was suggested to aid communication of revisions.

Capabilities in tracking revisions are anticipated in 1995 as computerized detailing software developers release new features. The technical part of revision tracking can be developed to a practical level if coordinated through a standard data exchange. The issue of files as official documents remains unresolved. Input from standards organizations and state regulatory boards will likely be required.

Bar Code, Tracking, Status

Tracking of steel pieces throughout the fabrication, shipping, erection, and invoicing cycle is greatly enhanced by the exchange of electronic data. Many fabricators are currently using bar coding or have plans to do so for obtaining up to the minute status of steel pieces.

One fabricator uses radio frequency bar code readers to transmit data from the field handlers to the computer data base for instant status and tracking. This information in well known DBase or Microsoft Access data file format can be transmitted by modem to the engineering office. Engineering, procurement and materials tracking personnel can use this data to suit their purposes. Development of color coded status on the computer model and on drawings is another concept under development.

Review of Details and Erection Drawings for Engineering Approval

In present practice, approval of details and erection drawings still involves large rolls of paper drawings sent through express mail each way. An engineer reviews and marks his comments on the drawings. The paper handling itself consumes much time in addition to mail delivery.

An engineer skilled in electronic viewing and annotating of graphic data could eliminate most of the paper handling during approval. One would need to be able to view and "redline" the details and erection sheets for approval and return the files electronically (or, at worst, by diskette in the mail.) There are several viewer and redline programs for this task on the market like MSreview and AutoView.

Pre-approval of a steel fabricator's computerized connection design assumptions, procedures, calculations, and use of mutually agreeable fabricator/engineering standards will speed up the approval process. Any cost savings for engineering approvals done using EDI and electronic documents will be most apparent on jobs with large tonnage of steel (i.e. hundreds of drawings).

Loads

Although Intergraph's structural modeling programs currently do not write the load packet data, the SDNF format accommodates specific connection design loads for all members in the model. Even so, only one load case is processed when the file is read into the existing detailing programs. This requires (as is the case when loads are shown on drawings) that the worst case loads be given. This is generally adequate except for complex connections with loads transferring through columns or beams to members on the other side of the connection. The use of accumulated worst case effects at these nodes causes uneconomical or impractical connection designs. The detailer consumes manhours struggling to design a practical connection without benefit of the results from the actual design cases analyzed by the engineer.

If the detailing program could cycle through multiple sets of load combinations (those occurring simultaneously in a rational analysis of the structure), and could determine resultant transfer forces, this would allow the engineer to supply a set of loads straight from his analysis. The connection design routines would have to be capable of resolving a 3D load combinations. Resulting design of column and beam reinforcement for through loads will be more economical. The engineer would spend less time reducing the loads to a single conservative worst case scenario. This would eliminate the time consuming rounds of communication between the detailer and the engineer as they try to determine a single combination of loads that control all of the connection components at a node.

AISC, European Efforts, STEP

A complete structural modeling, design, and fabrication data protocol standard is sorely needed to make electronic transmittal of a structure between engineering, fabrication, and construction a reality that is more efficient and beneficial to all parties. A more comprehensive data exchange protocol for steel design and construction is, in fact, making its way to an international standard (ISO 10303,STEP). A consortium of primarily European engineering companies is developing and performing tests on the standard application interfaces for steel. The PlantStep consortium is a US based effort to further understanding and implementation of STEP technology. The efforts of this group have been considerably less funded than the European STEP programs. The American Institute of Steel Construction, AISC, is currently considering participation in the STEP activities.

The STEP standards will include many application protocols (AP) for particular industries, product types, and technologies. Analogous to the SDNF would be a Building Structural Frame: Steelwork AP. Adoption of such an ISO standard AP by various software would guarantee that they would "talk" to each other.

Effects on the Engineering Process

In today's market, the possibilities of EDI has whetted the appetite for greater savings. EDI is focusing eyes on opportunities for productivity increases or savings in engineering such as in drawing production, namely elimination of drawings as the means to transmit structural data.

Drawings are typically manhour intensive. Even if views are extracted from a 3D model, much information is still added interactively such as dimension lines, detail references, special notes and symbols. Granted, information not drawn on 2D drawings must in some way be modeled as data. Data input requires manhours just as well, but data can be more easily transported and reused between many interrelated activities from conception, through design, fabrication, erection, and facility operation and maintenance. Resources can be brought to bear on modeling efficiency and data input and transfer. Elimination of information interpretation and input time associated with use of drawings will create savings. Additionally, use of electronic data provides an open door to new innovative methods.

Like any new process, the initial benefits are being balanced by other complicating issues. Steel detailing data is generated from 3 dimensional modeling. Many engineering projects produce only 2D drawings. Others are accustomed to 3D modeling of plant systems but are still tied to interactively extracted 2D views to verify and then transfer official information. Checks and balances have not yet evolved to create a high comfort level with purely electronic representation and official transmittal of structural designs. Data integrity and security are a concern to Engineers of Record and their companies. How will the concept of the PE stamp be applied to data? If the fabricator produces the erection drawings from the data file, how will special erection information from the engineering office be passed to construction?

To reduce reliance on paper drawings, we must be able to verify and document the correctness of the structure within the electronic media. In the PDS environment, all disciplines must be able to design on the electronic page (in one 3D data base exchangeable between AEC applications) where they can interact concurrently as the design evolves. A structural engineer must be able to see the status of process, electrical, piping, and architectural model data before he can realistically know if his supporting model data is valid in the facility. Lacking this knowledge creates more rework at later (and more expensive) stages. Knowing whether the equipment in and on your steel structure is preliminary, future, final, or has changes pending will effect how you design and issue steel data for purchase, fabrication and erection. Current practice relies on paper "check" drawings circulated through the disciplines to disseminate status information (noted for example as IFC, Hold, Preliminary). Electronic model reviews for interferences typically occur at intervals that insure rework and still do not always reveal true status of model data elements.

In the past, drawings have been the primary "worksheet" of the design process in engineering. Still persisting in this age of CAD/CAE, is the layout and design of facilities on paper followed by manual transfer into the electronic medium. Concurrent design cannot be truly done in such a system.

The human aspect of viewing graphic displays versus large paper drawings is not an insignificant issue. It is more difficult to visualize a structure through a small "window". One looses the peripheral view available on a big sheet of paper. Other disciplines such as piping find it even tougher to keep piping runs in mind with this small view. Use of computerized data verification, larger displays, dual displays, training, and experience will improve the ability to work comfortably and efficiently without sheets of paper.

Until we reach the level of computer use that eliminates the need for paper drawings, the engineers, suppliers, clients and others will still require them. For the near future, if we can eliminate the need for drawings as working documents during the design stage, we will save manhours by delaying drawing production to a one time task at the end of design when all information is complete. This could cut out the time creating and distributing iterations of the drawings between disciplines, managers, clients, et cetera.

What Needs To Be Done Next?

The single best enhancement to the process would be to adopt an internationally accepted application protocol such as forthcoming from STEP. Use of this more comprehensive data exchange standard will go far in satisfying the outstanding needs. Discussion on adopting the standard needs to be generated within the community of AEC software developers and the increasingly global engineering community.

Alternatively, enhance the current Intergraph SDNF: add curved (arc) elements, miscellaneous steel items like handrail, grating, bent plates, ladder data, treads, equipment holes, and other data to the eventual elimination of paper drawings.

Establish methods of verifying and controlling computerized structural data throughout the process to make it acceptable as official documents, thus eliminating dependence on production of 2D drawings and sketches. Establish PE stamp procedures for transmittal of electronic information.

Add capability to transmit the complete "As Fabricated" structural detail model back to the mainstream AEC software preferably using a national or international standard data protocol.(Some engineering companies are pursuing proprietary development in partnership with certain detailing software firms to add this capability for gusset plates for the purpose of interference checking.)

Acknowledgments

The writers would like to acknowledge the contributions to our understanding of EDI from the fabricators and detailing software firms mentioned as having participated in our pilot program. Through many meetings and conversations with their engineers, detailers, and management, we have a clearer picture of this emerging technology.

COMPUTER INTEGRATION THROUGHOUT AN
UNDERGRADUATE CIVIL ENGINEERING CURRICULUM

COL Thomas A. Lenox[1], Member, ASCE
LTC Robert J. O'Neill[2], Member, ASCE
LTC Norman D. Dennis, Jr.[3], Member, ASCE

ABSTRACT

This paper examines the authors' experiences in incorporating the personal computer into the civil engineering program at the United States Military Academy. It describes how use of the personal computer has permeated the math, science, and engineering courses in the civil engineering curriculum. It reviews how the personal computer is introduced to students during their freshman year and used in their basic science and math courses. The paper describes how civil engineering students encounter additional engineering computer applications as they move into their engineering science courses, and how they are required to use a wide variety of special purpose application software in their upper division analysis and design courses.

OVERVIEW OF THE USMA ACADEMIC PROGRAM FOR CE MAJORS

The purpose of the United States Military Academy is to provide the nation with leaders of character who serve the common defense. The four year undergraduate experience is built upon intellectual, military, and physical development programs. This academic program is deliberately broad in scope. It includes 31 core courses distributed almost evenly between mathematics, science, and engineering on one hand and the humanities and social sciences on the other.

[1]Professor of Civil Engineering; Director, Civil Engineering Program, Department of Civil & Mechanical Engineering, United States Military Academy, West Point, NY 10996.
[2]Associate Professor of Civil Engineering; Director, Civil Engineering Analysis Group, Department of Civil & Mechanical Engineering, USMA, West Point, NY 10996.
[3]Associate Professor of Civil Engineering; Director, Civil Engineering Design Group, Department of Civil & Mechanical Engineering, USMA, West Point, NY 10996.

The ABET-accredited civil engineering major, one of 17 majors available to cadets, requires twelve electives in addition to the 31 core courses. A typical academic program for the Civil Engineering major, excluding military science and physical education courses, is shown in Table 1 (*Academic* 1994-95).

TABLE 1. TYPICAL CIVIL ENGINEERING PROGRAM AT USMA

TERM 1	TERM 2	TERM 3	TERM 4
Discrete Dyn Sys	Calculus I	Calculus II	Prob & Stats
Chemistry I	Chemistry II	Physics I	Physics II
Intro to Computers	Psychology	Economics	Political Science
English Composition	English Literature	Foreign Language	Foreign Language
History	History	Philosophy	Earth Science
			Statics & Dynamics

TERM 5	TERM 6	TERM 7	TERM 8
Mech of Materials	Structural Analysis	Adv Struct Analysis	Dsgn of Struct Sys
Fluid Mcchanics	Hydro & Hydraulics	Dsgn of Steel Struct	Structural Mech
Electrical Engrg	Soil Mech & Found	Dsgn of Conc Struct	Intro to Env Engrg
Engrg Mathematics	Modern Physics	Vibration Engrg	Thermodynamics
Military History I	Military History II	Internat'nal Relations	Constitutional Law
Leader Psychology	Adv English Comp		

BRIEF HISTORY OF PERSONAL COMPUTER USE AT USMA

The availability and accessibility of the personal computer to undergraduate engineering students has changed many of the traditional methods of doing engineering tasks and learning engineering skills. This has certainly been the case at the United States Military Academy (USMA), the oldest engineering college in the United States.

In August 1986, the Academy began the computerization of its student body and faculty, clearly demonstrating its commitment to assuring the computer competency of all its graduates. Since this date, each freshman has been required to purchase an IBM-compatible personal computer and several common software items to be used throughout their four year academic program. By August 1989, ALL four undergraduate classes were equipped with their own personal computer and common software. Common software purchased by all cadets includes Windows, Microsoft Word for Windows, Quattro Pro, and Turbo Pascal. A personal computer is also furnished to every faculty member at the Academy.

Beginning in 1989, students began purchasing DERIVE from Soft Warehouse at the beginning of their freshman year. This DOS-based symbolic manipulator program intelligently applies the rules of algebra, trigonometry, calculus, and matrix algebra to solve a wide range of problems using a nonnumerical approach rather than approximate numerical techniques. DERIVE has given instructors and students the freedom to explore different approaches to problems -- to include "classical" approaches that might not be considered practical if they were being applied by hand.

In 1989, the Department of Mathematical Sciences articulated a requirement for an in-class computing instrument which freed cadets from the tedious and complex calculations which hindered the mastery of key mathematical concepts. Beginning in 1990, the HP-28S Advanced Scientific Calculator was selected as the common calculator for incoming students -- upgraded to the HP-48S in 1993. These calculators have a graphic display and symbolic capabilities, extremely beneficial in the instruction of calculus. Its capabilities include algebraic manipulation, derivatives, indefinite integrals, definite integrals, numerical solutions, plotting, statistics, matrix operations, complex math, and programming. While the HP-28S and HP-48S are advanced hand-held calculators and not computers, the computing capabilities of these instruments have made a tremendous impact on modernizing classroom instruction. Engineering faculty members are issued the same calculator as the students they instruct.

Each cadet's PC is networked (10base-T Ethernet) in their room, and connected to the campus wide Fiber (FDDI) ring. The FDDI ring also connects the staff and faculty, as well as many of the classrooms, laboratories, the library, the print plant, and the Internet. Over 50,000 electronic mail messages pass through our 6000 node TCP/IP multi-protocol routed network daily, in addition to hundreds of electronic bulletin board postings. Cadets share files, print to high speed laser printers, and run applications across over 100 different servers, including World Wide Web/Gopher access. Grades, evaluations, scheduling, course notes, lessons plans, CAI lessons, digital imagery, copies of slides, and reference material for courses are accessed on line, with hypertext help documents. Throughout the entire academic curriculum, the "Computer Thread" requires the use of their PCs in each and every course.

FRESHMAN INTRODUCTION TO COMPUTERS AND PROGRAMMING

Almost immediately upon receipt of their personal computer, freshman students are immersed in its use. In addition to using the computer as a word processing instrument in their core English, History, and Psychology courses; they make extensive use of the personal computer in all of their basic science and math courses. All freshman are required to take CS 105, Introduction to Computers and Pascal Programming. This course provides students with the fundamentals of the function, operation, programming, and use of digital computers. The course uses Turbo Pascal for Windows. Students learn a variety of problem solving methods and automate them through standard structured programming techniques.

The two chemistry core courses, CH 101-102, include required computer use in conjunction with laboratory exercises. Students perform and verify prelab exercises using software developed by the Chemistry Department faculty, and use graphing and curve fitting software to display data and results in lab reports. In addition, students are required to develop a number of their own spreadsheets to tabulate and analyze experimental data. Recently, chemistry students have begun to use interactive, tutorial software packages to study chemistry lessons on their personal computers.

In the core calculus courses, MA 103-104, cadets are required to use Quattro Pro to iterate solutions to discrete dynamical systems, create a table of function values, produce graphs, approximate definite integrals, and approximate the solution to differential equations. The mathematical assistant program, Derive, is used to solve systems of linear equations, solve the eigenvalue problem, compute limits, symbolically evaluate derivatives, find roots of equations, and plot direction fields of differential equations.

COMPUTER USE IN SOPHOMORE-LEVEL COURSES

Widespread use of the personal computer continues for all cadets throughout their second year. Students in the core probability and statistics course, MA 206, make extensive use of Quattro Pro and use MINITAB, a standard statistics package for PC's. In the core physics sequence, PH 201-202, cadets are required to use Quattro Pro to perform analysis of data acquired in the laboratory. Students are also required to use the "Physics Tutorial" software, developed in-house.

The core economics course, SS 201, requires each cadet to accomplish two "lab" exercises which provide extensive practice in the use of spreadsheet software. Students gain experience and skill in the use of formulas, file manipulation, graphing, functions, and statistics. In addition, second year students use the personal computer in their earth science course, EV 203, to generate climatic diagrams, compute time zones, and accomplish a multitude of other computer assisted instruction units.

In their statics and dynamics course, EM 302, sophomore engineering students exercise the full range of advanced capabilities of their programmable calculators. In addition, they use their personal computers to complete commercially-produced tutorials provided with their course texts.

APPLICATIONS IN ENGINEERING SCIENCE COURSES

Civil engineering students encounter additional computer applications as they advance into engineering science courses in their third year. In PH 365A, Modern Physics, a course required of all civil engineering students, computer analysis of lab data as well as original programming problems are included. Problems which have been assigned in the past include alpha particle decay simulation and hydrogen atom energy eigenvalue determination. Original programming is also required in EE 302, Introduction to Electrical Engineering, and EM 301A, Thermodynamics.

In EM 364A, Mechanics of Materials, extensive use is made of the spreadsheet to accomplish repetitive, iterative, trial and error, and simultaneous solutions as part of the laboratory and design problems that are integrated into the course. In many cases, cadets develop macros to accomplish the desired computer solution. Spreadsheet use continues in virtually all subsequent engineering science and design courses in the civil engineering program (Toomey and Lenox 1991). A summary of computer use in representative junior-level courses included in the civil engineering program is included in Table 2.

USE OF THE PC IN UPPER DIVISION CIVIL ENGINEERING COURSES

Civil engineering students are exposed to and required to use a wide variety of special purpose application software in the upper division courses. Specific programs include well known standard applications such as the HEC1 and HEC2 simulation programs in CE 380 (Hydrology and Hydraulic Design); reliable public

TABLE 2. COMPUTER USE IN ENGINEERING SCIENCE COURSES

EM364A   MECHANICS OF MATERIALS
Students prepare spreadsheet programs to design axial, torsion, and beam members.
Spreadsheet skills include data manipulation, table look-up, optimization, simultaneous
equations, and macros.

EE 302:   INTRODUCTION TO ELECTRICAL ENGINEERING
Uses MathCAD to determine currents & voltages in RLC circuits and verify manual
prelab calculations. Students write Turbo Pascal program to solve for node voltages
and loop currents in a network.

EM 301A   THERMODYNAMICS
Students use existing programs to analyze three different power cycle applications;
write original spreadsheet programs to determine the optimal design for a power cycle.

EM 362A   FLUID MECHANICS
Extensive design problem requires a spreadsheet solution to analyze varying flow
parameters and work towards an optimal solution.

MA 364   ENGINEERING MATHEMATICS
Derive is used extensively for algebraic manipulation, evaluating integrals, determining
eigenvalues and eigenvectors, operating on vectors, and graphing results.

PH 365   MODERN PHYSICS
Spreadsheet used for linear regression; requires original program to predict the half-life
of an alpha emitting nucleus from basic quantum principles.

EM 478   STRUCTURAL MECHANICS
Students use several "in-house" programs to transform tensors, perform the
characteristic value problem, and perform small finite element analysis problems.
Students use Derive or Mathcad in practically all homework assignments dealing with
various advanced strength of materials topics.

EM 486   VIBRATION ENGINEERING
Two extensive design problems require use of the personal computer. Students use
software applications programs available to them including mathematical problem
solvers (Mathcad or Derive) and spreadsheet programs. Design teams often use the
dynamic simulator, SIMULINK, in accomplishing vibration engineering design.

---

domain software such as the Corps of Engineers' CFRAME, a two-dimensional
stiffness-based frame analysis package used in CE 403 (Structural Analysis) and
CE 404 (Design of Steel Structures); and in-house developed software, such as
CONSOL, a finite difference-based program for predicting consolidation in cohesive
soils used in CE 370 (Soil Mechanics and Foundation Design). In CE 491
(Advanced Structural Analysis) virtually each of the 40 lessons of the course
requires the students to use the personal computer to demonstrate proficiency in
solving   analysis   problems   involving   the   direct   stiffness   method

(Lenox and Welch 1993), force method, energy methods (Lenox and Welch 1994), finite element method, and finite difference method.

In the capstone design course, CE492 (Design of Structural Systems), students in their final semester at the Academy are required to employ many of the analysis, design, and drafting software that they have used previously in the various required and elective courses of the civil engineering program (see Table 4).

MODERNIZATION

Each year, the Academy reexamines the standard personal computer package required of incoming students to insure that up-to-date hardware and software is being used by its engineering students. The goal is to provide sufficient computing capability to meet the future needs of the academic and engineering programs. As a result, each of the current classes at the Academy has different, yet relatively compatible, microcomputers as seen in Table 3.

TABLE 3. ANNUAL HARDWARE IMPROVEMENTS

CLASS	COMPANY	CPU	MEMORY	DRIVES	MONITOR	SPEED
1992	Zenith	286 LP	640K Base 256K EMS	720K 3.5" 720K 3.5"	Mono EGA	Smart
1993	Zenith	286 LP	640K Base 256K EMS	1.44MB 3.5" 1.44MB 3.5"	B&W VGA	Smart
1994	UNISYS	386 PW2	2MB	42MB Hard 1.44MB 3.5"	Color VGA	16 mhz
1995	UNISYS	386 SX	2MB	52MB Hard 1.44MB 3.5"	Color VGA	20 mhz
1996	Zenith	386 SX	4MB	80MB Hard 1.44MB 3.5"	Color VGA	20 mhz
1997	Zenith	486 SX	4MB	120MB Hard 1.44MB 3.5"	Color SVGA	25 mhz
1998	Zenith	486 DX	8MB	202MB Hard 1.44MB 3.5"	Color SVGA	33 mhx

TABLE 4. COMPUTER USE IN CIVIL ENGINEERING COURSES

CE 371 SOIL MECHANICS AND FOUNDATION ENGINEERING
Students use CONSOL, an "in-house" program to determine settlement and time rate of consolidation for layered soil systems; write spreadsheet programs to design retaining structures; use commercial software to prepare tables and graphs for lab exercises; and use public domain programs CBEAR, PHASE, FLOW, and GSD to master concepts.

CE 380 HYDROLOGY AND HYDRAULIC DESIGN
Programs HEC1 and HEC2 are used extensively in flood routing and plain design problems. Numerous applications of spreadsheets in storm sewer design and streamflow routing.

CE 403 STRUCTURAL ANALYSIS
A large analysis problem require use of a truss analysis program, CME-Truss, using a highly interactive graphical interface. A design problem requires use of CFRAME to analyze 2D frames using the stiffness method. Spreadsheets used throughout.

CE 404 DESIGN OF STEEL STRUCTURES
The matrix structural analysis programs CME-Truss and CFRAME are used throughout the course. An "in-house" program, LRFD92, is used to design beams, columns, or beam-columns. Spreadsheets are used for construction cost estimates.

CE 483 DESIGN OF CONCRETE STRUCTURES
Students create column interaction curves using both the program COLUMN and Mathcad in the analysis and design of columns. Students prepare spreadsheets and use Mathcad to iteractively design single and doubly reinforced beams (both rectangular and T-sections), floor slabs, and spread footings.. Other spreadsheet applications include shear and flexure design of beams and performance of mix design.

CE 491 ADVANCED STRUCTURAL ANALYSIS
CME-Truss and CFRAME used to check the student's manual calculations using the direct stiffness method. Students use spreadsheets throughout in accomplishing 11 homework sets dealing with the force method, direct stiffness method, flexibility to stiffness transformations, condensation techniques, finite difference method, and the finite element method. Students use program FEMCIVIL to solve 2D finite element problems using CST quadrilaterals. Students must master Robot V6, a state-of-the-art, three dimensional structural analysis program offering a variety of analysis options.

PHILOSOPHY OF COMPUTER EMPLOYMENT IN THE CE CURRICULUM

Within the civil engineering program, the faculty have been progressive, yet careful, in integrating computers and computer-based engineering into their courses (Lenox and Hand 1992). The civil engineering faculty is guided by a philosophy which maintains that the computer is an engineer's tool, not his surrogate. A good example of this is the widespread use of the spreadsheet in engineering computations. Initially taught to cadets in their core economics course during their second year, spreadsheet use continues in virtually all subsequent basic science, engineering science, and design courses in the civil engineering program. The spreadsheet is often found to be the most appropriate tool in solving a myriad of different types of "low-level" analyses (Toomey and Lenox 1991). The emphasis on spreadsheet use in the civil engineering courses reflects the faculty's conviction that spreadsheets are unsurpassed in their value and versatility as an engineering tool, a teaching tool, a developer of essential computer literacy/facility, and as an aid in developing orderly thinking and expression.

The civil engineering faculty attempts to use the computers in ways that will reinforce student understanding of physical behavior and engineering principles. Students are taught to be skeptical of computer-generated results; independent verification or validation is required before reliance on these results is permitted. The civil engineering faculty strongly believes that the undergraduate curriculum is the wrong place to train students to use "production" level design software, unless there is a bona fide pedagogical basis for it.

CONCLUSION

The philosophy of engineering education at West Point has remained essentially as it was before the advent of the personal computer. Engineering instruction plays a vital role in providing Academy graduates with the ability to use systematic, disciplined, and quantitative thinking to find practical solutions to complex and often ill-defined problems. The need for technically qualified officers to assume leadership roles in the technological environment of the "lean" Army of the future mandates that the Academy's civil engineering students acquire extensive computer skills, develop healthy computer habits, and employ computers effectively to help solve engineering problems.

APPENDIX. REFERENCES

Academic Program; Field Tables and Course Descriptions, AY 1994-1995. (1994).
Office of the Dean, United States Military Academy, West Point, New York, 129-131.

Lenox, T. A. and Hand, T. D. (1992). "Progressive Integration of the Personal Computer into an Undergraduate Civil Engineering Curriculum." *Proceedings of the Eighth Conference on Computing in Civil Engineering*, Dallas, TX, ASCE, 65-72.

Lenox, T. A. and Welch, R. W. (1993). "Opening the Black Box: Learning the Direct Stiffness Method Interactively Using the Microcomputer Spreadsheet." *Proceedings of the 1993 Annual Conference*, Urbana-Champaign, IL, American Society for Engineering Education, 1078-1087.

Lenox, T.A. and Welch, R. W. (1994). "Practical Solution of the Flexibility to Stiffness Transformation Problem With a Mathematical Assistant for the Personal Computer." *Proceeding of the First Congress on Computing in Civil Engineering*, Washington, DC, American Society of Civil Engineers, 1465-1472.

Toomey, C. J., and Lenox, T. A. (1991). "Introducing the Undergraduate Engineer to the Design Process." *Proceedings 1991 Annual Conference*, New Orleans, LA, American Society for Engineering Education, 370-374.

Not Just a Black Box:
Personal Computer Use in Undergraduate
Structural Analysis Courses.

LTC Robert J. O'Neill[1], PE, Member ASCE
CPT Scott R. Hamilton[2], PE, Member ASCE
COL Thomas A. Lenox[3], Member ASCE

Abstract

This paper describes the evolution of personal computer use in the two-course structural analysis sequence in the civil engineering program at the United States Military Academy. Specific uses of various software packages are addressed.

Introduction

The civil engineering majors in the structural engineering program at the Academy are required to take both of the structural analysis courses; however, many of our students have the opportunity to take the first structural analysis course ONLY. Over the last couple of years, the faculty have wrestled with the issue of how to properly incorporate computer-based methods in BOTH of these two courses to satisfy the needs of the different student populations (Lenox, Dennis and O'Neill, 1995).

This paper will show how changes have occurred in the two structural analysis courses over the last several years to properly introduce ALL undergraduate civil

[1]Associate Professor, Civil Engineering Division, Department of Civil and Mechanical Engineering, United States Military Academy, West Point, NY 10996.
[2]Instructor, Civil Engineering Division, Department of Civil and Mechanical Engineering, United States Military Academy, West Point, NY 10996.
[3]Professor, Civil Engineering Division, Department of Civil and Mechanical Engineering, United States Military Academy, West Point, NY 10996.

engineering students to the theory of matrix structural analysis in the FIRST course in structural analysis -- and more advanced structural analysis techniques in the second course. It will share the decision making experiences of the authors in deciding how to distribute topics between the two courses to properly integrate the use of the personal computer throughout the undergraduate structural analysis sequence. The paper will provide recommendations for incorporating computer methods in an undergraduate program that has only one structural analysis course, and also programs that have a follow-on advanced structural analysis course. The specific computer applications used in each of the two courses will be addressed. Lessons learned by the authors will be emphasized.

History

Prior to 1994, the first course (CE403: Structural Analysis) required the student to learn the classical methods of structural analysis; the second course (CE491: Advanced Structural Analysis) was the students' first exposure to the theory of matrix structural analysis and the finite element method. However, the students in the first course learned how to use two stiffness method computer programs in support of follow-on courses in steel and reinforced concrete design. While the students were successful in learning how to input data into the programs and in reading the output files during the first course, they used the programs as black boxes. The first course student had little or no appreciation for what was being done in the computational portion of the computer programs. Over time, the civil engineering faculty found this "black box" approach to structural analysis unacceptable.

Decisions

Since the "black box" approach was deemed unacceptable, the decision was made to incorporate some portion of the theory of matrix structural analysis in the first structural analysis course. This course now has 5 lessons that cover the direct stiffness method for truss analysis. These lessons were previously taught at the beginning of the second structural analysis course. The first course students are now exposed to the development of local element stiffness matrices, element transformation matrices and global element stiffness matrices. They are able to manually manipulate and assemble these matrices into a structure stiffness matrix in order to solve for displacements, reactions and member forces. With this knowledge the student has a better understanding and appreciation for the structural analysis computer programs used. In the second course this material is quickly reviewed and the course moves on to the direct stiffness method development for beams and frames. The next two sections will give a more detailed description of how the two structural analysis courses are currently organized, especially with respect to computer usage and software applications.

First Course in Structural Analysis

Table 1 shows the topics covered under the old and new curriculum. Analysis of this table shows the five new direct stiffness lessons were added by deleting lessons in influence lines, deflections, force methods and moment distribution. In this first course, students are exposed to the use of spreadsheets as aids to performing manual structural analysis using both classical and matrix methods. In addition, students use two structural analysis computer programs: CMETRUSS, a truss analysis program developed at the United States Military Academy, and CFRAME, a general frame analysis program developed by the U.S. Army Corps of Engineers. Table 2 provides a quick reference to the different computer applications the students are required to use and how they are used in the course. These applications are described in the following paragraphs.

Table 1: First Course Curriculum

Topic	Old Curriculum # Lessons	New Curriculum # Lessons
Basic Concepts	2	3
Truss Analysis	2	2
Shear & Moment	3	3
Influence Lines	4	3
Deflections	7	6
Force Methods	5	4
Approximate Analysis	2	2
Slope Deflection	4	4
Direct Stiffness	0	5
Moment Distribution	4	2
Compensatory Time	3	2
Examinations	2	2
Field Trip	1	1
Course Summary	1	1

The first computer requirement requires use of CMETRUSS. The students receive one lesson on its use, following one lesson reviewing determinate truss analysis. The student is then required to analyze a determinate truss using CMETRUSS and compare these results to the results determined from an earlier analysis using the method of joints and sections. Students are encouraged to use a spreadsheet to perform virtual work analysis of truss deflections. These are usually small graded homework assignments.

The first major graded assignment in this course is the engineer analysis project. Student teams of two or three are required to use CMETRUSS to generate influence lines for an indeterminate truss. The analysis teams then select initial member sizes for their truss. The second requirement is to solve for individual bar forces using the Force Method. In conjunction with their Force Method solution, students must use a spreadsheet to complete a virtual work table to do matrix operations to solve for the bar forces. As a third requirement the analysis teams return to CMETRUSS to verify their manual Force Method solution.

The students are next exposed to CFRAME. After using the program to solve a simple indeterminate rigid frame homework problem, they are given the next major graded exercise. This second major graded exercise in the course is the engineering design project. Once again student teams of two or three work together on the project. Each team performs an approximate analysis of a rigid frame and selects initial member sizes by hand. They then use CFRAME to perform an iterative design to select the lightest members. Finally they use the manual slope deflection method in conjunction with a spreadsheet to verify their CFRAME design.

The final computer requirement in this course is a direct stiffness truss analysis problem. This requires them to demonstrate how well they understand the material from the five lessons on the theory of matrix structural analysis. The students solve for the displacements, reactions and bar forces in an indeterminate truss using the direct stiffness method. The use of a spreadsheet to perform the matrix manipulations is required. The students are required to verify their solution using CMETRUSS.

Table 2: First Course Computer Usage

1. Use **CMETRUSS** to verify manual analysis of a determinate truss.
2. Use spreadsheet to perform virtual work analysis of truss deflections.
3. Use **CMETRUSS** to generate influence lines for an indeterminate truss. Using the force method with a spreadsheet, solve for the bar forces. Use **CMETRUSS** to verify force method solution.
4. Use **CFRAME** to perform an iterative design of a frame structure. Using the slope deflection method with a spreadsheet, verify the **CFRAME** solution.
5. Using the direct stiffness method and a spreadsheet, solve for the deflections, reactions and bar forces in a truss.

Second Course in Structural Analysis

Table 3 shows the topics covered under the old and new curriculum of the second course. Analysis of this table shows that six new lessons were added on ROBOT V6 by deleting lessons in direct stiffness method for trusses and Newmark's

Method. In this second course the use of the spreadsheet as a tool to assist in performing manual structural analysis calculations is reinforced. CMETRUSS and CFRAME are widely used throughout the course. In addition, the students are introduced to a state-of-the-art structural analysis/finite element analysis program, ROBOT V6. Table 4 provides a quick reference to the different computer applications the students are required to use and how they are used in the course.

Table 3: Second Course Curriculum

Topic	Old Curriculum # Lessons	New Curriculum # Lessons
Course Introduction	1	1
Deflection Methods Review	1	1
Force Method Review	2	2
Direct Stiffness Trusses	5	2
Displacement Methods Review	1	1
Direct Stiffness Beams	3	3
Direct Stiffness Frames	3	3
Direct Stiffness Method, Special Topics	3	3
Robot V6 for Structural Analysis	0	4
Energy Methods	3	4
Finite Element Method: 2D Elements	7	6
Robot V6 for Finite Element Analysis	0	2
Finite Difference Method	4	3
Newmark's Method	2	0
Cable Structures	2	2
Examinations	2	2
Course Summary	1	1

On the very first day of class in this second structural analysis course the students are given a homework assignment that requires them to review matrix math using all of the tools they have been taught in previous math, science, and engineering courses. This includes Quattro Pro, DERIVE (a symbolic mathematical manipulator), Mathcad and their HP28S calculator. They are required to write a series of five simultaneous equations in matrix form. The assignment requires them to multiply, invert and transpose a matrix. Finally they are required to solve the simultaneous equations using ALL of the above mentioned tools.

Since the first structural analysis course has already introduced the students to the direct stiffness method for trusses only a brief review is required. Included in this review is the introduction of a series of spreadsheet macros written to assist the student in creating global element stiffness matrices and assembling (stacking) these element matrices into the structure matrix (Lenox and Welch, 1993). These macros are named STRUSS, SBEAM and SFRAME, where the "S" stands for "stacking". With these macros the student is able to use the direct stiffness method and manually solve larger problems without spending a large amount of time in bookkeeping chores.

Throughout the first half of the course as the lessons proceed from trusses to beams to frames, the students are required to use spreadsheet macros for manually solving the problems assigned. In each case the student is also required to validate their solutions using CMETRUSS or CFRAME. In this way the student's dexterity in using all of these computer programs is greatly increased.

The next block of instruction is an introduction to a state-of-the-art structural analysis program. The one currently in use is ROBOT V6 from Metrosoft. The students spend four class periods in the computer lab going through guided instruction on the use of this program. They learn how to perform analyses on trusses and frames in 2D and 3D. The graphical user interface of the program makes the learning of this new software relatively easy for undergraduate engineers. Upon the completion of these lessons the students are required to analyze a frame structure using ROBOT V6 and then to verify the results by doing an analysis using CFRAME.

After completion of the direct stiffness method and ROBOT V6, the course moves on to Energy Methods, specifically the study of Castigliano's Second Theorem. The students are required to use either Mathcad or DERIVE to solve several problems symbolically. These include determining the 3x3 Flexibility matrix for a non-prismatic frame element, develop the 6x6 local element stiffness matrix for a non-prismatic frame element using the Flexibility to Stiffness Transformation, and determining the fixed end effects for a non-prismatic member under a uniform load. In order to accomplish this assignment the students become proficient with one of the two math assistance programs (Lenox and Welch, 1994).

The final portion of the course covers finite elements. Once the theory is covered the students are introduced to a simple finite element program, FEMCIVIL, developed at the United States Military Academy. It allows them to manually create a finite element mesh and solve relatively simple problems. An assignment requires them to analyze a uniformly loaded, simply supported beam to determine the maximum normal and shear stresses. They are required to develop a refined mesh in the area of interest and then compare their results to the "exact" (theoretical) results. At this point the students return to ROBOT V6 in order to solve more complex problems. Two lessons are spent in the computer lab learning how to create meshes, solving several examples problems and interpreting the results. Once again the students are given a small homework problem requiring the use of ROBOT V6 to solve a problem and then verifying the results using an independent procedure.

Table 4: Second Course Computer Usage

1.	Using a spreadsheet, **Mathcad**, **DERIVE** and HP28S calculator, multiply, invert and transpose matrices. Solve systems of simultaneous, linear equations.
2.	Program a spreadsheet macro to create a local element stiffness matrix, the transformation matrix and perform the matrix operations to create the global element stiffness matrix.
3.	Use spreadsheet macros (STRUSS, SBEAM and SFRAME) to assemble the structure stiffness matrix for a truss, continuous beam, and frame. Using the direct stiffness method with a spreadsheet, solve for displacements, reactions and member forces in a structure. Use **CMETRUSS** and/or **CFRAME** to verify manual calculations.
4.	Using **Mathcad** or **DERIVE**, perform symbolic calculations to determine the stiffness matrix and fixed end forces for a non-prismatic beam element.
5.	Analyze a 2D frame structure using **ROBOT V6**. Verify results using **CFRAME**.
6.	Using **FEMCIVIL**, perform a finite element analysis on a simply supported beam. Using **ROBOT V6**, verify your results and analyze more complex structures.

Conclusions, Lessons Learned

As previously mentioned, it was felt that all undergraduate civil engineering students should learn the theory of the direct stiffness method of structural analysis -- not just use the computer programs as black boxes. This has been accomplished by providing five lessons on the theory of matrix structural analysis in the first course in structural analysis. The students who recently had this exposure performed very well on their assigned projects and on the direct stiffness problem given on their final exam. The student taking the two course sequence receives a greater depth and a much broader understanding of the method.

Besides the inclusion of the study of the theory of the direct stiffness method into the first course, we have learned that it is very important to continuously provide the student with opportunities to hone their skills on the use of the various computer programs that are available to civil engineering students. By constantly requiring them to check a single answer by using two or more independent procedures, we counter their belief that "just because it came from a computer, it must be correct." In all cases the students have a better understanding of how structural analysis computer programs work, and some of their limitations. Applications become more than "black boxes" to the students.

REFERENCES:

1.   Lenox, T.A. and Welch, R.W., "Opening the Black Box, Learning Direct Stiffness Method Interactively Using the Microcomputer Spreadsheet," *Proceedings of the 1993 Annual Conference of the American Society for Engineering Education*, June 1993.

2.   Lenox, T.A. and Welch, R.W., "Practical Solution of the Flexibility to Stiffness Transformation Problem With a Mathematical Assistant for the Personal Computer," *Proceedings of the First Congress on Computing in Civil Engineering*, American Society of Civil Engineers, June 1994.

3.   Lenox, T.A., O'Neill, R.J. and Dennis, N.D., "Computer Integration Throughout an Undergraduate Civil Engineering Curriculum," Accepted for publication in *Proceedings of the Second Congress on Computing in Civil Engineering*, American Society of Civil Engineers, June 1995.

Using Spreadsheet Programming in an Undergraduate
Mechanics of Materials Course

Major Michael J. Gazzerro[1]
Lieutenant Colonel Norman D. Dennis, Jr., Member ASCE[2]
Captain Robert J. Carson[3]

Abstract

This paper examines the integration of computer spreadsheet programming into the Mechanics of Materials course at the United States Military Academy. It discusses specific advantages of spreadsheet programming over other programming formats. Detailed guidelines for how to incorporate spreadsheet programming into a mechanics of materials course are provided. The paper gives examples of how spreadsheet programming has been used in the West Point Mechanics of Materials course to provide an analysis tool for students in the laboratory, to conduct structural analysis, and to aid in the design of structural members. It details how personal computer use in the mechanics of materials course is an integral part of the computer thread in an engineering education. Successful implementation results in students who have an increased understanding of basic engineering principals.

Introduction

Fortunately, or maybe unfortunately, the microcomputer has found its way into the engineering classroom. Capable of producing massive amounts of visual data, computers have become far more indispensable than the slide rules, nomographs, and other design aids they have supplanted. There is hardly an engineering problem of any consequence that is not routinely solved using a

---

[1] Director of Public Works, Wiesbaden, Germany, and formerly, Assistant Professor of Civil Engineering, United States Military Academy, West Point, NY 10996

[2] Associate Professor of Civil Engineering, United States Military Academy, West Point, NY 10996

[3] Instructor of Civil Engineering, United States Military Academy, West Point, NY 10996

computer.   In fact, one of the toughest decisions an engineering firm or college engineering department has to make these days is what software package to select to solve a wide variety of engineering problems. There seems to be an endless array of computer software available to solve structural analysis and mechanics problems.

With all of the proliferation of computer software it is sometimes hard to keep in perspective the fact that any software is only as accurate as its users knowledge of the program's operation and underlying principles.   Computer programs require that the engineer using them understand the theory and the assumptions behind the program.   Unfortunately, production software often hides the theory and assumptions from the user.   This problem is especially significant when the user is a beginning engineering student who probably does not completely understand the theory, even if it were completely described within the program.

One obvious solution to this problem is to eschew the use of computers altogether in basic engineering courses such as introductory solid and fluid mechanics courses or thermodynamics. Of course, this solution disregards the integral place of the computer and fails to reinforce the notion that we should, in fact, make proper use of the tools at hand. The ability to evaluate computer output is an integral part of developing engineering judgment.   Furthermore, the Accreditation Board of Engineering and Technology (ABET) has stated that computer utilization is an integral part of a complete engineering education.

Another solution is to have students write their own software using a programming language such as FORTRAN or C++. This is appropriate in some higher level courses. At the basic level, however, the mechanics of writing a program can easily become overwhelming and cause the student to lose sight of the actual learning objective which is to apply and reinforce engineering theory.

At the United States Military Academy (USMA), students follow a third option--a kind of middle ground. Spreadsheet programming is used exclusively as a part of the mechanics of materials course.   Spreadsheet programming offers a highly effective way to integrate computers into basic engineering courses such as Mechanics of Materials. Spreadsheet use affords students the opportunity to apply solution logic without the input-output overhead associated with high level language programming. Students in the USMA Mechanics of Materials course use spreadsheets to prepare laboratory results, to conduct analysis of structural members, and as a tool in engineering design.

Advantages of Spreadsheet Programming in an Introductory Course

Within the context of an introductory engineering course, spreadsheet programming has many advantages over either commercial or "in-house" developed analysis programs. In many cases spreadsheets are superior to programming in a high level language because much of the input/output overhead is eliminated. Among the most important advantages are the cost, ease of learning, flexibility, the resemblance of the spreadsheet logic to "stubby pencil" logic, and excellent graphics capability.

Spreadsheets are relatively inexpensive. One of the biggest advantages that the USMA has, is that all cadets use the same spreadsheet package. Depending on the year they entered the academy, every cadet owns a version of Quattro ranging from Version 1.0 to Quattro Pro for Windows Version 5.0. The low cost of these spreadsheets packages means that every cadet has the software on his or her own machine. It also affords the instructor the ability to specify that certain assignments will be done with a spreadsheet.

Spreadsheets are relatively easy to learn. Once the user has mastered a few basic rules such as how to reference a cell in an equation, rudimentary spreadsheet programming is very straightforward. Yet, spreadsheets offer advanced users many tools to create truly sophisticated problem solving tools. Students can make the spreadsheet as simple or as intricate as their experience and training allow. More accomplished spreadsheet users can use advanced tools such as macros to allow branching operations or recursive calculations, while the less computer literate student can still write a simple spreadsheet that gets the job done.

Spreadsheets are flexible. Students need not buy a separate software program for each of the wide array of topics covered in an introductory course like mechanics of materials. For example, in the USMA Mechanics of Materials course students use their spreadsheet package to analyze and design structural members as well as to reduce and present laboratory data. A single software package allows students to examine axial loading, bending, torsion, and column behavior.

Spreadsheet logic closely resembles the logical flow of the engineering thought process. The equations a student would use to represent the engineering thought process can be directly replicated in the spreadsheet by transferring the set of equations to a series of rows or columns. Once the equations are entered into the appropriate cells, it is easy to check the validity of the logic because the calculations are immediate. As a result, there is an obvious relationship between the engineering thought process, the equations needed to represent that thought process, and the iteration through those equations to achieve an optimal solution. This notion is pictorially described in Fig. 1. Furthermore, there is a clear link between numerical data and any graphical output.

The current generation of spreadsheets provide outstanding graphics capabilities. The way in which spreadsheets tend to relate numbers to graphics is important. Since spreadsheets can easily generate graphs from a range of numerical values, it is easy for the student to generate a graphical depiction of a solution.

Furthermore, it is possible to directly alter the graphical output by changing the desired parameter(s).

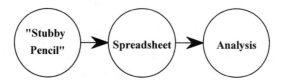

Figure 1. Spreadsheets have the advantage of directly replicating the equations one would use to solve an engineering problem by hand. It is straightforward, then, to use the output (which is frequently a graph) to analyze the results.

General Guidelines for Incorporating Spreadsheets

There are certainly as many ways to integrate computer programs in a course as there are combinations of engineering instructors and courses. The authors are offering an example of just one way to use the computer in a course. Of course, the ultimate goal of computer usage in any undergraduate course should be to aid in the learning of the actual course objectives. The instructor should avoid situations in which the effort required to learn a software package eclipses the student's effort devoted to learning the course objectives. Furthermore, the supporting engineering principles of a software package must be obvious to the user.

In the mechanics of materials course at USMA spreadsheets are used primarily to solve three design problems and to evaluate data taken in three separate laboratories. The authors have found the following guidelines to be appropriate when directing computer aided problem solving.

1. Specify that students will use a spreadsheet to solve the problem.
2. Require hand calculations to verify the output of the program.
3. Use the spreadsheet to generate graphic output.
4. Require minor extensions in spreadsheet proficiency for each succeeding exercise.
5. Constantly reinforce the notion that the computer is a tool, not a crutch.

Specify spreadsheet. Students must be directed to use a spreadsheet to perform iterative calculations. If they have done a good job of representing their solution process as a set of equations, then it will be easy to write a spreadsheet program. Recommend to students that they reference key variables such as; loads, stiffness, and certain geometric dimensions to a set of blocks which can be changed easily. This will allow them to use the spreadsheet as a tool to investigate the sensitivity of their design to various design parameters.

Hand calculations. Always demand hand calculations as a part of any solution. In the case of a design problem, calculations should be included for one complete iteration. Students should be told to do these hand calculations *before* doing any work on the computer. They should be pushed into the habit of outlining their solution thought process--a set of hand calculations. This is important for two reasons. First, it insures that every student can work through the theory. If hand calculations are not required, many students will attempt to simply transfer the equations directly from their head into the spreadsheet. This rarely works because students in basic courses have not developed a sophisticated enough level of engineering judgment to evaluate the reasonableness of their answers. Second, it will improve their chances of entering the correct equations into the program. Sample calculations are always included as an appendix of a design report.

Graphic Output. Insist that students generate some type of graphic output from their spreadsheets. This is important because it causes them to visualize the effect of their decisions. In some instances the graphical output is a direct map of the material's behavior. It is also useful to give the students a high quality example of graphical output so that they know what is expected of them.

Spreadsheet Knowledge. Students in the USMA Mechanics of Materials course come armed with fundamental spreadsheet programming skills. They can enter formulas in cells using both absolute and relative references. They can produce rudimentary graphs of columnar data. Throughout the course the students are required to implement more advanced features of the spreadsheet to include; table lookups, single and multi-variable optimization schemes, elementary macros to create decision logic, proper graphical annotation and presentation of tabular data. This is an evolutionary process, in that only one new advanced feature is developed for each graded exercise. At the conclusion of the six out of class graded exercises the students are capable spreadsheet programmers.

Computer as a Tool. Perhaps the most important thing that the instructor must do is to constantly reinforce the notion that it is the student--not the computer--who is doing the work and determining the answer.

Using Spreadsheet Programming in the Lab

One of the easiest ways to integrate spreadsheet programming into a mechanics of materials course is as a tool to evaluate lab data. In the USMA Mechanics of Materials course, three laboratory exercises are conducted--a simple tension test, a pure torsion test, and a simple four point elastic bending test. To successfully integrate the computer as a tool, the following steps are followed:

1. Collect lab data manually. While speed and accuracy would likely be increased by using some means of automatic data collection and processing as one would commonly do in a commercial or graduate level university lab, collecting the data by hand causes the student to see what is happening to the material as loads are increased.

2. Use the spreadsheet to analyze data. For example, in the simple tension test, the student uses the spreadsheet to convert load to stress and deformation to strain as illustrated in Fig. 2. The spreadsheet is used to calculate the modulus of elasticity and to plot the stress strain curve. Furthermore, the computer can be brought into the lab to provide immediate feedback to cadets.

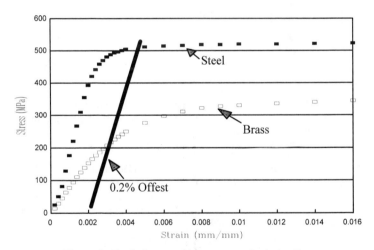

Figure 2. Typical normal stress-normal strain diagram for brass and steel, generated using Quattro Pro as part of a student lab report.

Using Spreadsheet Programming in Design and Analysis

The key to using a spreadsheet in design analysis is to constantly reinforce the notion that it is the engineer and not the computer who must eventually solve the problem. The computer is merely a tool, albeit a powerful tool.

A good example of how the personal computer is incorporated into design and analysis problems is the third design problem from a recent semester. The problem was to design the beam of the anchored bulkhead illustrated in Fig. 3. The beam was idealized as being fixed at one end, with a triangular distributed load along its length to represent the soil load, and a point load at the "free" end to represent the anchor. During the first phase of the problem--which was primarily an analysis phase--students were required to develop a spreadsheet to determine the shear and bending moment diagrams for the beam. The point load was made a function of the cable diameter and the distributed load was an algebraic function of the beam length. At this point the length of the beam and magnitude of the point load were unknowns. The intent was to create a generic spreadsheet.

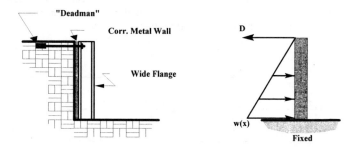

Figure 3. Soldier-beam type anchored bulkhead, idealized as a fixed cantilever beam with a hydrostatic and concentrated load applied.

During the initial phase of this problem each student created a spreadsheet to represent the shear and bending moment as a function of position on the beam. Once this was accomplished, it was easy to see the effect of the various unknown parameters on the behavior of the beam. For example, using the graphical representation of the moment function, as illustrated in Fig. 4, students could visually see how adjusting the point load could decrease the maximum moment.

The design phase of the project required the student to select the optimum combination of anchor cable and beam to minimize cost for a given wall height. A vertical look-up table was used to pick the appropriate wide flange beam properties and a two variable optimization table was used to pick the best combination. Students set up their solution logic by hand before and presented it in an in-progress-review before entering any equations into their computer.

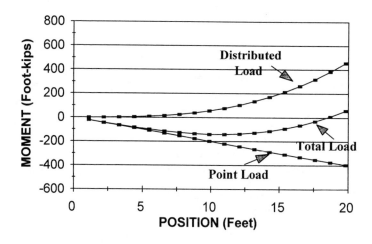

Figure 4. Moment diagram generated using Quattro Pro illustrating the use of superposition to determine the effects of combined loading.

Conclusions

Students in the mechanics of materials course have taken to spreadsheet programming as though the spreadsheet were an extension of their handheld calculator. If allowed to proceed unchecked, however, students would rarely develop their solution logic by hand before entering equations into the computer. Graduates of the USMA Mechanics of Materials course are able to create useful spreadsheets which have logical data entry and present solutions in a readily accessible manner. The most lasting impression instructors of the mechanics of materials course try to leave with the student is that, the results of the computer program are only as good as the programmer. He or she must critically evaluate their output. The graphical representation of data is extremely useful in this regard. Students should ask the questions: Can I replicate my spreadsheet calculations with a set of hand calculations? Do the results make sense? Is the effect of changing various parameters reasonable? Students must confidently verify that the answer to these questions is yes or the spreadsheet becomes a millstone around their neck.

A Tool But Not a Crutch:
Using Design Software in an Undergraduate Structural Steel Course

Captain Karl F. Meyer, Member, ASCE[1]
Major Stephen J. Ressler, Member, ASCE[2]

Abstract

The use of design software in an undergraduate structural steel course can significantly improve the quality of the educational experience if implemented correctly. By allowing its use only after proper mastery of course material, students learn to use design software as a tool which allows them to tackle steel design tasks more quickly and efficiently. Permitting its use too early results in a lack of understanding of basic skills and a dependence on design software as nothing more than a crutch.

Introduction

This paper describes the use of two design-oriented computer programs in an undergraduate structural steel course at the United States Military Academy at West Point. Both software packages were developed by members of the West Point civil engineering faculty, and both run on IBM-compatible personal computers. The authors have found that these programs, if used appropriately, can significantly enhance the quality of the students' educational experience. However, the authors also note that the use of design software in the classroom must be carefully controlled to ensure that students do not use the computer as a substitute for an understanding of fundamental concepts. The authors have come to believe that design software should only be made available to students after they have had ample opportunities to learn and practice the corresponding manual problem-solving skills.

---

[1] Instructor, Department of Civil and Mechanical Engineering, West Point, New York, 10996.
[2] Deputy Commander, New York District, U. S. Army Corps of Engineers, 26 Federal Plaza, New York, New York, 10278.

The Course

CE404, Design of Steel Structures, is an introductory steel course taken by all civil engineering majors in the Fall term of their senior year. The course uses the American Institute of Steel Construction Load and Resistance Factor Design (LRFD) methodology throughout. Five distinct blocks are covered within the course, including structural systems, tension members and tension connections, compression members, flexural members and beam-columns. For each block, students must perform two out-of-class written assignments: (1) a special problem (SP), worked individually, and (2) an engineering design problem (EDP), solved in groups of three. The SP involves analysis of an existing member or connection, while the EDP requires the design of several structural components or a simple structural system. Both the SP's and EDP's are based on a course project scenario, consisting of architectural drawings for a small commercial building. This scenario provides a thread of continuity throughout the course.

The two design-oriented software packages used in CE404 are **CME-Truss**, a truss analysis and design package, and **LRFD92**, a program for LRFD-based structural member design.

CME-Truss

The **CME-Truss** package actually consists of two different programs--the basic version used by students and an enhanced version, called **CME-Truss+LRFD LoadTest,** used by the instructor. Both are written in the Visual Basic for Windows programming language; both run on IBM-compatible personal computers with a 386 (or better) processor and a VGA (or better) color monitor.

CME-Truss is, for the most part, a conventional two-dimensional truss analysis program. It is based on the direct stiffness method of structural analysis and calculates member forces and displacements based on a user defined model. It does, however, have two unique features: it uses a highly interactive graphical interface, and it allows the user to assign cross-section properties by selecting from a list of AISC standard rolled shapes. Both features give students the capability to expedite the creation and alteration of structural models. Thus **CME-Truss** facilitates creative design by allowing students to explore many different alternatives in a relatively short period of time.

CME-Truss+LRFD LoadTest includes the same user interface and the same structural analysis module as the basic version of **CME-Truss**, but also has the capability to perform a comprehensive strength and serviceability evaluation of a steel truss.

Students are introduced to **CME-Truss** in the structural analysis course which precedes CE404; thus they are able to use the program early in CE404 to complete two special problems, with little or no additional training. In these special problems, students are given a steel truss with known dimensions and member sizes (from the course project scenario), and are required to evaluate its strength and serviceability. A number of design deficiencies are intentionally incorporated into the truss. In the process of discovering these deficiencies, students practice LRFD-based tension and compression member analysis, while also gaining proficiency with **CME-Truss**. That proficiency is put to the test in the EDP which follows.

In the EDP, students are asked to re-design the truss for better economy. The span, maximum depth, loads, steel grade, and connection configuration remain the same, and a limitation is placed on maximum top chord panel point spacing; but otherwise, the design is unconstrained. Students may use any truss configuration they desire. Their objective is to minimize the total cost of the truss without violating any LRFD limit states or exceeding a specified maximum service load deflection. Cost factors, which account for both material and fabrication costs, are provided by the instructor. The instructor provides no guidance concerning what an optimal truss design might look like or cost.

To stimulate interest in the project and to provide an incentive for superior performance, the EDP is administered as a design competition. Bonus credit is offered for the winning team.

As the students develop their designs, **CME-Truss** proves its worth. The simple graphical user interface facilitates the rapid creation of structural models; thus students are able to explore many alternative designs in a relatively short period of time. The instructor's design requirement specifies that students will investigate at least three alternative designs; yet in a recent competition, one design team developed over 20 different trusses, in an effort to achieve an optimal solution. **CME-Truss** also allows students to save their structural models on disk, in a specially formatted data file. The students are asked to submit a floppy disk with the saved data file, along with their completed design report.

The moment of truth arrives during the first class period following the EDP due date. During this class, the instructor uses **CME-Truss+LRFD LoadTest,** running on a PC in the classroom, to evaluate each of the student truss designs and determine the winner of the competition. Before class begins, the instructor loads a parameter file which specifies the magnitude of dead, live, snow and roof live loads which will be applied to the structure. These loads correspond to the ones specified to the students in the design requirement. With the competition underway, each student's truss design is loaded into the computer via the saved data file. Immediately a graphical image of the truss appears on the computer monitor, and a click of the mouse starts the simulated load test.

With each load test, **CME-Truss+LRFD LoadTest** performs the following actions:

- calculates joint loads in accordance with specified service loads and LRFD load case combinations,
- performs a structural analysis,
- calculates the strength of each member, with respect to all pertinent LRFD limit states,
- compares calculated member forces with corresponding strengths to determine if any members have failed,
- checks maximum service load deflection,
- calculates the cost of the truss, considering both materials and fabrication,
- checks that the length, depth, and support conditions of the truss are consistent with the design requirement,
- displays the results as an animated "load test" simulation,
- displays a summary screen showing the results of all strength and serviceability checks and the cost analysis,
- and provides a hard-copy printout containing the same information.

A "snapshot" of the simulated load test is shown below in Figure 1.

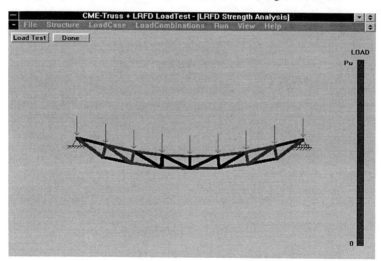

Figure 1. "Snapshot" of a simulated load test on a student truss design.

During the simulation, joint loads appear on the screen, and the truss deflects in response.   As load increases, members change from blue to red, in direct proportion to the ratio of member force to design strength. If that ratio exceeds one, the member disappears from the screen, indicating a failure.

During the load test, students display a level of enthusiasm normally reserved for the football field.   The level of excitement rises as the members of a truss change from blue to red.   Cheers of approval sound when a truss completes its load test and glows uniformly red in color--an indication that the structure is designed optimally, all of its members near their maximum capacity.   Failed trusses produce groans from their designers, cheers from their competitors.   When a truss fails, the designers often launch an impromptu forensic investigation, to discover the cause.

CME-Truss+LRFD LoadTest provides a scoreboard summarizing the results of each load test.   The program also creates a detailed output listing for each evaluation.   The listing shows the computed force, all design strengths, and the controlling limit state for each member.   By referring to this output, the instructor can provide immediate feedback to the student concerning the quality of his or her design. The number and character of member failures will almost invariably point to the specific design errors made by the student

The authors have found that the use of a computer-based design competition provides students with an incentive to investigate alternatives and thus enhances the quality of their design work; it also stimulates interest in the course and provides the instructor with an invaluable tool for objective and comprehensive assessment of student work.

LRFD92

LRFD92 is a program for optimal selection of steel beams, columns and beam-columns.   Given a set of user-specified design parameters (e.g., required strength, unbraced length, steel grade), **LRFD92** selects the lightest AISC rolled shape which satisfies all appropriate LRFD limit states.   Alternatively, for a given rolled shape, **LRFD92** calculates the design strength in axial compression, flexure, or combined flexure and axial loading.   In effect, the program does precisely what students are expected to be capable of doing manually, as a result of their participation in CE404.

In early iterations of CE404, the authors allowed students to use **LRFD92** for all three of their EDP submissions--the truss design, a roof framing system design, and rigid frame design.   The authors believed that students had ample opportunities to learn and practice manual problem solutions in class and on special problems. Having mastered the fundamentals, students could tackle larger and more comprehensive design projects with the aid of the automated design tool.

Unfortunately, this assessment proved to be incorrect. As soon as **LRFD92** became available, many students became excessively dependent on it. They tended to trust any output as correct, even if those results defied good judgment. In too many cases, the program became nothing more than a "black box." Manual problem-solving skills quickly atrophied, and students lost sight of the fundamental principles underlying those techniques. Exam scores suffered.

Based on this experience, the authors determined that the use of fully automated design software was *not* appropriate for routine use in an introductory structural steel course. In future iterations of the course, **LRFD92** was only introduced during the final phase of the course project--normally a rigid frame design. This change forced students to practice manual design throughout the course and produced a dramatic improvement in overall performance.

Through this trial-and-error experience, the authors learned a valuable lesson: the focus of an undergraduate structural steel design course should be on fundamental principles and problem-solving skills. The use of a fully automated design tool tends to detract from that focus and thus is inappropriate. The potential benefit of using such tools--the ability to assign design projects of much greater scope and complexity--is far outweighed by the inevitable decline in student performance.

The value of **LRFD92** is that it facilitates high-quality iterative design of very complex structural frames. During CE404, this is not an overriding consideration; what is important is the development of a sound understanding of basic member design skills. Thus **LRFD92** has found its most important application in the course which follows CE404--Design of Structural Systems, the ABET capstone design course for the West Point civil engineering program.

Conclusion

In the authors' experience, the use of design-oriented computer software can produce extremely valuable results. When used appropriately, programs like **CME-Truss** can greatly enhance students' design experience, without degrading their manual problem-solving skills. On the other hand, we find that fully automated design tools like **LRFD92** are inappropriate for use in an introductory steel design course. These programs tend to foster a "black box" mentality, which only serves to degrade student performance. Their use should be restricted primarily to more advanced design courses.

Computational Methods in Flood Plain Hydrology
in an Undergraduate Hydrology and Hydraulics Design Course

Major Edward C. Gully[1]
Lieutenant Colonel Norman D. Dennis, Jr., Member, ASCE[2]

Abstract

This paper discusses the benefits of the integration of a combination of commercial, public domain and "in-house" computer software in an undergraduate hydrology and hydraulics course. It describes how the careful implementation of computer-generated solutions has served to enhance the understanding of fundamental concepts. Various computer programs are used to predict water surface profiles, create and route hydrographs for reasonably complex watersheds and to size drainage structures. Computer methods are introduced only after underlying principles and manual solutions to problems are addressed and reinforced. Students then learn to use their personal computer and appropriate software as design tools in which key design parameters are easily varied so that efficient solutions can be identified.

Introduction

The use of computer software is an integral part of the undergraduate hydrology and hydraulics course taught at the United States Military Academy at West Point. All computer programs used by cadets in this course are microcomputer based and run on IBM-compatible personal computers. While the use of computer software is considered integral to the learning goals of the course, computer

---

[1] Assistant Professor, Department of Civil and Mechanical Engineering, West Point, New York, 10996.

[2] Associate Professor, Department of Civil and Mechanical Engineering, West Point, New York, 10996.

programs are only introduced after the theory and hand-generated solutions have been presented and reinforced. The authors firmly believe that students should understand the fundamental algorithms used in computer-based solutions before they ever attempt to use the software. The experiences with computer software in this course have not always been perfect. Numerous adjustments have been made to the syllabus, the method of presentation and the complexity of the problems assigned over the past several years to maximize the learning experience and minimize the frustration level of the student. As a result, the authors have found that judicious use of computer software can actually improve the students' understanding of the course concepts as well as illustrate the utility of the computer in engineering problem solving.

The Course

CE380, Hydrology and Hydraulic Design, is taken by all civil and environmental engineering majors in their junior year or senior year. CE380 is a somewhat unique course in that it combines hydrology and open channel hydraulics in a single course; whereas, most undergraduate engineering programs address the topics in separate courses. During the hydrology block of the course, the following topics are covered: descriptive hydrology, unit hydrograph theory, hydrologic routing, frequency analysis, and hydrologic design (rational method, storm sewer and detention design). In the hydraulics block of the course, the following topics are addressed: open channel flow, structures in open channels (weirs, culverts, channel transitions) and water surface profile generation for gradually varied flow conditions.

For each block, students must perform two out-of-class written assignments: an individually worked special problem and an engineering design problem solved in groups of three. All assignments are presented in realistic design scenarios requiring the student to filter a large quantity of material to arrive at one or more alternative solutions. The individual assignments consist of relatively simple problems that directly employ the theory presented in class and may be verified completely with hand calculations. The use of the computer at this stage serves only to introduce the student to the input requirements and the form of the output generated by the computer program. Ideally, the student's hand calculations and computer solution would agree completely. The design problem normally represents a scenario that is not easily completed by manual hand calculations. It is during this stage that the student learns the utility of the computer in engineer problem solving. Hopefully, by this point the student has a high level of confidence in the computer-generated solution and also has the manual tools to verify portions or all of the computer-generated solution.

The computer programs used in CE380 include HEC-1, HEC-2, CULVERT, and a standard spreadsheet package. Both HEC-1 and HEC-2 are the public domain versions developed by the Hydrologic Engineering Center in Davis, California.

CULVERT is a program developed in-house while the spreadsheet most commonly used by students is some version of Quattro Pro.

HEC-1

The HEC-1, Flood Hydrograph Package, is designed to simulate the rainfall-runoff process and generates hydrographs at specified locations. The program creates, routes and combines hydrographs by modeling the watershed as an interconnected system of components. The basic components of the model include surface runoff from sub-basins, stream channels and reservoirs (HEC-1 User's Manual).

In CE380 the HEC-1 Package is used as a tool to model the rainfall-runoff process and to evaluate the impact of increasing urbanization on a watershed. Prior to the introduction of HEC-1, the rainfall-runoff relationship is fully investigated in the classroom and reinforced with out-of-class assignments. HEC-1 is then presented and used to verify a manually solved classroom example which addresses the impact of urbanization on a simple watershed.

Students then use HEC-1 to solve an out-of-class assignment which focuses on the modeling of a similar yet more complex multi-basin watershed. The usefulness of the program is immediately evident to the student because of the complex routing and hydrograph combination scenario. During the exercise, the rainfall-runoff process is modeled using the Soil Conservation Service's (SCS) dimensionless unit hydrograph method. The students are required to use HEC-1 to conduct a sensitivity analysis of several key parameters (design storm, watershed area, base flow, SCS curve number, and SCS lag time). By performing this analysis, students develop a better understanding of the interaction of the various parameters impacting the rainfall-runoff process. They also quickly recognize the utility of the model when they are able to analyze the impact of increasing urbanization by adjusting only two parameters in the HEC-1 model (the SCS curve number and lag time). HEC-1 is then used to evaluate alternate strategies in controlling runoff. Throughout the process, the students are encouraged to perform an "engineering reality check" on all output.

The basic output from the program is rather crude by today's graphical standards, but it is complete, usable, and totally understandable by the students. The sophisticated presentation graphics of the Data Storage System (DDS) interface have not been incorporated into the course nor do the authors feel a need to do so at this time. HEC-1 was selected as the software of choice for this course over other, easier-to-use programs, because it represents the industry standard in modeling flood plain hydrology. The user interface of this program, while different from what the students are used to in a Windows environment, is indicative of a wide range of engineering programs currently used in everyday engineering practice.

HEC-2

The HEC-2, Water Surface Profile Program, is used to generate water surface profiles for steady, gradually varied flow in natural or man-made open channels. The effects of various obstructions such as bridges and culverts may be analyzed in the program. The computational procedure (the standard step method) is based upon the solution of the one-dimensional energy equation with energy loss evaluated with Manning's equation (HEC-2 User's Manual).

Prior to using HEC-2, students develop gradually varied flow water surface profiles using the direct step method, both by hand and through use of a computer spreadsheet solution. A classroom example which deals with a real life problem is first presented using a slide show to let the students visualize a stretch of open channel that may create a flooding condition in the vicinity of a public school. The water surface profile is first created by hand for this channel with simple geometry. The hand solution is then verified using HEC-2 in a follow-on classroom exercise. Aside from comparing results, the principal purpose of that exercise is to introduce the students to the input and output structure of the program. In a sequential scenario similar to that used with HEC-1, students are required to solve a similar out-of-class problem both manually and using HEC-2. The students are then required to investigate the impact of varying key parameters in the analysis (for example, Manning's roughness coefficient or the cross section geometry). By performing this sensitivity analysis, the students develop a better understanding of the hydraulic relationships involved in predicting a water surface profile. Students also use HEC-2 in a design scenario in which they evaluate a variety of flood control measures and recommend an optimal channel depth and geometry for a flood mitigation channel improvement project.

CULVERT

CULVERT is a program used to assist in the evaluation and design of culverts. CULVERT was originally developed by students at West Point in 1985 as a capstone project for a course in computer-aided design. The program was developed in FORTRAN, ran on a mainframe computer and was accessible via dumb terminals at that time. Later the program was refined by members of West Point's civil engineering faculty and re-compiled to run on a microcomputer. The program generates series of headwater versus discharge curves as a function of culvert slope and simultaneously plots a curve representing the transition between inlet and outlet control curves.

As in previous exercises, manual solution techniques for culvert analysis and design are initially presented. CULVERT is then used to demonstrate the impact of varying key design parameters (culvert shape, slope, and loss coefficients). A significant problem students often experience in culvert design involves the iteration through complicated orifice and open channel flow equations to determine if the culvert flow is governed by inlet or outlet control. The CULVERT program is very useful in depicting the differences between inlet and outlet control. An example of the CULVERT's graphical capabilities is provided in Figure 1. The design parameters can be easily varied, inlet or outlet control graphically ascertained, and

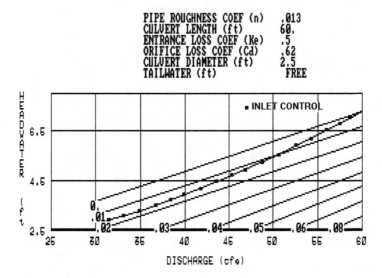

Figure 1 Graphical output from Program CULVERT illustrating the relationship between discharge and headwater elevation as a function of slope.

the culvert's capacity for a given set of conditions determined. If this program was not available, the sensitivity analysis for key design parameters would be very time consuming and calculation intensive. As a result, students would consume excessive time on repetitious calculations as opposed to developing an understanding of those properties that control the hydraulics of culvert flow.

SPREADSHEETS

Every cadet at the Military Academy owns a spreadsheet program of some type. Students are currently issued the latest version of Quattro Pro that is shipping during their freshman year. They are required to use the spreadsheet program in their freshman and sophomore math, science and economics courses. Throughout the program of instruction in CE380, the use of a standard spreadsheet program is integrated whenever feasible as a vehicle to improve student skills in engineer problem solving. Students are required to solve a variety of homework problems using only their spreadsheet package or their spreadsheet package in combination with hand calculations.   For example, the determination of Muskingum channel routing coefficients, reservoir routing and the direct step method for estimating elevations of water surface profiles are extremely conducive to spreadsheet solution. The graphical capabilities of spreadsheets are continually emphasized.

The Muskingum method is used for flow routing in the course.   In order to apply this method for a river reach, the Muskingum proportionally coefficient, K, and weighting factor, X, must be determined for observed inflow and outflow hydrographs.   Students are required to use their spreadsheet's graphical capabilities to solve this problem.   Students are required to assume values of X and develop a graph of the cumulative storage versus the weighted discharge, {X(Inflow)+(1-X)(Outflow)}.   The value of X that produces a loop closest to a single line is taken to be the best value for the reach.   An example of the graph required for solution of this problem is provided below in Figure 2.

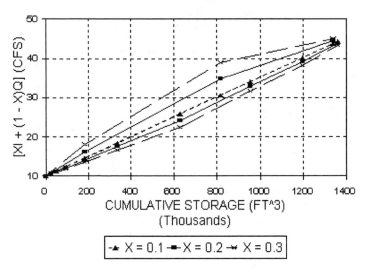

Figure 2   Graphical solution from Quattro Pro Spreadsheet illustrating the weighted discharge versus cumulative storage relationship.

The direct step method of estimating water surface profiles is also extremely conducive to spreadsheet solution. As before, the theory and method are presented in class. Both are illustrated with an in-class example. Students are then required to solve an out-of-class scenario based assignment using their spreadsheet package.

For example, students were required to investigate the behavior of an M2 water surface profile in the vicinity of an abrupt change in channel slope from a mild to a steep channel bottom slope category. The students were instructed to assume that the water surface passed through critical depth at the break in channel slope. They were to then investigate the flow's asymptotic return to approximately normal depth via the M2 water surface profile. The students were required to investigate the impact of the assumed step size as well as key design parameters such as Manning's roughness coefficient and the channel slope. They were also required to graph portions of the problem so they could visualize the associated changes in the water surface profile resulting from the variation of the key design parameters and distance upstream from the change in slope. An example of the required graphical output is provided below in Figure 3.

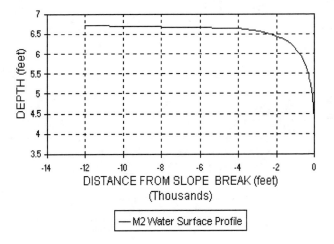

Figure 3   Graphical solution from Quattro Pro spreadsheet analysis illustrating the relationship between depth and distance upstream from a change in channel slope.

Conclusion

In the authors' experience, the use of computer software can produce extremely valuable results when used correctly. Computer usage should only be attempted after the underlying principles and methods of the solution are introduced and reinforced through manual problem-solving exercises. The software packages can then be used to enhance the focus on fundamental principles by providing an efficient means to illustrate the impacts of varying key design parameters. The programs are also useful as iterative design tools to identify efficient solutions. As with all learning exercises, the authors have wrestled with the level of complexity students should be exposed to in problem solving. The goal is not to overwhelm the student with the complexities or computational sophistication of the computer program. The focus should be on the students' interpretation of the results of the analysis and not the level of effort required to produce a solution. Therefore, the complete power of the computer programs used in CE380 is never fully tapped. Instead, the student is left with an appreciation of the capabilities of the program and has internalized the solution technique to the point that he or she can investigate the full capabilities of the program on their own.

References

1. United States Army Corps of Engineers, Hydrologic Engineering Center, HEC-1, Flood Hydrograph Package, User's Manual, 1990.

2. United States Army Corps of Engineers, Hydrologic Engineering Center, HEC-2, Water Surface Profiles, User's Manual, 1990.

Multiobjective Optimization of the Dial-a-Ride Problem Using Simulated Annealing

John W. Baugh Jr.,[1] Gopala Krishna Reddy Kakivaya,[2] and John R. Stone[3]

Abstract: Routing and scheduling in multi-vehicle dial-a-ride problems have attracted the attention of researchers in the last decade, and numerous techniques for generating approximate solutions have been proposed. While some of these techniques have mathematical foundations, it is often difficult to assess the global optimality of the generated solution due to their use of pure local improvement methods. In addition, most of these methods are based on a single objective, such as minimization of customer inconvenience, and cannot account for different or competing objectives that characterize the problem. We present a new approximate method based on simulated annealing, a stochastic approach to global optimization, that addresses these limitations.

1 Introduction

Vehicle routing and scheduling are key activities in many transit operations, including the demand-responsive systems that are sometimes referred to as "dial-a-ride." The basic elements of the dial-a-ride routing and scheduling problem include the following:

- *Service requirements* — customers place requests in advance to be picked up and dropped off at particular locations within time windows.

- *Provider capabilities* — a fleet of shared-ride vehicles is located at a central depot. Schedules are generated in advance.

The objective of the problem is to design a complete tour for each vehicle that services all the customers within their time windows. A description of a tour includes

---
[1]Associate Professor, Department of Civil Engineering, North Carolina State University, Raleigh, NC 27695-7908. E-mail: jwb@eos.ncsu.edu

[2]Graduate Research Assistant, Department of Civil Engineering, North Carolina State University, Raleigh, NC 27695-7908. E-mail: grkakiva@eos.ncsu.edu

[3]Associate Professor, Department of Civil Engineering, North Carolina State University, Raleigh, NC 27695-7908. E-mail: stone@eos.ncsu.edu

the 'route', or sequence of locations visited by each vehicle, and 'schedule', or times at which the vehicle should arrive at those locations. The problem is complicated by the presence of competing objectives such as minimization of the total distance traveled by the vehicles, customer inconvenience, number of vehicles used, etc.

In this paper we give an overview of related work and a concise description of the dial-a-ride problem that includes computational and tractability issues. A multiobjective approach using simulated annealing is then presented, followed by results obtained on a subset of real-world data. We conclude with some observations about the multiobjective nature of the dial-a-ride problem.

2 Related Work

Most of the early research efforts in vehicle routing and scheduling concentrated on the development of approximate methods for solution of the problem. The approximate approaches can be broadly classified into two types: *insertion heuristics* and *cluster-first and route-second heuristics*. In the first approach additional customers are inserted into an initial route in an optimal way. Thus the quality of the solution depends on the order in which the customers are selected. In the second approach the entire set of customers is partitioned into as many subsets as there are available vehicles, and routes are then developed for the individual vehicles. Thus the quality of the solution depends on the way this partitioning is done.

A solution to the dial-a-ride problem based on insertion heuristics has been proposed by many researchers (Jaw et al. 1986; Kikuchi and Rhee 1989; Kikuchi and Donelly 1992). The work of Jaw et al. was extended to include vehicle capacity constraints and was applied to a small-scale demand-response operation in Winnipeg (Alpha 1986).

The single vehicle dial-a-ride problem, although seldom occurring in practice, has been studied by a number of researchers because it occurs in the second part of cluster-first and route-second algorithms. An exact dynamic programming approach has been used to solve the single vehicle dial-a-ride problem (Psaraftis 1983). This work was extended by incorporating additional state elimination rules to reduce the computational requirements (Desrosiers et al. 1986). Other approaches to the single vehicle problem, including Benders decomposition (Sexton and Bodin 1985), have also been used. The scheduling component could be solved to optimality but the routing component defied rigorous optimization and hence space-time heuristics were used.

Other research (Solomon 1986) provides some insight into the worst case performance of well-known heuristics. An excellent overview of the algorithms and techniques used in vehicle routing and scheduling can be found in an article by Boden et al. (1983).

3 Problem Characteristics

The dial-a-ride problem (DARP) is characterized by the following:

- *Routing constraints* — each customer's origin and destination need to be visited exactly once.

- *Assignment constraints* — each customer must be assigned to a vehicle, and the number of available vehicles cannot be exceeded.

- *Timing constraints* — each customer's origin and destination can be visited only within the prescribed time windows.

- *Precedence constraints* — the same vehicle that picks up a customer at his origin should drop him at his destination.

- *Capacity constraints* — the capacity of the vehicle cannot be exceeded at any time.

- *Competing objectives* — various objectives should be minimized, including the distance traveled by all vehicles, customer inconvenience, and the number of vehicles used.

We show that DARP is NP-complete from two other NP-complete problems, viz., the Hamiltonian cycle problem and the traveling salesman problem with time windows (TSPTW). Informally, an input graph for a Hamiltonian cycle problem can be transformed in polynomial time to an input graph for TSPTW, and an input graph for TSPTW can be transformed in polynomial time to an input graph for DARP. Thus, any approach for solving DARP can be also be used to solve the Hamiltonian cycle problem, which is known to be NP-complete. The intractability of DARP justifies the development of approximate techniques.

4 Simulated Annealing

Since the dial-a-ride problem is NP-complete, it is unreasonable to expect globally optimal solutions in polynomial time in problem size. Various ad hoc methods mentioned in the literature are successful in solving realistic problems with minimal computational requirements but suffer from limitations concerning global optimality and the ability to account for different or competing objectives. Modern heuristic techniques, such as simulated annealing, tabu search, and genetic algorithms, are capable of efficiently searching the solution space for near-globally optimal solutions with respect to an appropriately defined objective function.

Our approach to the dial-a-ride problem is based on simulated annealing, a straightforward local improvement procedure with systematic randomness incorporated. The algorithm starts with an initial solution and temperature. At each step it generates a new solution that is accepted if it improves the objective; otherwise it is accepted with a probability $\exp(-\Delta C/t)$, where ΔC is the increase in the cost and t is the current annealing temperature. After a number of steps the annealing temperature is reduced and the process is repeated until the system "freezes" with a solution.

Of the modern heuristic techniques, simulated annealing appears to be most suitable for the dial-a-ride problem for the following reasons:

- The technique can easily be adopted for problems with a well-defined neighborhood structure.

- The technique has desirable theoretical convergence properties. Even though in the worst case it can take infinite amount of time to find the global optimum, given a suitable annealing schedule the algorithm has been shown to find near-globally optimal solutions in reasonable time.

- Other heuristics, especially tabu search, can easily be integrated into simulated annealing.

4.1 Cluster-First and Route-Second Strategy

The cluster-first and route-second strategy partitions the entire set of customers into clusters so that customers belonging to the same cluster are serviced together. Routes are then developed for the individual clusters. We approach the problem by using simulated annealing for clustering and a modified space-time nearest neighbor heuristic for developing the routes for the clusters. The reasons for this choice are given below.

- The most crucial decision for the DARP is the assignment of customers to available vehicles.

- Once customers have been clustered, very good heuristic algorithms, such as the Lin-Kernighan algorithm, are available for routing. Further, since the number of customers in a cluster is on the order of 10 or 20, even approximate heuristics like the space-time nearest neighbor heuristic give very good solutions with minimal computational requirements.

4.1.1 Clustering Strategy

Clustering is performed by the simulated annealing algorithm. The initial clustering is determined by a random assignment of customers to clusters. Routes for the clusters are developed using the modified space-time heuristic algorithm, and the cost of clustering is assessed.

Two operations are used to alter the clustering of customers at each stage of the simulated annealing algorithm.

- *Exchange* — Two customers are chosen randomly and their current clusters are identified; if both are in the same cluster, the customers are discarded and a new pair is generated. The cluster assignment of the customers is then exchanged, leaving the total number of clusters unchanged.

- *Swap* — A random customer and a random cluster are selected. The cluster of the customer is identified, and the customer is then swapped to the randomly generated cluster. Again, if the customer's cluster is the same as the randomly generated one, a new customer and cluster are randomly generated. The swap move has the potential to increase or decrease the total number of clusters by creating a new cluster for the chosen customer or by swapping out the only customer of a given cluster.

The neighborhood structure that results from these operations has the following desirable properties:

- It is simple.

- It is possible to generate any cluster from any other cluster by a series of transitions. This property is very important for the proper functioning of the simulated annealing algorithm.

- The exchange operation results in a smoother objective space, while the swap operation allows for the dynamic increase or decrease of the number of clusters.

At each iteration of the simulated annealing algorithm, the routing algorithm is invoked on the selected clusters. After developing routes for these clusters, the number of vehicles needed to service the developed routes is reassessed.[1] Finally, the objective function is evaluated by appropriately penalizing the following parameters of the solution: (a) the total distance traveled by all the vehicles, (b) the total disutility caused to the customers, and (c) the number of vehicles used.

If the new partitioning results in improvement, it is accepted; otherwise it is accepted with probability $\exp(-\Delta C/t)$, where ΔC is the increase in cost and t is the current temperature. The simulated annealing algorithm uses one type of transition operation (i.e., either an exchange or a swap) until a transition is rejected, at which point the alternate operation is used to generate new transitions.

4.1.2 Routing Strategy

After clusters have been generated, routing is performed by a greedy algorithm based on a space-time nearest neighbor heuristic. It starts a route by visiting the customer with the earliest pickup time. At each location the cost of visiting other locations is determined by estimating the cost of the next three succeeding moves if the location under consideration is visited. The algorithm identifies the succeeding moves by considering the space-time separation between locations and selecting the move with minimum space-time separation. The cost of a move is the summation of travel time between the two locations and the degree of violation of the time window. Of course, the costs are normalized by their associated penalties.

The implementation of this algorithm is realized by converting hard time windows to soft ones. A hard time window cannot be violated: in a given schedule, a vehicle must reach a destination within its time window, otherwise the solution is not feasible. In contrast, a soft window can be violated at a cost, and therefore may be regarded as a generalization of a hard time window. The advantages of converting hard windows to soft are as follows:

- In a real-world situation, windows are usually soft in the sense that there is a limit up to which a service provider will adhere to the time windows.

- Soft time windows are useful in evaluating the tradeoffs between service requirements and cost requirements. In addition, hard time windows can be modeled with soft ones by imposing large penalties on violated service requirements.

- Algorithms based on soft time windows are capable of finding solutions in cases where hard time windows would have failed. Thus, solutions obtained with soft windows indicate the degree of violation, allowing penalty methods to distinguish between a given pair of infeasible solutions (in attempting to find a feasible region).

---

[1]The number of vehicles needed to service a set of routes need not be same as the number of routes since a single vehicle can service routes that do not overlap in time.

4.2   Annealing Parameters

The success of the simulated annealing algorithm depends on the proper choice of the annealing parameters, viz., initial temperature, number of transitions carried out at each temperature, rate of cooling, and final temperature. In our current implementation, the initial temperature is chosen such that 90% of the proposed transitions are accepted. For a set of n customers the temperature is reduced after either $n(n-1)/2$ new transitions have been considered or n proposed transitions have been accepted, whichever occurs first. However, for problem sizes beyond 75 customers, the number of new transitions generated at each temperature is limited to $30n$ to reduce computation time. The reduction in temperature at each step is based on the current entropy of the system, for which we use the following equation (van Laarhoven 1988):

$$t_{k+1} = \frac{t_k}{1 + \frac{t_k \ln(1+\delta)}{3\sigma_k}}$$

where t_{k+1} is the new temperature, δ is a constant chosen to be 1 in the current implementation, and σ_k is the standard deviation of the cost distribution at temperature t_k. The ratio t_{k+1}/t_k is not allowed to drop below 0.8, and the stopping criterion is the occurrence of three successive temperatures that pass without an accepted transition.

To improve performance, a tabu list is used to give short-term memory to the annealing algorithm so that accepted transitions are not immediately reversed. That is, an accepted transition is retained for a fixed number of new state transitions before it is allowed to be reversed. This approach has the effect of allowing the newly accepted transitions some time to show their effectiveness.

5   Results

Although simulated annealing has been shown to converge to the global optimum, an infinite amount of computation time is required to prove optimality because of an extremely slow annealing schedule. However, less conservative (and less time-consuming) schedules have been shown to produce near-globally optimal solutions. A side-effect of any such practical annealing schedule is the variability of solutions in different runs. For a properly selected annealing schedule, however, this variability should be small.

The robustness of the present algorithm in consistently finding good optimal solutions is ascertained by running it on a problem with 25 customers with a different random seed for each run. The results are plotted as a histogram in Figure 1. The mean and standard deviation of the solutions are 421.46 and 18.38, respectively.

In the following example, over a hundred customer requests from the Winston-Salem Mobility Management National Demonstration Project are used. By varying the weights on objectives, we generate the tradeoff curve shown in Figure 2, where each circle represents a different schedule with an associated customer disutility and operation cost. The filled circles designate non-dominated solutions. Customer disutility is quantified by the amount of violation of time windows, and cost is quantified by the total travel time of all vehicles and the number of vehicles used. The tradeoff curve is interesting because it shows that no single-objective approach is adequate for this real-world data. For example, the minimum cost solution of 2675 has a disutility 188, but if one is willing to pay only 16 more cost units the customer disutility is nearly

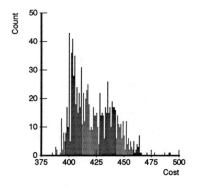

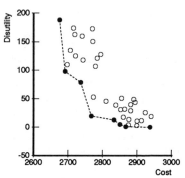

Figure 1: Histogram of SA Result   Figure 2: Cost vs. Disutility

halved. A similar argument can be made against the minimum disutility solution. Such tradeoff curves should help decision makers identify desirable operating strategies based on a realistic assessment of the various competing objectives.

6   Conclusion and Further Research

A multiobjective approach to the dial-a-ride problem allows us to investigate tradeoffs, particularly those between various measures of customer satisfaction and cost. Customer satisfaction is characterized by the degree of violation of desired time windows, and cost is characterized by the total travel time of all vehicles and the number of vehicles required.

We prove that the dial-a-ride problem is NP-complete, and therefore conclude that any practical approach to the problem must resort to heuristic techniques. The proposed approach, based on simulated annealing, gives near-globally optimal solutions, is robust, requires minimal user input in fine tuning the annealing parameters, and runs in time polynomial in input size.

There are many potential directions for subsequent study, including experimentation with the generated schedules in practice, extensions to combined demand-response and fixed-route systems, and an analysis of the effect of "no-shows" on schedules generated with various objectives. Because some paratransit operations have regular subscription trips, and there is some cost in running what may be completely different schedules each day, an interesting follow-on study would examine the reduction in cost obtained by rescheduling subscription trips daily as opposed to fixing subscription routes and adding non-subscription trips.

Acknowledgments

This research is supported in part by the Federal Transit Administration, the North Carolina Department of Transportation, and the Department of Civil Engineering at North Carolina State University.

References

Attahiru Sule Alfa (1986). "Scheduling of Vehicles for the Transportation of Elderly." *Transportation Planning and Technology*, Vol. 11, 203–212.

L. Bodin, B. Golden, A. Assad and M. Ball (1983). "The State of the Art in the Routing and Scheduling of Vehicles and Crews." *Computers and Operations Research*, Vol. 10, 63–211.

Jacques Desrosiers, Yvan Dumas and Francois Soumis (1986). "A Dynamic Programming Solution of the Large Scale Single Vehicle Dial-a-Ride Problem with Time Windows." *American Journal of Mathematical and Management Sciences*, Vol. 6, Nos. 3 & 4, 301–325.

Jang-Jei Jaw, Amedeo R. Odoni, Harilaos N. Psaraftis and Nigel H.M. Wilson (1986). "A Heuristic Algorithm for the Multi-Vehicle Advance Request Dial-a-Ride Problem with Time Windows." *Transportation Research-B*, Vol. 20B, No. 3, 243–257.

Shinya Kikuchi and Jong-Ho Rhee (1989). "Scheduling Method for Demand Responsive Transportation System." *Journal of Transportation Science*, Vol. 115, No. 6, 630–645.

Shinya Kikuchi and Robert A. Donnelly (1992). "Scheduling Demand-Responsive Transportation Vehicles using Fuzzy-Set Theory." *Journal of Transportation Science*, Vol. 118, No. 3, 391–409.

P.J.M. van Laarhoven (1988). *Theoretical and Computational Aspects of Simulated Annealing.* CWI Tract 51, Center for Mathematics and Computer Science, Amsterdam.

Harilaos N. Psaraftis (1983). "An Exact Algorithm for Single Vehicle Many-to-Many Dial-a-Ride Problem with Time Windows." *Transportation Science*, Vol. 17, No. 3, 351–357.

Thomas R. Sexton and Lawrence D. Bodin (1985). "Optimizing Single Vehicle Many-to-Many Operations with Desired Delivery Times: Scheduling and Routing." *Transportation Science*, Vol. 19, No. 4, 378–435.

Marius M. Solomon (1986). "On the Worst-Case Performance of Some Heuristics for the Vehicle Routing and Scheduling Problem with Time Windows." *Networks*, Vol. 16, 161–174.

Optimizing Roadway Design for Congestion Management: A Simulated Annealing Approach

Nicholas S. Flann[1]   John T. Taber[2]   William J. Grenney[2]

Abstract

With highway traffic increasing at the average rate of 2-3% per year, roadways are becoming increasingly congested. A major provision of recent surface transportation legislation was a mandate that states and local areas develop congestion management systems that are targeted at reducing roadway congestion. Since much of roadway congestion occurs at intersections, a system for designing cost-effective capital improvements at intersections will form a significant component of an overall congestion management system.

Designing *optimal* and *cost-effective* capacity improvements for at-grade intersections is a difficult and time-consuming task. Choosing appropriate values for parameters such as new lane-configurations and signal phasing and timing is difficult because of the vast number of legal combinations (approximately 10^{15}) that must be considered. Current methods rely on experienced engineers to suggest reasonable designs, which are often iteratively improved through the use of computer models. Such approaches require hours of an engineer's time and usually produce sub-optimal designs because only a few of the many possible alternatives are considered.

This paper introduces an intelligent computer assistant that automatically searches the space of possible capacity improvements for an intersection. Each combination of improvements (such as new lane construction or reallocation of lanes) is evaluated as to its cost and reduction in congestion (by optimizing for signal phasing and timing). Over 500 alternative designs are often considered for a single intersection. A rapid computer-based analysis of these alternatives produces a set of quantified and ranked projects, from which the engineer can choose the most cost effective design. The method *simulated annealing* is employed to search this space efficiently.

The computer tool is illustrated designing capacity improvement projects for two intersections in Utah, given current intersection geometry and projected movement volumes. In both cases, new designs where found that were less expensive and provided almost equivalent reduction in delay as those identified by engineers.

Key Words: Congestion management, Simulated annealing, combinatorial optimization, traffic signal timing, intersection design, capacity improvements, *IVHS*.

[1]Computer Science Department, Utah State University, Logan, UT 84322

[2]ASCE, Member, Civil Engineering Department, Utah State University, Logan, UT 84322

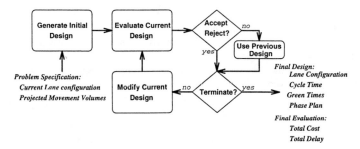

Figure 1: The design process employed to produce a cost-effective design for intersection capacity improvement.

Introduction

The current process used to identify a cost-effective capacity improvement design for an intersection is illustrated in Figure 1. This process is usually a cooperative one between an engineer and various computer analysis tools including expert systems. First, the engineer produces an initial design given the current lane configuration of the intersection and the projected traffic volumes. Here, the engineer chooses values for the design parameters, including a new lane configuration and an operational phasing and timing plan. The current design is then evaluated by employing some sort of highway intersection analysis software (such as the Highway Capacity Manual *HCM*, TRANSYT-7F or PASSER). Such an evaluation will determine the overall delay of vehicles through the intersection with the given capacity improvements and new signaling design. In addition, the cost of the new improvements will be calculated to determine if the improvements are cost effective.

Next, some modification is made (such as changing the lane configuration, timing or phase plan) and the new design is evaluated. Expert systems can assist the engineer at this stage by recommending certain design modifications. The engineer then enters an interactive process involving modification, evaluation and accept/reject stages. If the most recent change improves the design (by reducing its delay or reducing its cost), the new design is usually accepted, otherwise the new design is rejected and the previous unmodified design retained. The process of modification, evaluation then accept/reject now repeats while the engineer attempts to improve the design. The process terminates when an acceptable design has been achieved, usually based upon some pre-determined acceptable level of service and cost.

This process is time consuming and depends critically on the intuition and experience of traffic engineers. More importantly, the process is very likely to lead to a sub-optimal design because of the vast number of possible designs to consider. Simple counting arguments show that there are at least 10^{15} possible *legal* designs. Moreover, engineers usually employ a "greedy" strategy in developing designs, by only considering modifications that improve the current design. This strategy is very likely to lead to *locally optimal* designs.

This paper explores the use of an alternative technique that overcomes these limitations, while automating all the stages of design process given in Figure 1. It does away with the need for complex expert systems to generate initial designs or suggest modifications. Rather, the proposed method relies on an accurate and efficient generation and evaluation of candidate designs.

Previous Work

Expert decision support systems have been developed or suggested to assist engineers in the design of roadway intersections. In [10] an expert system was developed for determining cycle length and phase timing based on *HCM* delay formulas. Solutions were suggested from a set of heuristic rules. In [8] a knowledge-based system was developed that takes input of intersection geometry and movement volumes and develops a phase and timing plan. Rules for optimizing phase types and timings were based on interviews with traffic engineering experts and evaluated using procedures from the 1985 Highway Capacity Manual. In [9] a rule-based expert system was developed for suggesting low-cost capacity improvements for intersections. In [3] an expert system was developed to select traffic lane configurations using simplified item signalized intersection planning methodology rules.

Recently, direct optimization methods have been explored to set traffic signal green times dynamically [4]. A simplified simulation model was used to evaluate the alternative signaling designs and determine the overall average vehicle delay. A genetic algorithm—a method closely related to simulated annealing—was used to search the space of alternative signal designs. While this work found near-optimal signaling plans for day-to-day operation, it did not consider lane configuration alternatives for planning and congestion management. Indeed, it would be difficult to extend genetic algorithms to consider lane geometries because genetic algorithms represent design alternatives as strings of binary numbers (a 1 or 0). In [5] simulated annealing was applied to optimize signal phase sequencing and timing using TRANSYT-7F as its evaluation model. The technique successfully found near-optimal solutions for a 12-intersection corridor. Again, no consideration was given to alternative lane configurations.

Methodology

Simulated annealing is a method for solving combinatorial optimization problems that has been successfully applied to the physical design of computers [7], tours for a traveling sales person (TSP), scheduling bus routes and interpreting seismic waveforms, to name a few of many applications. One of the principal advantages of simulated annealing is that it can find near-optimal solutions to problems with very large numbers of possible solutions (for example, a 100 city TSP has over 10^{157} possible solutions). Another advantage is that little expert knowledge of the problem domain is needed. The principal information required is an efficient and accurate procedure for *evaluating* candidate solutions. For designing roadway intersection this procedure is readily available in the *HCM* manual or in *Traf-NetSim*, which contain procedures for computing the overall average vehicle delay, given a well specified design that includes the phase plan, green-times, cycle-time, lane configuration, additional factors and the volumes for each movement.

Simulated annealing is a variation of the iterative improvement method illustrated in Figure 1 that automates each stage. Below we describe in detail each step of the simulated annealing method:

Initial Design To produce an initial design, simulated annealing randomly generates a legal lane configuration, phase plan with length between 2 and 8 phases, a cycle-time between 60s and 120s and an assignment of green-times to each phase, such that each green-time has a minimum of 6s and the sum of the green-times plus the all-red-times is equal to the cycle-time. To produce a legal, yet random lane configuration a grammar is employed as a finite state machine that ensures that no lane crosses the path of an adjacent lane. Additional constraints are applied to limit the number of left, through and right lanes to reasonable values. To produce a random yet legal phase plan, the system exploits a transition table that gives all the legal transitions between each pair of phases. For a phase plan to be legal, each transition must be legal between all adjacent phases and each movement must have at least one permitted or protected phase.

Evaluation There are two criteria used to evaluate a design, its cost and the overall vehicle delay for the given movement volumes. The cost is evaluated by determining the construction changes needed to create the current lane configuration from the initial configuration, then summing a standard cost for each construction change. The overall delay time is determined by applying the procedure given in *Chap 9 Section III, Procedures for Application* of *HCM 1985*. This module returns a single numerical value giving the average delay/vehicle for the whole intersection.

Modification For simulated annealing to be effective, there are few constraints on the modification operation. The two key characteristics is that the operation be reversible and make only a small change. Six different modifications are currently performed. One is concerned with modifying the lane configuration, the other 5 modify the phase plan and timing. In all cases, the resulting modified solution is checked for legality. If the modification was found to generate an illegal solution, it is rejected and another attempted.

1. **Change a lane configuration.** A direction is chosen randomly and the current lane configuration is changed by the addition or reassignment of a lane.

2. **Change a green-time.** A phase is randomly selected and a random time generated between −10s and 10s. This time is then added to the time for that phase.

3. **Move some green-time.** Two distinct phases are randomly selected and a random time is generated between −10s and 10s. This time is added to the

time for one phase and subtracted from the time for the other phase.

4. **Substitute a phase.** A phase is randomly selected from the phase plan and replaced by a different phase.

5. **Insert a phase.** A position in the phase plan is randomly selected and a new phase is randomly generated and inserted into the phase plan at that position. A new random time between $6s$ and $15s$ is assigned to this new phase.

6. **Delete a phase.** A phase is randomly selected from the phase plan and deleted.

Accept? All modifications that reduce the overall delay time for the design are accepted. In this way the method makes progress towards a near-optimal solution. Since there is a danger of *only* accepting modifications that improve the design—allowing the system to get trapped in a local minimum—simulated annealing accepts modification that *increase* the delay non-deterministicly, with a probability dependent upon how many modifications the method has attempted. Early in the search, almost all "bad" modifications are accepted, but as the search progresses, the probability of accepting bad moves is gradually reduced to 0, when the method reduces to greedy search. It has been proved that this technique avoids all local minima and converges on the globally best solution if enough modification attempts are made [6].

Terminate? The method terminates and outputs the current best design when a given number of lane configurations have been considered. For the examples given here, approximately 500 designs were considered for a given intersection.

The method searches two distinct solution spaces to identify cost-effective capacity improvements. First, the method searches the space of lane configurations, controlled through the lane modification operator. Second, the method searches the space of signal phasing and timing plans. To accurately evaluate each alternative lane configuration, the best overall delay must be found by identifying a near-optimal signal timing plan. Hence, lane configurations must be searched at a slower rate than phasing and timing plans. Currently, one modification to the current lane configuration is made for every 20,000 phase and timing modifications. When a lane modification is made, the count of modifications employed in the **accept?** process is reset to 0 to ensure that a near-optimal signaling plan can be identified. Lane modifications are accepted based on the near-optimal signaling plan identified for the current lane configuration.

Evaluation of Approach

To evaluate the effectiveness of the simulated annealing method, a benchmark comparison was made of intersection designs and signal timing plans generated by the simulated annealing method versus those developed by experience engineers. Two actual design cases are evaluated:

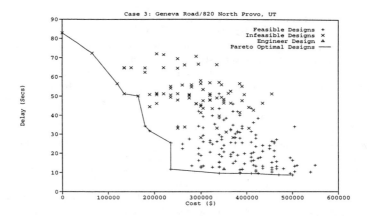

Figure 2: The space of lane configurations searched by simulated annealing for case 3 intersection. Each lane configuration considered is illustrated as a point on the graph. Pareto optimal designs can only be improved by spending more money.

The first case involves the intersection of a 2-lane state highway with a local 2-lane roadway. Recent development proposals project significantly higher traffic volumes on both roadways. Based on manual capacity analysis, it has been proposed that the State highway be expanded to a 5-lane highway (intersections would have 1 exclusive LT lane, 1 exclusive thru lane, 1 shared thru-right). The local roadway is proposed to have the westbound approach expanded to include 1 LT lane, 1 thru lane, and 1 RT lane. This design results in a delay of 12.7 seconds/vehicle and an estimated cost of $355,000.

The results of applying simulated annealing are shown in Figure 2. Notice that over 150 alternative designs were considered. Out of these, eight feasible pareto optimal designs are identified, which represent the choices offered to the engineer. The most cost effective design is clearly the one that yields an improved delay time of 11.2 seconds/vehicle with a savings of over $100,000 in construction costs compared to the original engineer's choice.

The second intersection considered is of two state highways adjacent to several shopping mall developments. Engineers proposed improvements to include in each direction: 2 LT lanes, 3 thru lanes, and 1 RT lane. The new level of service operates at overall average delay of 9.0 seconds/vehicle. The estimated cost of this improvement is approximately $800,000.

Projected improvements for this case using simulated annealing techniques produce a set of 10 pareto optimal designs that are less expensive, but with a slightly higher delay. The chosen design had a delay time of 11.0 seconds/vehicle with a savings of over $300,000 in construction costs.

Summary and Conclusions

A new method has been presented for generating and evaluating at-grade intersection improvements for congestion management and planning systems. Evaluation of designs includes cost, using a standard cost model, and delay, based on a near-optimal signal timing plan. The methodology is based on iterative improvement where hundreds of thousands of candidate designs are accurately evaluated at a rate of thousands per minute.

The results of the preliminary experiments described above suggest that simulated annealing can offer significant advantages over the current process for determining cost-effective capacity improvement projects for intersections. Current manual and computer assisted techniques do not ensure optimal results because they only consider a very small fraction of the possible alternatives in the search space. Moreover, current methods perform greedy searches thereby risking getting stuck in local minima. Sub-optimal decisions lead to poor utilization of road resources, unnecessary road construction costs, additional traffic delay and worsening air quality.

The simulated annealing process provides a set of qualified and ranked pareto optimal designs to choose from, thereby enabling the engineer to make more informed choices. The identification of pareto optimal designs ensures that the engineer only chooses cost-effective solutions (i.e., designs that can only be improved by spending more money). Moreover, by identifying all the reasonable alternatives, the computer system provides an clear justification for the engineer's final choice. Finally, the computer approach ensures that the engineer's time is more productively spent on considering reasonable alternatives, rather than having to generate and test many alternatives.

Future Research

The research contained in this paper was performed as part of a demonstration project of advanced machine intelligence and as part of continuing research in rural and urban highway planning and design. The success of this methodology indicates the need for expanding the research scope of applying simulated annealing methods to the following areas:

- Extend the use of optimization across networks of intersections;
- Incorporate safety and air pollution into the objective function;
- Evaluate alternative objective functions for calculating delay;
- Integrate adaptive controllers to better estimate delay;
- Use customized cost models that employ actual property values.

Acknowledgements

This research has in part been supported by continuing research in expert systems and artificial intelligence for rural highway design supported by a grant from the Mountain Plains Consortium of the University Transportation Centers program. Additional support is provided to Flann by the College of Science.

References

[1]   *Highway Capacity Manual*, TRB Special Report 209, Transportation Research Board, Washington, D.C., 1985.

[2]   Bielli,M., Ambrosino,G., Boero,M., and Mastretta,M. (1991). Artificial Intelligence Techniques for Urban Traffic Control, *Transportation Research A, Vol. 25A*, Pergammon Press, Great Britain, 1991, pg. 319–325.

[3]   Bryson,Jr.,D. and Stone,J., Intersection Advisor: An Expert System for Intersection Design, *Transportation Research Record 1145*, Transportation Research Board, Washington,D.C., 1987, pg.48-53.

[4]   Foy, M. D., Benekohal, R. F. & D. E. Goldberg., Signal Timing Determination using Genetic Algorithms, in *Transportation Research Record 1365*, Transportation Research Board, Washington, D.C. pg. 108–115.

[5]   Hadi, M. A. and Wallace, C. E. Optimization of signal phasing and timing using cauchy simulated annealing, Annual meeting of Transportation Research Board, Washington D.C. Jan. 1994.

[6]   Ingber L. & Rosen B. (1992). Genetic algorithms and very fast simulated reannealing: A comparison. In *Mathematical Computer Modeling*, Vol 16, No. 11, pp. 87–100.

[7]   Kirkpatrick S., Gelatt C. D. and Vecchi M. P. (1983). Optimization by simulated annealing, *Science, Vol 220, pp. 671–680, May 1993*.

[8]   Linkenheld,J., Benekohal,R., and Garrett,Jr.,J., Knowledge-Based System for Design of Signalized Intersections, *Journal of Transportation Engineering, Vol. 118, No. 2*, American Society of Civil Engineers, New York, 1992, pg. 241-257.

[9]   Morris,M. and Potgieter,L., Expert System for Aspects of the TSM Process, *Transportation Research Record 1280*, Transportation Research Board, Washington, D.C., pg 104-110.

[10]   Zozoya-Gorostiza,C. and Hendrickson,C. (1987). Expert System for Traffic Signal Setting Assistance, *Journal of Transportation Engineering*, American Society of Civil Engineers, New York, 1987, pg. 108-125.

The Role of Function–Behavior–Structure Models in Design

John S Gero[1]

Abstract

Function-behavior-structure (FBS) models have been developed as a powerful means of articulating a framework for design processes. This paper describes their role in design research and presents a recent application of their use in protocol studies of designers.

Introduction

Designing is concerned with the generation of design descriptions of potential artefacts in response to statements about the qualities to be exhibited by those artefacts (Coyne et al., 1990). These qualities are requirements, which arise from the client's and designer's intentions as well as those which appear as designing proceeds, and relate to functionality, performance, aesthetics, etc. How well the artefact satisfies these requirements can only be completely ascertained after the artefact's embodiment in the physical and social world. However, a prediction of the suitability of the design is possible if we can find a means of interpreting the design description in light of the requirements. Note we will use the word 'designing' for the act of producing a design and the word 'design' for the resultant.

Central to designing are mental models. These models capture the essence of the artefacts being designed and provide a repository of concepts which designers can appeal to for assistance. They provide a basis for design as designers can retrieve the models and utilise the knowledge encapsulated in them as they become relevant in a design process. This process is synonymous with the designer recalling and selecting relevant concepts that the designer uses during the process of developing a design. In routine design these models are sufficient to produce a design; the models act as generalised templates from which specific designs are derived. Retrieval and selection of suitable models, extraction of relevant information contained in the selected models, and finding suitable values for the variables are the main issues in operating these models in routine design. In non-routine design additional processes need to be utilised

[1] Professor of Design Science, Co-Director, Key Centre of Design Computing, Department of Architectural and Design Science, University of Sydney, NSW 2006, Australia, john@arch.su.edu.au

to modify the variables contained within these models in order to aid in the production of 'creative' designs.

Function-Behavior-Structure Models as Basis of a Framework Model

The use of the computer in engineering design has brought with it the need to organise engineering design in computable process terms. Whilst there are numerous individual processes it is appropriate and useful to develop a framework model for designs and designing. A useful characterisation of designing is through a framework model built around function, behavior and structure (Gero 1987). *Function* describes intended purposes and provides the teleology of the intended and resultant artefact. Function is a societal concept. Although, it appears counter-intuitive function is ascribed to designs rather than inhering in them, Figures 1 and 2.

Figure 1. Fixed structure used in function ascription (Finke 1990).

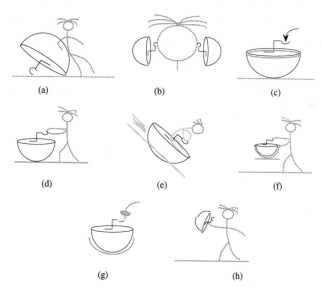

Figure 2. Various functions which could be ascribed to the structure in Figure 1, such as (a) lawn lounge (furniture), (b) global earrings (jewellery), (c) water weigher (scientific instruments), (d) portable agitator (appliance), (e) snow sled (transportation), (f) rotating masher (tools and utensils), (g) top or spinner (toys and games) and (h) slasher basher (weapons) (Finke 1990).

Behavior is derivable from structure and is the socio-technical means of articulating and achieving functions. *Structure* is the components and their relationships which go to make up the design, i.e. structure specifies what the artefact is composed of and how the components are connected. Figure 3 shows the function, behavior and structure subspaces and the mappings between them which constitute the space of designs.

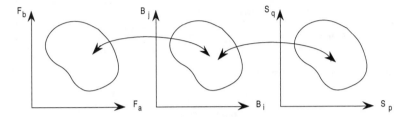

Figure 3. The three subspaces of function, *F*, behaviour, *B*, and structure, *S*, which constitute the state space of designs

Whilst there are transformations which map function to behavior and vice-versa and structure to behavior and vice-versa, there are no transformations which map function to structure directly. This is a version of the no-function-in-structure principle (de Kleer and Brown 1984; Gero 1990) where the teleology of an artefact is not found in its structure but is a contextual interpretation of its behavior. The corollary: no-structure-in-function also holds. This may, at first glance, also be counter-intuitive. The reason is that in human experience once a phenomenological connection between function and structure is made it is hard to unmake it. Thus, it appears that there is some sort of causal transformation between function and structure rather than a phenomenological one. The transformation which maps structure to behavior is a causal one.

These concepts of function, behavior and structure are sufficient to construct a framework model of design as a process, Figure 4. This model separates the behaviors expected of a design, B_e, from the actual behaviors, B_s, derived from a simulation of the design. For completeness, the design description, D, is included in this model.

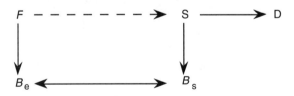

Figure 4. Framework model of design as a process.

where B_e = set of expected behaviors
B_s = set of actual behaviors
D = design description
F = set of functions
S = structure
—> = transformation
---> = occasional transformation
<—> = comparison

This framework model is sufficient to allow us to describe the primary processes which occur during designing:

formulation
synthesis
analysis
evaluation
reformulation
production of design description

These are described graphically in Figures 5 (a) to (f) (Gero 1990).

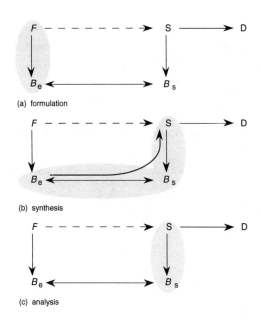

(a) formulation

(b) synthesis

(c) analysis

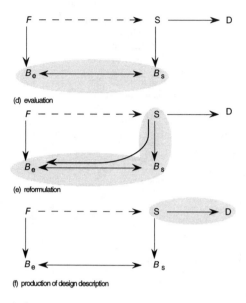

Figure 5. The generic processes which occur during designing mapped onto the
framework model (after Gero 1990).

Such a framework serves as a useful basis for design research since it provides a comprehensive foundation for design processes without requiring that the processes be uniquely specified (Tham 1991).

Function-Behavior-Structure Models as Basis of Protocol Models

Protocol studies are a means of extracting data from verbal utterances during intellectual activity (Ericsson and Simon 1984). They are beginning to be applied in design research as a means of developing an understanding of how human designers carry out the tasks of designing and as a foundation for computer-based design aids. A typical protocol study has the following activities:

(i)   taping a session
(ii)   transcribing the verbal utterances which occurred during that session,
(iii)   developing a coding scheme
(iv)   coding the transcript
(v)   analysing the result

In their use in design research protocol studies have been modified by the addition of an externally derived structure onto the coding development. The coding scheme we have developed (Purcell at al. 1994) is far richer than has been used in previous work. The danger in increasing the number of coding categories is that the results become too complex, potentially masking relationships and patterns in the data and can become too difficult to comprehend. Whilst a rich coding potentially presents these problems we have attempted to counter them through the use of the function-behavior-structure abstraction along with the structure that is present in the coding categories. The structure adopted segments the information into the following categories:

(i)   problem domain abstraction levels
(ii)   function-behavior-structure (F B S)
(iii)   analysis and evaluation
(iv)   synthesis micro-strategies
(v)   designing macro-strategies

Figure 6 shows the result of one such protocol analysis of a design session.

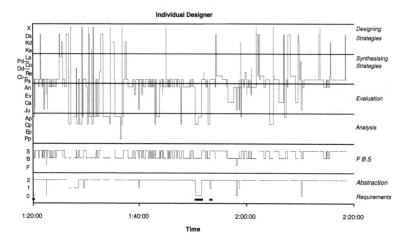

Figure 6. Result of a protocol analysis of a design session (after Purcell et al. 1994).

The use of the function-behavior-structure model allows an abstract representation to occur independent of the problem. Further it provides information not otherwise available, Figure 7.

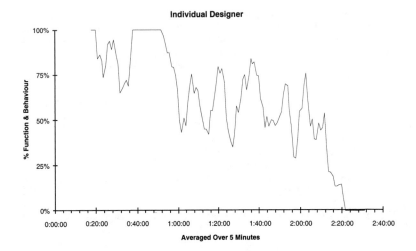

Figure 7. Percentage of time the designer spends on reasoning about function and behavior as opposed to structure (after Purcell et al. 1994).

Zero percent implies all the reasoning is about structure in that time interval, 50 percent implies an equal distribution of time, whilst 100 percent implies all the reasoning is about function and behavior and none about structure. Information derived from these framework models is valuable in helping in the development of theories of design.

Acknowledgments

This work has benefited significantly from discussions with members of the Key Centre of Design Computing, in particular Mike Rosenman, Mary Lou Maher and Kwok Wai Tham.

References

Coyne, R. D., Rosenman, M. A., Radford, A. D., Balachandran, M. and Gero, J. S. (1990) *Knowledge-Based Design Systems,* Addison Wesley, Reading.

de Kleer, J. and Brown, J. S. (1984) A qualitative physics based on confluences, *Artificial Intelligence* **24**: 411-436.

Ericsson, K. A. and Simon, H. A. (1984) *Protocol Analysis: Verbal Reports as Data,* MIT Press, Cambridge.

Finke, R. (1990) *Creative Imagery,* Lawrence Erlbaum, Hillsdale, NJ.

Gero, J. S. (1987) Prototypes: A new schema for knowledge-based design, *Working Paper,* Architectural Computing Unit, University of Sydney, Sydney.

Gero, J. S. (1990) Design prototypes: a knowledge representation schema for design, *AI Magazine* **11**(4): 26-36.

Purcell, A. T., Gero, J. S. and Edwards, H. (1994) The data in design protocols: the issue of data coding, data analysis in the development of models of the design process, *Preprints Analysing Design Activity—The Delft Protocols Workshop,* Technical University of Delft, Delft, pp. 169-187.

Tham, K. W. (1991) *A Model of Routine Design Using Design Prototypes,* Doctor of Philosophy dissertation (unpublished), University of Sydney, Sydney.

An Information Model for the Building Envelope

Hugues Rivard[1] , Claude Bédard[2], Kinh H. Ha[3], and Paul Fazio[4] M. ASCE.

Abstract

This paper presents an information model that enables the integration of the building envelope design process at the preliminary stage. Such an integrated design system facilitates data exchange between design tasks, improves communication between designers and supports the growth of data as the design process unfolds. The paper shows how the envelope design data can be organized into entities and relationships. These entities can then be decomposed into cohesive sets of data called primitives using the Primitive-Composite Approach. The paper concludes with a discussion of the advantages of storing data related to function and behaviour in addition to form in the context of building envelope design.

1. Introduction

The lack of information and communication among designers, during the building envelope design process, results in sub-optimal solutions leading to inadequate performance. To reverse this trend and to achieve better performance of the building envelope, professionals must be provided with computer integrated design tools which facilitate communications among participants and improve the transfer of data throughout the design process.

This paper presents an information model of the building envelope that can facilitate integration throughout the design process. The goal of such a model is to help analyze and organise the wealth of data used and generated during the process in an efficient manner. Such a model can be considered as a blueprint or plan which is used later for implementation of the integrated system. An information model developed for a design process can provide a logical representation to be shared by several users, thus facilitating the exchange of data.

The model consists primarily of building envelope entities and their relationships. Each envelope entity is decomposed into primitives which represent small cohesive sets of data. These primitives can be used like building blocks to form the basis of data integration. They represent specific views and concepts of an entity.

---

[1]Graduate student, Civil Engineering Dept., Carnegie Mellon University, Pittsburgh, PA.
[2]Associate Professor, [3]Professor, [4]Director, Centre for Building Studies, Concordia U., CANADA.

Finally, to improve the communication of design intent along with design results, primitives are added to store data related to function and behaviour. This is necessary to communicate the purpose of an object and its corresponding states.

2. Building Envelope Entities and Relationships

The data required during the design process is grouped by envelope entities. This method provides a structured data abstraction which can be readily understood by design participants. This section presents the entities and relationships that are necessary to describe the building envelope.

2.1. Building Envelope Entities

The building envelope can be viewed as an assembly of some basic entities. The building is enclosed by a number of planes which are called envelope planes. These planes are joined together at their edges through plane connections. Each plane can be subdivided into areas corresponding to different indoor spaces and envelope sections, and each plane can also be pierced by openings. The building envelope entities are depicted in Figure 1 and are explained below.

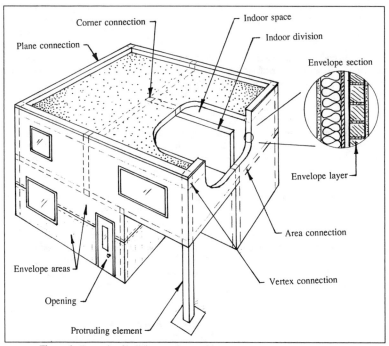

Figure 1. Example of building envelope entities.

The building volume is confined by a number of planes which are called **envelope planes**. Examples of envelope planes are: roofs, exterior walls, slab on grade and cantilevered floors.

Envelope surfaces which are curved with single or double curvature are a special case of envelope planes (e.g. conical roof and domes). Such surfaces can be decomposed into discrete envelope planes.

An envelope plane is partitioned into a number of non-overlapping **envelope areas**. An envelope area has only one type of envelope section and it corresponds to only one indoor space. Envelope areas are necessary to define specific regions of the plane which have different section characteristics and which are subject to distinct environmental indoor conditions. There is a need to partition the envelope plane into discrete areas corresponding to the different indoor spaces because the performance of a given envelope section depends on the indoor conditions (temperature and relative humidity) which can vary widely from one indoor space to another.

An **envelope section** is a sequence of construction products (or envelope layers) that is exposed when an envelope area is intersected by a cutting plane. A typical envelope section can be used in more than one envelope area.

An **envelope layer** corresponds to one construction product within an envelope section. Envelope layers may be classified according to their functions: cladding, coating, membrane, structure, panel, insulation, finishing and other. Examples of envelope layers include single-wythe brick veneer, expanded polystyrene, concrete blocks, polyethylene, gypsum board and oil base paint.

An **opening** is an element that pierces an envelope plane to allow light, views, access and ventilation. Examples of openings are windows, doors, skylights, HVAC exhaust and intake louvers. Openings are attached to envelope planes because an opening can span more than one envelope area. For instance, the door in Figure 1 is located in two envelope areas.

Each edge of a given plane is joined to another envelope plane through a **plane connection** entity. This entity represents the detailing at the junction of two planes. It may take several forms; the connection for two exterior walls may be a corner while the connection between an exterior wall and a roof may be a parapet or a cornice. Further detailing is required at the junction of three or more planes and is provided by **vertex connection** entities.

An opening is connected to an envelope plane through an entity **opening connection**. This entity represents the detailing at the junction between the plane and the opening on all its sides.

Envelope areas are connected at their edges with the entity **area connection**. This entity represents the detailing at the junction between two areas having different sections. Connection to floors and interior walls are also included in this type of connection since areas are delimited by the indoor spaces they enclose. Entity **corner connections** allow detailing at the junction of three or more envelope areas. These connections always occur within an envelope plane.

The volume of a building is divided into several indoor spaces by indoor divisions. An **indoor space** is characterised by an occupancy and environmental conditions to achieve a user-defined level of comfort. The volume of an indoor space is determined by enclosing envelope areas and indoor divisions. Examples of **indoor divisions** are floors and interior partitions.

The **building** entity includes all of the previous entities. It is a root entity which refers directly or indirectly to all the envelope entities. Furthermore, it characterises the building project being designed with general information such as address, owner and designers. The **city** entity defines the outdoor environment in which the building is located with characteristics that can affect the design of the envelope such as outdoor design temperature, relative humidity, wind pressures and seismic data.

A **protruding** entity is an element that extends from the building envelope, such as external columns and balconies. These entities are tied to the building through their connection with the envelope. They may be attached to either an area, a plane, a vertex or an opening connection entity. These entities often act as fins and dissipate significant amount of heat in cold climates.

2.2. Relationship Between Envelope Entities

The envelope entities have specific relationships with each other. Figure 2 shows the envelope entities and their relationships.

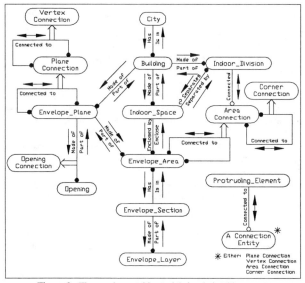

Figure 2. The envelope entities and their relationships.

A relationship defines an interaction between two entities and it is represented as a link between them. Each relationship is annotated with two notes describing the type of link. The notes have to be read in the direction of their arrows. Multiplicity, at the end of a link, designates how many entities of a given type may relate to a single associated entity at the other end of the link [Rumbaugh et al. 91]. Multiplicity is specified with the following symbols at the end of a link:

A black circle is for many, meaning zero or more.

An empty circle is for optional, meaning zero or one.

No symbol is for one-to-one relationship.

An integer indicates exact multiplicity.

Some relationships (the one representing connectivity) have special properties stored in an entity. This subsection describes the relationship between entities.

Many building entities can be located in a city entity. A building is made of several envelope planes, indoors spaces and indoor divisions. Envelope planes are connected to each other by plane connections and vertex connections. Envelope planes are themselves made up of openings and envelope areas. The opening connection entity could be tied to or integrated with the opening entity. An indoor division partitions two indoor spaces. Every envelope area has an envelope section and corresponds to one indoor space. The area connection entity connects two envelope areas and sometimes an indoor division. An envelope section is made of several envelope layers. The relationship between the protruding entity and the connection entities is shown at the bottom-right of the figure to avoid cluttering. A protruding entity may be tied to either an area, a plane, a vertex or an opening connection entity.

3. Decomposing the Envelope Entities Into Primitives

The entities presented in the previous section represent a logical manner for organizing the building envelope data and correspond to how designers see the envelope. These envelope entities can be stored as a set of attributes and relations in a database. To avoid the complexity of entities encompassing all possible representations and to support data integration, the entities are decomposed into primitive entities representing atomic concepts that are cohesive.

The Primitive-Composite Approach (or P-C Approach), developed by Phan and Howard (1993), is a data model and a structured methodology for modelling facility engineering processes and data to achieve integration. This approach addresses the problem of multiple views, schema evolution, data integration and the representation of form, function and behaviour. It is used here to decompose the envelope entities into primitives because it is flexible, extensible, and because the result can be customized. It is an approach which has much potential for integration over the entire building life-cycle. The primitives derived for a sub-system, such as structure or building envelope, can be merged with primitives derived from other fields of the building domain, thus augmenting the primitive schema. This approach allows different specialists to develop primitives in their own field of expertise and to merge them together in an integrated data model.

This section shows how the building envelope entities defined previously are broken down into primitive entities using the Domain Entities AnaLysis method (or DEAL) which is part of the Primitive-Composite approach. This method sets guidelines and procedures for decomposing entities of a given domain using a top-down approach into increasingly cohesive and reusable components. The final components represent the primitive entities of the domain [Phan and Howard 93]. A primitive entity is directly representable by an object within an object-oriented database system.

Cohesion is the only criterion used in decomposing the envelope entities and it is defined as a measurement that shows how closely the attributes of an entity relate to one another [Phan and Howard 93]. The general criterion of cohesion can be characterized into five specific criteria: the data attributes are stored in one location (access-cohesive), related to the same concept (concept-cohesive), not derived from each other (source-cohesive), instantiated and used at the same time (time- and use-cohesive). A functional and data flow analysis of the building envelope design process is needed to evaluate the cohesion of each envelope entity. This analysis was performed to identify the data required and the data generated throughout the design process and is presented in Rivard et al. (1994).

Figure 3 shows the decomposition of the entity envelope section into six cohesive primitives. Each decomposition is justified with an annotated arrow and a code on the resulting primitive. The annotated arrows show which criterion is used to decompose a leaf and the codes refer to corresponding items of the legend. The final primitives are shaded in grey and their data attributes are listed below.

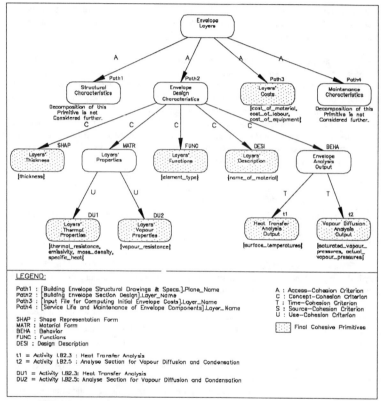

Figure 3. Decomposition of the envelope layer entity.

4. Form, Function and Behaviour

The primitives defining an entity can be grouped into three broad sets of data: form, function and behaviour. The form representation of an entity describes its physical characteristics, its actual design. This is the only representation found in typical building plans and specifications. The function representation of an entity includes its purpose and the requirements that have to be satisfied to realise that purpose. The behaviour representation of an entity contains its response to stimulations associated with design conditions. These three sets of data are necessary to completely define an entity in terms of what it is, what it is intended for and how it responds throughout working life.

The envelope layer decomposition shown in Figure 3 includes primitives about form (thickness, name, properties and cost), function (i.e. cladding, insulation, membrane) and behaviour (temperature and vapour pressures). The envelope requirements, such as minimum thermal resistance and maximum thickness, are defined at the envelope section level. Two sets of behaviour conditions are typically defined for each building envelope element as these are designed to operate under two extreme environmental conditions: cold and dry winter, hot and humid summer, as specified by codes such as ASHRAE (1989) and NBCC (1990).

Typically, only the form aspect is retained in construction technical documents (drawings and specs) whereas the information regarding function and behaviour is lost. Several advantages are gained by storing all three sets of data together in a common data repository. It can improve communication by allowing the purpose and the behaviour to be passed between designers along with the design description. It allows the investigation of what-if analysis, the reuse of design solutions in other projects and tutoring (necessary for case-based design). It can provide answers to requests about the intent or the status of an entity [Phan and Howard 93]. It finally represents a richer reference repository since the reason why a particular form was selected can be understood and justified from its function (requirements) and behaviour.

5. Conclusions

An information model of the building envelope has been developed to organise efficiently and logically the wealth of data used and generated in the preliminary stage of the design process. This data is grouped into envelope entities and relationships.

By using a systematic top-down approach (the Primitive-Composite Approach), each entity is decomposed into primitive entities which represent small sets of data attributes which are cohesive (i.e. closely related). The primitives found here can be used as the basis for data integration of the envelope design process. Specific primitives can be selected by a developer to present only the desired view of a complex entity. Furthermore, the primitive entities defined here for the envelope can be easily augmented by primitives derived for other building systems, thus increasing the data schema to support integration of several aspects of the building design process.

A database schema has been created based on the resulting information model and implemented in a commercial object-oriented database system O_2 to investigate feasibility. The adopted object-oriented approach presents many advantages: modelling capabilities, unification of programming language and database management system, reusability, modularity and the transparent use of persistent objects.

This central database represents the main repository of information for building envelope data. Testing of this implementation has shown the exchange of data to be workable along different preliminary design phases and among the various participants involved. It can support the growth of data throughout the process and store form, function and behaviour together whereas CAD systems currently support only the form aspects of objects.

6. References

ASHRAE Standard 90.1-1989, "Energy-Efficient Design of New Buildings Except Low-Rise Residential Buildings", American Society of Heating, Refrigerating and Air Conditioning Engineers, New-York, NY, 1989.

NBCC, "National Building Code of Canada" and the "Supplement to the National Building Code of Canada", National Research Council, Canada, 1990.

Phan, Douglas, and Craig Howard, "The Primitive-Composite (P-C) Approach - A Methodology for Developing Sharable Object-Oriented Data Representations For Facility Engineering Integration", Technical Report 85 (A and B), Center for Integrated Facility Engineering, Stanford University, August 1993.

Rivard, Hugues, Claude Bédard, Kinh H. Ha and Paul Fazio, "Functional Analysis of the Envelope Design Process for Integrated Building Design", ASCE, Proceedings of the 1st Congress of Computing in Civil Engineering, Washington D.C., June 20-22, 1994, pp. 71-78.

Rivard, Hugues, "Integration of the Building Envelope Design Process", Master Thesis, Centre for Building Studies, Concordia University, Montreal, Canada, 1994.

Rumbaugh, James, Michael Blaha, William Premerlani, Frederick Eddy and William Lorensen, "Object-Oriented Modeling and Design", Prentice Hall, 1991.

Behavior Follows Form Follows Function: A Theory of Design Evaluation

Mark J. Clayton[1] S.M., ASCE, Martin Fischer[2] A.M. ASCE, John C. Kunz[3] and Renate Fruchter[4], M. ASCE

Abstract

Within the established notion of design as a cycle of analysis, synthesis, and evaluation, we present a theory that effectively describes design evaluation. During evaluation of a design, *prediction* derives *behaviors* from the design *form*, describing the performance of the design in the given situations. *Assessment* compares those behaviors to the stated design *functions* that are the requirements in the design problem. Using these concepts, we studied examples of evaluation for three issues in building design: spatial evaluation, cost evaluation, and energy evaluation. We implemented a software prototype that works with a CAD representation of a building to provide these three evaluations. Our theory of design evaluation is useful for structuring design software to support automation of design processes, integration, and information sharing.

Background

In a widely accepted characterization, design in any domain consists of three activities (Asimow 1962):
- *analysis,* the process of achieving an understanding of the problem;
- *synthesis,* the process of generating solutions;
- *evaluation,* the process of judging and comparing alternative solutions.

[1] Research Assistant, Center for Integrated Facility Engineering, Stanford University, Stanford, CA 94305-4020

[2] Assistant Professor, Civil Engineering Department, Stanford University, Stanford, CA 94305-4020

[3] Senior Research Associate, Center for Integrated Facility Engineering, Stanford University, Stanford, CA 94305-4020

[4] Research Associate, Center for Integrated Facility Engineering, Stanford University, Stanford, CA 94305-4020

The activities are roughly sequential, although they are repeated in a cycle many times as the design alternatives converge to an acceptable solution. (More recent authors refer to this characterization and its broad acceptance, such as Coyne, et al., 1990).

Evaluation is a difficult and complex step that often addresses many different design issues. We define an *issue* as a coherent set of criteria and rationale by which a design may be evaluated. For instance, evaluation in building envelope design must address issues such as energy use, structural adequacy, cost estimating, and others (Rivard, et al., 1994). After a consideration of multiple issues, evaluation ultimately must support the selection of one alternative among many. Using concepts of form, function and behavior, we have formalized general principles that describe evaluation for diverse issues.

Form, Function and Behavior

Although he uses different terminology, the concepts of form, function and behavior appear in Herbert Simon's assertion that the design process is concerned with the relations among "the purpose or goal, the character of the artifact, and the environment in which the artifact performs" (Simon, 1969). In other work, the sum of the artifact's physical properties, including materials, geometry and spatial relations, is the artifact's form, while function is both the intended behavior and the actual performance (Flemming, et al., 1992). Other researchers have distinguished between function, as an assigned duty, and behavior, as a response of an artifact to specific circumstances (Luth, 1991). Similarly, a distinction has been drawn between the functional criteria or *requirements,* and the *performance* that is intended to satisfy a particular requirement (Carrara, et al., 1992). The concepts have been reified further in research on data modeling to support software integration and data sharing (Phan, 1993). Form objects are the physical components of the artifact. Function objects express the "purpose or intended role." Behavior objects describe "response to environmental stimuli."

Our definitions of the three concepts have been presented before (Kunz, 1994) but are recapitulated as follows:
- *Forms* are the geometry and materials of the artifact.
- *Functions* are the required and desired qualities of the artifact. Synonyms for function include *goals, intents, requirements,* and *purpose.*
- *Behavior* is the performance of the artifact under particular conditions.

In our definition, behaviors are any descriptors that must be derived from the form by means of reasoning. Consequently, we consider length, width, height, area, and volume to be geometric behaviors. We also consider spatial relations, such as adjacent-to and part-of, to be topological behaviors rather than part of the form, as they are derived by reasoning about the form.

These definitions relate well to Asimow's characterization of design. Analysis is the process of identifying and elaborating design functions. Synthesis is the production, selection, and arrangement of design forms. Evaluation is the prediction of the behaviors of the forms and the assessment of behaviors with respect to functions. To gain a better understanding of evaluation, we have studied form, function and behavior in three design issues.

Form, Function and Behavior in Three Design Issues

Our test case for this research has been the design of a suite of rooms to accommodate a medical cyclotron for a hospital. The addition is a small but interesting and realistic design problem that requires consideration of several issues. We have focused upon three design evaluation issues: spatial requirements for the cyclotron suite, construction costs, and heating and cooling energy loads. Table 1 illustrates forms, functions, and behaviors for the three issues for the test case.

Issues:	Space	Cost	Energy
Forms:	Geometry and materials	Geometry and materials	Geometry and materials
Functions:	required rooms, required dimensions, required floor area, required room adjacency, required room connectivity, required door size, required wall construction	project budget, budget per category	heating budget, cooling budget, energy flow budget per category
Behaviors:	existence of rooms, length, width, height, area, room adjacency, room connectivity, door size	length, width, height, area, volume, total cost, cost by category, item cost	surface area, volume, orientation, adjacency of spaces to walls, connectivity of windows to walls, energy flow through each item, temperatures,

Table 1: Form, function and behavior for three evaluation issues.

The spatial layout of the cyclotron suite must conform to requirements defined by the cyclotron manufacturer in specification documents. The forms mentioned in the documents include spatial entities such as the rooms; physical entities such as doors, hatches, walls and floors; and materials. The functions include requirements for particular rooms, the size of rooms, adjacency among rooms, access routes through doors, and the size of doors. Corresponding to each of these functions is a behavior of the forms, such as the designed size of rooms, adjacency of rooms, access routes, and size of doors.

For the construction cost issue, the functions are the project budget and cost objectives for various categories. The cost forms are the physical components of the building including their geometry and materials. The cost behaviors are the predicted cost for each item, per category, and for the building as a whole. Various geometric behaviors are used in computing item cost.

The energy issue is concerned with heating and cooling loads upon the building. The forms are primarily the geometric shapes and the materials of physical elements in the building envelope, and the geometry of the spaces. The functions are energy budgets for the maximum heating and cooling for the building, perhaps categorized by components and orientations. The behaviors are the predicted energy flows through the envelope and the temperatures in the spaces. These behaviors are derived from geometric behaviors such as surface area and volume. Behaviors might also include the costs for providing heating, ventilating, and air conditioning to the building.

Shared Forms, Functions and Behaviors

The depictions in the table reveal several form entities that are shared. The spaces are shared between the energy issue and the spatial issue. The interior doors are shared between the spatial issue and the construction cost issue. Exterior doors are shared in all three issues. The building envelope elements are shared in the energy and cost issues while the interior doors are shared in the spatial issue and the cost issue.

Behaviors are also widely shared. All three of the issues use geometric behaviors, such as length, width and area. Topological behaviors appear in the spatial issue and the energy issue as adjacency of rooms, adjacency of spaces to walls, connectivity of rooms through doors, and connectivity of windows to walls. Some behaviors are distinct, such as construction costs and energy flows.

There are conceptual similarities among many functions, although each issue defines distinct functions. The energy allocation and cost budget are conceptually similar, although they are itemized in different ways and use different units of measurement. Existence and adjacency functions used in the spatial issue are conceptually similar to functions in other issues that were not considered in this study, such as building codes and structural systems. As functions are more closely related to the special reasoning of an issue, they appear to be specialized for a particular issue to a greater extent than are forms and behaviors.

Predict! and Assess!

Using this characterization of form, function and behavior, evaluation of a design alternative for the hospital addition follows the same sequence of actions for each issue.

Given a description of the design alternative in terms of the form pertinent to the issue and the functions that must be satisfied, the evaluator must *predict* the behaviors. Having arrived at behaviors, the evaluator must *assess* the behaviors by how well they match the functions.

Prediction derives the behavior, or performance, of an artifact in a given situation. Many reasoning techniques may be used for prediction, including model-based simulation, heuristics, table look-up, and algebraic equations. Predictions may be exact and highly accurate or they may be probabilistic and approximate. They may be quantitative or qualitative. In these examples, length is predicted by rotating the form entity to align with the world coordinate system, getting its bounding box, calculating the dimension on each of the two horizontal axes, and returning the larger of the two. Connectivity is predicted by searching all candidate objects and performing interference checks on pairs of candidates. Cost is calculated by retrieving unit cost from a database and multiplying as needed by dimensional values.

Having predicted the behaviors, it is necessary to determine if the behaviors are appropriate with respect to functions. We refer to this action as *assessment*. Although one could introduce subtle degrees of satisfaction, assessment basically determines whether or not a function is satisfied. Assessment often consists of comparing a numeric value to a limit or a range, or merely determining that some aspect of the design exists. It might also involve probability and search.

The ideas of forms, functions, behaviors, prediction and assessment provide a basis for structuring integrated software. Behavior objects have a *predict!* method so that they can produce their own values. Function objects have an *assess!* method to compare the behaviors to the project requirements expressed by the function object.

Computer Model

To model these ideas, we have constructed a software prototype that is an enhancement of the SME software developed in other projects (Clayton, et al., 1994). As the new software has been developed with the Kappa knowledge engineering development tool, we refer to it as SME-K. It integrates three evaluation modules, addressing the design issues described above, with the AutoCAD design software. The concepts of form, function and behavior are reified as software objects and related by means of *interpretation objects* and *feature objects*. Additional object classes handle communication between the software and AutoCAD, user interfaces, saving and retrieving data, etc.

The reasoning and data for each design issue are encapsulated as a subclass of *interpretation object*. In SME-K, there is a cyclotron suite spatial interpretation, a cost estimating interpretation, and an energy interpretation. Each interpretation is responsible for extracting appropriate form and behavior information from the AutoCAD drawing and for providing an evaluation of the building design.

A *feature* is an object that relates particular form objects to function objects and to behavior objects, addressing some aspect of a design issue. Thus a feature object represents the geometry and materials of a design entity, the intent and purpose of the entity, and the performance of the entity. Each interpretation defines different and appropriate features. For example, the features for the cyclotron suite interpretation include: Cyclotron Room, Lab Room, Control Room, and Access Hatch. The Cyclotron Room feature associates the functions length requirement, width requirement, access requirements; and the behaviors length behavior, width behavior, and access behaviors. Each feature is responsible for evaluating the related part of the design and reporting the partial evaluation results to the interpretation.

Using SME-K, a designer can draw a building with standard AutoCAD commands and then rapidly and automatically evaluate the design for each issue. A graphic user interface allows the designer to create interpretations, instantiate features of each interpretation, and perform evaluations. Each interpretation produces an issue-specific evaluation in terms of the behaviors of the design and the satisfaction of functions.

An example evaluation from the spatial interpretation is shown in Figure 1. The **Critique Results** dialog box provides a list of functions. In the example, the selected function is concerned with the adjacency of two rooms, as indicated by the explanatory text in the dialog box. The **Feature Inspector,** accessed by clicking the **Form...** button,

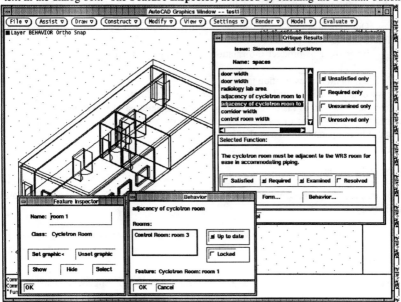

Fig. 1. Evaluation produced by the spaces interpretation.

allows one to highlight, show or hide the primary affected room in AutoCAD. The **Behavior** dialog box, accessed by clicking the **Behavior...** button, shows which rooms are adjacent to the primary room. Every function in the **Critique Results** list is associated with a form in AutoCAD and the behaviors that were used to assess the function.

Conclusions and Implications

The evidence gathered from this research supports a theory of the design evaluation process as follows:

- Design evaluation may be modeled as a process of relating the form, function and behavior using prediction and assessment.
- Prediction is the process of reasoning about design forms to produce models of behavior.
- Assessment is the process of determining the adequacy of behaviors with respect to functions.

In deference to Sullivan, we summarize this theory as "behavior follows form follows function follows behavior." Behavior is predicted from the form. Form is synthesized in response to the functions. Functions are assessed by comparing them to behaviors. The phrase neatly expresses the use of forms, functions and behaviors in the evaluation stage of the design cycle. By structuring three diverse design evaluation tools, the SME-K prototype provides evidence that the theory is general.

Our experience suggests that the initial steps achieved by this research may lead to future software that provides:

- integration of product information by sharing form, function and behavior objects; and
- automation of evaluation tasks.

We speculate that improved integration can ease coordination problems and increased automation can lead to faster design stages and consideration of more alternatives. The research may also lead to software development frameworks that allow rapid development of inter-operable design evaluation tools.

Acknowledgments

This research was funded by the Center for Integrated Facility Engineering at Stanford University.

AutoCAD is a trademark of Autodesk, Inc. Kappa is a trademark of Intellicorp, Inc.

Appendix. References

Asimow, M. (1962). *Introduction to Design*. Prentice-Hall, Inc., Englewood Cliffs, N.J.

Carrara, G., Kalay, Y. E., and Novembri, G. (1992). "Multi-Modal Representation of Design Knowledge." *Automation in Construction*, 1 (2), Elsevier Science Publishers, Amsterdam, the Netherlands, 111-122.

Clayton, M. J., Fruchter, R., Krawinkler, H., and Teicholz, P. (1994). "Interpretation Objects for Multi-Disciplinary Design." *Artificial Intelligence in Design '94*, Kluwer Academic Publishers, the Netherlands, 573-590.

Coyne, R. D., Rosenman, M. A., Radford, A. D., Balachandran, M., and Gero, J. S. (1990). *Knowledge-Based Design Systems*. Addison-Wesley Publishing Company, Reading, MA.

Flemming, U., Adams, J., Carlson, C., Coyne, R., Fenves, S., Finger, S., Ganeshan, R., Garrett, J., Gupta, A., Reich, Y., Siewiorek, D., Sturges, R., Thomas, D., and Woodbury, R. (1992). "Computational Models for Form-Functional Synthesis in Engineering Design." *EDRC 48-25-92*, Engineering Design Research Center, Carnegie-Mellon University, Pittsburgh, PA.

Kunz, J., Clayton, M., and Fischer, M. (1994). "Circle Integration." *Computing in Civil Engineering*, Vol. 1, ASCE, 55-62.

Luth, G. P., Krawinkler, H., and Law, K. (1991). "Representation and Reasoning for Integrated Structural Design." *Technical Report Number 55*. Center for Integrated Facility Engineering, Stanford, CA.

Phan, D. H., and Howard, H. C. (1993). "The Primitive-Composite (P-C) Approach -- A Methodology for Developing Sharable Object-Oriented Data Representations for Facility Engineering Integration." *Technical Report Number 85A and 85B*. Center for Integrated Facility Engineering, Stanford, CA.

Rivard, H., Bedard, C., Ha, K. H., and Fazio, P. (1994). "Functional Analysis of the Envelope Design Process for Integrated Building Design." *Computing in Civil Engineering*, Vol. 1, ASCE, 71-78.

Simon, Herbert A. (1969). *The Sciences of the Artificial*. The M.I.T. Press, Cambridge, MA.

Design Education Across the Disciplines

Janet L. Kolodner
and the EduTech Design Education Team[*]
EduTech Institute and College of Computing
Georgia Institute of Technology
Atlanta, GA 30332-0280

Abstract

Engineers, architects, and computer scientists all engage in design upon graduation. Yet, up to now, there has been little emphasis in post-secondary education in helping students learn the ins and outs of doing design successfully. Capstone projects provide too little design experience and come late in the curriculum. Freshman design experiences, when in place, tend to emphasize the design experience itself, often without helping students reflect on and understand what the experience is teaching them. Even when both components are in place, there is little connectivity between these early and late experiences.

Georgia Tech is taking the lead nationally in creating a design curriculum that helps students learn the skills involved in doing design and that is integrated throughout their undergraduate years. In EduTech's design focal group, faculty representing

[*] EduTech's design education team includes a variety of faculty who teach design, from a variety of engineering units (e.g., civil, mechanical, chemical, industrial), architecture, industrial design, software engineering, and rhetoric (document design?). It also includes faculty and students in the cognitive sciences, particularly those with background in case-based reasoning, design cognition, and learning; and faculty and students specializing in educational technology, interested particularly in software-realized scaffolding and software in support of collaborative learning. Shared by all is an interest in doing better at helping students learn the principles and skills behind good design and the conviction that collaboration with this broad range of participants is essential to a quality solution to that problem. EduTech Institute, which organizes this endeavor, has as its mission to use what we know about cognition to inform us as we design educational technology and the environments in which it will be used. Design education is a major focus of EduTech's current endeavors. Thanks to Farrokh Mistree for extensive comments on this paper.

the full variety of design disciplines are working along with cognitive scientists in an attempt to list the component skills involved in design, the kinds of activities and projects that promote such learning, and based on that, to devise curricular frameworks and guidelines for making design education work. We are working towards a core curriculum in design, one that can be shared by the full variety of design disciplines.

Our work is guided by the experience and needs of those in the design disciplines, the needs of the workplace, and by what cognitive scientists know about design cognition and about learning from experience (cases). Fortunately, all point in similar directions -- toward a curriculum that provides significant collaborative hands-on experience coupled with significant reflective activities. Our knowledge of case retrieval and indexing helps in guiding the sequencing of student activities, in suggesting several kinds of technology that might be useful both during learning and practice, and in informing us about the sorts of reflection to promote.

Design Education: What we are aiming for

Current approaches to education throughout the disciplines tend to emphasize individual work, learning of facts, and the analysis of artifacts. While engineering curricula also include one capstone design course in which students use a design project to integrate the concepts they have learned in individual courses and some include a freshman "design experience," few, if any, emphasize integrative activities as a fundamental part of the curriculum. Thus, too many of our students graduate knowing a lot of disjoint facts and concepts without knowing how or when to use those facts in responding to complex problems. Students are not yet ready at graduate time for the workplace, where use of knowledge for problem solving, design, and project management are essential. Nor are they prepared to collaborate well, to transfer their skills from one domain to another, or to learn and consider concepts beyond those learned in school.

Students currently use their first years in the workplace to integrate and learn to use the knowledge learned in college. We claim that if college education emphasized integration and use of knowledge, then necessary skills and the principles behind them would be learned by a broader cross-section of the workforce and in ways that would promote more effective use. Rather than educating our students for just one life-long career, we need to educate them to be instrumental in contributing to and managing technological change. This requires balanced experience in several areas: analysis and synthesis, science-based and process-based knowledge, communication, and human and societal issues. The body of knowledge that supports the practice of design is evolving from an art to a science, one that is different but complementary to other engineering sciences. Thus, we believe it is necessary to shift emphasis in design classes from lectures and puzzle-like problems to team activities in which participants learn by doing, acquire hands-on experience, and learn how to draw career-sustaining lessons from those experiences.

We are therefore aiming to put into place a core curriculum in design that will produce engineers, software designers, and architects with deep understanding of technical facts and the skills and foundational principles to be able to use them flexibly and well. We want to graduate students who can
- negotiate solutions to open problems;
- reflect on what they do and articulate critical decisions;
- represent, interpret, analyze, and communicate a problem;

- generate multiple alternatives in response to proposed problems;
- evaluate options and procedures both during and after determining appropriate responses;
- use mathematical and other technical resources effectively;
- manage the resources and skills needed to carry through on projects; and
- successfully work in teams; communicating at technical and professional levels.

These skills will allow our graduates to solve problems flexibly using the technological and scientific facts they are learning. But more than skill is needed to do these tasks well. Students will also need deep understanding of their own discipline, appreciation of what several related disciplines bring to their problems; and strong collaboration skills. This combination of deep knowledge and honed skills, we believe, will allow our graduates both to perform well and to learn from their individual experiences those things that will allow them to lead their companies into the future as they take on management and leadership positions.

Educational Philosophy

Recent educational research reveals several features of a learning environment that enhance learning. Students learn most effectively and permanently when they are active participants in the learning process (e.g., Chi et al., 1984). Particularly important is having the opportunity to explore the ins and outs of new concepts as they are learning them and to use newly-acquired knowledge to solve interesting problems (Bransford, Sherwood, Vye, & Reiser, 1986). Furthermore, there is much evidence that when students work in groups with others, their learning is enhanced (e.g., Brown et al., 1993). Experience shows that skills are best learned by practicing them and reflecting on that experience (Brown & Campion, 1994).

Problem-based learning (PBL), a scheme that includes all of these components, is being used substantially at medical and business schools (Barrows, 1985; Williams, 1993). In this scheme, students learn by solving authentic real-world problems. Because the problems are complex, students work in groups, where they pool their expertise and experience and together grapple with the complexities of the issues that must be considered. Coaches guide student reflection on these experiences, facilitating learning of the cognitive skills needed for problem solving, the full range of skills needed for collaboration and articulation, and the principles behind those skills. Because students are in charge of their learning, skills needed for life-long learning are also acquired, for they must manage what they need to learn and how it will be learned as they cope with the problems set for them. Experience in medical and business schools shows that problem-based curricula serve to enhance learning facts and concepts at a range of levels and to promote the learning of critical problem solving and collaboration skills needed for the workplace (Dolmans, 1994; Hmelo, 1994; Norman & Schmidt, 1992).

Based on these studies and experiences putting these programs into place in medical and business schools, the design core curriculum we are developing is collaborative and problem-based. We hypothesize that a problem-based curriculum that focuses on collaboratively solving authentic design problems will have the same benefits for our students that it has for medical students. It will allow them to better learn the factual and conceptual knowledge in the curriculum, to learn the applicability of that knowledge, to integrate facts learned across several courses, to become better designers, to learn and practice the skills they will need when they enter the workplace, and to learn the principles behind using those skills well. These principles, we believe, will allow our graduates to effectively transfer their knowledge and skills to solving new problems in new domains and that will allow them to creatively deal with the realities of noisy, ambiguous and incomplete data.

Putting it into Practice

But we can't simply transfer the problem-based curriculum of the medical schools to an undergraduate educational institution. In medical school, students focus full time on learning medicine. Medical students are older and more mature and already have at least some learning and study skills. Undergraduate education, on the other hand, is much broader, there is less focus, and students are still learning study skills. We need to borrow the best of problem-based learning as it is used in the medical schools, but to adapt it and shape it to the needs of undergraduate education.

The core we are developing cuts across a large and varied set of design disciplines (the range of engineering disciplines, architecture, and computing) and focuses on several fundamental skills and principles important to the practice of design in all of those disciplines. This interdisciplinary approach is intended to expose students to differing ways of approaching design problems beyond those prescribed by their home discipline. The courses will also be available and accessible to students outside of the design disciplines (e.g., management, public policy), providing opportunities for students who will work alongside designers to learn about design.

Our effort has three parts: (1) Development of a core curriculum that fosters integrative skills necessary for analysis, synthesis, evaluation, and contextual understanding of engineering, software, and architecture problems, with an aim toward emphasizing the learning of design skills. (2) Development of problem-oriented courses as part of this curriculum that address academic and professional requirements in tandem, helping students acquire collaboration skills, communication skills, life-long learning skills, and the principles behind those skills, in addition to factual knowledge, and helping them learn how to use the facts they are learning. (3) Integration of a software environment into these courses that promotes collaboration and reflection and that provides easy access to state-of-the-art computer technologies. The software environment is intended to promote transfer of knowledge skills learned in the design curriculum to other parts of the curriculum as well.

The Core Curriculum

Faculty in EduTech's design group agree that design curricula should begin in a non-disciplinary way, move forward in each discipline addressing relatively simple problems, and gradually move toward sophisticated interdisciplinary design projects. This gives us three kinds of courses: pre-disciplinary foundations courses that help students learn basic problem solving skills, articulation of decisions made, collaboration among people with different expertise, and so on; area-specific skills courses with different gradations of complexity in which students do more and more complex design using the analytic tools of their discipline for both synthesis and evaluation, building on the foundations courses and using design skills learned in those courses to learn disciplinary facts and skills; and interdisciplinary capstone courses in which students work in interdisciplinary teams on a problem of real-world complexity, integrating the expertise and analysis tools of several disciplines. The core curriculum will include the foundations courses and interdisciplinary courses. Each discipline, we hope, will gradually make its curriculum more design-oriented.

Key aims in the core curriculum are <u>continuity</u> and <u>transfer</u>. Continuity refers to introducing key skills early in the curriculum and revisiting them on a regular basis while adding more complexity to the situations in which they are carried out. Transfer refers to helping students learn to use these skills in a variety of different circumstances. Our analysis so far suggests a structuring of the core curriculum as shown in the figure below.

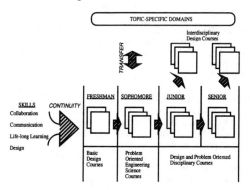

In the freshman year, we hope to have a sequence of two courses in which the essentials of design reasoning, collaboration, and communication are learned through work on non-trivial, but non-disciplinary, collaborative design projects. Problem-oriented "engineering science" and other courses, offered in the sophomore year and shared by the disciplines, will provide an opportunity to apply key design, collaboration, and communication skills as core concepts are learned. These courses, we hope, will replace existing fact-oriented science courses. In the following two years, design-oriented and problem-oriented disciplinary courses will help students cement the learning of key disciplinary concepts with design activities. Design experiences will be used to motivate reflection on problem solving, collaborative, and communicative skills used to successfully solve problems. In addition, we'd like to see interdisciplinary design-oriented courses and capstone options, in which students at several levels of expertise solve design problems together.

Throughout, several principles and values permeate our plans: active in-class involvement, education in both facts and reflection, continuous guidance, teamwork, systems-oriented problem solving, and learning by doing on authentic design problems. We want students to learn principles of design; we also want them to learn how to learn.

Central to the proposed curriculum is a software environment that promotes reflection and collaboration, supports design decision making, makes case studies available for perusal, makes internet resources available, and provides access to relevant discipline-specific software. The software will be used throughout the design curriculum and will play a key role in providing continuity and transfer across the many design projects students will carry out in the course of their undergraduate engineering education. We hypothesize that use of a tool that helps students make their decisions explicit will help them transfer the skills they learn early on into later and more complex projects; that use of a state-of-the-art design support environment will allow students to work on more complex and challenging design projects; and that use of a state-of-the-art collaboration support environment will help students to build upon each other's skills and knowledge and to leverage their own abilities.

The Freshman-Level Courses

The freshman experience, within our core, is comprised of two quarter-length courses. The first lays foundations for design; the second allows students to focus for ten weeks on a single project. In both courses, the majority of student time is spent collaboratively solving problems. Experiences in these courses are intended to form the foundation for carrying out design activities within their chosen discipline.

It is thus important for the teacher to find problems for students to solve that are representative of problems and issues that arise in the world, to promote the creation of general knowledge by sequencing problems such that comparing and contrasting of several different situations can be productive, and to provide help with group processes, reflection, and moving productively forward, aiming to insure that students are extracting appropriate skills from their experiences.

Course 1: Foundations of design I. This pre-disciplinary foundations course will give students their first hands-on experience with design. It comes at the beginning of freshman year. Students work in interdisciplinary teams on a series of small design projects that don't require significant hard-core disciplinary knowledge. They might design and build games of various kinds, plan and cook meals, or solve simple engineering problems and construct simple artifacts (e.g., variations on the classic egg-drop activity). Students discover, in this course, the ways in which they can effectively use problem-solving skills they already have to do design. They discover how to articulate the problem solving they already do.

In particular, they learn about the importance of prioritizing constraints and framing problems, the iterative nature of design activities, the types of decisions made during design, the role evaluation plays in discovering shortcomings of designs in progress, the ingenuity required to derive evaluation criteria beyond those in a specification, the necessity of generating and exploring several alternative solutions, and general-purpose strategies for doing each of these things. They also begin to learn how to negotiate their way productively in group activities. And they have their first experiences using the collaborative software environment. At the end of this course, we want students to have the vocabulary for discussing what they do when they do design, to understand the role several key skills from their discipline play in design, and to have an understanding of design that will allow them to appreciate, in later courses, the role the knowledge learned in those courses might play in carrying out complex problem solving and design.

Students will work on three projects in this course, a very short individual project (2 to 4 days), a group project of two-weeks duration or shorter, and a larger group project of 5 to six weeks duration. In this way, students will have the opportunity to "fail" softly early in the quarter, allowing them to learn from their mistakes and be more successful in later projects. The first project, a short individual one, is intended to introduce students to the concepts of understanding a problem, proposing multiple alternative solutions, and the need for evaluation. With this in hand, students will work in groups on a second project, this time using the project to point out some of the ins and outs of collaborating, the need for clear articulation and explanation, and the role software can play in aiding collaborative efforts. In the third project, groups will put these new collaboration principles and skills to work in designing and building a simple artifact. Our intention first time through is to have students design a souvenir for the Olympics.

Central to this class will be presentations of designs and the rationale behind them to the class as a whole and discussions about the different designs that promote recognition of the reasoning and skills used in understanding problems and coming up with solutions. Thus, much class time will be spent comparing and contrasting different solutions students come up with, discussing how different solutions were derived, and examining why different students and groups came up with different solutions. Other time in class will be spent with experts, who will model design reasoning for the students by solving problems and discussing their rationale for the students and by discussing the role design plays in the discipline or workplace the expert comes from. We hope to be able to present final projects to local industry as well as to the class and to gain feedback from industrial experts on a variety of design and manufacturing issues. This course will be offered for the first time in fall, 1995.

Course 2: Foundations of Design II. In this course, students will put what they've learned in the first course to use in collaboratively solving a larger-scale design problem, though still one that doesn't require detailed disciplinary knowledge. This pre-disciplinary foundations course will be modeled after the current ME3110 (Creative Decisions in Design), taught for many years now by Professors Farrokh Mistree, Janet Allen, and David Rosen.   Like ME3110, this course will reinforce the entire product realization process, from problem formulation through to design and to actually building a product, with special emphasis on planning the activities required to complete the full process. While ME3110 is taught to mechanical engineering juniors, the new course will be aimed at students in the variety of design disciplines and will be taken at the end of their freshman year. As this course is currently taught, students are introduced to an often-turbulent imaginary world. The inhabitants of that world must solve a serious problem that requires an engineered solution. Students design and build a solution to this problem. One quarter they may need to build an evacuation device, another quarter a device that will transport and drop sleeping potion in enemy territory. Important activities include problem formulation, design, construction, and testing. In this context, the students plan their activities for the quarter, allocate resources (cost, time, and so on), and have a device ready to compete near the quarter's end. Students experience the interplay inherent in meeting design requirements subject to resource constraints, selecting most-likely-to-succeed alternatives, and resolving trade-offs. Decisions are introduced as key engineering constructs to aid these tasks. Class time in this class will continue where the first class left off, helping students to further explore and understand cognitive, social, and technical skills and principles important to quality design. As in the first course, our software environment will play a key role.

Where is the Case-Based Reasoning?

This session is about case-based reasoning, but our discussion so far has neglected to mention it. CBR plays three important roles. First, an important part of the software support environment will be case-presentation software and case libraries. Case-presentation software will present the problems students will be solving, structuring the necessary background knowledge. Students will have access to libraries of cases that illustrate the issues and concepts they are exploring. Cases might provide suggestions about how to do things, but more importantly are intended to help students analyze problems to determine what their important characteristics are and where priorities should lie in devising solutions. The companion paper and presentation on Archie-2 will give more detail about this contribution.

The second contribution of case-based reasoning is more abstract. Our knowledge of case-based reasoning, especially its algorithms and heuristics for retrieval, adaptation, and indexing, suggest several insights into how we might induce learning in our students and the kinds of functionality and scaffolding our learning environments ought to have. Case-based reasoning tells us, for example, that learning happens as a result of interpreting and indexing experiences in ways that facilitate their retrieval at times when they might be useful. This suggests that we might induce reminding, and therefore transfer, by helping students to learn a vocabulary for describing design experiences and by helping them, before they are expert at using that vocabulary, to nonetheless interpret their experiences based on that vocabulary. Both the reflective discussions in class and the scaffolding provided in our collaboration environment will be designed to promote such learning and analysis. Based on case-based reasoning's premises, we are developing software for individual and collaborative use that is intended to induce skills transfer as students move through the curriculum in design-related disciplines (engineering, computing, and architecture). The paper and presentation on CaMILE explain the collaboration environment.

Finally, case-based reasoning suggests the importance of experience to learning, and the activities and materials being put together to support learning in our design courses reflect that insight. The paper and presentation on our sustainable technology course show how we are using insights from case-based reasoning and the case method of teaching to develop materials for design classes. While the sustainable technology course is not part of the design core, it is being developed with the same principles by people who are active members of EduTech's design education team.

References

1. Barrows, H.S. (1985). *How to design a problem-based curriculum for the preclinical years*. NY: Springer.
2. Bransford, J.D., Sherwood, R. S., Vye, N.J., & Rieser, J. (1986). Teaching thinking and problem solving: Research foundations. *American Psychologist, 41*, 1078-1089.
3 Brown, A.L., & Campione, J.C. (1994). Guided discovery in a community of learners. In K. McGilly (Eds.), *Classroom lessons: Integrating cognitive theory and classroom practice* (pp. 229-270). Cambridge, MA: MIT Press/Bradford Books.
4 Brown, A.L., Ash, D., Rutherford, M., Nakagawa, K., Gordon, A., & Campione, J.C. (1993). Distributed expertise in the classroom. In G. Salomon (Eds.), *Distributed Cognitions* (pp. 188-228). New York: Cambridge University Press.
5 Chi, M.T.H., Bassok, M., Lewis, M.W., Reimann, P., & Glaser, R. (1989). Self-explanations: How students study and use examples in learning to solve problems. *Cognitive Science, 13*, 145-182.
6. Dolmans, D. (1994). *How students learn in a problem-based curriculum.* Doctoral dissertation, University of Limburg, Maastricht, The Netherlands: University Pers Maastricht.
7. Mistree, F. & Muster, D (1985). A curriculum and paradigms for the science of design, *1985 ASEE Annual Conference Proceedings*, Atlanta, GA, pp. 101-107.
8. Williams, S.M. (1993). Putting case-based learning into context: Examples from legal, business, and medical education. *Journal of the Learning Sciences, 2*, 367-427.

A problem-based curriculum for sustainable technology,

Bert Bras (Mechanical Engineering),
Cindy Hmelo (EduTech),
Jim Mullholland (Civil and Environmental Engineering),
Matthew Realff (Chemical Engineering),
Jorge Vanegas [1](Civil and Environmental Engineering)

Georgia Institute of Technology, Atlanta, GA 30332.

Abstract

Sustainable technology has been defined as technology that provides for our current needs without sacrificing the ability of future populations to sustain themselves. Approaching the synthesis of sustainably engineered solutions requires weighing the qualities of different proposals from a variety of different perspectives and handling a variety of tradeoffs simultaneously and creatively over time horizons frequently far longer than normally considered for engineering projects. Practitioners of sustainable technology are well versed in their own discipline but don't necessarily know material from other disciplines that can contribute to these decisions.

Of necessity, these problems must be solved in multi-disciplinary groups. Students, therefore, need to learn not only what their own disciplines have to say about the issues, but they also need to be able to recognize the other kinds of issues that arise and to know which disciplines can contribute to their solutions. To further complicate things, this is a new way of addressing problems, and we can pinpoint only a small number of the issues that will arise in future problems. Thus, students need to learn how to recognize issues that have never been recognized before.

This state of affairs points to a need for students learn by working on cases in multi-disciplinary teams and to have access to previous cases that can help them identify what it is important to focus on and that they might adapt and merge to solve new problems. Learning in such an environment will provide students with cases that they can recall later in their careers. Equally important, however, is reflection on their experiences that will allow them to index those experiences appropriately for recall during later problem solving. Such reflection should also help them to assess new cases as they arise in the future. In this paper, we present several of the cases students are solving, reflecting on the content such cases need to have to promote effective learning.

---

[1] Author to whom correspondence should be addressed.

Background and Motivation

"Sustainable Development" does not have one definition or one set of concepts which it implies. In fact, part of the educational challenge is the ambiguity and imprecision of the term. As our point of departure, we take it to mean **"a process of change in which human attitudes and behaviors are modified so that, in endeavors to meet current needs, achieve aspirations and preserve options for future generations, individuals and communities will enhance and maintain their well being.", (Weston, 1994).** This is a far reaching concept that crosses disciplinary boundaries and requires students to be educated to continuously update their knowledge, attitudes and beliefs.

Under the sponsorship of the GE Foundation, the overall educational approach being adopted at Georgia Tech consists of starting with a three course sequence, and then migrating the ideas and concepts from these courses into the general curriculum. The first course is intended to expose the students to a framework in which questions about sustainability can be posed. This framework is based on four *dimensions of sustainability*, technology, economics, ecology and ethics, (Stan Carpenter, Personal Communication 1994). The second course in the sequence will employ this background to focus on a series of case studies using a problem-based-learning (PBL) approach (Barrows, 1985). The third course will involve more open-ended problem solving by an inter-disciplinary team of students. The core of the third course will be the solution of an open-ended problem by an inter-disciplinary team of students. The content, educational goals and methods of the second course, are the focus of the rest of this paper.

Problem-based Learning

New models of teaching and learning suggest that the emphasis of instruction needs to shift from teaching as knowledge transmission to less teacher-dependent learning. Learning needs to occur in problem-oriented situations if it is to be available for later use in those contexts (Bransford, Vye, Kinzer, & Risko, 1990). One approach to contextualized learning is through Problem-based learning. PBL was originally developed to help medical students learn the basic biomedical sciences (Barrows, 1985). PBL includes among its goals: 1) developing scientific understanding through real-world cases 2) developing reasoning strategies, and 3) developing self-directed learning strategies. Since its origin in medical education, PBL has been used in other settings such as engineering and architecure (e.g., Boud & Felletti, 1991).

Research and theorizing in cognitive science provides evidence that supports some of the general goals of PBL (Norman & Schmidt, 1992). As students articulate their knowledge in PBL, they develop more coherent understandings of science and its applications (Hmelo, 1994). In addition, the acquisition of prior examples that occurs during PBL should allow later problems to be solved by case-based reasoning (Kolodner, 1993). Finally, the active learning promoted in PBL should promote the self-directed learning strategies and attitudes needed for lifelong learning (Bereiter & Scardamalia, 1989). The self-directed learning objectives of PBL are particularly important in the sciences because PBL may endow individuals with the learning strategies necessary to stay informed in the face of rapid technological advances.

A PBL group consists of a multidisciplinary group of 5-7 students and a facilitator. The facilitator gives the students a small amount of information about the case and then the group's task is to evaluate and define different aspects of the problem and to gain insight into the underlying causes of the problem. This is done by questioning the facilitator, generating and evaluating hypotheses and learning issues. Learning issues are topics that the group has decided are relevant and which they need to learn more about. The group members divide the learning issues and research them, using both physical and human resources. The group members share the information and use it to understand the problem and propose alternative solutions. The students will then work in groups with other students in their own disciplines to use their technical knowledge for problem-solving. As in the real world, PBL groups are presented with incomplete information and they must identify what other information is needed about the problem as well as from other learning resources. The use of PBL in this setting requires collaboration. In addition, because this domain is so ill-defined, we want students to reflect on the nature of cases in sustainable technology. The next section of the paper will discuss technological tools to support the students' collaboration, reflection, and learning in this course.

Educational Technology

Collaborative learning is a key part of the PBL approach. As students articulate and reflect upon their knowledge, learning is enhanced (Brown & Palincsar, 1989). Group problem-solving allows students to tackle more complex problems than they could on their own. The major tool we will use to support collaboration is CaMile, an interactive, distributed, collaboration environment that scaffolds learning, reflection, access to materials, design, and problem-solving (Guzdial, Rappin, & Carlson, 1995). This environment supports collaboration through the NoteBase in which students enter comments of various types. They reflect on their thinking as they choose the type of note they enter. The students can attach documents of various forms to the their notes and can create links to the MediaBase, a multimedia repository of additional information.

The students will also use DesignMuse and ARCHIE-2 (Domeshek & Koldner, 1992) to assemble their own cases as they develop a case library that can be used for future reference. ARCHIE-2 provides a case library permits students to access the important parts of cases, understand their implications, and recognize the range of problems needing solving. These issues are important for students to consider as they assemble their own cases using DesignMuse, the case-authoring system.

Overview of Case Studies Developed

Drawing upon our collective experience in chemical systems optimization, chemical emission source, fate, and transport, mechanical design, and problem-based learning, ten case studies in sustainable development have been or are being developed (Table 1). Each case includes a factual case text (typically five to fifteen pages), supplementary material that provides case enrichment (e.g., reference text and video), and a teaching guide. Students are expected to expand the study of cases beyond these materials by identifying issues and exploring responses on their own. The role of the class instructor is a) to facilitate discussion around issues of technology, environment, economics, and ethics (the dimensions of sustainable development), and b) to encourage the use of fundamental principles and tools to address these issues.

A goal of each case is to expose students to a broader perspective of real-world problems while having them draw upon their disciplinary areas of knowledge to collaboratively develop solutions. Currently, ten cases have been developed (see Table 1). A common theme in these cases is that they involve the production and use of chemicals and their fate and transport in the environment. Application of principles of mass and material balancing, dynamic and steady state response, and life-cycle assessment is required in each case to assess issues of technology, economy, environment, and ethics. Cases 1 through 4 are broad in scope, addressing issues at the global, national, or industrial level. Cases 5 through 10 focus on particular sites. In all cases emphasis is given to the recovery and reuse of product, raw material, and/or energy. Topics addressed include the technical feasibility and economic tradeoffs of alternative processes and products; preservation and efficient use of material, energy, and space; and uncertainty in data in material properties, global, regional and local inventories, and health risk information.

Table 1. Case Studies and Issues in Sustainable Development.

Case:	Issues:
1. Renewable Energy.	Revisit course 1 issues; introduce theme of chemicals in the environment.
2. Automotive Recycling.	Product take-back legislation; multiple component recycling; closing the material loop; recycling processes.
3. Plastics Recycling.	Multiple levels of recycling; chemical processes.
4. The sustainability of Chlorine use - A Pulp and Paper Mill.	Overall approach to chemical use.
5. Automotive Body Repair.	Waste reduction; product maintenance industry.
6. Arcata Graphics: Solvent Recovery.	Recycling for profit; equipment selection.
7. Incinerator Air Emissions Estimate.	Alternative operation; byproduct emissions; energy recovery.
8. Learning from Industrial Accidents - Bhopal.	Accidental chemical releases; chemical toxicity.
9. Carpet Manufacturing: Textile Mill.	Capital equipment; alternative processes.
10. Sheet Molding Compound Manufacture.	Impact assessment and reduction; sustainability and economical trade-offs in industry.

Pilot Course

This course will be taught as a three credit hour elective cross-listed in Civil & Environmental Eng., Chemical Eng., and Mechanical Eng. The first two weeks will be used to acquaint students with the educational tools that will be used, and to introduce them to the concept of sustainable development and sustainability dimensions. These ideas are being developed in a course currently being offered. In the remaining eight weeks, three cases have been selected for study: Learning from Industrial Accidents - Bhopal, Premix' Sheet Molding Compound Manufacture, and Chlorine Use in a Pulp and Paper Mill.

Case 1: Learning from Industrial Accidents - Bhopal

Goals of the case study are: (1) to mine the field of human experience for information about sustainability that is otherwise not easily obtained, such as full-scale failure, human response, and health effects; (2) to consider what are appropriate levels of safety, with the associated economic and ethical questions addressed; and (3) to view the chemical system in question as a miniature model of sustainability, utilizing steady state and dynamic analyses to predict safe operating regimes. In addition to a factual case text, students will be encouraged to consider the perspective of a sociologist on Bhopal and hazard theory (Bogard, 1989), the accounts of individuals who were in Bhopal at the time of the accident (Kurzman, 1987), and lessons learned from other industrial accidents (Kletz, 1988).

The worst industrial accident in history occurred in the Union Carbide of India plant at Bhopal on the night of December 2, 1984. A leak of methyl isocyanate, being stored as an intermediate in the chemical synthesis of the insecticide carbaryl, spread beyond the plant boundary, killing thousands of people living in a shanty town nearby and injuring 200,000 more. Explanations from a variety of perspectives have been offered. These range from technical and human failures inside the plant, to corporate negligence, to regulatory failures on the part of the governments involved. In the case text, the political and economic setting in which the accident took place is described, the facts of the accident are reviewed, including an analysis of the runaway chemical reaction that resulted in the leak, and the aftermath of the tragedy is summarized.

The key question to be addressed by the students is as follows:

What can be done to prevent future Bhopals?

Students, having read the case text, will discuss facts and uncertainties about the case at the first class meeting. Four or five topics will be nominated for investigation. Groups will then be formed to generate specific questions that relate to these topics. Agreement will be reached regarding the format for subsequent class meetings, leaving time to synthesize findings in a case report. In addressing the key question above, possible study topics include: (1) alternative insecticides and agricultural practices; (2) alternative routes of carbaryl synthesis; (3) corporate ethics and multinational companies; (4) public policy and shanty towns.

Case 2: Sheet Molding Compound Manufacture

The goal of this case study it to make students aware of the sustainability issues of manufacturing half-products and plastics. Our specific pedagogical goal with this case is to enhance the students' ability to solve real-world problems which involve: 1) understanding of background and technology at hand; 2) Characterization of environmental impact; 3) Investigation of options to reduce or eliminate environmental impact and assessment of feasibility given a number of constraints; 4) Identification and implementation of a suitable solution to the problem at hand.

This case is not so much based on an event, but rather on a company, Premix, Inc., and its continuous strive towards sustainability within the economic, legal, and technical constraints. Premix is a manufacturer of sheet molding compound (SMC) products. SMC is a fiberglass-resin compound which is molded into products ranging from Jet-Ski hulls to valve covers and micro-wave dishes. SMC is manufactured by placing a layer of paste on a sheet of carrier film, dropping a layer of randomly oriented glass fibers, and adding a layer of paste on top of this. Clearly, the process uses natural resources. Not so clear is the fact that some hazardous chemicals are involved as well.

Premix' largest plant and corporate headquarters is based in North-Kingsville, Ohio. Premix is a good example of a medium-sized privately owned company which has to deal with environmental issues. Needless to say that Premix wants to stay in business as long as possible. The question arises whether they can sustain and even expand their current level of business of manufacturing materials and half-products in a technically, economically, ecologically, and ethically sound way. The key question we want the inter-disciplinary group of students to think about and answer is as follows:

What should Premix do in order to expand its facilities in an environmentally sound and sustainable way?

This question is very open-ended, like most real world problems. The case is set up to be open-ended and to form a bridge between Course 2 and 3. The students are given a large amount of (relatively) raw but relevant data and one pedagogical goal is to show them how to digest such amounts of material. The following case material will be distributed to the students: a short overview of SMC manufacturing; an interview with Dave Bonner, Premix's Chief Technical Officer and vice-president of research; a class report regarding the reduction of styrene emissions from the SMC process.; a copy of Premix's corporate business plan; reference material regarding emissions, processes, etc.; pictures of the actual manufacturing process; direct technical assistance from Premix.

The following exercises are part of answering the above question regarding Premix' expansion and help students to focus their thoughts: Create a flowchart of the SMC manufacture process; Identify the material and energy inputs and outputs at each stage in the flowchart; Perform an assessment of hazardous emissions and environmental impact; Identify technical, economical, ecological, and ethical constraints and goals relevant to enhancing sustainability from the attached case material; Identify and discuss major bottlenecks for Premix regarding sustainability; Identify a number of alternative ways to reduce the environmental impact and enhance sustainability. Two key issues to highlight in the case are that Premix is a real example of a larger group of companies and that short term solutions are often taken. The interview with Dr. Dave Bonner, Vice-President and Chief Technical Officer, provides good topics for discussion. For almost every question, one can talk about what Bonner's and Premix' reasons are for their choices and whether these are correct.

Case 3: The sustainability of Chlorine use - A Pulp and Paper Mill

The goals of this case are:

1. To explore the wider scale issues of chemical use through the example of the Great Lakes eco-economic system. 2. To explore the role of government and international policies in encouraging sustainable development. 3. To explore the possible responses of industry to potential regulatory changes.

There has been increasing concern about the role of chlorine containing compounds in the ecosystem. In addition to their toxic and carcinogenic risks there is emerging evidence of a subtle estrogenic effect on reproductive health. Chlorine and chlorine compounds are, however, a vitally important chemical to modern society. There are about 15,000 chlorinated compounds in commerce, including pesticides, disinfectants, and plastics. Almost 85% of the pharmaceuticals in the world use chlorine at some point in the reaction scheme,(Hileman, 1993).

The focus of the case will be a hypothetical pulp and paper mill situated in the Great Lakes area. Pulp and paper mills use chlorine and chlorine derivatives for bleaching pulp to make white paper. There is a significant amount of water used during the process, and even with emission controls some compounds can be released into plant effluents that reach the Lakes. The basis for the case will be that the International Joint Commission, the agency that act as an environmental watchdog for the Great Lakes, has proposed phasing out of chlorine use. The Province of Ontario has proposed reducing chlorine content of pulp and paper effluent streams to 1.5kg/ ton of pulp by 1995 and 0.8kg by 1999.

The key questions we want the inter-disciplinary group of students to think about and answer are as follows:

How should the mill respond, and how should government policy be designed to achieve these objectives and to approach zero emissions ?

The case materials will consist of the details of the particular setting of the mill, various reports on chlorine use and health effects, and sources for technical data on pulp and paper mill operations. The students will work in multi-disciplinary groups, one set of groups will be responsible for preparing the necessary policy strategy, and the other set for the company response. These groups will then present their findings to each other for critique. It is expected that the students will have to address the following types of questions, e.g.

1. Can the current levels of chlorine emissions from the mill be quantified and do solutions for reducing them exist ? 2. How much will it cost the mill to reduce chlorine emissions ? 3. Can the mill reduce emissions and still make product at a competitive price ? 4. Should we threaten closing the mill and taking the jobs elsewhere ? 5. Can we show that it isn't our effluents causing the problem ? 6. Will the government give tax incentives to comply with the new regulations ? 7. Can we design a tradeable permit scheme for pulp and paper mill emissions that will reach the overall target without closing down mills ? 8. How are we going to monitor compliance ?

Assessment and Evaluation

In examining the success of this program, we need to consider the students' learning ("assessment") and the effectiveness of the cases ("evaluation"). Assessment of the students' learning needs to be formative and summative. We will assess the working groups as well as the individuals. Group assessment will examine collaboration and how the groups consider the sustainability issues. This will be accomplished by videotaped analysis of the groups' activities. In addition, the students' collaborative discourse via CaMile will be considered. Some of the points that we will examine are group collaboration, individual contributions, discourse devoted to task-relevant activities, and learning strategies. In the research group, we will examine how these groups apply technical knowledge. For example, we will examine the extent to which students apply disciplinary principles such as mass and material balancing and life-cycle assessment.

In addition, we will use summative assessments of the individuals' learning and transfer of both the content they have learned and the values of sustainable technology to new problems. We will assess the extent to which students consider multiple perspectives in their problem-solving. We will use protocol analysis and written pre- and post- testing of the students' knowledge of sustainable technology to assess student learning.

Finally, we will assess the students' understanding of the nature of the issues in sustainable development by having them develop cases that can be used for future reference. These cases will be examined for evidence that the students collected information regarding the themes that were pervasive throughout the course: technology, ethics, economic, and ecology .

It will be important to evaluate the effectiveness of the cases used in this class. Some of the data gathered in the assessment of student learning will be reanalyzed using the case as the unit of analysis . By looking at problem-solving and collaboration as a function of the particular problem, we can look at the learning issues that students consider (i.e., do they match the objectives that the faculty had in mind when designing the cases) and if the students within each discipline apply the knowledge from their disciplines to the problem.

Future Issues

Our pilot implementation of this course will provide data that will inform subsequent instructional practice. We will use assessment data to refine the cases and to generate guidelines for the development of future cases. Future work will also be directed toward case enrichment and dissemination. As the cases are currently written, they are largely text with some pictorial information. In the future, we hope to add video and materials in other media to the existing cases. As these cases become rich multimedia resources, we plan to disseminate them via the World Wide Web to make them available to other educators. Finally, we hope to expand the information resources available to students by developing a case library that allows them to study both successes and failures in sustainable development in order to apply the lessons learned to their own problem-solving within the course and in future practice.

References

Barrows, H. S. (1985). *How to design a problem-based curriculum for the pre clinical years.* NY: Springer.

Bereiter, C., & Scardamalia, M. (1989). Intentional learning as a goal of instruction. In L. B. Resnick (Ed.), *Knowing, learning, and instruction: Essays in honor of Robert Glaser (pp. 361-392).* Hillsdale NJ: Erlbaum.

Bogard, W. (1989). *The Bhopal Tragedy: Language Logic, and Politics in the Production of a Hazard,* Westview Press, Boulder.

Boud, D., & Feletti., G. (1991). *The Challenge of problem based learning.* New York: St. Martin's Press.

Bransford, J. D., Vye, N., Kinzer, C., & Risko, R. (1990). Teaching thinking and content knowledge: Toward an integrated approach. *In* B. Jones &. L. Idol (Ed.s.), *Dimensions of thinking and cognitive instruction* (pp. 381-413). Hillsdale NJ:: Erlbaum.

Domeshek, E.A., & Kolodner, J.L. (1991). Toward a case-based aid for conceptual design. *International Journal of Expert Systems, 4(2): 201-220.*

Guzdial, M., Rappin, N., & Carlson, D. (1995). Collaborative and multimedia interactive learning environment for engineering education. *In ACM Symposium on Applied Computing 1995.* Accepted.

Hileman, B. (1993) Concerns Broaden over Chlorine and Chlorinated Hydrocarbons, *Chemical and Engineering News,* April 19, 1993.

Hmelo, C. E. (1994). Development of independent thinking and learning skills: A study of medical problem-solving and problem-based learning. *Unpublished doctoral dissertation,* Vanderbilt University.

Kletz, T. (1988). *Learning from Accidents in Industry,* Butterworths, London.

Kolodner, J. L. (1993) *Case Based Reasoning.* San Mateo, CA: Morgan Kaufmann Publishers.

Kurzman, D. (1987). *A Killing Wind: Inside Union Carbide and the Bhopal Disaster,* McGraw-Hill, New York.

Norman, G. R., & Schmidt, H.G. (1992). *The psychological basis of problem-based learning: A review of the evidence.* Academic Medicine, 67, 557-565.

Weston, R.F. (1993). Sustainable Development The Economic Model of the Future, Roy F. Weston Inc., Pa.

Supporting Collaboration and Reflection on Problem-Solving in a Project-Based Classroom

Mark Guzdial[1], Jorge Vanegas[2], Farrokh Mistree[3],
David Rosen[4], Janet Allen[5], Jennifer Turns[6], and David Carlson[7]

In this paper, we focus on technology to promote learning about design, particularly in situations where students are learning about design by *doing* design on significant projects. Two major learning needs arise in collaborative project-based classrooms. First, students need help in achieving their problem solving goals. Second, students need help reflecting on their experiences in such a way that they can get the most out of them and so that they can reuse their experiences in future situations.

CaMILE (Collaborative and Multimedia Interactive Learning Environment) [Guzdial, Rappin, & Carlson 1994] is a software support tool that facilitates and structures discussion among students (asynchronously and at diverse locations) as they are solving design problems. Using CaMILE, students comment on others' notes, explicitly identify the kind of comment being made (e.g., rebuttal, alternative, question), access libraries of information (including a case library), and annotate their comments with multimedia links, including links into a case base or network resources. CaMILE structures the discussion by defining the allowable types of comments and provides suggestions to students on what they might write in response to each type.

We are using CaMILE in two current contexts and planning to use it in a future context, all of which are relevant to this paper:
- In the first course in Mechanical Engineering Design at Georgia Institute of Technology, ME 3110, *Creative Decisions and Design*, as part of a *Design Learning Simulator*. The Simulator allows students to easily incorporate multimedia in their design activities, provides guidance in the design

[1]Assistant Professor, College of Computing, Georgia Institute of Technology, Atlanta, Georgia, 30332-0280

[2]Associate Professor, Associate Director Center for Sustainable Development and Technology, School of Civil and Environmental Engineering, Georgia Institute of Technology, Atlanta, Georgia, 30332-0350

[3]Professor, The George W. Woodruff School of Mechanical Engineering, Georgia Institute of Technology, Atlanta, Georgia, 30332-0405

[4]Assistant Professor, The George W. Woodruff School of Mechanical Engineering

[5]Senior Research Scientist, The George W. Woodruff School of Mechanical Engineering

[6]Graduate Research Assistant, School of Industrial and Systems Engineering, Georgia Insitute of Technology, Atlanta, Georgia, 30332-0205

[7]Graduate Research Assistant, College of Computing

learning process, and is Internet-based to allow students to collaborate on their assignments asynchronously and from any location. *Creative Decisions and Design* is the foundation course in Mechanical Engineering Design; eventually, we envision using the Simulator throughout the design sequence. In this course, learning about the science of design is orchestrated in the context of Decision-Based Design by using the Decision Support Problem Technique [Muster & Mistree 1985; Muster & Mistree 1988; Muster & Mistree 1989]. CaMILE's role in the Design Simulator is to scaffold collaboration and integration of resources in the Simulator, which facilitates learning in the computer-based Simulator.

• In the course sequence on sustainable development and technology being developed at Georgia Institute of Technology's Center for Sustainable Development and Technology [Vanegas & Guzdial 1995]. The objectives of the sustainable development and technology course sequence are to make students aware of the diverse issues to be considered in sustainable development decisions (e.g., ecological, economic, political, social, and technological) and to facilitate their decision-making and problem-solving processes with regards to sustainable development. The course sequence is particularly challenging because it's being offered campus-wide, crossing disciplinary boundaries. Thus, students taking the course may have little common background and a wide range of views and values on the issues of sustainable development. CaMILE's role in the sustainable development and technology course sequence is (1) to facilitate collaboration between students with diverse backgrounds and (2) to enable the integration of resources representing various views and constituencies.

• In the design education effort at Georgia Tech being spearheaded by Tech's EduTech Institute Director Janet Kolodner [Kolodner 1995]. The goal of the design education effort is to help undergraduates at Tech develop design skills and integrate design education experiences between a Freshman Design Experience and a Senior Capstone course. CaMILE's role in this context is to facilitate collaborative design and to provide continuity between the different design experiences that undergraduates at Tech have.

Of these three contexts, the greatest use has been in the Mechanical Engineering class, so we use that course as a detailed case study in this paper for how technology can support collaboration and reflection on problem-solving in a project-based classroom. In the following sections, we describe the course design philosophy and approach used in the Mechanical Engineering course; discuss the learning principles around which we have built the course; and how CaMILE (and other technologies) serve to support these approaches and principles.

Engineering Design Approach

Decisions about the format and content of education about engineering design should be guided by three factors: current perspectives on learning how to design as presented in the emerging design science literature, cognitive science theories about learning, and classroom and/or pragmatic considerations. A design course thus needs to ensure that it reflects well-accepted views of the design process including those by Pahl and Beitz [Pahl & Beitz 1984], Dixon [Dixon 1991], and others. A design course should also embody our best understanding of the constructivist theories of how students learn. Finally, any course needs to operate within the time, space, and other constraints imposed by external sources.

Decision-Based Design is a term coined to emphasize a different perspective from which to develop methods for design. The principal role of a designer, in Decision-Based Design, is to make decisions. In general, decisions are characterized by information from many sources (and disciplines) and may have wide ranging repercussions. In Decision-Based Design, decisions serve as markers to identify the progression of a design from initiation to implementation to termination. The implementation of Decision-Based Design can take different forms. Our approach is called the *Decision Support Problem (DSP) Technique* which consists of three principal components: a design philosophy, an approach for identifying and formulating Decision Support Problems (DSPs), and software.

Decision Support Problems provide a means for modeling decisions encountered in design. Formulation and solution of DSPs provide a means for making the following types of decisions:

- **Selection** - the indication of a preference, based on multiple attributes, for one among several alternatives.
- **Compromise** - the improvement of an alternative through modification.
- **Coupled or hierarchical** - decisions that are linked together; selection/selection, compromise/compromise and selection/compromise decisions may be coupled.

Applications of DSPs include the design of ships, damage tolerant structural and mechanical systems, the design of aircraft, mechanisms, thermal energy systems, design using composite materials and data compression.

Learning Principles

Engineering design education based on constructivist theories of cognitive science suggests that effective learning occurs in a project-based situation that is at once (that is, in an integrated manner) authentic, supportive of the learning process, and scaffolded. Authenticity is a factor in increasing student motivation and to increase the likelihood of transfer between the classroom situations and real world situations [Koschman, Myers, Feltovich, & Barrows 1994]. Students should be asked to solve problems that are similar in complexity and in components to problems that they will be facing in their post-graduation experience as designers.

In order to support the learning process, students need opportunities to identify and articulate problems, reflect on these problems until they reach a solution, and then articulate their solution and what they learned. In engineering design classrooms, articulation opportunities are usually limited to essay testing or interpretation of homework assignments. In the ME course, students are encouraged to learn *about* learning and design through additional activities such as written "What I learned" essays and classroom reflection exercises. Articulation and reflection occur frequently in collaborative environments in which knowledge sharing is occurring. The audience in such environments can provide feedback, offer contrasting views, ask for clarification, or extend ideas – all of which improves student understanding and learning through increased reflection and often additional articulation. In most classroom situations, the instructor is the only audience for the students' work, so there is limited opportunity for this interaction with an audience.

The third implication of a constructivist view of design education is that learning opportunities should be *scaffolded* [Collins, Brown, & Newman 1989; Guzdial 1994]. Scaffolding is generally defined as support for (1) doing an activity and (2) learning through and about that activity. We can envision providing scaffolding for the many activities going on in a project-based learning environment:

- *Scaffolding for the activity of design* itself which can include modeling the process of design through introduction of case studies; providing tools which communicate and facilitate good design process; creating opportunities for students to interact with one another and thus elicit articulation and reflection about their designs and design process; facilitating or guiding design activities.
- *Scaffolding for collaboration* which is an activity that students are often weak at. Scaffolding for collaboration can include facilitating interaction (e.g., encouraging use of productive phrases, discouraging use of unproductive comments); suggesting useful ways to collaborate; and pointing out opportunities for collaboration.
- *Scaffolding for case interpretation* may play an important role in gaining transfer of design learning. By learning to interpret a case abstractly, students may be able to recognize more easily how two different design problems are related, which will facilitate reuse of knowledge from one problem in another. Scaffolding for case interpretation may include prompting a student to describe a design problem; coaching them about how to abstractly interpret a design problem; and modeling the process through example interpretations of design problems.

In the sense that the DSP Technique identifies components of a design process and describes how to participate in that process, the DSP Technique is scaffolding by modeling the activity of a design process. More directly and explicitly, the instructors provide scaffolding through their modeling of process, working with and coaching students, and encouraging students to articulate and reflect. Through use of computer-based tools, we can more actively provide scaffolding for students' design activities while they are engaged in those activities.

The challenge for providing all of these supportive facets of an effective design learning environment is to provide an integrated course, not a disconnected collection of useful exercises (which destroys authenticity and thus motivation). While scaffolding can be provided by an instructor, and is currently provided by the instructors of ME3110, it is difficult to keep all of the pieces tightly integrated. Often, the result is that students feel that they have many separate things to do and learn as opposed to learning a single skill (design) which has integrated components. For this reason, we are creating a computer-based design simulator (a simulation in contrast to a complex, multi-month, high-pressure industrial design experience) which provides an integrated, scaffolded design experience with access to a wide variety of resources.

Technological Supports for Problem-Based Design Education

CaMILE extends and integrates the existing resources in ME 3110. CaMILE is a network-based (Internet) collaboration environment which provides a forum for sharing, discussing, and reflecting as well as integrating several information bases. What makes CaMILE different from other network resource browsers is the emphasis on (1) having students actively post material to the network and relate resources they find there to their own discussions and activities and (2) guiding the students in this activity. Guidance and technological support can facilitate both the doing and learning of design [Guzdial, Weingrad, Boyle, & Soloway 1992].

The core of CaMILE is a collaborative NoteBase where students can post multimedia notes in group discussions. The NoteBase builds on work previously done in the area of student created multimedia authoring [Hay, Guzdial, Jackson, Boyle, & Soloway 1994]. One solution is for the student to create a document in which the text is on one side and all of the multimedia annotations are placed on the other side of the document, called the multimedia margin. In order to maintain the relationship between the multimedia annotations and the text, the annotations are positioned in the margin so that the relationship is readily apparent. Figure 1 shows CaMILE open to a group discussion and one of the students' notes in that discussion.

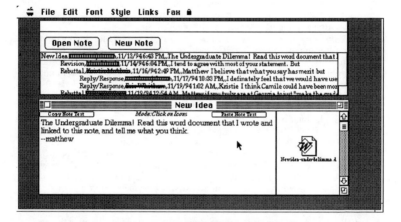

Figure 1: CaMILE Discussion and Note

CaMILE's NoteBase allows for several types of annotations, including the following: documents (including word processing and spreadsheets), pictures, sounds, videos, links to the World Wide Web [Berners-Lee, Cailliau, Luotonen, Nielsen, & Secret 1994], links into the multimedia databases (discussed below), and links into a case-based design aid [Domeshek & Kolodner 1992]. CaMILE's NoteBase is different from other kinds of collaborative learning environments (such as CSILE [Scardamalia, Bereiter, McLean, Swallow, & Woodruff 1989]) and multimedia communication environments (such as the World Wide Web or Usenet newsgroups) in three respects.
- First, the CaMILE NoteBase is focused on integrating diverse media: From any kind of document on the desktop, to multimedia databases, to case-bases. Integration supports students in discovering relationships between various kinds of problems, views, and media. By recognizing abstracted relationships which highlight similarities between problems and solutions, we hope to encourage students in developing skills that will lead to enhanced transfer between subject areas.
- Second, while other collaboration environments have been focused on collaborative discourse, we are focusing on supporting collaborative design,

of which collaborative discourse is a subset. Using the NoteBase to support collaborative design has placed several additional demands on the environment. The range of interactions between students is different: setting up meetings, collaboratively (yet asynchronously) creating final documents, and having discussions about design issues. By providing for integration of design tools, CaMILE has essentially become a window into a set of tools that the students can use to support design.

• Third, like CSILE but unlike the World Wide Web or newsgroups, the NoteBase provides support (scaffolding) to aid the process of collaboration.

As seen in Figure 2 (left side), CaMILE asks students creating new notes to identify the *kind* of note that they are creating in terms of the contribution's role in the overall discussion (e.g., raising a *New Idea*, or making a *Rebuttal*, or identifying a *Question*). Based on the kind of note identified, CaMILE then makes *suggestions* about what are good things to say in this kind of note (Figure 2 right side). This *scaffolding for collaboration* in CaMILE is based on the *procedural facilitation* used in CSILE [Scardamalia et al., 1989].

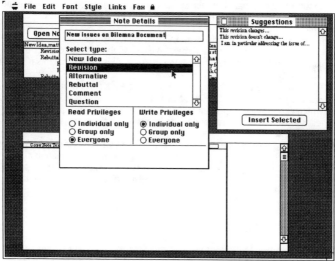

Figure 2: Prompts in CaMILE for kind of note and suggestions

Along with the NoteBase, CaMILE provides multimedia databases which students can review for more information on a topic or to gather evidence to be linked to a note.

• One of these resource collections is a multimedia database, MediaBase, which collects a wide variety of media into nodes (text with multimedia annotations, like a note from the NoteBase) and relates nodes into collections called *paths* [Rappin, Guzdial, & Vanegas 1994]. The MediaBase provides a non-traditional medium for presenting information to students. Its advantages include its ability to integrate all types of information into the notes, the flexibility in linking these notes together (demonstrating to the students the tight interconnections between the ideas), and its feature whereby several interpretations (e.g., a philosophical, economic, native American Indian, etc.) on a single note can be stored.

- CaMILE also provides a facility to create and use Electronic books. Both the MediaBase and the Electronic books allow students to view material in non-linear manner and allow the students to create annotations (e.g., margin notes, underlined pages, links from sections of texts to notes in the CaMILE NoteBase) and otherwise personalize the environment with their own thoughts and ideas. Textbooks available in word processing documents can be easily converted into the electronic book form. The advantages of this format include the ability of students to integrate the material into their discussions through annotations, to copy and paste the text material into their documents, and to make personal annotations and share these annotations without defacing the text. These two resource collections provide students with alternative ways of interacting with the course content.

Having students use CaMILE to support their design process provides us with unique opportunities to study both the student design process (not just the product) as well as the student use of technology and media. Our assessment of CaMILE focuses on three kinds of data:

- CaMILE currently records, in addition to the notes that students create, extensive log files of their activities. We can analyze these to study the usability of the software and the process strategies of the students [Guzdial, Walton, Konemann, & Soloway 1993].
- We also have the NoteBase itself. The Notes are a form of *intermediate artifact* [Blumenfeld, Soloway, Marx, Krajcik, Guzdial, & Palincsar 1991] which can provide important insight into the students' designing and learning activity. We are conducting linguistic and content analysis of the Notes.
- Finally, we do have the students' artifacts, the end-product designs of their process. By comparing the Notes of CaMILE-using students, their products, and the products of non-CaMILE-using students, we are seeking insight into the role that the collaboration environment is playing in the students' designs.

Current Status and Directions

CaMILE has now been in use for two quarters in the Mechanical Engineering course and one quarter in the Sustainable Development course sequence. The students using CaMILE in ME, many of whom had rarely used a compute prior to the course, volunteered to participate in a one credit laboratory, explicity as an experiment but also as an opportunity for students to learn more about technological support for design and learning. The students were required to attend a laboratory session for three hours each week, but were encouraged to use CaMILE on their own time at any computer facility on campus. After about two weeks of activities designed to teach them how to use CaMILE as well as word processing software, drawing packages, and electronic mail, they started using CaMILE outside of the classroom. Additional activities were created to demonstrate other uses of CaMILE to the students (distance collaboration to answer questions, creation and participation in discussions designed to promote articulation and reflection on issues related to design, etc.).

Outside of the in-class activities, students used CaMILE on their own in support of their design activities and in several novel ways. They used CaMILE to integrate their design work, to communicate with each other, and to create histories of their design decisions. In the first quarter of use, the 14 students, TA, and software support person generated well over 400 Notes with over 100 links to various kinds of resources.

In addition, the content of students' notes suggests that students are (1) recognizing the strengths of the environment (both on its own and in relation to other technologies) and (2) the importance of a collaborative approach to design, as seen in the following quotes:

- It could be that a situation arose six months ago that influenced your decision that people have long forgotten about. With CaMILE you can go back and see exactly what went on and what you as an individual and a group were thinking.
- WWW - I think the www is probably most effective for communicating information to a large number of people, most of which you are not looking for a response from.
  CaMILE - I think CaMILE is most effective for communicating information to a small group of people who you want responses from and want to carry on a conversation with, but don't want everyone interrupting.
- To sum it up. I guess I'm learning a lot about what the most effective means of communicating with people is. I used to think that calling everyone is the easiest, but now I'm realizing that e-mail and CaMILE are much more effective use of my time for communicating information to our group.
- The more I think about how we have used CaMILE the more I realize how much more we can learn from talking with others. I have always been the type to try and figure things out myself using what I know and my own point of view. It is very valuable to actually talk to others and find out what they think and what they know.

That the students gained these kinds of insights is notable, because using CaMILE and integrating it into the existing technological infrastructure has been a significant challenge. Students had to deal with networks on the same campus which ran incompatible operating system versions, with computers running the smallest possible monitors and memory configurations, and with conflicting and oftentimes buggy software.

While we feel that CaMILE has been a successful intervention thus far, we also recognize the limitations of our approach. The most critical are probably the following three:

- While CaMILE provides *scaffolding for collaboration*, it is an environment that supports other kinds of activities which are *not* scaffolded. For example, students using CaMILE integrate a wide variety of resources, including a recent development of links to a case base. However, the activity of *finding relevant resources* is a complex one, well worth scaffolding to support the learning and doing. Further, collaboration is much more complex than simply rhetorical roles and suggested phrases. Thus, CaMILE's scaffolding is incomplete for the activities it supports.

- There is much about learning to design and to learn in problem-based classrooms which is not at all supported in CaMILE. For example, the activities of reflection, articulation, and design itself are unsupported in CaMILE. While CaMILE provides *opportunities* for these activities, there is no support to enhance student learning and doing in these activities.
- Finally, CaMILE is a relatively generic environment right now. Much could be done to deeply integrate CaMILE into the ME 3110 class, the sustainable development classes, and the EduTech Design Education Initiative. For example, CaMILE could be made aware of the curriculum being used and it might adjust the kinds of prompts and suggestions to reflect the developing complexity of the students' understanding of the course domain. Further, the design of CaMILE does not yet reflect an understanding of the problem-based learning process. We envision CaMILE becoming more integrated with course-specific materials, so that collaboration becomes part-and-parcel of study and design in these classrooms.

In sum, we view CaMILE as a natural extension of the kinds of activities already ongoing at Georgia Tech. CaMILE integrates well with the design and learning philosophies on which ME 3110 is based. It provides an important integrative component and provides for enhanced learning through collaboration as part of a design learning simulator. However, we also note that we are in the midst of an ongoing research project, to understand and improve design education through technological supports to support learning in a problem-based classroom.

References

Berners-Lee, T., Cailliau, R., Luotonen, A., Nielsen, H. F., & Secret, A. (1994). The world-wide web. Communications of the ACM, 37(8), 76-82.

Blumenfeld, P. C., Soloway, E., Marx, R. W., Krajcik, J. S., Guzdial, M., & Palincsar, A. (1991). Motivating project-based learning: Sustaining the doing, supporting the learning. Educational Psychologist, 26(3 & 4), 369-398.

Collins, A., Brown, J. S., & Newman, S. E. (1989). Cognitive apprenticeship: Teaching the craft of reading, writing, and mathematics. In L. B. Resnick (Ed.), Knowing, Learning, and Instruction: Essays in Honor of Robert Glaser Hillsdale, NJ: Lawrence Erlbaum and Associates.

Dixon, J. R. (1991). The State of Education, Parts I & II. Mechanical Engineering(February and March), 64-67 (February), 56-62 March.

Domeshek, E. A., & Kolodner, J. L. (1992). A case-based design aid for architecture. In J. Gero (Ed.), Proceedings of the Second International Conference on Artificial Intelligence and Design

Guzdial, M. (1994). Software-realized scaffolding to facilitate programming for science learning. Interactive Learning Environments, In Press.

Guzdial, M., Rappin, N., & Carlson, D. (1994). Collaborative and multimedia interactive learning environment for engineering education. In ACM Symposium on Applied Computing 1995 (In press.).

Guzdial, M., Walton, C., Konemann, M., & Soloway, E. (1993). Characterizing process change using log file data (GVU Center Technical Report No. 93-44). Georgia Institute of Technology.

Guzdial, M., Weingrad, P., Boyle, R., & Soloway, E. (1992). Design support environment for end users. In B. A. Myers (Ed.), Languages for developing user interfaces (pp. 57-78). Boston, MA: Jones and Bartlett.

Hay, K. E., Guzdial, M., Jackson, S., Boyle, R. A., & Soloway, E. (1994). Student composition of multimedia documents: A preliminary study. Computers and Education.

Kolodner, J. (1995). Design Education Across the Disciplines. In Second Congress on Computing in Civil Engineering. Atlanta, GA: American Society of Civil Engineers.

Koschman, T. D., Myers, A. C., Feltovich, P. J., & Barrows, H. S. (1994). Using technology to assist in realizing effective learning and instruction: A principled approach to the use of computers in collaborative learning. Journal of the Learning Sciences, 3(3), 225-262.

Muster, D., & Mistree, F. (1985). A curriculum and paradigms for the science of design. Annual Conference Proceedings of the American Society for Engineering Education.

Muster, D., & Mistree, F. (1988). The decision-support problem technique in engineering design. International Journal of Applied Education, 4(1), 23-33.

Muster, D., & Mistree, F. (1989). Engineering design as it moves from an art toward a science: Its impact on the education process. International Journal of Applied Engineering Education, 5(2), 239-246.

Pahl, G., & Beitz, W. (1984). Engineering Design (A. Pomerans, Trans.). London/Berlin: The Design Council/Springer-Verlag.

Rappin, N., Guzdial, M., & Vanegas, J. A. (1994). Supporting distinct roles in a multimedia database. InConference Proceedings of the Third Annual Conference on Multimedia in Education & Industry (pp. 202-204). Charleston, SC: Association for Applied Interactive Multimedia.

Scardamalia, M., Bereiter, C., McLean, R., Swallow, J., & Woodruff, E. (1989). Computer-supported intentional learning environments. Journal of Educational Computing Research, 5(1), 51-68.

Vanegas, J., & Guzdial, M. (1995). A Collaborative and Multimedia Interactive Learning Environment for Engineering Education in Sustainable Development and Technology. In Second Congress on Computing in Civil Engineering. Atlanta, GA: American Society of Civil Engineers.

Improving the Management Decision Process at Fort Stewart, Georgia

Bruce C. Goettel[1]

Abstract

A changing Army mission, troop restructuring, and reduced funding necessitates optimizing decision-making processes currently facing the Directorate of Public Works (DPW) at Fort Stewart, Georgia. While many Army installations are seeing reduced military strength, Fort Stewart is anticipating an increase in military personnel. However, at the same time the civilian work force and installation budget are facing funding cutbacks. The DPW is responsible for maintaining the base infrastructure at a Army installation, and has the responsibility to perform many engineering functions such as maintenance, repair and construction, engineering design and services, facilities operation, and land management. Given the constrained funding available for these functions, management and the installation engineers needs better tools to create optimal facility investments.

This paper describes a study of the information requirements at the DPW at Fort Stewart, Georgia which was undertaken by the U. S. Army Corps of Engineers Construction Engineering Research Laboratory (USACERL). This paper describes the process USACERL used in determining the information requirements of the DPW, the resulting steps used to prioritize the requirements, the recommended solutions, and the phasing needed to develop the solutions.

With most of the requirements involving the two engineering divisions, interconnectivity became the primary goal and the development of a DPW network infrastructure became the top priority. However, simply installing

---

[1]Principal Investigator, U. S. Army Construction Engineering Research Laboratories, P.O. Box 9005, Champaign, IL 61826-9005

344

a bare bones network was not a realistic answer, some functionality needed to be provided, as well as a game plan for future action. This paper describes the solutions, both long and short range, developed to give the engineering division both connectivity and functionality with the constraints of a limited budget along with a uncertain potential for additional funding.

Introduction

A changing Army mission, troop restructuring, and reduced funding necessitates optimizing decision-making processes currently facing the Directorate of Public Works (DPW) at Fort Stewart, Georgia. While many Army installations are seeing reduced military strength, Fort Stewart is anticipating an increase in military personnel. However, at the same time the civilian work force and installation budget are facing funding cutbacks. The DPW is responsible for maintaining the base infrastructure at a Army installation, and has the responsibility to perform many engineering functions such as maintenance, repair and construction, engineering design and services, facilities operation, and land management. Given the constrained funding available for these functions, management and the installation engineers requirements, better tools to create optimal facility investments.

The DPW engineer must live with the consequences of declining budgets and the need for improved information management. The information may exist but is on different systems, in different locations and not easily accessible. Much time is spent in information collection and reporting to others in the organization. Data bases need to be designed and made accessible to many organizational elements. A variety of automation hardware and software exists in the organization, but limited financial resources increase the need for sharing.

With most of the requirements involving the two engineering divisions, interconnectivity became the primary goal and the development of a DPW network infrastructure became the top priority. However, simply installing a bare bones network was not a realistic answer, some functionality needed to be provided, as well as a game plan for future action. The study identified solutions, both long and short range, developed to give the engineering division both connectivity and functionality with the constraints of a limited budget along with a uncertain potential for additional funding.

Fort Stewart proved to be an excellent site for this study. Unlike many Army installations, Fort Stewart is expecting a increase in the numbers of military personnel, but this is coupled with a reduction in the numbers of the civilian workforce which maintains the infrastructure necessitating an enhancement in the current way operations are performed. Also, the

Colonel in charge of the project and the division heads and managers were very supportive of automation efforts and all worked to support our efforts.

Objective

The objective of this effort was to develop an overall information plan for the Fort Stewart DPW. The plan would include cost estimates for each action and a plan for acquisition, development, and deployment which would be prioritized by return on investment and any necessary precedent relationships. In addition, the following goals for the first year implementation were established:

o   Provide a balance between functionality, connectivity, savings, and cost

o   Support expendability in direction of a long-range configuration

o   Maintain consistency with Army-wide automation systems of present and those under development

o   Maintain consistency with the existing automation standards at Fort Stewart

o   Require no additional funding to keep systems operational (other than maintenance) once development was complete

Approach

A six step approach was followed in developing an overall information systems plan for the Fort Stewart DPW.

Analysis of Requirements. The first step in this process was to determine the information requirements of Fort Stewart. Key Fort Stewart DPW personnel were interviewed along with personnel from other parts of the organization to define their requirements. As a result of these interviews, an assessment of the existing hardware, software and communications configuration was developed. A analysis of these interviews also generated a list of over fifty separate functional and connectivity initiatives for inclusion in the plan.

Development costs were developed for each initiative by estimating equipment, software and programming costs. The most costly initiative being creation of a local area network infrastructure.

Return-On-Investment Analysis. A return-on-investment analysis was performed on selected initiatives. For each selected initiative, a description of the baseline (existing) and proposed methods of operation were presented and non-quantifiable benefits identified. Calculations were developed which estimated the cost savings.

Funding Guidance. Costs estimates were also developed for an 8-year system life cycle and documented in a report which was distributed to Fort Stewart before the prioritization meeting.

Prioritization. The functional and connectivity items were presented to the Fort Stewart DPW Information Steering Committee which consisted of representatives of all of the DPW divisions. The report was presented and each item was quickly described. After each item was discussed, each committee member rated it on a 1-5 scale. Overnight, the results were tabulated and the prioritized list was presented to the committee. After a couple of minor changes, the list was divided into grouping of the most mission-critical (Category 1) and those of lessor mission critically (Categories 2 and 3).

Figure 1 is an example of the top rated functions from the prioritized list. This list is displayed by their old number. Not all functions had Life-Cycle costs calculated, so those fields are blank.

Analysis of Technology. Alternative available technologies were considered to support the DPW requirements. The technologies considered included USACERL developed systems, interfaces to existing Army-wide systems, commercially available systems, and custom solutions.

Analysis of Alternative Implementation Strategies. With a limited amount of funding available, various implementation strategies were considered so that a balance among Functionality, Connectivity, Cost and Savings could be achieved. The implementation process could not be accomplished by simply selecting the first n initiatives until the money ran out. Several functions required another function to be implemented first or equipment to exist before they could be developed. Others could be partially satisfied with a portion of the estimated requirements (data storage for example). Most all the initiatives required the existence of a local area network before they could occur.

Analysis of Technology. Alternative available technologies were considered to support the DPW requirements. The technologies considered included USACERL developed systems, interfaces to existing Army-wide systems, commercially available systems, and custom solutions.

LIST OF PRIORITIZED FUNCTIONS

Number	Description	Life-Cycle Labor Savings	Savings/ Investment Ratio	Total Cost	Initial DEH-Wide Mission Priority	Category
1	Housing Phone Mail	$577,266	16.5	$35,000	5.0	1
2	Work Services Status Dial In System	$83,210	2.4	$35,000	4.8	1
52	Network Infrastructure			$1,488,950	4.8	1
50	Training for the Network			$300,000	4.7	1
4	Inventory Bar Coding for the DEH Supply Operation	$2,021,096	33.2	$60,000	4.6	1
45	Network Security Software and Menu System			$80,000	4.6	1
31	Complete Service Order in the Pick Up Trucks	$2,708,615	54.2	$50,000	4.6	1
30	Print Service Orders to the Maintenance Shops and in pick up trucks	$1,525,522	4.6	$330,000	4.5	1

Figure 1 Example of Functions List in DPW Priority Order

<u>Analysis of Alternative Implementation Strategies.</u> With a limited amount of funding available, various implementation strategies were considered so that a balance among Functionality, Connectivity, Cost and Savings could be achieved. The implementation process could not be accomplished by simply selecting the first n initiatives until the money ran out. Several functions required another function to be implemented first or equipment to exist before they could be developed. Others could be partially satisfied with a portion of the estimated requirements (data storage for example). Most all the initiatives required the existence of a local area network before they could occur.

Table 1 shows the balance that was achieved among the Functionality, Connectivity, Cost and Savings strategies. Under the connectivity consideration, the table shows we were able to connect 13 out of 17 buildings and connect 83 out of 122 workstations. The connection would be made by either a fiber link between the buildings or by a remote connection via a modem.

For functionality, we were able to totally or partially satisfy 13 out of 23 for the Category 1 items, 2 out of 14 Category 2 items, and 3 out of 14 for Category 3 items. Several Category 2 and 3 items were included because of the ability to complete these items at a minimal cost since they used the same automation solution for a Category 1 item. "Partially satisfy" means that (1) not all of the DPW staff would have access to the function,

Table 1: Assessment of First-Year Automation Strategy

Criteria	Type	Satisfaction Number / Possible
CONNECTIVITY	Repository Interbuilding Intrabuilding	1 / 2 13 / 17 83 /122
FUNCTIONALITY	Category 1 Category 2 Category 3	13 / 23 2 / 14 3 / 14
POTENTIAL SAVINGS		$ 5.1M / $10.1M
TOTAL COST		$ 1.2M / $ 5.2M

or (2) the hardware/software provided would not address the total capacity for the function.

The plan projected a potential savings estimate of $5.1 million out of a total of $10.1 if we had covered everything. The total cost was estimated at $ 1.2 million out of a possible $5.2 million if we have covered everything. Overall, for approximately 1/4 the total Life-Cycle cost, a savings of approximately 1/2 of the Potential Life-Cycle was realized.

Project Implementation

Several minor changes, reflecting Fort Stewart's concern for future funding implications, were made to the scope of the project and approval was given for the implementation of the first years work.

Building the automation infrastructure was the first priority of the first phase as so many of the functional initiatives depended upon a LAN for connectivity. Novell Netware 3.11 was selected for the network software, Hubs from Cabletron were selected for the network hardware. DPW personnel installed the network cabling and CERL personnel installed the Novell software. Several network applications are being developed. In time, as the new users become cognizant of the potential of networks, additional capabilities which were identified by the study for Phase two can be added. The Fort Stewart study is meant to be revisited each year, reviewed and updated to provide for changes in requirements and increased network usage.

Summary

The automation study and the initial implementation has contributed to a constructive beginning in developing solutions to Fort Stewart's automation needs. At the time this paper is being written, the network has been functional for about six months. Additional buildings are being considered for connection to the network. How well were the needs anticipated, the solutions developed, how did the implementation proceed, and what problems were incurred, are all questions that need to be studied at a future time. USACERL has developed a good relationship with Fort Stewart and hopes to continue to provide assist them in their automation efforts.

Integrated Computer Information System in Facility Service Department of A School District

Yi Liu[1], Matt Syal[2], Mike Spearnak[3]

ABSTRACT: Integrated computer information systems represent the latest trend in the architecture, engineering and construction (AEC) industry because they provide fast, accurate and consistent information flow by representing and transferring information effectively. Several issues are studied in this paper that includes the existing problems in construction project management practice, how the applications of integrated computer information system affect the production and productivity of the real world construction industry, the limitations of the system, and the role of integrated computer information system as a key tool in the whole project management system.

Introduction

In today's environment of large projects with short schedules, a slim margin of profits and increasingly complex management requirements, better control of project costs and construction duration can be maintained because a great deal of software is available to deal with design drafting, construction scheduling, and detailed cost estimating of construction projects.

In the past few years, initial progresses have been achieved in computer integration of individual application packages, while in the construction industry, the net impact on the construction project management decision-making process has not fully kept pace with the glowing promises of new integrated computer systems. Many construction companies have computerized their operations to a certain extent, but most small to medium size construction companies do not have

---

1. Graduate Assistant, Construction Management and Civil Engineering, 224A Guggenheim Hall, Colorado State University, Fort Collins, CO. 80523
2. Associate Professor, Department of Agricultural Engineering, 203 Farrall hall, Michigan State University, East Lansing, MI 48824-1323
3. Construction Project Manager, Poudre School District R-1, 2407 LaPorte Avenue, Fort Collins, CO. 80521-2297

such integrated computer systems, and still use several different software packages separately for a single project.

Poudre School District R-1 has 42 elementary, junior and senior high schools. The total cost estimate of construction projects is about $118,431,000 during 1991-1995 based on the estimated cost on November 30, 1994. As of November 1994, seventeen building projects were completed, eight projects were under construction and, six projects were in design or planning stages. As owner's representative and construction project manager, the facility service department of the Poudre School District R-1 (brief as PR-1 in the following text) provides services for the school district's new or remodel construction, from conceptual planning through all the construction processes to facility maintenance. Due to a lack of enough experienced professionals, the need for establishing computer-aided information system becomes more important since existing computer facilities are only available to perform routine financial accounting and daily office word processing.

Although the PR-1 is not a professional construction company, its problems are common to many small to medium construction companies. They do not have enough professional personnel, and they have difficulties in investing time and money on computerization. But they would like to make a greater commitment to the long term benefits from computerization.

Since some integrated construction estimating/scheduling, resource and facility management (FM) commercial packages are available, PR-1 started the setting up of the integrated computer information system at the facility service department and, its final goal is to modernize its management process gradually by integrating accounting, job cost, construction estimating and scheduling, facility management system and the historical database, to control projects from conceptual planning through facility maintenance.

This paper discusses the projected benefits and limitations of the computer applications being implemented in this school district. It shows the use of computer information system as possible solutions to the problems being faced by the facility service department of the school district. The paper concludes with the discussion of the role of integrated computer information system for the school district in the next 5-10 years.

Computer Hardware and Software
        Following steps were considered for setting up a computer integrated information system:
        set up necessary computer hardware and software
        scan blue drawings and convert them to the AutoCAD format
        incorporate drawing standards including names, sizes, scales, symbols, etc.

develop user-friendly archive system and backup standards
develop system for training PR-1 staff in system use
integrate AutoCAD drawings with FM system
setup a computer-based historical database for design, technical
    specification, facility management,   financial accounting, primary
    contractor performance records and all other related information.

The information flow for a typical construction project at PR-1 is shown in figure 1.

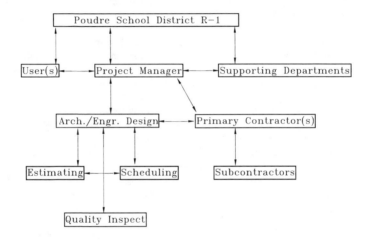

Figure 1. Information Flow of A Typical Project at PR-1

The basic computer hardware and software setup is as follows:

Hardware: 20" Compaq QVision color monitor; HP Vectra LE486/66 (8MB RAM, 1000 MB fast SCSI disk); HP DesignJet 650 Color Plotter; HP 5GB DAT Tape Drive; HP Laserjet 4 Printer; Summagraphics 12x12 Tablet;

Software: AutoCAD for Windows   R12; ASG   for Windows Version 6; Microsoft Project for Windows Version 3; EXCEL for Windows Version 4.0; Microsoft Word for Windows Version 6.0; Archibus (to be procured)

Problems & Computer-Based Solutions
1.Planning Stage: Without a computer based construction project database, PR-1 estimated the construction budget and duration of a project based on previous experience and on whatever information was available at its conceptual planning

stages. It led to certain differences between the approved budget and preliminary estimating cost from designers at the later stage. Once funding was available, the only source for further evaluating project cost and schedule used to be pre-bid estimating from the architects, who were in charge of the project design and most of them usually used a spreadsheet for the preparation of their estimating for the school district. With the help of a computer integrated system and taking advantage of the historical database, PR-1 can do a much better job at planning stages, by performing estimating and scheduling in a more consistent way. Besides for reducing repetitive data exchange, the computer integrated information system has great advantages of keeping potential information for using in the subsequent stages and create a situation of project transparency.

2.Design and Specifications: Since there were almost no automation and standards for construction project management at PR-1, different architects, or sometimes even the same architect, used different symbols and conventions in construction drawings. This led to confusion and conflicts among project participants. During the detailed design stage, PR-1 needs to provide technical specifications to the designers. Although the electrical and mechanical specifications are available at PR-1, there was no general specification data. This took PR-1 staff a lot of repetitive efforts and time without a construction project historical database to maintain such information. Therefore, outside professionals had to be consulted, and this led to differentiation from one project to another. If standards are developed and required for all construction drawings, technical specifications can be traced systematically from the database, problems at construction procurement stages will be reduced greatly at design phases.

3.Monitoring Construction Process: PR-1 places heavy emphasis on monitoring the construction process because most of the projects are time driven. Checking field reports and job sites are the major measurements for monitoring projects at PR-1. For large, complex projects or multi- small/medium projects, PR-1 hires professional construction managers or consultants. These construction management or consulting companies normally have some computer-aided tools for supporting their work. Without a computer integrated information utilization system at PR-1, it is almost impossible for PR-1 construction managers to catch real-time information since the physical monitoring in real time is getting more difficult. The computer integrated information system can provide a powerful tool and convenient environment to PR-1 construction managers with timely information. So PR-1 construction managers may compare the actual project progress with the planned schedule, even trace the actual cost of labor, materials, equipment and subcontract.

4. Reporting System: The facility service department is supposed to maintain all the school facility drawings and related paperwork. Years after construction, it is usually found that the original blueprints are almost useless because these drawings

are either incomplete or in good conditions. As part of the project, these blueprints are being scanned and converted to the AutoCAD format. Since many schools are kept remodeling, PR-1 is going to integrate all these addition plans for tracking update information. All these files are organized in a way to be compatible with the PR-1 internal accounting code system and backed on tape regularly.

Limitations and areas of Future Research

The processes of computer-integrated systems have their primary value in creating an organized system for planning, scheduling, and budgeting which in turn provide an information framework for project control purposes at PR-1. On the other hand, there are certain obstacles to the development and application of computer-integrated system in construction project management.

1. A major problem in computer applications for construction which is currently being faced at PR-1 is that of systems and software integration. There are many good software packages available for the construction industry. But most tend to provide limited integration between design drawing and estimating, estimating and scheduling, i.e., they are basically a one-way low-level data transfer between modules and are not quite flexible. So two-way communication is one of the areas of current interest in computer integration.

2. Most software packages are far from user-friendly. They require some kind of training and many are difficult to be customized. But construction managers demonstrate a great diversity in approach because they use different techniques and methods in solving real world problems, so it is a painful process for PR-1 to train its staff in using computer information system.

3. Even with the current realm of sophisticated software, one very important aspect of project control is "coding intelligence". But most of those packages are lack of some kind of human intelligence. Many of the critical decisions in construction management are still made based on experience, intuition and judgement. Therefore, more efforts needed to ensure that the computer information system can help human being in their decision-making process by stimulating human knowledge and experience. The lack of such capability can be an obstacle for an experienced professional to accept such a computer integrated information system as a practical tool.

In summary, a lack of consistent information flow, incomplete standards and/or general specification, no well-organized database, without an effective tool for real-time information tracing, etc., all these existing problems in construction practice are common in AEC industry. Therefore, integrated computer information systems not only serve as a powerful tool to facilitate data exchange between different project stages, but more importantly, brake the information communication barrier among various disciplines and smooth the information flow

·in project cycle. No doubt, this kind of integration system is very important and helpful in the construction management decision-making process.

Practice at PR-1 has shown that computers may play an even more important role by adding capabilities to construction management system. For example, computer integration starts from planning, design, estimating, scheduling, job cost, reporting through maintenance, that is, the whole construction cycle. An immediate instance is the migration of network-based computer software programs from the home office mainframe to job site PCs, which have created a great opportunity for construction project managers to control the project with first hand contemporary information. Taking advantages of the existing local network, the PR-1 construction project managers can transfer project documents in a quick, more convenient and accurate way. That is the goal PR-1 heads for, and an ideal information flow for a typical project at PR-1 is shown as figure 2.

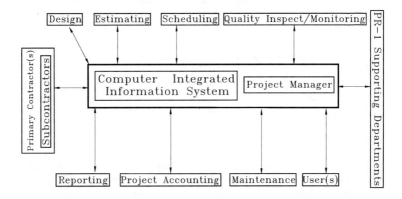

Figure 2.   Ideal Information Flow for A Typical Project at PR-1

Conclusions

With the increasing of market competition and regulations, it is crucial to improve production and productivity of the construction project management process. A few construction companies have implemented integrated information management system, to meet the needs of complex situations and to handle multi-projects more skillfully. The application of the computer integrated information system at PR-1 may be a good example for small and medium construction companies to stay with or ahead of the competition in the construction industry.

Acknowledgements

The authors are thankful to the Poudre School District R-1 for providing funding for the project, and access to these applications.

References

[Coles 1994]   *Bruce C.Coles, Kenneth F.Reinschmit*, "Computer-Integrated Construction", Civil Engineering, ASCE June 1994

[Ed Ott 1994]   *Ed Ott, Matt Syal and Gary Gehrig*, "Computer-Integrated Construction and CM Firms: A Profitable Combination", Computing in Civil Engineering, Proceedings of the First Congress held in conjunction with A/E/C Systems '94

[Parfitt 1993]   *M. K. Parfitt, M. G. Syal, M. Khalvati and S. Bhatia*, "Computer-Integrated Design Drawings and Construction Project Plans", Journal of Construction Engineering and Management, 119(4), December 1993

[Poudre R-1]   *Poudre R-1 Construction Services*, "Building Fund Project Overview 1991-1995", November 30, 1994

[Syal 1991]   *M. G. Syal, M. K. Parfitt, J. H. Willenbrock*, "Computer Integrated Design Drawing, Cost Estimating, and Construction Scheduling", Report No. 11, NAHB/NRC Designated Housing Research Center at Penn State, Sept. 1991.

[Wagter 1990]   *H. Wagter* ed, "Computer Integrated Construction", Proceedings of The Second CIB W78 & W74 Joint Seminar on Computer Integrated Construction, Tokyo, Japan, 17-19 September, 1990.

[Westney 1992]   *R. Westney*, "Computerized Management of Multiple Small Projects", Marcel Dekker, Inc. 1992.

Process-Oriented Intelligent Planning for the Delivery of Reinforcing Bars

Md. Salim[1], Member, ASCE

Abstract

The need to increase productivity, quality, and safety in construction challenges researchers to seek and develop bold innovative changes. This paper provides an insight into some central aspects of automated process planning, a key element of computer-integrated construction (CIC). The work discussed in this paper includes the establishment of a feature-based and process oriented planning framework to allow the integration of computer aided design (CAD) and artificial intelligence (AI) for rebar placement planning. The paper also discusses a prototype system for feature-based and process oriented rebar placement planning. The prototype system is implemented on a PC-486 by integrating AutoCAD, dBASE and LEVEL5 OBJECT software. Finally, implementation of the planning model for the delivery of rebar in a six-story building project and the results of the study are discussed.

Introduction and Background

Process planning is traditionally considered as the planning functions that define and sequence the tasks and machines necessary to convert a CAD image into a physical unit. The goal of automated process planning is the integration of design and production data for generating usable process plans. Two main factors contribute to the need for automated process planning. The first stems

----

Asst. Prof., Construction Management, Dept. of Industrial Tech., Univ. of Northern Iowa, Cedar Falls, IA 50614-0178

from the limitations of CAD to represent knowledge. The simple images of reinforcement for concrete slabs or columns in CAD do not provide any information on how those design objects could be fabricated or assembled. In order to bridge the gap between CAD and CIC, the intelligence of what to do with a CAD image has to be integrated with the computer representation of the item. This intelligence includes properties or operational instructions on how to construct/manufacture what is represented as points, lines or shapes (Bernold and Reinhart 1990). While engineers might associate a picture of mould with the processes of how to manufacture it, the CAD image is totally ignorant of such matters. The CAD system needs to be provided with the required "smarts" so that it can be used to give automatic computer commands to a manufacturing or construction system. One way of accomplishing such a task is the use of artificial intelligence (AI) techniques, and in particular the knowledge based ones which is capable of expressing the heuristics that are commonly used by the human experts. Furthermore, the structure of knowledge based system is flexible enough for easy updating of planning processes which change in a dynamic environment.

The use of reinforced concrete constitutes a major portion of U.S. construction; and the assembly of (reinforcing bars) rebar is an integral part of this construction. The efficiency of the installation or placement of cut and bent rebar depends on how effectively the delivery is organized. A basic requirement for the placement-oriented delivery of rebar is to have a placement plan before the start of shop fabrication and site delivery. Automatic evaluation of CAD represented design with a linked process planning system using AI could be the key to provide a detailed placement plan for the rebar. This plan could subsequently be used for organizing the delivery of rebar integrated with the actual placement.

<u>Intelligent Process Planning</u>

Process planning entails the preparation of a detailed plan for the production of a piece, part or assembly. "Process plans consist of sequential lists of individual manufacturing operations with all relevant associated information necessary to produce a part in a certain manufacturing facility" (Ansaldi et al. 1989). Traditionally, process planning functions are performed by a highly skilled individual who is intimately familiar with all aspects of the respective manufacturing or assembly operation. With the advent of computers, continuous attempts are being made to enhance the process planning functions by augmenting the abilities of human expert with those of computer. According to Ansaldi et al. (1989), developments in manufacturing industry aim at an intelligent process planning system that would integrate design and production data for generating usable process plans. A CAD representation of a part or object is used to provide input data to the process planning system, eliminating the need of human intervention for translating the drawing into a suitable form for automatic processing. Modern solid modeling CAD systems are being used by the designer in order to import a list of shapes to the screen. In this case, adding of features to the CAD image is associated with simultaneous development of production code.

Traditional Rebar Design and Delivery

Design drawings for rebar form the basis for fabrication and delivery. Rebar need to be placed in the concrete forms accurately as required by the design drawings, and adequately tied and secured before the placement of concrete. Most commonly, the general contractor orders rebar from a fabricator and a separate subcontractor is engaged for receiving, handling and placing the fabricated rebar. Therefore, a distinct sequence of events evolves for designing and placement of rebar as presented in Fig.1. (Bernold et al. 1993).

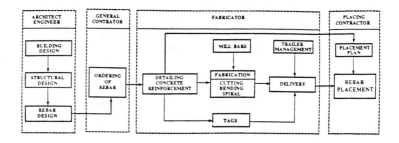

Figure 1. Flow Diagram for Rebar in Construction

As can be seen from Fig. 1, the engineering drawings for rebar are prepared by the Architect/Engineer (A/E). Upon award of the contract to the general contractor, the assembly or installation of rebar in the concrete form is accomplished in five distinct steps: 1) ordering, 2) detailing (preparation of shop drawings), 3) fabrication, 4) delivery, and 5) placement (Fig.1). Traditionally, the fabricator is responsible for the delivery of rebar as ordered by the general contractor based upon the specifications.

According to the traditional contract between the general contractor and the rebar fabricator in the eastern part of the U.S., detailing, fabrication, and delivery are the responsibilities of the fabricator. In the western part, however, this contract includes the placement of rebar in addition to detailing, fabrication, and delivery. As described in the ACI Detailing Manual-1988 (ACI Committee 315), detailing consists of the preparation of the placing drawings which may comprise bar lists, rebar schedules, bending details, placing details, and placing plans or elevations. Placing drawings (commonly known as shop drawings) are based on the rather generic engineering drawings, and are completed by adding additional information to facilitate the fabrication, delivery, and placement. Placing drawings are used by the job superintendent and the foreman for actual placing or assembly operations.

The following section discusses the development of a process-oriented planning system for construction.

A CAD-Integrated Process Planning System (CPPS)

Computer aided design (CAD) is one of the most advanced tools to develop electronically represented models of physical objects. Because of their importance in the industry, it was decided to use the well accepted CAD software package AutoCAD as the basic design tool. The CPPS was implemented by integrating LEVEL5 OBJECT (an object-oriented database with a rule base) running under Microsoft Windows, AutoCAD (Release 11) and dBASE IV. The implementation of the CPPS is discussed with rebar placement planning function in construction as an example. One of the processes in reinforced concrete construction is the assembly or placement of rebar. Detailing in CPPS is supported by a feature library within AutoCAD which contains standard rebar configurations. Each rebar configuration is further characterized by the size(s) and length(s) of different bar sections. As indicated by the name of the library, the configuration or shape of each rebar (e.g., closed stirrup) is a feature that is directly related to the sequence with which the rebar is placed into the concrete form.

In the CPPS, the designer is able to pick any standard bar from the library and insert it into the design drawing. Frame based attribute lists enable the designer to specify size, lengths of different sections, etc. Upon completion of the feature-oriented rebar design, CAD data is made available to a data base management system (DBMS). Instead of sending all the detail design data to the intelligent process planner, the data is massaged and reorganized within the DBMS and required data for the process planner is stored in a separate database file to use as input to the planner. Placement of rebar is based on hard and soft operational rules which heavily depend on the features of rebar and the concrete elements. The knowledge base of the CPPS uses a forward chaining rule based approach to generate rebar placement plans for a building structure.

Field Testing of the Planning Model

A six story office building project having beam and joist floor system entirely of reinforced concrete was selected for the testing of the planning model. Rebar placement operation for each floor was accomplished in two phases, namely, north and south pours. The main goal of the field testing was to allow comparative measurements of crew performance. The placing of rebar was planned in conjunction with the formwork construction. Since the formwork construction was done in two phases, namely, north and south pours, placing of rebar for each pour was done separately. The four identical floors of the building project provided the opportunity to collect significant amount of data to make valid comparison of crew performance.

Bundling, delivery, staging, and placement of rebar for the beams of 2nd and 3rd floors were based on the traditional method. Detailed placing drawings based upon the engineering and approved shop drawings were prepared by the placing contractor. Rebar placing operation for these floors were closely observed. Bundling, delivery, staging, and placement for the beams of the 4th floor were based on the concept of placement planning. Rebar placement sequences were utilized in creating micro (smaller bundles within a master bundle) and master

bundles as well as assigning staging areas to master
bundles. For the 5th floor, however, traditional bundles
were used. Staging assignments of master bundles for the
5th floor were based on consideration of the sequential
placement of different sets (micro bundles) within the
master bundle, final locations of the sets of bars, and
weight of individual micro bundle.

One of the effective ways of measuring crew
performance is productivity ratings. For details of the
productivity ratings, refer to Oglesby et al. (1989).
For productivity ratings, the main categories of crew
activities are: 1) Direct or effective work; 2) essential
contributory work; and 3) ineffective work. Effective
work involves the activities of the crew that are
directly involved in the actual process of putting
together or adding to a unit being constructed (Oglesby
et al. 1989). Essential contributory work includes all
elements that are required to complete a work unit,
although not adding to the unit being constructed.
Ineffective work includes doing nothing or doing
something not required to complete the work unit. The
results of the direct or effective work for all the four
floors are presented in Fig. 2.

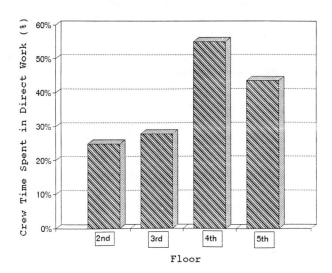

Figure 2. Results of Productivity Ratings (Direct Work)

As can be seen from Fig. 2, direct work for the second and third floors accounted for 24.8% and 27.8% of crew time respectively, with a difference of 3.0%. Direct work for fourth and fifth floors accounted for 54.9% and 43.5% respectively.

Conclusions

Automated process planning plays a very important role in construction. This paper reviewed the basic concepts of intelligent process planning. It was demonstrated how such approaches can be applied to construction processes. Implementation of the process planning model resulted in a drastic productivity improvement for rebar assembly. Productivity decline and global competitiveness in construction will demand for more process planning services in future. Automatic generation of a process plan for construction processes could be the key for productivity improvement in construction.

APPENDIX. REFERENCES

Ansaldi, S., Boato, L., Canto, M, Fusconi, F., and Giannini, F. (1989) "Integration of AI techniques and CAD solid modeling for process planning applications," Computer applications in production and engineering, Elsevier Science Publishers B.V. North Holland.

ACI Detailing Manual-1988. ACI Committee 315, Publication SP-66(88), American Concrete Institute, Detroit, MI.

Bernold L.E., Salim, M., and Dunston, P.S. (1993) "Automation in design and fabrication of concrete reinforcement," Proceedings of the Third International Conference on Fossil Plant Construction, Palm Beach, FL.

Bernold L.E., and Reinhart, D.B. (1990). "Process planning for automated stone cutting," J. Computing in Civil Engineering, ASCE, Vol. 4, No. 2.

Oglesby, C.H., Parker, H.W., and Howell G.A. (1989). Productivity Improvement in Construction, McGraw-Hill, NY.

Salim, M., and Bernold, L.E. (1994). "Effects of design-integrated process planning on productivity in rebar placement," J. Construction Engineering and Management, ASCE, Vol. 120, No.4.

Development of a practical knowledge based adviser
to diagnose defects in flat roofs.

R.J.Allwood[1] & A.Goodier[2]

Abstract

The opportunity of creating a large practical expert
system to diagnose faults in flat roofs has been used to
develop some new techniques particularly in deriving a
full set of goals, in structuring the rules to enhance
comprehensibility, in imposing a control strategy over
the inferencing process and in providing distributed
help. Extensive field trials and blind testing are also
reported.

Introduction

There is a great deal more to creating an effective
diagnostic expert system than eliciting knowledge and
converting it into rules. This paper will report on some
of the following tasks:-

* Deriving a complete set of goals,
* Structuring the rules in the knowledge base,
* Achieving a truly comprehensible set of rules,
* Imposing control over the inferencing process,
* Distributing additional knowledge,
* Graphics screens to find faults at details,
* Testing and field trials,
* Assessing blind tests.

The completion recently of a substantial system to
diagnose defects in flat roofs (Allwood & Goodier 1995) -
with 276 goals and over 1000 rules, financially sponsored

----

[1] Reader, [2] Research Assistant, Department of Civil &
Building Engineering, University of Technology,
Loughborough, Leics., LE11 3TU, UK.

equally by UK Industry and Government, has provided an opportunity to work on many important facets of expert system creation. The industrial backers, who also provided expertise, were:- Laing Technology Group, Sainsbury plc., Royal Institute of Chartered Surveyors, British Flat Roofing Council. The Building Research Establishment also helped sponsor the work and provided expertise and test material. The system was created using the XiPlus shell from Inference Corporation.

Deriving goals

Diagnosis of defects is an excellent application for the backward chaining inference process used by expert systems and which is such an efficient method for selecting just those goals of a pre-defined set for which the associated rules are true for the observed symptoms (Allwood, 1989). The goals of such a system are all the possible defects that can be suffered and it is desirable for a full list of these to be developed as soon in the development as possible. The scale of the system can then be ascertained and the knowledge elicitation conducted in a structured fashion by discussing with experts one goal at a time. However, many experts do not find it easy to list goals since they represent the end results of their thinking whereas a consultation will start with symptoms. The defects for the flat roof diagnostic system proved especially difficult to draw up due to the very considerable number, 276 in the completed system, representing a range of problems experienced by many experts but outside the scope of any one. By defect, we mean the fundamental cause of distress rather than the symptoms the defect creates. For example moisture entrapped during construction is a defect leading to symptoms such as blistering.

It became clear early in the development that our group of experts would need some assistance to achieve a reasonable full set of defects. We first identified the basic components of flat roof constructions and their common combinations, commonly known as cold, warm and inverted decks. By focusing firstly on defects of each layer, eg protective layer, insulation, etc. then on defects of the roof system as a whole and finally on defects in the many common details, particularly in roof penetrations, our industrial experts generated a good starting list.

At a later stage the list was amplified by using each defect as a 'seed' for lateral thinking in which we were able to use a computer to generate new possible defects from combinations of elements of the starting defect. As an illustration, an early defect concerned the effect of thermal expansion of a particular insulant on a waterproof membrane bonded to the insulant. The basic phenomenon behind this defect could clearly apply to other combinations of a material weak in tension, bonded to a material which expanded or contracted for any reason, not necessarily thermal. The computer-aided 'lateral thinking' started from a data file of several hundred records created from the permutations of all possible pairs of adjacent materials. This file was then examined by a set of production rules to select feasible combinations of materials known to be weak in tension or likely to expand or contract. The resulting set was then considered by experts to determine whether new and practical defects had been found. A detailed paper on this technique is being prepared.

The final set of defects consists of 70 defects for the roof layers and the roof system plus 206 for defects in details.

Structuring rules for defects

The task of eliciting a full set of rules for each defect can be greatly aided by adopting a consistent and logical structure into which all rules can fall. For the defects in flat roofs it became clear that all rules could be consistently split at a high level into two parts. This came from the concept that each defect will occur when a set of practical conditions exist which define a 'situation' where the defect could occur and a set of 'symptoms' in the right locality which confirm that the defect has indeed developed. By making the top level rule for a defect into one which requires both the situation to exist and the symptoms to be seen, knowledge elicitation was immediately simplified by directing the expert to focus on each component.

We consider as an example, a defect which occurred only too often - a built-up membrane spanning a movement joint in the roof deck without having an appropriate movement detail in the membrane itself. The 'situation' for this defect consists of a membrane of low-grade roofing felt, a roof deck with a movement joint and the membrane having no movement joint. If this situation exists, the membrane will be subjected to considerable tension and release and will develop splits along the line of the joint. Thus the 'symptoms' will be splits

in the roofing felt, the pattern of the splits will be
linear and the orientation of the splits will be parallel
to the deck movement joint.   The resulting two rules
together with the top level rule for this defect are
shown in Figure 1 taken directly from the knowledge base
prepared for the XiPlus shell.

---

if situation allows cause c3.5
and symptoms indicate cause c3.5
then causes include c3.5 built-up roofing improperly spanning
expansion joint

if type of membrane is built-up roofing
and built-up membrane has base of low grade
and roof deck includes movement joint
and membrane does not include movement detail
then situation allows cause c3.5

if presence of splits is yes
and orientation of splits include in-line with movement joint
and location of splits include at movement joint or offset
slightly from movement joint
then symptoms indicate cause c3.5

Figure 1.   Rules for one defect showing division of
knowledge into 'situation' and 'symptoms'.

Comprehensible rules

     An early claim of expert systems, although not often
achieved, is that the rules forming the knowledge base
can be read by a non-computer person.   Apart from
providing, by a display of the relevant rule, a simple
way of answering the user's questions of 'How' an answer
was obtained or 'Why' a question is being asked, having
a knowledge base that can be understood by anyone
provides an important aid to verification.   If the
expert(s) supplying the raw knowledge can read and
understand the way it has been represented their
agreement to its verity is an important step in the final
testing stage.

     The rules for the flat roof system make much
reference to observable symptoms and at the outset we
drew up a global list of symptoms and associated
appearances just as we had for the defects.   This
ensured a consistent style and the inadvertent use of
alternative words for a symptom, eg 'rips' instead of
'splits'.   It also led to the recognition that symptoms
such as splits in felt would have several features or

attributes eg orientation, spacing, pattern and extent.
Where possible, a consistent set of attributes was
developed and used for many symptoms, eg pattern of
splits, of blister, of ridges, etc.   The common style of
rule clause was of the form 'attribute' of 'object'
'relation' 'value' as seen in the rule shown above in
Figure 1.   (Experienced users of expert system shells
will recognise the notion of 'slots' in 'frames'.)

Assertions were used where a single fact needed to
be determined and the only difficulty in composing
suitable words was the desirability of avoiding the
relation words 'is', 'are', etc or the software would not
recognise the phrase as an assertion. If this could not
be avoided, an underscore character was inserted to
disguise the relation word, eg 'if the roof_is
permanently shaded'. These precepts, together with the
rule structuring set out above and careful choice of
words yielded a knowledge base readily understood by our
experts.

Need for overall and dynamic strategy

With a backward-chaining inferencing method, there is
strictly no need to interfere with the searching software
but just leave the system to ask questions as needed.
This leads to an almost random sequence of unconnected
questions being asked of a user generating the real risk
of losing the user's confidence right from the start.
Additionally, consideration must be given to how the
expert system should stop, particularly when it is
possible that it may find several feasible alternative
solutions to a problem, as in many building defect
diagnoses.   Choosing and implementing an appropriate
strategy is therefore another important task in creating
an expert system (Allwood & Goodier, 1993).

The strategy incorporated into the flat roof
diagnostic system is as follows:- the user is asked some
preliminary questions and this information is used by a
special set of rules to establish which of five 'types'
of water may be the source of the problem.   These are:
rain penetration, condensation, built-in water, leakage
of services pipes or defects without water ingress such
as ponding.   On the basis of the result, the groups of
defects are ordered to present the most likely groups
first.   For example, if the source of water is rain
penetration, the first group of defects considered are
potentially faulty details such as an internal outlet
with a badly dressed membrane.   After that group, the

system considers defects in the overall roof such as splits in a membrane caused by unauthorised traffic, and then problems with pipes, built-in water, condensation, etc. until all defects have been examined or the user chooses to stop the investigation.

Opportunities to incorporate knowledge and help

An important element in any expert system is one of providing help to the user, if needed, when a question is being asked. It is often not appreciated just how much knowledge can be incorporated in additional text and diagrams for this 'help' function. Another opportunity to embed knowledge lies simply in the choice of text for the questions. These do not have to be bald one-liners. A second paragraph, amplifying the question can convey extra information to be read or ignored according to the user's need. All types of 'help' must clearly be formulated with the target user in mind and ours is a building professional not particularly experienced in flat roof work. The text of every question has been composed for that level of user to understand and, if judged necessary, an extra sentence or two added for clarification. Figure 2 shows one question about the need for a spreader if a load is applied to mastic asphalt illustrating that style.

Is there a spreader present between an item on the roof and the mastic asphalt membrane that reduces the load?

Though ideally a plinth should be taken from the deck through the membrane to support any items or plant on the roof, commonly this is not practicable. When an item is placed upon a mastic asphalt roof a spreader should be placed between the item and the membrane to reduce the pressure.

Figure 2 Knowledge included in a question

More comprehensive help is given by separate displays of text or graphics optionally shown before answering a question. Such text can be quite extensive. We have, for example, a three page note on the 14 types of insulation allowed by the system when answering a question about the make-up of the roof. Shorter texts are more common and most questions have a 'help' screen of one page. Some questions require a diagram to assist the user in giving an accurate answer. We have 70 built-in diagrams for display, mainly concerned with details. These have been taken from a current flat roof design guide (BRFC/CIRIA, 1993).

Testing and field trials

The 'clean' knowledge was read through by the research team and, selectively, in the presence of the industrial experts. Case study material had been gathered at the start of the project and these problems were used for early trials of the complete system. Visits were then made to nearby organisations with flat roof problems. Each appointment was preceded with requests to recall past problems of flat roofs for which solutions had been found. During the visit the system was explained and then the local problems were run through with the request that the solution be concealed until the system had produced its answers. Despite the general lack of formal documentation, most problems were fully and enthusiastically described.

The systematic and thorough nature of the questions asked by the adviser was noted and well received by virtually all participants. Seventy nine tests were made and all were re-run through the system in its final condition, producing a near 100% accuracy except for problems outside the scope of the system. It was usual for several defects to be found for each problem and this coincided with the experts own conclusions.

Assessing blind tests

One technique of verifying knowledge bases, often discussed but rarely performed, is by submitting a set of test problems to human experts and to an expert system with the various answers compared anonymously by a panel - hence 'blind' testing. The flat roof adviser testing was concluded by such a test. Twenty test problems were selected from the field trials supplemented by some from early Case Studies. The data given to the computer system in the form of answers to questions was supplied to several experts along with the full list of defects embodied in the system for them to select from. Each problem was considered by two experts as well as by the computer.

Since the 'answer' to each problem was not usually one defect but a set of up to 6 possible defects, those selected by the experts and the computer were compared against the 'correct' set as selected by the building inspector. A simple scoring system to judge the accuracy of the results was devised by assigning a weight to each defect, since some were more important to find than others, and then calculating the sum of the weights

of the correct defects found. The ratio of the sum
obtained by each expert or the computer against the sum
of the inspector's defects, gives an unbiased comparison.
The results as seen in Figure 3 showed that the computer
scored as well or better than the human experts.

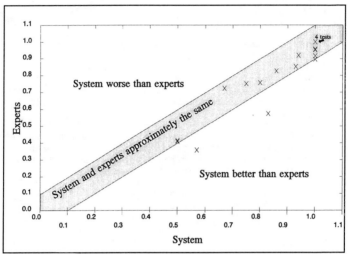

Figure 3 Result of blind testing.

Appendix

R.J.Allwood, "Techniques and Applications of Expert
Systems in the Construction Industry", Ellis Horwood,
Jan. 1989 pp143.

R.J.Allwood & A.Goodier, "Overall strategies for
diagnostic expert systems", CIVIL-COMP93, Artificial
Intelligence CIVIL-COMP93, 17th-19th August, 1993,
Edinburgh, pp73-78

"Flat roofing: Design & Good Practice", BFRC/CIRIA,
London, 1993.

R.J.Allwood & A.Goodier, 'A knowledge based adviser to
diagnose faults in flat roofing systems', Chartered
Surveyor Monthly, RICS, London, V.4 (6), April, 1995.

Of Tools and Toys:
Software for Water Resources Decision Support Systems (DSS)

René F. Reitsma[1], Carol A. Marra[1], Stephen C. Wehrend[1]

Abstract

The arsenal of tools for developing environmental decision support systems has evolved quickly over the last ten years or so. In this paper we briefly discuss a few of the most significant changes and most recent developments[2].

Databases

Perhaps the most important difference between today's and yesterday's DSS, concerns the way in which information on the state of an environmental system is maintained in terms of the storage, retrieval and overall management of a system's data. Where modern systems are built on a *data-centered architecture* with a comprehensive *geo-relational database* at their center, early DSS developers did not have such comprehensive tools for data management available. Instead, they developed their own data storage facilities and their own data storage and retrieval functions. With the recent advance of relational data base management and GIS systems, however, DSS developers have been given a number of most effective tools for the management of their, often very complex, data.

The power of, for instance, the *relational model* for representing attribute data is remarkable, especially in light of its simplicity. In relational databases, data is represented as tables, not unlike traditional data matrices, i.e. variables as columns, observations as rows. However, when different tables share one or more columns, complex search or join operations can be executed. Structured query languages such as SQL (Date, 1989) provide the tools for conducting these types of operations from both command-based SQL interpreters and from inside programs (embedded SQL). The associated low-level data storage, searching and computing are carried out by the relational database software.

---

1. Center for Advanced Decision Support Water and Environmental Systems (CADSWES), Dept. of Civil, Environmental and Architectural Engineering, University of Colorado at Boulder.

---

2. A more complete presentation of this material is published in Reitsma et al. (forthcoming) Decision Support Systems for Water Resources Management; in: Mays, L. (ed.) Handbook of Water Resources; McGraw-Hill Publishing Company.

Geographic Information Systems (GIS)

Whereas relational database management systems increased both the productivity and amenities of a developer's life regarding various types of *attribute data*, *GIS* have supported the management and manipulation of complex *spatial data* sets. GIS are database management systems for the storage, retrieval and management of spatial data (Burrough, 1986; Scholten and Stillwell, 1990; Star and Estes, 1990). Unlike most attribute data, spatial data have a well-defined structure and are composed of simple building blocks such as points, lines and polygons. Furthermore, spatial data are limited enough in their dimensional extent to be made subject to equally well-defined mathematical manipulations. This implies that with relatively simple spatial building blocks complex structures and manipulations of those structures can be achieved. Examples of more complex structures are complex surfaces such as digital elevation models, stream networks or finite element meshes. Examples of spatial computations are the calculation of area, the length of a route through a topological network, the intersection of a group of superimposed polygons or the aggregation of values from irregularly shaped surfaces into a single composite value.

By coupling both relational and spatial data models, today's DSS developers have the availability of powerful geo-relational data models which serve as data managers for complex DSS applications.

Object Orientation

The advance of object-oriented programming (OOP) (Ellis and Stroustrup, 1990; Lippman, 1991; Coplien, 1992) caused another revolution in both the modeling and coding of complex computer-based systems. Object-oriented techniques discretize a modeling or programming problem in self-contained domains of action or objects. Objects maintain both data and exhibit behavior in a true stimulus-response manner; i.e., an object's state changes as a function of the application of a dynamic. The dynamic itself is triggered as a result of receiving a message from another object or as the result of a (previous) state change. Applied to a model of a river basin, objects could be reservoirs, power plants, diversions or reaches, each of which maintains its own state by managing its own data and applying its own dynamics whenever necessary. The objects communicate with each other by sending each other water or by requesting water. Although each of the objects acts autonomously, collectively their behavior models the behavior of the system as a whole.

With these new object-oriented programming techniques, it became easier to address many of the programming problems in ways that approximate coders' tasks and objectives (true modularity, encapsulation of data, class hierarchies and inheritance, etc.). By the same token, new compilers and new complete programming environments facilitated the management and coding of large complicated programs and eliminated a lot of tedious and often difficult code testing. New and continuing improvements in language standards keep supporting this process.

Graphical Standards and Interface Tools

Another major leap in the development of DSS was caused by the coming of age of graphics development tools such as graphics libraries, user interface builders, graphics standards and of late, platform independent development libraries. Although it is still possible for developers to write their own graphics libraries and, if so desired, their own complete windowing systems, writing graphics applications has become increasingly more efficient and accessible through the availability of these high-level libraries.

Advances in graphics systems have occurred hand-in-hand with the emergence of windowing systems and mouse-driven interface systems. Both tools have dramatically increased the opportunities for both data visualization and the development of user interfaces.

Platform-independent Development Libraries

Recently, various software manufacturers have made available so-called platform-independent development environments. These systems provide extensive function libraries for many types of applications. Examples are libraries for file manipulation, text and string manipulation, graphics routines, map manipulation, client-server applications, database interaction or hypertext facilities. These libraries are implemented "on top of" the native platform specific system software, one implementation for each different platform. As a result, code written using these libraries will compile and execute on every platform for which the library is available (provided that auxiliary code is also platform-independent).

Networks

Productivity and comfort levels of DSS developers were further boosted by the emergence and rapid growth of national and international computer networks. World-wide networks, together with such institutions as the Open Software Foundation (OSF) and clever constructs such as *share-ware*, have created a very large market where it is a pleasure to shop for other people's intellectual gems.

Networks do not only facilitate the work of DSS developers, they also open new opportunities for putting DSS technology to work. In the case of the Tennessee Valley Authority DSS, for instance, people throughout the TVA system can simultaneously log onto the DSS' database server using the corporate Wide-Area Network (WAN). The DSS itself is equipped with data collection programs which, at specified times, use the network to see if new data for incorporation is available at various sites within the organization. If so, the data is copied over, checked for operational constraint violations and incorporated into the database. Violations are logged on a bulletin board and individual users throughout the organization are notified about the new postings to their bulletin board.

Interpreters

A number of less dramatic innovations have further strengthened a DSS developer's arsenal of tools and toys. A very recent one, for instance, is the (re)emergence of interpreters, written on top of conventional programming languages such as C, which allow for the creation of complex, interpreted code segments. Some of these languages such as "Tcl" (Ousterhout, 1990) or Tcl-based tools such as "expect" (Libes, 1991), allow for the modification and extension of DSS without having to re-compile the system's source codes. As such, developers can field a system while continuing to work on, for instance, the library of visualization tools. Every time a new tool is completed it can be added to the system without having to go through the often complicated process of version release and control. Similar benefits are associated with interpreter-based plotting tools such as ACE/Xmgr (Turner, 1993) or statistical data analysis and visualization packages such as S-PLUS (StatSci, 1991; Becker, Chambers and Wilks, 1988).

References

Becker, R.A., Chambers, J.M., Wilks, A.R. (1991) The New S Language; Wadsworth & Brooks/Cole Computer Science Series.

Burrough, P.A. (1986) Principles of Geographical Information Systems for Land Resources Assessment; Oxford Science Publications.

Coplien, J. (1992) Advanced C++. Programming Styles and Idioms; Addison-Wesley Publishing Company.

Date, C.J. (1989) The SQL Standard; Second Edition; Addison-Wesley Publishing Company.

Ellis, M.A., Stroustrup, B. (1990) The Annotated C++ Reference Manual; Addison-Wesley Publishing Company.

Libes, D. (1991) expect: Scripts for Controlling Interactive Processes; Computing Systems; vol. 4.

Lippman, S.B. (1991) C++ Primer; Addison-Wesley Publishing Company.

Ousterhout, J. (1990) "Tcl": An Embeddable Command Language; Proceedings of the Winter 1990 USENIX Conference; Washington D.C., Jan. 22-26.

Scholten, H.J., Stillwell, J.C.H. (1990) Geographical Information Systems for Urban and Regional Planning; Kluwer Academic Publishers; Dordrecht.

Star, J., Estes, J. (1990) Geographic Information Systems: An Introduction; Prentice Hall; Englewood Cliffs, NJ.

StatSci (1991) S-PLUS User Manual, Vol. 1; Statistical Sciences Inc., Seattle, WA.

Turner, P. (1993) ACE/gr User's Manual: Graphics for Exploratory Data Analysis; Software Documentation Series, SDS3, 91-3.

An Architecture for an Extensible Modelling Environment for Water Resources Management

Terri Betancourt[1], Bill Oakley[2], Tim Hart[3]

Abstract

PRSYM architecture was designed with several groups in mind: end users, Civil Engineering programmers, and Software Engineers. By focusing on the requirements for flexibility, extensibility, and modularity for each of these groups, PRSYM has evolved into a robust modelling environment for river basin management. In this paper we examine some of the key features underlying the PRSYM architecture.

Introduction

PRSYM is the third generation of an application architecture long under development at CADSWES. The fundamental simulation behavior underlying PRSYM was implemented in the first generation architecture. The second generation focused on an object-oriented architecture in support of simulation. Now, in its third generation, PRSYM architecture has focused on three common software objectives: flexibility, extensibility, and modularity.

The major thrust of PRSYM flexibility was defined to be a graphical application which provides the end user with the ability to interactively construct and model any river basin. In addition to the non-basin-specific requirement, PRSYM was charged with building a modelling framework for activities ranging from power operations to water allocation planning.

Three of the key extensibility features identified were 1) to allow Civil Engineers to develop multiple solution methods for their problems and to provide a mechanism for the end user to select among the various solution techniques; 2) to provide switching between various PRSYM controllers including Discrete Timestep Simulation, Rule-based Simulation, and Goal Programming Optimization; 3) to provide a data management interface which allows the user to easily load a variety of input data.

---

1. Research Scientist, Center for Advanced Decision Support, Water and Environmental Systems (CADSWES), Univ. of Colorado, Campus Box 421, Boulder, CO 80309
2. Research Scientist, CADSWES.
3. Research Scientist, CADSWES.

The major role of PRSYM modularity was to define a software structure which provides a development environment for programmers of varying skill levels. To meet this requirement, PRSYM software development has proceeded along three major libraries: GUI (the graphical user interface), SimLib (the simulation behavior), and PrsymObj (the engineering algorithms). The lines of division for these three libraries follows the programming skill sets of PRSYM developers. GUI developers tend to be computer scientists, often with some art background, who specialize in graphical layout and user interaction. SimLib developers tend to be computer scientists who focus on model framework and interaction between modelling components. PrsymObjs developers are Civil Engineers with special knowledge in many aspects of river basin modelling.

GUI

PRSYM's graphical user interface (GUI) is built upon the portable, cross-platform development tool Galaxy by Visix. This tool allows for rapid prototypes and minimizes development efforts for future migration across platforms. Through a callback mechanism, the GUI communicates with the simulation library in a one-way dependency which allows for modifications to the interface without affecting the simulation or PrsymObj libraries. This communication link results in an architecture well structured for interchangeable front ends. For example, a second PRSYM front end currently under design is a non-graphical batch driver for un-assisted run management and system testing.

The PRSYM user interface gives a user tremendous flexibility when building and simulating their river basin. They may:

- add objects to the model from a palette of engineering objects. This feature allows users to construct a model representation for any river basin;
- delete engineering objects from the model;
- create links to control the information flow between objects. For example, an upstream reservoir's outflow may be linked to a downstream reach's inflow such that the modelling results from the reservoir will automatically be sent to the reach (or vice versa);
- select the technique to be used to solve a particular category of problem. For example, select the routing technique to use for reaches;
- select model run parameters such as the start and end dates, the timestep, and the type of model run (simulation, optimization, etc.);
- get an in-depth view of the basin model by opening views of individual components;
- get a concise view of the basin model by creating a Simulation Control Table, or SCT;

SimLib

The simulation library (SimLib) embodies the framework for PRSYM modelling. SimLib contains all the building blocks from which Civil Engineering programmers construct PRSYM's engineering behavior. From an engineering perspective, the three major building blocks of SimLib are SimObj, Slots, and Methods.

The SimObj class is composed of a collection of Slots and Methods, and every engineering object is a specialized type of SimObj. For example, a Reach is a SimObj with four Slots (inflow, outflow, loss coefficient, and lag time) and several Methods (solveForOutflow, solveForInflow, timeLagRouting, and impulseResponseRouting). Slots can be thought of as variables which hold data on physical characteristics. Methods are simply engineering functions (or subroutines) and fall into two categories: dispatch Methods (solveForOutflow and solveForInflow) and user-selectable Methods (timeLagRouting and impulseResponseRouting). Dispatch Methods vary according to what variables are being solved. User-selectable Methods vary according to the solution technique used.

The relationship between Slots and dispatch Methods is implemented in SimLib as a *callback* mechanism such that when enough Slot values are known, the dispatch Method is invoked. In the Reach example, when Inflow is known (and Outflow is unknown), the Method solveForOutflow is invoked. The engineering objects (in the PrsymObj library) are responsible for identifying all dispatch conditions, i.e., the combinations of known slots, unknown slots, and the associated solution methods. In turn, the SimLib manages all simulation state changes and invokes the appropriate Method when a dispatch condition is met.

User-selectable Methods represent multiple engineering solutions to a single problem, for example, multiple routing techniques. Engineering objects may provide as many user-selectable Methods as is appropriate. Through the GUI the user may choose which solution technique is most appropriate for his/her model. SimLib is responsible for insuring that the correct technique (the one selected by the user) is used in the engineering solution.

PrsymObj

From SimLib components -- SimObj, Slots, and Methods -- the library of engineering objects (PrsymObj library) is constructed. This library currently contains definitions for objects such as Reach, Confluence, Canal, StorageReservoir, PowerReservoir, and PumpedStorageReservoir. However, the ability to easily populate this library with additional engineering objects is one of the major driving forces in the current PRSYM architecture. The process of defining a new engineering object includes:

- identifying the physical characteristics which will be modelled by the object;
- defining the algorithms needed to model these physical characteristics;
- declaring the type of slot appropriate for the data required in the algorithms;
- registering dispatch conditions with their associated dispatch Method;
- programming each dispatch Method (algorithm) registered;
- registering user-selectable Methods;
- programming each user-selectable Method (algorithm) registered.

From this simple framework new engineering objects are added to PRSYM, made available to the end-user, and managed within the modelling environment.

Data Types

The Slot is the fundamental data type within PRSYM. Common attributes among the slots include units, scale, formatting (floating point, integer, or scientific), display precision, minimum and maximum limits, and convergence criteria. These attributes can be modified by the end-user on a per-slot basis as appropriate to the available data.

PRSYM stores all values and performs all calculations in internal, or standard, units. A standard unit is defined for each unit type (Length, Area, Flow, etc.). However, PRSYM's user interface allows the user to specify different units (other than internal units) for importing, exporting, and displaying data. Furthermore, PRSYM allows the user to define their own units, provided that they supply a conversion factor between the standard unit and their unit.

PRSYM contains default values for all of the common Slot attributes. For example, there are default units, scale, formatting, convergence criteria, and so on. Although these default configuration parameters can be changed through the user interface,. PRSYM's Resource Database (RDB) provides the added capability to override the defaults on a site-specific basis.

In addition to the common Slot features, there are four Slot types which provide specialized features and behavior: Time Series, Tabular, Scalar, and Expression.

1) Time Series

   Time Series data - data which are on a "per-timestep" basis - are stored in SeriesSlots. Examples of Time Series data include inflows, outflows, and pool elevation, all of which vary over time.

   A SeriesSlot has a start date, an end date, and a fixed timestep which together determine the number of values in the series. The currently supported timesteps are hourly, 6-hourly, daily, weekly, and monthly. Each value may be either input (a forcing function) or output (calculated) on a per-timestep basis.

2) Tabular

   Tabular data are stored in TableSlots. A TableSlot is M columns by N rows, and is "column major." The number of columns as well as the column attributes are defined by the engineering objects; however, the number of rows is variable, and may be changed by the end-user. A primary use of TableSlots is for rating curves. To this end, the TableSlot contains many methods related to table lookup and table interpolation.

3) Scalar

   Scalar data is stored in ScalarSlots. Examples of scalar data include maximum pool elevation and Reach lag coefficient.

4) Expression

   Expression data - data which are represented as an arithmetic expression of other data - are stored in ExpressionSlots. ExpressionSlots are specialized SeriesSlots which add an arithmetic expression capability to the SeriesSlot. The user interface includes an expression editor which assists the user in defining the expressions. The expression itself allows for standard arithmetic operators, as well as special "summation" operators, for example, "sum the energy in storage across all LevelPowerReserviors."

Data Management Interface

For PRSYM to be used operationally, it is necessary to update the Time Series data on a regular basis. For instance, a daily operational model would, on a daily basis: move the Time Series forward (by setting the start date and end date); and import observed data to replace previously calculated values. Also, it might be necessary to calculate values at the end of a model run for further analysis or reporting. Typically the source of the data being imported or the destination of the data being exported will be a database, although this isn't always the case.

The PRSYM user interface provides a means of performing these tasks, but to do so on a regular basis would be time consuming and error prone. For this reason the Data Management Interface, or DMI, was developed. The DMI enables the user to accomplish the aforementioned tasks with "the click of a button". There were several requirements which influenced the DMI design:

- that PRSYM not require the user's data source or destination to be in a particular format (e.g. a particular relation database schema);
- that PRSYM neither know, nor care, the user's data source or destination format;
- that the user be able to use existing data extraction tools, with little or no modification.

To this end, the DMI uses ASCII files as intermediate data files. There are two files which the user provides for the DMI:

- a control file, which maps PRSYM object and Slot pairs to the data files they import or export, and which contains data file attributes (such as units and scale). The user interface includes an editor which assists the user in creating the control file;
- an executable program which either creates the intermediate data files from the data source, or writes the intermediate data files to the data destination.

The user creates a DMI scenario, which includes the control file, the executable, and user defined options. The user defined options enable the user to parameterize differences between scenarios (for example, between a low hydrology assumption and a high hydrology assumption), which in turn allows the them to use the same control file and executable for a variety of scenarios.

For an input DMI, PRSYM will invoke the executable (which creates the data files from the data source), and then import the data files. For an output DMI PRSYM will export the data, and then invoke the executable (which writes the data files to the data destination).

Multiple Controllers

In addition to its data management facilities, PRSYM has a modelling management facility as well -- the Controller. The first controller developed for PRSYM was Discrete Timestep Simulation where mass balance is calculated for timestep units defined by the end user. For example, a user may want to run a mass balance simulation for one year at monthly timesteps. An extension to mass balance simulation also provides water quality modelling.

A second PRSYM controller, the Rulebased Simulation, is an extension of the Discrete Timestep Simulation. Rulebased Simulation provides the user with the additional flexibility of defining policy rules within the river basin which alter the simulation. One example of a policy might be "flood control" which specifies maximum reservoir storages in the spring. Any particular rulebase may contain multiple policies with relative priorities associated with each for multi-criteria evaluation. Rulebased simulation allows policies to be modelled for their effects on river basin management.

Optimization is a third controller currently available within PRSYM. The optimization technique used is goal programming. PRSYM relies on CPLEX as the optimization engine. In this mode of execution, the end user may specify goals for optimization such as "minimize spill for all reservoirs."

In order to provide an extensible framework for supporting multiple controllers, a model management facility was developed based on state transition. The model management approach allows for process events such as START, STEP, PAUSE, and ABORT to be handled appropriately regardless of the specific controller in execution.

The requirements for developing a new controller are simply a matter of providing a body of code which support seven functions: beginningOfRun, beginningOfBlock, executeBlock, endOfBlock, advanceBlock, endOfRun, and percentDone. The interpretation and specific implementation of each of these functions is up to the developer of the controller. As an example, in PRSYM's Discrete Timestep Simulation the seven functions take on the following behavior:

beginningOfRun	initialize water objects for new simulation run
beginningOfBlock	notify water objects of current timestep
executeBlock	solve mass balance for current timestep
endOfBlock	(no special behavior)
endOfRun	check for underdetermined solution
advanceBlock	advance clock one timestep
percentDone	calculate percentage of timesteps completed

Any controlling mechanism which can be expressed using these seven functions can be readily integrated into the PRSYM architecture.

Conclusions

From the end-user's perspective, PRSYM provides a flexible environment for river basin modelling through a variety of modelling facilities. The user begins by building the river basin representation from a palette of common modelling components (reservoirs, reaches, power plants, canals, etc.). Next, to assist in populating the model with available data, PRSYM's DMI facility allows for automated connections to existing data sources. The data are imported and viewed in user-selectable units, scale, and precision. Through the GUI, the user examines as little or as much modelling detail as necessary. The user runs mass balance simulations with or without water quality, with or without prioritized rules, or optimizes against policy constraints. Within each of these modelling approaches, the user selects from among a variety of solution techniques. Finally, the user saves complete model results or select results to various output formats.

For the Civil Engineering programmer and modeler, PRSYM provides yet another level of flexibility and extensibility. At this level new engineering objects may be added to the PRSYM palette, for example, a weir. New engineering methods may be added to existing objects, for example, a special tailwater calculation method. Even new controllers may be added for additional modelling capability, such as water rights allocation.

Software Engineers involved in PRSYM development as most assisted by its modular architecture. Both the object-oriented implementation and the clean separation of the GUI, SimLib, and Engineering libraries eases the process of software change and integration. In addition, the widespread use of non-commercial C++ libraries such as LEDA and development tools such as GNU make software portability much easier to achieve.

As a software application PRSYM is built on an architecture which provides flexibility, extensibility, and modularity for the end-user, Civil Engineering programmers, and Software Engineers alike.

References

CPLEX is a registered trademark of CPLEX Optimization, Inc., Suite 279, 930 Tahoe Blvd., Bldg. 802, Incline Village, NV 89451-9436.

Galaxy Application Environment is a registered trademark of Visix Software Inc., 11440 Commerce Park Drive, Reston, VA 22091.

GNU software is available from Free Software Foundation, 675 Massachusetts Avenue, Cambridge, MA 02139-3309.

LEDA is a C++ "Library of Efficient Data Types and Algorithms" available from the Max-Planck-Institut f"ur Informatik, Im Stadtwald, 6600 Saarbr"ucken, FRG.

Automatic Object Oriented Generation of Goal Programming Models for Multi-Reservoir Management

Elizabeth A. Eschenbach[1], Evan R. Zweifel[2], Timothy M. Magee[3], Carl F. Grinstead[3], and Edith A. Zagona[4]

Abstract

PRSYM (Power Reservoir System Model) has been developed as a part of the INTEGRAL Project. (See other papers in this proceedings for a description of the INTEGRAL Project). PRSYM is an object oriented software tool that allows water resources engineers to both simulate and optimize the management of a system of hydropower reservoirs. This paper focuses on how PRSYM optimization automatically generates the goal program formulation of a reservoir system, given the user input of: 1) physical characteristics of the reservoir system, 2) the prioritized policy goals for optimization, and 3) parameters for constraint linearization. The input of this information is facilitated using various PRSYM Graphical User Interfaces (GUI).

Introduction

The Integral Project, a collaborative effort between EPRI (Electric Power Research Institute), TVA (Tennessee Valley Authority), WAPA (Western Area Power Administration) and USBR (U.S. Bureau of Reclamation), has focused on developing software tools for managing hydropower reservoir systems. Part of the Integral project is PRSYM (Power Reservoir System Model), an object oriented software tool that has both simulation and optimization capabilities. This paper will focus on the automatic generation of the goal program formulation for an LP (Linear Programming) solver. (For more information on the Integral project and PRSYM simulation, see related papers in this proceedings).

Goal programming and LP are well established methods for solving water resources problems[1]. The innovative aspect of PRSYM optimization is that

1. Asst. Professor, Environmental Resources Engineering, Humbolt State University, Arcata, CA 95521, ASCE Member
2. Postdoctoral Research Associate, Center for Advanced Decision Support for Water and Environmental Systems (CADSWES), University of Colorado, Boulder, CO 80309-0421
3. Professional Research Assistant, CADSWES, University of Colorado, Boulder, CO 80309-0421
4. Research Associate, CADSWES, University of Colorado, Boulder, CO 80309-0421, ASCE Member

the software is built so that it can automatically generate the appropriate goal program formulation, given user input. The following input is needed from the user for PRSYM optimization: a physical model of the reservoir system, a list of prioritized policy goals associated with the system, and selection of parameters for constraint linearization methods. The organization of this paper is as follows. First, a brief overview of the optimization formulation is described. Then, an introduction to the user interface for describing the physical system is presented. Then, the graphical user interface for entering the prioritized goals is described, and the user selected linearization methods are listed. Lastly, the algorithm for automatically generating the goal program using the user's prioritized goals is outlined.

Goal Programming

PRSYM optimization utilizes pre-emptive goal programming, using linear programming (LP) as an engine to optimize each of the prioritized goals input by the user. The optimal solution of a higher priority goal is not sacrificed in order to optimize a lower priority goal.

The goals are indexed by i where $i = 1, ..., P$ and are represented as linear functions, $F_i(\mathbf{x})$, with a desired target value, T_i. The goal program seeks to minimize the positive and/or negative deviations, d_i^+ and d_i^- respectively, of each goal with respect to its target. Traditional multi-objective goal programming translates the goals to a single objective function minimizing the weighted sum of deviations.

$$\text{Minimize } \sum_{i=1}^{P} W_i \left(d_i^+ + d_i^- \right) \qquad \text{(Eq. 1)}$$

subject to

$$F_i(x) - d_i^+ + d_i^- = T_i \qquad i = 1, 2, ..., P \qquad \text{(Eq. 2)}$$

$$x \in X \qquad \text{(Eq. 3)}$$

$$d_i^+, d_i^- \geq 0 \qquad i = 1, 2, ..., P \qquad \text{(Eq. 4)}$$

The user inputs the constraints represented by Equation 2, but the constraints represented by Equation 3 are automatically generated by PRSYM optimization. Equation 3 represents the set of physical constraints that define the feasible region for the goal program and include the physical constraints for the topology of the hydropower reservoir system. For example, a power reservoir object automatically generates its mass balance equation and the equation that describes how the turbine capacity changes with elevation.

Unlike the formulation presented in Equation 1, PRSYM optimization allows a number of constraints to be associated with a given priority i. For each constraint k (k=1,..., N) associated with priority i (i = 1,..., P), there is a deviation $d_{i,k}$ that is minimized. PRSYM optimization allows the user to choose one of three formulations to compute the minimum deviation: Minimax, Summation and Objective.

Minimax goals minimize the maximum deviation of one of the k constraints associated with a given priority i. The Minimax method formulates the LP for priority p to compute the following.

$$\text{Minimize} \left\{ \begin{array}{c} \text{Maximum} \\ k = 1, ..., N \end{array} \left(d_k^+, d_k^- \right) \right\} \tag{Eq. 5}$$

The Minimax method requires several optimizations in practice. After the original expession has been minimized, the largest deviation is "frozen" at its minimal value and is removed from the objective. The optimization successively minimizes the remaining deviations, freezes the largest deviation and minimizes the remaining deviations.

Summation goals minimize the sum of all deviations of the k constraints associated with a given priority i. The summation method formulates the LP for priority i in the following way.

$$\text{Minimize} \left\{ \sum_{k=1}^{N} \left(d_k^+ + d_k^- \right) \right\} \tag{Eq. 6}$$

Objective goals are of a much different form. The user creates an expression of the decision variables **x** and requests that the goal program either minimize or maximize that expression. For example, the user could associate priority i with an expression for the cumulative spill over time periods t (t = 1,...,T) at reservoir Watauga and require PRSYM to minimize the expression.

$$\text{Minimize} \left\{ \sum_{t=1}^{T} \text{Watauga.Spill} [t] \right\} \tag{Eq. 7}$$

Physical Model Input

Before PRSYM optimization can be run, the user inputs a physical model of the reservoir system using the PRSYM GUI. Figure 1 shows the GUI that allows the user to build the physical model.

The user builds the reservoir system using a palette of engineering objects. (reservoirs, canals, reaches, etc.) These objects are then linked together using a link editor. For instance, the outflow from the upstream reservoir is linked to the inflow of the down stream reservoir. Data is input to each object by opening the object and inputting the appropriate data in the appropriate slot. (See related papers in this proceedings for more information on building models using PRSYM.)

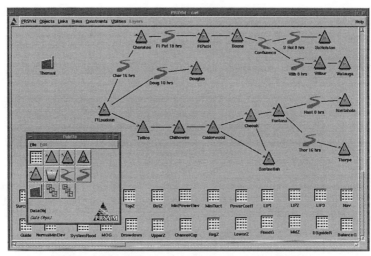

Figure 1. Building a reservoir system using PRSYM

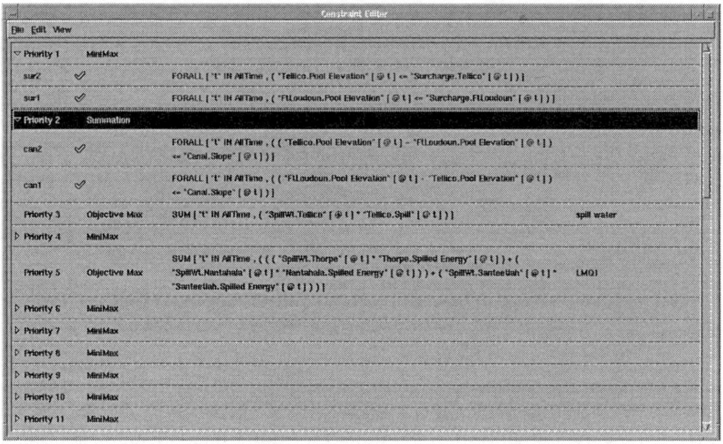

Figure 2. PRSYM Optimization Goal Editor

Policy Constraint and Expression Editors

Figure 2 provides an example of the Constraint Editor showing that a number of different constraints can be associated with a single priority (Goals 1 and 2). Goal 1 is solved using Minimax and Goal 2 is solved using Summation. Goal

3 has only one expression associated with it and it is an Objective goal. All the information in the Constraint Editor is input by the user. In order to write a constraint, the user selects the goal to add the constraint to, then selects the "new constraint" option from the Edit menu.

Figure 3 shows an example of the syntax directed expression editor, which

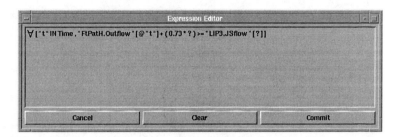

Figure 3. PRSYM Optimization Constraint Editor

assists the user in writing syntactically correct policy constraints. The editor will only allow the user to input expressions that contain variables that are accepted by PRSYM optimization. Question marks represent parts of the equation which have yet to be built by the user. The editor works by simultaneously manipulating both the text and a context free grammar describing the expression language. The user can use the symbols Σ, Π, $\bar{x}$, and $\forall$ to represent summation, product, average, and forall operations. For example, the user can write an objective that will minimize the sum the spill of all reservoirs in the following manner:

$$\text{Objective Minimum} \sum [\text{R in Reservoir}, \sum [\text{t in Time, R.Spill[t]}]] \qquad \text{(Eq. 8)}$$

Automatic Constraint Linearization

In addition to inputting the policy constraints, the user must parameterize the user selected linearization methods before PRSYM optimization can run.

The optimization model uses two types of constraints: physical constraints and policy constraints. In addition to automatically generating the physical constraints, PRSYM automatically generates the mathematical description of each of the constraints. Part of the automatic generation includes linearizing any nonlinear constraints, using the user's preferred method for linearization.

For example, a user could write the following nonlinear policy constraint.

$$\text{Watauga.Pool Elevation[t]} \leq \text{FloodGuide.Watauga[t]} \qquad \text{(Eq. 9)}$$

where FloodGuide is a data object containing the elevation levels for Watauga Reservoir's flood guide curve. This constraint (Equation 9) can be linearized by expressing Pool Elevation at Watauga in terms of Storage. The Watauga reservoir object contains a data table that records the corresponding storage value for a given pool elevation. Conditioned on the convexity of the constraint and the number of terms in the constraint, the user may choose how to represent a linear relationship between pool elevation and storage using one of the following:

-a line tangent to the data at a user specified point,
-a line going through the data at two user specified points,
-a piecewise line going through the data at n user specified points,
-a direct substitution of the constraint.

Automatic Goal Program Formulation

Once the user has built the physical model, input the policy constraints, and chosen and parameterized the linearization methods, the user selects RUN and the PRSYM optimization controller formulates a sequence of problems for an LP solver (CPLEX [2]). The optimization controller formulation process implements the following steps:

1) Build a dictionary of variables acceptable for PRSYM optimization, differentiating between variables that make constraints nonlinear, and those that do not.

2) Collect the physical constraints from each engineering object in the work space.

3) Collect the policy constraints and the goals for the goal program that were input by the user.

4) Linearize all nonlinear physical or policy constraints.

5) For each priority in the constraint set do the following

   a) Expand each constraint for this priority for all appropriate time steps.

   b) Add in the appropriate deviation variable for this priority to each policy constraint.

   c) Give the problem to the LP solver.

6) Propagate the solution from the LP solver back to the PRSYM interface.

Steps 5a and 5b require further explanation. Suppose the user has requested that the optimization be run for 2 weeks at a 6 hour time step, (indicating a total of 56 time periods). Physical constraints that are similar for all time periods are expressed to the controller only once. For example, the mass balance equation for a reservoir named Watauga is expressed as:

$$\text{Watauga.Spill } [t] - \text{Watauga.Storage } [t-1] + (\text{Watauga.Outflow } [t-1] - \text{Watauga.Inflow } [t-1]) \Delta t = 0 \qquad \text{(Eq. 10)}$$

In Step 5a, the controller expands Equation 10, into 56 separate equality constraints; one mass balance equation for each time period t (t=1,..., 56). The expansion of this constraint includes setting the initial conditions when t=1.

In step 5b, the optimization controller takes the user's prioritized policy constraints from the constraint editor and adds in deviation variables Z_i for the goal program formulation. For example, say the user selects Minimax for priority i and states a policy limiting the spill at Watauga.

$$Watauga.Spill[t] \leq SpillLim.Watauga[t] \qquad \text{(Eq. 11)}$$

The optimization controller uses a dummy variable, Z_i to evenly enforce the policy constraint across time periods. When $Z_i = 0$ spill is limited to the appicability limit, and when $Z_i = 1$ the stronger policy limit is fully satisfied. The controller formulates:

$$\text{Maximize } Z_i \qquad \text{(Eq. 12)}$$

$$0 \leq Z_i \leq 1 \qquad \text{(Eq. 13)}$$

$$Watauga.Spill[t] - (SpillLim.Watauga[t] - Watauga.Spill.applim) \cdot Z_i \qquad \text{(Eq. 14)}$$
$$\leq Watauga.Spill.applim \qquad t = 1, ..., T$$

where T is the number of time periods and WatuguaSpill.applim is the physical upper limit of spill possible. This applicability limit is part of the physical model input by the user. SpillLim.Wataugua is the policy bound on spill at Watauga. This data resides in the SpillLim data object.

If the user had chosen summation for the constraint in Equation 11, the controller would create additional dummy variables, Z_{it}, for each time period. The Z_{it} replaces Z_i in equation 14:

$$Watauga.Spill[t] - (SpillLim.Watauga[t] - Watauga.Spill.applim) \cdot Z_{it} \qquad \text{(Eq. 15)}$$
$$\leq Watauga.Spill.applim \qquad t = 1, ..., T$$

The controller completes the formulation by defining Z_i to be the sum of deviations:

$$Z_i = \sum_{t=1}^{T} Z_{it} \qquad \text{(Eq. 16)}$$

$$0 \leq Z_{it} \leq 1 \qquad t = 1, ..., T \qquad \text{(Eq. 17)}$$

and Z_i is still maximized.

If the user chooses to use an Objective Goal to minimized spill at Watauga, then the following constraint is formulated by the controller and the goal program attempts to minimize Z_i.

$$Z_i = \sum_{t=1}^{T} Watauga.Spill[t] \qquad \text{(Eq. 18)}$$

Finally, after all priorities have been processed and optimized, the results are propagated from CPLEX back into the model for viewing by the PRSYM GUI.

Summary

A brief introduction to PRSYM optimization has been presented here. PRSYM optimization allows reservoir systems managers to solve a complicated optimization problem, by specifying prioritized policy goals, linearization parameters and a physical model of the system. The specification of this information is facilitated by a user friendly GUI. Unlike previous water resources systems applications, the optimization formulation is automatically generated, using the user input. The problem is then sent to an LP solver and returns the solution to the user via the GUI.

References

[1] Shane, Richard M. and Boston, William T., TVA's Weekly Scheduling Model: Overview and Implementation, WATERPOWER '83 -- International Conference on Hydropower, Sept. 1983.

[2] CPLEX, CPLEX Optimization, Inc. Suite 279, 930 Tahoe Blvd., Bldg. 802, Incline Village, NV 89451, 1994.

[3] Shane, Richard M. and Gilbert, Kenneth C., Weekly Time Step Reservoir System Scheduling Model, Tennessee Valley Authority, Office of Natural Resources, Division of Water Resources, Water Systems Development Branch. Norris, Tennessee, June 1980.

[4] CADSWES (1993) Power and Reservoir System Model (PRSYM) Design Document for 1993 Implementation; CADSWES, University of Colorado at Boulder, CO.

[5] Gilbert, Kenneth C. Management Science In Reservoir Operation, Applications of Management Science, Vol 4, p. 107-130.

A Rule Language to Express Policy in a River Basin Simulator

Stephen C. Wehrend[1] and René F. Reitsma[1]

Abstract

The need to express policies for a generic river basin simulator can be met by attaching an interpreted language to the simulator. This allows the policies to be modified between runs, without requiring the simulator to be changed or rebuilt. In this paper, we briefly discuss the need for such a language, the nature of this language and how it fits into a generic river basin simulator.

Background

Power and Reservoir Simulation System (PRSYM) [CADSWES, 1993] is a generic river basin simulator. In PRSYM, a basin to be modelled consists of a set of objects, such as reservoirs, power plants and reaches, which are linked together to form a network. Each of these objects is autonomous and given the attributes necessary to accurately model an object of its type. For instance, a reservoir object will have, among other things, a current elevation, an inflow, a release and a method for computing evaporation. The links, which connect the objects, are used to pass information from one object to another. An example of a link is a connection between the outflow of an upstream reservoir and the inflow of a downstream reservoir.

PRSYM has a number of methods for solving the river systems it models. The two main methods are mass balance and optimization. For more information on how optimization is used in PRSYM, see the accompanying paper by Eschenbach, et al. in these proceedings. In PRSYM, mass balance is a method whereby each of the objects in the network uses its own internal state and information received from its neighbors to balance itself. When all of the objects are balanced, the network is balanced.

While mass balance in PRSYM provides the ability to simulate different scenarios on river basins by allowing state variables on objects to be changed, it does not allow a simulation to make conditional decisions. Conditional decisions are often necessary to the formation of policies which are used by operators of river systems. For example, the following simple conditional policy could be used for flood control:

---

1. Center for Advanced Decision Support for Water and Environmental Systems (CADSWES), Dept. of Civil, Environmental and Architectural Engineering, University of Colorado at Boulder, CO.

```
IF  Mead's elevation > 1229.0 THEN
    Mead's release = Mead's inflow
ENDIF
```

Here, if Mead's elevation is greater than 1229.0 feet, its release is set equal to its inflow. Otherwise, its release will be computed by another policy, if one applies, or through mass balance.

Policy in a River Basin Simulator

Expressing policy in simulation programs is not new. Dedicated models, such as the Colorado River System Simulation (CRSS) [Schuster, 1987], combine a basin-specific simulator with basin-specific policy. In these systems, much of the information, from the configuration of the basin to the policies which control the flow of water through the basin, are built into the program. As such, if changes to the policies are needed, the code which contains this policy must be changed and the program rebuilt. This approach is generally useful when information about the basin and its policies is unlikely to change.

Another approach is to require the form of policies to be built into the program and to allow the values that the policies rely on to be changed at run-time. For example, the policy mentioned above might be built into the program using the following form:

```
IF  Mead's inflow > value THEN
    Mead's release = Mead's inflow
ENDIF
```

The user would supply *value* at run-time. This approach has the advantage over the previous approach that some changes to the policies can be made without rebuilding the program. It has the disadvantage that the policies expressed are likely to be very specific and any changes to the form of a policy will require the program to be rebuilt.

A last approach is to allow policies to be completely modifiable without requiring the underlying system to be rebuilt. This can be accomplished by attaching a language interpreter to the program and requiring the policies to be written in the accompanying language. Policies are interpreted at run-time and changes to them do not require the program to be rebuilt. While this approach provides a great deal of flexibility, it comes with a price. Since the policies are interpreted at run-time, the program will tend to be slower than programs that use either of the other two approaches.

Rule Language

There exist at least two options for the integration of an interpreted language into a simulator: develop a custom language or use an existing language. A custom language has the advantage that it can be developed to exactly meet the needs of the users. It has the disadvantage that its development and maintenance can be time-consuming and expensive, especially given that the users' needs, and therefore the language, will likely evolve over time.

Using an existing language has the advantage that there are many well tested and documented interpreted languages in existence. These languages are typically full-featured and extensible. In addition, some of them are intended to be embedded into existing applications. And since they have a predefined language definition and built-in interpreter, time can be spent constructing support facilities, such as an editor and debugger, rather than developing domain-specific language. The disadvantages of this approach include the possibility that the language may not have the exact features that the users need and its interpreter may be slow.

For PRSYM, an existing language was selected: the Tool Command Language (Tcl) [Ousterhout, 1994]. Tcl is extensible, can be easily embedded into existing C programs and has a large number of language primitives. While Tcl has many advantages, there are some concerns about the speed of its interpreter and its occasionally clumsy syntax. It is possible that these concerns will motivate a language switch. Figure 1 shows a simple rule which uses Tcl.

```
RULE_NAME:              Mead Flood Control
RULE_PRIORITY:          1
RULE_DEPENDENCIES:      mead.elevation, mead.inflow
RULE_INCLUDES

    source MeadFloodControl_Procs.TCL

RULE_BEGIN

    set mead_elevation    [GetValue mead.elevation 1.0 ft]

    if {$mead_elevation >= 1229.0} {
        [SetValue mead.release [GetValue mead.inflow 1.0 cfs] 1.0 cfs]
    }

RULE_END
```

Figure 1. Simple rule using Tcl.

Integration with PRSYM

Rules are used in PRSYM in a manner that is similar to the way rules are used in a traditional, forward-chaining expert system [Brownston, et al., 1985]. Like an expert system, a list of rules which are available for execution is kept. This list, called an agenda, is updated as the state of the system is changed. When a rule is needed for execution, the rule with the highest priority is taken from the agenda and executed. Rules are added to the agenda when it is determined that their execution is likely to change the state of the system. This is determined by associating with each rule a list of state variables that the rule either reads or writes. If any of these state variables changes since the last execution of the rule, the rule is added to the agenda.

The current design calls for the execution of a rule whenever PRSYM reaches a state that it can no longer mass balance, regardless of whether the network is fully solved. Thus, PRSYM will mass balance without using rules until it has either completely solved the system or it does not have enough information to continue. At this point, it will execute a rule, which will more than likely change the state of the system. If, after the rule has executed, PRSYM can continue mass balancing, it does so. Otherwise it continues to execute rules until either it can continue mass balancing or it has exhausted all its rules. When all rules are exhausted, the simulation terminates. If the network is fully solved, the system is fully-determined; otherwise, the system is under-determined. Figure 2 contains a representation of this process.

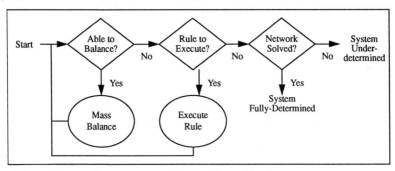

Figure 2. Flow of control in mass balance solver with rules.

One likely result of adding rules to a system is that objects will become over-determined. This can be illustrated using the first policy example given. If Mead is already mass balanced when the above rule tries to set its release, Mead will become over-determined. In order for Mead to mass balance given the new release, one of the state variables it uses in its mass balance must be overwritten.

A method to determine the state variable to overwrite, which relies on the priorities of the rules, was developed. As part of this method, each state variable on an object is assigned a priority when it is set. This priority depends on who or what set the state variable. If the state variable is set explicitly by the user, the priority is the highest possible. If the state variable is set by a rule, the priority is set to the priority of the rule. In either case, where setting a state variable on an object would result in an over-determined object, the priorities of the state variables are used to determine the state variable to overwrite. In all cases, the state variable with the lowest priority is forfeit. Thus, given a conflict between two rules, the rule with the highest priority will take precedence. This has the effect that a rule with a low priority may be prohibited from executing.

References

CADSWES (1993) Power and Reservoir System Model (PRSYM) Design Document for 1993 Implementation; CADSWES, University of Colorado at Boulder, CO.

Schuster, R.J. (1987) Colorado River Simulation System Documentation. System Overview; U.S. Department of the Interior, United States Bureau of Reclamation, Denver, CO.

Ousterhout, J. (1994) Tcl and the Tk Toolkit; Addison-Wesley Publishing Company; Reading, MA.

Brownston L., Farrell R., Kant E., Martin N. (1985) Programming Expert Systems in OPS5: An Introduction to Rule-Based Programming; Addison-Wesley Publishing Company; Reading, MA.

AN IMPROVED MODEL FOR BEAMS ON ELASTIC FOUNDATIONS

Y. C. Das[1] and C. V. G. Vallabhan[2]

INTRODUCTION

The concept of beams and slabs on elastic foundations has been extensively used by geotechnical, pavement and railroad engineers for the analysis and design of foundations. The majority of work in this area has been done by Hetenyi (1946) and Westergaard (1948) using the classical Winkler model where a coefficient k called the modulus of subgrade reaction of the foundation, is employed. The use of Winkler model involves a major problem of determining the modulus of subgrade reaction k representing the soil continuum and a behavioral inconsistency that a beam or a slab carrying a uniformly distributed load will produce a rigid body displacement. To eradicate this inconsistency in the Winkler concept, Vlasov and Leont'ev (1966) developed a unique two parameter model using a variational approach. An extensive study of the beams on elastic foundation has been given by Kerr (1964) and Scott (1981). The Vlasov model, in spite of its conceptual elegance, required the estimation of a parameter γ, which controls the decay of stress distribution within the foundation. Jones (1977) established a relationship between γ and the displacement characteristics, but did not actually determine the value. Vallabhan and Das (1987) developed a simple iterative technique to uniquely determine the γ-parameter for uniformly distributed loading condition. Later, Vallabhan and Das (1991), using finite difference method, solved many problems of beams on elastic foundations for a wide variety of loading conditions. The authors called this method "Modified Vlasov Model". When the results -- from this model were compared to those obtained from more

---

[1]Member, ASCE, Department of Civil Engineering, McNeese State University, Lake Charles, LA.
[2]Fellow, ASCE, Department of Civil Engineering, Texas Tech University, Lubbock, TX.

sophisticated finite element solutions, it was found that
the Modified Vlasov Model yielded satisfactory solutions
when the beams were carrying uniformly distributed loads.
However, when the beams carried concentrated loads,
especially at the ends, the solutions deviated more from
the corresponding finite element solutions. The
inability of the Modified Vlasov Model to adequately
represent the elastic foundation led the authors to
modify the assumed vertical displacement function in the
elastic continuum. The authors introduced a series of
vertical displacement functions instead of one function,
and this modification is the theme of this paper.

Using a series of vertical displacement functions,
the total potential energy of the beam-foundation system
is minimized by variational principle and the resulting
equations of beam-foundation system are solved by finite
difference method along with an iterative technique. The
results obtained from this "Improved Model" are compared
with those obtained from "Modified Vlasov Model" and the
finite element models. The results indicated substantial
improvements in the accuracy of the solutions, especially
when the beams carry concentrated loads. The beam
displacements, shears and moments have been computed as
functions of beam-foundation parameters and the results
are presented graphically and these will be of great use
to design engineers. The method involved in this
"Improved Model" can be easily programmed on an IBM-PC
compatible computer and can be used effectively by
engineers for quick analysis of the problems involved
with beams on elastic foundations.

IMPROVED FOUNDATION MODEL

Fig. 1 illustrates a uniform beam resting on an
elastic foundation of thickness H with a rigid base at
the bottom. The elastic foundation is assumed to be in
a plane strain condition. The minimum potential energy
theorem is employed here to derive the governing
differential equations. The total potential energy
function for the beam-foundation system is given by

$$\pi = \frac{1}{2} \int_{\frac{1}{2}}^{\frac{1}{2}} E_b I_b \left(\frac{d^2w}{dx^2}\right)^2 dx + \frac{b}{2} \int_{-\infty}^{\infty} \int_0^H (\sigma_x \epsilon_x + \sigma_z \epsilon_z + \tau_{xz} \gamma_{xz})\, dx\, dz - \int_{\frac{-1}{2}}^{\frac{1}{2}} qw\, dx$$

(1)

where E_b - Young's modulus of elasticity, I_b = moment of
inertia of the cross section of the beam, w = the lateral
displacement of the beam, $\sigma_x, \sigma_z, \tau_{xz}$ = the components of
stress in the elastic foundation, $\epsilon_x, \epsilon_z, \gamma_{xz}$ = the
corresponding components of strain in the elastic

continuum, q = the applied lateral force on the beam, b,l = the width and length of the beam, and H = the height of the elastic foundation.

ASSUMPTIONS

The vertical displacements at any point (x,z) in the elastic foundation are $u(x,z)$ and $w(x,z)$ in the x- and z-directions respectively. It is assumed that

1. the vertical displacement

$$\overline{w}(x,z) = \sum_{i=1,2}^{n} w_i(x)\ \phi_i(z)$$

(2)

such that

$$\phi_i(0) = 1;\ \phi_i(H) = 0\ (i = 1,2,\ldots n)$$

2. the horizontal displacement $u(x,z) = 0$ (3)
everywhere in the elastic continuum.

Here the $\Phi_i(z)$, $(i = 1, 2,..n)$ functions describe the decay of the vertical displacement $w(x,z)$ in the z-direction. Incorporating the strain-displacement equations from the linear elasticity theory, the constitutive relationship and minimizing the total potential energy function and taking variations in w_i and Φ_i respectively, the following governing equations are obtained in the matrix form.

Field Equations for the Beam: $(-1/2 < x < 1/2)$

$$[El]\{W^{IV}\} - [T]\{W^{11}\} + [K]\{W\} = \{Q\}$$ (4)

where the elements of the matrices [El], [T] and [K] are

$$[EI]_{ij} = E_b I_b\ \phi_i(0)\ \phi_j(0)$$

$$K_{ij} = \int_0^H (\lambda + 2G)\ \frac{d\phi_i}{dz} \cdot \frac{d\phi_j}{dz}\ dz$$

$$T_{ij} = \int_0^H G\ \phi_i\ \phi_j\ dz \text{ for } i = j = 1,2,3\ldots n$$

(5)

For the vectors $\{W\}$ and $\{Q\}$, the column elements are

$$W_i(x) \text{ and } q_i(x) = q(x)\ \Phi_i\ [0]$$ (6)

The superscripts "II" and "IV" on W denote differentiation with respect to x, twice and four times respectively. The boundary conditions at x = ± 1/2 are:

$$[EI]\{W^{111}\} - [T]\{W^1\} = \{F_{net}\} \tag{7}$$

or prescribed displacements. The other set of boundary conditions are prescribed moments

$$[EI]\{W^{11}\} = \{M_{net}\} \tag{8}$$

or prescribed slopes at x = ± 1/2.

Field Equations Outside the Beam: (-∞ < x < 1/2 and 1/2 < x < ∞)

$$-[T]\{W^{11}\} + [K]\{W\} = \{0\} \tag{9}$$

with boundary conditions,

x → ± ∞, {W} = {0}, and
x = ± 1/2, [T]{W^1} + {F} = {0} (10)

Field Equations in the Elastic Foundation: Using variations of Φ_i, we get

$$[M]\left\{\frac{d^2}{dz^2}\Phi\right\} - [N]\{\Phi\} = \{0\} \tag{11}$$

where the elements of matrices M and N are:

$$M_{ij} = \int_{-\infty}^{\infty} (\lambda + 2G) w_i \, w_j \, dx \tag{12}$$

and

$$N_{ij} = \int_{-\infty}^{\infty} G \frac{dw_i}{dx} \cdot \frac{dw_j}{dx} \, dx \tag{13}$$

Here the elements of the vector $\{\Phi\}$ are: Φ_i (z) (i = 1,2,..n)

The boundary conditions for $\Phi_i(z)$ are given in Eq. (2).

SOLUTION OF FIELD EQUATIONS

Solution of Equations in the Elastic Foundation

Since Eq. (11) is a coupled homogeneous differential equation, its solution can be achieved by assuming

$$\Phi_i(z) = a_i e^{rz} \text{ for } i = 1,2,...n \tag{14}$$

where a_is are constants. Here γ is a foundation parameter. Substituting Eq. (14), we have:

$$\gamma^2[M]\{a\} = [N]\{a\} \qquad (15)$$

This is a linear algebraic eigenvalue problem. There are n eigenvalues γ_i^2 (i = 1,2,..n) and corresponding eigenvectors $\{a\}_i$. Since M and N matrices are symmetric and eigenvalues are distinct. The eigenvectors are orthogonal.

Using the orthogonal property solutions to satisfy the boundary conditions given in Eq. (2) can be written as:

$$\Phi_j(z) = \sum_{i=1,2}^{n} a_{ji} A_i \Phi_i(z)$$

where

$$A_i = \frac{\{a\}_i^T [M]\{1\}}{\{a\}_i^T [M]\{a\}_i}$$

$$\Phi_i(z) = \frac{\sinh \gamma_i (H-Z)}{\sinh \gamma_i H}$$

(16)

(17)

Using $\Phi_i(z)$, we now can evaluate the coefficients of the T and K matrices.

Solution of Equations Outside the Beam (1/2 < x < ∞)

Here also (see Eq. (9)), we have coupled homogeneous linear differential equations. We assume

$$w_i(x) = b_i e^{sx} \text{ for } i = 1,2,..n \qquad (18)$$

where b_is are constants. Substituting Eq. (18) in Eq. (9), we have

$$s^2[T]\{b\} = [K]\{b\} \qquad (19)$$

which is another linear algebraic eigenvalue problem. Here we obtain n eigenvalues s_i^2 (i = 1,2,..n) and corresponding eigenvectors $\{b\}_i$. These eigenvectors are orthogonal to each other since T and K matrices are symmetric. Using the boundary conditions at x = 1/2 and x → ∞, we get

$$w_j(x) = \sum_{i=1,2}^{n} b_{ji} \, B_i \, e^{-s_i(x-1/2)}$$

(20)

where

$$B_i = \frac{\{b\}_j^T [T] \{W^*\}}{\{b\}_j^T [T] \{b\}_i}$$

(21)

Here W* is displacements of the beam at x = 1/2. Similar analysis can be done for the region $-\infty < x < -1/2$.

Solution Technique

Only two terms are considered as an example. The method starts with assumed values of the two γ-parameters as 1 and 2 respectively. The following are the steps used in the iterative procedure.

1. Assume value of γ_1 and γ_2.
2. Calculate coefficients of matrices K and T.
3. Determine eigenvalues and eigenvectors of K and T.
4. Solve Eq. (5) for w_1 and w_2.
5. Calculate coefficients of matrices M and N.
6. Determine eigenvalues of γ_1 and γ_2 and corresponding eigenvectors.
7. Compare new and old values of γ-parameters. Repeat from step 2, if convergence of γ is not satisfactory.

The convergence criterion is achieved using a norm of the γ-parameters, equal to

$$\|\gamma\| = \sqrt{\gamma_1^2 + \gamma_2^2}$$

(22)

and the iteration is continued until the new and old norms of the γ differ by 0.001. Not more than six iterations were required for the example problem shown below to get the required convergence.

NUMERICAL EXAMPLES

The examples selected here are the same as those solved by Vallabhan and Das (1987), where they compared the results obtained from the modified Vlasov method and finite element methods.

As numerical example, a beam of length 100 ft, width 1 ft and depth 5 ft with modulus of elasticity 4.32 x 10^6 psf is considered to be supported by foundation with depth of soil stratum 100 ft, modulus of elasticity 4.32 x 10^6 psf and Poisson's ratio 0.2. Three loads, a uniformly distributed load (5 k/ft), a concentrated load in the middle (100 k) and concentrated loads at the ends (each 100 k) are selected. The vertical displacements of the beams are compared with other solutions and they are illustrated in Figs. 2, 3 and 4.

CONCLUSIONS

1. If the loads are uniformly distributed on the beam, the improved foundation model does not increase the accuracy of the solution substantially.
2. For the case of concentrated loads, especially at or near the ends, the improved foundation model is much superior to the simple modified Vlasov model. The displacements obtained are parallel to the finite element displacements.
3. The method can be easily programmed on an IBM-PC compatible computer and can be used effectively by the engineer for quick analysis of such problems.

REFERENCES

1. Hetenyi, M.   (1946).   "Beams   on   Elastic Foundation", Uni. of Michigan Press, Ann Arbor, Mich.

2. Jones, R. (1977). "The Vlasov Foundation Model", Int. J. Mech. Sci., Pergamon Press, Elmsford, N.Y.

3. Kerr, A. D.   (1964).   "Elastic and Viscoelastic Foundation Models", J. Appl. Mech., Sept.

4. Scott, R. F.   (1981).   "Foundation Analysis", Prentice Hall, New Jersey.

5. Vallabhan, C. V. G. and Das, Y. C.   (1987).   "A Note on Elastic Foundations", IDR Report, Dept. of Civ. Engrg., Texas Tech Univ., Lubbock, Tex.

6. Vallabhan, C. V. G. and Das, Y. C.   (1991). "Modified Vlasov Model for Beams on Elastic Foundations", J. of Geotech. Engrg., Vol. 117, No. 6, June.

7. Vlasov, V. Z. and Leont'ev, N.N. (1966). "Beams, Plates and Shells on Elastic Foundations", Translated from Russian by Israel Program for Scientific Translations, NTIS Accession No. N67-14238.

8. Westergaard, H. M. (1948). "New Formulas for Stresses in Concrete Pavements of Airfields", Trans., ASCE, 113.

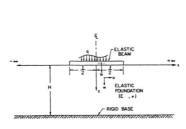

FIG. 1. BEAM ON ELASTIC FOUNDATION (VLASOV MODEL)

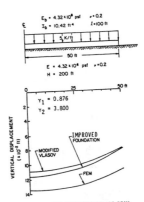

FIG. 2. VERTICAL DISPLACEMENTS OF BEAM DUE TO UNIFORMLY DISTRIBUTED LOAD (GEOMETRY 2)

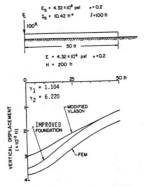

FIG. 3. VERTICAL DISPLACEMENTS OF BEAM DUE TO A CONCENTRATED LOAD IN THE MIDDLE (GEOMETRY 2)

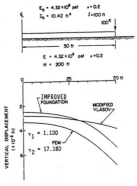

FIG. 4. VERTICAL DISPLACEMENTS OF BEAM DUE TO A CONCENTRATED EDGE LOAD (GEOMETRY 2)

SI Conversion: ft = 0.3048m
                psf = 47.88Pa

Three-Dimensional Stress Analysis of the First Molar

Elfatih M. Ahmed, P.E.[1] and Robert E. Efimba, Sc.D., P.E.[2]

1. Abstract

Three-dimensional solid finite element models are used to study the behavior of the human first molar, as a follow-up to a study reported by Efimba and Ahmed (1994) in which 2-D models were used for the maxillary central incisor. Some researchers have long realized that two-dimensional models are not suitable for the molar, and have resorted to an axisymmetric model, as reported by Farah and Craig (1974) and, more recently, by Asaoka (1994). However, the authors believe that a 3-D model that takes advantage of two-fold approximate symmetry is a more accurate representation of the molar, and have used such a model with isoparametric quadratic solid (IPQS) "brick" elements as implemented in the computer code GTSTRUDL (1985). As in the previous study, convergence of displacement studies led to the actual mesh pattern used to determine the respective stresses due to mechanical and thermal loads applied to a molar, with and without a full gold crown preparation. The maximum principal normal and shearing stresses caused by a 100°C change in temperature are found to be much smaller than those due to a mechanical load of only 50N in each of the four quadrants, but are significant because of the distortions that they can cause in dental restorations.

[1]Chief Bridge Engineer, Widmer Engineering, Beaver Falls, PA 15010

[2]Member, ASCE; Associate Professor of Civil Engineering, Howard University, Washington, DC 20059

2. Introduction

In this study, the first molar is modeled as a 3-D solid using the geometry shown in Figure 2, with dimensions from the axisymmetric model presented by Farah and Craig (1974). Three different models are used to study the convergence of the results, with 8, 16, and 23 of the isoparametric quadratic solid 'IPQS' elements, respectively. A natural first molar is first analyzed under different loading conditions and the results are compared. Then a first molar with full crown preparation using gold is examined, and the results are compared with those obtained in the axisymmetric idealization study by Farah and Craig (1974).

3. Materials and Methods

For the natural first molar, the elastic modulus, Poisson's ratio, and the coefficient of thermal expansion are shown in Table 1 for the different materials. The material properties shown in Table 1 encompass a range consistent with values given by Selna, et al. (1975), Wilner and Pattensbargen (1942), and by Farah and Craig (1974), reflecting differences in material properties given in the dental literature. The pulp was not considered and was treated as a void. Consideration of the pulp requires modelling its biological behavior which is out of the scope of this study.

Table 1. Material and Physical Properties

Material	Elastic[**] Modulus (GPa)	Proportion-al[*] Limit (MPa)	Poisson's Ratio[**]	Coefficient[***] of Thermal Expansion per °C
Enamel	35.0	224	0.25	11.4×10^{-6}
Dentin	14.0 - 18.6	148	0.25 - 0.31	8.3×10^{-6}
Gold	89.5	379-534	0.33	14.0×10^{-6}

[*] Phillips (1982); [**] Farah and Craig (1974); Selna, et al. (1975);
[***] Wilner and Pattensbargen (1942); Cutnell and Johnson (1992)

Three models are examined in this study. Model 1 consists of 94 nodes and 8 isoparametric quadratic solid 'IPQS' elements of the type shown in Figure 1. The 'IPQS' selected from the GTSTRUDL (1985) element library, is six-sided with six quadrilateral faces and 20 nodes per element, and has an expansion function:

$$
\begin{aligned}
f = \ & \alpha_1 + \alpha_2 e + \alpha_3 \eta + \alpha_4 \xi + \alpha_5 e \eta + \alpha_6 e \xi + \alpha_7 \eta \xi \\
& + \alpha_8 e \eta \xi + \alpha_9 e^2 + \alpha_{10} \eta^2 + \alpha_{11} \xi^2 + \alpha_{12} e^2 \eta + \alpha_{13} e^2 \xi \\
& + \alpha_{14} e \eta^2 + \alpha_{15} \eta^2 \xi + \alpha_{16} e \xi^2 + \alpha_{17} \eta \xi^2 + \alpha_{18} e^2 \eta \xi \\
& + \alpha_{19} e \eta^2 \xi + \alpha_{20} e \eta \xi^2
\end{aligned}
\tag{1}
$$

where α_1 to α_{20} are constants; and e, η, ξ are dimensionless coordinates.

This expansion yields a displacement field that is quadratic within the element and quadratic along the edges, so it is suitable to model structures with irregular geometry like the first molar. This model has 94 nodes. Model 2 has 16 and 149 nodes. Model 3, the final refined mesh, has 23 'IPQS' elements and 202 nodes. Convergence of solutions is investigated based on the results of these 3 models. Then a first molar with full crown preparation is analyzed in a fourth Model 4, which is the same as Model 3, except for the different material properties.

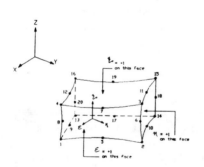

Figure 1. Isoparametric Quadratic Solid (IPQS) Element

This fourth model is used to compare results with those obtained by Farah and Craig (1974). The same material properties and loading used by Farah and Craig are used in this model. The enamel is assumed to be completely removed in this study.

4. Loading

To study the convergence of the results a load of (50N) has been applied at joint 15, (see Figure 2) in the three models 1, 2, and 3. The 50N in the one quadrant is equivalent to a total load of 200N applied to the molar. Model 3 is then subjected to a temperature increase of 100°C at all the joints in the surface above level AB. The thermal load is applied at surface joints because only surface joints experience immediate temperature changes that might occur in the mouth. Stresses due to each loading condition are compared in Table 2. An edge force is applied to Model 4 which is close enough to the axisymmetric load of Farah and Craig (1974), and the results of the two studies are compared.

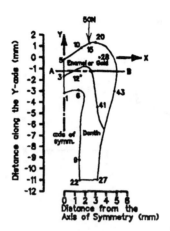

Figure 2. A 3-D Finite Element Model of the First Molar with Some Nodes

5. Results

When displacements in y, and x directions for models 1, 2 and 3 were compared, the percentage difference between the two refined models 2 and 3 are within 17.4%, 7.9% and 11%, respectively. Monotonic convergence of displacements is observed at most points. Maximum principal stresses and maximum shear stresses due to a change of temperature of 100°C at some points in the surface are compared to those due to an assumed concentrated load of 50N applied vertically downward at joint 15 (Figure 2). The stresses due to concentrated loads are higher than stresses produced by temperature change (Table 2). However, the thermal load produced a maximum principal stress of 43 MPa which might cause significant distortion of the restorations, a fact that reflects the importance of thermal stress analysis.

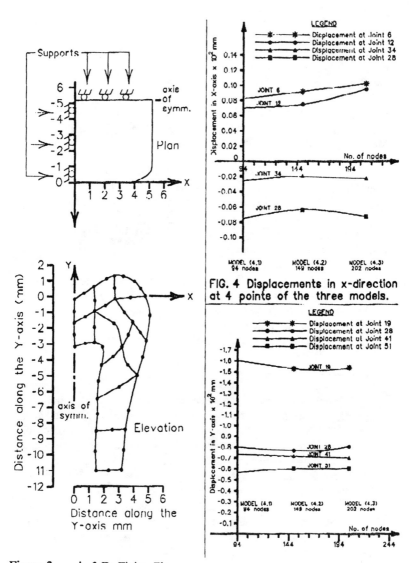

FIG. 4 Displacements in x-direction at 4 points of the three models.

Figure 3.   A 3-D Finite Element Model of the First Molar with Some Elements

Figure 5.   Displacements in y-direction at 4 points of the three models

Table 2. Maximum Stresses Due to Concentrated and Thermal Loads

	Max. Principal Stress (MPa)		Max. Shear Stress (MPa)	
Joint	Concentrated Load	Thermal Load	Concentrated Load	Thermal Load
5	- 24.3	- 3.11	25.5	20.04
6	- 1.1	- 0.42	5.89	3.25
10	64.1	- 3.55	40.3	17.27
15	-171.3	- 40.5	108.3	20.10
20	- 86.8	- 32.84	76.36	16.28
27	- 12.3	1.18	4.4	2.24
41	- 8.0	- 23.65	7.1	12.79

Results obtained from Model 4 are compared to those obtained by Farah and Craig (1974) in Table 3. The results compared are within 20%.

Table 3. Comparison of Stresses With Those Obtained by Farah and Craig

Case Description	Model 4 (MPa)	* (MPa)	% Difference
Max. shear stress at gold dentin interface	2.8	3.36	20
Max. shear stress in gold crown	7.6	6.8	10.5
Min. shear stress at gold dentin interface	0.87	0.95	9.2

*Farah and Craig (1974)

6.   Discussion

The factors which may account for differences in stresses obtained from the two studies include the type, size and number of elements and to a lesser extent, the type of solver used in each study. In the earlier study by Efimba and Ahmed (1994), it was found that modeling the base as a set of horizontal and vertical non-linear springs with different properties did not significantly change the results from those obtained assuming a fixed base. Therefore, the usual assumption of a fixed base is retained in this study.

The maximum principal stress of 43 MPa produced by a temperature increase of 100°C at some surface points underscores the importance of finite element dental thermal stress analysis. The stresses obtained from model 4 and stresses reported by Farah and Craig (1974) are within 20%. In Farah and Craig's study, the first molar was idealized as an axisymmetric solid. This idealization is close, but is not the exact representation of the geometry of the first molar. Furthermore, only special types of loading could be considered using axisymmetric models, for which some loading conditions, such as concentrated loads, cannot be considered.

7.   Conclusions and Recommendations

In this study, only a few elements are used to obtain satisfactory results, thanks to the efficiency of the 'IPQS' elements with a high degree polynomial in the displacement expansion function.

Experimental studies of the base could be used to determine its actual behavior. The results and conclusions are based on some assumptions for the behavior of the base. Investigation of the actual behavior and how it affects the results obtained in this research are suggested for further work. Analysis of the whole jaw-bone, including its dynamics and taking into consideration the effect of the interaction between one tooth and another, is called for, and characterization of the biomechanical properties of the pulp should be used to further improve current finite element models.

8. Acknowledgements

The authors acknowledge the support from the Department of Civil Engineering at Howard University. The skillful preparation of the manuscript by Ms. Patricia Williams is greatly appreciated.

9. Appendix I - References

Ahmed, E.M. (1988) "The Application of the Finite Element Method to the Analysis of Dental Structures," Master of Engineering Thesis, Department of Civil Engineering, Howard University, Washington, DC 20059

Asaoka, K. (1994) "Effects of Creep Value on Occlusal Force on Marginal Adaptation of Amalgam Filling," J. Dent. Res. 73(9) 1539-1545.

Cutnell, J.D. and K.W. Johnson (1992) Physics, 2nd Ed., Wiley, p. 332, Table 12.2.

Efimba, R.E. and E.M. Ahmed (1994) "Extensions of Finite Element Dental Stress Analysis," Proc. First ASCE Congress on Computing in Civil Engineering, Washington, DC, pp. 1928-1935.

Farah, J.W. and R.G. Craig (1974) "Finite Element Stress Analysis of a Restored Axisymmetric First Molar," J. Dent. Res, 53:859-866.

GTSTRUDL User's Manual (1985) GTICES Systems Laboratory, Georgia Institute of Technology, Atlanta, GA.

Phillips, R.W. (1982) Skinner's Science of Dental Materials, 8th Ed., W.B. Saunders Co., Philadelphia, PA.

Selna, L.G., Shillingburg, H.T. and P.A. Kerr (1975) "Finite Element Analysis of Dental Structures, Axisymmetric and Plane Stress Idealization," J. Biomed. Mater. Res., 9:237-252.

Wilner, Sounder and Pattensbargen, George C. (1942) "Physical Properties of Dental Materials," National Bureau of Standards, (now the National Institute for Standards and Technology), No. C433, U.S. Government Printing Office, Washington, DC.

Simplification Errors in the Classic Truss Model: Are They Still Tolerable in This Age of Microcomputers?

Primus V. Mtenga,[1] P.E., Member ASCE and Edward Lloyd[2]

Abstract

Recent developments in computational abilities plus the availability of powerful low-priced microcomputers have driven the need for more efficient analysis of trusses. The classical truss model provides no recognition of semirigid behavior of truss joints. A model to predict axial forces and moments in trusses which deviate from the classic truss model assumptions is presented. Results show significant reduction in moments in the continuous chords of a truss as the rigidity of the truss member joints increases.

Key Words: semirigid joints, joint slippage, continuous chords, spring coefficients, classical truss model

Introduction and Background

The classical truss model is a structure composed of individual members joined together to form a series of triangles. To analyze the truss structure the following simplifying assumptions are made:

a) The truss members are connected together by frictionless pins. This assumption implies that as the members change their lengths due to axial forces, they will be free to rotate without causing any bending in the truss members;

b) Truss members are straight, and thus there is no bending moments due to P-Δ effect;

---

[1] Asst. Prof., Dept. of Civil Engr., Florida A&M Univ./Florida State Univ. College of Engr. (FAMU/FSU CoE), Tallahassee, FL 32316-2175.
[2] Graduate Student, FAMU/FSU CoE Tallahassee, FL 32316

c) The structural system is arranged in such a manner that all loads are transmitted to the joints only.

In real life, however, there are deviations from these assumptions. For example, multiple bolt joints and welded joints are far away from the frictionless pin assumption. Therefore, bending moments and associated shear forces will arise, whenever these rigid or semi-rigid joints resist the rotation of the members as they change in length due to axial forces. The fact that some of the truss members (such as the top and bottom chords) are continuous is a deviation from the ideal truss assumption. This deviation in addition to the fact that not all loads are nodal, such as the top chord trusses with sheathing attached (both floor and roof trusses), will cause bending moments in the truss members. The bending of the compression members is particularly critical since it will lead to the P-Δ effect.

The aforementioned simplifications were very important in the days when our computational abilities were limited. However, recent developments in computer applications in the field of structural analysis have made it very easy to operate with models which are much closer to real life situations. It will take some extra effort, on the part of software developer, to include semi-rigid joints as well as the modeling of continuous chords in his analysis. However, it will take almost no extra effort for the software user to incorporate the extra data required.

A number of methods have been used in an attempt to account for the deviations from the classical model. For example, the Truss Plate Institute(TPI) (Truss.. 85) Specification for the design of metal plate connected wood trusses requires the consideration of moments, in addition to the axial forces obtained from the classical truss model. Models dealing with these types of trusses have been proposed. Such models include the Purdue Plane Structures Analyzer (PPSA) (Suddarth and Wolfe, 1984), in which fictitious analog members are introduced to model the rigidity of the joints. On the other hand the Wisconsin Model (Cramer et. al. 1993), uses a more refined connector model based on connector behavior proposed by Foschi (1977). All these models, which are closer to real life situation than the classical model, rely heavily on recent developments in computation capabilities.

In recent times a number of methods of dealing with semirigid joints have been proposed. These include studies by Seif et. al. (1981), Wang (1983), and Maraghechi and Itani (1984). In these studies the semirigid effect was accounted for by introducing springs with coefficients varying from zero to one. Study on the effect of bolt slippage is reported by Kitipornchai et. al. (1994). Results of this study illustrate that joint slippage has significant effects on deflections.

Model

A member with semi-rigid connections can be decomposed into two parts as shown in Fig.1; a beam element and a truss element.

The local element stiffness matrix will then be:

$$[k_e] = [k_{bl}] + [k_{tl}] \tag{1}$$

where

$[k_e]$ = local element stiffness matrix

$[k_{bl}]$ = local stiffness matrix for the beam part of the element

$[k_{tl}]$ = local stiffness matrix for the truss part of the element

The modified element local stiffness matrix for the beam part (Fig.1 b) can be derived from what has been proposed by Wang (1993) as follows:

$$[k_{bl}] = [A] [S] [A^T] \tag{2}$$

where

$$[A] = \begin{bmatrix} \dfrac{1}{L} & \dfrac{1}{L} \\ 1 & 0 \\ -\dfrac{1}{L} & -\dfrac{1}{L} \\ 0 & 1 \end{bmatrix} \tag{3}$$

$$[S] = \begin{bmatrix} s_{ii}\dfrac{EI}{L} & s_{ij}\dfrac{EI}{L} \\ s_{ji}\dfrac{EI}{L} & s_{jj}\dfrac{EI}{L} \end{bmatrix} \tag{4}$$

$$s_{ii} = \frac{12 \, p_i}{4 - p_i \, p_j} \, , \quad s_{ij} = s_{ji} = \frac{6 \, p_i \, p_j}{4 - p_i \, p_j} \quad \text{and} \quad s_{jj} = \frac{12 \, p_j}{4 - p_i \, p_j} \quad (5)$$

p_i and p_j are the fixity factors at member ends i and j respectively. These factors are zero for hinged connections and equal 1 for rigid connections.

For the truss part of the element (Fig.1a), there is a possibility of connection slippage along the axis of the member. Slippage can occur in two possible ways: 1) gradual slippage increasing in proportion to the axial load, or 2) sudden slippage taking place after a load sufficient to overcome the friction between the connected members is reached. Presented in Fig.2 are possible truss member load-elongation curves for the two slippage mechanisms described above. Stresses occurring at the connectors are always higher than those at other locations of a truss member. Because of these higher stresses local deformations around the connectors will be higher than what is considered to be the deformation of a bar in axial loading. In such a scenario, the gradual slippage model will be a better approximation to reality. The gradual slippage model was used in this study.

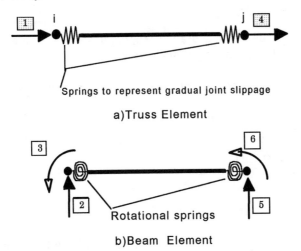

Springs to represent gradual joint slippage

a)Truss Element

Rotational springs

b)Beam Element

Fig.1. Model of a Member in a Truss With Semi-rigid Joints

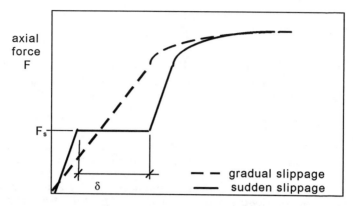

Fig.2. Truss Member Joint Slippage Characteristics

Following is the derivation of the truss element stiffness characteristics under the assumption of gradual connection slippage.

Let axial stiffness of the member be

$$\alpha_m = \frac{EA}{L} \tag{6}$$

and the stiffness of the gradual slippage springs at member ends i and j be α_i and α_j respectively.

The equivalent stiffness, α_e, for these three springs in series can then be written as:

$$\alpha_e = \frac{\alpha_i \alpha_j \alpha_m}{\alpha_i \alpha_m + \alpha_j \alpha_m + \alpha_i \alpha_j} \tag{7}$$

The stiffness of the joints can be expressed as multiples of the member axial stiffness, as follows:

$$\alpha_i = \xi_i \alpha_m \quad \text{and} \quad \alpha_j = \xi_j \alpha_m \tag{8}$$

where the factors ξ_i and ξ_i will have a value of infinity for joints with no axial flexibility. With these factors Eq.7 will be of the form

$$\alpha_e = \frac{\xi_i \xi_j \alpha_m}{\xi_i + \xi_j + \xi_i \xi_j} \tag{9}$$

Defining the axial fixity of the member as

$$p_a = \frac{1}{1 + \dfrac{1}{\xi_j} + \dfrac{1}{\xi_i}} \tag{10}$$

then the axial stiffness of the member will be

$$\alpha_e = p_a \frac{EA}{L} \tag{11}$$

The local element stiffness matrix for the truss part will therefore be:

$$[k_{tl}] = \quad \begin{array}{|c|c|} \hline p_a \dfrac{EA}{L} & -p_a \dfrac{EA}{L} \\ \hline -p_a \dfrac{EA}{L} & p_a \dfrac{EA}{L} \\ \hline \end{array} \tag{12}$$

Presented in Fig. 3 is a sketch of a truss analyzed using the model described above. This truss has a span of 8.4 m and a slope of 1 to 2. It was analyzed at a load of 1.44 KN/m (2.4 KN/m^2 applied to trusses placed at 600 mm on center), which is the design load for a wood truss of this configuration.

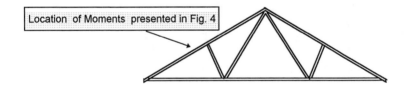

Location of Moments presented in Fig. 4

Fig. 3 A Sketch of one of the Trusses Used in this Study.

The axial force and moments at the top-chord at a location indicated in Fig. 3 are presented and discussed below.

Results and Discussions

Presented in Fig.4 are the variations of axial forces and moments with respect to the rotational rigidity coefficients in the truss members. It can be seen that as the rigidity coefficients increase, both the axial force and the moment at the chosen location in the top-chord decrease. This may be explained by the fact that for non-zero rigidity coefficients, the truss members take part in restraining the rotation of the joints. Consequently, the overall rotations of the joints are less, and thus there is less bending in the top-chord.

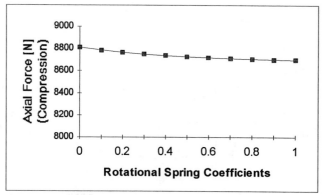

a)Axial force variation with respect to rotational spring coefficients

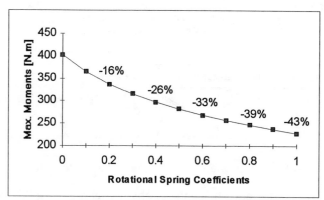

b)Relationship between moments and rotational spring coefficients

Fig. 4 Variation of Axial Force and Moments in a Top-Chord of a Fink Truss (shown in Fig. 3)

In Fig.4 b) the percent moment reduction at selected rigidity coefficients are computed. From these values it can be seen that even at low rigidity coefficients there are significant reduction in bending moments. For example at a coefficient of 0.2, a value that may be quite conservative for heavy bolted connections or connections with gusset plates, the moment reduction is more than 15 percent.

Conclusion

The results presented in this paper suggest that there is need for including the semi-rigid nature of "truss" joints in the analysis of trusses. The classical truss model gives results that are quite different from reality. There is need for software developers and researchers to take advantage of recent developments in computer applications in the field of structural analysis by developing codes and databases that incorporate the semi-rigid nature of joints in the analysis of trusses. It will take little extra effort on the part of a software developer to include these effects in the code, however, it will take almost no extra effort for the software user to incorporate the extra data required.

References

Cramer, S. M., Shrestha, D. K., and Mtenga, P. V. (1993) Computation of Member Forces in Metal-Plate Connected Wood Trusses. *Struc. Engr. Review, August 1993.*

Foschi, R. O. (1977) Analysis of wood diaphragms and trusses. Part II: Truss-Plate connections. *Can. J. Civil. Eng.,* 4(3), 353-62.

Kitipornchai, S., Al-Bermani, F. G. A., and Peyrot, A. H. (1994) Effect of Bolt Slippage on Ultimate Behavior of Lattice Structures. *J of Stru. Engr., August 1994.*

Maraghechi, K., and Itani, R. Y. (1984) Influence of truss plate connectors on the analysis of light frame structures. *Wood and Fiber Sci.,* 16(3). 306-322.

Seif, S., Vanderbilt, M. D., and Goodman, J. R. (1981) Analysis of Composite Wood Trusses. *Struc. Research Report No. 38,* Civ. Engr. Dept., Colorado State Univ., Fort Collins Co.

Suddarth, S. K., and Wolfe, R. W. (1984) Purdue Plane structures Analyzer II: A computerized wood engineering system. *General Technical Report FPL 40,*USDA Forest Service, FPL, Madison, W!.

Truss Plate Institute 1985. *Design specification for metal plate connected wood trusses,* Truss Plate Institute, Madison, WI.

Wolfe, R. W., La Bissoniere, T, and Traver, R. (1988) Roof system study number II: Progress Report #2., Forest Prod. Lab., Madison, WI.

Information Technology and Education:
Towards The Virtual Integrated
Architecture/Engineering/Construction Environment

Hossam El-Bibany[1], A. M. ASCE, Matt Vande[2], Poonam Gowda[2], Ben Branch[3],
John Groth[3] and Katharine Schuld[3]

Abstract

This paper describes a continuous effort to design and test educational concepts of integrated education through the use of information technology. The concepts have been used and tested for five semesters of undergraduate education. The paper starts by describing the goals of using information technology in integrated education. It then proceeds to briefly describe two interrelated example projects.

Introduction

The last decade has witnessed various research efforts in computational tools towards the Virtual Integrated Architecture/Engineering/Construction Environment. This environment, however, cannot be achieved without integration of knowledge in the participating human minds. Education, therefore, plays a major role.

Education, however, has always followed a model of specialization from organizational behavior research. This model has proven inadequate, in some cases due to the nature of a specific industry like construction, and in others due to economic pressure on organizations. Integrated education has therefore emerged as a need for the twenty first century. But what is integrated education? and what is the role of information technology in integrated education?

As there is no general answer for these questions, the continuous quality improvement effort reported in this paper tries to study the role of information technology in the virtual integrated Architecture/Engineering/Construction educational environment. The paper describes a continuous effort to design and test educational concepts of integrated education both on the courseware level and the curriculum level.

---

[1] Assistant Professor, Department of Architectural Engineering, PennState University, University Park, PA 16802-1416.

[2] Graduate Research and Teaching Assistant, Department of Architectural Engineering, PennState University, University Park, PA 16802-1416.

[3] Undergraduate Research and Teaching Assistant, Department of Architectural Engineering, PennState University, University Park, PA 16802-1416.

The effort has started at various levels for the last five semesters of undergraduate education. One basic concept underway include the use of information sciences and technology like interactive multimedia knowledge transfer and object-oriented simulation and modeling tools to advance educational and training systems. Another concept looks at the process of courseware integration through curriculum design and the use of information technology.

Goals

The **objective** of the effort reported in this paper is to study the **strategies** of gradually introducing various educational concepts using information technology and evaluating its success based on students input and evaluation within the special academic needs of the integrated design/construction educational environment. These needs could be summarized as follows:

Integrated Knowledge and Educational Model. The various fields of design, construction and management education have been always treated as separate in the last decades based on theories of specialization. Courses generally followed a model that divides knowledge top-down into its lower level theoretical elements and start teaching from the theoretical up. The interfaces between the various courses have not always been well designed to let the student glide through the knowledge environment. Furthermore, in the construction industry, due to the product and process nature, the integrated body of knowledge has proven difficult to formalize due to the difficulty of generalization. Therefore, the integrated knowledge in the educational environment needs to be developed and refined.

Educational Methods. Some fields have proven difficult to teach in the current academic setting. Construction knowledge (e.g., equipment and methods) is particularly difficult to teach adequately in the classroom, even though this is the typical way it is presented. Textbook pictures, movies and slides help to convey some of the scale and complexity of the subject, but they are passive rather than active media from the student's perspective. Construction education could be enhanced greatly through hands-on experimentation, yet owing to cost, safety and accessibility issues, few students have a chance for such experience [El-Bibany and Paulson 91].

Teaching Efficiency. Teaching methods predominantly remain based on lectures where a classroom instructor presents material to students in a linear sequence. All students in a class passively receive the identical body of topics selected to represent the subject being taught, and they receive it at the same pace regardless of abilities. Courses with a large student attendance necessitate the use of more than one instructor, usually in the form of teaching assistants, with different experience and educational backgrounds. This often results in students complaining of the inconsistency of the knowledge transferred to them, specifically in low level undergraduate courses.

The use of information technology to create an the integrated design/construction environment is also recommended to be performed within the following processes:

1. The Redesign and Development of Undergraduate and Graduate Curricula

Investment in information technology is expensive and time consuming. Its introduction needs, therefore, to be done within a new curriculum design that takes into account how the various educational processes will change with the new technology.

2. The Active Participation of Students in the Design and Development of the Various Education Projects

The success of information technology in education is partly based on its fit to the students needs. Participation of students helps ensure that these needs are considered. It reinforces leadership skills in students and creates a better environment for student acceptance of the technology needed.

Example Projects

As an example of the use of information technology in integrated education, one effort at PennState University may be considered. *AE221—Construction Materials and Methods* and *AE222—Working Drawings* are two courses that were redesigned to emphasize integrated design/construction knowledge. The teaching tools and methodology were also revised to take advantage of the most suitable technologies like multimedia educational material, and organizing construction site visits. The process of integration of the two courses follow the Total Quality Management (TQM) recommendations of the Construction industry. Important aspects of such recommendations include the provision of real construction experience to engineering students, and training students to perform constructability analyses in generating the final construction drawings.

The following paragraphs describe two pilot projects underway.

1.Project DeCON – a project to create an interactive multimedia educational environment depicting the design and construction stages of a commercial building

The goals of this project are to:

• capture the nature and organization of the Architectural/Engineering/Construction industry,

• illustrate the interaction between the design and construction team members,

• stress the importance of and describe various foundational elements in design/construction integration, and

• illustrate the basic concepts of decision making related to materials, systems and construction processes.

[El-Bibany and Schuld 95] describes the initial project stages of multimedia knowledge collection, training and preparation of the development team, and various concepts, refers to strategies of instructional design in multimedia environment and summarizes the long term benefits of such multimedia projects. The design and development effort is based on the experiences gained in the work reported in [El-Bibany and Paulson 91, El-Bibany et al 92, and El-Bibany 95].

2. Project 3-D – a project to use 3-D modeling tools for better integration of design requirements into the construction working drawings

Our construction industry interaction and research indicates that there is a high need to incorporate constructability knowledge and perform alternative economic analyses in the design stage in the generation of the construction working drawings. This has proven very difficult under the current setting manual drawing in AE222. In this course, students generate hand drawings while having major problems with basic visualization, lack of knowledge of building details, and practically impossible to perform design changes. This project, therefore, attempts to solve this problem using the right computer tools.

Market research and studies have identified the requirement for the best suitable computer tools for the building industry as follows:

* Able to help students perform design in 3-D, then subsequently generate the working drawings.
* Be object oriented, in which students generate and manipulate building objects as opposed to lines and points.
* Able to attach to or collect information from any building object for utilization in further analyses.
* Able to perform major engineering functions like architectural design, structural design, project estimation and construction simulation.
* Have a shallow learning curve. The students should spend most of their times learning about and analyzing buildings and not struggling with the software capabilities.
* Have a potential to become an every day tool for students to use in any undergraduate and graduate classes in the Architecture, Architectural Engineering and Civil Engineering Departments.

ArchiCAD™ was chosen and used for two semesters of education in combination with manual drawing techniques. Based on student feedback and evaluation, the project has proven successful in achieving the goals set of this stage.

Course Evolution

The two projects described in the previous section were introduced to the integrated courses based on the understanding of the student needs. As project DeCON is still under development, only the video material depicting a complete construction stage of a commercial building has been used in conjunction with site visits. Basing the course example on one continuous real construction process has raised the level of student understanding of the integrated knowledge environment.

Project 3-D completed the normal hand drawing processes historically used in these courses. Our goal was to understand how the two processes completed each other. It also served as a context for student introduction to group work with shared responsibilities.

Based on the progress in these projects, we can conclude that their success is mainly based on the value added to the courses from the students point of view.

Conclusions

The process of introducing information technology in education should be design within the large picture of integrated education. It should be evolutionary rather than revolutionary. Various technologies could complement the current educational methods. The success of the process depends on the target student input and involvement in the design and development stages.

References

El-Bibany, H., 1995, *MutliMedia Knowledge Transfer in Construction Education and Training,* Engineer Degree Thesis, Civil Engineering Department, Stanford University, Stanford, CA, June.

El-Bibany, H., and Paulson, B.C., 1991, "Microcomputer/Videodisc System for Construction Education," *Microcomputers in Civil Engineering*, Vol. 6 N0. 2.

El-Bibany, H., Manavazhi, M., Paulson, B.C., and Ryssdal A., 1992, *MultiTool: A Toolkit for the Development of Computer-Based Multimedia Courseware*, Software Manual, Civil Engineering Department, Stanford University, CA 94305-4020, June, 1992.

El-Bibany, H., and Schuld, K., 1995, "Design/Construction Integrated Education Through Project-based Interactive Multimedia," *Proceedings of the Second ASCE Congress on Computing in Civil Engineering*, June, 1995, Atlanta, Georgia.

A Collaborative and Multimedia Interactive Learning Environment for Engineering Education in Sustainable Development and Technology

Jorge A. Vanegas, Ph.D.,[1] Associate Member, ASCE
Mark Guzdial, Ph.D.[2]

Abstract

As our society, and indeed all societies of the world, head into the 21st century, this concept of sustainable development will likely become ever more important. The United Nations World Commission on Environment and Development has defined sustainable development as "meeting the needs of the present without compromising the ability of future generations to meet their own needs." This definition translates into many complex challenges, with serious implications for all professions for the next century, especially engineering. Sustainable development also creates a need to educate a new generation of engineering professionals who can effectively face the challenges and contribute to the finding of solutions to the problems we need to address to achieve sustainability.

This paper describes a tool that supports engineering education in sustainable development and technology: Collaborative and Multimedia Interactive Learning Environment (CaMILE). This system will be initially used in an introductory course sequence in sustainable development and technology, currently being developed at Georgia Tech. This sequence will introduce major curricular changes, not only in the way each course is designed and developed, but more importantly, in the way students learn and are taught within a problem–based, case–based and collaborative learning and reasoning environment. CaMILE uses state–of–the–art computer technologies to (1) provide instructors and students with the necessary tools to enhance the teaching and learning processes; (2) provide a mechanism to effectively integrate the academic, industry and societal requirements of sustainable development within the academic environment; and (3) assist in developing and strengthening students' integrative skills in analysis, synthesis and contextual understanding of sustainable development problems.

---

[1] Associate Professor, School of Civil Engineering, Georgia Institute of Technology, Atlanta, GA 30332-0355; Tel. (404) 894-9881, jvanegas@ce.gatech.edu

[2] Assistant Professor, College of Computing, Georgia Institute of Technology, Atlanta, GA 30332-0280; Tel. (404) 893-9387, guzdial@terminus.cc.gatech.edu

Introduction

The conflicts and complex interrelationships that exist between the economic development needs of both developed and developing nations, and the environmental problems resulting from development efforts, are a growing concern for the world today. To study this issue, the United Nations formed in 1984 the *World Commission on Environment and Development* (WCED). This Commission, in its report Our Common Future (1987), defined sustainable development as "meeting the needs of the present without compromising the ability of future generations to meet their own needs." This deceivingly simple statement translates into many complex challenges, with serious implications for all professions, especially engineering.

Sustainable development challenges institutions at every level, from global to local, to create new avenues for the design of products and the use of natural resources. New ways to market, produce, deliver and dispose of products are needed, which may lead to the development of new services and even new infrastructures. This is not an easy task given the magnitude of the political, social, economic, technological and cultural differences in countries around the world. At the basis of the vision of sustainable development is a new paradigm of economic growth that is in harmony with the environment. This requires the education of engineers, technological professionals, and decision/policy makers with an integrated view of technologies and their applications, and a sensitivity to the complexity and diversity of the cultural, natural, and societal environment.

The American Association of Engineering Societies (AAES) has suggested a conceptual framework for sustainable development for engineers (AAES, 1994). First, engineers must be trained and engaged more actively in political, economic, technical and social discussions and processes to help set a new direction for the world and its development. Second, engineers need to use environmentally sensitive and responsive economic tools, in order to integrate environment and social conditions into market economics. Third, in planning for sustainable economic development, engineering should become a unifying, not a partitioning, discipline, and engineers need to look at systems as a whole, as opposed to looking at fragmented or single parts. Fourth, engineers and scientists must work together to adapt existing technologies and create and disseminate new technologies that will facilitate the practice of sustainable engineering and meet societal needs. Fifth, the knowledge, skills and insights of the physical as well as the social sciences, together with all engineering disciplines must be brought together in a new collaborative partnership. Finally, engineers must cultivate an understanding of environmental issues, problems, risks and potential impacts of what they do. Specifically, engineering education must instill in its students an early respect and ethical awareness for sustainable development, including an understanding and appreciation of cultural and social characteristics and differences among various world communities. In addition, students need to acquire the analytical tools to assess risks and impacts, to perform life cycle analyses, and to solve technical problems, cognizant of and taking into consideration the economic, socio–political and environmental implications.

Current undergraduate engineering education is not preparing students to work within this framework. Students have a limited exposure to the multi–disciplinary nature of real–world engineering projects, and students do not have the opportunity to participate in the complete process of design conception, analysis, synthesis, implementation; they are normally exposed to only one part of the process at a time. They also lack a contextual understanding of the issues, challenges and problems of achieving development in a global context. Furthermore, solutions to complex, real–world problems are increasingly requiring the use of advanced technologies for numerical simulation, visualization, communication, and for acquisition, storage, retrieval and manipulation of massive amounts of data. Students, due to the limitations imposed by the costs of these advanced technologies and the structure of classroom instruction, are receiving limited exposure to such technologies.

A current effort at Georgia Tech to overcome these problems is the development of a carefully planned and designed engineering curriculum in sustainable development and technology, within a multi– and interdisciplinary learning environment that incorporates the latest advances in cognitive science and computer–aided instruction and learning. One of the tools within this learning environment is **CaMILE, Collaborative and Multimedia Interactive Learning Environment. Interactive Learning Environment**.

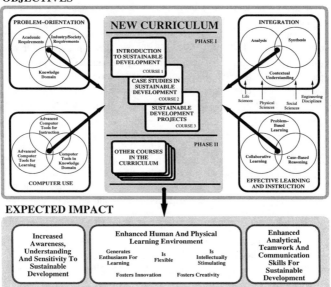

Figure 1.– Objectives and Impact of the Curriculum Development Initiative

Drivers

CaMILE is a direct result of a multi-disciplinary curriculum development effort in sustainable development and technology at Georgia Tech. Unlike many of the subjects in traditional engineering, sustainable development does not lend itself to the creation of a set of rules and procedures for creating optimized solutions. Rather, sustainable development requires the balancing of many different issues: ecological, economic, cultural, technology capabilities, design life cycle, and so on. Thus, an important objective of the curriculum development effort was to create a course sequence where this diversity would be met head-on, within which an integrated, collaborative, and multimedia environment would help facilitate (1) the development of educational materials to support teaching and learning of fundamental issues and concepts surrounding sustainability; (2) the teaching of these issues and concepts; and (3) the learning of these issues and concepts.

In addition, the development of CaMILE was driven by a number of specific problems or difficulties that engineering education faces today. The crux of the difficulty is that the engineering educators prepare students for the engineering needs of industry in a completely different environment — the classroom, which is organized and structured in very different ways than industry. Courses in engineering present information structured in terms of single disciplines. As a result of this structure, students often miss connections between various data and techniques that do not fit into that particular structure. Industry problems are usually met with cross-disciplinary techniques. In addition, to encourage general learning, engineering concepts are often taught in the abstract. Students often have difficulty understanding these abstract concepts as divorced from concrete examples [Kolodner, 1993]. Finally, graduates are expected to work as a member of a team while classroom work is primarily individual Consequently, in order to meet the needs of modern industry, engineering education has the following requirements placed upon it: (1) effective presentation of information which has cross-discipline connections; (2) bridge the gap between the abstract and the concrete; and (3) prepare students for work in teams.

CaMILE

CaMILE provides a forum for sharing, discussing, and reflecting, as well as multimedia information bases to aid students in locating and making sense of multimedia information. It emphasizes (1) having students actively post material to the network and relate resources they find there to their own discussions and activities, and (2) guiding students in their participation, from prompts while making text comments to guides which direct students through a multimedia database. These emphases are informed by cognitive science research on how learning is best facilitated. Guided student participation is critical since it places students in the role of active learners [Collins et al., 1989; Farnham, 1990]. The results to date have exceeded the original scope and expectations.

CaMILE is based on an extended notion of multimedia which encompasses students' own work. Rather than simply defining multimedia as unstructured and homogeneous types (e.g., sound, video, and graphics), CaMILE also supports links to structured documents (e.g., spreadsheet, word-processing, and CAD files) and into multimedia databases (e.g., a single node in a multimedia database or a single page in an electronic book). The extended definition of multimedia used in CaMILE encourages students to see their own work as media which are as reasonable to reference in the collaborative exchange as any media found across the network.

CaMILE has three important advantages over traditional teaching and technology in use in engineering classrooms, and particularly in addressing complex topics as sustainable development and technology. First, it integrates various kinds of media and perspectives to produce a complex and multifaceted view of knowledge, and provides guidance to comprehend this view. Second, it encourages sharing of student work. By relating abstract concepts to concrete examples (student work and other media), CaMILE supports greater depth of understanding. Finally, CaMILE encourages collaborative learning with students working in teams.

CaMILE consists of three components, NoteBase, MediaBase and Electronic Books, each carrying through the themes of participation and guidance, i.e., each component provides a means for students to comment, modify, and interact–to actively participate, and furthermore, each component provides a structure or explicit guidance that helps in directing the student's participation in fruitful ways.

NoteBase

The core of CaMILE is a collaborative **NoteBase,** where students read and post text commentary with multimedia annotations in group discussions, and are guided through a form of procedural facilitation [Scardamalia et al., 1984; Scardamalia et al., 1989]. The NoteBase provides a structure for the notes such that each new note is a response to something already said in a discussion. When a student reading through a discussion finds a note that she wishes to respond to, she may create a new note that he/she must specifically identify as a Question, a Comment, a Rebuttal, an Alternative, or so on, as shown in Figure 2.

Figure 2.– Types of Question Prompt

As shown in Figure 3, notes in CaMILE look like MediaText documents, which have been shown to be a particularly useful form for student-based multimedia composition [Hay et al., 1995]. These notes are text-based, but can contain multimedia annotations in the margin. Students can create links that refer to pictures, sound, spreadsheets, word-processors, a URL (i.e., an addressing form used with Mosaic and the World Wide Web), a page in the electronic books, a node in the MediaBase — any kind of document on the student's own disk or available somewhere on the network. While editing a note, a floating Suggestions window provides phrases that the student is encouraged to use in a note of the selected type.

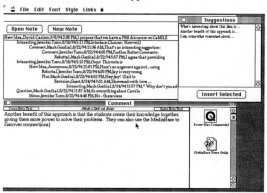

Figure 3.– Screen Shot of CaMILE Notebase

MediaBase

The **MediaBase** [Rappin et al., 1994], is a multimedia database that serves as a resource which students can review for more information on a topic or to gather evidence to add to a discussion. In the MediaBase, a content expert has previously placed information into *nodes*, which are collections of text with multimedia annotations, similar to the notes in the NoteBase. These nodes are organized into collections called *paths*. Each path contains nodes that are related to a common topic or theme.

Students can participate in the MediaBase on a variety of levels. At the lowest level, they can traverse paths that have already been laid out for them, exploring along a single theme. Students can participate more actively in the MediaBase by jumping along the hypermedia network formed when nodes are on more than one path. By traversing the MediaBase in this more self-directed manner, the student can construct their own meaning for the information presented. If the student sees a connection between nodes that is not already part of the MediaBase, the student can create their own path through the data, and can post the path to the NoteBase as a trail for other students to explore. In addition, students can create multimedia annotations to the nodes, which can be saved, shared with other students explicitly, or posted to the NoteBase. These annotations can be retrieved by the student even when he or she is not connected to the MediaBase.

Students are guided in the MediaBase by anthropomorphic agents appropriately called *guides*. A MediaBase guide presents a specific point of view on the presented data.

Guides have been shown to help students make sense of complex domains that can be approached from several different angles [Oren et al., 1990]. A guide presents an opinion on a given node, and points to a path, called an *agenda*, with more information about the guide's subject. A node can have any number of guides presenting different points of view on the node. For example, a MediaBase node about the population explosion would contain facts about how quickly the population is growing. Guides for this node might include the perspective of an industrialist from a developed country, the leader of a developing country, an environmental group, a religious organization, and so on.

The MediaBase encourages students to explore a topic in multiple media guides them toward alternative and often conflicting perspectives on a subject. Learning to explore these different viewpoints is an important problem-solving skill for students.

Electronic books

The third component of CaMILE is an Electronic Book (*Ebook*) format, based on the AutoFlow electronic book system developed at Dartmouth by Matthew Williams. Much of the information available to undergraduates is in the form of text. While in its paper form, text provides information, but it cannot be easily linked with other related resources, commented upon, nor shared. Existing word-processing files can easily be converted into the CaMILE EBook format which supports linking of the text to external multimedia resources, allowing students to annotate the book in useful ways (e.g., writing notes in the margin, underlining key phrases, adding bookmarks, creating links from relevant phrases in the book to their own documents), and sharing of students' notes with one another. Each student's annotations are stored in a separate Notes file. By opening their books using another student's Notes file for annotations, students can gain other students' perspectives on the text.

Electronic books provide media in a familiar, structured form. The free-form, hypertext structure of the NoteBase and MediaBase is less familiar to students, and thus requires new kinds of guidance. Electronic books extend the traditional book with links to multiple media and the opportunity for shared annotations. Such annotations provide yet another opportunity for students to write about and relate course material to their own understanding.

Future Directions

Comprehensive assessment and evaluation plans are currently in place. In addition, to integrate the diverse activities on an engineering campus and to allow widespread collaboration, CaMILE will move beyond its current single platform base, to work across the common platforms in use in engineering schools: Apple Macintosh, Microsoft Windows, and UNIX X-Windows based systems. The research team currently exploring development options for re-implementing CaMILE cross-platform.

Also, a CaMILE implementation on top of the world-wide web (WWW) using the new HTML+ language and corresponding protocol is being considered [Berners-Lee et al., 1994]. HTML+ allows for truly interactive multimedia documents across platforms.

Acknowledgments

CaMILE was developed in the Graphics, Visualization, and Usability Laboratory of Georgia Tech, with support from the EduTech Institute at Georgia Tech, the GE Foundation, and the National Science Foundation. In addition, various collaborators from different disciplines including Computing, Civil and Environmental Engineering, Mechanical Engineering, Industrial and Systems Engineering and Public Policy at Georgia Tech were excellent sources of ideas and advice.

References

American Association of Engineering Societies (1994). *""The Role of the Engineer in Sustainable Development."* American Association of Engineering Societies, Washington, D.C.

Berners-Lee, Tim, Robert Cailliau, Ari Luotonen, Henrik Frystyk Nielsen, and Arther Secret (1994). "The world-wide web." Communications of the ACM 37: 76-82.

Collins, Allan, John Seely Brown, and Susan E. Newman (1989). "Cognitive apprenticeship: Teaching the craft of reading, writing, and mathematics." In Knowing, Learning, and Instruction: Essays in Honor of Robert Glaser, ed. Lauren B. Resnick. Hillsdale, NJ: Lawrence Erlbaum and Associates, .

Farnham-Diggory, Sylvia (1990). Schooling. The Developing Child, eds. Jerome Bruner, Michael Cole, and Barbara Lloyd. Cambridge, MA: Harvard University Press.

Hay, Ken E., Mark Guzdial, Shari Jackson, Robert A. Boyle, and Elliot Soloway (1995). "Student composition of multimedia documents: A preliminary study." Computers and Education, In Press.

Kolodner, Janet (1993). Case Based Reasoning. San Mateo, CA: Morgan Kaufmann Publishers.

Oren, Tim, Gitta Salomon, K. Kreitman, and Abbe Don (1990). "Guides: Characterizing the interface." In The Art of Human-Computer Interface Design, ed. Brenda Laurel. Reading, MA: Addison-Wesley.

Rappin, Noel, Mark Guzdial, and Jorge A. Vanegas (1994). "Supporting distinct roles in a multimedia database." In Conference Proceedings of the Third Annual Conference on Multimedia in Education & Industry, 202-204. Charleston, SC: Association for Applied Interactive Multimedia.

Scardamalia, Marlene, Carl Bereiter, R. McLean, J. Swallow, and E. Woodruff (1989). "Computer-supported intentional learning environments." Journal of Educational Computing Research 5: 51-68.

Scardamalia, Marlene, Carl Bereiter, and R. Steinbach (1984). "Teachability of reflective processes in written composition." Cognitive Science 8: 173-190.

World Commission on Environment and Development (1987). *Our Common Future.* Great Britain: Oxford University Press.

An Interdisciplinary Course in Engineering Synthesis

Steven J. Fenves,[1]Hon. M. ASCE

Abstract

The course *Synthesis of Engineering Systems* is one of three interdisciplinary courses offered by the Engineering Design Research Center at Carnegie Mellon University. Students are exposed to two major approaches for engineering design synthesis: mathematical programming and knowledge-based systems. An integral part of the course is a term project in which the students apply both synthesis approaches to a design problem in their domain. This paper summarizes the philosophy and contents of the course, illustrates a representative student project, and discusses future research and tool developments needs and possible educational approaches.

1. Introduction

The Engineering Design Research Center (EDRC) is one of the interdisciplinary Engineering Research Centers sponsored by the National Science Foundation. EDRC's primary mission is the development and dissemination of computer-based technologies to serve the competitive needs for reduced product development cycle, improved product quality, and reduced product costs. The four principal goals of EDRC are to:

1. make significant contributions to a science base of design in the form of methodologies, computational tools, and environments for engineering design;
2. educate a new generation of engineering design practitioners, educators, and researchers for industry and academia;
3. infuse the engineering curriculum with engineering design textbooks and other course materials; and
4. collaborate with industry to support improved design practice by exchanging knowledge, people, and tools.

EDRC is organized into three laboratories: Synthesis, Design for Manufacturing, and Design Systems. The Synthesis Laboratory deals with methods which expand or elaborate objectives and constraints at one level into structure and behavior descriptions at the next level. The Design for Manufacturing Laboratory addresses two kinds of information flow: from the

---

[1]Sun Company U iversity, Professor of Civil and Environmental Engineering, Carnegie Mellon University, Pittsbui , PA

design to the manufacturing phases, attempting to integrate the two; and from manufacturing to design, developing abstraction mechanisms that assist designers in anticipating downstream consequences. The Design Systems Laboratory (abolished in late 1993) deals with the integration of designers, tools and information into environments supporting the overall design process.

Each laboratory is responsible for three to five thrusts, which change over the years as the Center's strategic plan develops. A thrust typically consists of two to six projects conducted by the Center faculty and their students. There are approximately 35 affiliated faculty, representing the Chemical, Civil, Electrical and Computer, and Mechanical Engineering departments in the College of Engineering and Architecture, Design, Industrial Administration, and Computer Science outside the College. The approximately 60 graduate students per year in the Center are enrolled in their respective academic departments. An overview of EDRC is presented in [Demes et al., 1993].

One of EDRC's prime concerns is that the concepts and methods developed in the Center's projects be applicable and useful across the widest range of design disciplines. The educational component of EDRC is one of its mechanisms for disseminating its research results across the engineering college.

The course *Synthesis of Engineering Systems* is a first year graduate level course offered by EDRC. It is one of three interdisciplinary courses taught by members of the EDRC. The courses are administered by the college of engineering and accepted by all engineering departments. The other two courses are: *Design, Manufacturing and Marketing of New Products*, addressing on a case-study basis the full cycle of product development and innovation; and *Engineering Design: The Creation of Products and Processes*, intended to provide a fundamental understanding of the design process. The overall motivation of the suite of courses is twofold:

- to provide a rapid introduction for new graduate students and research assistants to the common methodologies underlying EDRC's research activities; and
- to foster an interdisciplinary setting for problem-solving that reflects the multidisciplinary nature of engineering design.

Specifically, the synthesis course is intended to present, compare, and contrast a set of generic, domain-independent synthesis approaches and to provide an opportunity for students to apply and integrate these approaches on design problems in their domain of specialization.

The purpose of this paper is to present a brief overview of the course, in terms of its basic philosophy, major components and the role of student projects. Research and educational needs to further improve the course are discussed. This paper is an abridged and updated version of the material presented in [Fenves and Grossmann, 1992].

2. Course Overview

For the purposes of the course, synthesis is defined as the generation, selection, and integration of systems, subsystems, and components satisfying global objectives and requirements, as well as constraints due to the physical laws governing the behavior of the systems. Synthesis is thus viewed as a generic design activity in which an abstract description of the artifact (objectives and requirements) are *elaborated* into a more detailed description (structure and behavior).

Synthesis is viewed as a design task where the designer has to select from a space of alternatives the topology (i.e., the hierarchical structure and interconnection of subsystems or components) and parameters that define the structure of a system. The behavior of a synthesized system is then determined by an analysis step followed by an evaluation. The results are fed back to the synthesis step in order to make a new choice of the topology and/or parameters with the aim of satisfying the global objectives and design requirements. The three tasks of synthesis, analysis and evaluation may be integrated into one process or performed sequentially. Methodologies for synthesis must have the capability to systematically generate alternatives and to perform decisions to identify improved designs. Mathematical programming and knowledge-based systems are two approaches that can, in principle, address some of these issues. The former is based on a quantitative framework that emphasizes simultaneous optimization and handling of constraints, whereas the latter is based on a symbolic framework that emphasizes heuristic reasoning and decomposition. As quantitative and qualitative sources of knowledge are invariably involved in engineering synthesis, there is a need for combining and integrating the two approaches. The potential benefits of doing this have been recognized by a number of researchers ([Glover, 1986]; [Simon, 1987)], and are starting to be explored in various areas of applications [Kusiak, 1991].

A major goal of the course is to promote the synergism between mathematical programming and knowledge-based approaches to synthesis. The course exposes students to the two approaches and explores their integration and combination through a term project in a problem of the student's domain. The two major components of the course are first summarized, followed by the description of the student projects.

Mathematical Programming Component. This component presents optimization concepts and algorithms arising from Operations Research that can be applied to the synthesis of engineering systems [Levary, 1988]. These techniques are appropriate when synthesis problems can be posed quantitatively as well-defined problems that require satisfaction of constraints and determination of optimal trade-offs, and when synthesis, analysis and evaluation can be integrated in one process.

For the application of mathematical programming techniques, the following design methodology is followed. First, a representation of alternatives is developed for the choices of topologies and parameters; this representation, denoted as a superstructure, is then modeled with discrete and continuous variables. The former are used to model the selection of the topology, while the latter are used to model the selection of design parameters and state variables in the system. The performance of the system and the design specifications are represented through equations, and the criterion for the selection through an objective function. Finally, the design is obtained by solving the corresponding mathematical program, which, in general, is a mixed-integer-nonlinear programming problem [Grossmann, 1990]. The mixed-integer nonlinear programming (MINLP) problem may reduce to any of the following optimization problems: linear programming (LP), mixed-integer linear programming (MILP), or nonlinear programming (NLP).

In order to apply the above design methodology, emphasis is first placed on the development of the superstructure of alternatives which may be represented in the form of a tree or, more generally, a network. As there is no formal theory for developing these representations, they are illustrated through example problems. Basic concepts of optimization are then covered to first present NLP and LP techniques, in order to address design problems with only continuous variables, followed by MILP and MINLP techniques.

Students are given homework assignments in which they have to model and solve small optimization problems in various domains. Students face several difficulties. First, some of the students may have had no, or very little, previous exposure to optimization. We do not assume previous knowledge, except for basic background in calculus and linear algebra. The second difficulty, which is perhaps the most serious one, is that students experience difficulty in formulating optimization problems. We devote some lectures to modeling, and present examples derived from research work in the EDRC. Finally, although the tools we provide are powerful platforms for optimization, they require some effort and time to learn the syntax. We try to overcome this problem by providing sample problems. **Knowledge-Based Component.** This component presents synthesis methodologies arising from the fields of Artificial Intelligence and knowledge-based systems. Due to the relative recency of the field, the applicable methodologies and approaches cannot be categorized and formalized to the same extent as the techniques of mathematical programming. The general model of synthesis presented above is used, with the following limitations:

1. qualitative synthesis tends to emphasize topology and may be less concerned with parameter value selection;
2. analysis may be nonexistent or may only involve a weak quantitative model; and
3. evaluation tends to emphasize *satisficing* rather than optimizing.

Two methodologies are discussed in depth: rule-based expert systems and hierarchical decomposition. The rule-based formalism is used to decompose the design problem into a hierarchy of goals. Heuristic classification is implemented by sets of rules for the identification, selection, and evaluation or elimination of alternatives for each goal element [Clancey, 1985]. Hierarchical decomposition is a further formalization of this approach, where the knowledge base represents a hierarchy of system, subsystem, or component alternatives available at each level of the artifact decomposition; problem-specific facts and heuristic constraints among alternatives at various levels represented in the knowledge base are used to systematically generate feasible alternate configurations[McDermott, 1981]. Other knowledge-based synthesis strategies, including case- and prototype-based reasoning, analogical reasoning, and model-based reasoning are treated in lesser detail and illustrated with engineering synthesis applications.

Students are given homework assignments in which they develop small expert systems using the tools provided: a PC-based shell, such as VP-Expert; and EDESYN, a domain-independent shell specifically geared to synthesis by hierarchical decomposition [Maher, 1988]. A single application, understandable to students with varying technical backgrounds, is chosen for implementation, such as the design a screw for connecting two parts of wood subject to various constraints. The students' difficulties manifest themselves at two levels. First, students are unfamiliar with rule-based programming and attempt to program with strict procedural control structures. Second, the students have difficulty in collecting, organizing, and coding domain knowledge, the traditional "bottle-neck" of expert systems development. We try to overcome these problems by giving several lectures on knowledge acquisition, and directing students to handbooks on technical data. Some students have interviewed craftsmen and hardware store salesmen to collect domain knowledge.

3. The Student Project

As stated above, homework assignments are given in both components of the course, illustrating and elaborating the lecture material on assigned problems. The problem sets are either generic (i.e., independent of any specific engineering domain) or drawn from specific disciplines (e.g., chemical, civil, or mechanical engineering). In order to complement these assignments, each student completes a term project incorporating the two methodologies on a synthesis problem of the student's choice. The student projects serve two functions:

1. to tie the two components of the course together by exploring the potential for combining and/or integrating the two approaches; and
2. to provide an opportunity of applying the methodologies presented to a problem in the student's specific domain of specialization.

The first function helps to motivate the evolution of the course towards closer integration, as well as to generate new research questions and ideas. The second function is crucial since students come into the course with a variety of engineering backgrounds, and as graduate students they intend to further specialize in their respective engineering disciplines. It is the theme of this and other EDRC-sponsored courses that the best way to "open the student's eyes" to interdisciplinary methodologies, transcending their discipline-based education, is to provide a setting in which they can explore, apply, and evaluate such methodologies in solving a "familiar" problem. Our experience shows that such an approach yields better understanding of the role of the methodologies presented than any set of instructor-assigned problems.

Each student is given the choice of selecting a synthesis problem of his/her domain.

In a representative project, a first-year graduate student in Civil Engineering investigated the optimum design of truss structures and the use of heuristics for selecting truss types and preliminary proportions. The mathematical optimization component formulated truss design as an LP problem subject to equilibrium and stress constraints. The geometry of the truss was fixed, and all potential truss members were included in the formulation of the superstructure. Under a single loading condition, the LP converged to a statically determine configuration, as expected. This is a simple example of topology or configuration optimization. The knowledge-based component consisted of sets of rules for selecting the layout of the truss, the preliminary dimensions, and the preferred configuration from among six standard truss types. The heuristics were extracted from old (circa 1920) structural design textbooks. The student commented that these rules may not be realistic for modern bridge design practices. The student proposed two integration schemes: the KBES as a preprocessor supplying layout and dimensions to the LP program; or the KBES as a postprocessor evaluating the optimal design in terms of suitability for its intended use.

4. Generalizations

The experience in teaching the course, and particularly the feedback from student projects, clearly supports our premise that engineering synthesis can significantly benefit from the creative integration of quantitative, mathematical optimization-based methods with qualitative, AI-based methods including knowledge-based systems methodologies. On the other hand, as the above student project demonstrates, the tools implementing these two methods presently form two unconnected "islands of automation" with no direct linkages between them. This latter observation suggests an agenda for considerable future work. This agenda is briefly discussed below in the major categories of research, tool development, and education.

Research. A basic question that has emerged from the experience with the projects in this course, and with other much larger research projects in the EDRC, is how to develop a common framework for combining or integrating the mathematical programming and knowledge-based approaches for the synthesis of engineering systems. In general, the two major alternative are:

1. the quantitative and qualitative models act as two *interfaced* knowledge sources under a common coordination scheme; or

2. the quantitative and qualitative models are tightly *integrated* in a common framework.

The first alternative would build on the emerging methodologies of distributed problem-solving, cooperative problem-solving, and concurrent design [Durfee et al., 1989]. In such systems, the overall problem-solving task is distributed to a number of heterogeneous agents, each of which is capable of contributing some portion to the overall solution process. Communication, coordination, and control of agents is exercised in a variety of ways, of which the blackboard architecture is the most popular. Major research issues that need to be addressed include the development of symbolic problem representations that can support the full range of quantitative and qualitative approaches and the development of domain-independent representations of the heuristics and problem superstructures involved.

It is understood and accepted that interfaced systems of the class sketched cannot aim at producing global optima over all the variables and constraints. However, they may lead to increasingly comprehensive optimization problems or, at least, provide optima over broader ranges of problems considerations than possible with quantitative optimization models.

The second alternative is the one where both qualitative and quantitative knowledge sources are activated simultaneously. This alternative would be particularly relevant for expediting the search in large combinatorial optimization problems. Here one possible integration framework is to convert the heuristics and logic of the knowledge base into an MILP model [Raman and Grossmann, 1991]. In this approach, the qualitative MILP model is obtained by stating the logic and heuristics in conjunctive normal form which is then translated into inequalities with 0-1 variables. Boolean variables are also assigned to the possible violation of heuristics whose weighted sum is minimized in the objective function.

Tool Development. The last decade has witnessed the development of a number of modeling tools for optimization which represent a major improvement over the low-level access and formulation techniques of the past. Although the current modeling systems have greatly facilitated the use of optimization, the fact remains that they still require significant learning and coding effort. Furthermore, except for a few LP modeling systems, the transfer and preparation of data can be a major task. It would be clearly desirable to develop optimization tools that require little expertise by the users, which can handle symbolic information and be accessed in a variety of forms, and which can be easily interfaced to databases or spreadsheets.

In a similar vein, tools for knowledge-based synthesis are still rudimentary. The available shells are largely geared to diagnostic problems. EDESYN [Maher, 1988] is the first attempt to develop a domain-independent synthesis shell comparable in scope to the available diagnosis frameworks, but is still quite limited. Entirely new generations of knowledge-based synthesis frameworks suitable for broad classes of engineering synthesis problems are needed.

Education. Despite the fact that synthesis lies at the core of engineering design, there has been slow progress in incorporating synthesis in the undergraduate and graduate engineering curricula. Since decisions at the synthesis level have a major impact in the cost and quality of the designed artifacts, there is a clear need for synthesis to permeate engineering education. Experience with our course has shown that synthesis is teachable and that there are benefits from doing this in an interdisciplinary setting. This is, of course, not to say that discipline-based synthesis courses are not needed. These are certainly required to exploit the physical insight and knowledge of a particular discipline. On the other hand, an interdisciplinary course can emphasize basic design methodologies and provide students with a broader perspective.

There are several major question that need to be addressed for effectively teaching synthesis as an interdisciplinary graduate course. The first is how to present synthesis concepts that emphasize strategies and formulations without mathematical optimization becoming a "black box". The second is how to provide "synthetic" knowledge-bases to supplement students' missing or limited domain knowledge. The final question is how to effectively present the combination and relation of optimization and knowledge-based approaches to synthesis. Undoubtedly, answers to these questions will require both basic research as well as additional experience in teaching courses such as the one described.

5. Summary and Conclusions

This paper has reported the experience in teaching an interdisciplinary graduate course on synthesis in the Engineering Design Research Center at Carnegie Mellon. The course has been described, emphasizing the experience with projects where students are asked to combine optimization and AI-based techniques in a variety of engineering domains. While projects provide interesting results and insights, the tools implementing these two methods presently form two unconnected "islands of automation" with no direct linkages. Based on this experience, some possible directions for future research, tool development and education have been pointed out.

Acknowledgements

This work has been supported by an NSF grant to the Engineering Design Research Center. The collaboration of colleagues Ignacio Grossmann and Mary Lou Maher in developing and teaching the course is gratefully acknowledged, as is the contribution of all the students who have participated in the course.

References

[Clancey, 1985] Clancey, W.J., "Heuristic Classification", *Artificial Intelligence*, Vol. 27, pp. 289-350.

[Demes et al., 1993] Demes, G.H., S.J. Fenves, I.E. Grossmann, C.T. Hendrickson, T.M. Mitchell, F.B. Prinz, D.P. Siewiorek, E. Subrahmanian, S. Talukdar and A.W. Westerberg, "The Engineering Design Center at Carnegie Mellon University" *Proceedings of the IEEE*, Vol. 81, No. 1, pp. 10-24.

[Durfee et al., 1989] Durfee, E.H., V.R. Lesser and D.D. Corkill "Trends in Cooperative Distributed Problem Solving", *IEEE Transactions on Knowledge and Data Engineering*, Vol. KDE-1:1.

[Fenves & Grossmann, 1992] Fenves, S.J. and I.E. Grossmann, "An Interdisciplinary Course in Engineering Synthesis" *Research in Engineering Design*, Vol. 3, pp. 223-231.

[Glover, 1986] Glover, F, "Future Paths for Integer Programming and Links to Artificial Intelligence", *Computers and Operations Research*, Vol. 13, pp. 533-549.

[Grossmann, 1990] Grossmann, I.E., "Mixed-Integer Nonlinear Programming Techniques for the Synthesis of Engineering Systems", *Research in Engineering Design*, Vol. 1, pp. 205-228.

[Kusiak, 1991] Kusiak, A. "Process Planning: A Knowledge-Based and Optimization Perspective", *IEEE Transactions on Robotics and Automation*, Vol. 7, pp. 257-266.

[Levary, 1988] Levary, R.R. (Ed) "Engineering Design-Better Results Through Operations Research Methods", North Holland, Amsterdam.

[Maher, 1988] Maher, M.L. "Engineering Design Synthesis: A Domain

Independent Representation", *Artificial Intelligence for Engineering Design, Analysis and Manufacturing*, Vol. 1, pp. 207-213.

[McDermott, 1981]   McDermott, J. "R1: The Formative Years", *AI Magazine*, Vol. 2.

[Simon, 1987]   Simon, H.A. "Two Heads Are Better Than One: The Collaboration Between AI and OR", *Interfaces*, Vol. 17, pp. 8-15.

[Raman, 1991]   Raman, R. and I.E. Grossmann, "Relation Between MILP Modelling and Logical Inference for Chemical Process Synthesis], *Computers and Chemical Engineering*, Vol. 15, pp. 73-84.

A/E/C Teamwork

Renate Fruchter[1] and Helmut Krawinkler[2]

Abstract

This paper describes an on-going effort, initiated by Stanford's Civil Engineering Department in collaboration with the Architecture Department at UC Berkeley, on the development and testing of a new and innovative *computer integrated Architecture/Engineering/Construction* (A/E/C) teaching environment. In this computer integrated A/E/C environment a new generation of architecture, structural engineering, and construction management students learn how to team up with other disciplines and take advantage of emerging information technologies for collaborative work in order to design and build higher quality buildings faster and more economical. The paper presents the vision, along with the research, development, and strategies undertaken to develop the environment and build an infrastructure towards its realization.

1 Background

Communication is and will be critical in achieving better cooperation and coordination among professionals and across organizations in tomorrow's business world of global markets and virtual corporations. The emerging technologies will provide the means to bridge the gap among professionals and organizations, and to overcome the limitations of both geography and time. However, technology by itself, without improved teamwork, will fail. The virtual corporation, as presented by Davidow [Davidow 1992], will thrive in an environment of teamwork, in which owners, designers, contractors and business people all work together to achieve common goals. The vision of the virtual corporation describes a team of professionals, which will form and re-form as needed around a project, and will use the emerging technologies which support team work. In the A/E/C industry, the owner's requirements and the different issues related to a building project determine the team's composition over the life cycle of the project. However, the involvement of the owner in the process and participatory computer-aided, integrated collaborative work are not yet in place in the A/E/C industry.

Because the corporation of the future will be built on information, it means that it will be necessary to educate professionals about the tools that control and manipulate

---

[1] Res. Assoc., Center for Integrated Facility Engineering, Stanford University, Stanford, CA 94305.

[2] Prof., Department of Civil Engineering, Stanford University, Stanford, CA 94305.

information, and support collaborative and concurrent work. And because teamwork will be the primary work mode, it means that training in consensus building, group dynamics, and problem solving by using numerous and diverse technology advances will be essential.
Our thesis is that education is a key to improved communication. Teamwork is needed, is hard to teach, however, it is possible to teach. This paper presents a preliminary prototype environment, a *computer integrated A/E/C* course and laboratory in which teams of architecture, structural engineering, and construction management students work together in A/E/C teams on building projects. The course is aimed to increase the number of students who will:

- understand how the three disciplines - architectural design, structural design, and construction - impact each other,
- learn how new technologies can provide support for collaborative and concurrent A/E/C teamwork, and
- understand how concurrent engineering and collaboration technology can be modeled and simulated from an organization point of view.

2 Motivation

The distinction among A/E/C professionals, which is emphasized by divergent education, is today's status quo. It is our believe that some of the reasons of the current poor coordination and communication among professionals (e.g., architects, structural engineers, construction managers) in the fragmented A/E/C industry and among project phases (e.g., conceptual design, detailed design, planning) are rooted in the way education is structured today, by discipline, and the sequential "waterfall" paradigm in which projects are executed, by phases. The combination of education by discipline and project execution by phases leads to the fragmentation of the A/E/C community into loosely coupled "operational islands of knowledge" among professionals, project stages, and organizations (Figure 1, adapted from [Richard 1986]). The divergence in curricula and specialization are major factors that magnify this fragmentation.

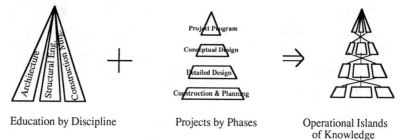

Education by Discipline   Projects by Phases   Operational Islands
                                                              of Knowledge

Figure 1: Poor Communication and Coordination in the Fragmented A/E/C Industry -- A Result of Specialization and Divergence in Curricula

Although architects, structural engineers, and contractors work together to achieve a common goal, that is "build a building," they think differently about the same product, the building. Each mode of professional thinking could enrich the other, and each would profit from an interchange of viewpoints. However, difficulty in communicating design concepts and decisions among professionals often hinders such knowledge sharing. Communication problems in the A/E/C industry are rooted in the differences in interests and modes of thought of professionals, which are fostered, magnified, and made more rigid by the divergence in their education. Some of the salient observations of the state-of-the-art in A/E/C education are:

- Students in architecture, structural engineering, and construction management programs have little opportunity to attend classes in the other two programs.
- Architectural design, structural design, and construction management classes are taught by discipline, and few courses exists that present the relationship among disciplines.
- Engineers generally use the abstract language of mathematics to develop their designs, whereas architects employ visual language and graphic notation.
- Architects and engineers are trained to see different things when looking at the same object.
- Architecture and engineering students are being made aware that their professional activities will have different and sometimes conflicting goals.

What transpires in conflict situations is that the A/E/C professionals not only have different ways of thinking and, hence, different approaches to the same problem, but also have real in communication difficulties since they talk different languages.

In addition, very limited curriculum-development effort is devoted to the introduction of modeling and analysis concepts for the management of human and technology integration from an organizational point of view. Architecture and engineering students lack the global understanding and the means to evaluate and appreciate the impacts of rapid changes in today's technologies on the organizational structures of the teams in which they work.

3 Teamwork in the Computer Integrated A/E/C Course

The experimental computer integrated A/E/C course is a collaborative effort of the Civil Engineering Department, at Stanford, together with the Architecture and Civil Engineering Department, at UC Berkeley. The course is intended to provide an environment and an environment for inter-disciplinary apprenticeship, enable teamwork, sharing of knowledge and information, and overcome the limitations of both geography and time. The computer integrated A/E/C course engages faculty members and practitioners from A/E/C disciplines, researchers who develop new computer technologies, and students, and tries to tie the diverse computer-aided engineering innovations together into a single cohesive course and laboratory environment. The course presents basic principles and concepts of each discipline, focusing on:

- how the disciplines are linked together,
- what information and knowledge is shared,
- how information/data of one discipline impacts the other disciplines.

Emphasis is placed upon team work in a project that integrates these principles and concepts. Teams of three students each (i.e., each student having a background in one of the three disciplines), are involved in hands-on assignments to model, implement, refine, and document the design; and analyze their team's organization. The course (1) gives the students a general perspective of the purpose of the course, focusing on integration issues and reasons for modeling organizations, and (2) identifies the need for software tools, the way these tools can support disciplinary and interdisciplinary tasks, the way the tools interact in order to support collaborative and concurrent engineering, and the way the tools support the systematic design of organization structures as information-processing and communication systems to evaluate the efficiency of multidisciplinary engineering design teams.

Roles in the Computer Integrated A/E/C Environment. In a collaborative effort faculty members and research associates who have been involved in teaching the A/E/C disciplines and developing new computer-based tools, play the role of "master builders." They expose in this course a new generation of students to teamwork, integration issues, modeling and analyzing organization structures. Undergraduate and graduate students with a background in one of the three disciplines, play the roles of apprentice and journeyman, respectively. They develop a holistic view of A/E/C industry, learn to work in inter-disciplinary teams, and gain a better understanding of how the three disciplines can be integrated in terms of the product and the process, and how computer technology can be used to support tight collaborative and distributed concurrent engineering.

Teamwork and activities in the computer integrated A/E/C course. Teamwork is a process of reaching a shared understanding of the design and construction domains, the requirements, the building to be built, the design process itself and the commitments it entails. The understanding emerges over time as each team member develops an understanding of his/her own part of the project and provides information that allows others to progress. The process involves communication, negotiation, and team learning. The activities in the computer integrated A/E/C course may focus, for instance, on three phases -- the collaborative conceptual design phase, the concurrent detailed design phase, and the organization modeling phase. Teams of students are involved in a multi-disciplinary building project in which they model, refine and document the design product, the process, and their implementation. The students learn (1) to regroup as the different discipline issues become central problems and impact other discipline, (2) to use computer tools supporting specific discipline tasks and (3) to navigate among the different computer integrated environments supporting different degrees of collaboration. The projects progress from conceptual design to a computer model of the building and a final report. As in the real world, the teams have tight deadlines, have to engage in design reviews, and negotiate modifications. As the project progresses, information and knowledge is gathered, sorted, modeled, reorganized, and implemented. A team's cross-disciplinary understanding evolves over the life of the project. It starts with individual discipline information and knowledge, definition of concepts, and redefinition of issues, using each discipline's natural idioms. As the project progresses a number of things happen: (1) the concepts are transformed into models, (2) the models become more detailed, (3) discipline models are linked, and (4) information is reorganized so that it can be shared among the participants.

The first experimental course was offered last Spring Quarter 1994, to three A/E/C student teams from Stanford. This academic year, we offer the course to Stanford and UC Berkeley students. The course is taking place in Winter Quarter 1995 (at the time this paper is written), and involves five A/E/C teams which are composed of both Stanford and Berkeley students. In addition, each team has an apprentice represented by an undergraduate student from Stanford. The mixed Stanford & Berkeley student teams adds the "real-world" collaboration complexity to the experiment, which includes space and time coordination and cooperation needs. The results of the second experiment will be reported in a future paper.

Emerging technologies in the Computer Integrated A/E/C Course. Visionary cooperative information systems, such as the Electronic Superhighway, will manage and access large amounts of information and computing services. They will support individual and collaborative work. This will facilitate the computer integrated A/E/C course to be any-where-any-time supported by an open, heterogeneous network-oriented environment. Lectures and meetings will be face-to-face, tele- or video-conferenced. Communication and sharing of information among team members will be done using centralized or distributed integrated and interoperable computer environments, depending on the degree of collaboration at various stages of a project. Modeling, visualizing, sharing, storing design models will be enabled by tools such as shared 3D graphic CAD, electronic white-boards, multi-media, component catalog and handbook data-bases. Database systems will contribute information management techniques for distributed, heterogeneous databases. Computation will be conducted concurrently over the network by use of tools ranging from conventional to advanced application systems (e.g., structural analysis, knowledge-based critique). One of the challenges of the computer integrated A/E/C course model is to effectively combine the emerging technologies and their contributions to meet the diverse requirements of teamwork.

In recent years growing attention has been given at some campuses to the issues of integration, collaborative work, concurrent engineering, and modeling of organizations. [Fenves 1990] [Sriram 1991] [Pohl 1992] [Levitt 1992] [Fruchter 1993] [Khedro 1993] The thesis of this paper is that in addition to cutting edge research and technology, education is a key to improved communication and the evolution of today's A/E/C industry into tomorrow's virtual corporations.

Computational environment. The computer integrated A/E/C course focuses on integration, information, and organization modeling which support the life cycle of a facility. Internet mediated design communication, integration and organization frameworks, groupware technology and multimedia are used in the experimental course-lab:

• *World Wide Web* (Mosaic) is used in a team building exercise and as a medium to disseminate and share conceptual design solutions of the design teams. The "team building on the Web" exercise, is based on generic skill definitions of the A/E/C students and hypothetical project calls for bids posted on the Web. Students have to identify the specific project they can work on among the different calls for bids and publish on the Web their skills, project preference, and request for collaborators from the other disciplines. This exercise exposes students to the Web and one of its future potential commerce and business applications.

• *Interdisciplinary Communication Medium, ICM* [Fruchter 1993], is used as an integration environment to support the graphic representation in AutoCAD and symbolic modeling and reasoning of A/E/C in conceptual collaborative design. This shared environment supports concurrent engineering by improving communication among members of interdisciplinary teams. The purpose is to let students explore the different cross-disciplinary issues among architectural and structural form modeling, and constructibility. ICM implements an iterative "Propose-Interpret-Critique-Explain" (PICE) model as a communication cycle in collaborative conceptual design. ICM integrates a shared graphical modeling environment and network based services. ICM graphics includes 3D models of evolving designs, and network based services may include knowledge-based reasoning tools that critique the performance or cost of a proposed design. The

key technical concept of ICM is that a graphical design environment (such as AutoCAD) can also serve as the central interface among designers (human-to-human) and as the gateway to tools/services (human-to-machine) in support of interdisciplinary design. This computer based graphical environment enables designers to share and explore designs in the following ways:
   ° define and view definitions of objects and functions at different levels of abstraction;
   ° directly engage students in network distributed design/analysis/critique which use tools/services to define geometric, spatial, graphic, and symbolic information models;
   ° interlink different discipline interests via shared graphics.

• *Federation of Collaborative Design Agents, FCDA* [Khedro 1993], is used as an integration environment among the distributed, heterogeneous tools that support the reasoning of A/E/C in the detailed design. The suite of heterogeneous tools are research and commercial software applications running on different software and hardware platforms. FCDA aims at integrating heterogeneous applications (i.e., applications which have been implemented in different languages and run on different hardware platforms), called design agents, by allowing better software interoperability. The design agents exchange design information and knowledge by using a formal language Agent Communication Language as a communication standard. The communication of design information among linked design agents is coordinated via system programs called facilitators in a federation architecture. The federation architecture specifies the way design agents and facilitator communicate in an integrated environment. The facilitators are also the repository of a shared vocabulary (ontology). The ontology consists of those words that refer to objects, functions, and relation which describe the design domain disciplines. The design agents characterize their role in the integrated environment by specifying: (1) interests, which consist of the design information about which the agents desire to be informed, (2) perspectives, which define the attributes and formats in which the agent desires to receive and send information, and (3) behaviors, which describe the agents' activities in the environment. The intended use of FCDA is for distributed design performed in a partially independent fashion by design agents and professionals.

• *Virtual Design Team, VDT* [Levitt 1992], is used to understand how organization structures, project tasks, and collaboration technologies can be modeled and simulated in order to analyze the impact of communication technologies on organization behavior and performance. VDT is used to model the teams' organization and work process, given various organization structures and communication tools. The purpose of the experiment is to expose students to organizational and management issues, in addition to technical issues, involved in multidisciplinary projects.

• *Video conferencing* capabilities linked with computers running AutoCAD provide the necessary medium for the A/E/C design teams and "owners," who are in two different geographic locations, to discuss the design solution in real time.

• This year (Winter 95) we will also test the use of *groupware*, such as Timbuktu, xmx, Xshare, and xmove, which will enable collaborative teamwork for the geographically dispersed teams, that is, between Berkeley and Stanford students. These groupware tools enable the team members to share and work concurrently on the same desktop, Xwindow, or application (e.g., AutoCAD).

4 Expected Significance and Impact

The experimental computer integrated A/E/C course explores the integration of *people*, *knowledge*, *building systems* , and *technology* in a cohesive environment. Its vision is to create:

- *An exciting environment* which will motivate students for a life long odyssey in creative facility design and construction management. The course is aimed to significantly increase the number of students who will become agile, innovative, flexible, and adaptable professionals.
- *A test-bed* to explore and assess the use of research prototype tools which become robust and mature, and invite integration and testing in a controlled environment such as a course-lab.
- *A new culture* which brings together undergraduate and graduate students, faculty members, and researchers, and in the future involve also industry representatives from the different disciplines.
- *A technology transfer bridge* between the research laboratories in the university and the industry. Even though information represents a competitive advantage for industry, education, and research, it is perceived and used differently in the three environments. For industry, competitive advantage is represented by secured information. For education, competitive advantage is represented by open information exchange. For research, competitive advantage is represented by open information creation. These different perspectives offer this course an excellent opportunity to provide a two way technology transfer bridge between the research laboratories, which develop new methodologies and tools, and industry organizations, which provide the real-world experience and future business needs.

Educational significance and impact.

- The *computer integrated A/E/C course* model will affect the way students will view their disciplinary tasks based on a holistic understanding of the cross-disciplinary impacts and constraints, and will prepare them to adapt to the rapid changes in today's technologies.

- The effects of the *computer integrated A/E/C course* on institutions in which *computer integrated A/E/C course* models will be explored and tested, will include:

  ° an increase in interdepartmental efforts to close the gap between architects and civil engineers,
  ° a framework for cross-disciplinary collaboration and exchange of research and teaching ideas related to this educational model, and
  ° an integrated computer environment that is flexible, modular and extensible, enabling the addition of new and exciting spin-off research prototypes developed at those institutions.

Academic significance and impact.
The computer integrated A/E/C course will provide:

- A methodology for adapting research prototypes to "hands-on" computer lab exercises.
- A test-bed for evaluation of the emerging technologies.

- Observations on the way students adapt and improve their performance by using different computer tools which support teamwork.
- Case study observations (i.e., student teams) regarding the modeling and analysis of organizations.

Practical significance and impact. The development and implementation of the *computer integrated A/E/C course* model will provide a new generation of professionals educated to perform more effectively in tomorrow's virtual corporations.

Acknowledgments. This research is funded in 94/95 by the NSF Synthesis Coalition. The first experiment was partially supported in 93/94 by the Center for Integrated Facility Engineering (CIFE), at Stanford University. We thank the first generation of pioneering students who worked with us in the first experimental course in 93/94. We express our thanks to R. Levitt, P. Teicholz, M. Fischer, T. Khedro, Y. Jin, from Stanford University, Y. Kalay, M. Martin, from the Architecture Department at UC Berkeley, and L. Demsetz, C. Thewalt, from the Civil Engineering Department at UC Berkeley for their participation in this course. We thank the industry support group of the *"Computer Integrated A/E/C"* course: Autodesk, Sun Microsystems, and IntelliCorp for providing the necessary software and hardware.

World Wide Web presentation
For a World Wide Web presentation of the experimental *"Computer Integrated A/E/C"* course, see:
> **URL: http://cdr.stanford.edu/html/ICM/AEC.html**

References

[Davidow, 92] Davidow, W.H. and Malone, M.S., "The Virtual Corporation," HarperBusiness a Division of HarperCollins Publisher, 1992.

[Fenves, 90] Fenves, S.J. et al., "Integrated Software Environment for Building Design and Construction," *Computer-Aided Design*, 1990, Vol. 22, No. 1, 27-36.

[Fruchter, 93] Fruchter, R., Clayton, M., Krawinkler, H., Kunz, J., Teicholz, P. "Interdisciplinary Communication Medium for Collaborative Design". *Proc. of the 3rd International Conference on A I in Civil Engineering,* Edinburgh, 1993.

[Khedro, 93] Khedro, T., Genesereth, M., Teicholz, P. "FCDA: A Framework for collaborative distributed multidisciplinary design", *AI in Collaborative Design Workshop Notes.* Menlo Park, CA, AAAI, 1993.

[Levitt, 92] Levitt, R., Cohen, G., Kunz, J., Nass, C., "The Virtual Design Team: Using computers to Model Information Processing and Communication in Organizations," *Center for Integrated Facility Engineering*, Stanford University, CIFE Working Paper 016, 1992.

[Pohl, 92] Pohl, J., et al. "A Computer-Based Design Environment: Implemented and Planned Extensions of the ICADS Model," *Design Institute Report, CADRU - 06-92, School of Architecture and Environmental Design, California Polytechnic State University,* San Luis Obispo, CA, 1992.

[Richard, 86] Richard D. Rush, ed., "The Building System Integration Handbook." John Wiley & Sons, San Luis Obispo, CA, 1986.

[Sriram, 91] Sriram, D., "Computer Aided Collaborative Product Development," Research Report R91-14, *Intelligent Engineering Systems Laboratory, Department of Civil Engineering, MIT,* Cambridge, MA, 1991.

Educating Future Master Builders

David R. Riley[1] and Victor E. Sanvido[2]

As industry innovators experiment with design-build delivery methods, partnering, and interdisciplinary design integration, the roles of individuals and organizations in the AEC industry become less specialized and more dynamic. As a result, the differences between planning, design, and construction phases in a project become blurred. The skills required to orchestrate a project team thus become more diverse and technically challenging. To be successful participants in this ever-changing industry, young professionals must understand the entire process of providing a constructed facility, from initial idea through design, construction and operation.

Research efforts in computer integrated construction continue to improve our understanding of the information required to drive a project. Technological advances in the form of electronic information transfer provide promise of an "information highway," where project participants are offered opportunities to communicate with each other in ways limited only by their paradigms. Before students are released into the industry they need to be provided a solid, broad fundamental education. Secondly they should be exposed to the electronic tools of their trade and shown how to use them effectively.

This paper describes an effort to provide a balanced educational program for future "Master Builders," and to prepare them to enter the industry as innovators and

---

[1] Post-Doctoral Scholar, CIC Research Lab, Dept. of Architectural Engineering, The Pennsylvania State Univ., 104 Eng. Unit A, University Park, PA 16802.

[2] Associate Prof., CIC Research Lab, Dept. of Architectural Engr., The Pennsylvania State Univ., 104 Eng. Unit A, Univ. Park, PA.

managers. Academic programs in building construction face a variety of new challenges in educating successful engineering professionals. As engineers, a working knowledge of the technical systems of modern facilities must be provided. Although a student may initially be responsible for designing or building only one specialty system for a facility, an ignorance of the relationships between this and other system components will often result in design conflicts. In some cases the resulting project plans and designs do not allow the building to be an integrated functional facility. As professionals, individual project roles must be placed in the context of an overall project outlook. Individuals cannot operate on an island, isolated from the risks and responsibilities of other project players. An appreciation for the contributions of all participants must be developed to establish working relationships and functional project teams. In response to these needs, a unique five year curriculum that educates future AEC leaders with a strong technical background and a complete project perspective is presented.

First, an introduction to the building design and the construction industry is incorporated into the first year of core engineering courses. This stimulates an early interest from freshmen engineering students who are otherwise laden with a demanding technical course load. As a result, the most creative and gifted students are often inspired to pursue a unique five-year curriculum in architectural engineering. Beginning in the second year, core courses are complimented with architectural theory and introductory systems design courses in structural, mechanical, lighting, plumbing, and fire protection disciplines. During this year, students are also introduced to building materials through an architectural drawing course which emphasizes the development of integrated building details and connections. Students develop functional floor layouts and design interfaces between architectural, structural, and technical building systems. Complete building plans and details are first developed by hand, then enhanced through computer modeling. Practical depth on the fundamentals of engineering mechanics is provided through a laboratory course in structural analysis. Here students explore the loads imposed on buildings and how these are absorbed, distributed, and transmitted to foundations. The affects of compressive, tensile and shear stress are explored through laboratory experiments that challenge students to design and build various structures and test them destruction.

The third year curriculum is dominated by a series of specialty courses which present detailed criteria and analysis methods for major building systems. This series makes up the backbone of a broad technical background for each student. The focal point of this year is an introduction to the building industry through a course in construction project management. Here students are exposed to the concept of the master builder, a team leader who must recognize the roles of all project players and plan and manage the entire process of providing a facility. In each course, students are required to use computers as time saving tools which provide information for decisions, but not to replace their technical skills. Teamwork and leadership traits are cultivated through group assignments that necessitate coordinated efforts. A demanding load of independent work develops self discipline and time management skills. At the conclusion of the third year, students use their experiences in breadth courses as a basis for the selecting a specialty track in construction management. Students with a career interest in engineering design can elect to specialize in structural, mechanical, or lighting/electrical design for the remainder of their curriculum.

Fourth year coursework in the construction option allows students to round out their technical skills and develop traits of a successful professional. Progressive courses in construction management strive to integrate to roles of managers at the project level. Several courses address the means and methods for the construction of architectural, structural, and mechanical systems. The project management functions during site exploration, design analysis, constructability, estimating and scheduling are explored in increasing levels of complexity. Rather than separating these elements into specialty courses, they are integrating into full semester projects, placing students into the role of a "real" project manager for the first time. On each of these projects, industry standard computing tools are utilized to develop estimates, schedules, and logistics plans. Concurrently, the integration of form and function is practiced in an architectural design studio to stimulate creativity and clear expression of ideas. Perhaps more importantly, students develop an understanding of the architect's perspective and the goals and values inherent to a successful architectural design process. To complete the fourth year, essential courses in accounting, management, and economics provide awareness of modern business practices.

In the fifth year students are asked to take the background they have been given and to apply it to current practice. Each select a large scale construction project to investigate for the first semester. A comprehensive study is performed of the building systems, project players, construction methods, project control, and professional services provided by the managing entity. Students are required to seek information from all key players on the project, including owners, architects, construction managers, and subcontractors. This gives them hands-on experience interacting with the different personalities and attitudes assembled on a single project. They are asked to then assess the systems they find, critique the performance of the construction manager, and provide recommendations for improvement. As a basis for their recommendations, students must take advantage of a select team of industry professional guest lecturers and advice from experts on the subjects under investigation. They also compare their findings during group meetings set in a professional atmosphere. This builds and understanding of what types of project management functions are necessary for varying types of facilities and delivery systems. To complete this semester's work, a final technical report is assembled which describes their findings and summarizes their recommendations for the project.

In their final year, students also perform independent thesis research. Here they are challenged to recognize problems as opportunities for change and improvement, and then synthesize solutions utilizing both their technical skills and creativity. During the fall semester, students attend a roundtable meeting with leading industry professionals to discuss pertinent issues facing the building industry. At the conclusion of this meeting, students are able to identify relevant topics for their research in the spring, and establish industry contacts to assist them. They are required to define their scope of work, conduct independent research, and apply their findings back to the industry. Elements of this include a formal topic proposal, a comprehensive review of literature and current practices, the development and refinement of a solution to the problem, the application of the solution to the building industry, and the final oral presentation of their findings. This presentation is made at a professional seminar attended by industry members. This professional-level interaction fosters sharing of ideas and awareness of current issues facing the industry.

Throughout the curriculum, an outlook is maintained on the dynamic role of professionals in the evolving AEC industry. This diverse program builds a mutual awareness and appreciation for the skills of all project players. A fundamental emphasis is placed on education first. Computing tools are viewed as enhancements to a solid technical background. As they enter the AEC industry, graduates are prepared to accept the challenge of managing the design and construction of modern facilities.

A COMPUTER BASED MODEL FOR INTEGRATED ASSISTANCE TO THE PLANNING OF BUILDING PROJECTS

José Tiberio Hernández, PhD[1]
Associate Professor and Head, Computer Science Dept.
Diego Echeverry, PhD, Assoc. Member ASCE
Associate Professor, Civil Engineering Dept.
Gloria Cortés, MSc
Assistant Professor, Computer Science Dept.

Universidad de Los Andes, Bogotá, Colombia

ABSTRACT

The research project described in this paper has the objetive of designing an integrated plataform to support the decision-making process at the feseability stage of a building project. The model is based on a feature-oriented, multi-view representation of a building project, with emphasis on the relations between diferent points of view of the project (financial, technical, legal, etc). A prototype system was developed which was fed cartographic information about the city of Bogotá. In addition, the prototype integrates other functions such as time and cost estimates.

INTRODUCTION

The present model is based on the conceptual models proposed by Yoshikawa regarding the desirable characteristics of intelligent computer aided design tools [Yos92]. A number of different proposals for design knowledge structuring, such as classifications and rules[KRA89], evolve from these works. An obvious problem that appears here is the difficulty for finding an appropriate language to express design decisions and to deal with the different points of view of an object in the process of being designed. On the other hand, in the metal-mechanical sector and headed by

---

[1]   Hernández, Head , Computer Science Department, Universidad de Los Andes, Carrera 1E # 18A
    - 10, Santa Fe de Bogotá, Colombia. e-mail: jhernand@uniandes.edu.co
    Echeverry, Associate Professor , Civil Engineering Department, Universidad de Los Andes,
    Carrera 1E#18A - 10, Santa Fe de Bogotá, Colombia. e-mail: dechever@cdcnet.uniandes.edu.co
    Cortés, Assistant Professor , Computer Science Department, Universidad de Los Andes, Carrera
    1E # 18A - 10, Santa Fe de Bogotá, Colombia. e-mail: gcortes@uniandes.edu.co
This work is sponsored by CDM - Apple Colombia.

CAM-i, proposals that deal with object representation centered on the object's interesting aspects and not only on its geometry were originated. Feature based representations gave a great impulse to a new generation of CAD/CAPP/CAM systems in this field [Cam-i89] [HER94a] [SHA89].

Recent works attempt to bring together these lines of work. We find, among others, the works of Tomiyama, in which the use of the object-oriented paradigms for structuring knowledge is proposed and Mantyla[Man90] who proposes a scheme of feature representations with uncertain dimensions.

Our work fits into the works of this last group and proposes formal representation of features, "rough representations" and designs that permit the characterization of specific domains and, furthermore, it permit the formulation of specific purpose design aid systems (CADE's) in a systematic and structured manner [Car92][Her94b].

Our model is based of a formal representation of sketches (it is important to point out that by sketch we mean partially instantiated specifications that can be satisfied by more than one product) that can in turn be "seen" from one or more points of view thanks to a feature definition over the sketches. The design is, therefore, the process of decision making in which the partial information of the sketches is completed with decisions that are taken from different points of view, until a specification which is only satisfied by one product is obtained.

This model has the great advantage of allowing the formal definition of Views, the decision propagation of a view in the sketch representation, and the explicit representation of the design process.

This article deals with the idea explained above in the context of the construction industry, and presents a prototype which takes into consideration the feasibility stage of a project.

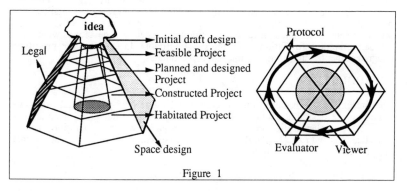

Figure 1

We can see in Figure 1 the structure of the process as a pyramid on whose peak we find the gestation of a project and, in each one of its sides, a perspective of the project from the point of view of one of the experts participating in it. The first representation is an initial sketch which, as we gain information with the decisions of the various experts, improves the specification of the project.

In the case of the feasibility stage, we find that the points of view are, among others, localization, spaces, time, and costs. As is shown in the same figure each one of the views must count on a visualization of the project, a scheme of evaluation of the state of the project, an aid to decision making and a protocol that allows the propagation the decisions made to the other views by the project representation.

This requires a computer architecture that favors this proposal and effectively supports a process of concurrent engineering in the construction industry. Our proposal is based on:

• A feature based representation where the features are not necessarily completely instantiated (e.g. sketches) and are grouped in Views according to the aspects of interest for decision making in the project.

• A knowledge representation based on sketch classification and distributed in Views.

• A project representation manager responsible for maintaining the model's coherence and for propagating decisions correctly.

• A Client/Server scheme for achieving efficient support to decision making.

Following this methodological scheme, the feasibility stage (the stage that covers the time period beginning with the gestation of the project's initial idea up till the decision of economical and of technical feasibility) was modeled, and a prototype that integrates the different tools that support the model was developed.

AN INTEGRATED MODEL FOR THE FEASIBILITY STAGE OF A BUILDING PROJECT

At the planning stage of a building project, considerations must be made to develop the appropiate type of facility. These considerations involve the interpretation of data associated with the geographical location of the building site. Data such as site size and shape, access roads, access to utilities, city regulations, local market characteristics etc. should be aviable to the decision makers at planning stage. Another fundamental aspect in the decision making is the cost and time estimation.

Representation

Taking into consideration the information needs in the feasibility stage of a construction project, the following views were defined:

• Localization View
• Product View (Architectural Design, Structural design, hydraulic design, etc.)
• Financial View (Costs)
• Activity planning view (Time)

The representation used for these views can be seen in Figure 2.

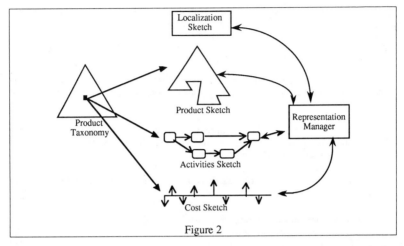

Figure 2

The project's sketch is composed of the localization sketch, the product sketch, the activities sketch, and the cost sketch. The representation manager has functions that permit it to communicate the changes performed on the Views to the sketches that should be modified.

Localization Sketch

A localization sketch is made up of two components: the location sketch and the plot sketch. Depending of the level of specification, this structure takes different values. An initial sketch of localization can be the general location of the plot, city zone where it is found, and the average size of plots in that zone. A sketch that is more specific contains the exact localization information, cartographic location and the plot's perimeter. It is, in this sketch, where city building regulations are taken into account.

Product Sketch

The product sketch's definition is an n-ary tree the root of which contains the minimum information of a product that can be possessed. The sub-trees associated with the root represent the structure of the product's components. To descend a level in a branch of a tree entails that a decision has been made or that information has been gained regarding the components of a part of the product. Summarizing, the product sketch is a tree of components. The information associated with the nodes of the tree should include data pertaining to cost and time of manufacturing, installation and completion.

A complete tree describes a specific product. The leaves make reference to specific parts (i.e., a "MadeFlex528" door). Information regarding cost and time of manufacturing, installation and completion of a given part of the product is fopund on the leaves as weel as restrictions over these same variables. For this purpose, component taxonomies could be used. An incomplete tree describes a stage of the product's design. If the information of the tree's leaves is "read", the state of the design should be "seen".

Figure 3 shows the diagram of a part of a project sketch. Note that not all branches of the tree are specified and that may have different levels of definition.

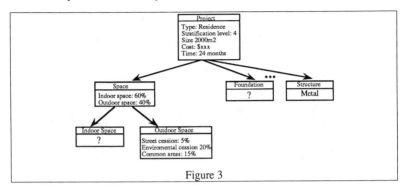

Figure 3

Activities Sketch

An activities sketch is a directed activities graph (CPM). The information associated with the nodes of the graph indicate time indexes. The arcs in the graph indicate precedence between activities. The initial activities sketch has a single node that only contains information regarding the estimated time it will take to complete the project. A sketch which has been further specified can be a graph with a node for each of the principal activities involved in the project (i.e. construction and selling) and the time that will be taken by each one.

Cost Sketch

A cost sketch is a list of costs and income ordered by date (cash flow). The initial cost sketch is a list with a single element which contains the estimated cost for the whole project. A more refined sketch can be a cash flow with costs and income grouped by activities. (i.e. cost for construction, income for sales and financing).

Initially, to define a project, we have a classification of the of the business' products. This taxonomy is a tree where each node contains information about the type of product. Each node has associated with, or permits the generation of a project sketch: in other words, a product sketch and an activities sketch and a cost sketch. As the product is specified (as we descend in the classification tree), the sketches will be more complete..

Views

As mentioned in the introduction, a View is the set of features of the project that interests one of the experts that participate in it. In our model, each of the Views has the following:

- An associated sketch, in other words, each view manages, with the help of the representation manager, the information regarding the sketch that corresponds to it.

- An evaluator that permits "seeing" the state of the project from the expert's point of view at any time.

- A viewer that permits "seeing" the sketch associated with the view.

Bellow, the components of each view for the feasibility stage are described.

View	Evaluator	Viewer
Localization	In charge of: given the localization sketch, applying the city regulation and sending data that affect other sketches (i.e. maximum number of floors) to the representation manager.	Allows seeing in the city map the location of the project, depending on how well defined the sketch is. For example if the exact location is still unknown only the city zone is shown.
Product	In charge of: given the product sketch, producing its visualization. For example if the location, the shape of the lot and the number of floors are known, the associated polyhedron is generated.	In this case, it is the same as the evaluator
Activities	In charge of given the activities sketch compute how long the project will take.	Allows viewing the project's activities graph. The graph's level of detail depends on the sketch's level of specification.
Financial	In charge of given the financial (or cost) sketch, compute the cost effectiveness of the project.	Allows viewing the project's cash flow. The cash flow's level of detail depends on the sketch's level of specification.

Representation Manager

The representation manager is in charge of maintaining coherence between the different sketches' information. The sketches notify the manager of any change in the information (this may happen if an expert, interacting with a view, makes a decision that adds information to the specification). The manager must decide which view should be notified of the change. The Views receive the changes of information, use the viewers to reflect them, and evaluate the project to obtain a new state.

MODEL IMPLEMENTATION FOR THE FEASIBILITY STAGE

To test the model, a prototype of the application was implemented on a Macintosh platform integrating different software tools. Some of these are tools that specialize in dealing with views: MapInfo, a GIS system to manage location views, Excel to manage the financial view, and a planning and control program to manage the activities view. The other tools were constructed for the specific case of the feasibility stage: an image viewer was adapted for the product View and a representation manager was constructed. These last two were implemented using the C programming language.

The View managers were constructed using the facilities offered by specialized tools: MapBasic for the localization view and macros for the financial view. On the other hand, the representation manager communicates with the view managers using

AppleEvents. This scheme allows considering implementing the prototype on more than one machine.

Bellow the typical screen of the application is shown.

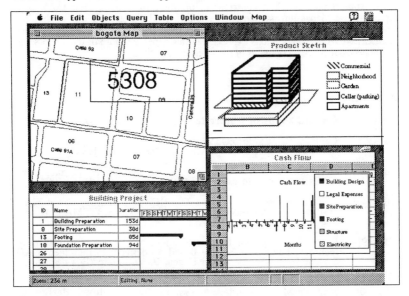

CONCLUSION

The initial result obtained with the prototype described in this paper are very interesting due to the integration obtained and to the fact that an appropriate language was offered to each expert participating in the complex process of the construction industry. However, it is necessary to remain prudent, due to the complexity that appears as the project's information grows. We believe that the great conceptual challenge in our ambitious project lies in finding a way to represent decision propagation in incomplete representations.

On the other hand, the distributed and open architecture that appears naturally in the project favors the view managers' growth and specialization in the context of a particular construction company.

ACKNOWLEDGMENTS

The participation of the graduate students Eliana Navarro and Ricardo Hoyos is gratefully acknowledged. The authors are also grateful for the funding provided by CDM-Apple Colombia. The contribution of MapInfo through their Colombian representative Mapas Informaticos was acknowledged. Parcial funding for the time dedicated by D. Echeverry was provided by the Colombian Institute of Science (COLCIENCIAS). The opinions expressed in this paper are those of the authors and do not necessarily reflect opinions of the sponsoring institutions.

REFERENCES

[CAM-i89] Current status of feature technology. CAM-i 1989.

[CAR92] Cardoso, R. , Cardoso,O,Hernández,J, Ramos O. : Especificación y modelaje de diseños en CAD. CLEI'92. España. 1992

[HER94a] Hernández José T. "Rasgos y Clasificaciones como base para CADs de propósito especifico." Reporte interno DFAC. 1994

[HER94b] Hernández José T. , Morales J." PICAS: a feature-based CAD system for quasi-axisymmetrical parts". SISTEMAS, N. 58. March 1994

[KRA89] Krause F. Vosgeram F., Yaramanoglu N. . "Implementation of technical rules in a feature Based Modeller". In Intelligent CAD System II, V. Akman, P ten Hagen, P. Veerkamp (Eds.) Springer-Verlag 1989

[MAN90] Mantyla,M. A modeling system for top-down design of assembled products. IBM Journal of research and development, Vol34 No.5 Sep.1990.

[SHA89] J. Sha et al. "Functional Requirements for feature based modeler". CAM-I, Arlington Texas, 1989

[YOS92] Toshikawa et al. "An Intelligent Integrated Interactive CAD. A preliminary Report". in Intelligent Computer Aided Design. D.C. Brown, M.B.Waldron, H. Yoshikawa (Eds.) North-Holland 1992

An Integrated Transportation Planning Process for System Development Utilizing Geographic Information Systems (GIS)

Thomas Andriola[1]

Abstract

Geographic information systems (GIS) are commonly used in the evaluation of future transportation projects and the choice between alternatives. However, the potential utilization of GIS has yet to be fully realized. As the criteria for making investments in transportation infrastructure become more stringent the use of GIS must expand beyond the role of being strictly an analysis tool and provide a framework for making decisions.

This paper documents the application of GIS for a decision-support process in evaluating alternatives for the site of a magnetic levitation (maglev) system between Baltimore and Washington D.C. GIS functions were two-fold. First, GIS was used as an analytical tool to collect, organize, and analyze information for environmental impacts, right-of-way (ROW) takings, and other land use senstivities. These are the traditional uses. The second used GIS as a means to link analysis tools and create an integrated and iterative process for evaluating alternatives through information sharing.

Development of Evaluation Criteria

The evaluation of alternatives explored the use of existing rights-of-way (ROW) for siting the guideway and minimizing the amount of land required to construct a maglev system. Several factors were identified for developing and evaluating corridor alternatives. The objective was to rank alternatives that would offer the greatest potential to meet the goals of the system, which were stated in the ISTEA legislation. Table 1 contains the criteria set used.

---

[1] Transportation Planner, Frederic R. Harris, Inc., 4115 East Fowler Avenue, Tampa, FL, 33617.

Table 1 - Evaluation Criteria

- High Speed Operation
- Right-of-Way Utilization
- Intermodal Connections
- Ridership and Revenues
- Costs (capital and operating)

- Safety
- Environmental Impacts
- Energy Consumption
- Cost Effectiveness
- Joint Development Opportunities

Use of GIS

GIS was used in two ways. First, it was used to perform analysis of alignment location, curvature determination, environmental assessment, and ROW utilization. These are the traditional analyses for GIS which are widely understood and well-documented. Second, GIS was used as a means to link transportation-analysis tools together so they could take advantage of the analysis results in other study components and incorporate them into their own analysis. These tools received necessary data from GIS to perform other vital elements of the study, including maglev vehicle performance, ridership estimates, operational concepts, and life-cycle costs. Building these links created the opportunity to pass information freely between analysis tools and provided two major benefits.

The first benefit derived from using GIS was that it allowed for a greater number of alternatives to be analyzed. By automating much of the process the amount of time spent in setup was greatly reduced, which freed more time for analysis. The second benefit was the ability to integrate the study components. This allowed the analysis to consider the interdependencies between study components and share analysis results to identify the best alternatives. This is a current deficiency in many planning projects and will be discussed in greater detail in the discussion on the integrated planning approach.

Data Types

Several data types were used in the analysis of alternatives. Data was gathered from various sources and entered into the GIS. Common formats and data conversion programs were built to facilitate the transfer of data. Table 2 lists some of the different data sets used in the study.

Table 2 - Data Sets

- Wetlands and Floodplains
- Hazardous Waste Sites
- Scanned Topographic Map Images
- Land-Use and Socio-economic Data

- Parks and Wildlife Sanctuaries
- Historical Landmarks
- Right-of-Way Boundaries
- Hydrographic Features

The rest of this paper will be dedicated to explaining the analysis functions provided by GIS in determining the best alternatives and the concept of integrated transportation planning.

Guideway Alignments and Station Locations

Maglev guideway alignments were created using the Ground Transportation Analysis System (GTAS), a Macintosh-based GIS and transportation planning tool. GTAS created and modified alignments and stations using a combination of vector and raster data and algorithms for alignment curvatures, spiral lengths, and other maglev system requirements. Scanned United States Geological Survey (USGS) quadrangle maps provided information on location of highways, roads, railroads, buildings, park lands, and other features. Knowledge of their locations enabled alignments to be specified to minimize the relocation of existing structures. Digital Line Graph (DLG) files were used to verify highway and railroad locations and create ROW boundaries to examine ROW utilization. A digital terrain model, specifically the USGS Digital Elevation Model (DEM), was used to develop vertical profile information.

Environmental Impact Assessment

ARCINFO was used to establish an efficient and accurate method for assessing the impact of the proposed maglev system on various environmental features within the region. Many of the data sets listed in Table 2 were incoporated into ARCINFO for the environmental assessment. The various interrelationships among map features were analyzed and cross-tabulated. By using map overlay techniques, specific areas were calculated for each feature lying within the boundary of a sixty-foot wide maglev guideway corridor. The use of ARCINFO made this task much more efficient than traditional manual techniques and provided the opportunity to include data from GTAS and other planning tools.

The environmental screening was performed by generating buffers of the base centerline DLG data using average widths identified on existing maps and plans. By overlaying the ROW polygon for alignment corridors on the various environmental map layers, a "polygon clip" process was applied to calculate the areas of each feature that fell within each polygon boundary. The results of this analysis were compiled and tabulated so that comparisons of the range of impacts could be made.

Assessment of Rights-of-Way Usage

The assessment of existing ROW utilization and the need for additional ROW was a major thrust in the study. By digitizing existing ROW boundaries from paper maps, ROW plats and as-built plans, information was obtained to determine existing utilization and overall ROW requirements. Because this information was made readily available in electronic format the analysis also could be performed using GIS.

ROW boundaries were digitized and geographically corrected to the project coordinate base, NAD27 Maryland State Plane. This information was then used as a baseline for the analysis of various alignments. Alignments were imported into ARCINFO directly from GTAS to identify those portions that fell outside existing ROW boundaries. The data resulting from this analysis was used to identify usable portions of the ROW and determine additional ROW requirements for the cost analysis.

Maglev Vehicle Performance

Maglev operating speeds and corridor travel times were necessary for all alignments. Alignments were subjected to a rigorous analysis to examine system performance and its ability to compete with other modes in the corridor. Proposed alignments needed to generate sufficent ridership to warrant initial construction and future commercial operation. The vehicle performance model (VPM) within GTAS calculated performance results and expressed them in speed-vs-distance charts, such as is shown in Figure 1. Results were also exported to support analyses for environmental impacts, ROW utilization, and operating requirements.

Figure 1 - Vehicle Performance Model Output

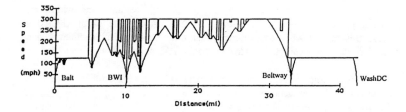

Performance analyses calculated vehicle speeds and travel times based on the system banking characteristics, passenger comfort standards, and vehicle acceleration characteristics. Conversely, speeds calculated by the VPM were used to generate designs for high-speed transition spirals necessary for maglev system operation.

Operations

A system operations plan was developed to determine operating requirements including fleet size, operating schedules, minimum headways, spare parts requirements, energy consumption, and staffing levels. These requirements were based on inputs provided from other study components including ridership estimates, operating speeds, travel times, and level-of-service requirements.

Two models were used in developing the operations plan: the Fleet Planning Model (FPM) and the Train Performance Calculator (TPC). FPM used ridership demand and system capacity to determine optimum sizes, staffing, and schedules. TPC used vehicle speeds (received from the VPM) and detailed knowledge of the propulsion system to calculate energy requirements and optimum operating speeds.

The Integrated Planning Approach

As stated earlier, an integrated planning approach was used to capture the interdependencies between of study components. As transportation systems development becomes more complex and the scrutiny over spending tax dollars increases the need for systematic techniques to evaluate alternatives becomes essential. In the case of a maglev system alternatives were multi-dimensional and needed to analyze all combinations of alignments, station locations, technologies, operating scenarios, ridership, and fare structures. This created many potential alternatives which by traditional methods could not have been done because of the time and effort requirements.

From the inception of the study, relationships between study components were identified. Significant improvements in productivity and accuracy were thought achievable by recognizing that relationships between study components could be iterative in nature (feedback loops) rather than simply causal. Figure 2 depicts the interrelationships present in the study.

By using GIS, data collection and preparation time was reduced through automation and the ability to create the necessary file formats. Since many of the analyses performed were GIS-based this allowed for the quick and easy transfer of data and provided the most current data to be available. Likewise, for those tools that were not GIS-based, it created file formats that could be directly read into or out of the tools to greatly facilitate analysis. This created a seamless transfer of data thanks to the compatible formats and the modern-day modem. This data processing structure is analagous to a client-server system in the business process environment. GIS acts as the server allowing many to access information; the specific analysis tools act as the clients and request information as needed.

Figure 2 - Interrelated Components

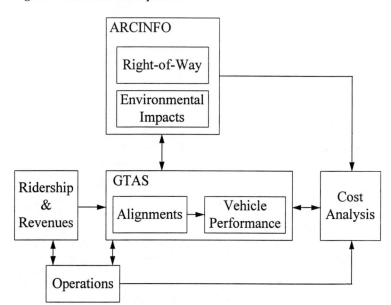

Summary

The Baltimore-Washington Maglev Study offered an opportunity to expand the role of GIS in the planning process. By recognizing that transportation systems development contains interdependencies an integrated and iterative approach was developed to assess alternatives. To date most system development efforts have separated the functions of design and analysis. However, this paper shows that it is possible to combine these functions by recognizing interrelationships, designing a process to exploit them, using today's advanced processing power to make them faster, and using technologies such as GIS to coordinate them.

The key to successful system development lies in integrating planning tools. Modern technology, such as GIS, exists to access information and tackle complex problems in an efficient and effective manner. By doing so better decisions can be made and tax dollars leveraged. For the Baltimore-Washington study the proposed approach made it posssible to evaluate several alternatives with a high degree of fidelity and eliminated much of the wasted time and effort of independent analysis.

References

"Baltimore-Washington Corridor Magnetic Levitation Feasibility Study: Executive Summary," prepared by KCI Technologies, Inc. et al., June 1994.

"Baltimore-Washington Corridor Magnetic Levitation Feasibility Study: Working Paper 8.1 - Evaluation Criteria," prepared by KCI Technologies, Inc. et al., June 1993.

"Baltimore-Washington Corridor Magnetic Levitation Feasibility Study: Working Paper 4.2 - Interim Operations Plan," prepared by KCI Technologies, Inc. et al., June 1993.

"Baltimore-Washington Corridor Magnetic Levitation Feasibility Study: Working Paper 3.1.5 - Market Analysis Ridership Projections," prepared by KCI Technologies, Inc. et al., June 1993.

"The Use of GIS in the Study of the Location and Feasibility of a Maglev System within the Baltimore-Washington Corridor," Thomas Andriola & Steven Beck, Fourth Annual TRB Transportation Planning Methods Applications Conference, Daytona Beach, FL, May 1993.

JOINING DYNAMICALLY SEGMENTED INFRASTRUCTURE DATA IN RELATIONAL DATABASE SYSTEMS

C.W. Schwartz[1] and N.L. Bentahar[2]

INTRODUCTION

The maintenance and rehabilitation of large civil infrastructure networks is increasingly planned and engineered with the aid of computer-based infrastructure management systems (IMS). A key IMS component is the database of infrastructure **attributes**--e.g., geometric and other inventory data, demand estimates, condition indices, etc. For linear infrastructure networks such as roads, rail lines, sewers, and water and gas pipelines these attributes are most naturally associated with network **segments**--e.g., the grade of a sanitary sewer between two stations, the capacity of a water main between two valve locations, or the average roughness of an arterial between two milepoints. Definition of a segmentation scheme is a fundamental issue that must be addressed early during IMS development because of its important implications for how the associated attribute data will be stored and accessed in the IMS database. Since different categories of data--e.g., inventory vs. demand vs. condition--will generally vary in dissimilar ways over the length of the network, different segmentations may be appropriate for the different data categories.

Unfortunately, not all IMS segmentation schemes are fully compatible with "standard" relational database management systems (RDBMS), the most common vehicle for IMS implementations. However, in order for these operations to work successfully for segmented IMS data, the data tables being joined must have *identical* segmentation. This requirement is not satisfied by **dynamic segmentation**, a comparatively new but increasingly common scheme for segmenting IMS data.

The difficulties of joining dynamically segmented data tables becomes even more severe in distributed client/server database environments, where all database queries must usually be formulated using SQL (Structured Query Language). The logical and mathematical constructs in "standard" SQL are very limited and not well suited to the complex operations required to join dynamically segmented infrastructure data.

Our purpose in this paper is to describe a new procedure for joining dynamically segmented infrastructure data that is compatible with SQL-based client/server database architectures. We begin with a brief review of the segmentation issues and the basic principles

---

[1]Associate Professor of Civil Engineering, University of Maryland, College Park, MD 20742

[2]Senior Systems Engineer, PCS/Law Engineering, Beltsville, MD 20705

of join operations in RDBMS. We conclude with a description of a new **overlap inner join** algorithm, which is our generalization of the standard inner join operation to the case of dynamically segmented infrastructure data.

DYNAMIC SEGMENTATION

For linear infrastructure networks, data attributes are most naturally associated with network segments. This is best illustrated using an example. We will consider a single road with location references defined by milepoints (this can be easily generalized to a road network, pipelines, etc.). In order to keep the example as simple as possible, we will consider only two attribute categories and only one attribute within each category: inventory data, with pavement surface type (AC/PCC) as the attribute; and condition data, with PSI (Present Serviceability Index, scale of 0 to 5) as the attribute. The hypothetical variations of the data attributes along the five mile length of the example road are summarized in Table 1. Note that there is a gap in the PSI condition data between milepoints 3.0 and 3.5; incomplete data is the rule rather than the exception in the real world.

Attribute	Beginning Milepoint	Ending Milepoint	Attribute Value
Surface Type	0.0	2.0	AC
	2.0	3.0	PCC
	3.0	5.0	AC
PSI	0.0	1.5	3.0
	1.5	2.0	3.8
	2.0	3.0	2.3
	3.5	5.0	3.6

Table 1. Values of Attributes Along Example Roadway

The simplest segmentation scheme is **fixed segmentation**. In fixed segmentation, each road is divided *a priori* into a set of segments. The segments along a road may all have the same length (e.g., a tenth of a mile) or they may have different lengths (e.g., the distance between adjacent intersections). The essential feature, however, is that the segments are the same for *all* data attributes. The values of the attributes associated with each segment represent average (or predominant or typical) values for that segment--e.g., length-weighted average PSI for the segment. Thus, to some extent the segmentation influences the attribute values.

In our simple example problem, we will use 1 mile long fixed segments and two attribute tables to organize the road data. Segments can be conveniently identified by sequential numbers (SegID), as defined in Table 2; note that in the fixed segmentation scheme there are no overlaps among segments. The data tables as constructed using the fixed segmentation scheme are summarized in Table 3; note that both tables are segmented identically. There is some redundancy in the inventory table (e.g., the first two rows in Table 3a) because the fixed segmentation scheme requires identical segmentation for *all* data tables regardless of whether the fixed segment breaks are necessary for any individual data table. There are also some complications in the condition table: the PSI value for segment 2 is an average value since PSI changes within the segment, and there is no row corresponding to segment 4 because PSI is unknown for a portion of this segment.

SegID	BeginMP	EndMP
1	0.0	1.0
2	1.0	2.0
3	2.0	3.0
4	3.0	4.0
5	4.0	5.0

Table 2. Segment List for Fixed Segmentation

SegID	SurfType		SegID	PSI
1	AC		1	3.0
2	AC		2	3.4
3	PCC		3	2.3
4	AC		5	3.6
5	AC			

(a) Inventory   *(c) Condition*

Table 3. Data Tables for Fixed Segmentation

A more elegant alternative to fixed segmentation is **dynamic segmentation** (Nyerges, 1990). The road is no longer divided into segments *a priori*; instead, the segments are defined by the attributes--or, more specifically, segment *breaks* are defined by *changes* in attribute values. A change in the surface type, for example, defines a boundary between two inventory segments. The essential feature here is that the segments are defined by the attribute values (or changes in these values) rather than predefined simply by milepoint coordinates. An important consequence of this is that the road segmentation will generally be different for different attributes (e.g., inventory and condition). In short, the attribute values control the segmentation.

The segments can again be identified by sequential numbers (SegID), as defined in Table 4. Each data table has its own distinct segmentation. Segments may now overlap (e.g., SegID 1 and 4) and the same segment may appear in multiple data tables (SegID 2) although this will generally be the exception rather than the rule. The data tables constructed using the dynamic segmentation scheme are summarized in Table 5. Note that there is no longer any redundancy within the tables because of the table-specific segmentation. In addition, there no longer is any need to compute an average PSI value over any segment, since condition segment breaks are defined any time PSI changes.

SegID	BeginMP	EndMP
1	0.0	2.0 (inventory segment)
2	2.0	3.0 (inventory and condition segment)
3	3.0	5.0 (inventory segment)
4	0.0	1.5 (condition segment)
5	1.5	2.0 " "
6	3.5	5.0 " "

Table 4. Segment List for Dynamic Segmentation

SegID	SurfType		SegID	PSI
1	AC		4	3.0
2	PCC		5	3.8
3	AC		2	2.3
			6	3.6

(a) Inventory   *(c) Condition*

Table 5. Data Tables for Dynamic Segmentation

Although this example database is very simple and small, it still illustrates some of the advantages of dynamic segmentation over fixed segmentation. First, dynamic segmentation eliminates the problem of defining an "average" or "predominant" value for a segment attribute. The actual values of the attributes are associated with each segment, and changes in these values define segment breaks. Second, dynamic segmentation represents a more "natural" structure for the attribute data. The segmentation is governed by the data, which is the item of inherent interest, rather than by some arbitrary *a priori* subdivision procedure. Third, dynamic segmentation is more compatible with many automated field collection methods that record continuous streams of attribute data (e.g,, roughness); these streams of data can be dynamically segmented "on the fly". And finally, dynamic segmentation eliminates the linkage between database structure and specific applications that is inherent in fixed segmentation. This is increasingly important as infrastructure databases become more centralized and shared among multiple applications. The optimal fixed segmentation for a pavement management system, for example, will not necessarily be the same as the optimal fixed segmentation for a traffic planning model. Dynamic segmentation in this case represents a "neutral" scheme; each application can take the dynamically segmented data and resegment it as most appropriate for its own purposes.

CONVENTIONAL TABLE JOINS

Table joins are fundamental operations in all RDBMS (Date, 1995). The basic requirement is that the tables being joined contain a common column (attribute). In an inner join operation (the only case considered here), rows having the same attribute value in the common column are combined to form a composite table, the output from the join operation.

The conventional RDBMS inner join operation is completely compatible with the fixed segmentation scheme for linear infrastructure data. In our simple road example, the inventory and condition data tables (Table 3) all contain a SegID column for the segment identifier, so this becomes the common column needed for the join operation. The row having a SegID value of "3" in the inventory table corresponds--i.e., is physically congruent--to the segment having a SegID value of "3" in the condition table; the attributes in these two rows can therefore be combined to form the corresponding row in the composite inventory-condition table shown in Table 6. All rows having the same SegID value correspond to the same physical length of road, and thus the table join operation makes physical sense. Note that in Table 6 there is no row for segment 4, because segment 4 is not present in the condition table and thus there is no match.

SegID	SurfType	PSI
1	AC	3.0
2	AC	3.4
3	PCC	2.3
5	AC	3.6

Table 6.   Composite Data Table for Fixed Segmentation
(Inner Join)

In dynamic segmentation, the conventional inner join operation unfortunately breaks down. If we attempt to join the dynamically segmented data tables in Table 5, the result will be the single row shown in Table 7. Although both input tables still share a common SegID column, there is only one row pair in the two tables that has matching SegID values. In physical terms, only one of the pairs of inventory and condition segments in Table 5 represents *congruent* lengths of road. Even though none of the other segment pairs in Table 5 are congruent, nonetheless several of these other segment pairs do *overlap*, and for these segments we should theoretically be able to determine the composite inventory and condition attributes along the overlapped common length. However, since these pairs do not represent identical segments and consequently have differend SegID values, they cannot be combined using the conventional join operation. Our desired database operation therefore becomes the more complicated problem of determining the overlaps between segments and the corresponding multi-table attribute values along these overlaps.

SegID	SurfType	PSI
2	PCC	2.3

Table 7.   Composite Data Table for Dynamic Segmentation
(Inner Join)

The overlaps between dynamically segmented linear infrastructure data cannot be determined easily using "standard" RDBMS manipulations. It becomes even more difficult in state-of-the-art client/server database architectures, which increasingly are the target environment for infrastructure management systems (Ullman, 1993). In the client/server architecture, the engineering application (e.g., a road management system) executes on a local client workstation (e.g., a PC) and issues queries (i.e., requests for data) across a high speed communications network to a remote specialized database server (e.g., another PC). The database server receives the query, efficiently extracts the appropriate data from its associated database tables, and then sends the requested data back over the network to the client. The principle advantages of the client/server database architecture are: (a) fast query processing, since the database server is dedicated to this single task and thus can be highly optimized; and (b) reduced network traffic, since only data satisfying the query (as opposed to all of the data in the tables) is transmitted from the server to the client.

In the client/server database environment, queries are formulated using the "neutral" SQL query language (Date and Darwen, 1994). Unfortunately, although SQL is adequate for most conventional RDBMS requests, including the conventional table join operations, standard SQL has inadequate logical and mathematical manipulation capabilities for directly determining the overlaps between dynamically segmented data. Implementations of dynamically segmented data tables in client/server database environments thus must forfeit the main advantages of the database server. Instead of relying upon the database server for the bulk of the query processing, the individual data tables are simply transmitted back to the client workstation, where the dynamic segmentation operations are performed locally. The optimized query processing capabilities of the database server are bypassed and network traffic is greatly increased.

THE OVERLAP JOIN

To eliminate the inherent incompatibility between dynamic segmentation and RDBMS capabilities, particularly as implemented in a client/server database architecture, we have developed an algorithm that determines whether two segments from two different data tables overlap, and if so, extracts the beginning and ending length coordinates of the overlap and the multi-table attributes along the overlap segment. We term this algorithm the *overlap join* operation. There are two variations: the *inner overlap join*, which produces only the overlapping segments; and the *outer overlap join*, which produces the overlapping segments plus any nonoverlapping segments that occur in either of the tables. Because of space limitations, we will only present the simpler inner overlap join here.

The overlap join problem can be described in terms of six fundamental cases, as illustrated in Figure 1. Pairs of segments consisting of one segment each from the two tables being joined are evaluated against these six cases in sequence. Cases 1 and 2 represent the two nonoverlapping conditions, while cases 3 through 6 represent all possible overlap conditions. The formal definition for each case and the corresponding beginning and ending length coordinates for the overlap segment are also summarized in Figure 1. Note that the overlap definitions are specified very carefully to eliminate certain overlap conditions from satisfying more than one case. For example, congruent segments (i.e., B1=B2 and E1=E2) could easily be interpreted as a valid special condition of any of cases 3 through 6. However, close examination of the case definitions in Figure 1 show that congruent segments satisfy case 3, are explicitly excluded from case 4 by the 'AND NOT' clause, and implicitly excluded from cases 5 and 6 through careful use of the '<' rather than '<=' operator. It is important that each segment pair be evaluated against the six cases in the sequence shown to ensure that all special conditions are treated properly.

The ease with which the logic embodied in Figure 1 can be implemented in a RDBMS query depends directly upon the logical and mathematical capabilities available in each specific RDBMS. For example, Microsoft Access has a rich set of these capabilities, including the ability to incorporate user-defined Access Basic function procedures in database queries. Using these function procedures, the overlap join algorithm can be directly and easily incorporated into the standard query framework.

Implementation of the overlap join algorithm becomes more difficult in a distributed client/server database environment. There is no direct way to perform the overlap join operation using a single SQL query comparable to that for the conventional join operation. However, the logic embedded in the six cases in Figure 1 can be implemented as a *series* of SQL queries. Each of these queries can be processed directly on the database server with intermediate results stored as temporary data tables on the server. Only the final composite table generated at the end of the SQL query series is returned over the network to the requesting client workstation. The principle advantages of the client/server database architecture are maintained: query processing is performed efficiently on the optimized database server, and only data satisfying the query are transmitted over the network from the server to the client.

In order to perform the overlap join using standard SQL statements, it is more convenient to define the dynamic segments in terms of their beginning and ending length coordinates rather than in terms of an arbitrary segment identifier. The SegID segment identifier in our example problem can be replaced by beginning and ending milepoints either by performing a conventional inner join between the segment list table (Table 4) and the attribute tables (Table 5) or by simply defining the attribute tables in terms of beginning and ending milepoints from the start. In either case, the revised inventory (INV) and condition (COND) attribute tables have the form given in Table 8.

BeginMP	EndMP	SurfType
0.0	2.0	AC
2.0	3.0	PCC
3.0	5.0	AC

BeginMP	EndMP	PSI
0.0	1.5	3.0
1.5	2.0	3.8
2.0	3.0	2.3
3.5	5.0	3.6

(a) Inventory (INV)   *(c) Condition (COND)*

Table 8. Revised Data Tables for Dynamic Segmentation

The SQL statements required to join the milepoint-attributed dynamically segmented inventory and condition data tables (Table 8) for our example road problem are listed in Figure 2. The code in Figure 2 is written in the Watcom SQL (Powersoft, 1994), which generally conforms to the SQL/89 "lowest common denominator" standard for current SQL dialects. Section {1} of the algorithm simply creates a temporary table *temp* that is used to store intermediate results during processing; columns in this table contain the two sets of beginning and ending milepoints for each pair of overlapping inventory and condition segments plus the corresponding inventory and condition attributes. Section {2} creates a data table *join2* used to store the final results from the overlap join operation; columns in this table contain the beginning and ending milepoints for the overlap and the corresponding inventory and condition attributes. Section {3} finds all pairs of segments that overlap and stores their milepoints and attributes in the previously-defined *temp* table. The 'where' clause in section {3} automatically excludes all nonoverlapping segments, thereby automatically incorporating cases 1 and 2 of Figure 1 and simultaneously minimizing the size of the temporary table. Sections {4} through {7} sequentially execute the logic in the overlap cases 3 through 6 of Figure 2. Each overlapping segment satisfies the 'where' clause of one and only one of sections {4} through {7}; the appropriate beginning and ending milepoint values for the overlap are then extracted from the *temp* table along with the corresponding inventory and condition attributes and used to create a new segment (row) in the output *join2* table.

The results from the overlap join operation for our example problem are summarized in Table 9. All physical lengths of road for which both inventory and condition attributes are known are included in this table, and the attribute values represent the actual values along each segment. As before, there is no segment in Table 9 for the length of road between milepoints 3.0 and 3.5 because the PSI value is unknown. An outer overlap join algorithm would include this segment in the table with a null value for the unknown PSI, but discussion of this algorithm is beyond the scope of this paper.

BeginMP	EndMP	SurfType	PSI
0.0	1.5	AC	3.0
1.5	2.0	AC	3.8
2.0	3.0	PCC	2.3
3.5	5.0	AC	3.6

Table 9. *Results from Inner Overlap Join Operation for Example Problem*

It is important to note that the overlap join algorithm described in Figure 2 is fully compatible with the underlying philosophy of the client/server database architecture. All of the operations are performed on the database server, and only the records satisfying the overlap join operation are stored in the *join2* output table for transfer back over the network to the requesting client application (the *join2* table could be dropped after the transfer, if desired). Network traffic is thus minimized. The computational load for the database server is increased for the overlap join operation as compared to a conventional join, but this is to be expected given the much greater complexity of logic required. The bulk of the time for the overlap join operation is consumed in creating the initial *temp* table of overlapping segment pairs; the amount of time required for this operation depends principally on the ultimate size of this table.

SUMMARY

Dynamic segmentation of attribute data for linear infrastructure networks is an attractive and increasingly common approach for organizing databases in infrastructure management systems (IMS). To our knowledge, no one has previously developed an algorithm for joining dynamically segmented data tables that is syntactically and philosophically compatible with SQL-compliant client/server database architectures. These types of manipulations are increasingly vital in state-of-the-art IMS and computing environments. The technique presented in this paper for the inner overlap join operation provides a valuable tool for developers of infrastructure management systems. This and other similar techniques such as the outer overlap join are also relevant for geographic information systems, which share many of the same segmentation and data access problems.

ACKNOWLEDGMENTS

Many of the ideas in this paper were formulated during the development of a statewide pavement management system for the Delaware Department of Transportation. The contributions of Mr. Eugene Abbott and Mr. David Matsen, Director and Assistant Director of Planning, are gratefully acknowledged.

REFERENCES

Date, C.J. (1995) *An Introduction to Database Systems (6th Ed.)*, Addison-Wesley, New York, NY.

Date, C.J., and Darwen, H. (1994) *A Guide to the SQL Standard (3rd Ed.)*, Addison-Wesley, New York, NY.

Nyerges, T.L. (1990) "Location Referencing and Highway Segmentation in a Geographic Information System," *ITE Journal*, March, pp. 27-31.

Powersoft (1994) *Watcom SQL Version 3.0 Reference Manual*, Powersoft Corp., Burlington, MA.

Ullman, E. (1993) "Client/Server Frees Data," *Byte*, Vol. 18, No. 7, June, pp. 96-106.

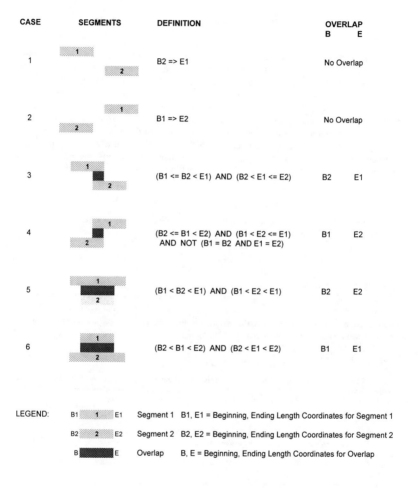

CASE	SEGMENTS	DEFINITION	OVERLAP B E
1		B2 => E1	No Overlap
2		B1 => E2	No Overlap
3		(B1 <= B2 < E1) AND (B2 < E1 <= E2)	B2 E1
4		(B2 <= B1 < E2) AND (B1 < E2 <= E1) AND NOT (B1 = B2 AND E1 = E2)	B1 E2
5		(B1 < B2 < E1) AND (B1 < E2 < E1)	B2 E2
6		(B2 < B1 < E2) AND (B2 < E1 < E2)	B1 E1

LEGEND:
B1 ▨1▨ E1   Segment 1   B1, E1 = Beginning, Ending Length Coordinates for Segment 1
B2 ▨2▨ E2   Segment 2   B2, E2 = Beginning, Ending Length Coordinates for Segment 2
B ▬▬ E   Overlap   B, E = Beginning, Ending Length Coordinates for Overlap

Figure 1. Six Cases for Segment Overlap Problem

*Note** SQL Code:*

```
{1}   create table temp (beginmp1 numeric(3,1), endmp1 numeric(3,1),
                          beginmp2 numeric(3,1), endmp2 numeric(3,1),
                          surftype char(3), psi numeric(3,1));

{2}   create table join2 (beginmp numeric(3,1), endmp numeric(3,1),
                           surftype char(3), psi numeric(3,1));

{3}   insert into temp
         select inv.beginmp, inv.endmp, cond.beginmp, cond.endmp,
                inv.surftype, cond.psi
         from inv, cond
         where not(cond.beginmp >= inv.endmp or
                   inv.beginmp >= cond.endmp);

{4}   insert into join2
         select beginmp2, endmp1, surftype, psi
         from temp
         where (beginmp1 <= beginmp2 and beginmp2 < endmp1) and
               (beginmp2 < endmp1 and endmp1 <= endmp2);

{5}   insert into join2
         select beginmp1, endmp2, surftype, psi
         from temp
         where (beginmp2 <= beginmp1 and beginmp1 < endmp2) and
               (beginmp1 < endmp2 and endmp2 <= endmp1) and
               not (beginmp1 = beginmp2 and endmp1 = endmp2);

{6}   insert into join2
         select beginmp2, endmp2, surftype, psi
         from temp
         where (beginmp1 < beginmp2 and beginmp2 < endmp1) and
               (beginmp1 < endmp2 and endmp2 < endmp1);

{7}   insert into join2
         select beginmp1, endmp1, surftype, psi
         from temp
         where (beginmp2 < beginmp1 and beginmp1 < endmp2) and
               (beginmp2 < endmp1 and endmp1 < endmp2);

{8}   drop table temp;
```

*See text for descriptions of each section of code.

Figure 2. SQL Statements for Inner Overlap Join for Example Problem

Routing Applications Developed in a GIS Environment

by

Carl E. Kurt[1]
Li Qiang[2]
Yanning Zhu[2]

Introduction

There are many commercial desktop mapping software packages available to engineers and planners. In addition, there are a few very sophisticated, and expensive, GIS software packages available which have routing applications built into them. These built in routing applications are rather difficult to use and may not be available to the casual user. This paper describes the procedures used to develop a series of routing functions. In addition, several applications built from these routing functions will be described.

Basic Information and Limitations

Most desktop mapping and GIS platforms contain the information required to develop a routing application. This information includes line type and geometry (x and y coordinates at the start and end of each street segment, segment length, and other data information). For polyline type segments, x and y coordinates of all intermediate, or shape, points can usually be found. Data, such as average travel speed and one way street information, can be attached to each street segment. This additional information is very useful when the user wants to optimize a "least cost" function based on this data attached to a segment. This functionality permits the user to find the shortest time or least cost, based on distance travelled and time to travel that distance.

---

[1]Professor, Department of Civil Engineering, University of Kansas, Lawrence, KS 66045

[2]Research Assistant, Kansas University Transportation Center, Lawrence, KS 66045

Most platforms, which do not support routing applications, lack the capability to link adjacent street segments together. To provide this functionality, the software must assign to each map segment a starting node ID and an ending node ID. Through queries, the software can determine that a certain group of streets have a common name, but it does not know which segments are adjacent to any given segment. This connectivity information is critical when developing a network routing application.

Another crucial ingredient is a development environment. In the microcomputer/Windows environment, some of the packages take direct advantage of Visual Basic or Visual C++ development environments. Other packages have developed their own development environments.

The method for developing the development environment is not as critical as the capabilities provided the user or developer. It is imperative that the user be able to interrogate the map database. For example, for each segment in the map database, the user must find the coordinates of the start and end points of the line. If the software automatically has the functionality to directly give the user segment length, it is very convenient. If not, the user will need to calculate segment length from node geometry.

With this background, the development of a network analysis capability requires the user to convert an electronic map with unconnected street, or line, segments to an electronic map with unique nodes and links. Graphically, one can see in Figure 1, the consequences of this action.

One of the software packages that meets the criteria discussed is MapInfo. It has a development environment, MapBASIC, street maps readily available and functionality to connect to other development environments through the support of Dynamic Data Exchange (DDE) and Dynamic Link Libraries(DLL).

Development of Link, Node and Routing Tables

As previously discussed, most desktop mapping software packages have much of the information stored in the map database. However, what they lack is the information required to define connectivity between the individual map segments.

After considerable investigation, the authors decided to create three separate tables using information found in the map databases and additional information developed with the software. The first table is the link table. Because street segments are stored in a fixed order, their assigned link ID number is the order stored in the map database. Also, the start and end points of each link was assigned a unique Node ID. Finally, the barrier status of the street was stored. One way street information and other street data is also stored in this table. The support of two different types of street barriers will be discussed later.

The assignment of node IDs is a two step process. In the first pass, node IDs are assigned in numerical order at the ends of each link. Obviously, there are several node IDs at the intersection of two or more links. In the second pass, all duplicate nodes are eliminated.

In the Node table, some very critical information is stored. For each node, results of the network analysis are stored. The optimum cost function (or length) from each node to the source node, the next adjacent link, and node ID on the optimum path are stored.

The last table created is the Route Table. Because the user has no, or little, control over the street map, there is a critical need to keep data about how the network is arranged. Theoretically, it is possible for an infinite number of links to be attached to a given node. If this concept was supported, the required route table size is n x n for a network with n nodes. However, for most transportation networks, each node is connected to a finite number of links. The maximum number of links was set at eight which significantly minimized the memory requirements for large networks.

Street Barriers

Two different barriers are found in practice. The first type of barrier is a link barrier. This barrier type is representative of a complete road reconstruction. That is, the entire road is closed. A second barrier type is a point barrier. This type of barrier could be represented as a pipe line construction through a road. Unless the road is a one way road, users may enter the link from either end but they can not go directly from one end to the other. This barrier information is stored in a barrier table. In this table, the link ID, type of barrier and location, if required, of the point barrier is stored.

One and Two Way Streets

In practice, one and two way streets occur in network analysis. Each segment has a direction when drawn in the electronic map. If the direction of travel is in the direction of the segment, a code is assigned to the link. Different codes are provided when the travel direction is in the other direction and for two way streets. This information is stored in the link table.

"Least Cost" Criteria

The simplest and classical function used to conduct a network analysis is distance. However, if one expands the definition of least distance to least cost, a broader range of applications can be conducted. For example, if average speed is added to a map database, one can calculate time by dividing the segment length by the average speed. Then, the network analysis could work on finding least time, rather than least distance.

This least cost function could also be a more complex function. For example, transportation costs could be separated into a distance component and a time component. With the least cost concept, the network could be evaluated based on the total transportation costs.

Network Analysis

Once the network files, link, node and routing tables are generated and street barrier and one/two way street information updated, it is time to conduct the network analysis. The network analysis was conducted in two environments. In both cases, the Dijkstra (1959) Method was selected to find the shortest path from every node in the network to the source node. The first time through, the Dijkstra Method was programmed in MapBasic. While this was relatively easy, we found the execution of the routine to be very slow.

Taking advantage of the DLL capabilities built into MapBasic and Windows, the network analysis was rewritten in Visual C++ and compiled as a DLL. Values of critical network information were passed to the Analysis DLL through arrays. This approach greatly improved the performance of the entire package.

Several different networks were evaluated to compare performance standards. One network with approximately 3,750 links took approximately 13 hours to analyze with the MapBasic version. It took approximately 6 minutes in the first DLL version. Subsequent improvements in software development have reduced the time to analyze this network of 3,750 links to 3 seconds. Thus, moving to the C++ environment was a major step to improve performance.

Network Based Applications

With the basic network connectivity information and an efficient analysis capability developed, it was time to develop specific applications. The first application calculates and draws the shortest path between a destination point and the source node. Three methods for defining the destination point are supported. They are: typing in a street address, pointing to a selected node, or pointing to any point close to a network link. In each case, the software snaps to the closest link in the network. With the data found in the link and node tables, the software can calculate the shortest path from the destination point to the source node.

A typical example is shown in Figure 1. While it may look like this shortest path is not correct, it really is. Since the street in front of the destination point is a one way street in the wrong direction, it is not possible to take the "shortest" path.

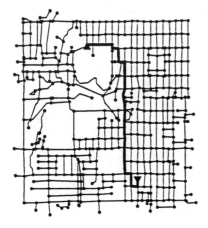

Street	Cost	Length
11th St	0.251	0.251
Mississippi St	0.035	0.035
11th St	0.059	0.059
Indiana St	0.124	0.124
W 12th St	0.061	0.061
Louisiana St	0.256	0.256
Alumni Pl	0.079	0.079
Louisiana St	0.735	0.735
W 21st St	0.059	0.059
Ohio St	0.117	0.117
W 22nd St	0.065	0.065
Tennessee St	0.063	0.063
TOTAL COST	1.904	1.904

Shortest Path From Source to Destination

▲Source
▼Destination

Figure 1. Shortest Path

As part of this option, the user is provided a map of the shortest path. One should also note that a table is provided to the user. This routing table permits the user to have access to the street names, costs assigned to the street and travel distance.

Another classical problem using network analysis is the travelling salesman problem. In this problem, the user is asked to travel through one or more points while returning to the start point. To use this approach, it is necessary to calculate the path and least cost or distance from each pick up point to all other pick up points. Since one way streets are supported, the least cost from point A to B will not necessarily be the least cost from point B to A. After all combinations of least costs and paths are calculated, the nearest neighbor method was chosen to determine the final path (Lawler, et al - 1985).

A typical example of the shortest path solution for 5 points is shown in Figure 2. Because of the variances found in the nearest neighbor method, we have found that the "least cost" path found has been a function of the start and end node. While this is somewhat bothersome, we have found that while the two paths may look very dissimilar, the total cost between the two options is very similar (in the 1-2% range).

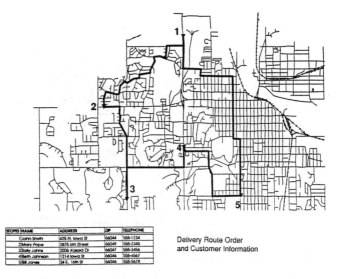

STOP ID	NAME	ADDRESS	ZIP	TELEPHONE
1	John Smith	425 N. Iowa St	66044	555-1234
2	Mary Pope	3875 6th Street	66049	556-2345
3	Sally Johns	2006 Kasold Dr	66047	555-3456
4	Beth Johnson	1214 Iowa St	66046	555-4567
5	Bill Jones	34 E. 16th St	66046	555-5678

Delivery Route Order
and Customer Information

Figure 2. Traveling Salesman Problem Connecting Five Stops

A take off of the classical travelling salesman problem is a class of problems similar to routing paratransit applications. In these cases, the unit may start at a depot, pick up one or more clients and take them to the first drop off point. Then, there may be a requirement to pick up several other clients and drop them off to a second drop off point. This procedure could be repeated for several hours, or a day. Now, what is the most efficient path to pick up these clients and drop them off at the appropriate site? While the authors have not developed a working algorithm for this problem, they have done enough research into the solution to know that by modifying the travelling salesman solution, it is possible to calculate the optimum path and to display this path on a map.

To date, classical examples of network analysis have been demonstrated. However, what other applications can be developed with this basic information? One such example is a path dependent buffer function. For many site selection and health care analyze, decisions are made based on the data within a circle drawn a certain distance about the site. The authors asked a different question. What would be the coverage area if the distance was measured along a path? A typical example is shown in Figure 3 where the criteria was set at 0.75 miles. As you may note, the geometry looks a little lop sided. Again, this is the influence of one way streets.

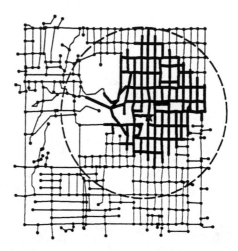

Figure 3. Comparison Circle vs. Path Dependent Buffer Criteria (Note Effect of One Way Streets)

In this case a boundary was developed by connecting points that just met the criteria. In this case, the area of this boundary was approximately 40% of the circle area.

Since MapBasic permits the user to interrogate the entire map database, you may notice at the edge of the area within the coverage area that many of segments lie only partially within the criteria. For those segments where one node is within the criteria and one node is outside the criteria, the software goes to the segment geometry and calculates the point that just meets the criteria.

What about the area enclosed by the street segments within the criteria? While each street configuration is unique, areas within the selection criteria of a path dependent buffer are many times only one-half the area of a circle with the same criteria. This is especially true when streets are not layed out in a regular pattern or when natural barriers are present. The authors have seen cases where patients are assigned to the closest hospital using a straight line criteria when, in practice, they must drive three times farther and right past an alternate hospital.

The last example is the capability to develop maps based on street data, such as maps outlining carrier routes. Since each street in the United States is assigned a specific carrier route, a query is made selecting all streets for a given carrier route. After a network is generated from this query, an algorithm was developed to find all segments on the edge of the network. A sample of this feature is shown in Figure 4.

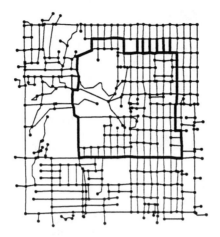

Figure 4. Typical Carrier Route Boundary

Summary and Conclusions

The case has been made for developing a network analysis capability using GIS and desktop mapping software which does not directly support network analysis. In the windows environment, the case has been made to make use of DDE capability and the DLL feature in Windows. Significant improvements in performance and capabilities were observed.

Once the basic network tables are generated, applications do not necessarily have to be transportation related. For example, the generation of areas for site selection and health care management are easy to develop. Also the generation of specific boundaries for cases such as carrier route boundary files is possible. As the user base for these new tools becomes more sophisticated and experienced, new and other interesting applications can be supported.

Appendix

Dijkstra, E. (1959), "A note on two problems in connexion with graphs". Numeriche Mathematics **1**, 269-271.

Lawler, E. L., Lenstra, J. K., Rinnooy Kan, A. H. G. and D. B. Shmonys (1985). *The Traveling Salesman Problem*. John Wiley & Sons, New York, NY.

Design Challenges For Contractors' Integrated Management Systems

Alan Russell, Member, ASCE[1], and Thomas Froese, Member, ASCE[2]

Abstract

This paper examines integrated computer systems for the construction industry, particularly medium-sized contractors. It addresses the practical organizational and computing context in which such systems will be used, discusses the potential benefits and barriers facing such systems, and outlines some central challenges and opportunities facing the research community in this area.

Introduction

"Integration" has become a watchword for much current research on computer tools for the construction industry. Motivated by substantial potential benefits of information exchange among sophisticated computer-aided design and management software, many solutions are offered by researchers and systems developers. While differing in approach, all involve pervasive computer use throughout the construction project life cycle. In contrast, the construction industry itself remains slow in embracing even well-established computer tools. In particular, medium-sized contractors, who perform the majority of building construction work, do not seem ready to adopt high-end computer integrated management systems. This paper examines the construction and project views context in which such systems must operate, discusses the potential benefits and impediments of integrated systems, and outlines design challenges involved in creating practical integrated systems for medium-sized contractors.

Our focus is on firms that construct projects which range from a few million to tens of millions of dollars—the vast majority of building and civil engineering projects and firms. These firms, which we refer to as *medium-sized contractors*, generally do not offer design services nor possess the resources and expertise required of the sophisticated systems used by EPC firms. The way in which these firms staff and run projects and internally manage project information services has important implications for the design of management systems.

Figure 1 identifies some of the distinct views that describe a project, all of which are united by the actual project itself. In this paper, we discuss the design of computer systems that address process views (how, who, when, where) and as-built views (what

---

[1] Professor, adr@civil.ubc.ca; [2]Assistant Professor, tfroese@civil.ubc.ca;
Dept. of Civil Eng., Univ. of British Columbia, Vancouver, B.C., Canada, V6T 1Z4

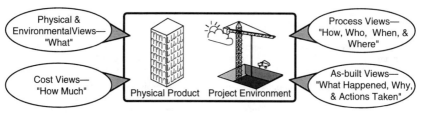

Figure 1. Project Views

happened, why and actions taken), although we recognize the areas of overlap with other views.

Integrated systems in this paper refer to systems that offer one-stop shopping through a single software system or, more likely, through sets of separate application programs which share common process models and common databases. Thus, we differentiate between *integrated systems* and *communication among systems*. For the latter, we mean mechanisms for exchanging data among different programs which may or may not share common databases, interfaces, product models, or project views.

The Construction Context

An assessment of the potential for integrated systems for medium-sized contractors must be founded in an understanding of the industry and working environment in which they perform. There are several characteristics of this environment that place unusually difficult constraints on the feasibility and practicality of integrated computing systems. While the physical environment itself is challenging, the organizational and human resource environments provide the most significant challenges. Like other large business ventures, construction projects are carried out by large and complex organizations (hundreds of individuals). Unlike many other ventures, however, this large organization is made up of many much smaller companies (several dozen) whose involvement in the project changes rapidly over the relatively short project duration (one to three years). The responsibility assignments, coordination interfaces, and communication paths among the companies are based on long-standing traditional roles. But these roles are often difficult to express in the form of standard procedures and formats that are acceptable to most firms and industry personnel, and hence are difficult to capture in computer applications that have broad appeal. Also, the cut-throat nature of construction leads to low profit margins and, correspondingly, to small management teams, low investment in computing and other forms of management support tools (it is still common, for example, to have *no* computers in construction site offices), and little investment in training. Additionally, the typical background of contractors' project management personnel is from field activities, and computing expertise, if it exists, resides generally with the more junior personnel.

To illustrate the typical situation we have encountered with medium-sized contractors (also see Jagbeck, 1994), general contractor site staff typically consist of a project manager, a site superintendent, maybe a foreman, a safety officer, a site secretary, and a project engineer or coordinator. Additional support is offered through head office dealing with estimating, cost accounting, client relations, and so forth. [Cost accounting, part of the cost views of a project and a head office function, is the one computerized project view that is used and enforced without fail within a firm, and is generally adequately staffed. It also has the advantages of the existence of standards (some government imposed) and relatively comprehensive software systems, which cannot be said for any other computerized view.] Nevertheless, most site offices are set up to be as self sufficient as possible, with as many support services as possible being charged directly to the project rather than to general head office overhead. When computers are used on site, invariably they are not linked using a LAN nor are they linked with head office using a WAN or even a modem. This exacerbates the slow turn-around time of job costing information and feedback to project personnel.

The most prevalent computer tools used directly by project personnel at the site or in head office are word processors, spreadsheets, and to a lesser extent, database and CAD software. The value of these tools lies in their relative ease of use, advances in integration and common interfaces, and most of all, the flexibility they offer to the user in crafting procedures and formats that respond to the project manager's view of how he or she wants things done.

In terms of project management systems used—most of which focus on planning, scheduling and associated activity-based functions such as resource planning and leveling, cash flow forecasting, etc. (few of which are used)—adherence to a single system is seldom done. The desire to do so, if it exists, is lessened for at least two reasons. First, clients may contractually specify the use of different systems. Second, the restricted set of project management functions treated by these systems means that many of the essential duties performed by project management personnel are little affected by the choice made. Thus, the selection of a project management system can depend on what system the most junior, computer literate member of the project team is familiar with, as this person will be responsible for preparing and maintaining the schedule, with some input at the beginning from the project manager and project superintendent. Further, varying degrees of reliance are placed on the project plan and schedule and the system used to generate it. The schedule is often seen as something more for use by the owner or his agents than by the contractor himself. Consequently, the plan and schedule often lack the detail essential to be useful road maps and strategic tools for guiding the project (e.g., procurement, inspection, and commissioning activities may be missing, logic may be incompletely defined yielding unrealistic float values, etc.), lessening dependency on the system used to generate them. The contractor will often rely mostly on "throw away" short-cycle schedules that are produced either manually, using a spreadsheet, or using a generic planning and scheduling package as opposed to a more complex system capable of supporting a number of project management functions.

We have observed that planning, scheduling, and supporting functions are seldom accorded a prominent role in the firm. Few, if any, resources are provided for formal training in project management software and related functions, and money is rarely allocated to the development of custom software. Those filling the role on a transient basis are either junior engineers or building technologists who have acquired basic planning and scheduling skills and some exposure to current project management software while in school. But they soon realize that there is no future in specializing in planning and scheduling, and inevitably they aspire to become project coordinators, project managers, or occasionally, project superintendents. Thus, there is a lack of continuity in staffing for computerized support of project management functions (the same cannot be said for cost accounting or estimating which are seen as the life blood of the firm). In some cases, this leads to a contracting out of selected project management functions, especially planning and scheduling, resulting in a loss of expertise internal to the firm, ownership of project schedules, and the ability to modify plans and schedules on an ongoing basis in order to respond to actual project environments. (As soon as the service is contracted out, one of the objectives becomes to minimize consulting costs, leading to fewer schedule revisions and updates.)

Project Views

Figure 2 categorizes construction related software according to its role in supporting four different but interrelated project views. The figure does not show word processing, spreadsheet and database software, though their importance, as previously noted, should not be underestimated. The physical and environmental view of the project describes what is to built in terms of geometry, topology, physical systems, materials, etc., and the physical, economic and socio-political environment in which the project will proceed. The process view describes how the project will be constructed, who is responsible for different aspects of the work, when it will be done, and where. The cost view deals with the cost structure of individual parts as well as the overall project from various perspectives (subtrades, general, owner), and involves initial cost estimates and cost tracking throughout the construction phase. The as-built view describes what happened during the journey, why, and what actions were taken.

As noted in figure 2, some redundancy exists in the software and data used to support the different views. Such redundancy is one indicator of where beneficial integration or communication strategies may be pursued. Currently, much of the input required for use of many of these software applications is derived from other applications, but no interface exists between the applications.

Figure 2 can be used to classify ongoing research work in terms of its contribution to improving existing tools, developing new ones, developing standards and supporting structures for them, and pursuing integration and communication among functions. Our focus here is on the last three items, although we note that considerable research is still directed at improving existing tools (resource modeling, activity modeling, simulation, etc.). Examples of research in support of process views include Alkass et al. (1993), Hornaday et al. (1993), Tommelein and Zouein (1993), Ioannou and Liu (1993), Kamarthi et al. (1992), and Mahoney and Tatum (1994). Examples of new applications in support of as-built views include Russell and Fayek (1994), Yates (1993), Russell (1993), Abu-Hijleh and Ibbs (1993), Diekmann and Gjertsen (1992), and Roth and Hendrickson (1991). Work on integration, communication, and product modeling in support of these strategies includes Froese (1995), Teicholz and Fischer (1994), Parfitt et al. (1993), East and Kim (1993), Carr (1993), Miyatake and Kangari (1993), Björk (1994), Evt et al. (1992), and Rasdorf and Abudayyeh (1991). Much of this work explicitly or implicitly assumes an application to large projects with highly skilled users and high-end hardware environments.

Figure 2 can also be used to help in identifying data flows and overlaps among functions. This analysis can be used to assist in determining which components of the different views could be beneficially integrated (e.g., much of the process and as-built views), what communication links should be developed, and where standard models should be formulated to assist the integration and communication tasks.

The Potential of Computer-Integrated Management

While integration of construction computing systems is currently an active research area, we believe that the exact role that integration has to play in the industry and the specific benefits that it has to offer have not been well established. In this section, we briefly reflect on the potential benefits from the viewpoints of researchers, system developers, and end-users.

The benefits of integration are often presented by the research community in terms of increased *"information sharing"*. Computer integration, it is argued, will lead to improved communication among computer systems which will, in turn, lead to better information sharing among project participants. Not only will this reduce errors and

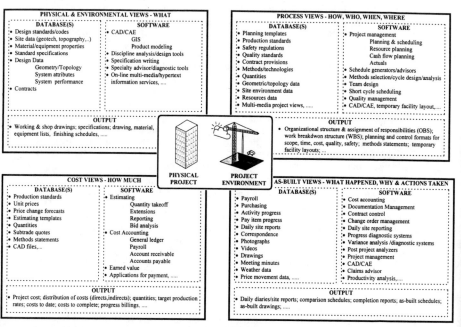

PHYSICAL & ENVIRONMENTAL VIEWS - WHAT

DATABASE(S)
- Design standards/codes
- Site data (geotech, topography,..)
- Material/equipment properties
- Standard specifications
- Design Data
  - Geometry/Topology
  - System attributes
  - System performance
- Contracts

SOFTWARE
- CAD/CAE
  - GIS
  - Product modeling
- Discipline analysis/design tools
- Specification writing
- Specialty advisor/diagnostic tools
- On-line multi-media/hypertext
  information services, ...

OUTPUT
- Working & shop drawings; specifications; drawing, material,
  equipment lists, finishing schedules, ...

PROCESS VIEWS - HOW, WHO, WHEN, WHERE

DATABASE(S)
- Planning templates
- Production standards
- Safety regulations
- Quality standards
- Contract provisions
- Methods/technologies
- Quantities
- Geometric/topology data
- Site environment data
- Resources data
- Multi-media project views, ...

SOFTWARE
- Project management
  - Planning & scheduling
  - Resource planning
  - Cash flow planning
  - Actuals
- Schedule generators/advisors
- Methods selection/cycle design/analysis
- Team design
- Short cycle scheduling
- Quality management
- CAD/CAE, temporary facility layout,....

OUTPUT
- Organizational structure & assignment of responsibilities (OBS);
  work breakdwon structure (WBS); planning and control formats for
  scope, time, cost, quality, safety; methods statements; temporary
  facility layouts; ...

PHYSICAL PROJECT   **PROJECT ENVIRONMENT**

COST VIEWS - HOW MUCH

DATABASE(S)
- Production standards
- Unit prices
- Price change forecasts
- Estimating templates
- Quantities
- Subtrade quotes
- Methods statements
- CAD files,...

SOFTWARE
- Estimating
  - Quantity takeoff
  - Extensions
  - Reporting
  - Bid analysis
- Cost Accounting
  - General ledger
  - Payroll
  - Account receivable
  - Accounts payable
- Earned value
- Applications for payment, ...

OUTPUT
- Project cost; distribution of costs (directs,indirects); quantities; target production
  rates; costs to date; costs to complete; progress billings. ....

AS-BUILT VIEWS - WHAT HAPPENED, WHY & ACTIONS TAKEN

DATABASE(S)
- Payroll
- Purchasing
- Activity progress
- Pay item progress
- Daily site reports
- Correspondence
- Photographs
- Videos
- Drawings
- Meeting minutes
- Weather data
- Price movement data, ...

SOFTWARE
- Cost accounting
- Documentation Management
- Contract control
- Change order management
- Daily site reporting
- Progress diagnostic systems
- Variance analysis /diagnostic systems
- Post project analyzers
- Project management
- CAD/CAE
- Claims advisor
- Productivity analysis,....

OUTPUT
- Daily diaries/site reports; comparison schedules; completion reports; as-built schedules;
  as-built drawings; .....

Figure 2. Contractor Project Views

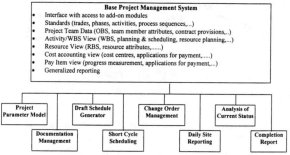

Base Project Management System
- Interface with access to add-on modules
- Standards (trades, phases, activities, process sequences,...)
- Project Team Data (OBS, team member attributes, contract provisions,..)
- Activity/WBS View (WBS, planning & scheduling, resource planning,...)
- Resource View (RBS, resource attributes,......)
- Cost accounting view (cost centres, applications for payment,.....)
- Pay Item view (progress measurement, applications for payment,...)
- Generalized reporting

Project Parameter Model	Draft Schedule Generator	Change Order Management	Analysis of Current Status

Documentation Management	Short Cycle Scheduling	Daily Site Reporting	Completion Report

Figure 3 Schematic of Integrated System Components

inefficiencies resulting from inaccurate, untimely, or missing information, but it will help foster better coordination and cooperation of the highly-fragmented construction participants. In our view, this argument is at least partially borne out. Figure 2 clearly shows that many of the construction management processes rely heavily on information produced by others in separate phases or views of the project, and the quality, efficiency, and timeliness of these processes must partially depend upon the quality, efficiency, and timeliness of their required information inputs. From the perspective of the project client, any increase in the quality and quantity of information available to all participants *can only increase* the ability of the participants to serve the clients needs.

However, providing, and even using, this information is not without cost to participants. Information produced within one company for use by others must be of higher quality and rigor than information intended for internal use only since the company no longer has control over how the information will be used. Under current project relationships, a company gains no added value or profit from sharing information, yet it maintains full liability for the information's use or misuse. Similarly, a company that uses information shared from another company will not, without extra effort, have as much knowledge of the appropriateness and accuracy of the information as it would for information generated internally. Even internal information sharing creates increased dependencies among company personnel and departments, and prerogatives for selecting systems to use, especially for site personnel, may be diminished or lost.

From a system developer perspective, a case can be made that the more specialized to one audience (e.g., the construction industry) and integrated (e.g., function and feature rich for a specific industry) that software becomes, the less viable it becomes commercially (the market gets significantly smaller and the low price mentality created by "shrink-wrapped" software precludes the high prices implied by a small market). This is an impediment to realizing the potential benefits offered from ongoing research on specialized advisory and planning and control tools for the construction industry. Our view is that integration strategies must be pursued that reflect the realities of the market place while seeking the benefits of full integration. Thus, a modular development strategy is desirable—create a core, generic system that has broad appeal to users inside and outside of the construction industry, and then develop specialist modules for different target markets that are an integral part of the user interface and that employ seamlessly shared product models and databases.

From an end-user or construction firm perspective, motivations for integration must come from demonstrable cost savings at the level of the firm (e.g., less personnel); development of a competitive advantage (e.g., increase speed of delivery, preferential pricing from subtrades because of better coordination and management of a project); the capture and use of expertise from past and current employees (resulting in fewer mistakes, oversights, etc.); the creation of new opportunities (e.g., the ability to offer reimbursable pretendering services in an attempt to acquire negotiated as opposed to lump sum work); enhancement of the company's image because it is seen to be on the cutting edge of technology; or the ability to support a much broader range of time-consuming project management functions.

Design Challenges for Computer-Integrated Management

In this section we outline briefly our thoughts on some of the research and design challenges that must be resolved in order to develop integrated systems that respond to the realities of the working environment in which medium-sized contractors find themselves. What should be in the system, how should the system be organized, and how should integration be effected are touched upon.

The first challenge is to identify the function set and supporting information that should be included within an integrated system. Because reference to the physical characteristics of the project on an ongoing basis is essential to construction of the project, information from the physical and environmental views of a project should be accessible. Passive representations (viewing only) of the project should be available, in multi-media form. These would include scanned in site photographs, drawings, finishing schedules and key contract clauses as well as parameter models which describe project scale and physical system characteristics. An ability to associate these representations (photos, building parameters) with other representations (e.g. activity and pay item) of the project is essential.

High priority should be given to integrating the following process, cost and as-built views with current planning, scheduling and control process software: initial schedule generation; short cycle scheduling; pay item reporting and applications for payment; documentation management; change order management; daily site reporting; automated analysis of current project status; and completion report generation. Challenges arising from these priorities deal with definition and support of other project representations, mappings to connect representations, and the ability to work with knowledge in large chunks. Current process management software facilitates the description of a project in terms of activity, organization and resource hierarchies (work breakdown structure, organization breakdown structure and resource breakdown structure). However, estimating, cost accounting and/or pay item representations used by head office or the client are not derivable from these hierarchical structures, and yet project staff must analyze performance and report using them, as well as others such as physical and as-built documentation records (photos, correspondence, drawing list, etc.). Thus, generalized mapping schema in the form of many-to-many relationships are essential to broadening the function set supported. Enhanced usability of management systems could also arise from the ability to incorporate standards and previous experience in systems so that information can be manipulated in large chunks, coding can be standardized, and automated analysis schemes devised. Examples include standard trade, phase, activity, cost code, and problem classification lists, subproject networks that can be scaled up or down depending on project parameter values, etc.

A second major challenge is to define an architecture and interface for an integrated system. A schematic of the organizational schema and components to be included is depicted in figure 3. The base module mirrors much of what is available in current project management software systems, except that it would support cost accounting and pay item views with appropriate mapping tools, and all of the optional add-on modules shown below would appear in the base system interface. Design of the add-on modules would be based on the product models used for the various representations supported by the base system.

A third and fundamental challenge deals with the definition of the various representations supported by the integrated system. This requires rich and flexible formulations of product and process models for building form and systems, activities, resources, organization entities, cost accounts, pay items, documents, etc., along with the ability to create associations among them. The creation of standards for these models to facilitate general integration and communication is also a major challenge.

If these challenges are met, the construction community has much more incentive to adopt integrated project management systems, because they support a much fuller spectrum of the activities carried out by project management staff on a day-to-day basis.

Acknowledgments: We gratefully acknowledge financial support for this work by the Natural Sciences and Engineering Research Council of Canada.

References

Abu-Hijleh, S. and Ibbs, W. (1993). "Systematic Automated Management Exception Reporting". *ASCE J. of Construction Eng. & Management,* 119(1), 87-104.

Alkass, S., Aronian, A. and Moselhi, O. (1993). "Computer-Aided Equipment Selection for Tranporting and Placing Concrete", *ASCE J. of Construction Eng. & Management,* 119(3), 445-465

Bjork, B. (1994). "RATAS Project - Developing an Infrastructure for Computer-Integrated Construction", *ASCE J. of Computing in Civil Eng.,* 8(4), 401-419.

Carr, R., (1993). "Cost, Schedule, and Time Variances and Integration", *ASCE J. of Construction Eng. & Management,* 119(2), 245-265.

Diekmann, J. and Gjertsen, K. (1992). "Site Event Advisor: Expert System for Contract Claims". *ASCE J. of Computing in Civil Eng.,* 6(4), 472-479.

East, E. and Kim, S., (1993). "Standardizing Scheduling Data Exchange", *ASCE J. of Construction Eng. & Management,* 119(2), 215-225.

Evt, S., Khayyal, S. and Sanvido, V. (1992). "Representing Building Product Information Using Hypermedia", *ASCE J. of Computing in Civil Eng.,* 6(1), 3-18

Froese, T. (1995). "Models of Construction Process Information", *Computing in Civil Eng.: Proc. of the Second Congress,* ASCE.

Hornaday, W., Haas, C., O'Connor, J., and Wen, J. (1993) "Computer-Aided Planning for Heavy Lifts", *J. of Construction Eng. & Management,* 119(3), 498-515.

Ioannou, P. and Liu, L. (1993). "Advanced Construction Technology System - ACTS", *ASCE J. of Construction Eng. & Management,* 119(2), 288-306.

Jagbeck, A. (1994). "MDA Planner: Interactive Planning Tool Using Product Models and Construction Methods", *ASCE J. of Computing in Civil Eng.,* 8(4), 536-554.

Kamarthi, S., Sanvido, V. & Kumara, S. (1992). "Neuroform-Neural Network System for Vertical Formwork Selection", *ASCE J. of Comp. in Civil Eng.,* 6(2), 178-199.

Mahoney, J. and Tatum, C. (1994) "Construction Site Applications of CAD", *ASCE J. of Construction Eng. & Management,* 120(3), 617-631.

Miyatake, Y. and Kangari, R., (1993). "Experiencing Computer Integrated Construction", *ASCE J. of Construction Eng. & Management,* 119(2), 307-322.

Moselhi, O., Hegazy, T. amd Fazio, P. (1993). "DBID: Analogy-Based DSS for Bidding in Construction", *J. of Construction Eng. & Management,* 119(3), 466-479.

Oloufa, A., Eltahan, A. and Papacostas, C. (1994). "Integrated GIS for Construction Site Investigation", *ASCE J. of Construction Eng. & Management,* 120(1), 211-222.

Parfitt, M., Syal, M., Khalvati, M., and Bhatia, S. (1993) "Computer-Integrated Design Drawings and Construction Project Plans", *ASCE J. of Construction Eng. & Management,* 119(4), 729-742.

Rasdorf, W. and Abudayyeh, O. (1991). "Cost- and Schedule-Control Integration: Issues and Needs", *ASCE J. of Construction Eng. & Management,* 117(3), 486-502.

Roth, S. and Hendrickson, C. (1991). "Computer-Generated Explanations in Project Management Systems", *ASCE J. of Computing in Civil Eng.,* 5(2), 231-244.

Rowings. J. (1991). "Project-Controls Systems Opportunities", *ASCE J. of Construction Eng. & Management,* 117(4), 691-697.

Russell, A. (1993). "Computerized Daily Site Reporting", *ASCE J. of Construction Eng. & Management,* 119(2), 385-402.

Russell, A. D. and Fayek, A. (1994). "Automated Corrective Action Selection Assistant", *ASCE J. of Construction Eng. & Management,* 120(1), 11-33

Teicholz, P. and Fischer, M. (1994). "Strategy for Computer Integrated Construction Technology", *ASCE J. of Construction Eng. & Management,* 120(1), 117-131.

Tommelein, I. and Zouein, P. (1993). "Interactive Dynamic Layout Planning", *ASCE J. of Construction Eng. & Management,* 119(2), 266-287.

Yates, J., (1993). "Construction Decision Support System for Delay Analysis", *ASCE J. of Construction Eng. & Management,* 119(2), 226-244.

Object-Oriented Scheduling Model for Integrated Schedule Information from Design to Construction Management

Kenji Ito, Member AIJ and JSAI[1]

Abstract

As clients' requests are becoming more diverse and extensive than ever, it is necessary to concentrate all the company's efforts and to fortify cooperation among different divisions in the A/E/C industry. In the past, development of information systems was basically focused on increasing efficiency of particular domain tasks, and was not enough to respond current problems. Therefore, a new method of developing collaborative and integrated information systems is required to synthesize resources in order to achieve total efficiency. However, it is difficult to generate, share and maintain project data during the various phases of the A/E/C project life cycle from design to construction. Especially, the schedule information needs to be stored, retrieved, manipulated, and updated by many project participants, each with his/her own view of the information and several kind of project constraint, such as site area, uses of facility, cost of material, structural and construction methods, temporary facility, labor cost and so on. Therefore, the author has been proposing an object-oriented schedule model supporting the multiple participants in A/E/C projects, which is very important for the integration of the A/E/C industry.

The scope of this research is to establish an object-oriented scheduling model shared by various A/E/C project participants and provide several types of schedule information by computer-based systems according to the A/E/C project process. The objective is the eventual development of an integrated system which includes activities from design, estimation, construction planning and construction management. The object-oriented scheduling model is

---

[1] Research Engineer and System Engineer, Information Systems Dept., Information Systems Div., Shimizu Corporation. No. 1-2-3, Shibaura, Minato-Ku, Tokyo 105-07, Japan.

intended to link several kind of knowledge-based expert systems, planning system, scheduling system and database.

This paper describes the development, present status and future directions for object-oriented scheduling model to facilitate schedule information sharing along the paths through which work flows between project stages or specialists.

The work described in this paper is still in process. The paper will conclude by sketching out of future plans for extensions and generalizations to the object-oriented scheduling model.

1. Introduction

A number of computer-based systems have been developed for the A/E/C industry. However many of these systems can only be utilized within narrow application domains. Efforts have been attempted to integrate these application software by linking systems and providing data transfer interface so that a richer communication among applications of different domains can be achieved. A project model that can properly describe a facility and is accessible by multiple participants of different disciplines is a very important ingredient for integration for the A/E/C industry.

Therefore many researchers have been attempting to develop a product model or a project model using object-oriented methodology. On the other hand, some researchers have been attempting to realize a integrated system using a product model or a project model. However, there are few results to support a global product or project model which is shared by the various participants or many applications of an A/E/C process.

Therefore, establishing of project model which supports the multiple participants needs is one of the key issues for the computer integrated construction and the author has been proposing the object-oriented project model called PMAPM [Ito 90, 91, 93, 94] which supports multiple participants of A/E/C project.

2. PMAPM

Each participant of an A/E/C project has his/her own view of the information about a constructed facility. In general, each participant has his/her own viewpoint, own data requirement, own data format and own computer-based applications, and all participants are expecting more information by less data input about the constructed facility. On the other side, most of researches on product model or system integration are considering only local information sharing or domain specific integration. Because it is difficult to develop the common product model which supports the various participants of project from domain specific product model, and theses product models will be not a kernel of the integrated system environment in the future. That is because the most important roles of product model or project model are:

• How to store project information and common building elements information with their properties as a project model and a product model.

• How to provide the domain specific sub model for any participants or computer-based applications from project model and common product model .

Then for the integrated system environment, information sharing mechanism for any participants is the most important issue. This mechanism will be consisted of the product model which does not depend on the specific participants view and the process model which includes several kinds of methods in order to retrieve and update the project information from product model according to any participants requirements. Therefore, in PMAPM (Object-Oriented Project Model for A/E/C Process with Multiple-Views), there are many kinds of building elements defined as an object with properties and there are many kinds of views defined as an object which are according to the view points of each participant of the project.

3. Object-Oriented Scheduling Model

Research on PMAPM, the author has been trying to provide the common process and product model for any type of A/E/C project which supports all the project's participants. However, it is very difficult to apply this prototype model to the actual project, because there are many existing computer-based applications in the industry and most of these applications have no objet-oriented data structure or interface to the object-oriented database.

On the other side, we have to prove the ability of object-oriented project model applying to the real world in order to show the future direction and capability of integrated information environment. Therefore, we decided to define the target domain to prove our concept with conventional applications and that was integration of schedule information.

There are many research results about the integration systems related to schedule information such as integration between CAD systems and planning systems or integration between planning systems and scheduling systems. PMAPM also achieved such a schedule related integration as follows:

• Three planning and scheduling systems are integrated by PMAPM using the domain specific sub model.

• Information for construction planig are created from CAD information through PMAPM.

However, these integration are still prototype environment and it is not enough to apply this environment to the real project in the industry. On the other side, in Shimizu, we have several kinds of schedule related computer-based systems for each participants as follows:

3.1 PREPLAN

PREPLAN is a knowledge-based expert system for evaluating the preliminary construction period at the project planning stage or early design stage. This system evaluate the

- Underground construction work duration
- Super structure construction work duration
- Finishes work duration
- Total construction period

from following information:

- Total construction area
- Main structure of facility
- Number of stories
- Ground condition of construction site
- Usage of facility

Figure-1 shows the user interface of PREPLAN. The result of evaluation store into the own ASCII format file.

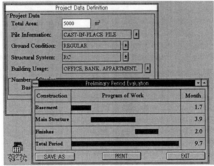

Figure-1: User Interface of PREPLAN

3.2 STDPLAN

STDPLAN is a knowledge-based expert system for evaluating the standard construction term at the end of detail design stage, estimation stage and early construction planning stage. This system evaluate the following construction work duration

- Preparation work
- Pile driving and Excavation work
- Super structure construction work

- • Finishes work
- • External work
- • Fabrication work(Steel, Precast member)
- • Total construction period

from following information:

- • Site area and construction area
- • Structure of facility
- • Number of stories
- • Ground condition of construction site
- • Usage of facility
- • Structural and construction method(Industrialization)
- • Holiday information, weather information
- • Labor information
- • Fabrication information

Figure-2 shows the user interface of STDPLAN. The result of evaluation store into the own ASCII format file.

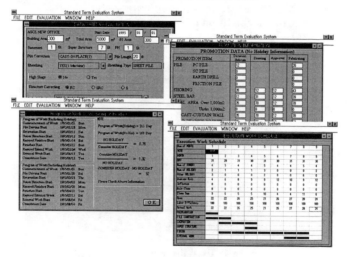

Figure-2: User Interface of STDPLAN

3.3 NETMAIN

NETMAIN is a conventional network diagram system for construction planning and construction management stage. This system provides the graphical user interface like a drawing software in order to define the activities and resources for each activities and show the result of resource analysis and critical path evaluation. Figure-3 shows the user interface of NETMAIN. The result of evaluation and drawing store into the own Binary format file.

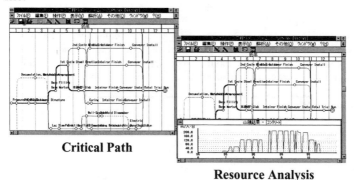

Critical Path

Resource Analysis

Figure-3: User Interface of NETMAIN

3.4 PMNG

PMNG is a conventional bar chart diagram system for construction management stage for managing the program and progress of work at the construction site. This system provides the graphical user interface like a drawing software in order to define the detail activities as a weekly or monthly schedule and also input the actual results of each work. Basically, the purpose of this system is drawing not a simulation, however, this system has a capability to define the relationship between the activities. Figure-4 shows the user interface of PMNG. The result of drawing store into the own Binary format file.

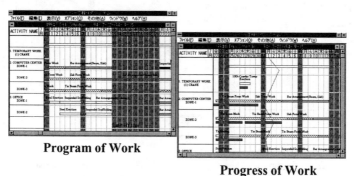

Program of Work

Progress of Work

Figure-4: User Interface of PMNG

3.5 Object-Oriented Scheduling Model

These four schedule related applications were used in the separated domain or stage in Shimizu, and there were no information sharing each other. Then the author proposed the object-oriented scheduling model as a domain specific sub model of PMAPM in order to share the schedule related information among them. Figure-5 shows the basic structure of scheduling model and schematic image of the integrated system environment.

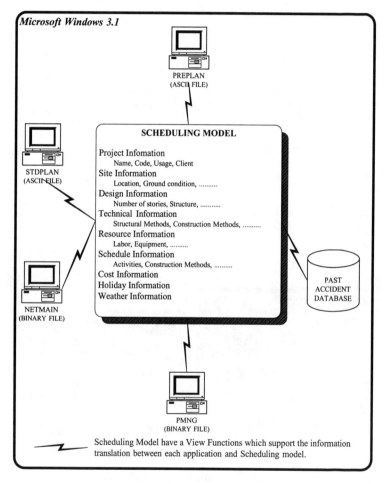

Figure-5: Schematic Image of Sceduling Model

4. Conclusion and Discussion

This paper described an object-oriented scheduling model that supports multiple views for building projects and this model creates the many kind of schedule related sub model according to the user needs or existing application needs in the Shimizu. Currently, this scheduling model has been implemented to the some construction sites as the test case of the information integration with applications as described in this paper and the past accident database[Ito 95a].

During the course of developing this scheduling model, the advantage and disadvantage of global project or product model and domain specific model in the A/E/C industry have been found and discussed. [Ito 95b]

Finally, the purpose of this work is to support the integrated environment for schedule related information using an object-oriented scheduling model to be referenced by multiple disciplines throughout the A/E/C process. In order to achieve this purpose, continuing effort is needed to evaluate the result of test at the construction sites and to increase the connecting applications which create or use the schedule related information.

References

[Ito 90]   K. Ito, K. Law and R. Levitt, "PMAPM: An Object Oriented Project Model for A/E/C Process with Multiple Views," *CIFE Technical Report*, No. 34, Stanford University, July 1990.

[Ito 91]   K. Ito, "Design and Construction Integration Using Object Oriented Project Model with Multiple Views," *Proceedings on Construction Congress II*, ASCE, pp. 336-341, Boston, U.S.A., April 1991.

[Ito 93]   K. Ito, "Constraint Management for Concurrent Design and Construction Using an Object-Oriented Project Model," *Proceedings of 5th International Conference on Computing in Civil and Building Engineering*, ASCE, pp. 1588-1591, Anaheim, U.S.A., June 1993.

[Ito 94]   K. Ito, "Integrated System Environment Using An Object-Oriented Project Model with Multiple-Views" *Proceedings of The First ASCE Congress on Computing in Civil Engineering*, ASCE, pp. 1204-1211, Washington D.C., U.S.A., June 1994.

[Ito 95a]   K. Ito, "An Object Model for Integrated Construction Planning and Safety Prevention Database," *Proceedings of The Second ASCE Congress on Computing in Civil Engineering*, ASCE, Atlanta, U.S.A., June 1995. (will be appeared)

[Ito 95b]   K. Ito, "General Product Model and Domain Specific Product Model in the A/E/C Industry," *Proceedings of The Second ASCE Congress on Computing in Civil Engineering*, ASCE, Atlanta, U.S.A., June 1995. (will be appeared)

Integration in Building Engineering Design with Databases

Claude Bédard[1] and Hugues Rivard[2]

Abstract

Design in building engineering is presented as a complex process for which integration is a necessity, i.e. to reconcile a variety of information elements representing disciplinary viewpoints and building components through time. A comprehensive information framework is proposed to 'anchor' such elements of information in a unique and systematic fashion so that interactions can be clearly identified. From this framework an information system is developed for the preliminary envelope design process and a OODBMS implementation is reported.

Introduction

The field of Building Engineering (also called Architectural Engineering in the US) can be characterized as one that cuts across the traditional technical disciplines related to buildings: architecture, civil engineering (structures and foundations), mechanical engineering (heating, ventilation, air conditioning), construction engineering and management etc. The ultimate objective of building engineering design is to deliver the best possible building over the entire life-cycle, under any foreseeable condition of use and occupancy. To be successful, building engineering design must therefore **integrate**, i.e. reconcile a multitude of different, sometimes divergent, viewpoints and characteristics, as well as interrelations among them, in order to produce the best possible building. As it is impractical to change traditional disciplines, the only alternative to enforce integration in the building design process is by means of computerization, first through the development of a rich information system that can accommodate this multitude of aspects. After describing a building information framework that forms the basis of integrated design

[1] Associate Professor, Centre for Building Studies, Concordia University, 1455 de Maisonneuve W, Montréal H3G 1M8 Canada.
[2] Graduate student, Civil Engineering Dept., Carnegie Mellon University, Pittsburgh, PA.

applications, an example implementation is presented for the preliminary design of the building envelope by means of an object-oriented database management system (OODBMS).

A building information framework: the 3P model

The results of poor decisions made at the building design stage are everywhere: leaking roofs, air infiltration through walls and around openings, interference between elements that goes undetected until construction, cost overruns. The cause of such deficiencies in design decisions are profound and can be linked to the fragmented nature of the construction industry, and to the complexity of buildings which constitute an intricate assemblage of disparate components. The resulting problems are generally costly, occur during construction or service life and reflect communication breakdown during the design phase as well as the fact that traditional disciplines do not cover critical aspects that are not central to their own field of interest, e.g. the building envelope which is not designed currently by any of the traditional disciplines. In response to this situation, building engineering has emerged as a discipline that encompasses all technical aspects and focuses on getting the overall system right, i.e. the entire building, rather than any isolated component. To be successful, design in building engineering must therefore **integrate**, i.e. reconcile a multitude of different, sometimes divergent, viewpoints and characteristics, as well as interrelations among them, in order to produce better buildings than before.

In order to support integrated design applications, an information system must first be developed that can address the multiplicity of aspects required for building design. First, buildings must be considered as complex systems made of numerous components. These interact with each other in a number of ways: such interactions must be determined as completely as possible at the design stage. Second, these interactions change with time depending on the design, construction or operation phase considered. Third, a multitude of participants contribute to the design and ensuing construction processes, including different teams of professionals and non-professionals, like trade people, manufacturers and suppliers. Thus to attempt building design in integrated fashion, various elements of information need to be accessible at any one time. Moreover, to be able to manipulate the right elements at the right time, an underlying information **framework** must be agreed upon by all participants, one hopefully that follows some 'natural order' to facilitate reference and to enable the study of interactions among distinct elements. There is really no common ground between a beam (structural component), a ceiling diffuser (mechanical component) and a window (envelope component) other than they belong to the same building and may interact with each other. The development of an information framework that can support the description of building elements in terms of form, function and behaviour thus requires the definition of very general 'ranges' to accommodate this multitude of aspects. Such building information framework is therefore required to account for all factors and actors in the design process, as well as their interrelations.

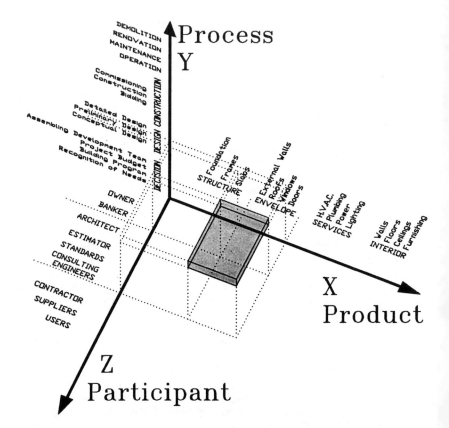

Figure 1. 3P model representation of the preliminary envelope design process

The so-called multidimensional model of buildings, or **3P model** as shown in Figure 1, has been developed for this purpose as a hierarchical information framework that can refer any element of information with respect to three major axes, the X or Product axis, the Y or Process axis, and the Z or Participant axis. The 3P model can support a comprehensive description of building design with its many facets and more importantly, account for the intricate interactions among these facets (Bédard 1989). In essence, the X axis of the 3P model represents the physical description and associated properties of a building: geometry (usual 3D description of CAD programs), cost, materials, details of connections, responses to different conditions. The Y axis covers the succession of events taking place through time while the Z axis accounts for the variety of participants involved in the building delivery process, directly or indirectly, through their requirements, reactions, different means of interventions. The main advantage of such an information framework for building design is that it can be used as a guide for the development of a central repository of information about any building task, thus saving time-consuming and error-prone data entry and exchange. It also enables the evaluation of several alternatives by moving along different branches of the information tree. Integration issues can also be addressed systematically by relating elements of information to the three main axes of the 3P model.

An OODBMS implementation is presented below for the integrated design of the building envelope system at the preliminary stage. The 3P model representation in Figure 1 clearly circumscribes the ranges of information considered as well as the solution space that will be explored to develop an overall best design solution. Following this general framework, an information system is put together by means of process and product modelling to evaluate simple concepts that can be easily translated into computing objects. An implementation of this information system is finally realised in a DBMS to facilitate communications among disciplines while avoiding duplications about each building component, irrespective of the specific viewpoint.

A database schema for designing the building envelope

The data generated and used for this implementation was analyzed and first cast in the form of a process model (Rivard et al. 94). Then, an information model was developed to organize efficiently and logically the wealth of data generated. The data is grouped into entities and relationships: the building envelope is seen as an aggregation of entities such as envelope planes, envelope sections, envelope layers, connections and openings (Rivard et al. 95). These entities are then decomposed into cohesive sets of attributes called primitives using the Primitive-Composite approach (Phan et al. 93). The resulting information model is implemented as a central database in an object-oriented database management system to investigate integration.

Each building envelope entity is modeled as a class. The primitives describing a given entity are implemented either as a method or as a class attached to the entity. Figure 2 shows the implementation of the envelope section entity using the Object Modeling Technique notation (Rumbaugh et al. 91). The Envelope_Section class contains the design description as attributes (name of section and list of envelope layers) and it refers to its primitives: the requirements of the section, the properties of the section and the behavior of the section. The cost, thickness and properties primitives are implemented as methods since they represent derived attributes: comp_cost, compt_thickness, comp_thermal_R and comp_vapor_R. All methods save their computed values in the class Section_Properties, thus explaining the presence of thickness and cost with the attributes thermal and vapor resistances. Three methods are used to manage the design of the section which is a list of envelope layers (i.e. add_layer, del_layer and

insert_layer). The last method, check_charact, is used to compare the resulting
behavior and properties with the requirements.

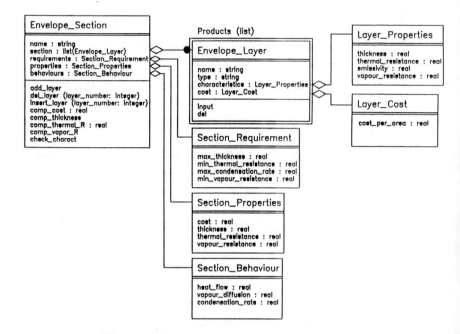

Figure 2. Database schema of the envelope section entity

The way this entity is defined allows the user to concentrate on one of its
views at a time. To illustrate this, the steps taken to design an envelope section are
described based on the building envelope process model (Rivard 94).

1. Envelope sections are instantiated by the architect and assigned to regions
   of the envelope plane. A name is assigned to each Envelope_Section
   object.
2. The envelope designer specifies the requirements of each section using
   instances of the Section_Requirement class.
3. The envelope designer design the sections by assigning a list of envelope
   layers in the Envelope_Section object.
4. The section properties are computed based on the list of envelope layers
   comprising it and stored in an instance of the Section_Properties class.
5. The behavior of each section with respect to condensation is stored in an
   instance of the class Section_Behaviour.

It can be seen from this design description that only a portion of the attributes describing a section were used at any given time. Dividing the attributes into several objects allows the display of only the needed attributes for each step. It is less confusing and overwhelming than an object showing the whole list of attributes of a given entity.

The information model is implemented in a commercially available object-oriented database system called O_2. The following advantages are realized through this experimentation:
1. Exchange of data between different steps of the preliminary design process is made possible.
2. Transfer of data among design participants involved is facilitated.
3. Growth of data is supported as the design unfolds.
4. Existing computer applications can interact with the database.
5. Form, function and behavior are stored together in the same data repository: this contrasts with existing CAD systems which only store the form aspects of objects.
6. A specific view of an entity can be supported through the use of primitives.

The capability of the object-oriented paradigm to capture the building envelope design process shows that OODBMS appear to be the best available database technology to integrate the building design process. The following advantages specific to OODBMS are identified:
1. The modeling capabilities of the problem domain are improved because of the semantic richness of object-oriented languages. The various building envelope entities and their complex relationships can be effectively stored.
2. Inheritance allows the reusability of common data structure and behavior.
3. Application programs can either be outside the OODBMS or stored within the database as methods attached to classes.
4. The database could be queried in an ad hoc manner.
5. Object-oriented programming allows the same approach to be used for the database and non-database segments of an application, thus avoiding the need for data transformation between the program application and the database structure.
6. Encapsulation enhance code reliability and modularity.

Conclusion

An integrated design approach is necessary in building engineering if better buildings than before are to be delivered. Given the inherent complexity of buildings and the multiplicity of viewpoints to support, an information framework is proposed that allows the complete description of actors and factors intervening through the entire building life-cycle. An example implementation is also presented for the preliminary design of the building envelope by means of a DBMS. The object-oriented implementation supports a logical growth of the database following the design process and it provides different views of envelope entities. The implementation demonstrates the feasibility of accounting for integration in the design process as the resulting database supports data creation, maintenance and communication at various stages, thus showing that applications can be developed to interact with the database to extract and store data.

References

Bédard C. (1989), "Research Directions for AI in Design: a Building Engineer's View of Design Research", Workshop on Research Directions for AI in Design, Stanford University, edited by J.S. Gero, University of Sydney, pp. 39-42.

Phan D.H.D., and Howard H.C. (1993), "The Primitive-Composite (P-C) Approach - A Methodology for Developing Sharable Object-Oriented Data Representations for Facility Engineering Integration", Technical Report 85 (A and B), Center for Integrated Facility Engineering, Stanford University.

Rivard H., Bédard C., Ha K.H. and Fazio P. (1994), "Functional Analysis of the Envelope Design Process for Integrated Building Design", Proceedings of the First Congress of Computing in Civil Engineering, Washington, pp.71-78.

Rivard H. (1994), "Integration of the Building Envelope Design Process", Master Thesis, Centre for Building Studies, Concordia University, Montréal, Canada.

Rivard H., Bédard C., Ha K.H. and Fazio P. (1995), "An Information Model for the Building Envelope", Proceedings of the Second Congress of Computing in Civil Engineering, Atlanta.

Rumbaugh J., Blaha M., Premerlani W., Eddy F. and Lorensen W. (1991), "Object-Oriented Modeling and Design", Prentice Hall.

Organizational Knowledge Creation in Integrated Construction Planning System

Yusuke Yamazaki[1]

Abstract

This paper attempts to clarify organizational knowledge creation in building construction, which requires dynamic circulations and interactions among implicit knowledge and explicit knowledge. Frameworks on organizational knowledge creation are briefly described. Three cases of organizational knowledge creation in building construction are also presented: 1) analysis on utilization of newly developed construction technologies, 2) estimation of standard construction period, and 3) introduction of integrated construction drawings by CAD. Through the analysis, a mechanism which can transmit implicit knowledge to explicit knowledge should at first be established. The mechanism requires metaphors which are identified as conceptual models with object-oriented paradigm when information technology is applied to solve the problems. Also, an integrated construction planning system (ICPS) is described from a viewpoint of organizational knowledge creation.

1. Introduction

In recent years, integration of design and construction has been a challenging issue to innovate construction industry. In spite of an increasing demand for integration of design and construction functions, the shortage in the skilled and knowledgeable designers, engineers, planners and managers, and the ageing of such people have come to present serious problems. It has been pointed out that the background for such a situation is that construction activities require wide-spread knowledge in various domains with long-term experiences acquired through participation in actual building projects. The serious problems faced by the construction industry today are to eliminate defects caused by shortage in the skilled and knowledgeable designers, engineers, planners and managers, and to improve design, engineering, planning and management productivity by introducing information technology.

Research focus here is a practical application of organizational knowledge creation model to building construction. Organizational knowledge creation is defined as dynamic circulations and interactions among implicit knowledge and explicit knowledge through different levels of organizations (person, group, organization, and inter-organization). Usually, knowledge used in design and construction planning is divided into two types: 1) Implicit knowledge such as evaluation and application procedures of building systems and construction systems, which are acquired through experience and are possessed by individual designer and planner; and 2) Explicit knowledge such as composition and functions of building systems and construction systems, which are shared as technical standards among groups, departments and organizations and are usually stored in the form of documents.

[1] Manager, Intelligent Engineering Systems Department, Technology Division, Shimizu Corporation, No.1-2-3, Shibaura, Minato-Ku, Tokyo 105-07, Japan

Reviewing successfully developed construction technologies, design, engineering, planning and management knowledge associated with the developed technologies are recognized as implicit knowledge at the beginning of development. According to the exploration of development, such knowledge is transformed into explicit knowledge such as technical standards with application procedures in the form of documents. The explicit knowledge is applied to actual building projects, and then the technologies require improvement and sophistication to be a more innovative technology development, such as hybrid building systems, automated construction systems, and advanced computer applications. These development need integrations and interactions among knowledge with broader technological areas. Consequently, new implicit knowledge in design, engineering and construction are created through integrated and interactive investigations by development organizations. Thus organizational knowledge creation is performed as dynamic circulations and interactions among implicit knowledge and explicit knowledge through different levels of development stages.

One reason why the impact of the organizational knowledge creation approach on the building construction will be crucial is that it supports a high level of coordination between human and computer. A second strength of this approach is that it provides a well-organized environment for transmission of information and knowledge, a prerequisite for the use of an object-oriented approach in the integration of information and knowledge. This has removed poorly coordinated transfers of information and knowledge between the various organizations. Also, the environment provides a flexible decision making process in a global concept that integrates all critical design and planning processes. Thus organizational knowledge creation in building construction is relevant.

2. Frameworks of Organizational Knowledge Creation

In manufacturing, a series of researches with regard to organizational knowledge creation have been performed by Nonaka [Nonaka 90]. Our research started from the application of framework presented by Nonaka to building construction. The integrated organizational knowledge creation model illustrated by Nonaka presents dynamic interactions as four types of activities: socialization, externalization, combination, and internalization. When the concept is applied to building construction, transmission of expert skills and implicit knowledge through actual jobs from a skilled engineer to an unskilled engineer is socialization. Documentation and standardization of personal or group know-how is externalization. Systematization of documented knowledge on design, engineering, planning and management by computer programs is combination. Knowledge acquisition through development of new technology based on established technologies is internalization. Our focuses are on internalization and externalization, which are caused by interactions between implicit knowledge and explicit knowledge. Since the organizational knowledge creation does not necessary take an up-stream exploration starting from person level, a creative organization should have a system or a structure which can explore linkages between four modes of knowledge creation.

In construction industry, researches on constructability presented similar results. A research pointed out a need for increasing understanding of the type of knowledge involved in experiences from engineering and construction projects.[Tatum 87]. Constructability improvement is recognized as evolutions of the four interaction modes of knowledge creation based on opportunities to better capture and transfer experiences and knowledge between design and construction. As constructability researches took analytical and formulated approaches, utilization of exogenous source of knowledge and combination of formalized knowledge were relevant. Consequently, emphasis were posed on documentation and computer utilization to establish database and knowledge-base with constructability. The activities were viewed as combination and externalization. Reflecting these researches, our focus were posed on dynamics of organizational knowledge creation, especially a mechanism for building construction.

3.Cases of Organizational Knowledge Creation in Building Construction
Three cases in which organizational knowledge creation is identified are presented.they indicate key issues to efficiently adopt information technology to building construction.

Analysis on Utilization of Construction Technologies
To efficiently utilize newly developed construction technologies, a survey was performed in 1984 with the following items: 1) how widely technologies are understood and utilized by designers, engineers, planners and managers; and 2) what are barriers for utilization and what kind of knowledge is required to efficiently utilize them. Through analysis, the followings were found.
Average of understanding was approximately less than 30% and utilization was approximately less than 10%. Between understanding and utilization of a technology has a high correlation (the coefficient of correlation is 0.91). Through this analysis, three unique technologies were found: No.48 Micco joist slab system, No. 68 Tile-preset form system, and No.76 Asbestos spray robot. Micco joist slab system and tile-preset form system were well utilized at that period, because they improve quality and reduce skilled construction work forces. These technologies have been substituted by newly developed construction systems, such as composite precast concrete slab and tile-preset precast concrete form system, which can further reduce labor force and improve quality.
Except cost problem, major barriers were identified as both organizational and technological: 1) insufficient application procedure, 2) non-systematic evaluation method, 3) insufficient supporting functions, and 4) inflexibility in technology itself. These problems are often identified in transferring technology, knowledge and information. On the other hand, efficiently utilized technologies involve common features: 1) technical standard, design standard and construction procedures were established, 2) applicable conditions, application procedure, issues to be investigated are clearly defined, and 3) effects by introducing the technology are also evaluated and stored as information or data of productivity, cost and quality.
These analysis indicate that transmission of implicit knowledge to explicit knowledge is important to efficiently utilize newly developed technologies as well as established technologies. Also, systematized technologies, such as integrated construction system and hybrid building system were difficult to be adopted and evaluated in actual building projects. Because, comprehensive knowledge to perform a sequence of design, engineering, planning and management were necessary, which were sometimes shared only by development group or acquired only by expert planners as their implicit know-how. A stress should be put on a mechanism to transmit these implicit knowledge to explicit knowledge to be shared at organization level.

Estimation of Standard Construction Period
To estimate appropriate construction period for a project is a key to success in project management. There have been several surveys and proposals in this area.The first research was performed by Asakura through analysis of construction period data with 320 building projects during 1974 to 1976 [Asakura 79].The analysis applied statical method extensively. Over 50 forms of formula were at first assumed reflecting planning procedures in conceptual scheduling. Then these formulas were analyzed through multiple-correlation analysis to be in acceptable forms by scheduling knowledge of planners, resulting in four formula of estimating standard construction period for piling/underground, superstructure, finishing, and overall project. The

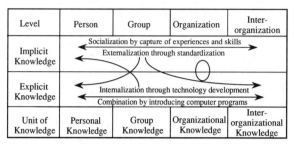

Level	Person	Group	Organization	Inter-organization
Implicit Knowledge	Socialization by capture of experiences and skills → ← Externalization through standardization			
Explicit Knowledge	← Internalization through technology development → ← Combination by introducing computer programs			
Unit of Knowledge	Personal Knowledge	Group Knowledge	Organizational Knowledge	Inter-organizational Knowledge

Figure.1 Organizational Knowledge Creation Model for Building Construction
(Modified based on Nonaka's model)

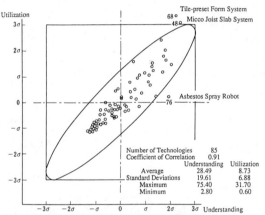

Number of Technologies 85
Coefficient of Correlation 0.91

	Understanding	Utilization
Average	28.49	8.73
Standard Deviations	19.61	6.88
Maximum	75.40	31.70
Minimum	2.80	0.60

Figure.2 Correlation of Understanding and Utilization with Technologies

Table.1 Formulas for Estimating Standard Construction Period

Examples of Formulas (1977)

Piling/Underground Construction Period
$$[\,0.6 + 0.02 \times \sqrt{S_F} \times a + 1.1 \times N_U\,] \times b$$

Superstructure Construction Period
$$[\,0.6 + 0.02 \times \sqrt{S_T} + 0.35 \times \{\,N_U + N_S + 0.3 \times N_P\,\}\,] \times c$$

Time scale : month (including unworkable days)

a: adjustment by type of piles, b: adjustment by soil condition, c: adjustment by type of structure

Examples of Formulas (1989)

Piling Period
$$[\,0.25 + a_1 \times \sqrt{(S_F \times L_P)}\,] \times f_1$$

Underground Construction Period
$$[\,0.2 \times \sqrt{S_F} + \{\,0.5 + 0.2 \times b_1 \times b_2 \times \sqrt{S_U \times N_U}\,\} + 5 \times \{\,N_U + b_3\,\} + 12\,] \times f_2$$

Superstructure Construction Period
$$[\,\{\,45 + 0.9 \times \sqrt{S_T} + 16 \times N_U + 7 \times N_S + 4 \times N_P\,\} \times c \times i\,] \times f_3$$

Time scale : day (workable days)

a1: adjustment by type of piles and number of piling machines,

b1: adjustment by type of shoring, b2: adjustment by type of sheeting, b3: adjustment by temporary access road,

c: adjustment by type of structure, i: adjustment by industrialization, fn: adjustment by perspectives,

St : Total floor areas, Sf: Standard floor areas, Su: Standard floor areas (underground), Lp: Length of Piles

Ns : Number of stories, Nu: Number of stories(underground), Np: Number of stories (penthouse)

formulas used a time scale as month. These formulas can be viewed as explicit knowledge extracted from implicit knowledge with conceptual scheduling through externalization. The formulas had been widely utilized in actual building projects. In 1989, the formulas were required to be up-dated, due to labor shortage and increase of holidays caused by changes in social environment, and intensive use of prefabricated and systematized building/construction systems supported by intensive technology developments. Formulas were also established through analysis of about 200 building projects. The process of analysis followed the procedure taken by Asakura. As shown in figure 4, the major forms of new formulas are almost same as the former formula, except piling/underground part. Major factors which caused modification in the form are recognized as effects accompanied with development of new piling systems and systematized underground construction systems. Also, we introduced industrialization ratio as representative of labor reduction to associate conceptual scheduling and plannings of prefabricated and systematized systems to superstructure of a building. These formulation indicate that new implicit knowledge is created from explicit knowledge through internalization. A mechanism dynamically cause interactive creations of implicit knowledge and explicit knowledge is important to maintain effectiveness of planning knowledge at an organizational level.

Introduction of Integrated Construction Drawings by CAD

Recently, integrated construction drawings are focused as interface to negotiate and coordinate design and construction planning at product design stage. CAD systems have been introduced in product design, to effectively produce various kind of construction drawings and to efficiently reuse drawing data by defining layers. According to the application of CAD systems, implicit knowledge possessed by skilled engineers required to be extracted and represented as CAD objects and layers in CAD. These CAD objects and layers can be viewed as explicit knowledge created through combination. The background are identified as followings: 1) personal improvement in technique for information management using CAD system through establishing CAD object library; 2) a need to transfer technical knowledge with building construction to CAD operators who substitute drawing engineers who are knowledgeable both in design and construction; and 3) establishment of standards for data exchange format and layers at construction industry level. Current focus on organizational knowledge creation by CAD is how extract implicit knowledge with decision making processes and transmit them to explicit knowledge. Generally, a drawing is viewed as a combination of a field and objects. A field is recognized as a function to perform intended investigation [Hirasawa and Yamazaki 94]. For example, in a drawing for construction facility planning, objects are a crane, steel elements and site layout, and functions are to identify the position of crane based on planning conditions and then to modify the site layout. More attention should be paid on these functions of each drawing as well as representation of objects. These functions imply a need for domain specific models.

Reviewing organizational knowledge creation in building construction, standardization and formalization of knowledge sometimes obstruct dynamic interaction between implicit knowledge and explicit knowledge. However creation and exploration of knowledge are necessary in all levels of organization, dynamic circulation of knowledge is stopped when major portion of knowledge is once formalized and objectively structured. This also implies a requirement for establishing a mechanism of organizational knowledge creation in building construction based on information technology especially by introducing modelling techniques.

4. Introduction of Organizational Knowledge Creation to ICPS

Current researches on product models and process models focusing on standardization and formalization of information and knowledge are assumed to involve mechanisms for organizational knowledge creation. Because, product model and process model essentially include mechanisms such as abstraction, generalization, formalization, classification and standardization. We assumed product model and process model as metaphors to transmit implicit knowledge to explicit knowledge.

Metaphors are important, because they allow flexibility in representing unformulated knowledge as well as formulated knowledge. Industrialization ratio settled through analysis of standard construction period, planning process models extracted from investigation of construction planning, and standard detail models identified through analysis of integrated construction drawings may be recognized as metaphors.

Our current attempt is to involve mechanisms of organizational knowledge creation in an integrated construction planning system (ICPS) depending on metaphors. ICPS is originally a knowledge-based system which represents and formalizes different types of knowledge among building system planning, construction system planning, scheduling and construction facility planning. The ICPS has introduced object-oriented paradigm extensively to introduce metaphors as intermediate planning objects [Yamazaki 94]. The system has been developed in UNIX environment based on a commercial knowledge-base shell (XPT2) and a relational database (ORACLE). The system is currently being transferred to WINDOWS environment .

5. Integrated Building and Construction Planning for Slab

As an example, a slab system planning module which is a sub-module of ICPS is described. Object composition of slab planning system is shown in figure 3.

A project can be subdivided into several planning areas based on project planning perspective of a planner or a planning team. Structural systems present major scopes to subdivide planning areas. The project is estimated and evaluated through summary of these areas.

A slab system planning area is a class object , which controls several sets of slab assembly models by sending messages to building planning system manager and construction system planning manager to retrieve required data from objects, to actuate estimation, and to accuses database. Building system manager controls building assembly models, and construction system manager controls construction systems by combining building component models and construction system models. Building assembly models are produced as instances and control message passings among building component models (consist of slab, joist, beam, wall, and column) by linking them. Current ICPS defines 14 slab assembly models, which are instantiated based on a structural system assumed in a slab planning area.

A building component model represents column, beam, joist, slab and wall with entities of size, structure, composition and method to calculate amount of volumes in several ways. Each building component model sends message via building assembly models. A construction system model is dynamically instantiated by a building component, which defines an estimation procedure and contains entities of construction methods, such as formwork method, reinforcing method, and concrete placing method. These models are viewed as representatives of explicit knowledge.

To evaluate instantiated plannings as combination of these models, appropriate index such as unit prices should also be estimated. Estimation of unit prices are well depend on planning scopes for a area, such as optimization of prefabrication and site works, systematization of available resources through work scheduling, and minimizing of temporary equipments and facilities. These planning scopes are viewed as implicit knowledge possessed by planners, which are defined as planning models in ICPS to reflect planner's implicit knowledge to evaluate plannings. Thus object-oriented

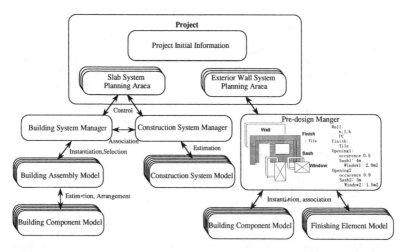

Figure.3 A Conceptual View of Classes and Objects in ICPS for WINDOWS

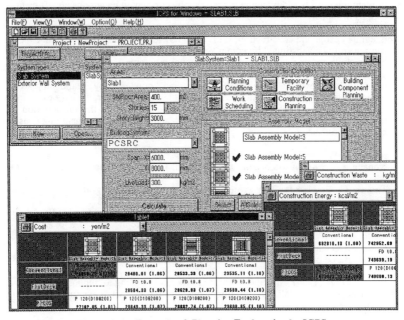

Figure.4 An Example of Planning Exploration by ICPS

paradigm and modelling technique are applied to ICPS, which aims to present a viewpoint of organizational knowledge creation to planners. Through the implementation, we found several issues to be improved in future. Usually, new planning knowledge is at first recognized as implicit knowledge to settle appropriate assumptions of planning scope to the system, and then the knowledge is evaluated to be established as a general planning model. Consequently, the system needs a module to easily instantiate user-defined planning models as well as general planning models to efficiently apply implicit knowledge to plannings. Also, user-interface will be improved by introducing a graphical modeler, which is viewed as an abstraction tool for implicit knowledge of users. These problems will be solved in an implementation of exterior wall planning module, which is developed as a sub-module of the ICPS.

Summary and discussions

Organizational knowledge creation should be established as a mechanism and be involved in design , engineering , planning and management in building construction to transmit and to share both explicit and implicit knowledge through different levels of organizations, such as person, group, organization and inter-organization. Especially when introducing information technology, product model and process model should be applied to clarify functions and objects assumed as metaphors of implicit knowledge and explicit knowledge. Through analysis of three cases, industrialization ratio object, planning models and standard detail models were identified as metaphors. In ICPS several types of planning models were viewed as metaphors to utilize implicit knowledge as well as explicit knowledge possessed by a planner. Through modelling of planning objects and planning processes, object-oriented paradigms are efficiently applied to support modularity of different planning functions.

However the introduction of object-oriented paradigms and modelling techniques to integrate design and construction, skills and knowledge possessed by human experts can not fully be substituted by computer programs. If we pursuit full substitution of human skills and human knowledge by computer programs, it will require tremendous investment for computer hardware and software. On the other hand, if we pursuit to maintain high level of human skills and human knowledge by establishing well-organized learning environment, it will also cost highly. Therefor, we have to balance computer hardware/software with human knowledgeware in the introduction of information technology to design, engineering, planning and management functions in building construction.

References

Nonaka, I. (1990), A Theory of Organizational knowledge Creation. (in Japanese), Nihon Keizai Shinbunsya, Tokyo, 1990.

Tatum, C. B. (1987), The Project Manager's Role in Integrating Design and Construction, Project Management Journal Vol. XVIII No.2, 1987.

Asakura, T., et al (1979), An Estimation Method for Standard Construction Period. (in Japanese), Summaries of Technical Papers of Annual Meeting, pp 355-356, AIJ, 1979.

Hirasawa, G. and Yamazaki, Y.(1994), A Frame-Based Drawing Management System Based on Building Object Model , Computing in Civil Engineering, pp 998-1005, ASCE, 1994.

Yamazaki, Y. (1994), An Object-Oriented Process Modeling to Support Cooperative Planning in Industrialized Building Construction Project, Computing in Civil Engineering, pp 998-1005, ASCE, 1994.

A DATA MODEL AND DATABASE SUPPORTING INTEGRITY MANAGEMENT

by C. Eastman, M.S. Cho, T.S. Jeng and H.H. Assal[1]

Abstract

The need for integrity management in design is reviewed. The features that support integrity management are presented for EDM-2, a data model and database for product modeling. The features include constraints that check the status of rules, specific structures for representing the knowledge embedded in applications and automatic constraint management logic. An example is included from concrete design.

I. Introduction

Architecture and building construction is a knowledge rich area. Many different technologies exist for construction: for example, for structural systems, cladding, HVAC systems and interior partitions. Each type of construction involves its own set of components, rules of composition and often their own methods of analysis. Similarly, there is a rich vocabulary of building types, such as hospitals, airports and schools, and for individual spaces, such as auditoriums and cafeterias. Each building type and type of space use has its own set of activities, elements to be composed, relevant furniture and criteria for good design. Today, most of the knowledge about both construction technologies and building types are packaged into reference books and the knowledge is applied manually. Soon, such knowledge will be embedded in computer applications and applied interactively or automatically.

Architectural design is challenged with the creative application of construction technologies and the generation of designs responsive to contextual conditions and to generate high levels of performance. It does so by applying complex mixtures of technology to diverse mixtures of building type and space use. The combinatorial richness of knowledge to be applied to different building projects prohibits widespread use of any single fixed building model. Rather an open-ended model that can support dynamic interfacing to new applications, is becoming widely recognized

---

[1] Center for Design and Computation and Dept. of Architecture, University of California, Los Angeles, 90024-1467

as the only practical long term approach for developing intelligent tools for computer aided building design (Augenbroe, 1992), (Eastman, 1993), (Galle, 1994).

Secondly, design decision making is iterative and convergent, certainly not linear. Design processes must respond to many different goals and performance criteria that are often conflicting. Buildings are designed by multiple specialists. Decisions in each technical area have complex dependencies and these dependencies cross area boundaries, resulting in complex interactions among decisions. Because of this, one iterated decision may invalidate much existing data.

One need only consider a team of designers working on a project from remote locations, using a shared server. As one team defines the basic design, others use the data generated to develop their parts and detail them. Later some of the initial data changes, for example the spacing of the structural grid. Data based on the old grid spacing is no longer valid. Later, consider a point where the building and its facade have been laid out, the mechanical system and structural system sized. Based on a client request, the windows in one area of the building are made bigger. Consider what data has been automatically made invalid by this action. The structural loads are changed where the windows were altered. The energy loads where the windows were changed are modified, possibly requiring the mechanical system to be re-evaluated. A different size mechanical system may result in different loads on the structure. The interior partitions may line up poorly with the windows, ... and so on.

From these two examples, it is clear that data in a building model may be valid or invalid, based on the criteria by which it has been judged. *Integrity* is the relation between data and physical reality. Integrity of a model is always only partial, specific to certain relations of interest. Integrity can be defined by rules, and expressed in terms of whether they are satisfied within a model (Ullman, 1988).

In design, integrity is built up incrementally. Consistency of a model in relation to reality is small at the beginning and is added as design proceeds, so as to be complete for the rules being modeled at the end (Eastman and Kutay, 1991). Design operations typically involve integrity checking or taking actions that change the integrity of the design. It is very easy for such changes -- that are now managed manually -- to be missed, especially as design members become distributed on a network. It is imperative that designers know the integrity of data available within a building model.

In this paper, we report on EDM-2, a prototype backend database, with its own language and data model developed for use in product modeling. EDM-2 is a second generation model based on the earlier EDM (Eastman etal, 1993.). EDM-2 is a client-server database system that supports both schema evolution and dynamic linking of applications. It also supports explicit and automatic integrity management.

In this paper, we present the constructs and methods for integrity management used within EDM-2. In section 2 is presented the concepts and approach used for integrity management. In section 3, the language constructs used to represent semantic integrity rules are given and an outline of their implementation. In section 4 is given the logic methods used to maintain integrity. We incorporate an example, using concrete design.

2. Integrity in a Product Model Database

The logical correctness and consistency of data in a product model can be determined by a set of integrity rules. These rules can be roughly categorized:

(1) *definitional rules:* e.g., a line has two endpoints, a quadrilateral has four points; the base dimensions of a grid are monotonic;

(2) *well-formedness rules:* the graph of a structural system is connected; all building elements with weight must have a path transferring that weight to the ground; a polygon used for area calculations cannot be self-intersecting;

(3) *rules corresponding to physical laws:* solid objects cannot overlap in space-time; materials obey stress, electrical, acoustic, thermal and magnetic behavior models;

(4) *standards and codes:* agreed upon good practices in response to public safety;

(5) *design goals:* that the structural safety factor is greater than a given number, that a certain set of activity spaces are provided.

Such rules are organized in a rough hierarchical order, where the lower level rules must be satisfied before the higher level rules are meaningfully evaluated. That is, a path for structural transfer can be evaluated only if the lines representing it are correctly defined; stress calculations can be meaningfully evaluated only if the structural graph is well-formed; design goals can be evaluated only if the physical performance of materials and composition can be meaningfully defined. Obviously, a growing number of these rules are satisfied by application programs.

(a) correct data is available to evaluate the rule and it is True or False;
(b) the data is not available to determine if the rule is satisfied;
(c) the data or rule has been changed since it was evaluated;
(d) a rule required to be true to allow evaluation of this one is not satisfied.

FIGURE ONE: Integrity rules can have several states, which may change over time according to the conditions shown.

EDM-2 uses *constraints* to represent integrity rules, including the five categories of rules above (and possibly others). Constraints are of two kinds: Variant and Invariant. *Invariant constraints* are like most constraint systems implemented in databases; they are automatically evaluated after every action, and must be True or the action is rejected. Invariant constraints are appropriate for definitional rules. Variant constraints are the second type, and are evaluated upon user command. They may evaluate to True or False. Both Invariant and Variant constraints may also evaluate to Undefined, meaning that there is incomplete data to evaluate them. In addition, Variant constraints may have a value of NULL, which means that their value is unknown, possibly because of conditions (c) or (d) in Figure One. Variant constraints are appropriate for well-formedness rules, physical laws, standards and codes and design goals.

In addition to these rule definition capabilities, EDM-2 includes automatic constraint management capabilities. Specifically, it has built in capabilities to guarantee that constraints that have been set to True or False are reset to NULL under conditions (c) and (d) of Figure One. These capabilities will be described in more detail later.

	CONSTRAINT VALUES	
Condition	**Variant**	**Invariant**
Ccall evaluates to True	True= 1	True = 1
Ccall evaluates to False	False = 0	(Not allowed transaction rejected)
parameters or constraint body missing	Undefined = -1	Undefined = - 1
condition unknown	(blank) = NULL	(blank) = Null

FIGURE TWO: There are four possible states for Variant constraint calls and two possible values for Invariant Constraint calls.

3. EDM-2

EDM-2 is an object-based database management system that includes objects, which can be defined using multiple inheritance of other objects. Objects are called *Design Entities (DEs)*. All attributes of an object are also objects. That is, they have a global definition and if that definition is modified, then all objects with those attributes are also modified. Object attributes defined with a single value are called *Domains*, complex objects are DEs. Attributes may be variable length SEQUENCEs or SETs. Domains and DEs are defined as classes, then multiple instances defined of those classes.

EDM-2 relies on two types of object constructions. The first is the *specialization* lattice, defined by inheritance. The other is *composition*. Composition is the aggregation of heterogeneous DEs to define another DE. The aggregation DE is called the *target* and the components are called *parts*. There may be multiple compositions describing the same target, for example one set of finite elements to define the thermal performance of a shape and another to define the structural performance. The same part may be in multiple compositions. For example a pump may be part of the fluid flow composition and also the electrical composition. Compositions may be defined top-down or bottom-up.

Dynamic schema evolution allows new DEs and Domains to be created at any time, or existing ones to be modified. In addition, DE instances may be migrated up or down a specialization lattice, with the effect that new attributes may be added to or deleted from existing DE instances.

In EDM-2, the construct for defining integrity rules is the *Constraint*. Constraints are associated with either a DE or Composition class and apply to all instances of them. (Domain constraints, defining optional bounds, are built in.) Constraints are defined in two parts: the Constraint is the rule specification; the constraint call or *Ccall* is the specific invocation site. The Constraint specifies the argument types and when it exists, the constraint body. The Ccall provides the specific arguments at a call site.

```
/*layout of construction grid */
CREATE DOMAIN dist:REAL
  LBOUND 0.0
  DESC "distance in meters;
CREATE DOMAIN coord:REAL
  DESC "coordinates in meters;
/* create a general gridline and
   two specializations */
CREATE gen_gridline
  ATTR (edge:dist);
CREATE DE v_grid SHARED KEYNAME
  ATTR (e:gen_gridline);
CREATE DE h_grid SHARED KEYNAME
  ATTR (e:gen_gridline);
/* define grid as system */
CREATE DE grid KEYNAME
  ATTR (
  v_edge :SEQUENCE OF v_grid,
  h_edge :SEQUENCE OF h_grid,
  DESC "the aggregate structure
       for a complete grid";
/* constraints to make grid
   lines ascending */
CREATE CONSTRAINT mono
  (SEQUENCE OF *,char(1))
  IMPL <filename>;
  DESC "checks that a sequence
  is monotonic and ascending(A)
  or descending(D)";
```

```
CREATE CCALL monotonic_v
  CONSTRAINT mono
  (v_edge,"A"):I;
CREATE CCALL monotonic_h
  CONSTRAINT mono
  (h_edge,"A"):I;
MODIFY DE grid
  ADD CCALL
  (monotonic_v,monotonic_h);
/* a possible example layout */
:v1=INSERT INTO DE v_grid v1
      (e.edge=124.0);
:v2=INSERT INTO DE v_grid v2
      (e.edge=128.2);
:v3=INSERT INTO DE v_grid v3
      (e.edge=132. 5);
:v4=INSERT INTO DE v_grid v4
      (e.edge=137.0);
:h1=INSERT INTO DE h_grid h1
      (e.edge=11.2);
:h2=INSERT INTO DE h_grid h2
      (e.edge=15.2);
:h3=INSERT INTO DE h_grid h3
      (e.edge=19.4);
INSERT INTO DE grid grid1
(v_edge[1]=:v1,v_edge[2]=:v2,
v_edge[3]=:v3,v_edge[4]=:v4,
h_edge[1]=:h1,h_edge[2]=:h2,
      h_edge[3]=:h3);
```

FIGURE THREE: The EDM-2 schema definition of a grid system for use in a concrete structure. The syntax of EDM-2 is given in (EDM Group, 1994).

Ccalls are either Variant or Invariant. Their truth tables are shown in Figure Two. EDM-2 does not provide a specific language for writing Constraint bodies. They are implemented externally (in C or C++), defined in a file and dynamically linked. They use an external application library for accessing and writing database values and some issues of automatic management. Invariant Ccalls are always guaranteed to be True oe Undefined at the end of a transaction. Each Variant and Invariant CCall instance has its own value -- one of those defined in the truth table. Over time a large library of Constraints is assumed to become available.

An example EDM-2 schema that defines a structural grid is begun in Figure Three. The grid is defined as a general gridline, specialized into horizontal and vertical gridlines. Two invariant Ccalls, monotonic_h and monotonic_v, guarantee that the gridlines are monotonically increasing. Sequences of the gridlines are referenced by a grid DE. A 2x3 grid is then instantiated. Here, EDM-2 local variables are used identified by a name starting with a colon (:). C code defining the Constraint bodies is not included.

```
/* floor structural system */        CREATE CON slab_s ( )
 CREATE DE floor_struc                 IMPL <filename> ;
  ATTR ( mesh:grid,                   CREATE CCALL slab_size
  floor_no:INT)                        CON slab_s ( ):V;
  DESC "floor structure              MODIFY DE slab_geom
  aggregation";                        ADD CCALL slab_size;
 /* define a slab */                 CREATE COMP to_floor_struc
 CREATE DE slab_geom                   TARGET floor_struc
 ATTR (left_edge:v_grid,               PART ( SEQUENCE OF slab_geom);
  right_edge:v_grid,                 /* a constraint to make
  top_edge:h_grid,                     slab layout well-formed */
  bot_edge:h_grid,                   CREATE CONSTRAINT slab_lap
  thickness:dist,                      (SEQUENCE OF slab_geom)
  width_in_x:dist,                     IMPL <filename>
  length_in_y:dist );                DESC "checks no slabs overlap";
  DESC "slab defined by 4 grids      CREATE CCALL slab_overlap
  and thickness";                    CONSTRAINT slab_lap
 /*constraint checks slab size */     (part_slab_geom):V;
```

FIGURE FOUR: The EDM-2 definition of a floor structure composed of slabs.

Given this grid definition, slabs may be defined to compose a floor structural system, as shown in Figure Four. The floor_struc DE holds a grid system and an identifier as attributes. It is the target for the to_floor_struc Composition. The Composition parts are slabs (a fuller example would include beams and girders). A slab references the top, bottom, left and right gridlines that border it. Some grid cells may remain unfilled, for example for an elevator. Slabs may span more than one gridline. However, it is not allowed that they overlap and an invariant Ccall in the Composition checks for this condition. Slab also has a depth, length and width attribute. The width and length have associated Ccalls that check they are consistent with the panel borders. These Ccalls are Variant.

4. Constraint Management

CCalls in the EDM-2 environment may be assigned values in two ways. They may be assigned by the body of a Constraint; they may also be assigned by an external application, during the write-back at the application's completion. In this way, the application can identify the conditions that it has satisfied.

Constraint management in EDM-2 responds to conditions (c) and (d) in Figure One. First, consider the case where the values accessed by a Ccall are changed. The EDM-2 C application interface library has database access functions that set flags on all variables and structures accessed within a Ccall. These flags identify all Variant Ccall instances that access a variable or structure. When any variable or structure is modified, EDM's modification operators set to NULL all the Ccall instances that access it. These flags are created and managed automatically, largely unknown to the user.

For defining the relation of constraints used in an application, EDM-2 incorporates a structure called an Accumulation. *Accumulations* carry the following information: (1) a set of pre-condition constraints that define the rules required for an application to be executed, (2) a set of post condition constraints, identifying the rules satisfied by the application, (3) and a Composition or DE that the Accumulation is associated with. An Accumulation is a class structure corresponding to an application program. Since an application may be executed multiple times, each execution corresponds to an Accumulation instance. An Accumulation instance identifies in addition for each invocation the DE instances that are written by the application. The integrity logic implemented within EDM is that if the pre-condition constraints for an operation become not True, then the post-condition constraints that the operation satisfied are also no longer guaranteed and must be set to NULL. This can result in a forward chain, setting to NULL all data derived from the changed values.

An example is presented in Figure Five. It is assumed that an external application exists to define `slab_geom` instances, represented by the Accumulation define_slabs. For each slab generated, it will satisfy all the associated constraints. Two slab instances are created the first time the application is called. An Accumulation instances records this event. An external application is used to calculate the slab thickness (and reinforcing, from loads, not presented here). The application is represented by an Accumulation called `slab_thickness`. It has two pre-condition constraints, `slab_size` and `loads` The Accumulation instance for two slabs S1 and S2 are shown. `slab_size` has a post-condition constraint, `con_slab_th`, corresponding to the application.

```
/*define floor_struc instance */        /* insert accum. instances as
INSERT INTO DE floor_struc fs1              composition parts */
   (mesh=grid1,floor_no=1);             INSERT (s1,s2)INTO ACCUM
/*accum. for slab layout appl.*/          define_slabs;
CREATE ACCUM define_slabs              UPDATE fs1 OF COMP floor_struc
   REF to_floor_struc                    ADD PART (slab_geom:s1,s2);
   PRE ( )                             /*constraint on slab thickness*/
   POST (slab_overlap)                CREATE CON sl_t
/* result of slab application */         (loads,slab_geom);
INSERT INTO DE slab_geom s1           CREATE CCALL conc_slab_th
   (left_edge=:v1,                      CONSTRAINT sl_t
   right_edge=:v2,                       (loads,targ):V;
   top_edge=:h1,                      /*accum. for thick.application*/
   bot_edge=:h2,                      CREATE ACCUM slab_thickness
   width_inx=4.2,                       REF slab_geom
   length_iny=4.0  );                   PRE (slab_size,loads)
INSERT INTO DE slab_geom s2             POST (conc_slab_th)
   (left_edge=:v2,                    /* result of 2 runs of thickness
   right_edge=:v3,                       application */
   top_edge=:h1,                      INSERT (s1)INTO ACCUM
   bot_edge=:h2,                         slab_thickness;
   width_inx=3.8,                     INSERT (s2)INTO ACCUM
   length_iny=4.0  );                 slab_thickness;
```

FIGURE FIVE: The database operations corresponding to the definition of two slabs and calculation of their thickness.

Now let us suppose that some data is changed. For example, the vertical gridline whose keyname is v1 might be moved. The following constraint checking takes place:

(1) the invariant Ccall monotonic_v checks that the change does not violate the ascending order rule; if it does, the action is rejected; we assume it is okay here;

(2) since the Variant CCall instance for slab_geom s1 accesses gridline v1, the changed value automatically sets this instance of slab_size to NULL, flagging it to be rechecked. If other slabs rely on gridline v1, they too would be flagged.

Let us now consider the second kind of effect. slab_size is a pre-condition to the first slab_thickness Accumulation instance. Since its pre-conditions are no longer satisfied, its post-condition constraint is set to NULL, requiring it to be re-evaluated. That is, the effects of the change have been captured. The full logic for these two kinds of integrity management and constraint propagation are presented elsewhere (Eastman, 1995).

5. Conclusion

Databases for engineering and design, to be useful in practice, must be able to guarantee the integrity condition of the data carried. The methods presented provide such a guarantee and can deal with a wide range of engineering relations. NOTE: This work was supported by NSF grant, No. IRI-9319982.

REFERENCES

H. Assal, C.Cho, and C. Eastman, [1991], "An EDM Model of Concrete Structures", Design and Computation research report, Graduate School of Architecture and Urban Planning, U.C.L.A., Los Angeles.

Augenbroe,G. [1992] "Integrated Building Performance Evaluation in the Early Design Stages" Building and Environment., 27, pp.149-161.

Eastman, C.M., and A. Kutay, [1991] "Transaction management in design databases", in D. Sriram, etal, Concurrency in Engineering Data Management, Springer-Verlag, N.Y.

Eastman, C.M., S. Chase and H. Assal, [1993], "System architecture for computer integration of design and construction knowledge", Automation in Construction, 2:2, (July), pp. 95-108.

Eastman, C.M. [1995] "Managing Integrity in Design Information Flows", Research Report, Center for Design and Computation., University of California, Los Angeles, CA. 90024.

EDM Group, [1994] EDM-2 Reference Manual, Research Report, Center for Design and Computation., University of California, Los Angeles, CA. 90024

Galle, Per, [1994] "Towards integrated 'intelligent' and compliant computer modeling of buildings", Report of Computer Aided Building Design Unit, Techical University of Denmark, Lyngby, Denmark.

Ullman, Jeffrey D. [1988], Principles of Database and Knowledge-Base Systems, Vol.1, Computer Science Press, Rockville, MD.

Improving Design Problem Formulations Using Machine Learning

John S Gero[1]

Abstract

This paper applies a machine learning approach to the improvement of the design formulation. It utilises an evolutionary technique to learn better problem reformulations in order to improve the formulation. This results in both a near optimal problem formulation and an improvement in the solution synthesised from that formulation.

Introduction

Most of machine learning in design has focussed on learning generalisations in order to predict some future behavior of the system under consideration. Such approaches have been applied primarily to the analysis and synthesis stages of designing. There has been very little work done relating to the formulation stage. This paper applies a particular machine learning approach to the improvement of the formal description of the design formulation. It applies an evolutionary technique to the problem reformulation in order to improve the formulation. This results in both a near optimal problem formulation and an improvement in the solution synthesised from that formulation.

Design can be considered a purposeful, constrained, decision-making, exploration and learning activity. Decision-making implies a set of variables, the values of which need to be determined. Search is the common process by which values for variables are determined. Exploration is the process by which the variables which are utilised to describe the design problem are decided. The effect of exploration is to define the problem space within which decision making occurs. Learning implies the restructuring of knowledge as opposed to restructuring of facts.

Searching in design is concerned with restructuring the facts, Figure 1, while exploration in design is a class of learning processes that restructures the knowledge used to define the space within which search occurs, Figure 2. Thus, the result of exploration is the equivalent of reformulating the design problem.

---

[1] Professor of Design Science, Co-Director, Key Centre of Design Computing, Department of Architectural and Design Science, University of Sydney, NSW 2006, Australia, john@arch.su.edu.au

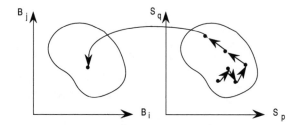

Figure 1. Search is carried out within a fixed design space of structure space guided by performances in behavior space.

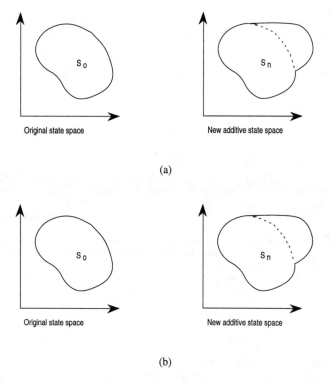

Figure 2. Exploration modifies the design space; design spaces can be: (a) added to or (b) substituted for.

Machine learning is comprised of four principal paradigms (Carbonell 1990; Mitchell et al. 1986): the inductive paradigm, the analytic paradigm, the genetic paradigm and the connectionist paradigm. All of these paradigms are employed within a shared view of the role of machine learning in design; namely that of "learning to perform old tasks better using available tools". This, the main emphasis is on improving existing tasks. Inductive learning is an example of this. The system is provided with a set of examples and it produces generalisation of those examples under a set of concepts that are used to describe those examples. These concepts are then used to predict the classification of future examples.

The type of learning that we wish to describe here is more concerned with " learning to restructure knowledge to produce results which could not have been achieved using current knowledge" (Gero et al 1994).

Learning in Design

Learning is always concerned with improving the quality of knowledge, traditionally from examples. A number of different teleologies drive machine learning not all of which are directly applicable to design. The foci in machine learning in design include: concept formation (Maher and Li 1992: 1993), learning high level relationships (Rao and Lu 1993; Gunaratnam and Gero 1993) and increasing the efficacy of available knowledge (Arciszewski et al 1987). The approach adopted here is to improve the quality of the knowledge. It is concerned with restructuring design knowledge which is already in the form of generalisations.

The focus of the application of machine learning in design has been largely in the areas of design analysis and design synthesis. The work reported here is concerned with that aspect of design called design formulation. Design formulation deals with the early part of designing when the design problem is being described with variables and relationships between those variables. This area is not well understood although there is a nexus between the formulation of a design problem and its solution. We are all familiar with the aphorism: "if you are not part of the solution you are part of the problem".

Learning Improved Design Formulations

Normally the formulation of a design is fixed. For example, in design optimization the variables are specified as are the constraints which define relationships between them. This would appear to rule out any opportunity to learn better formulations. However, if we replace the fixed variables in the formulation by a richer representation such as a state transition grammar then there are opportunities to learn grammars which are better representations than the initial grammar which maps onto the fixed variables in the original formulation.

A typical state transition grammar has the following form:

LHS —> RHS

where   **LHS** is a symbolic left hand side found in the current state
           RHS is a symbolic right hand side which is substituted for the LHS to
                produce a new state.

For example:

          A —> A+B

where   **A** could have any semantic meaning in a context such as, say, a beam
           B could have any semantic meaning in a context such as, say, a column.

This rule in the grammar would state that whenever there is a beam by itself, replace it
with the beam and a column, thus defining a relationship between the beam and the
column.

With a fixed grammar the order of the execution of the grammar rules is what is
to be determined. This then becomes the optimization problem in design. The optimal
design is produced by the optimal order of execution of the grammar rules so as to
optimize a set of defined objectives.

If we allow the rules to change so that not only is the order of execution to be
determined but the 'best' rules which can be found then we can expect better results.
By allowing the rules to change we are changing the formulation of the design problem. How do we allow a system to learn better rules? One simple approach is to treat
the rules of the grammar as the genes in an evolutionary system driven by a genetic
algorithm (Goldberg 1989). With the rules as the genes we need to allow crossover to
occur in such a way that new rules are created from existing rules, Figure 3.

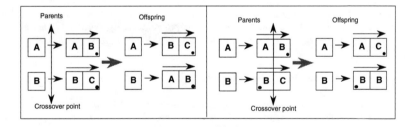

Figure 3. Generating new grammar rules by crossover; different crossover points
within a rule produce different potential offspring rules.

These new rules then produce different designs than the original rules were capable of
producing. The system learns which of these new rules produce better results and adopts
those in succeeding generations. Each generation is capable of creating new rules and
each set of rules is a learned grammar which corresponds to a reformulation of the
design problem using the same set of initial variables but with different relationships

between them. In this way it is possible to begin to address the issue of improving problem formulations in order to improve the resulting designs.

Acknowledgments

This work is supported by a grant from the Australian Research Council.

References

Arciszewski, T., Mustafa, Z. and Ziarko, W. (1987) A methodology of design knowledge acquisition for use in learning expert systems, *Man-Machine Studies* **27**: 23–32.
Carbonell, J. G. (1990) Paradigms for machine learning, *in* J. Carbonell ed., *Machine Learning Paradigms and Methods*, MIT/Elsevier, Cambridge, MA. pp. 1–10.
Gero, J. S., Louis, S. and Kundu, S. (1994) Evolutionary learning of novel grammars for design improvement, *AIEDAM* **8**: 83–94.
Goldberg, D. (1989) *Genetic Algorithms in Search, Optimization and Machine Learning*, Addison-Wesley, Reading, MA.
Gunaratnam, D. and Gero, J. S. (1993) Neural network learning in structural engineering applications, *in* L. F. Cohen ed., *Computing in Civil and Building Engineering*, ASCE, New York. pp.1448–1455.
Maher, M. L. and Li, H. (1992) Automatically learning preliminary design knowledge from design examples, *Microcomputers in Civil Engineering* **7**: 73–80.
Maher, M. L. and Li, H. (1993) Adapting conceptual clustering for preliminary structural design, *in* L. F. Cohen ed., *Computing in Civil and Building Engineering*, ASCE, New York. pp.1432–1439.
Mitchell, T, Carbonell, J. and Michalski, R. (eds) (1986) *Machine Learning A Guide to Current Research*, Kluwer, Boston.
Rao, R. and Lu, S. (1993) A knowledge-based equation discovery system for engineering domains, *IEEE Expert* **8**(4): 37–42.

A Probabilistic Machine Learning Model for Supporting Collaborative Design

Nenad Ivezic[1] and James H. Garrett, Jr.[2]

Abstract

Decision support systems that support early collaborative design processes must assist designers from different design perspectives: (1) in synthesizing and analyzing parts of a solution; (2) in identifying conflicts among the different perspectives; and (3) in managing design tradeoffs. A probabilistic machine learning–based model can be used to support such activities. This model provides a basis for interaction among designers during their decision making and for applying machine learning to acquire and integrate design knowledge. A framework in which an experimental evaluation of this model–based approach can be conducted is discussed.

1 Introduction

Decision support systems that support an early collaborative design process must assist designers from different design perspectives: (1) in synthesizing and analyzing parts of a solution; (2) in identifying conflicts among the different perspectives; and (3) in managing design tradeoffs. Requirements for such systems can be seen from three different perspectives: the structural perspective, the functional perspective, and the dynamic perspective. From the *structural* perspective, such decision support systems must manage a distributed network of designers (actors), each with a possibly different design perspective and responsible for a set of decisions appropriate for their perspective. To be able to contribute, designers must have a shared understanding of the progress being made during the design process. Hence, a shared description of the evolving solution must be maintained and managed by a decision support system. From the *functional* perspective, decision support systems must support decision making at a high level of abstraction so that key decisions are identified and the decision process is manageable. As a result, decision support systems must deal with uncertainty in the design process. Such systems must also support bi–directional inferences: performance estimation (i.e., analysis) and search for alternatives (synthesis). From the *dynamic* perspective, decision support systems must support concurrent and asynchronous decision making among a diverse, interacting set of designers. Inevitably, conflicts arise when these independent decision making processes are integrated. Thus, identification and management of these conflicts must be supported.

We are developing and testing a three stage methodology for building and using such decision support systems. In the first stage, machine learning is used to extract and compile relationships

---

[1] Research Assistant, Dept. of Civil and Env. Eng., Carnegie Mellon University
[2] Associate Professor, Dept. of Civil and Env. Eng., Carnegie Mellon University

among the elements of a design decision problem from existing design knowledge resources (i.e., design tools and databases.) In the second stage, such compiled design relationships are used to quickly estimate performances and to search for alternative design solutions. In the third stage, the identified relationships are used to identify possible conflicts among the multiple design perspectives in the design process and to provide guidance in resolving these conflicts.

In the remainder of this document, we describe in more detail these three stages of our methodology for building and using decision support systems. In Section 2, we review the proposed model of early design decision making processes—the behavior–evaluation model. This model provides the basis on which design relationships are extracted using machine learning approaches. In Section 3, we discuss the alternative machine learning approaches that are being used to build the components of the behavior–evaluation model. In Section 4, we present the setup with which we will experimentally evaluate this methodology.

2 Behavior–Evaluation Model of Collaborative Decision Making

Two goals have been posed for the development of a model of early design decision making—the behavior–evaluation model. First, the model needs to provide for a formal, yet intuitive, representation of design decision making problems. Second, the model needs to allow for management of complexities when applying machine learning to learn design knowledge.

Two types of actors participate in the collaborative design process using the proposed decision support systems: (1) *design agents* representing a specific design discipline and perspective involved in the design process; and (2) a *project manager* representing the overall concerns of the product design process and who may be thought of as an owner/customer representative.

Each design agent is exclusively responsible for making a collection of decisions about the form of the product that belongs to that agent's discipline. On the other hand, the interest of the design agents may extend to decisions made in a number of disciplines. In summary, the authority of agents are exclusive, while the interests are overlapping. The project manager's responsibility and interests are to effectively enforce the overall objectives for the design problem. The project manager has no interest in making decisions about the form of the product except to enforce specifications, such as using a specific material.

Figure 1 illustrates the behavior–evaluation model (Ivezic 94a). This model is based on an idealization of the decision making process as consisting of three sub–processes: (1) selecting some or all attributes of a design; (2) determining behavior of the design; and (2) evaluating behavior of the design in terms of the objectives of the design process. We make a further assumption that these three sub–processes can be effectively captured within two mappings: (1) the behavior model; and (2) the evaluation model. Consider, for example, the structural designer shown in Figure 1. The mappings (i.e., the behavior and evaluation models) that underlie the structural decision making process are contained in a module which plays a role of an intelligent design assistant, or IDA (Anderson 94). The behavior model within this IDA relates structural decisions to behaviors of the product design from a structural perspective: structural strength and deflection. The evaluation model within the same structural IDA relates the behaviors of the design to the objectives of structural design: structural safety and serviceability. On the other hand, the project manager has an associated IDA that contains only an evaluation model. This model relates the behaviors of a product design from relevant disciplines to the global objectives.

A decision support system addressing a specific collaborative product design problem may include numerous models of behavior and evaluation. Figure 1 illustrates conceptually an integral model of such a system for the collaborative design of a bridge structure. The model consists of a collection of IDAs, each containing a unique set of behavior and evaluation models and support-

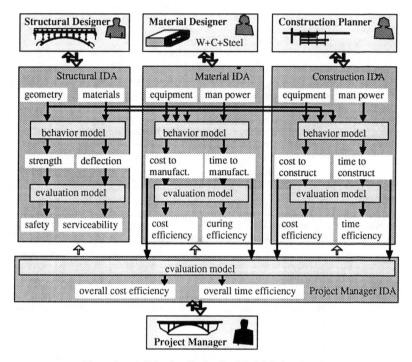

Figure 1 — A Behavior–Evaluation Model Example

ing a participant from a specific design discipline. Integration of these models into a collaborative design decision support system is achieved at two levels: (1) at the decision level, which allows propagation of decisions from each discipline to all other affected disciplines and enables understanding of impacts of decision making across disciplines; and (2) at the objectives level, where relating the global objectives (indirectly through behaviors) to local objectives provides a means by which to understand the problem–specific design tradeoffs and to control the design process by imposing policies on design tradeoffs.

By using the proposed behavior–evaluation model, the overall requirements for decision support systems, introduced in Section 1, can be met. From the structural perspective, the distributed and integral behavior–evaluation model allows for implementation of a distributed network of designers and decision variables. This model also allows for the building of a shared design description (described in detail in (Ivezic 94a)). From the functional perspective, this model provides a basis for the abstraction process. Management of uncertainty is provided by viewing the behavior–evaluation model in a probabilistic framework. Treating this model probabilistically also provides for bi–directional inferences. Finally, from the dynamic perspective, the distributed probabilistic behavior–evaluation model allows concurrent and asynchronous processing of parts of this model while the bi–directional inferences allow for conflict management.

We now turn to the user requirements for such decision support systems. Five basic functionalities must be supported by such systems: (1) estimation and evaluation of design performances; (2) search for alternative design solutions; (3) conflict identification; (4) conflict resolution; and (5) performing design tradeoffs. Figure 2 presents a probabilistic view of a portion of the behavior–evaluation model that we use to show how these functionalities are provided.

The user of a decision support system is able to specify ranges of values for decision variables. User specifies constraints both on form variables (e.g., product material and geometry) and performance variables (e.g., strength, cost, duration of manufacturing process). A probabilistic behavior–evaluation model provides a window onto the state of compatibility of constraints imposed on these decision variables. Figure 2 illustrates the manner in which the behavior–evaluation model is used: each variable of the model has an associated estimate of probability distribution (computed using the probabilistic behavior–evaluation model) which is interpreted as the *likelihood of values* that may be assumed by the variable conditioned on all specified constraints on variable value ranges. The constraints on the value ranges may be specified for any variable in the model, allowing effectively for (bi–directional) inferences of likelihoods of these values.

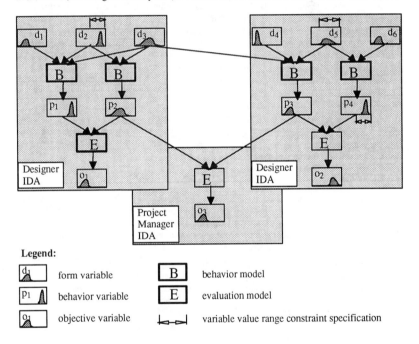

Figure 2 — Probabilistic Behavior–Evaluation Model

Once the constraints are specified for the decision variable ranges, the user is capable of using the decision support system to obtain estimates of design performances and to perform the search for alternative decisions about form variables. These two uses are possible because the probabilistic model can be used to estimate probability distributions on *all* variables. Identification of conflicts is straightforward: if there exists an incompatible collection of design specifications (specified in the form of constraints on variable values), the probability distributions for all variables will not be computable. This is the consequence of the procedures used to estimate the probability distributions based on Monte Carlo simulations.

In order to achieve support for conflict resolution, we make use of differentiable probabilistic models (e.g., multi-layer perceptrons), which allow us to determine sensitivities of objectives and performances with respect to decisions (Klimasauskas 91). Then, in the case of a conflict, we can use the probabilistic model to determine which decisions need to be altered in order to achieve specified constraints on behaviors and objectives. Finally, to support design tradeoffs, we use the evaluation model to specify preferences, in terms of explicit objectives, on design behaviors.

3   Machine Learning Approaches to Building Behavior Models

The intent of this research is to provide automated knowledge acquisition approaches to build, use, and maintain behavior models of product and process designs. These models are one of the principal components in decision support systems for early stages of a collaborative design. On the other hand, means for manual construction of relatively simple models of evaluation knowledge for a product will be provided. The rationale for this approach to building decision support systems has two parts: (1) knowledge of product behavior is typically complex while evaluation knowledge in many cases of early stages of design consists of simple relationships that compare specific behaviors of design to target behavior values; and (2) behavior knowledge is relatively invariant for different design contexts while evaluation knowledge is volatile and subject to many contextual conditions.

In this section we focus on building behavior models by applying machine learning approaches to transform and operationalize existing design knowledge resources. It is beneficial to view this process in terms of two sub-goals: (1) structure discovery of the behavior model; and (2) estimation of the behavior model parameters.

3.1   Structure Discovery of Behavior Models

The underlying probabilistic structure of a behavior model may be conveniently represented as a probabilistic graphical model. Learning of graphical structures is researched in a number of camps within the fields of of Artificial Intelligence, Machine Learning, and Statistics. An approach, which we are considering in our work, is based on the theory developed by Spirtes *et al.* in (Spirtes 93). At the heart of this approach lies the notion of causality.

The concept of causality emerges in a number of problems. For example, in *the collapsibility problem*, one is faced with the following question: when can the same conclusions about the influence of variable A on variable B be obtained by analyzing a reduced set of variables (containing A and B). If one knew the causal structure of the problem, one would be in a position to disregard those variables which influence variable B only indirectly through, say, variable C, given that variable C is known. The other problem is that of *prediction under forced actions*: what effects due to forced manipulations on the model variables may be predicted and how. The answer to these questions does not follow from the estimate of an actual probability distribution but rather from the causal interpretation of the probability relationships.

To address the issue of causality within the probabilistic framework, Spirtes *et al.* propose an axiomatic approach which effectively ties the notion of causality to the probabilistic graphical structures by two axioms which have a "natural" interpretation. A very practical result of this approach is that traditional statistical approaches to prediction under forced actions are considerably suboptimal. In other words, the theory has resulted in algorithms which lead to considerably improved predictions compared to those produced with the traditional statistical approaches.

Behavior and evaluation relationships in the behavior–evaluation model must be treated as causal relationships; thus, we apply Spirtes' approaches to building such models. Two uses of these algorithms are planned: (1) discovery and validation of the proposed causal structure of the behavior models; and (2) validation of the learning data sample with respect to the causal structure. The algorithms are presently used on a class of probabilistic graphical models with restrictive assumptions. We look at adaptations of these algorithms to problems where these assumptions may be relaxed and these algorithms applied to more general structure discovery problems such as those that appear in our research.

3.2 Estimation of Behavior Model Parameters

In many cases of learning mappings, the outcome is a non–convex mapping where not only many solutions exist for a single input but also traditional function approximation would lead to solutions outside the allowable region. A number of recent papers have considered this problem and, more generally, the problem of learning parameter estimates of mappings that are inherently non-deterministic (Ghahramani 94, Bishop 94). Use of mixture models (i.e., models built from a combination of relatively simple probabilistic models) is common for all these approaches and is essential for learning general non–deterministic mappings (Buntine 94). In the approach described in (Ghahramani 94) no distinction is made between input and output variables as density estimation is performed over all problem variables. The density estimation is based on maximizing the likelihood of a parametric mixture model using the expectation–maximization algorithm (i.e., algorithm that finds the parameters of a model that maximize likelihood of given data). A supervised learning formulation is posed for the problem. The benefit of this approach is that it provides a single framework for real, discrete, or mixed data and generalizes to data analysis tasks with arbitrary missing data patterns. The problem involving the prediction of continuous variables is described in (Bishop 94). The problem described in this paper is that the conditional averages provided by standard machine learning approaches are very limited descriptions of target variables. In order to obtain a more complete description of the data, for the purposes of predicting the outputs corresponding to new input vectors, a model of conditional probability distribution for the target data is needed.

The importance of the above referenced work is that it represents one of the first attempts to address the need of learning of multi–valued, non–functional mappings in the machine learning community. For knowledge acquisition in the context of early collaborative design, multi–valued, non–functional mappings are frequently encountered. Mixture models and other approaches to estimating parameters that lead to probability density estimates are used to estimate behavior model parameters in our methodology for building decision support systems.

4 Experimental Evaluation of Methodology

The primary component of future work in this project involves the evaluation of our methodology for creating decision support systems. In this section, we describe the experimental evaluation of the methodology. First, we describe the aspects of our evaluation approach, and then we give the overall strategy for this approach.

4.1 Aspects of the Evaluation Approach

Figure 3 shows the four aspects of our experimental evaluation approach and the considerations within these aspects. The first aspect is evaluation criteria: (1) usability; and (2) effectiveness. The second aspect is required functionality: (1) support of evaluation of design performances; (2)

search for alternative design solutions; (3) support of conflict identification; (4) support of conflict resolution; and (5) support for performing design tradeoffs. The third aspect is design problem complexity: (1) parametric design; (2) configuration design; and (3) time–dependent design. The fourth aspect is the mode of use: (1) single–user scenario; and (2) collaborative scenario.

Evaluation Aspects	Considerations				
Evaluation Criteria	usability	effectiveness			
Functionality	performance estimation	search for alternatives	conflict identification	conflict resolution	design tradeoffs
Problem Complexity	parametric	configuration	time– dependent		
Mode of Use	single– user	collaborative			

Figure 3 — Aspects and Dimensions of the Experimental Evaluation

4.2 Overall Strategy of the Evaluation Approach

The key issue in developing an evaluation approach is dealing with the intractable nature of realistic design problems. In general, it is not possible to use a deductive approach in evaluating problem solving approaches for these problems. Faced with this issue, we intend to concentrate on solving specific problems that are realistic and representative of a specific class of design problems. Then, we hope to be able to generalize from our findings and apply our approach to other design problems. Further, in performing our experiments, we intend to first reduce the complexity of the problem by constructing a simpler version of the problem and understand the nature of the problem and the capabilities of a specific machine learning approach by solving a series of problems of increasing complexities. An instance of such an approach was presented in (Ivezic 94b), where an artificial design problem was used to compare alternative machine learning approaches.

Ultimately, we are interested in evaluating the usability and effectiveness of the proposed methodology to build decision support systems for early stages of collaborative design. The usability criterion will evaluate the users' attitudes towards the decision support system framework. The effectiveness criterion will evaluate the capabilities of the adopted machine learning and other supporting computational approaches within the decision support system framework. To help in the evaluation of these two criteria, we have identified sub–criteria and, ultimately, metrics.

For usability, we identify three sub–criteria: comprehensiveness, expressiveness, and interpretability. The metrics we intend to use for these sub–criteria will be measurements of the time needed to learn and demonstrate understanding of the methodology, the number of a specific decision problem constructs that are able to be expressed in the language of behavior–evaluation model, and the frequency of correct interpretations of outputs of the decision support system, respectively. For effectiveness, we will create experiments to measure the accuracy and precision of the machine learning approaches in supporting the functionalities shown in Figure 3.

We are interested in how the decision support system methodology performs on representative parametric, configuration, and time–dependent design problems, with varying numbers of collaborating design agents. To perform these measurements, we are in the process of preparing two experiments. The first experiment will take place in the classroom. We will allow students who are familiar with using traditional analysis and simulation tools to use a decision support system for a relatively simple design problem. With this experiment, we intend to primarily measure the usability of the decision support system. A second experiment is planned for a realistic design setting where a number of tools and databases will be used to create a decision support system. We will use this experiment to confirm our findings about usability of the approach and perform further measurement of effectiveness of the methodology. The results of these experiments will be reported on in a future publication.

5 Summary

We presented a probabilistic machine learning model for support of collaborative design processes in terms of two sub–processes: behavior estimation and behavior evaluation. By casting this behavior–evaluation model in a probabilistic framework, we are able to provide five user–specific functionalities needed to support collaborative design processes: (1) support of evaluation of design performances; (2) search for alternative design solutions; (3) support of conflict identification; (4) support of conflict resolution; and (5) support for performing design tradeoffs. For each variable of the behavior–evaluation model, we are able to estimate a probability distribution given specifications for value ranges of some of the variables. These probability distributions can be thought of as the likelihood of values for a variable conditioned on all specified variable constraints. Several machine learning approaches will be used in constructing a behavior–evaluation model for a given design problem. Probabilistic machine learning algorithms will be used for both acquiring model structure and for learning model parameters. A set of experiments are planned to evaluate the usability and effectiveness of this approach.

6 Acknowledgements

This work is supported by the Engineering Design Research Center, an NSF Engineering Research Center.

7 References

Anderson, T. (1994). "Intelligent Design Assistants: The Need for Supporting User Augmentation of Knowledge." In *Proceedings of the International Workshop on the Future Directions of Computer–Aided Engineering*, D. Rehak, ed., Carnegie Mellon University, pp. 173–176.

Bishop, C. M. (1994). *Mixture Density Networks*. Neural Computing Research Group Report NCRG/4288, Department of Computer Science, Aston University.

Buntine, W. L. (1994). "Learning and Probabilities." Tutorial in *1994 MLnet Summer School on Machine Learning and Knowledge Acquisition.*

Ghahramani, Z. (1994). "Solving Inverse Problems Using an EM Approach to Density Estimation." In *Proceedings of the 1993 Connectionist Models Summer School*, Mozer, M. C., P. Smolensky, D. S. Touretzky, J. L. Elman, and A. S. Weigend (eds), Hillsdale, NJ, Erlbaum Associates, pp. 316–323.

Ivezic N. (1994a). *Machine Learning Support of Collaborative Design*. Unpublished PhD Proposal, Civil and Environmental Engineering Department, Carnegie Mellon University, Pittsburgh, PA, 62 pgs.

Ivezic, N. and J. H. Garrett, Jr. (1994b) "A Neural Network–based Machine Learning Approach for Supporting Synthesis." *Artificial Intelligence for Engineering Design, Analysis, and Manufacturing*, vol. 8, pp. 143–161.

Klimasauskas, C. C. (1991). "Neural Nets Tell Why." *Dr. Dobb's Journal*, vol. 16, no. 4, April 1991, pp. 16–24.

Spirtes, P., Glymour, C., and Scheines, R. (1993)., *Causation, Prediction, and Search*. Springer Verlag, 1993.

Preparation of Examples for Learning Systems:
A Strategy for Computer-Generated Examples

Mohamad Mustafa[1], Member ASCE

Abstract

This paper proposes a new strategy for the preparation of examples for acquiring conceptual structural design knowledge through the use of inductive learning systems. The utilization of computer-generated examples in knowledge acquisition has several advantages over the traditional human experts prepared examples. It provides the user with the flexibility to generate as many examples as desired to complete the learning process in a relatively short time and it allows the full control over the number and character of examples generated. The paper discusses the stages in the preparation of examples using computer-generated example's strategy. It also describes the process of acquiring engineering design knowledge using AQ15 learning system. The final results of learning include selected decision rules from the domain of wind bracing systems of tall buildings. The acquired knowledge is verified using the leave-one-out empirical error rate method. The paper includes a comparison of results obtained in earlier studies using the human-expert-prepared examples.

Introduction

Structural design process can be divided into two design stages: conceptual and detailed design stages. In the conceptual design stage, constraints and available knowledge are used to produce a class of abstract descriptions of the structural system being designed; such as one, two, or three-bay rigid frame or truss wind bracing system when designing a three-bay skeleton structure. The selection of such structural design systems is time consuming and greatly depends on the designer's experience. It is usually conducted in an informal way. In the detailed design stage, the selected structural systems are analyzed, designed, and optimized to develop the final structural system. In this stage, available structural design software are heavily utilized by engineers to detail products. The use of these software is an acute problem particularly when inexperienced designers use them.

---

[1]Assistant Professor, Department of Engineering Technology, Department of Civil Engineering Technology, Savannah State College, Savannah, GA 31404

Their lack of experience in the form of abstract knowledge may lead to designing structural systems which could be improved through simple changes in the structural configuration.

Recently, a number of research projects on automated knowledge acquisition were initiated to improve the current practice of acquiring abstract descriptions of structural systems in the conceptual design stage. For example, a genetic algorithm was used by Grierson and Pak in a system for the optimization of configurations of skeleton structures (Grierson and Pak, 1992). The same algorithm was also utilized by Maher to develop a learning system for the general purpose of knowledge acquisition about conceptual design (Maher, 1992). Arciszewski et al. used constructive induction for learning design rules for wind bracing systems in tall buildings (Arciszewski et al., 1994). The author used AQ15[2] and Datalogic[3] learning systems to develop the methodology of inductive learning for structural engineering applications (Mustafa, 1994). In that particular study, the author indicated that one of the crucial elements for acquiring structural design knowledge using learning systems is the strategy for the preparation of design examples. It was found that the traditional approach to the preparation of structural engineering design examples which involves the consultation of domain experts is not sufficient to complete the learning process and obtain sufficiently accurate decision rules. For this reason, and the availability of many structural optimization design and analysis software, the strategy for computer-generated examples is developed.

Strategy for Computer-Generated Examples

The proposed strategy is related to a more general knowledge acquisition process proposed in (Arciszewski et al., 1994).

An example is understood in this study as a feasible combination of nominal attributes and their values which describes a given engineering system (here - a wind bracing system) under investigation.

Four major stages are distinguished in the strategy for computer-generated examples as shown in Figure 1 below: 1) *Development of Mathematical Models*. In this stage, the available domain knowledge is used to develop a system of mathematical models describing the given engineering system under consideration. The mathematical models are next implemented in computer programs. 2) *Model Implementations*. This stage involves the development of computer software using the developed system of mathematical models for the general domain under investigation. As an example of this stage in the area of steel design structures, Professor Donald Grierson of Waterloo University, Canada used the available general steel structural design knowledge to develop a system of mathematical models for the analysis, design, and optimization of steel structures under static loading which he later implemented into a computer package called SODA[4]. 3) *Development of Representation Space*. In this stage, the available general domain knowledge is used to identify a number of relevant nominal attributes and their feasible values to describe a given

---

[2]Developed by the Machine Learning and Inference Center, George Mason University, Fairfax, VA

[3]Developed by Reduct Systems Inc., Regina, Canada.

[4]SODA is a Structural Optimization Design & Analysis Software for Structural Engineering, Developed by Waterloo Engineering Software, Waterloo, Ontario, Canada.

engineering system under investigation. 4) *Model-Based Example Generation.* This stage involves the preparation of a collection of examples using the developed software and the prepared attributes and their values for the specific engineering system under consideration.

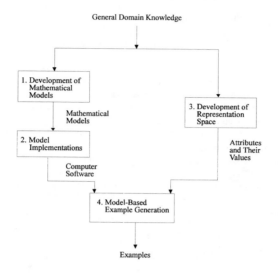

General Domain Knowledge

1. Development of Mathematical Models

Mathematical Models

2. Model Implementations

Computer Software

3. Development of Representation Space

Attributes and Their Values

4. Model-Based Example Generation

Examples

Figure 1. Preparation of Examples: Computer-Generation

Using the process described above, the computer-generated examples strategy is defined as examples generated using analytical, design and optimization software developed and implemented for the specific domain under consideration.

Preparation of Wind Bracing Examples

The general domain knowledge of wind bracing for steel skeleton structures was initially used by Arciszewski (1985) to develop a general representation space for all types of wind bracing systems and it was developed next by the author for the purpose of his project. This representation space consists of eight attributes, which include nominal characteristics with symbolic values, such as the attribute "D. Static character of joints," and numerical attributes of discrete character, such as the attribute "A. Number of stories" as shown in Table 1. The development of these attributes and their feasible values are based on the following assumptions: 1) The number of stories of a three bay steel skeleton structure ranges from six to thirty stories with variation of six. Therefore attribute "A. number of stories" has five feasible values of 6, 12, 18, 24, and 30 stories. has five feasible values of 6, 12, 18, 24, and 30 stories. 2) Each story is assumed to be 13 feet high. This value was selected to limit the ratio of height to least horizontal dimension of the building to less than 5 (Taranath, 1988). 3) Attribute "B. bay length" is assumed to have two feasible values of 20 and 30 feet length. They are the most commonly used values by architects for office and residential buildings. 4) It is assumed that the primary occupancy of the building is one in

Attributes	Attribute	Values			
	1	2	3	4	5
A. Number of stories	6	12	18	24	30
B. Bay length (feet)	20	30			
C. Importance factor (I)	1.07	1.11			
D. Static character of joints	Rigid	Hinged	Rigid & Hinged		
E. Number of bays entirely occupied by bracing	1	2	3		
F. Number of vertical trusses	0	1	2	3	
G. Number of Horizontal trusses	0	1	2	3	
H. Steel unit weight	High	Medium	Low	Infeasible	

Table 1. Representation Space for Computer-Generated Wind Bracing Examples

which more than 300 people congregate in one area. Therefore the attribute "C. importance factor for wind", which is needed for wind pressure calculation, has two feasible values. The first value is 1.07 when the building is 100 miles from hurricane ocean line and in other areas. The second value is 1.11 when it is at hurricane ocean line. 5). The material to be used is steel and has a yield stress of 36 ksi. 7) Design constraints are based on the American Institute of Steel Construction manual, ninth edition, stress and deflection constraints. 8) Attributes "D. static character of joints" through "G. number of horizontal trusses" are used to describe the structural components of rigid frames and truss wind bracing systems. Their feasible values are as shown in Table 1. 9) The gravity pressure is assumed to be constant and its values are as follow: ROOF: Dead Load (DL) = Live Load (LL) = 30.0 psf. FLOOR: DL = 60.0 psf, Superimposed Dead Load (SDL) = 28.0 psf, LL = 80.0 psf.

Using the above eight described attributes with their feasible values and the SODA software, 336 compatible combinations of a minimum weight design of a wind bracing system were prepared and used as examples for the learning process. Eighty four examples are for each of the six, twelve, and eighteen story buildings. Only forty two examples are for each of the twenty four and thirty story buildings because attribute "B. bay length" cannot be equal to the 20 feet value in order to satisfy the third assumption described above.

Since the 336 examples were generated using analytical and design software, the obtained values of the attribute "H. Unit steel weight" have numerical character. These values are converted then to nominal values. For each tall building height considered in this study, the actual unit steel weight range of variation is determined and the unit weights of individual wind bracing normalized. Table 2 below shows the calculated unit steel weight range of variation for each individual heights of tall building considered.

Number of stories	Steel unit weight range of variation (pcf)	Normalized range
6	≥ 0.058 and ≤ 0.090	0 to 1
12	≥ 0.068 and ≤ 0.196	0 to 1
18	≥ 0.078 and ≤ 0.261	0 to 1
24	≥ 0.088 and ≤ 0.303	0 to 1
30	≥ 0.100 and ≤ 0.423	0 to 1

Table 2. Unit Steel Weight Range of Variations

Next, the normalized (0,1) range was divided into three equal sub ranges and nominal values assigned to these sub ranges: low, medium and high. For example, Low means the low relative unit steel weight which is in the range of variation (0 - 0.33), while High means the high relative unit steel weight of a given wind bracing system under a set of design requirements, which is in the range of variation (0.66 - 1.0). The medium relative unit steel weight is in the range of variation (o.33 - 0.66).

Learning Process

The learning was conducted in accordance with the multi-stage learning strategy proposed in (Arciszewski and Mustafa, 1989) and seven learning were used. For this reason, the set of 336 examples was divided into 7 subsets, each subset contained an equal number of examples (i.e. 48 examples). Then the 336 examples were numbered and their sequence for individual subsets was randomly determined by the learning system.

The AQ15 learning system (Michalski et al., 1986) was then used to generate decision rules at individual learning stages. The first subset of examples (i.e. 48 examples) was entered into the learning system to be analyzed in the first learning stage. Then, the second subset of examples was added to the first subset. This new set of examples (96 examples) was entered into the learning system to be analyzed in the second learning stage. This process was continued until all seven subsets of examples were exhausted.

The acquired structural design knowledge wass verified using the leave-one-out overall empirical error rate method (Weiss and Kulikowski, 1991; Arciszewski and Dybala, 1992). In the leave-one-out method, the learning system randomly selects (N-1) examples, from a given set of N examples for learning and the remaining one example is used for testing. This process is repeated by the system N times; i.e., each example is used as a testing example. For example, in the first learning stage the learning system randomly selects 47 examples for learning and one example is for testing. This process is repeated 48 times and the average overall empirical error rate is calculated and recorded. The progress of learning is then monitored using inductive learning curves. In this case, the learning curve is a graphical representation of the relationship between a given overall empirical error rate method and individual stages of the learning process.

Results and Conclusions

Only selected decision rules, which describe the structural design knowledge acquisition for optimal wind bracing design, are presented with their structural interpretation.

Decision Rule 1:
IF: No. of stories = 30 stories, Static character of joints = rigid and hinged connections, No. of bays entirely occupied by bracing = 2 bays, and No. of horizontal trusses = 1 or 2 horizontal trusses
THEN: Steel unit weight = high

Structural Interpretation: When designing a building in the height range of 30 stories the following two configurations of structural components of a wind bracing system should be avoided: 1) a two bay wind bracing system of rigid frames and one horizontal truss. 2) a two bay wind bracing system of rigid frames and two horizontal trusses.

This decision rule is very useful in the conceptual design stage of wind bracing design. It limits the options of wind bracing systems that should be considered for the final analysis and reduces the costly time of analyzing and designing these systems. Since the consideration of these wind bracing systems will yield a high unit steel weight, this decision rule will serve the structural designer to justify his or her final selection of the wind bracing system.

Decision Rule 2:
IF: No. of stories = 30 stories, No. of vertical trusses = 2 vertical trusses, and No. of horizontal trusses = none
THEN: Steel unit weight = medium

Structural Interpretation: When designing a building in the height range of 30 stories without horizontal trusses, the following configurations of structural components of a wind bracing system, which yields an acceptable unit steel weight value, should be considered: a two-bay wind bracing system of vertical trusses.

Such a rule is useful when there are limitations imposed by architects or clients on the wind bracing selection process. It will also prevent analyzing wind bracing systems that yield a high unit steel weight of the structure under consideration.

The constructed learning curve for this study, Figure 2, clearly demonstrate the progress of learning. The acquired structural knowledge, using computer-generated examples, at the seventh learning stage is expected to produce an empirical error rate of 3.9% when used for learning about the unit steel weight optimization of a new wind bracing system for given structural design requirements.

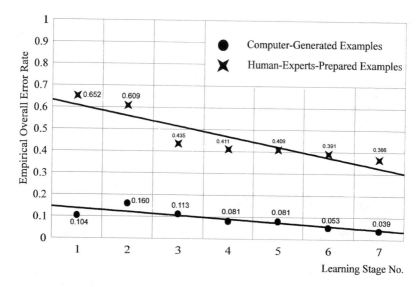

Figure 2. Learning Curves

By comparing results obtained using the human-expert-prepared examples (Mustafa and Arciszewski, 1992) and computer-generated examples, the constructed learning curves shown in Figure 2 clearly indicate that the learning process using human-expert-prepared examples is far from completion. A learning process is considered to be close to completion when the empirical error rates converge toward zero which is the case for computer-generated examples. The gap between the two regression lines can be attributed to the inconsistency of the human-expert-prepared examples. It should be emphasized here that this result does not necessarily mean that the computer-generated examples are "better" than those examples prepared by human experts. They are simply more consistent.

The computer-generated examples strategy clearly shows its feasibility in the process of learning. However its feasibility is demonstrated only in the domain of this study. The entire process should be investigated and tested for other structural design engineering domain. Further study is yet needed to investigate a new strategy which combines the computer-generated examples with the human-experts-prepared examples.

Acknowledgments

The author gratefuly acknowledges the cooperation of Waterloo Engineering Software, Ontario, Canada, which provided SODA, the software used in the preparation of examples. Also, the cooperation of the Machine Learning and Inference Center at George Mason

University, Fairfax, VA, is acknowledged. The Center provided a learning system based on AQ15 learning algorithm and participated in the analysis of results.

Appendix-References

Arciszewski T., (1985). "Decision making parameters and their computer-aided analysis for wind bracings in steel skeleton structures," Advances in Tall buildings, Van Nostrand Publishing Co., New York, N.Y.

Arciszewski T. and Mustafa, M., (1989). "Inductive Learning Process: The User Perspective," chapter in Machine Learning, edited by R. Forsyth, Chapmon and Hall.

Arciszewski, T., Dybala, T., and Wnek, J., (1992). "A Method for Evaluation of Learning Systems," Journal of Knowledge Engineering Heuristics, Vol. 5, No. 4, pp. 22-31.

Arciszewski, T., Bloedorn, E., Michalski, R. S., Mustafa, M., Wnek, J., (1994). "Learning Design Rules for Wind Bracings in Tall Buildings," Vol. 8, No. 3, July 1994.

Grierson, D. E., Pak, W. H., (1992). "Discrete Optimal Design Using a Genetic Algorithm," Proceeding of NATO Conference on Topology Design of Structures, Portugal.

Maher, M. L., (1992). "Automated Knowledge Acquisition of Preliminary Design Concepts," Proceedings of the ASCE Conference for Computing in Civil Engineering, Dallas, Texas, 975-982.

Michalski, R., Mozetic, I., Hong, J., and Lavrac, N., (1986). "The AQ15 Inductive Learning System: An Overview and Experiments," Report, Intelligent Systems Group, Department of Computer Science, University of Illinois at Urbana-Champaign.

Mustafa, M. and Arciszewski, T., (1992). "Inductive Learning of Wind Bracing Design for Tall Buildings," in Knowledge Acquisition in Civil Engineering, T. Arciszewski and L. Rossman (Eds.), the American Society of Civil Engineers.

Mustafa, M., (1994). "Methodology of Inductive Learning: Structural Engineering Application," Ph.D. Dissertation, Civil Engineering Department, Wayne State University, Detroit, MI 48202.

Taranath, B. S., (1988). Structural Analysis and Design of Tall Buildings, McGraw-Hill Book Company, New York.

Weiss, S. M. and Kulikowski, C. A., (1991). Computers that Learn, Morgan Kaufman Publishers, San Mateo, California, 1991.

DEDUCTIVE CAD

Sivand Lakmazaheri[*] A.M. ASCE

I. Introduction

Conventional CAD systems store and display geometric information, and perform computations according to the procedures that are embedded in them. When presented with unanticipated questions or tasks, such systems fail to perform adequately. For example, although a conventional CAD system can be used to model a bridge structure, the system cannot be used to answer questions such as "How many barrier rails are used in the bridge?" or "What is the relationship between a load and the size of a pile?" or "Is the generated geometric model of the bridge correct?" In general, conventional CAD systems neither can understand nor have the necessary computational capability to answer such questions. This limitation can be overcome by incorporating a deductive mechanism (a mechanism capable of drawing logical inferences) in conventional CAD systems. Such a mechanism makes it possible to create and manipulate geometric objects, and to answer complex queries about objects and relations among objects via deductive reasoning.

Deductive reasoning is the process of inferring the "truth value" of a statement (referred to as goal) from available knowledge. When augmented with the ability to enforce the truth and falsity of goal statements, deductive reasoning becomes a powerful tool for creating and manipulating objects and for reasoning about them [1].

A CAD system which supports deductive reasoning is called a *Deductive CAD* (DCAD) system. In such a system one can describe objects and relations declaratively, and generate and manipulate objects deductively. This paper describes the elements of a typical DCAD system and exemplifies its use.

II. What is a Deductive CAD System?

Deductive CAD systems are generalization of conventional CAD systems. DCAD systems support deductive definition and manipulation of objects (geometric and non-geometric) and relations (arithmetic and non-arithmetic) among objects. In deductive

---

[*] Civil Engineering Department, The Catholic University of America, Washington, D.C. 20064
Lakmazaheri@cua.edu

CAD objects and their inter-relationships are represented declaratively using the language of predicate logic [2]. Two types of objects can be defined in DCAD systems. These are: simple objects and derived objects. If the attributes of an object are stored explicitly, that object is called a simple object. An object whose attributes can be deduced from simple objects is called a derived object. For example, the lines constituting a rectangle can be viewed as simple objects while the rectangle itself can be viewed as a derived object. Derived objects are said to be represented implicitly in terms of simple objects. Rather than storing derived objects, in DCAD systems, one stores their definitions as rules. The attributes of derived objects are computed, when needed, based on their definitions.

A DCAD system consists of a deductive mechanism and a conventional CAD component (see Figure 1). The conventional CAD system is composed of a CAD engine for creating and manipulating simple geometric objects and a CAD file for storing simple objects. The deductive mechanism has a rule base for storing deductive object definitions; it uses the CAD file for storing simple objects; and it employs an inference engine for performing logical inferences. In a DCAD system procedural attachments are used for accessing the CAD engine. These attachments appear as built-in predicates in rules. Through the use of built-in predicates, the deductive mechanism can access the geometry creation and manipulation functions supported by the CAD engine.

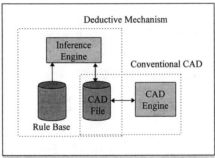

Figure 1: Components of a Deductive CAD System

Deductive CAD systems differ from typical knowledge-based CAD systems [3]. DCAD systems are based on predicate logic whereas most knowledge-based CAD systems are based on propositional logic. There are two major differences between the two logics. Propositional logic uses pattern matching whereas predicate logic uses resolution for drawing logical inferences. As a result, in DCAD systems relations among objects can be manipulated non-deterministically. This enables DCAD systems to respond to unanticipated questions. Furthermore, propositional logic does not support constraint-based computations whereas constraints can be easily handled in predicate logic. The ability to handle constraints is needed for solving many engineering problems.

III. Modeling in DCAD

Figure 2 shows the hierarchical decomposition of a standard type of bridge structure. The geometry of such a bridge can be defined deductively using 19 rules. Three of these rules are given below.

$bridge(SL,RW,PL,PW,M,PT) \Leftarrow spanS(0,0,N,SL,RW) \wedge$
$\qquad bentS(0,0,N,SL,RW,PL,PW) \wedge abutmentS(0,0,N,SL,RW,M,PT).$

$spanS(X,Y,N,SL,RW) \Leftarrow span(X,Y,SL,RW) \wedge spanS(X+SL,Y,N-1,SL,RW).$

$span(X,Y,SL,RW) \Leftarrow deck(X,Y,SL,RW) \wedge barrier(X,Y,SL,RW).$

The first rule defines object bridge in terms of objects spans, bents and abutments. The second rule recursively defines object spans in terms of individual span objects. The last rule defines object span in terms of objects deck and barrier rail.

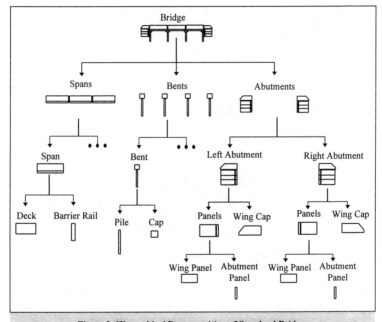

Figure 2: Hierarchical Decomposition of Standard Bridges

Suppose we are given a bridge geometry and we are asked to verify its correctness. That is, we are to determine whether or not the given model adheres to the rule-based bridge definition. This is done by determining the truth value of the goal statement that describes the modeled bridge. The inference mechanism, which makes this determination, returns *true* if the bridge model is consistent with the bridge definition.

Otherwise, it returns *false*. The goal statement that describes a bridge with the following specifications is { bridge(5,10,7,8,300,3,'A-1') }.

Bridge Specification

Number of Spans:	5
Span Length:	10 m
Roadway Width:	7 m
Pile Length:	8 m
Pile Size:	300 mm
Number of Abutment Panel rows:	3
Abutment Panels type:	A-1

The rule-based bridge definition can also be used to create bridge models. This is done by enforcing the truth of the goal statement that describes the bridge.

IV. Discussion

Deductive reasoning, when incorporated into conventional CAD systems, yields a general computational framework for engineering problem solving and decision making. In such a framework geometric and non-geometric information (objects) can be created, manipulated, and reasoned about effectively.

Natural language is the primary means of communication among engineers. A CAD environment that supports communication via natural language, therefore, is a highly useful tool in engineering practice. The DCAD framework facilitates natural language communication. Objects and relations can be defined using suitable natural language statements rather than logical expressions. Such natural language statements can be transformed into logical statements which, then, can be manipulated by the system. For example, bridges can be defined using natural language statements such as {*Bridge consists of superstructure and substructure. Superstructure consists of N spans. Each span consists of five deck units and two barrier rails ... *}. These statements, which can be communicated in written or spoken form, can be translated and then used by the system to generate bridge models. Obviously, such a natural language should be expressive enough so that complex definitions can be clearly stated. And, it should be natural enough so that the user can formulate the needed statements with ease.

Significant advances have been made in the area of speech recognition and synthesis. Speech, natural language, and DCAD technologies provide a practical foundation for developing conversational CAD systems. Such systems will have the potential to revolutionize the use of computers in engineering practice.

V. References

[1] Lakmazaheri, S. (1995) "A Logic-Based Mechanism for Integrity Maintenance of Engineering Databases," *Engineering with Computers*, 11: 46-57.

[2] Lakmazaheri, S. (1995) "Logic Programming for Structural Engineering: An Introduction and Overview," Proceedings of Structures Congress XIII, Boston, MA, April 3-5.

[3] Oxman, R.M. and Oxman, R.E. (1991) "Formal Knowledge in Knowledge-Based CAD," Building & Environment, 26: 35-40.

Constructing Simulation Models Around Resources Through
Intelligent Interfaces

Simaan M. AbouRizk[1], A.M., ASCE

and

Jingsheng Shi[2]

Abstract

This paper presents automatic model generation using the resource-based
modeling (RBM) methodology (Shi and AbouRizk 1994). First, basic concepts and
general overview of the RBM environment is introduced. Then, various direct and
indirect linking structures are addressed for coupling purpose. Finally an
earthmoving example is used to illustrate the entire modeling process.

Introduction

Construction simulation was initiated by Halpin (1973) with the development
of the CYCLONE modeling methodology. It has become the basis for a number of
construction simulation systems including INSIGHT (Paulson 1978), RESQUE
(Chang 1987), UM-CYCLONE (Ioannou 1989), and COOPS (Liu and Ioannou
1992). Unfortunately, construction simulation is presently limited to academic
research with limited successful applications in the industry. The major obstacles to
its use by industry are the complexities involved in constructing a model and the
resultant time requirement--the technique is not yet cost effective.

Researchers have attempted different ways to simplify the modeling process
including: 1) model resuability (Bortscheller and Saulnier 1992), 2) computer-aided
modeling approach (Balci and Nance 1992), and 3) hierarchical and modular concept
(Zeigler 1987). Paul notes that it is impossible to produce an all purpose simulation
modeling system that can handle any problem that one might wish to model (Paul
1992). Major researches in simplifying construction modeling include graphical

----

[1]Alberta Construction Industry Professor, Department of Civil Engineerimg,
University of Alberta, Edmonton, Alberta, Canada T6G 2G7.

[2]Ph.D. Candidate and Research Associate, Department of Civil Engineerimg,
University of Alberta, Edmonton, Alberta, Canada T6G 2G7.

modeling interfaces (Liu and Ioannou 1992, Huang et. al. 1994) and model resuability (Halpin et. al. 1990, and McCahill and Bernold 1993).

The objective of this research was to develop, test and implement a Resource-Based Modeling (RBM) methodology that enables the modeler to build accurate construction simulation models using resources as the basic building blocks. The whole modeling process will be automatic with the modeler specifying resources and site (project) conditions.

Resource-based Modeling Methodology

Hierarchical and modular modeling concepts were first presented by Zeigler (1976, and 1987). They have been useful in simplifying the constructing process of simulation models, particularly for large and complex systems (Luna 1992). The basic components of the concepts include the 'atomic model', the 'model library', and 'coupling'. An atomic model is a basic and unique description of a particular process. A model library consists of numerous atomic models, which are to be used in various combinations to construct a high level model. Coupling is the act of combining related atomic models.

In general terms, the following tasks and issues must be addressed to apply the hierarchical approach to model construction projects:

1. The atomic models that are to be included in the model library must be defined and designed.

2. Coupling procedures that address the actual requirements of construction projects must be developed. Zeigler (1987) suggested creating new models by combining two or more atomic models through model input and output ports. The authors' early experimentation with this approach showed that two or more atomic models cannot always be directly linked through simple input/output ports. Various types of linking structures must be defined and implemented to facilitate the coupling process.

3. A means of integrating the attributes and boundaries of the physical system environment (e.g. project site information) must be incorporated into the modeling process.

Many real-world systems are characterized by dynamic resource interactions (e.g. different types of equipment in earthmoving construction). Although each construction system is unique, the operating processes of its component resources are usually somewhat generic. They can be pre-defined as atomic models and stored in a model library. Atomic models from the library can then be surveyed by project-associated data and modified to form project-specific atomic models. The environment will then identify the appropriate linking structures and assemble the working model.

Overall Structure of a Resource-based Modeling Environment

Conventionally a user must understand both the simulation theory and the selected simulation language, then he/she can construct a simulation model through the simulation language provided interface.   The RBM environment is a pre-processor for simulation languages.   Through this pre-processor the user can construct a simulation model by simply specifying required resources and project-related information.   This environment should include enough information and expertise in order to automate the modeling process as much as possible.   The environment is composed of eight basic components:

1. a database to store resource attributes;

2. an atomic model library which includes all types of resources for a specific type of construction project;

3. a user interface that allows the user to specify required resources, project-related resource attributes, and other project information;

4. a module (knowledge-based if necessary) which can convert physical site conditions to simulation information, for example, computing the duration of a work task of a construction process from given physical project site conditions;

5. an atomic model generation module which can combine resource attributes and project-related information with atomic models in the library to produce project-specific atomic models;

6. a knowledge-based module which can identify and generate proper linking structures to suite the atomic models;

7. a module which can assemble all atomic models with linking structures to generate a working simulation model;

8. an interface which can call the selected simulation language and allow the user to experiment with the generated model.

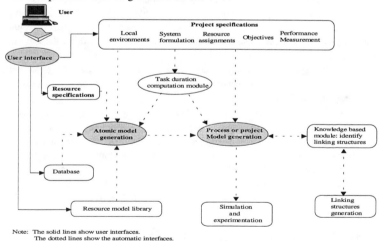

Figure 1   Overall system structure

Figure 1 illustrates the structure of the RBM environment and the interaction of the eight basic components. When a user is using this environment to construct a simulation model, three steps are involved: 1) resource specification, 2) project specification, and 3) model generation. The first two steps have to be completed by the user through provided interfaces. The third step does not need direct involvement from the user. The environment will automatically associate the user-specified resource and project information with atomic models in the library to generate a simulation model. The model generation can be divided into atomic model and entire model generation. Resource and project specifications were briefly introduced by Shi and AbouRizk (1994). This paper focuses on the model generation.

Resource Specification

The user can specify all resources required by a project through the provided user interface. The attributes of these resources can be obtained from embedded databases and user specification.

Project specification

Project specification requires detailing of the physical features of the project. Five aspects of information are required: 1) system specification -- to define the makeup of a construction system by basic resource processes; 2) resource assignment -- to associate the user-specified resources with resource processes to achieve planned operations; 3) site conditions; 4) measurement -- to define the production per cycle of each resource process; and 5) system objectives -- to define the termination of simulation operations.

Atomic Model Generation

Atomic models in the library only describe the basic logic structures of corresponding resource operations. No actual data are included in these models. The atomic model generation module constructs project-specific atomic models from library-resident models for each specified resource process in accordance with specified resource attributes, site conditions and resource process objectives. These generated models become the bases for the entire model generation.

Entire Model Generation

The entire simulation model is the final output of the modeling process. It is obtained by combining generated resource process models through linking structures. Atomic models are like building blocks. Required building can be generated by properly linking the appropriate blocks. Linking structures in the resource-based modeling methodology are used to link atomic models into one entire simulation model. A linking structure is a submodel which receives input from one atomic model and transfers it to another atomic model. The purpose of a linking structure is to correctly assemble related atomic models according to their characteristics.

Simulation entity is defined as information flowing through a simulation model (Pritsker, 1986). An entity is any object, resource, unit of information, or combination thereof which can define or can alter the state of the system. Simulation entities traverse through a simulation model as the system's status change. They can also traverse from one atomic model to another through communication ports. In some cases, the entities in two models are dimensionally equivalent. Therefore, an entity released from a model can be directly routed to its following model. A "direct link" can be defined to model this situation. However, the communication between models sometimes can not be directly carried on because the entities in these models are not dimensionally compatible. For instance, where a backhoe excavator can load 8 m³ soil into a truck in each bucket, and the truck can hold 50 m³ soil, six entities in loading model equal to one entity in hauling model. An "indirect link" which balances the difference between entities in different models can be defined. The selection of the type of linking structures depends upon the characteristics of atomic models and is achieved through embedded knowledge base.

Direct linking structures

Direct linking structures will not alter simulation entities during the transfer process. Various scenarios of direct linking structures can be defined to cope with different requirements.

1. One-one link

A one-one link is the simplest scenario. The output of one model is required as the sole input to another model. Using arrows to define the coupling process, only one arrow is required to link the two models.

2. One-multiple link with all branch releases

Where multiple models require the single output of a preceding model, a "*continue*" function node, which releases simulation entities to each of the following models, can be added to the linking structure.

3. One-multiple link with one branch release

Where multiple models follow a single model, and only one of them can be released, a function "*select*" node can be added to the linking structure as shown in Figure 2. The selection rule could be "cyclic"or "priority" depending on the selected simulation language.

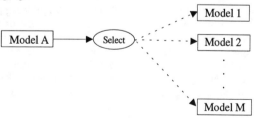

Figure 2 One-multiple link

4. Multiple-one with multiple releases

If multiple models are followed by a single model and all outputs are to be routed to the following model, the treatment is similar to the one-one situation. For each preceding model, an arrow routes its output to the following model.

5. Multiple-one link with one release

Where multiple models precede a single model, and one output is required from each of them to release the following model, a function "Consolidate" node can be added to the structure.

Indirect linking structures

Indirect linking structures are used to couple models that are not dimensionally compatible in simulation entities. A batch, an unbatch function node and a queue node are needed to implement various indirect linking structures.

1. One-one indirect link

Similar to a direct one-one link, a one-one indirect link implies a single model is followed by another single model. In this case, the entity released from model A is unbatched into multiple units which are routed to a QUEUE node, then multiple entities released from this QUEUE node are batched into one entity which is routed to Model B as shown in Figure 3.

Figure 3   One-one indirect link

2. One-multiple indirect link

If multiple models require the output of a single preceding model, the first part of linking structure will be identical the one-one scenario. After QUEUE node, each following model should have a separate batch node.

3. Multiple-one indirect link

If multiple models are followed by a single model, all entities released from proceeding models are to be unbatched and routed to a QUEUE node. Then entities released from the QUEUE node are batched and routed to model B.

An earthmoving example

To illustrate the RBM concept, consider an earthmoving project with five resource processes: soil preparation, loading at location 1, loading at location 2, hauling, and spreading. The sequence of the five resource processes are specified as shown in Figure 4. Five resources have been specified: a CAT-D8 dozer, a UH-501 backhoe excavator, an EX-300 excavator, five CAT-777 trucks, and a CAT-7 dozer. The *resource assignments* are: CAT-D8 for soil preparation, UH-501 for loading at location 1 ; EX-300 for loading at location 2, CAT-777 trucks for hauling, and CAT-D7 for spreading. *Local environments* are: hard clay ground for CAT-D8 (hard pushing), loose stockpiled clay (easy cut) for UH-501 and EX-300; a haul route consisting of a section of 10% inclining grade and a section of busy traffic for the trucks, and loose stockpiled clay (easy pushing) for CAT-D7. *The system objective* is to spread 100,000 m³ of excavated soil at the dump area.

This information is specified by the user through provided interfaces, and will be incorporated into corresponding atomic models in the library to generate project-specific atomic models. The linking structure required are: one-multiple indirect link between dirt preparation and loading, multiple-one indirect link between loading and hauling, and one-one indirect link between haling and spreading. By identifying appropriate linking structures, the entire project simulation model is constructed by assembling the atomic models by following the logic sequence in system specification.

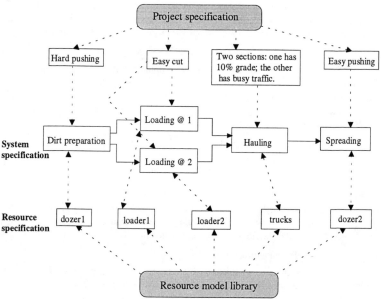

Figure 4 An earthmoving construction modeling example

Conclusions

The resource-based modeling methodology unitizes generic programming tools with input provided from a user specification interface, automatically construct and configure a project-specific simulation model. The user must be familiar with the operations and resources that will be required for the project, and with the project site conditions, but no simulation language expertise is required. This approach may significantly improve the cost-effectiveness of construction simulation, allowing the transition of construction simulation from institutions to industry.

Aknowledgement: This project is funded by NSERC-CRD Grant # 166483.

References

AbouRizk, S. M. (1994). Technical Discussion on: "Modeling Operational Activities in Object-Oriented Simulation." by A. A. Oloufa. *J. of Computing in Civil Engineering*, ASCE.

Balci, O. and R. E. Nance (1992). "The Simulation Model Development Environment: An Overview." *P. of the 1992 Winter Simulation Conference*, 762-736.

Bortscheller, B. J. and E. T. Saulnier (1992). "Model Reusability in A Graphical Simulation Package." *P. of the 1992 Winter Simulation Conference*, 764-772.

Halpin, D. W. (1990). "MicroCYCLONE User's Manual." Division of Construction Engineering and Management, Purdue University, West Lafayette, Indiana.

Huang, R. Y, A. M. Grigoriadis, and D. W. Halpin (1994). "Simulation of Cable-Styed Bridges Using DISCO." *P. of the 1994 Winter Simulation Conference*, 1130-1136.

Ibbs, C.W. (1987). "Future direction for computerized construction research." *J. of Constr. Engrg. and Mgmt.*, ASCE. 112(3), 326-345.

Liu, L. Y. and P. G. Ioannou (1992). "Graphical Object-Oriented Discrete-Event Simulation System." Proceedings of Winter Simulation Conference, 1285-1291.

Luna, J. (1992). "Hierarchical, Modular Concepts Applied to an Object-Oriented Simulation Model Development Environment." Proceedings of Winter Simulation Conference, 694-699.

McCahill, D.F. and L. E. Bernold (1993). "Resource-Oriented Modeling and Simulation in Construction." *J. of Constr. Engrg. and Mgmt.*, ASCE, 119(3), 590-606.

Paul, R. J. (1992). "The Computer Aided Simulation Modeling Environment: An Overview." *P. of the 1992 Winter Simulation Conference,* 737-746.

Paulson, B.C. Jr., (1978). "Interactive Graphics for Simulating Construction Operations." *J. of Constr. Div.*, ASCE, 104(1), 69-76.

Shi, J. and S. M. AbouRizk (1994). "A Resource-Based Simulation Approach with Application in Earthmoving/Strip Mining." *P. of the 1994 Winter Simulation Conference,* 1124-1129.

Zeigler, B.P. (1976). "Theory of Modelling and Simulation." Wiley, N.Y. (Reissued by Krieger Pub. Co., Malabar, Fla. 1985).

Zeigler, B.P. (1987). "Hierarchical, modular discrete-event modeling in an object-oriented environment." *Simulation*, 49(5), 219-230.

Recent Advancements in Simulation User Interface - A Description of the DISCO Environment

Rong-Yau Huang, Ph.D.[1], Associate Member, ASCE
Daniel W. Halpin, Ph.D.[2], Member, ASCE

Abstract

Interest in the application of computer simulation for planning and analysis of construction operations has been generated since the introduction of CYCLONE by Halpin in 1973. Numerous examples of construction process models have been developed in many publications. Nevertheless, practicing engineers have been slow to utilize process modeling as a regular planning and analysis tool. Partially, this is due to the unfamiliarity of construction personnel with simulation techniques. A graphically-based simulation interface called DISCO (Dynamic Interface for Simulation of Construction Operations) has been recently developed to take advantage of modern computer graphics technologies. DISCO allows the building of a schematic model diagram graphically on the screen. It displays the model dynamics by employing a "node coloring" mechanism and by continuously updating the node information during the simulation run. It reports node statistics information graphically and in tabular form. The DISCO interface creates a graphical modeling and simulation environment and facilitates the application of computer simulation for planning and analysis of construction operations.

Introduction

Computer simulation gives the analyst an insight into resource interaction and is well suited to the study of resource driven processes. Computer simulation for planning and analysis of construction operations has been an active area of development since the introduction of CYCLONE modeling methodology [Halpin

---

[1]Research Associate, The Mortenson Center for Construction Innovation, International Market Square, 275 Market St., Suite 545, Minneapolis, MN 55405; Tel. (612) 338-1818, Fax (612) 338-1311, E-Mail: rhuang@mortcenter.com

[2]Professor and Head, Div. of Construction and Management, Purdue University, West Lafayette, Indiana 47907; Tel. (317) 494-2240, Fax (317) 494-0644, E-Mail: halpin@ecn.purdue.edu

1973]. Numerous examples of construction process models have been developed in Halpin [1990b] and Halpin and Riggs [1992]. These include earth-moving, pavement, tunneling, segmental construction of an elevated structure, and concrete placement in a high-rise building. The CYCLONE methodology offers a simple graphical method for the modeling and analysis of construction operations and is by far the most widely used modeling methodology for simulation in the construction research community [Huang 1994].

Although the CYCLONE modeling environment provides a simple set of graphical elements which can be used to model construction operations, practicing engineers have been slow to utilize process modeling as a regular planning and analysis tool. Managers typically leave the design of construction operations to field personnel who rely upon a "cut and try" approach grounded in previous experience with similar processes. This typically leads to very inefficient work sites characterized by poor materials handling systems and high levels of idleness for key resources.

Recently, continuing breakthroughs in both the hardware and software aspects of computer technology have greatly improved the environment within which man and machine interact. The Windows™, OS/2™, Macintosh™, X-window™ operating systems make possible the creation of more friendly and effective user interfaces. The DISCO (Dynamic Interface for Simulation of Construction Operations) program which operates under Windows environment was developed by the authors [Huang & Halpin 1993] to integrate with the CYCLONE system. It takes advantage of modern computer graphical technologies and creates a graphical modeling and simulation environment to facilitate construction process simulation. This paper introduces the DISCO program and examines how the program can assist in planning and analysis of construction operations. A dry batch delivery and concrete placement operation is used for demonstration.

DISCO Program

The DISCO program is a Windows based application written in the Visual Basic language. The DISCO user interface, as shown in Figure 1, employs the conventional Windows' format -- maximizing & minimizing buttons, title bar, menu bar, scroll bars, and so on. The bulk blank area in the middle of the window is the area for model drawing. The CYCLONE methodology is employed for abstract model building and its set of standard elements constitutes the basic icons in the DISCO graphical menu bar. Readers are referred to Halpin and Riggs [1992] for details regarding the CYCLONE modeling methodology.

Basically, the DISCO program serves as both a preprocessor and post processor for the MicroCYCLONE program [Halpin 1990]. DISCO generates the

MicroCYCLONE input file and takes the chronological list file generated by running the MicroCYCLONE program as its input. It then recreates the entire course of simulation dynamically on the computer screen with the schematic model diagram used as the display mechanism. The schematic CYCLONE diagram is made dynamic by the introduction of color marking which reflects the movement of resources on the process diagram. This approach relies upon the concept of a dynamic schematic diagram such as those that are used in the control rooms of various plants (e.g. refineries, power stations, etc.) to reflect the dynamic status of the plant at any given time. Control room consoles use colors, blinking lights, and audible alarms to keep the operator updated regarding the operations in progress. A similar technique is used in DISCO.

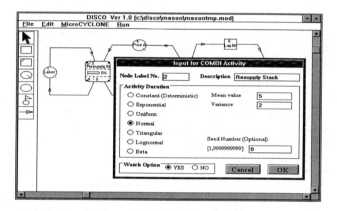

Fig. 1. DISCO User Interface and Input Dialogue Box

The DISCO program performs three major functions as follows:
(1). It builds a schematic model diagram graphically on the screen and generates the MicroCYCLONE input file.
(2). It displays the model dynamics by continuously updating the associated information for each node during the simulation run.
(3). It reports node statistics information graphically and in tabular form at any simulation event time point.

Building the Schematic Model Diagram

DISCO allows the creation of the CYCLONE schematic model diagram graphically on the screen. The user can create nodes or connectors between nodes by simply clicking on the appropriate icon from the DISCO graphical menu bar (Figure 1) and dragging it over to the drawing area. A dialogue box will pop up asking for the input of information associated with that particular node, such as label number, node description, distribution type of activity time duration and its corresponding parameters, and so on. Users can always double click on any node to recall the dialogue box for editing the node information. The "Watch Option" (Yes or No) information in each node determines whether the simulation run-time information associated with a particular node such as % time idle, will be "watched" (shown) later during the simulation or not. Nodes with "YES" in the "Watch Option" box have a bigger node size in order to display simulation run-time information (e.g., Node 2 in Figure 1). Again, users can go back and forth to change the option except for the COUNTER node for which the "Watch Option" is automatically invoked. This mechanism provides the user opportunities to watch nodes alternatively and enables the user to focus on watching certain nodes while not being distracted by too many "watched" nodes during the simulation.

Once the model formulation is completed, DISCO generates the MicroCYCLONE input file automatically. The user is exempted from the necessity of being familiar with MicroCYCLONE's Problem-Oriented Language (POL) which is required for translating a CYCLONE model into a MicroCYCLONE input file. This file is then used to run the MicroCYCLONE program. In addition, DISCO allows the user to run the MicroCYCLONE program from within the DISCO environment. Users do not have to access and exit the DISCO program in order to run MicroCYCLONE program since the two programs are integrated.

Display of Model Dynamics

After the schematic model diagram is drawn on the DISCO screen and the chronological list file (.DIS) is generated by running the MicroCYCLONE program, the user can trigger playback of the simulation. DISCO recreates the entire course of simulation dynamically on the screen. A simulation clock is shown on the screen to indicate the updated simulation time point T-NOW. As the value of simulation time advances, run-time information for those nodes being "watched" is continuously updated throughout the simulation. The display format of the run-time information contained in "watched" nodes is as shown in Figure 2.

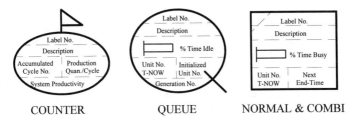

Fig. 2. Format of DISCO "watched" Nodes

The DISCO program attempts to create the impression of "animation" during the simulation run by employing colors. Each type of node is assigned a different color. The color goes on if a unit "enters" an empty node. The entering unit can be a truck returning to an empty queue, or the processing of a task. Similarly, the color goes off when the last unit "exits" a node. Once the simulation starts, the user is able to see colors flashing on and off on the screen, along with continuously updated run-time information. This mechanism allows the user to experience system dynamics during the simulation run.

Report Node Statistics

Throughout the simulation, the DISCO program keeps track of all the run-time information associated with each node. The user can request statistical reports on a particular node at any simulation time by double clicking on it. A report box will appear on the screen and the execution of the simulation will be interrupted. The user has to select the type of report to be viewed and the report is then displayed both graphically and in tabular format. Table I lists the types of reports provided by the DISCO program.

Table I. DISCO Report Types

	NORMAL	COMBI	QUEUE	CONSOLIDATE	COUNTER
Unit No. Profile	✓	✓	✓	✓	
% Time Idle (Busy)	✓	✓	✓		
Activity End Time	✓	✓			
System Productivity					✓

The execution of the simulation resumes automatically once the user has completed viewing the reports. The overall statistical information of each node can be obtained in the same manner at the end of simulation.

A Dry Batch Delivery and Concrete Placement Model

Model Description

Figure 3 shows the model diagram of a dry batch delivery and concrete placement model in DISCO format. This model basically involves the interaction of five types of resources -- crane, bucket, crew, truck, and concrete mixer. Hauling trucks carry dry concrete batches from batch plant to the paving site and dump them to the skip of the mixer. After the mixer mixes the concrete, wet batches are dumped into a concrete bucket and lifted by a crane to the placement location where they are dumped, spread and finished by a labor crew. Readers are referred to Halpin and Riggs [1992] p.197 for this model.

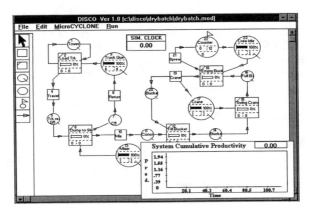

Fig. 3. Dry Batch Delivery and Concrete Placement Model in DISCO

In order to monitor the performance of the trucks, mixer, crane and labor crew in this particular simulation run, Nodes 9, 13, 17, and 23 are set to be "watched." In addition, selective work tasks such as "Load Truck" (Node 2), "Dump to Skip" (Node 6), "Fill Bucket" (Node 12), etc., are "watched" so as to better understand the system dynamics. As shown in Figure 3, prior to the beginning of the simulation, the simulation clock is set to 0.00. Green color (not visible in the figure) in all the QUEUE nodes with initialized resources (e.g., crane, bucket, trucks, etc.) signals the availability of a resource. The % Idle time in these nodes is set to 100 initially. A productivity report window shows up which allows the monitoring of the development of the system productivity through the simulation run. It is immediately apparent that "Load Truck" (Node 2) will be the only task to be scheduled at simulation clock (T-NOW) 0 due to the availability of both required resources -- loading tower and hauling truck.

DISCO Run

Once the simulation starts, the node color of "Load Truck" (blue) signals the commencement of this task and the unit no. at the left node bottom is updated to 1. The next-end-time at the right node bottom is updated to 12.00 minutes to indicate the time loading of the first truck will be completed. It takes the tower 12 minutes to load a concrete truck. Meanwhile, the node color of "Tower" goes off since it is busy loading the truck. In addition, the unit no. in the "Truck Queue" (Node 9) is decreased to 3 meaning that three trucks are now idle at the truck queue.

T-NOW is advanced to 12 minutes at which time the loading of the first truck is finished. The node color of "Load Truck" goes off. The first truck carries dry concrete batch and travels to the site for mixing. The node color of "Travel"

(Node 4) flashes on and the next-event-time at the right node bottom is updated to 22.00 (12.00+10.00) minutes. It takes the truck 10 minutes to travel to the site. The tower is now available (node color on) for loading another truck. Since three trucks are waiting in the truck queue, another "Load Truck" task can be scheduled at this time point. Again, the node color of "Tower" and "Load Truck" flashes off and on respectively and the next-event time of "Load Truck" is updated to 24.00 (12.00+12.00) minutes. Meanwhile, all the run-time information contained in these nodes is updated accordingly.

As the simulation continues, the user can view the loaded trucks as they arrive at job site and dump dry concrete to the mixer skip. The mixer mixes the concrete and fills the crane bucket with wet concrete. The crane then lifts and swings the bucket to the paving site and a labor crew empties the bucket, spreads and finishes the concrete while the crane swings back for lifting another bucket.

Report Analysis

A truck utilization profile (Figure 4) is obtained by double-clicking on the "Truck Que" node (Node 9) at simulation clock time 90.85 minutes. It records the unit number of trucks in queue through time, waiting at the tower to load concrete. The step like profile indicates that the number of trucks waiting in queue decreases from 3 to 2 at T-NOW 12.00 minutes; from 2 to 1 at 24.00, and from 1 to 0 at 36.00. No truck is idle after T-NOW 36.00 minutes. This reflects that after initial start-up, the trucks are not delayed in loading at the tower. Similar reports on the crane, the mixer, and the labor crew at different simulation clock times can be obtained in the same manner.

In addition, the updated cumulative system productivity is shown in the productivity report window (Figure 4). The productivity curve continues to climb and the operation has not yet reached a steady state of performance at the observed simulation clock time of 90.85 minutes (This is a deterministic model.) The current productivity is 1.59 cubic yards per minute or 95.4 cubic yards per hour for a crew of three labors. More detailed information regarding the system performance (productivity) can be obtained by double clicking on the "Counter" node (Node 22).

Conclusions

The DISCO interface allows a simulation experiment to be conducted in a more intuitive and relatively easy fashion. Its capability of graphical modeling reduces the learning efforts and greatly improves the accessibility of simulation as an analysis tool. This capability also reduces enormously the efforts needed for model modification and increases the model reusability. System dynamics are better understood through the use of the "node coloring" mechanism and the continuous updating of the node information. In addition, DISCO graphical and tabular reports allow monitoring of chronological information on each node. This further facilitates the use of the CYCLONE modeling format and the DISCO program interface as a process simulation base for planning and analysis of construction operations.

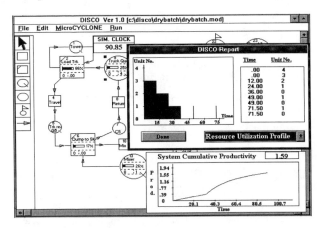

Fig. 4. DISCO Report: Concrete Truck Utilization Profile

References

AbouRizk, S., Halpin, D. and Lutz, J. 1992. State of The Art in Construction Simulation. *Proceedings of the 1992 Winter Simulation Conference*, Ed. J. Swain, D. Goldsman, R. Crain and J. Wilson., 1271-1277

Halpin, D.W. 1973. An investigation of the use of simulation networks for modeling construction operations. Ph.D. Thesis, U. of Illinois, Urbana-Champaign, IL.

Halpin, D.W. & Riggs, L.S. 1992. *Planning and Analysis of Construction Operations*. John Wiley and Sons, Inc., New York, NY.

Halpin, D.W. 1990a *MicroCYCLONE User's Manual*. Learning System Inc., West Lafayette, IN.

Halpin, D.W. 1990b *MicroCYCLONE System Manual*, Division of Construction Engineering and Management, Purdue University, West Lafayette, IN.

Huang, Rong-Yau and Halpin, Daniel W. 1993. *DISCO User's Guide*, Division of Construction Engr. and Mgmt., Purdue University, W. Lafayette, IN.

Huang, Rong-Yau and Halpin, Daniel W. 1994. Visual Construction Operation Simulation - The DISCO Approach. *Journal of Microcomputers in Civil Engineering*, 9 (1994), pp.175-184.

Huang, Rong-Yau. 1994. A Graphical-Based Method for Transient Evaluation of Construction Operations. Ph.D. Thesis, Purdue University, West Lafayette, IN.

Virtual Reality Environment for Design and Analysis of Automated Construction Equipment

Augusto Op den Bosch[1] and Makarand Hastak[1], AM-ASCE

Abstract

This paper describes a methodology to develop virtual prototypes of Automated Construction Equipment (ACE) design. Virtual reality not only provides a platform for testing and validation of the original equipment design but also allows the equipment designer to test different design alternatives with respect to the productivity and effectiveness of the equipment design. This kind of visual validation is invaluable for promoting the development and application of automated construction equipment. The validity and effectiveness of virtual reality for equipment design is illustrated with the help of a virtual prototype of a new Automated Pipe Laying Equipment (APLE) recently developed at Purdue University.

Introduction

People in the construction industry tend to be very skeptical when it comes to the automation of conventional construction processes. This skepticism is justified since it is difficult to analyze the specific benefits and drawbacks of using new Automated Construction Equipment (ACE) in place of a conventional process. To promote the use of ACE, it is important to develop techniques that can allow us to visualize the consequences of using automation on a specific construction site without having to develop ACE prototypes. In addition, it should also allow us to validate new ACE designs and perform a cost/benefit analysis.

Virtual environments can provide an ideal scenario to develop and test ACE. With the recent innovations in computer graphics technology, complex simulations of dynamic interdependent objects can be performed in real time. This factor allows the interactive manipulation of virtual objects in computer generated environments. Since ACE prototypes are expensive to develop, virtual reality provides an inexpensive alternative to develop and test the validity of new equipment designs. All the important information needed to approve production of new equipment can be found in a virtual environment for a fraction of cost involved in developing real prototypes. A methodology to improve the way machine behavior is modeled in virtual environments has been developed and tested in the Interactive Visualizer Plus Plus (IV++) at Georgia Tech.

---

1. Post Doctoral Fellows, School of Civil and Environmental Engineering, Georgia Institute of Technology, Atlanta, Georgia 30332-0355, Tel. (404) 894-2390, Fax (404) 894-2278

The IV++ tool allows ACE designers to visualize not only the geometric constraints of designs but the operating constraints as well. Task duration and other variables can be found during the virtual simulation and used in further stochastic analysis to determine the benefits and the cost effectiveness of new equipment design. Furthermore, the new methodology supports concurrent design by allowing access to the virtual environment from multiple workstations with a wide range of user interfaces. The different groups of a construction team can access the same environment and obtain the information that is relevant to the particular group while a simulation takes place.

This paper will illustrate the application of virtual reality for design and analysis of new automated construction equipment. In addition this paper will describe a new methodology developed to improve the modeling of virtual machines in computer generated environments and the various benefits derived from it. An example will illustrate the implementation of a virtual prototype of a new automated pipe laying equipment design.

Interactive Visualizer (IV++) - A Virtual Reality Environment for Engineers

Spatial engineering problems are usually very difficult to model and visualize because they involve a large number of components that interact in complex ways. Very often, analytical modeling is not enough because either the models are not accurate enough to capture all the relevant behaviors of physical systems in sufficient detail or there is not enough knowledge to create such detailed models. Virtual environment tools have been used to help engineers solve these kinds of problems. Such tools allow engineers to visualize relevant features of the problem and enable them to get early feedback on design decisions. Such feedback would be too costly or extremely difficult to obtain from the real world and is very valuable because it accelerates the design process.

A framework for representing spatial engineering problems in a simulated virtual environment and an authoring tool, the Interactive Visualizer (IV++), that implements this framework were developed at Georgia Institute of Technology (OpdenBosch 1994). The framework supports detailed modeling of the problem's components and their behaviors without sacrificing clarity of representation or accuracy. It also allows visualization of relevant features of the problem under study and assists the engineer in obtaining early and useful feedback. IV++ is an instantiation of the framework that is implemented in the C programming language and takes advantage of the Silicon Graphics platform.

Conceptually, the framework consists of entities that represent physical or abstract objects in the real world. Engineers can model a system by specifying the entities that constitute such system and by describing how these entities behave and interact with each other. There are four basic types of entities that can be used for modeling a system: *machines, cameras, lights,* and the *medium* in which they exist. Machines correspond to objects in the real world having general behaviors, which can be directly specified by the engineers or emerge from the interaction of the machine components. Cameras and lights metaphorically represent tools that allow visualization of the problem under study. These tools provide the visual interface between the engineers and the computer. A medium entity such as terrain, water, or space represents containment and defines the spatial limits of the environment. For example, a terrain is a medium that constrains entities by gravity to travel on a ground surface.

Interactive Visualizer (IV++) is a program that implements the framework presented above allowing engineers to use the entities to manipulate, display and visualize the modeled task in a virtual environment. The program is capable of supporting any combination of existing machines and provides the user with an extensive library to aid in the creation of new ones, following and object-oriented paradigm. Users can combine pre-defined machines with other entities (i.e. cameras, lights, and medium) on-line to create interactive environments where they can operate and test these entities in a truly interactive fashion.

The versatility of the Interactive Visualizer (IV++) presents a powerful medium for design and analysis of innovative automated construction equipment. Under normal circumstances, it is very expensive and time consuming to develop and test a prototype of a new equipment design. However, with the help of virtual reality and systems such as Interactive Visualizer (IV++) the time and cost involvement are considerably reduced. Additionally, it is substantially easy to implement and test design changes in a virtual environment rather than in a scaled prototype. The virtual reality paradigm and the Interactive Visualizer were applied to a conceptual equipment design recently developed at Purdue University for automating the pipe laying operation, called APLE (Hastak and Skibniewski 1993). Figure 1 shows the system layout.

Figure 1

Illustration of the Virtual Protype Environment

Automated Pipe Laying Equipment (APLE)

The primary objective of developing APLE (Automated Pipe Laying Equipment) was to eliminate the need for a worker inside the trench, while achieving increased productivity and reduction in operation costs. Often the complete automation of a construction operation requires extensive equipment and process modifications. In developing APLE these factors were taken into account and minimum changes in the process and equipment design were recommended (Hastak and Skibniewski 1993).

APLE has been conceived to include an excavator with a specially designed manipulator attachment for the pipe lifting, pipe lowering, positioning, and fixing operations (refer to Figure 2). The entire system includes an excavator with a three link boom and a bucket arrangement, a modified second link of the excavator to resemble a box truss (similar to the boom of a tower crane), and the specially designed manipulator attached under the second link of the excavator boom. Apart from the five degrees of freedom of the excavator, the manipulator is designed to have two suitably modified Stewart platforms (described in the following paragraph) with six degrees of freedom each (refer to Figure 2). Under the Stewart platforms two end effectors are attached, with each having three degrees of freedom at the wrist and a gripper arrangement. In addition to this, one degree of freedom (a translation motion along the x-axis) is available at the joints between the excavator boom (second link) and the manipulator attachments. The equipment is designed to be track mounted with two sets of rigid legs to help in excavation and pipe laying operation (see Figure 2).

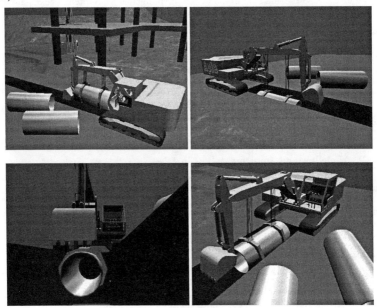

Figure 2
APLE Prototype Operating in the Virtual Environment IV++

Manipulator and the Stewart Platforms

The Stewart platform is composed of two platforms, the upper one is called the base platform, the lower one the payload platform. The shape of the platforms varies from an octahedron to a triangular shape, contingent to the desired operation. The payload and the base platforms are connected by six links (refer to Hastak and Skibniewski 1993). Usually the links are six linear actuators composed of DC analog motors and ball screws (Smith and Nguyen 1991). In this case a modification in the links is suggested, i.e., instead of using actuators, wire ropes (cables) should be used for links. These links would be controlled by a motor driven pulley arrangement provided in the box truss type second link of the excavator boom. This modification is desired because considerable manipulator reach is required in the pipe laying operation for which the actuators would not be adequate. Pulley arrangement is desirable in order to retract and lower the manipulator in the trench.

Operational Steps

There are 16 steps followed by APLE in performing the pipe laying operation: (1) excavate the trench by using the excavator and the bucket, (2) execute preliminary grading with the excavator bucket, (3) lower grading material into the trench, (4) grade the trench base with the help of the bucket and the laser. A laser field generator would be provided over the trench to act as a reference. A laser controlled probe attached to the excavator boom is suggested to assist in grading the trench base, (5) pull the trench box in position to avoid a trench cave-in (optional), (6) concurrently apply sealant to the pipe pre-laid next to the trench (performed manually by human), (7) grab the pipe pre-laid next to the trench (perpendicular to the trench center line) with the end effector gripper attached to the manipulator, (8) align the manipulator on top of the trench using the rotational motion (about z-axis) of the excavator, (9) align the second link of the excavator horizontally over the trench. The lower link of the excavator should point towards the trench so as to make use of the camera vision, (10) lower the pipe to the base of the trench using the Stewart platforms (i.e., translation motion along z-axis), (11) connect the pipe to the previous section using the translation motion (along x-axis) available at the joint of the manipulator and the second link of the excavator boom and the three degrees of freedom of the grippers, (12) release the grippers, (13) retract the Stewart platforms, (14) lower the bedding material with the help of a chute (performed by human on the surface), (15) move forward along the trench and go to step 1, and (16) end of cycle.

ACE Modeling and Implementation in IV++

This section will describe the implementation of APLE (a type of ACE) as a virtual machine in IV++. Even though IV++ has an extensive library of machines, it has to be taught how to handle new cases. This procedure involves both the solid modeling of a new machine using CAD and the implementation of new source code within IV++ to handle new machine behavior. The CAD modeling step is the easiest one and can be simplified in many cases if one can get a hold of an existing CAD representation of the machine in question or even some of its sub components. The implementation of behavior is more involved and requires knowledge of C++. It is important to note, though, that IV++ provides not only a modeling methodology to make the implementation process easier but also a library of functions to support code development. The following sections describe some of the details associated with both CAD modeling and behavior definition of IV++ machines.

Representation of Machine Geometry using CAD applications

The basic unit of a machine's geometric file is a 3-D point. Points are used to define polygon vertices and polygons are used to represent the surfaces of all machine objects. This hierarchical structure is commonly used in virtual environments but very few CAD packages represent objects this way. A common output format for CAD applications is a list of all entity polygons. The polygons are defined as a list of vertices with some additional attributes. By taking advantage of some of these attributes (e.g., color, border thickness, group, level, layer, etc.) one can incorporate hierarchical information in CAD files and then convert them to a standard object-polygon-point format. The standard machine representation format "VH" in IV++ is a file that contains a list of points followed by a list of polygons followed by a list objects. At the object level there is information about hierarchy and dependency as well as some physical attributes of the machine's object such as color, material, and shading conditions.

The following list includes the basic requirements to define a useful machine that can be used as a virtual prototype:

-convert all entities of the machine model to polygons.

-group all the polygons that belong to the same object together

-give each object an attribute that can be traced later to its dependency to other objects. For example: If an arm (object#2) is attached to the main body (object#1), it can be given an attribute that can later relate it to "object#1". In CADKEY ® for instance, the arm can be placed in "Level" #1 since the level information is one of the attributes contained in the CDL output file .

One of the advantages of having a hierarchical representation is that it makes it easier for developers to program the behavior of virtual machines. The next section will provide some insight into this procedure.

Representation of Machine Behavior within IV++

In IV++ machines are handled as variables of a type called "MACHINE" which contains a series of functions to help developers put together new instances of this type of variable. The actual implementation is divided into two modules: Controls and Dynamics. The controls module contains the handling instruction of input devices (e.g., keyboard, mouse, spaceball, position trackers, additional input gadgets, etc.). In essence the input devices alter the variables that define the current state of the machine. The Dynamic module interprets the state variables and uses them in kinematic and dynamic equations which also update the state of the machine. Usually these equations change the state variables associated with the orientation and location of the machine in the environment.

The following example illustrates the implementation of one aspect of the APLE machine; i.e., the motion of the bucket when the bucket control is activated.

Controls Code:

```
machine(Apple).object(Bucket).state(Valve)= read_gadget( bucket_control )
```

Dynamics Code:

```
machine(Apple).object(Scoop).compute_rotation(machine(Apple).object(Scoop).state)
```

The function "compute_rotation" contains the equations that will determine the final rotation angle of the bucket by using all the relevant information including the "valve" status.

The design of the user interface of a new machine can be accomplished either during the behavior definition cycle or after its completion. The real machine interface (controls, handles, buttons, etc.) can be approximated with the computer keyboard and mouse for debugging purposes during the behavior definition stage. Real controls and gadgets can be connected and used in IV++ using analog to serial converters. This allows the ACE designers to put different interface options to the test. The last step before the virtual ACE prototype testing can take place is the definition of the virtual environment layout. This step involves the placement of all relevant machines and objects in the environment including cameras, lights, terrain topography, and additional objects like the pipes that will be manipulated by APLE. IV++ provides tools that allow the users to design an environment interactively. Figure 3 illustrates all the steps involved in the implementation and testing of a new ACE. The insight that the system provides can be used by designers to change the original ACE specifications and its user interface.

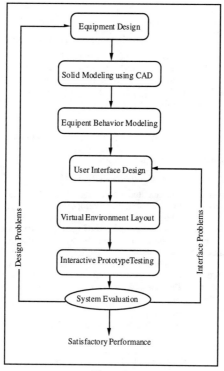

Figure 3
System Implementation Flowchart

Benefits and Conclusion

Virtual reality and IV++ were presented as viable medium for testing and validating new Automated Construction Equipment (ACE) designs. One of the significant advantages of virtual environment prototypes is that they provide the means to simulate the operation of new Automated Construction Equipment (ACE) at considerably lower costs as compared to conventional equipment prototypes. Virtual prototypes can be used to verify not only the mechanical validity of the original design, but also the effectiveness and efficiency of new ACE's in performing specific tasks.

The virtual environment IV++ proved to be not only a notable platform for conducting the tests and analysis, but a powerful tool that can help programmers define and implement virtual prototypes. The modular methodology and the object-oriented framework of IV++ simplified the complex task of implementing the Automated Construction Equipment (ACE) prototypes. The applicability of the concept was illustrated with the help of a virtual prototype of the Automated Pipe Laying Equipment (APLE) design.

References

Hastak, Makarand and Miroslaw J. Skibniewski. (1993). "Automation Potential of Pipe Laying Operations." International Journal of *Automation in Construction*, Vol. 2, pp. 65-79.

Opdenbosch, A.(1994), "Design/Construction Processes Simulation in Real-time Object-Oriented Environments", Ph.D. Thesis, Georgia Institute of Technology.

Paulson, B. C. (1985). "Automation and robotics for construction." *Journal of Construction Engineering and Management.* ASCE, 111(2), 190-207.

Smith, W. F. and C. C. Nguyen. (1991). "On the mechanical design of a Stewart platform-based robotic end-effector." *IEEE Proceedings of the SOUTHEASTCON.* IEEE. v 2, 875-879.

Tatum, C. B. and A. T. Funke. (1988). "Partially automated grading: Construction process innovation." *Journal of Construction Engineering and Management.* ASCE, 114(1), 19-35.

Surface Wave Testing Inversion by Neural Networks

Nenad Gucunski[1], Trefor P. Williams[2] and Vedrana Krstić[3]

Introduction

During the past ten years significant interest has been generated towards application of nondestructive methods in monitoring of infrastructure systems. Major areas of interest include quality control, detection of distress precursors and prediction of deterioration processes. The Spectral-Analysis-of-Surface-Waves (SASW) method is a seismic nondestructive technique for evaluation of elastic moduli and layer thicknesses of layered systems, like pavements. The SASW method, in conjuction with other seismic techniques (Gucunski and Maher, 1995), can be utilized for the purposes described above. The objective of the test is to determine the Rayleigh wave dispersion curve, i.e. velocity of a Rayleigh wave as a function of frequency, and then through a process of inversion to define the elastic modulus profile.

The SASW test has been fully automated (Nazarian et al., 1993), except in the inversion phase where automation is limited to certain classes of problems. The complexity of the inversion process results from a number of reasons: interference of surface and body waves, "near-field" effects, significant contribution of higher Rayleigh modes, etc. Significant effort is being oriented towards the development of a general automated inversion procedure. Neural networks were used with high success in a number of civil engineering applications where the complexity of a problem, or limitations of available data, did not allow development of a direct or iterative solution procedure. The paper presents development and application of neural networks for an automated inversion of SASW test data. The neural model herein is for asphalt concrete (AC) pavements, but similar models can be developed for other types of layered systems.

---

[1]Assistant Professor, Department of Civil and Environmental Engineering, Rutgers University, Piscataway, NJ 08855-0909
[2]Associate Professor, Department of Civil and Environmental Engineering, Rutgers University, Piscataway, NJ 08855-0909
[3]Graduate Student, Department of Civil and Environmental Engineering, Rutgers University, Piscataway, NJ 08855-0909

Generation of SASW Dispersions Curves

The dispersion curves used in the development of the neural network were developed by numerical simulation of the SASW test (Gucunski and Woods, 1992). In the first phase of the simulation displacements, a response due to a circular vertical source, are evaluated at a number of distances from the source. These displacements correspond to displacements, usually recorded as velocities or accelerations, during the actual SASW test. In the second phase the dispersion curves for several receiver spacings and the average dispersion curve are calculated in the way identical to that for a real test (Nazarian, 1984). A commonly used assumption is that wave propagation is dominated by a single surface (Rayleigh) wave. So, a number of inversion procedures developed utilize this assumption to simplify the procedure. The numerical simulation used in this work considers a fact that the field dispersion curve is a result of the overall wave propagation, i.e. superposition of body and surface waves of a number of modes.

Neural Network Model

Back-propagation neural network models with jump connections were found to be the most effective models in solving the complex problem of inversion of the SASW test (Williams and Gucunski, 1995). Three- and five-layer models with jump connections produced the best predictions of an asphalt-concrete (AC) pavement profile (Figure 1). The objective of this study is to improve the obtained models by increasing a number of patterns in training and test sets, and to examine the influence of the number of neurons in hidden layers on the network learning process.

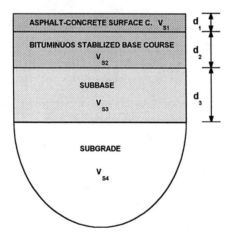

Figure 1. Assumed Layered Profile.

The total number of patterns used in the study was 188. Each pattern consisted of a dispersion curve and an associated profile defined by five dimensionless parameters: d_2/d_1, d_3/d_1, V_{s2}/V_{s1}, V_{s3}/V_{s1} and V_{s4}/V_{s1}. The profile is described by only these five dimensionless parameters because the shear wave velocity, V_{s1}, and the thickness, d_1, of the surface layer, in most cases, can be easily obtained (Roesset et al., 1989). Figure 2 illustrates a typical dispersion curve from SASW testing. The thickness of the surface course matches the wavelength at a point of the largest curvature, while velocity V_{s1} is equal to roughly 1.1 phase velocity of the initial, almost flat, portion of the curve. Ranges of values of the parameters were selected in a way so that they cover road profiles typical for the State of New Jersey (Gucunski and Maher, 1993). The dispersion curves used contain 78 dimensionless phase velocities as a function of a dimensionless frequency. The dimensionless phase velocity is defined as a ratio of the phase velocity and shear wave velocity V_{s1}, while the dimensionless frequency as fd_1/V_{s1}, where f is frequency in Hz. The range of the dimensionless frequency used is 0.007 to 0.833.

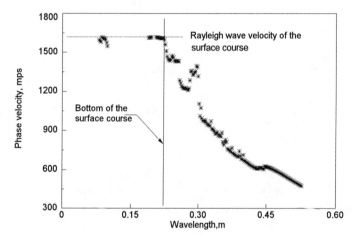

Figure 2. Average Dispersion Curve for Location 1 of SHRP Site near Syracuse
(Maher and Gucunski, 1992).

The training set consisted of 152 patterns, while the test set of 36 patterns. Selected patterns can be grouped into four groups based on the shape of the dispersion curve (Figure 3). A parameter that controls the shape of the curve is V_{s2}/V_{s1} ratio. Each of the groups included approximately the same number of patterns. To prevent network overtraining two criteria were considered: number of events since minimal average error (set to > than 40,000 for the test set) (Gucunski and Williams, 1995) and average error for the test set (set to < 0.001). The neural networks were developed using the Microsoft Windows version of the Neuroshell 2 neural network development program and a PC 486-50 MHz with 8 MB of RAM.

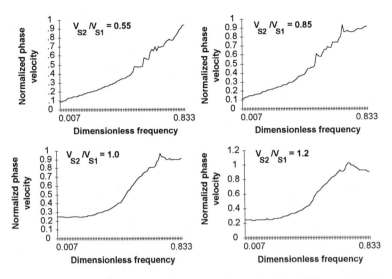

Figure 3. Typical Shapes of Dispersion Curves Covered by Neural Network Model.

Results

The model was first trained under the same conditions like the old one with 98 patterns with overtraining criterion of 40,000 events after the minimal average error. This time the learning period was halved for both 3-layer and 5-layer models. This can be explained to be a result of a more careful selection of a test set. On the other hand, prediction of the profile was somewhat worse. To improve matching with the actual data, a number of neurons in hidden layers was varied and overtraining criterion was changed to the minimal average error for the test set. The obtained results are given in Table 1 for 3-layer networks and in Table 2 for 5-layer networks.

Table 1. Mean Squared Error for 3-layer Network with Jump Connections

Number of neurons in hidden layer	Mean squared error				
	d_2/d_1	d_3/d_1	V_{s2}/V_{s1}	V_{s3}/V_{s1}	V_{s4}/V_{s1}
54	0.011	1.902	0	0	0
60	0.005	1.188	0	0	0
66	0.007	1.602	0	0	0
70	0.006	1.054	0	0	0
74	0.007	1.439	0	0	0
78	0.011	1.972	0	0	0
82	0.005	1.156	0	0	0

Table 2. Mean Squared Error for 5-layer Network with Jump Connections

Total number of neurons in hidden layers	Mean squared error				
	d_2/d_1	d_3/d_1	V_{s2}/V_{s1}	V_{s3}/V_{s1}	V_{s4}/V_{s1}
54	0.008	1.547	0	0	0
60	0.005	1.418	0	0	0
66	0.004	0.820	0	0	0
69	0.007	1.460	0	0	0
75	0.008	1.548	0	0	0
81	0.005	0.932	0	0	0

An almost perfect match between input and learned data for output variables d_2/d_1, V_{s2}/V_{s1}, V_{s3}/V_{s1} and V_{s4}/V_{s1} can be observed in the tables and in Figure 4. At the same time, the network was not able to completely learn the d_3/d_1 ratio pattern. This

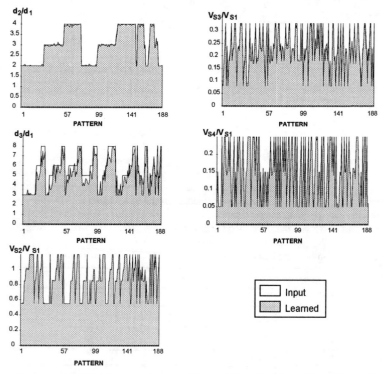

Figure 4. Matching Between Input and Learned Data for 5-layer Neural Network Model with 66 Neurons in Hidden Layers.

is in agreement with previous observations (Gucunski and Maher, 1993) that accurate evaluation of properties of a layer immediately beneath asphalt concrete courses represents the most difficult task in profile definition.

Influence of the number of hidden neurons on the mean squared error for the d_3/d_1 ratio is given in Figure 5. The best result is obtained with a 5-layer network with jump connections and with a total number of 66 neurons in the hidden layers.

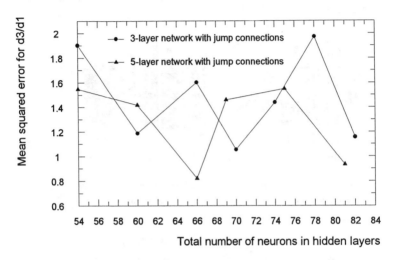

Figure 5. Effect of Hidden Neurons Number Variation on Mean Squared Error for d_3/d_1 Ratio.

The learning time for all models was approximately 24 hours. Almost all of them showed better progress if learning was occasionally interrupted. Also, during the learning period overtraining criterion was never reached, but training was considered over when a particular model showed the same mean squared error after two consecutive interruptions. It is reasonable to assume that a longer learning time and a larger number of input patterns would lead to even better matching.

Variation of other neural network parameters, like momentum, learning rate and initial weights, or switching between random or rotation pattern selections, did not affect the accuracy of the obtained models.

Contribution factors for a 5-layer network, shown in Figure 6, are significantly higher for lower frequencies. This is again in agreement with practical experience and theoretical results (Gucunski and Maher, 1993) on the sensitivity of the dispersion curve with variation of properties of different layers of a pavement.

Variation of properties of the upper pavement courses affect the shape of the dispersion curve in a high frequency range far stronger than variation of properties of deeper layers affects it in a low frequency range. This is not observed in a 3-layer model, which leads to a conclusion that the learning process of a 5-layer model is a more appropriate one (Figure 7).

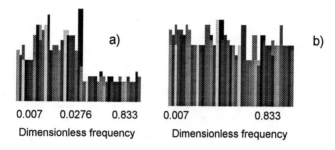

0.007   0.0276   0.833   0.007   0.833
Dimensionless frequency   Dimensionless frequency

Figure 6. Contribution Factors for a) 5-layer Network with 66 Neurons in Hidden Layers and b) 3-layer Network with 70 Neurons in Hidden Layer.

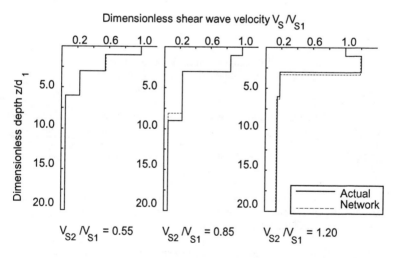

Figure 7. Comparison between the Actual Output and 5-layer Neural Network Model with Jump Connections and 66 Neurons in Hidden Layers.

Summary and Conclusions

Neural network models for inversion of dispersion curves from the SASW test on a four-layer AC pavement were presented. The best developed model is a five-layer back-propagation model with jumps. The model almost perfectly predicts thicknesses and shear wave velocity for all the layers, except the thickness of the subbase layer. While results indicate that the error can be reduced by variation of a number of neurons in a hidden layer, no rules could be defined.

References

Gucunski, N. and Maher, M. H. (1995). "Application of Nondestructive Wave Propagation Methods in Pavement Condition Assessment," to appear in the Proceedings of 1995 Transportation Congress on *Transportation for a Sustainable World Economy*, ASCE, San Diego, Calif., October 1995, 12 pp.

Gucunski, N. and Maher, M. H. (1993). "Database of Dispersion Curves for Surface Wave Testing," *Advances in Site Characterization: Data Acquisition, Data Management and Data Interpretation*, Geotechnical Special Publication No. 37, ASCE, 1-12.

Gucunski, N. and Woods, R. D. (1992). "Numerical Simulation of the SASW Test," *Soil Dynamics and Earthquake Engineering Journal*, 11(4)213-227.

Maher, M. H. and Gucunski, N. (1992), "Comparative Evaluation of Deflection and Wave Propagation Nondestructive Testing Methods for Pavements: Implications for Implementation at State and Local Level," Cycle V Report, University Transportation Research Center, Region II, City University of New York, N.Y.

Nazarian, S. (1984). "In Situ Determination of Elastic Moduli of Soil Deposits and Pavement System by Spectral-Analysis-of-Surface-Waves Method, Ph.D. Dissertation, Department of Civil Engineering, University of Texas at Austin.

Nazarian, S., Baker, M. R. and Crain, K. (1993), "Development and Testing of a Seismic Pavement Analyses," Report SHRP-H-375, Strategic highway Research Program, National Research Council, Washington, D.C.

Neuroshell 2 manual (1993). Ward Systems Group, Inc., Frederick, MA.

Roesset, J. M., Chang, D. -W., Stokoe, K. H. II, and Aouad, M. (1989) "Modulus and Thickness of the Pavement Surface Layer from SASW Tests," *Transportation Research Record*, No. 1260, 53-63.

Williams, T. P. and Gucunski, N. (1995) "Neural Network Models for Backcalculation of Moduli from SASW Test," *Journal of Computing in Civil Engineering*, ASCE, Vol. 9, No. 1, 1-8.

AN EXPERT SYSTEM FOR THE IDENTIFICATION OF CAUSES OF FAILURE OF ASPHALT CONCRETE PAVEMENT

John A. Kuprenas[1], Member, ASCE, Ricardo Salazar[2], and Rod Posada[3]

ABSTRACT

Incorrect assessment of the cause of pavement failures often leads to the performance of non-required or inappropriate pavement work. A microcomputer based expert system, PAVE, is introduced to find the most common causes of failures on asphalt concrete pavements. The top down design, forward chaining method of inference, and formulation of rules for the new system are described. The use of confidence factors within PAVE to model uncertainty is reviewed. Conclusions summarize future refinement and expansion possibilities for PAVE.

INTRODUCTION

Federal, state, and local governments spend billions of dollars annually in the maintenance and rehabilitation of existing roads and highways. The determination of the causes of pavement failures on these roads and highways is of great importance in order to improve future pavement designs and to determine the best of course of action to repair existing, damaged pavements. Incorrect assessment of the cause of pavement failures will lead to waste of tax dollars in the performance of non-required or inappropriate pavement work. This paper describes a microcomputer based expert system, **PAVE**, for the determination of the root cause for failures in asphalt concrete pavements.

THE PAVE SYSTEM

The use of expert systems in transportation and roadways has been well established. Current expert systems focus on pavement management and pavement

[1]Const. Mgr., Vanir Const. Mgmt., 3435 Wilshire Blvd., Los Angeles, CA 90010
Lecturer, Dept. of Civ. Engrg., California State Univ., Long Beach, CA 90840
[2]Grad. Student, Dept. of Civ. Engrg., Calif. State Univ., Long Beach, CA 90840
[3]Grad. Student, Dept. of Civ. Engrg., Calif. State Univ., Long Beach, CA 90840

evaluation (Allez 1988, Hendrickson and Janson 1987, Ritchie 1987, Shen and Grivas 1993). An expert system to determine failure causes and select rehabilitation technique jointed reinforced concrete, jointed plain concrete, and continuously reinforced concrete is particularly impressive (Hall et al. 1988).

The new PAVE expert system assists technicians with basic knowledge of pavements, in identifying the possible causes of the failure of a pavement, that in turn will help in determining the most appropriate rehabilitation strategy for a deteriorated pavement. This new approach will standardize the decision-making, will help in the planning of the streets of most needed rehabilitation, and will lead to a more rational distribution of the tax dollars. The most likely users for PAVE would be personnel working for the public works department of public agencies. The intended audience for this work would include not only engineers involved in pavement design, but also engineers and researchers interested in microcomputer expert systems. Researchers interested in computer modelling of uncertainty, particularly with confidence factors, would also be interested in this work.

PAVE was created in the expert system shell called VP Expert by WordTech Systems. The inference engine, user interface, knowledge base editor are all contained within VP expert (Luce 1992). The knowledge base consisting of facts, rules, and heuristic are supplied through programming. PAVE was created at the California State University, Long Beach. The design of **PAVE** was researched using existing textbooks on the failure of pavement mixes, complemented with the personal experience of this project team members.

Structure

PAVE was created using a top down design. Top down design divides a large problem into progressively smaller problems. Some sub-problems must be further subdivided in order to be solved, while some do not require further refinement. This top down design allows an expert system to be systematically expanded through a top down development while maintaining an emphasis on the major problem. The top down design diagram for the PAVE expert system is shown in figure 1.

As seen in figure 1, the top down design first called for the identification of failure types. Based on expert knowledge, three failure types were identified. The distress or failure in the pavement caused by vehicle loads or by material forces generated by changes in moisture, sunlight, and other environmental consequences, visibly manifests itself as a surface defect, as pavement deformation, or as cracking. These are the three potential failure types for asphalt pavement. Using a top down approach, each of these three failure types was then subdivided into different failure manifestations. For example, how can surface defects be manifested in asphalt pavement? Expert knowledge and research were used to answer each of these questions. Table 1 shows the possible manifestations for each of these three types of failures.

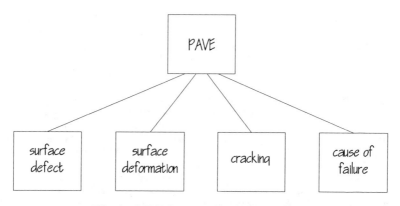

FIG. 1. PAVE System Top Down Design

Rules

Expert knowledge is again used to identify root causes for pavement failures. The PAVE expert system uses information related to the types of failure (surface defects, surface deformation, and cracking) to assess the cause for the failure. The system is based on a data driven search, hence the method of inference is forward chaining. Using a forward chaining inference method means that forward rules attempt to prove goals (Badiru 1992, Parasaye and Chignall 1988). If-then rules start with facts and determine conclusions that can be reached from these known facts. This inference method suits pavement assessment since, pavement failures are **known facts**, while knowledge supplied by experts to diagnose root causes for these failures are reached through **conclusions**.

All three types of distress are used to indicate the one root cause for failure. The number of rules required within PAVE is, therefore, determined by the different number of combinations of the failure manifestations. Therefore, the number of rules, N_{rules}, for PAVE can be expressed as

$$N_{rules} = N_{defect} \times N_{deform} \times N_{cracking} \dots \dots \dots \dots \dots \dots \dots \dots \dots (1)$$

with N_{defect} is the number of possible manifestations for a surface defect failure, N_{deform} is the number of possible manifestations of surface deformation failure, and $N_{cracking}$ is the number of possible manifestations of cracking failures. Given the information supplied in Table 1 that $N_{defect} = 4$, $N_{deform} = 5$, and $N_{cracking} = 4$, and using equation (1), one sees that the number of rules for PAVE, $N_{rules\ PAVE}$, can be calculated as

$$N_{rules\ PAVE} = 4 \times 5 \times 4 = 80 \dots \dots \dots \dots \dots \dots \dots \dots \dots \dots \dots \dots \dots (2)$$

assuming three types of failure as indicated in Table 1.

TABLE 1. Manifestations of Asphalt Pavement Failures

Failure manifestation (1)	No. of outcomes, N (2)	Type of failure (3)
Surface Defects	4	Gravel Loss Ravelling Flushing None
Surface Deformations	5	Ripping Shoving Rutting Distortion None
Cracking	4	Longitudinal Transversal Alligator None

VP expert allows induction of rules from a spreadsheet file. Table 2 is a portion of a spreadsheet table that depicts the knowledge encoded into the first twenty rules of PAVE before the rules were induced. The three first headers are the type of failures or distresses experienced; the fourth header is the most likely cause for the pavement failure. The spreadsheet in Table 2 shows all combinations of data for a surface defect failure type manifested as "gravel loss". The number of rules required to represent this knowledge can be calculated using equation (1), with N_{defect} = 1 ("gravel loss"), N_{deform} = 5, and $N_{cracking}$ = 4. The number of rules used to represent surface defect failure type manifested as "gravel loss", $N_{rules \ "gravel \ loss"}$, can be calculated as

$$N_{rules \ "gravel"} = 1 \times 5 \times 4 = 20 \dots \dots \dots \dots \dots \dots \dots \dots (3)$$

The spreadsheet used to induce the rules for PAVE included the remaining sixty rows of data for all remaining failure manifestation combinations. A sample rule from the actual expert system is shown in Figure 2. The rule represents the rule induced from the last line of the spreadsheet in Table 2.

Uncertainty

Since pavement failure assessments are inherently imprecise, the PAVE system makes use of confidence factors. Confidence factors are a common representation of heuristic weights not based on any mathematical or logical statistical representations. In the PAVE system, the relative degree of confidence the user has in a fact is assigned a confidence factor number. This number is between zero and

TABLE 2. Data for a Surface Defect Failure Type - "Gravel Loss"

Surface defect (1)	Surface deformation (2)	Type of cracking (3)	Cause of failure (4)
Gravel loss	None	None	No bond
Gravel loss	Ripping	None	No bond & thick tack coat
Gravel loss	Shoving	None	No bond & lub tack coat
Gravel loss	Rutting	None	No bond & comp on AC
Gravel loss	Distortion	None	No bond & base comp
Gravel loss	None	Longitudinal	No bond & thermal shrink
Gravel loss	None	Transversal	No bond & ref cracks
Gravel loss	None	Alligator	No bond & pbs
Gravel loss	Ripping	Longitudinal	No bd & thk tack coat & thrm shk
Gravel loss	Ripping	Transversal	No bd & thk tack coat & ref crcks
Gravel loss	Ripping	Alligator	No bond & thick tack coat & pbs
Gravel loss	Shoving	Longitudinal	No bd & lub tack coat & thrm shk
Gravel loss	Shoving	Transversal	No bd & lub tack coat & ref crcks
Gravel loss	Shoving	Alligator	No bond & lub tack coat & pbs
Gravel loss	Rutting	Longitudinal	No bd & comp on AC & thrm shk
Gravel loss	Rutting	Transversal	No bd & comp on AC & ref crcks
Gravel loss	Rutting	Alligator	No bond & comp on AC & pbs
Gravel loss	Distortion	Longitudinal	No bd & base comp & therm shk
Gravel loss	Distortion	Transversal	No bd & base comp & ref cracks
Gravel loss	Distortion	Alligator	No bond & base comp & pbs

one hundred. An assignment of one hundred indicates the user has absolute confidence that the fact is true (Luce 1992).

Procedures for performing operations on certainty factors are not based on any rigid theory. In the expert system shell used to develop PAVE, the following formula is used to calculate the final confidence factor for a certainty of a rule, CNF_{final}, is defined as

$$CNF_{final} = \min(CNF_{condition}) \times CNF_{conclusion} \div 100 \quad \ldots \ldots \ldots \ldots \ldots (4)$$

where $CNF_{condition}$ is the certainty factor for all conditions of the rule, and $CNF_{conclusion}$ is the coded certainty factor for the rule conclusion (Luce 1992).

The current programming of PAVE has set the CNF of conclusion to be 100 for all rules in the expert system. The user is asked to input the certainty factor for each conditioning element to the rule, $CNF_{condition}$, through the user interface. When the user indicates failure manifestations through the interface questions, the user is given the option of supplying a certainty factor to any of the manifestations selected.

```
RULE 19
IF        Surf_Defects=Gravel_loss AND
          Surf_Deform=Distortion AND
          Cracking=Alligator
THEN      Cause_of_Failure=No_bond_&_base_comp_&_pbs;
```

FIG. 2. Example Rule from PAVE

For example, using the rule depicted in figure 2, the user may define the certainty factor for the surface defect failure manifestation of "gravel loss", $CNF_{condition, defect}$, to be 90, the certainty factor for the surface deformation failure manifestation of "distortion", $CNF_{condition, deform}$, to be 60, the certainty factor for the cracking failure manifestation of "alligator", $CNF_{condition, cracking}$, to be 80. The final confidence factor is then calculated as

$$CNF_{final, rule\ 19} = min(90, 60, 80) \times 100 \div 100 = 60 \dots \dots (4)$$

where $CNF_{conclusion}$ is set to be 100.

CONCLUSIONS

PAVE presents a practical, simple tool for assisting technicians with basic knowledge of pavements, in identifying the possible causes of the failure of a pavement. This tool will therefore help in determining the most appropriate rehabilitation strategy for a deteriorated pavement. The advantages of PAVE are its the simplicity of the knowledge base structure, relative ease of identification of asphalt pavement failure types, and the ease of use of the system itself. One disadvantage of PAVE is that the use of confidence factors depends on the sophistication of the users. Another disadvantage is the limited scope of the system since it currently only is limited to assessments only for asphalt pavements.

Future expansion of PAVE can lead in two directions, refinement and expansion. Research can further refine the current three failure modes to make more precise, which should in theory refine the identification for the root cause of the failure. A second potential refinement would be to improve ability of the system to model uncertainty. Perhaps the use of fuzzy logic would be simpler to use than certainty factors. A more practical expansion of PAVE would be to expand scope of the system. The most next logical step of such an expansion would be increase the system to include other pavement types other than asphalt. Another possible expansion would be to recommend repair techniques based on failure assessments and root causes of failures. This type of expansion, however, would profoundly increase complexity of the system since the repair technique is dependent on both the state to the existing pavement and the cause for the initial pavement failure.

APPENDIX. REFERENCES

Allez, F. (1988). "ERASME: An expert system for pavement maintenance." *Transportation Research Record 1205*, Transportation Research Board, Washington D. C., 1-5.

Badiru, Adedeji B. (1992). *Expert Systems Applications in Engineering and Manufacturing*, Prentice Hall, Inc., Englewood Cliffs, N. J. p.106, 107.

Hall K. T., Conner, J. M., Darter, M. I., and Carpenter, S. H. (1988). "Expert system for concrete pavement evaluation and rehabilitation," *Transportation Research Record 1207*, Transportation Research Board, Washington D. C., 21-29.

Hendrickson, C. T., and Janson, B. N. (1987). "Expert systems and pavement management." *Proc. 2nd North Am. Conf. on Managing Pavements*. Vol. 2, Ministry of Transportation, Toronto, Ontario, Canada, 255-266.

Luce, Thom (1992). *Using VP-Expert in Business*, Mitchell McGraw-Hill, Inc., Watsonville, California, p. 120-121

Parasaye, Kamran and Chignall, Mark (1988). *Expert Systems for Experts*, John Wiley and Sons, New York, New York, p. 95-96

Ritchie, D. S. (1987). "Application of expert systems in managing pavements." *Proc. 2nd North Am. Conf. on Managing Pavements*. Vol. 2, Ministry of Transportation, Toronto, Ontario, Canada, 277-288.

Shen, Y. C. and Grivas D. A. (1993). "Use of knowledge graphs to formalize decisions in preserving pavements." *J. Comp. in Civ. Engrg.*, ASCE, 7(4), 475-494.

FEASIBILITY OF APPLYING EMBEDDED NEURAL NET CHIP TO IMPROVE PAVEMENT SURFACE IMAGE PROCESSING

Kelvin C.P. Wang[1] Member, ASCE

ABSTRACT: Due to limitations in processing speed, algorithm's accuracy, and implementation costs, there is no widely accepted system for automated pavement surface distress survey. This paper presents the feasibility of using a specially designed and programmable neural net chip in a microcomputer to conduct pavement surface image processing. This neural net chip, Ni1000, has massive parallel processing capabilities. With the help of high speed bus and a state of the art microcomputer, the resulting system will be able to provide real-time processing at traveling speed (55 mph). This paper presents the potentials of the Ni1000 and the research plan to use the chip for pavement surface distress survey. The results from this study will not only provide know-how for the highway industry in automating the survey of pavement surface distress, but also demonstrate the feasibility of integrating low-cost, extremely powerful components into a useful vision system.

INTRODUCTION

The categorization and quantification of the type, severity, and extent of surface distress is a primary method for assessing the condition of highway pavements. The most widely used method to conduct such surveys is based on human observation. This approach is extremely labor intensive, prone to errors and poses traffic hazard. In the last few years, methodologies were developed to automate pavement distress surveys. However, none of these applications is widely accepted, due to implementation costs, low speed, and accuracy limitations. Recent technological innovations in computer hardware and pattern recognition techniques provide opportunities to explore a new methodology for the automation of pavement distress survey in the most cost-effective way. Ritchie (1990) presented some background information on automated approaches.

The goal of this paper is to study the feasibility of using an efficient and low-cost approach to process video images of pavement surface, which would detect, classify, and quantify pavement surface distresses by using a microcomputer with a neural net based add-on board, and a digital camera. The focus of the research is on developing a new cracking recognition and classification approach for project level pavement surface distress survey, including the improvement of image interpretation quality. The specially designed and manufactured neural net chip will be used to perform parallel computation at an order of magnitude faster speed than the fastest microprocessor available today, and at a lower price.

[1] Assistant Professor, Department of Civil Engineering, University of Arkansas, Fayetteville, Arkansas 72701, Telephone: (501)-575-8425, Fax: (501)-575-7168

BACKGROUND

An important element in pavement management is the determination of pavement condition. The assessment of pavement condition is usually conducted through pavement distress survey, at either project level or network level. Visual surveys are still the most common technique used to collect data. The concerns of safety, cost and human error provide an impetus to automate the distress survey process. Successful automation will reduce the overall cost of performing pavement surface distress survey and to provide more objective and standardized results. The survey automation involves two major areas: data collection and data interpretation.

The system concept for the automation of pavement distress survey is shown in Figure 1. There are two distinct subsystems: an image acquisition subsystem and an image display and interpretation subsystem. The image acquisition subsystem is becoming a standard technology. More recent development involves high resolution video cameras and laser recording technologies. Image data in most cases are recorded in analog format. The analog video images can then be digitized by a "frame grabbing" process. Each frame of analog image data base is converted to digital format. In the recent years, much research and development have been aimed at the computerized processing of images in order to augment the computer's power with some "human-like" visual sensing capability. This technology, often termed computer vision or machine vision, is related with the second subsystem of image interpretation of collected pavement surface data.

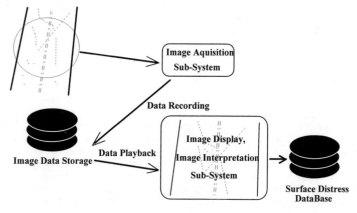

FIG. 1. System Concept in the Pavement Surface Distress Survey

Many academic and industrial studies have attempted to automate the evaluation of pavement surface distress. The core of these systems relies on the processing algorithms which are analytical models used to filter image noise and identify the type and severity of various cracks. There is no single system developed so far that received widespread acceptance from the highway industry. There are a number of reasons why those efforts did not produce useful systems. First, the implementations employ different image processing algorithms and different hardware design, which resulted in non-comparable survey data from these equipment. Second, image processing for pavement surface distress survey at any practical speed requires high performance computing equipment. When a compromise was made in respect of equipment performance, data quality may have been affected. Third, there is no standard indexes to quantitatively define the types, severity, and extent of pavement surface distress. However, efforts are underway to initialize a set of standards. Fourth, image processing as whole is a relatively new filed. Only in recent years were the major applications of automated pavement surface distress survey implemented.

Most of the developed systems include vehicle equipped with proper video gear traveling at highway speeds. Pavement surface images are collected into analog storage devices through camera(s) mounted on the vehicle. A post processing is conducted on the collected images. Dougherty and Giardina (1987) proposed a summary of the main steps used in image processing, which is adapted for pavement image processing shown in Figure 2. The image process begins with collecting the raw data in analog format. Digitized images is defined on a grid or array within the field of observation. The set of points comprising the grid is selected through the sampling process, for example 450 points by 450 points. Quantization is carried out concerning the discrete scaling of the illumination levels in the image to the gray scale values chosen for analysis. Image restoration involves cleaning up the image, such as filtering out noise and sensor bias. An image can be enhanced before further processing so that particular features, such as edges of objects, are more distinctive. Segmentation can be used to identify regions where the color or brightness levels are relatively uniform. In feature selection stage, parameters are obtained geometrically, statistically, or through the use of transform techniques, to identify features that could be useful for image characterization and classification. Image registration is included to account for sensor attributes that could impact the image, such as velocity and attitude. Finally, image classification attempts to match vectors of features against existing data to achieve a match and a detection decision for the original image or object. The following discussions cover two developed applications.

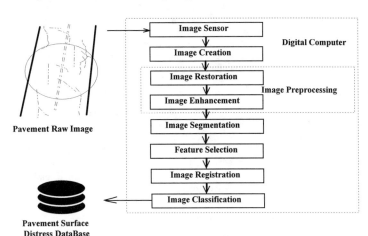

FIG. 2. Steps in Pavement Image Processing

Komatsu System

In the late 1980s, the Japanese consortium Komatsu built an automated-pavement-distress-survey system (Fukuhara, et al. 1990), comprising a survey vehicle and data-processing system on board to simultaneously measure cracking, rutting, and longitudinal profile. Maximum resolution of 2048 x 2048 is obtained at the speed of 10 km/h. The Komatsu system works only at night to control lighting conditions. Digital image processing techniques are applied to crack image data (representing 4 m by 16 m images) in a post processing mode. Parallel processing is used in two stages to determine cracking parameters such as the number, width, and length of cracks, which are then stored in the pavement data bank. In the first processing stage (image segmentation), a massive 64 MC68020 parallel microprocessors are used (a MC68020 is equivalent to an entry level 386 chip). The system design permits up to 512 32-bit microprocessors to be used in parallel to further improve performance. In the second stage (crack classification), seven T800 transputers are used in parallel. The Komatsu system represents an implementation of the most sophisticated hardware technologies at that time. However, it does not output the types of cracking. Another barrier to implement the Komatsu system is that it virtually requires the power of multiple super-computers to carry out the two stage analyses.

Triple Vision's NCHRP Project

In the Triple Vision project completed in 1991, sample video data based on PASCO 35 mm films were transferred to a video disk (Fundakowski, et al. 1991). The data acquisition was conducted by PASCO using 35mm film camera at highway speed. The algorithm for the video image processing were distributed across two computer systems. The bulk of the image processing operations (i.e. image preprocessing and segmentation) were performed in a DSP-1000 image processing system (product of Datacube, inc., Peabody, MA.). The feature extraction and classification stages of the video image processing system were implemented in a standard 386 PC. The DSP-1000 incorporates several image processing boards, a digitizer/frame grabber and a display board. In the report produced in this project, the comparison of the machine generated cracking recognition, and the data based on the two experts on pavement engineering is quite poor. In addition, the applicability of the system for actual highway use is questionable. An upgraded system with a 33 MHz 486 computer and the Data Cube processor can only process 1 frame of image per minute. That is equivalent to 29 hours of processing time per one lane-mile.

RESEARCH APPROACH

Pattern recognition techniques in most of the current pavement image processing applications do not imitate the pattern recognition that takes place in the neural systems of pavement engineering experts. The results from existing applications show that their industrial applicability is limited due to the unacceptable deviations of their recognition results from experts opinions. In addition, their requirement of hardware equipment is astronomical for near-real-time processing, which is required for industrial application. Therefore, there is a need to use new technologies to improve the quality of survey data, increase the processing speed and reduce the implementation costs.

Pattern Recognition with Neural Nets

A neural network consists of many processing elements, which are usually organized into a sequence of layers with full or partial connections between successive layers. There is an input buffer where data are presented to the network, and an output buffer, which holds the response of the network to a given input. Other intermediate layers are known as hidden layers. In most neural networks, the combined input to a processing element in any layer is obtained through a weighted summation of the outputs of elements in the previous layer. Next, an output is computed from the combined input through a transfer function, typically a sigmoid function, which modulates the combined input so that the output approaches one when the input gets larger and approaches zero when the input gets smaller.

Neural networks have the ability to learn from examples. This ability is acquired through a learning process where connecting weights are adjusted in response to stimuli presented at the input buffer, and also at the output buffer. After a learning phase, a recall phase may be activated to process a stimulus presented at the input buffer of the trained network and create a response at its output buffer. The versatile feature of the computing memory of a neural network, together with its ability to learn further as more information is gathered, is likely to be useful in improving the image recognition in the automation of pavement surface distress survey. Kaseko et al. (1993) conducted research employing neural net techniques for pavement surface distress survey, which has some promising capabilities. However, the neural net system in that study was realized in software. Therefore, speed was limited due to the parallel structure of neural nets which requires parallel processing for high speed computation.

Pavement Surface Distress Analysis with the Ni1000 Neural Net Chip

Collaborative research between Intel Corporation and Nestor Inc. produced a high performance recognition accelerator, the Ni1000, for imaging and signal processing applications. For a typical signal processing application, the speed is estimated to increase 100 to 1000 times than a host microprocessors and digital signal processor (DSP). Ni1000 contains 3.7 million transistors and is manufactured using Intel's 0.8u CMOS FLASH EPROM (Erasable Programmable Read Only Memory) technology. At 40 MHz, Ni1000 performs 20 billion five bit integer subtract and accumulate operations per second, and 160 MFLOPS (Park, et al. 1993). Ni1000 can perform pattern recognition at 40,000 patterns per second. Input patterns of 256 dimensions by 5 bits are transferred from the host to the Ni1000 and compared with the chip's memory of 1024 stored reference patterns, in parallel. The 'memories' are stored on chip in a 1.3 Mb FLASH EPROM. Therefore, no battery backup is required. A custom 16 bit on-chip micro controller runs at 20 MIPS and controls all the programming and algorithm functions. Once programmed, the Ni1000 computes in a highly pipelined fashion, implementing the statistical algorithms which use Radial Basis Functions (RBFs). RBFs are considered an advancement over traditional template matching algorithms and back propagation (Park, et al. 1993).

Assume that pavement surface width is 12 feet. A digital picture of a pavement surface covers an area of 12 feet (transverse) by 10 feet (longitudinal) with a minimum resolution of 100 dots per inch. Also assume that there are a total of 1000 patterns for various pavement surface distresses. Based on the full speed of 40,000 patterns per second, a total of 40 pictures of the size 12 feet by 10 feet can be processed. The translating speed is 400 feet per second or 273 mph. Even though there most likely will be input/output bandwidth and other overhead restrictions imposed on the sub-systems, the reduced processing speed will still be adequate for project level study for pavement survey distress survey. For real-time processing where 55 mph is a typical travel speed for surveying vehicles, a fast I/O and data capturing speed matching the Ni1000 pattern rate are required. In addition, Ni1000 is capable of processing patterns at the dimension of 256. Pavement surface image is captured in two dimensions. The Ni1000's capabilities of processing large number of dimensions can be converted to additional power to accelerate the two dimension processing.

RESEARCH STEPS

There are three major components of the system's hardware: a data collection equipment, a high performance microcomputer, and the Ni1000 based add-on boards. Figure 3 demonstrates the data flow in the proposed image processing system. The Eastman Kodak still digital camera, Model DCS 420C, uses a conventional Nikon N90 as the base capturing device and saves the captured images to an attached hard drive with a resolution of 1,500 x 1,000. A Pentium (100 MHz) computer with PCI local bus is to be used. The Ni1000 development kit includes the PCI compatible Ni1000 board and the development software modules. The PCI bus provides a burst data rate of 155 M bytes per second. It is expected that a PCI Ni1000 board will provide image processing performances comparable to a super computer.

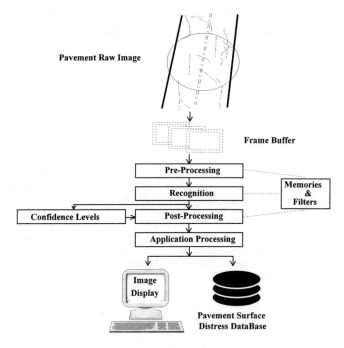

FIG. 3. Data Flow in the Proposed Research (Based on Nestor, Inc.)

The DCS 420C camera can be used to capture the pavement surface image with the dimension of 12 feet by 8 feet. The resulting resolution is approximate 120 dots per foot.

As this research is related to project level and data collection speed is not a critical factor, a number of sample images can be shot with the digital still camera and the image can be conveniently transferred into the microcomputer from the hard drive in the digital camera.

There are four phases to develop the system: software development. neural net training, application, and evaluation. A feedback process is used to improve the software development during the three phases of training, application, and evaluation. Intel and Nestor provide a complete design package for the Intel-based PC platform. Many of the modules provided with the Ni1000 kit are accessible from MS Windows based host system software, NestorACCESS, or run from the DOS prompt. A high level C language applications programming interface (API), the same interface used by NestorACCESS GUI, permit to call NestorACCESS functions from your application program. NestorACCESS provides a step by step method of generating 'memories' by describing a sequence of 'job' to run. There are several standard recognition algorithms that have been optimized to be used on the chip. Custom code can also be written by using the Ni1000 micro code. As Ni1000 uses EPROM, the customized code can be downloaded to the chip fairly easily.

The identification and quantification of surface distress will concentrate on cracking on asphalt concrete surface. The procedure and definitions used in the *Distress Identification Manual for the Long-Term Pavement Performance Project* (1993) will be followed. A total of six crack types will be covered: (1). Fatigue cracking, (2). Block cracking, (3). Edge Cracking, (4). Wheel Path/Non-Wheel Path longitudinal Cracking, (5). Reflection Cracking, (6). Transverse Cracking.

CONCLUSION

Based on the feasibility study, it was determined that the Ni1000 chip is capable of providing real-time processing for pavement surface distress survey. Research is currently underway to use the development kit to develop a fully functional project level distress survey system. It is very hopeful that a low-cost automated system for network level distress survey can be built with the Ni1000.

APPENDIX I. REFERENCES

Dougherty, E. R. and Giardina, C. R. (1987). *Image Processing - Continuous to Discrete, volume 1*, Prentice-Hall

Fukuhara, T., et al. (1990). "Automatic Pavement-Distress-Survey System," ASCE Journal of Transportation Engineering, May/June.

Fundakowski, R. A., et al. (1991). *Video Image Processing for Evaluating Pavement Surface Distress*, Final Report, National Cooperative Highway Research Program 1-27, September.

Kaseko, M. S. and Ritchie, S. G. (1993). "A Neural Network based Methodology for Pavement Crack Detection and Classification." Transportation Research, Vol. 1C, No. 4, December.

Park, C. et al. (1993). "A Radial Basis Function Neural Network with On-Chip Learning." An Intel-Nestor Paper.

Ritchie, S. (1990). "Digital Imaging Concepts and Applications in Pavement Management," ASCE Journal of Transportation Engineering, May/June.

A New Method of Truss Analysis for Large Deformations and Snap-Through

Apostolos Fafitis[1] and Feng Luo[2]

1. INTRODUCTION

This paper presents a numerical approach to the analysis of trusses, different from the classical structural analysis approach (e.g. the Stiffness or the flexibility method). The presented method is more general and it can handle in a unified way such problems as arbitrarily large deformations, sudden change in geometry as it happens in snap-through or buckling, inelastic nonlinear materials and non-proportional loading. The analysis of linearly elastic structures subjected to small deformations, is a special case of the proposed method.

2. DESCRIPTION OF THE METHOD

The structural problem is reduced to the mathematical problem of solving n nonlinear equations with n unknowns. The equations to be solved are the equilibrium equations of the external and the internal forces acting at each degree of freedom of the structure. The unknowns are the displacements at those degrees of freedom.

Let i be a typical joint connected with n_i members of the truss as shown in Fig. 1. The internal force exerted by member m at the α-th degree of freedom (say at joint i in the x-direction) is

$$P_{ij} = A_m \sigma_m \qquad (1)$$

- - - - - - - - - - - - - -

1. Associate Professor of Civil Engineering, Member ASCE, Civil Engineering Department, Arizona State University, Tempe, AZ85287-5306.

2. Graduate Student, Civil Engineering Department, Arizona State University, Tempe, AZ85287-5306.

where σ_m is the stress of member m and it is a function of the strain of this member. The function relating the stress with the strain is the constitutive relationship of the material.

$$\sigma_m = f_m \left(\epsilon_m \right) \tag{2}$$

The strain is defined as

$$\epsilon_m = \frac{l_m^2 - l_{mo}^2}{l_{mo}^2} \tag{3}$$

where l_{mo} is the original length of the member and l_m is its strained (current) length.

$$l_{mo}^2 = (X_j - X_i)^2 + (Y_j - Y_i)^2 \tag{4}$$

$$l_m^2 = [(X_j - U_j) - (X_i - U_i)]^2 + [(Y_j + W_j) - (Y_i + W_i)]^2 \tag{5}$$

in which Xi, Yi, Xj and Yj are the coordinates of the undeformed structure, and Ui, Wi, Uj and Wj are the displacements of joint i and j respectively.

The internal force in the α-th degree of freedom exerted by the member m is then

$$P_{ijx} = P_{ij} \cos\theta_{ij} = g_{ijx} (U_1 U_2 .. U_r W_1 W_2 .. W_s) \tag{6}$$

where θ_{ij} is the angle of member ij with the x-axis.

$$\theta_{ij} = atn [(Y_j - Y_i)/(X_j - X_i)] \tag{7}$$

The equilibrium equation is then

$$\sum^{r_i} P_{ijx} - F_{ix} = 0 \tag{8}$$

or

$$g_\alpha (U_1 U_2 .. U_r W_1 W_2 .. W_s) = F_{ix} \quad (i = 1,2..n)$$

(9)

where F_{ix} is the x-component of the external force at joint i, and $g_\alpha (U_1, U_2, .. U_v, W_1, W_2, .. W_s)$ is a nonlinear function of the displacements of the joints. Thus eq. 9 is the equilibrium equation at the α-th degree of freedom. The number of equations like this is n = r+s, where r and s are the numbers of unrestrained degrees of freedom in the x- and y-direction respectively.

The external force F_{ix} is applied incrementally at the α-th degree of freedom and the load history (e.g. rate of load variation) does not have to be the same for all degrees of freedom, therefore the load can be non proportional.

The system of the above n nonlinear equations is solved for the current external load numerically by the Newton-Raphson method. The Jacobian of the equilibrium equations, (eq. 9) is calculated numerically by giving small variations to the current joint coordinates in the direction of the unrestrained degrees of freedom, one at the time, and calculating the rate of change of the internal force at each degree of freedom. Note that, as it is apparent from eq. 9, the physical meaning of element J_{ij} of the Jacobian is the rate of change of the internal force g_α as the displacements (U's and W's) vary.

The initial values of the displacements at each cycle of iteration, that is at each load increment, are taken equal to the displacements that have been calculated at the previous load increment. The very first values are taken equal to zero (displacements of the undeformed structure). Each cycle of iterations is continued until equilibrium of the internal and the current external forces is reached.

3. NUMERICAL EXAMPLES

The method is applied in three numerical examples. In the first example, a cantilever truss (Fig.2) is subjected to an imposed horizontal displacement (Displacement

Control). In the other two examples two roof trusses are subjected to displacement and load control. The truss configuration is shown in Fig.3. One of these trusses is a rather shallow (with rise equal to 5 units i.e. x=3) and the other deeper (with 13 units rise i.e. x=11).

3.1 Cantilever Truss

The truss shown in Fig. 2 is subjected to an imposed horizontal displacement of joint 9. The joint is free in the y-direction. The Load-Displacement curve at five characteristic configurations of the structure are shown in Fig.4. The first configuration is a large deformation(horizontal and vertical displacement equal to 11.30 and 14.92 respectively). The second and third configurations capture a sudden transition (snap). By increasing the displacement the configuration changes suddenly again to the one in the fourth row of Fig.4. Note that at this configuration the force at joint 9 is negative (push). Finally the reversed structure of row 5 in Fig.4 is subjected to eccentric axial tensile force as the displacement of joint 9 increases. The variations of stress in member 1 (first from left, lower horizontal member) are shown in the same figure.

3.2 Elastic Response of Roof Trusses

The shallow roof truss of Fig.3 was analyzed by the proposed method for an imposed displacement (displacement-control) at the apex joint . Because of symmetry, only half of the structure was considered.

The displacement-control response with the configuration of the structure are shown in Fig.5. The joint is displaced to 8 units downwards (loading) and then it is restored to its original position (unloading). The stress in member 1 during this loading-unloading is shown in the same figure.

The deep truss of Fig.3 was subjected to a vertical load at the apex joint and the response is shown in Fig.6. Note that, because this roof has a large rise (13 units compared with the 5 units of the previous truss), there is no snap-through upon unloading but the structure remains "blocked" as shown in Fig.6 (row 5).

3.3 Elastic-Plastic Response of Roof Truss

The response of the shallow truss of Fig.3, with elastoplastic members is shown in Fig.7. The loading in this case was displacement-control. Note that when the controlled apex joint is restored to its original position, the structure is not restored to its original configurations but there is a permanent deformation, as shown in Fig.7(row 5). The members of the structure after a full cycle of loading are left with residual forces.

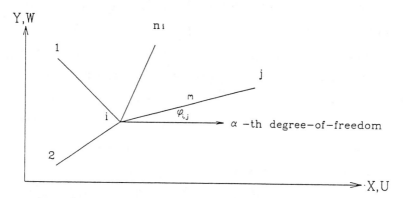

Fig.1 Typical Truss Joint

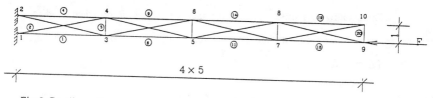

Fig.2 Cantilever Truss

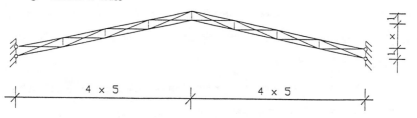

Fig.3 Roof Truss (Shallow Truss: x = 3; Deep Truss: x = 11)

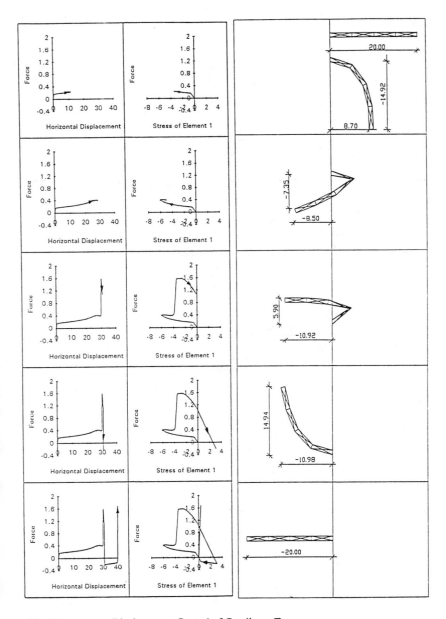

Fig.4 Response to Displacement Control of Cantilever Truss

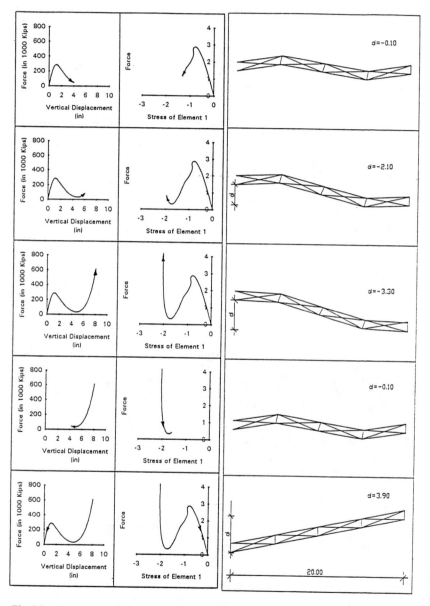

Fig.5 Response to Displacement Control of Roof Truss with 5 Rise

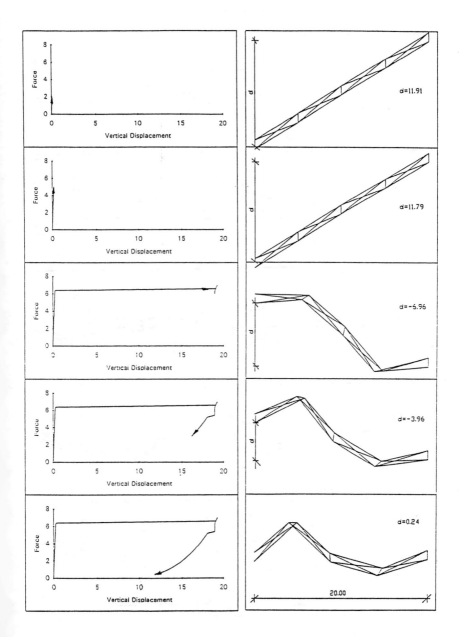

Fig.6 Response to Load Control of Roof Truss with 13 Rise

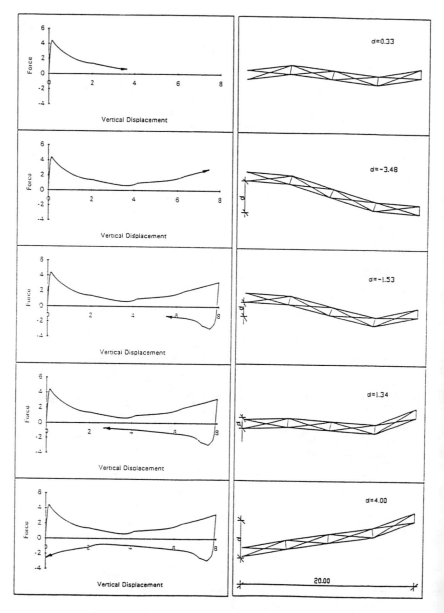

Fig.7 Response to Displacement Control of Roof Truss with 5 Rise

SPREAD-PLASTIC MEMBER MODELS
FOR INELASTIC ANALYSIS OF STRUCTURES

Balram Gupta[1] and Sashi Kunnath[2]

Abstract

The modeling of spread plasticity formulations for structural members and its application in inelastic structural analysis is examined. Three primary and practical approaches to stiffness formulation using distributed flexibility are explored in this study with a view to assessing their suitability in nonlinear analysis of structures. The member models are developed from considerations of plastic hinge-length and moment variation across the element. The models are applied to the analysis of a model frame subjected to both static and dynamic loads. A discussion of the implications, both practical and numerical, of using each of the models in inelastic structural analysis is also included.

Introduction

The basic objective in developing a member model for nonlinear analysis is to adequately characterize the varying stiffness properties of the element during inelastic action. A number of member models have been proposed in the past, ranging from simple multicomponent representations (Clough et al., 1965) with bilinear force-deformation hysteresis to more complex fiber model (Spacone and Filippou, 1994) and finite element discretizations wherein material behavior is prescribed at the constitutive level.

Spread-plastic member models are those which take into consideration the phenomena of yield penetration (steel and concrete) and crack propagation (primarily concrete). The modeling of yield penetration (finite plastic zones) and propagating cracks is accomplished through distribution of the flexibility (or stiffness) along the member. This paper investigates the performance of three types of spread-plastic member models for use in nonlinear analysis of structural systems.

----

[1]Graduate Student, and [2]Assistant Professor, Department of Civil Engineering, University of Central Florida, Orlando, FL 32816.

Description of Member Models Used in Evaluation

Spread plastic member models use flexibility formulations to develop the stiffness matrix. The typical moment-rotation relationship for a typical beam element shown in Figure 1 will assume the following form:

$$\begin{Bmatrix} \theta_A \\ \theta_B \end{Bmatrix} = L_n \begin{bmatrix} f_{11} & f_{12} \\ f_{21} & f_{22} \end{bmatrix} \begin{Bmatrix} M_A \\ M_B \end{Bmatrix} \tag{1}$$

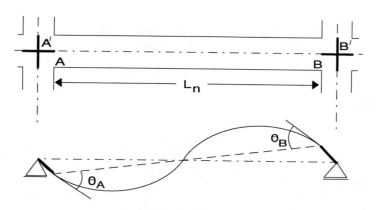

Figure 1. Typical Beam Element with Rigid End Zones

Three variations in the representation of spread plasticity are evaluated in this study. Based on a linear moment distribution, the stiffness variation across the element for each of the three models under consideration is shown in Figure 2. The first model is based on using a finite plastic zone and will be referred to as the "Hinge-Length Model" (Model 1). The second model considers a linear distribution of flexibility and is called the "Linearly Distributed Model" (Model 2). The third model investigated combines critical aspects of both the above models and is referred to as the "Hybrid Model" (Model 3). In Models 1 and 3, the length of the plastic zone is established from a pre-determined moment-curvature or moment-rotation envelope for the member. In the initial elastic phase, the hinge lengths are zero and a primarily elastic response is obtained. Yielding at either end of the member results in a new stiffness matrix, which is then constantly updated as the hinge length increases. The hinge length is a function of the previous maximum moment and does not change until the "zone" is exceeded by additional inelastic excursions. The flexibility coefficients for Model 1 assume the following form:

$$f_{11} = (3\alpha - 3\alpha^2 + \alpha^3)/(3(EI)_a + ((1-\alpha)^3 - \beta^3)/3(EI)_o + \beta^3/3(EI)_b \quad (2)$$

$$f_{12} = f_{21} = (3\alpha^2 - 2\alpha^3)/6(EI)_a + (1 - 3\beta^2 + 2\beta^3 - 3\alpha^2 + 2\alpha^3)/6(EI)_o$$
$$+ (3\beta^2 - 2\beta^3)/6(EI)_b \quad (3)$$

$$f_{22} = (3\beta - 3\beta^2 + \beta^3)/3(EI)_b + ((1-\beta)^3 - \alpha^3)/3(EI)_o + \alpha^3/3(EI)_a \quad (4)$$

where : $\alpha = (M_a - M_y)/(M_a - M_b)$; $\beta = (M_b - M_y)/(M_a - M_b)$

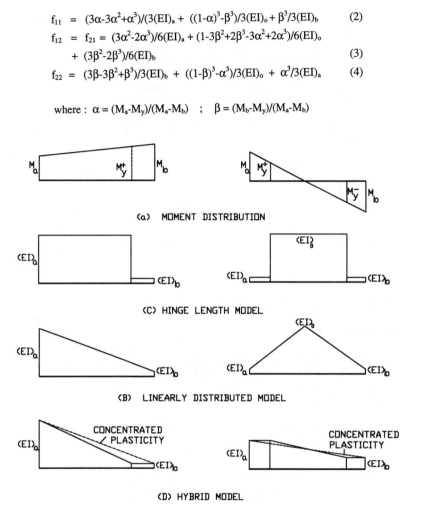

(a) MOMENT DISTRIBUTION

(C) HINGE LENGTH MODEL

(B) LINEARLY DISTRIBUTED MODEL

(D) HYBRID MODEL

Figure 2. Spread-Plastic Member Models Evaluated in this Study

In the above expressions, the end moments are functions of time and change with each step of the analysis. The yield moments are predetermined and remain fixed throughout the analysis. Care must be taken to ensure that the yield moment has been reached before computing the plastic hinge lengths.

In Model 2, a linear variation of flexibility is assumed from each end of the member to the contraflexure point. The contraflexure point, which may lie within or outside the member, may be fixed or allowed to vary. In either case, the resulting stiffness matrix is symmetric. The coefficients of the flexibility matrix of Eq.(1) will be as follows:

$$f_{11} = (6\alpha-4\alpha^2+\alpha^3)/12(EI)_a + (1-3\alpha+3\alpha^2-\alpha^3)/12(EI)_b$$
$$+ (3-3\alpha+\alpha^2)/12(EI)_o \tag{5}$$

$$f_{12} = f_{21} = (-2\alpha+\alpha^3)/12(EI)_a + (-1+\alpha+\alpha^2-\alpha^3)/12(EI)_b$$
$$+ (-1-\alpha+\alpha^2)/12(EI)_o \tag{6}$$

$$f_{22} = \alpha^3/12(EI)_a + (3-\alpha-\alpha^2-\alpha^3)/12(EI)_b + (1+\alpha+\alpha^2)/12(EI)_o \tag{7}$$

where : $\alpha=M_a/(M_a-M_b)$

The above closed-form expressions for the linearly distributed model are valid when the contraflexure point is within the member. A simpler subset can be derived when the contraflexure point lies outside the member.

Model 3 is a combination of the above two models. The length of the plastic zones on either side is established from the moment-curvature envelope (as in Model 1) while a linear variation of flexibility is assumed for the remaining length. The coefficients of the flexibility matrix are:

$$f_{11} = [R(6\beta^2+4\beta R+R^2)+4\beta(3-3\beta+\beta^2)]/12(EI)_b$$
$$+ [R(12\beta^2+8R\beta+3R^2)+4\alpha(3-3\alpha+\alpha^2)]/12(EI)_a \tag{8}$$

$$f_{12} = f_{21} = [R(6\alpha-6\alpha^2+2R-4\alpha R-R^2)+\alpha^2(6-4\alpha)]/12(EI)_a$$
$$+ [R(6\alpha-6\alpha^2+4R-8\alpha R-3R^2) + \beta^2(6-4\beta)]/12(EI)_b \tag{9}$$

$$f_{22} = [R(6\alpha^2+4\alpha R+R^2)+4\alpha(3-3\alpha+\alpha^2)]/12(EI)_a$$
$$+ [R(12\alpha^2+8R\alpha+3R^2)+4\beta(3-3\beta+\beta^2)]/12(EI)_b \tag{10}$$

where : $R = (1-\alpha-\beta)$; $\alpha = (M_a-M_y)/(M_a-M_b)$; $\beta = (M_b-M_y)/(M_a-M_b)$

The above expressions are valid when point of inflexion lies within the member. A simpler subset is possible if the point of inflexion lies outside the element.

Evaluation of Spread-Plastic Member Models

The evaluation of the three member models described in the previous section will be examined by applying them in a series of analyses of a single storey single bay plane frame subjected to both static and dynamic loads. The member models were implemented in the computer program IDARC (Kunnath et al., 1992). Details of the frame used in the evaluation are presented in Figure 3.

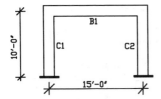

ELEM	My	EI	EA	GA
B1	650.0	5E6	4E5	1.5E5
C1	1800.0	7.5E6	6E5	2.5E5
C2	1800.0	7.5E6	6E5	2.5E5

(ALL UNITS Kip-inch)

Figure 3. Configuration and Member Properties of Model Frame
Used in Evaluation

Three loading cases, each simulating different levels of inelasticity, were considered. In the first load case, the applied lateral load was limited to produce an elastic response. The magnitude of the load was then increased to produce yielding in the beam member only. Finally, the load was increased substantially to yield both the beam and column elements. Results of the evaluation are presented in Table 1.

Table 1. Results for Incremental Nonlinear Analysis under Static Loads

MODELS	LOAD (kips)	25	35	50
MODEL 1	M col	954.0	1427.4	2156.5
	M beam	546.0	672.6	843.5
	δ	0.439	0.704	3.149
MODEL 2	M col	954.0	1432.0	2114.8
	M beam	546.0	668.6	885.3
	δ	0.439	0.707	3.368
MODEL 3	M col	954.0	1431.9	2147.7
	M beam	546.0	668.1	852.3
	δ	0.439	0.708	3.924

While the overall member forces computed using all three models are similar (within 5% of each other), it can be observed that the Hybrid Model results in story displacements that are almost 25% in excess of those produced by the hinge-length model indicating that it is more conservative for drift-based design.

The second subset of numerical studies involved seismic ground motions. The 1940 El Centro record was used in the evalautions. The peak ground accelerations (PGA) were adjusted to produce the similar inelastic levels as the earlier study involving static loads. It was found that a PGA of 0.5 g resulted in an elastic system response, 1.5 g produced yielding in the beams only, while a PGA of 2.5 g yielded both beams and columns. Results are given in Table 2.

Table 2. Results for Incremental Dynamic Analysis under Earthquake Loads

MODELS	PGA (g)	0.5	1.5	2.5
MODEL 1	M col	528.0	1445.0	1812.0
	M beam	302.2	673.7	709.1
	δ	0.243	0.715	1.022
MODEL 2	M col	528.0	1441.0	1810.0
	M beam	302.2	669.2	698.2
	δ	0.243	0.714	1.016
MODEL 3	M col	528.0	1441.0	1806.0
	M beam	302.2	668.6	692.0
	δ	0.243	0.714	0.979

Here, the maximum displacements were produced using the hinge length model. It must be pointed out that the hinge length was fixed at 10% of the member length and not allowed to vary as a function of the moment. The Hybrid model, however, permitted variation of the hinge length. In general, it was found that allowing the hinge length to vary from zero resulted in stiffer member models.

Concluding Remarks

It can be stated, in general, that there are two primary approaches in spread-plastic modeling of members that are suitable for use in analysis of large structural systems. The first uses the concept of a hinge length, while the other accounts for the distribution of flexibility across the member using some pre-determined variation (a linear variation was used for the model evaluated in this paper). The basic intention of all models is to account for the spread of inelastic deformations into the member. While these models appear to be relatively simple models, the problems associated with their implementation in computer programs for nonlinear analysis and their use in practical structural evaluation presents a host of problems.

The hinge-length model is dependent on two critical parameters: the post-yield stiffness and the maximum length of the inelastic zone. While the two are related, they can also independently influence the overall response of the member. The higher the post-yield stiffness, the larger the potential plastic zone. The following factors must be taken into consideration in using the hinge-length model:

1. Since the hinge length becomes non-zero only after the onset of yielding, this model does not lend itself to trilinear force-deformation envelopes which distinguishes cracking from yielding.

2. The plastic hinge length can extend across the entire member length if allowed to grow unbounded. For this reason, Soleimani et al.(1979) and Filippou and Issa (1988) suggest limiting values of the hinge length. The effect of restricting the hinge length is to impose indirect constraints on the post-yield stiffness.

The linearly distributed model, on the other hand, has some distinct advantages over the hinge-length model with respect to the above issues, since it can accommodate either bilinear or trilinear envelopes and is independent of the plastic hinge zone.

Other issues related to the use of these models have to do with numerical stability and implemenation in a computer code. The flexibility matrix for a member needs to be updated for any or all of the following reasons: (a) a transition in stiffness as prescribed by the hysteretic force-deformation model (all models); (b) a change in the plastic hinge length (Models 1&3); and (c) a shift in the contraflexure point (Model 2 only). All such changes lead to unbalanced forces between two solution steps. Item (a) can be dealt with using an event-to-event strategy which can be extremely time-intensive. Changes in plastic hinge length require special treatment in the solution algorithm since no event transition is defined. The difficulties associated with a varying contraflexure point is also not associated with any predefined event change. Hence, an iterative approach to ensure stability of the final solution is necessary. In general, for all models, it is advisable to use predetermined fixed contraflexure points (not necessarily at the center of the member) and a fixed hinge length to ensure stability of the solution.

Acknowledgements

This study is part of a more comprehensive effort to develop analytical models for use in nonductile concrete buildings. Financial support provided by the National Center for Earthquake Engineering Research (NCEER), which in turn is supported by the National Science Foundation and the State of New York, is gratefully acknowledged.

References

Clough, R.W., Benuska, K.L. and Wilson, E.L. (1965). "Inelastic Earthquake Response of Tall Buildings", Proceedings of the 3rd World Conference on Earthquake Engineering, New Zealand, Vol.II, pp. 68-89.

Filippou, F.C. and Issa, A. (1988). "Nonlinear Analysis of Reinforced Concrete Frames Under Cyclic Load Reversals", Report No. UCB/EERC/88/12, University of California, Berkeley.

Kunnath, S.K., Reinhorn, A.M. and Lobo, R.F. (1992). "IDARC - Version 3.0: A Program for Inelastic Damage Analysis of RC Structures", Technical Report NCEER-92-0022, National Center for Earthquake Engrg., Buffalo, New York.

Spacone, E. and Filippou, F.C. (1994). "RC Beam-Column Element for Nonlinear Frame Analysis", Proc., 11th Conference on Analysis and Computation, Atlanta, ASCE, New York.

Soleimani, D., Popov, E.P. and Bertero, V.V. (1979). "Nonlinear Beam Model for RC Frame Analysis", Proceedings of the 7th ASCE Conference on Electronic Compution, St. Louis, Missouri.

Non-Linear Dynamic Analysis of a Safety Wall

Cornelis van der Veen[1] and Johan Blaauwendraad[1]

Abstract

Nowadays gases are in general preferably stored in liquid form at atmospheric pressure. Therefore gases are cooled to their boiling point, which for most gases is below 0°C. For safety reasons a double-walled tank is normally used for the bulk storage of large quantities of L.N.G. Such a tank comprises an inner tank made of nickel steel and an outer wall constructed of prestressed concrete. This paper describes a finite element analysis of the concrete outer wall which is loaded by a pressure wave which results from a gas cloud explosion at some distance of the wall. The dynamic analysis is done in two ways. First, for linear-elastic material behaviour. Second, for non-linear material behaviour duo to cracking and/or plasticity. The results are compared and discussed in detail. In both approaches we can speak of a real dynamic analysis. In some situations a complex exact dynamic analysis is not yet needed. It can be sufficient in a design phase to accept a simple approximation with less demand on time and money. Another reason for a simpler computation is the validation of the results of a complex exact dynamic analysis. The more complex the analysis, the higher the risk for errors. It then is of great value if the order of magnitude can be estimated in an independent parallel way, therefore much attention has been paid to an approximate analysis. The results are compared with the computed results.

Introduction

A cylindrical wall of structural concrete with circular ground-plan protects a steel inner tank for mass LNG storage. The concrete wall is loaded by a passing pressure wave which results from a gas cloud explosion at some distance of the wall. The dynamic study is done in two ways. First, for linear-elastic material behaviour. Second, for non-linear material behaviour due to cracking and or plasticity. The finite element package `DIANA' has been used to study more in detail the response of the concrete wall. The dynamic analysis has been performed by direct integration of the equations of motion.

----
[1] Delft University of Technology, Faculty of Civil Engineering
     Stevinweg 1, 2628 CN Delft, the Netherlands

This approach is suitable for both linear and non-linear analysis. In some situations a complex exact dynamic analysis is not yet needed. It can be sufficient in a design phase to accept a simple approximation with less demand on time and money. For a preliminary design no big precision is wanted, but rather speed and insight. An approximate analysis is based on the idea that the real complex structure can be seen as a simple one-degree-of-freedom mass-spring-system. The approximate analysis consists of four steps: 1) a linear static analysis, 2) computation of the period T for the lowest eigen mode, 3) determination of the dynamic load factor (DLF), 4) multiplication of the static results by DLF.

The final goal in the present non-linear dynamic analysis will be approached in 5 distinct steps.

1. Linear static analysis
2. Determination of eigenfrequencies and eigenmodes
3. Approximated dynamic response by determination of a dynamic load factor (DLF)
4. Linear dynamic analysis by direct integration
5. Non-linear dynamic analysis by direct integration

The static linear analysis is done for a pressure loading which causes the maximum horizontal resultant in x-direction. The results of the static analysis give a first impression of the behaviour of the structure. The absolute values do not have any meaning for the dynamic analysis, but the distribution of forces is already revealed more or less. Above that, the finite element mesh and several input data can be checked adequately. Finally the displacement field of the static run can be used in the Rayleigh-method to estimate the lowest frequency.

<u>Input data</u>

The concrete cylindrical shell has a radius R=27.2 m, a wall thickness h=0.5 m, and a height H=32.4 m. The shell is open at the upper end (no roof) and has a pile foundation at the bottom. The pile distance is small with regard to the shell circumference. Therefore, the ring of piles can be considered as a uniformly distributed line-spring. In the present analysis the springs are linear-elastic for both horizontal and vertical forces. A pin connection is assumed between the shell and the springs, so rotation of the shell can occur freely in the bottom edge. The vertical spring stiffness per m^1 of the shell is k_v=50,000 kN/m. Horizontally the stiffness is k_h=2000 kN/m, both in radial and in tangential direction.

For the damping ratio a value ξ=0,05 has been chosen, which can be justified for rather heavily loaded concrete. The use of the parameter ξ presupposes proportional damping. Possible additional damping due to the foundation is not considered in the present study.

The gas cloud explosion (deflagration) causes a pressure wave which propagates with a velocity v in a terrain with obstacles. A good approximation for v is the propagation velocity of sound in air at atmospheric circumstances, v=340 m/s. Here we mention the additional pressure above the atmospheric, the *effective pressure*. When we measure the effective pressure at the passing of an undisturbed pressure wave in a terrain without obstacles a pressure-time diagram is recorded. In this example a diagram is used such that the effective pressure $p(t)$ increases linearly from zero to a value p in a time interval t_d *(duration time)*. After t_d seconds the effective pressure immediately falls down to zero. If the cylindrical shell is being passed, the effective pressure will change due to reflections by the obstacle. In this study a reflection coefficient r is used which depends on the cylinder coordinate φ. The formula $r(\varphi)$ has been derived by the Prins Maurits Laboratory of TNO in the Netherlands.

$$r(\varphi) = 4.333(1 + 0.0368 \cos \varphi)^7 - 3.333$$

The resulting effective pressure is always normal to the shell surface. The effective pressure is constant over the height of the shell, but has a continuously changing distribution in time in circumferential direction. To give an impression of the value of $r(\varphi)$, it has been calculated for three salient positions along the shell:

- front ($\varphi = 0$)   : $r(\varphi) = 2.25$

- side ($\varphi = \frac{\pi}{2}$)   : $r(\varphi) = 1$

- back ($\varphi = \pi$)   : $r(\varphi) = 0$

Figure 1 gives an overview of the input data.

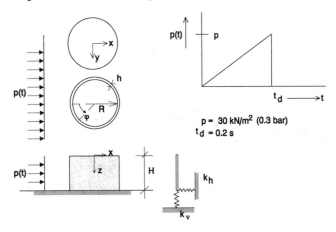

$$p = 30 \text{ kN/m}^2 \text{ (0.3 bar)}$$
$$t_d = 0.2 \text{ s}$$

Figure 1. Overview of the input data

A curved shell element has been chosen to model the concrete wall. The element has been derived from a solid element on basis of a parabolic displacement field. This means that a linear distribution of strains occurs. Hence, a linear distribution of moments, shear forces and membrane forces is shown per element.
Only half of the cylinder is modelled. In circumferential direction 15 elements are chosen. In hight direction an equidistant mesh of 8 to 10 elements would be fine. However, a finer mesh is needed for the bending disturbance edge zone in the lower region of the shell. Consequently, the mesh shown in figure 2 has been chosen.

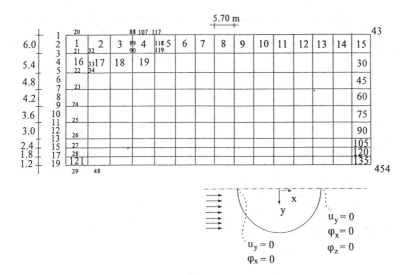

Figure 2. Element mesh and boundary conditions

Approximate dynamic response

First, a static analysis is carried out. Second, an eigenfrequency analysis is performed to estimate the dynamic load factor (DLF) of the shell. We first consider a simple-mass-spring system with one degree of freedom. The system has a period T and is loaded by a force which increases linearly in t_d seconds from zero to F. The maximum dynamic displacement $u_{max.d}$ is related to the maximum static displacement $u_{max,s}$ for a static load F by the dynamic load factor: $u_{max.d}/u_{max.s}$.

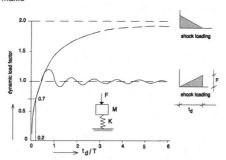

Figure 3. Dynamic load factor versus t_d/T

Figure 3 shows that the DLF depends on the quotient t_d/T. If $t_d/T > 1$ the value of DLF oscillates around the value 1. The duration t_d of the force F is relatively long compared to the period T of the system, so not much dynamic effect occurs. For $0.4 < t_d/T < 1.0$ the dynamic load is more severe than the static load. At maximum an increase of about 25% will be seen. For values of $t_d/T < 0.4$ the dynamic response is less than static behaviour. Now the displacement is reduced because $DLF < 1$.
In the present example a value $t_d/T=0.2$ was found ($T=1s$; $t_d=0.2s$). From figure 3 we read: $DLF=0.7$. The value of the DLF is applied to the results of the static analysis. This yields a design estimate for the dynamic behaviour of the structure.
Instead of making a DIANA run for the computation of the eigenfrequencies, one also can approximate the lowest frequency by the Rayleigh method. We reduce the shell structure to a simple one degree of freedom mass-spring-system. This is done by assuming a displacement field with one control displacement as fundamental unknown. We choose the displacements for the static run and adopt the horizontal displacement u_{x1} in node 1 in the upper edge ($\varphi=0$; $x=R$) as the control displacement. The distributed mass in the shell is replaced by an equivalent mass M in node 1. The kinetic energy must be represented correctly in the replacing mass-spring-system.
The stiffness K of the mass-spring-system must be approximated as well. This is done by computing the force F which is equivalent to the distributed pressure. The wanted stiffness is then computed from $K=F/\mu_{x1}$. In this way the eigenfrequency can be found from $\omega^2=K/M$ and from this the period T. A period of $T=0.90s$ was found which is somewhat lower than the computer calculation ($T=1.0s$).

Linear Dynamic Analysis

The explosion loading is specified per column (vertical series) of elements. The choice of the time step is influenced by the correct description of the system response and the load-time diagram. As for the load-time diagram, the time step Δt must be sufficiently small in comparison with the load duration time $t_d(=0.2s)$. A more precise criterion for Δt is the condition that not more than one column of elements is unloaded in one time step.

Table 1. Comparison of static and dynamic response

	dynamic	static	dyn. factor
u_x node 1	0.25	0.45	0.56
u_x node 19	0.19	0.36	0.53
m_{yy} node 1	-120	-237	0.51
n_{yy} node 1	-1250	-1633	0.77
m_{xx} node 15	-310	-577	0.54
n_{xx} node 15	697	900	0.77
n_{xx} node 19	818	1123	0.73
vert.reaction node 19	818	1121	0.73
horz.reaction node 19	351	675	0.52

Furthermore, a correct reproduction of the system response poses another condition on the time slep. Preferably 20 time steps must be applied in the (small) T_i of the highest mode which is activated. For the present shell structure only the first 15 modes are of importance. Consequently, a time step $\Delta t=0.002\ s$ has been used.

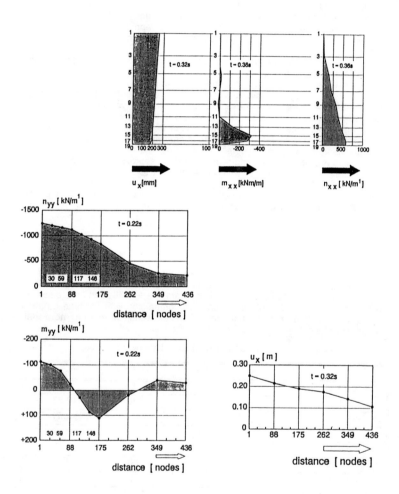

Figure 4. Maximum value during the linear dynamic response

The results of the dynamic analysis are compared with the results of the static analysis and are collected in Table 1. A *DLF* of *0.70* is calculated with the aid of the approximate dynamic analysis. A dynamic load factor in the range of *0.51* to *0.77* was calculated. Thus the *DFL* corresponds reasonably well with the value of the *DFL* calculated with the approximate analysis.

Figure 4 shows the most important design quantities which are plotted along important lines/cuts of the shell. The time is chosen at which these quantities are maximum.

Non-linear Dynamic analysis

The non-linear dynamic analysis differs from the linear dynamic analysis as for the additional material data and loading cases. For the non-linear dynamic analysis more integration points over the thickness are necessary in order to simulate a non-linear material behaviour over the thickness e.g. cracking with softening and plasticity. In this case 7 integration points are used over the thickness and a Simpson-integration over the thickness is performed.

Comparable results are found with the nonlinear dynamic analysis. Differences are found for the moment m_{xx} and the vertical normal force n_{xx} in node 15, when we look more in detail. At the instant of cracking the moment and normal force reduce significantly, see Fig. 5. Cracks occur in the wall only at the front (inside) and rear side of the wall.

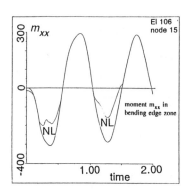 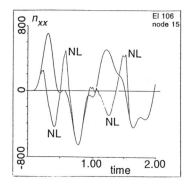

Figure 5. Comparison of the linear and nonlinear dynamic results

Note that at the front side (where the pressure wave hits the wall for the first time) five elements which are connected to the foundation are cracked completely through the thickness of the wall, see Fig. 6,. This phenomenon affects very much the nonlinear response of the normal force, see Fig. 5.

It was found that no plasticity occurs for the reinforcement and prestressing steel. However, we have to bear in mind that the *DFL* was smaller than one (*0.7 to 0.5*).

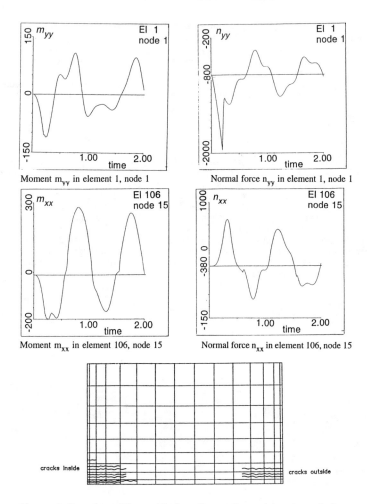

Figure. 6. Overview of the results from the nonlinear dynamic analysis

Conclusion

The general conclusion may be drawn that the global behaviour of the wall is not significantly affected by the nonlinear dynamic analysis. There is, however, a distinct difference in the local behaviour as a result of excessive cracking.

Computer-Aided Design in Foundation Engineering
Chandra S. Brahma, F. ASCE[1]

Abstract

The basic objective of the paper is to illustrate with the software "Foundation" to both practitioners and students the advantage of using a computer to perform a design process that is manually lengthy and tedious. The "Foundation" software package is an interactive program developed for the complete design of shallow foundations. The program uses recent design practices, computerizes manual methods of design to achieve improved speed as well as reliability of calculations, and runs on IBM-PC or other compatible microcomputer hardware using a DOS operating system. The program follows a minimum learning curve strategy and achieves a high degree of user friendliness through an easily followed format and highly explicit data prompts.

The software assesses effects of complex as well as nonuniform soil conditions and varying structural loading conditions in order to achieve the safest and most economical design. It considers effects of footing shape, depths to footing, eccentric as well as inclined loadings, base and ground inclinations, groundwater table, and layered soils in computing the maximum allowable supporting capacity of founding soils at several depths below ground surface. While both immediate and long-term settlements, incorporating effects of layered systems and lateral strains, are calculated for fine-grained soils, the distortion settlements of cohesionless soils are ascertained with the semi-empirical procedures adjusted to fit experimental and in-situ findings. For a predetermined amount of maximum tolerable settlement, based on the designer's experience and judgement, the software which is capable of ascertaining the magnitude of total as well as long-term settlements at each specified depth to footing, is developed for proportioning adequate footings. For each depth to footing below ground surface, the structural design of a reinforced concrete footing provides a designer with the required thickness of footing as well as several choices of flexural reinforcements, based on the requirements of the American Concrete Institute (ACI) Code.

[1]Professor of Civil Engineering, California State University - Fresno, 2320 East San Ramon Avenue, Fresno, California 93740-0094

Introduction

Besides considering a number of adverse environmental factors, a designed footing must not transmit stresses in founding soils which exceed their maximum allowable supporting capacities, must be safe against structural failures of reinforced concrete, must not exceed tolerable displacements for the type of structure and its function, and must represent the most acceptable compromise between performance and cost. In order to meet these requirements, the designer, based on general site conditions and structural constraints, must select a trial footing size and carry out the analysis, using the appropriate analytical tools. The result of such analysis yields initial values for comparison with specific characteristics of the design. Footing sizes are thereafter modified many times until subsequent iterations result in the optimum design consistent with the particular method of analysis and the given range of soil response. In the absence to date of a convenient source of codification of detailed hand calculational routines, the quantitative analyses of the process leading to economical design of even simple footings require prodigious amounts of an engineer's time which is better spent in selecting the most appropriate investigative and analytical techniques and in studying the effect of varying geotechnical and/or structural parameters against the engineer's own framework of organized experience.

Allowable Bearing Capacity

The foundation is required to transfer the load from the superstructure to the underlying soil deposits. In so doing, it must neither overstress the soil nor cause shear failures of the soil. A safe foundation design, therefore, incorporates a predetermined factor of safety against the soil's net limiting shear resistance (q_n). While various equations have been in use for calculating limiting shear resistance, the following basic equations (Brahma, 1992; Meyerhoff, 1963; and Vesic, 1970) for the allowable bearing capacity (q_a) considers the shape and depth of the footing, and the inclination of the loading, the base of the footing and the ground surface.

$$q_a = q_n/FS = [cN_c\xi_c\xi_{ic}\xi_{dc}\xi_{bc}\xi_{gc} + \gamma D_f(N_q-1)\xi_q\xi_{iq}\xi_{dq}\xi_{bq}\xi_{gq} + .5\gamma_c B'N_\gamma\xi_\gamma\xi_{i\gamma}\xi_{d\gamma}\xi_{b\gamma}\xi_{g\gamma}]/F.S.$$

where,   c = cohesion of soil
  FS = factor of safety equal to 3.0 against a bearing capacity failure
  γ = unit weight of overburden soil at footing level
  D_f = Depth of the base of the footing below ground
  γ_c = average unit weight of soil below foundation level
  B' = effective width of the footing
  N_c, N_q, N_γ = bearing capacity factors
  ξ_c, ξ_q, ξ_γ = shape factors
  ξ_{ic}, ξ_{iq}, $\xi_{i\gamma}$ = load inclination factors
  ξ_{dc}, ξ_{dq}, $\xi_{d\gamma}$ = depth factors
  ξ_{bc}, ξ_{dq}, $\xi_{d\gamma}$ = base inclination factors
  ξ_{gc}, ξ_{gq}, $\xi_{g\gamma}$ = ground inclination factors

The software takes account of eccentricity by introducing a fictitious effective width $B' = B - 2e_x$ and a fictitious effective length $L' = L - 2e_L$ of the footing, instead of its actual width and length. The load thereby acts at the geometric center of the reduced area of dimension $B' \times L'$ of the footing. The various above-mentioned factors incorporated in the software are as follows:

Bearing Capacity Factors:
$$N_q = \tan^2(\pi/4 + \phi/2) \exp(\pi \tan \phi)$$
$$N_c = (N_q - 1) \cot \phi$$
$$N_\gamma = 2 (N_q + 1) \tan \phi$$
where, ϕ = angle of internal friction of soil

Shape factors:
$$\xi_c = 1 + (B'/L') (N_q/N_c)$$
$$\xi_q = 1 + (B'/L') \tan \phi$$
$$\xi_\gamma = 1 - 0.4 (B'/L')$$

Inclination factors:
$$\xi_{iq} = [1 - P / (Q + B' L' c \cot \phi)]^{\,m}$$
$$\xi_{ic} = \xi_{iq} - (1 - \xi_{iq}) / (N_c \tan \phi)$$
$$\xi_{i\gamma} = [1 - P / (Q + B' L' c \cot \phi)]^{\,m+1}$$
$$\xi_{ic} = 1 - (mP / (B'L'cN_c)) \qquad \text{for } \phi = 0$$

where, P and Q are the horizontal and vertical components of the footing reaction

For inclinations of the load in directions of the width B and the longer dimension L, the respective exponents m (m_B and m_L) are given by
$$m_B = (2 + B/L) / (1 + B/L) \quad \text{and} \quad m_L = (2 + L/B) / (1 + L/B)$$

For the load inclined in the direction A, making an angle ψ_A with the direction of the long dimension L, the exponent m_A may be interpolated between exponents m_B and m_L such that
$$m_A = m_L \cos^2 \psi_A + m_B \sin^2 \psi_A$$

Depth factors:
$$\xi_{dc} = 1 + 0.2 (D_f/B') \tan (45 + \phi / 2)$$
$$\xi_{dq} = \xi_{d\gamma} = 1 + 0.1 (D_f/B') \tan (45 + \phi / 2)$$
$$\xi_{dq} = \xi_{d\gamma} = 1 \qquad \text{for } \phi = 0$$

Base inclination factors:
$$\xi_{bq} = \xi_{b\gamma} = (1 - \alpha \tan \phi)^2$$
$$\xi_{bc} = \xi_{ic}$$
$$\xi_{bc} = 1 - [2\alpha/(\pi + 2)] \qquad \text{for } \phi = 0$$

where, α = the angle of base inclination, positive downward

Ground inclination factors:
$$\xi_{gq} = \xi_{g\gamma} = [1 - \tan \omega]^2$$
$$\xi_{gc} = \xi_{ic}$$
$$\xi_{gc} = 1 - [2\omega/(\pi + 2)] \qquad \text{for } \phi = 0$$

where, ω = the angle of ground surface slope, positive downward

When the groundwater level is within depth B' below the footing level, the average effective unit weight (γ_c) of the soil below the base of the footing may be calculated using the relationship:
$$\gamma_c = \gamma' + (d_w/B') (\gamma - \gamma')$$
where, γ_m, γ', and d_w are respectively moist unit weight of soil above water table, submerged unit weight of soil below water table, and the depth to groundwater

level below the footing.

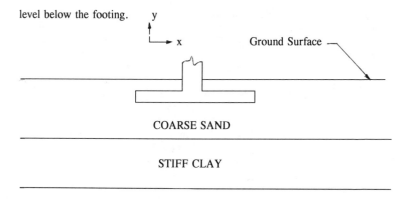

FIGURE 1: RECTANGULAR FOOTING

    Input data for footings include an initial depth to footing below the ground surface, dead as well as live loads, magnitude of tolerable total or long-term settlement, eccentricities and angles of inclination of load in vertical and long directions, dimensions of column or wall, concrete strength of supporting as well as supported member, length to width ratio of the desired footing, and yield stress of reinforcement. Other input related to subsurface profile include number and types (cohesive or cohesionless) of soil layers beneath the footing, depth to water table below ground surface, thickness of each layer, unit weight, angle of internal friction, and cohesion. For cohesive soils, pore pressure parameter A, undrained shear strength, Poisson's ratio, compression and recompression indices, void ratio, and maximum preconsolidation pressure and overconsolidation ratio, if any, are required. In case of predominantly cohesionless soils, further required inputs are standard penetration test blows per 0.3m and detailed descriptions of soil.

    Using the above-mentioned relationships, the software calculates the allowable bearing capacity of all specified components, resistant or otherwise, of the soil deposits below the footing based on the effective strength parameters for highly permeable granular materials and the total strength parameters from undrained tests for predominantly fine-grained soils. In proportioning square, rectangular or continuous wall footings to the nearest of 76 mm at various depths at an interval of 0.5m below the ground surface, the maximum pressure due to structural loadings at the upper surface of each stratum or lenticular element is not allowed to exceed the allowable bearing capacity. For eccentrically loaded footings, the dimensions of the footing are designed such that the resulting reaction passes through the middle third of the footing dimension in each direction of eccentricity.

Consolidation and Distortion Settlements

Time-dependent consolidation settlements arise from the compression of the soil skeleton, and, therefore, are likely to be significant in general for mostly fine-grained soils. Based on results from field curves wherein effective stress (σ') is plotted to a logarithmic scale against void ratio (e), the maximum one-dimensional consolidation settlement (s_c) is given by,

$$s_c = \Sigma(H/1+e_0)[C_e \log(\sigma_c/\sigma_0' + C_c \log\{(\sigma_0'+\Delta\sigma)/\sigma_c\}] \quad \text{for } \sigma_0' < \sigma_c < \sigma_0' + \Delta\sigma$$
$$= \Sigma(H/1+e_0) \ C_c \log\{(\sigma_0'+\Delta\sigma)/\sigma_0'\} \quad \text{for } \sigma_0' + \Delta\sigma < \sigma_c$$
$$= \Sigma(H/1+e_0) \ C_c \log\{(\sigma_0'+\Delta\sigma)/\sigma_0'\} \quad \text{for normally consolidated clay } \sigma_0' = \sigma_c$$

Where,   H   is the thickness of cohesive layer
e_0 is the initial void ratio
C_e is the recompression index
C_c is the compression index
σ_c is the maximum past pressure
σ_0' is the average effective overburden pressure
$\Delta\sigma$ is the vertical pressure increase

Under site conditions where the footing area is not large compared with the thickness of the cohesive layer, the three-dimensional nature of the problem will invariably influence the magnitude of settlement. The maximum one-dimensional consolidation settlement (s_c) is corrected [$s_{corr} = \lambda s_c$] to allow for the variation in excess pore pressure due to lateral strain (Skempton and Bjerrum, 1957). The coefficient λ is a function of porewater pressure parameter A and a specified boundary loading.

The immediate distortion settlement of a footing supported by a fine-grained soil layer of finite thickness is expressed as,

$$s_{fd} = c_d \ p_{net} \ B \ (1 - v^2) \ / \ E$$

where, c_d is an influence factor (Harr, 1966) which accounts for the shape of the loaded area and the position of the point for which settlement is being calculated,

E is Young's modulus and v is Poisson's ratio for the elastic medium,

p_{net} is the magnitude of net uniform vertical pressure increase,

and B is a characteristic dimension of the footing.

The semi-empirical methods (Schmertmann, 1970; Schmertmann et al, 1978) provide a practical rational solution to the problem of predicting distortion settlement of shallow footings on cohesionless masses. The settlement of a footing subjected to a net uniform pressure of p_{net} is

$$s_{cd} = \Sigma \ \epsilon_z \ \Delta z = C_1 \ C_2 \ p_{net} \ \Sigma \ (I_z \ \Delta z) \ / \ E_d$$

where, C_1 is a correction factor incorporating the effect of strain relief due to embedment $= 1 - 0.5 \ (\sigma' \ / \ p_{net}) \ \geq \ 0.5$

C_2 is a correction factor incorporating continued settlements due to creep over a period of $t_{year} = 1 + 0.2 \log (t_{year} \ / \ 0.1)$

E_d is the appropriate deformation modulus at the middle of each layer of thickness Δz, σ' is the effective in-situ overburden pressure at the footing depth, and I_z is a strain influence factor

The peak value of strain infuence factor I_{zm} increases with the magnitude of the net footing pressure and is given as,

$$I_{zm} = 0.5 + 0.1(p_{net}/\sigma'_m)^{0.5}$$

where, σ'_m is the effective vertical pressure at a depth of 0.5B below the base of square (L/B = 1) footings or B below the base of strip (L/B $\geqslant$ 10) footings.

For square footing, the strain influence distribution diagrams has ordinates of 0.1, I_{zm}, and 0 at depths of 0, 0.5B, and 2B respectively below the footing. The diagram for strip footing has ordinates of 0.2, I_{zm}, and 0 at depths of 0, B and 4B respectively below the footing. The deformation modulus for square or long strip footings used in the software is obtained from the suggested empirical relationships in terms of either the static cone resistance q_c or standard penetration test N-values (in blows/0.3m). The user of the software may also input any deformation modulus deemed appropriate for the site conditions being analyzed.

Once the footing is adequately proportioned from the standpoint of supporting capacity, the software calculates distortion settlement in each cohesionless layer or immediate distortion as well as long-term settlements in each cohesive layer as applicable. The surface subsidence is then verified against the specified total or long-term tolerable settlement. Should the surface subsidence exceed the specified tolerable settlement, the footing dimension in each lateral direction is changed in increments of 76 mm until the designed footing meets criteria relating to allowable supporting capacity and settlement.

Structural Design

The factored loads and moments for the controlling loading condition is then ascertained and the required nominal resisting values are calculated by dividing the factored loads and moments by the applicable strength reduction factors. By trial and adjustment, the required effective depth of the footing section which has adequate one-way beam shear as well as two-way punching shear capacities, is determined to the nearest 12.7 mm. Subsequently, a total reinforcement area and several alternate sizes and spacings of flexural reinforcements in long and short directions meeting the development length required by the ACI Code, based on nominal moment in each direction and the applicable effective depth of footing, are selected. For footings having an eccentricity in one or two directions, the applicable shear and moment are calculated at appropriate sections, based on non-uniform contact pressure distributions. Based on the bearing stresses on the footing and the column or wall at their interface, the number and size of the dowel bars that transfer the load from the supported to the supporting member or that are required by the ACI Code are determined.

Computer Software

The foregoing sequence of operations are then repeated for design of footings at each desired depth to footing below ground surface in increments of 0.5m from that initially specified. The use of the developed software is demonstrated hereinafter with the design of a rectangular footing subject to axial as well as moment loadings.

Design Example:

A 30.5 cm X 30.5 cm reinforced concrete column (Figure 1) carries an axial dead load of 533.8 KN and an axial live load of 355.8 KN. The column is subjected to dead load and live load moments of 54.23 and 13.56 KN-m respectively in x direction and 81.35 and 54.23 KN-m respectively in y direction. Assuming that the strength of column as well as footing concrete is 27.58 MN/m^2 and that the yield stress of reinforcement is 413.64 MN/m^2, a reinforced concrete rectangular (length to width ratio of 2) footing in a three-layered system is to be proportioned in accordance with the ACI specifications for ultimate strength design such that the depth to the footing below ground surface D_f is $\geq$ 0.6 meter and tolerable long-term settlement of footing is $\leq$ 1.0 cm. Details of each soil layer are as follows:

Coarse Sand: thickness = 2.43m; cone penetration resistance q_c = 11970 KN/m^2; ϕ = 35°; depth to GWT = 1.52m; γ_m = 19.64 KN/m^3; and γ_{sat} = 20.81 KN/m^3

Stiff Clay: thickness = 1.52m; undrained strength q_u = 287.3 KN/m^2; ϕ_u = 0; C_o = 0.02; C_e = 0.005; γ_{sat} = 20.01 KN/m^3; v = 0.4; σ_c = 383.04 KN/m^2; e_0 = 0.9; and porewater pressure parameter A = 0.8

Soft Clay: thickness = 40.0m; undrained strength q_u = 95.76 KN/m^2; ϕ_u = 0; γ_{sat} = 17.66 KN/m^3; v = 0.4; e_0 = 1.0; C_c = 0.05; C_e = 0.001; and porewater pressure parameter A = 1.0

Computer output for design of the footing, placed at a depth of 0.61m below ground surface, are presented as follows:
Depth to footing below ground: 0.61m; Long-term Settlement: 0.795 cm;
Maximum Allowable Bearing Capacity: 339.06 KN/m^2;
Footing Size: 1.35 m X 2.7 m; effective thickness of concrete: 45.77 cm;
Steel Area Required For Bearing: 465.13 mm^2; Recommended steel bars; 4 #4;
Total Flexural Steel Area Required (Short Direction): 2224.41 mm^2

Suggested Steel Area (mm^2)	NO. of Bars	Bar Size #	No. of Bars in Band Width B	Spacing of Bars in Band Width B (mm)
2277.79	32	3	22	60.0
2280.18	18	4	12	120.0
2376.69	12	5	8	190.0

Total Flexural Steel Area Required (Long Direction): 2715.37 mm²;

Steel Area Suggested (mm²)	Number of Bars	Bar Size (#)	Spacing (mm)
2776.06	39	3	30.0
2786.89	22	4	55.0
2772.80	14	5	90.0
2850.23	10	6	130.0
3102.19	8	7	170.0
3040.25	6	8	235.0
3223.36	5	9	295.0

Conclusion

Besides considering a number of adverse environmental factors, a designed footing must not transmit stresses in founding soils exceeding their maximum allowable supporting capacities, must be safe against structural failures of reinforced concrete, must not undergo displacements beyond those deemed tolerable for the type of structure and its function, and must represent the most acceptable compromise between performance and cost. In order to meet these requirements, the software "Foundation" yields several detailed designs of footings placed at intervals of 0.5m below the initial depth to footing, resulting in the optimum design consistent with the given range of soil response and structural contraints.

Appendix - References

1. Brahma, C. S., "Computer-Aided Design of Shallow Foundation", International Conference on Geotechnical Engineering, Johor Bahru, Malaysia, 1992

2. Meyerhoff, G. G., "Some recent research on bearing capacity of foundations", Canadian Geotechnical Journal, Volume 1, No. 1, pp 16 - 26, 1963

3. Schmertmann, J. H., "Static cone to compute settlement over sand", Journal of the Soil Mechanics and Foundations Division, ASCE, Vol. 96, No. SM3, pp 1011 - 1043, 1970

4. Schmertmann, J. H., Hartman, J. P. and Brown, P. R., "Improved Strain Influence Factor Diagrams", Journal of the Geotechnical Engineering Division, ASCE, Vol. 104, No. GT8, pp 1131 - 1135, 1978

5. Skempton, A. W. and Bjerrum, L., "A contribution to settlement analysis of foundation in clay", Geotechnique, London, Volume 7, 1957

6. Vesic, A. S., Research on Bearing Capacity of Soils, Unpublished Report, 1970

NON-LINEAR LIMIT ANALYSIS FOR THE BEARING CAPACITY OF HIGHLY FRICTIONAL SOILS

Steven W. Perkins[1], Member, ASCE

Abstract

Highly angular granular soils exhibiting elevated levels of friction and a significant degree of non-linearity in the material failure envelope pose significant difficulties when attempting to use conventional solutions to compute bearing capacity. Theories of limit analysis can be formulated to include a non-linear failure criterion. This formulation has been applied to the problem of bearing capacity. Analytical results are compared to results from centrifuge experiments on a very dense, highly angular, silty sand.

Introduction

Dense granular soils commonly exhibit a degree of non-linearity in the material failure envelope. This non-linear behavior becomes significant when the secant measure of the friction angle becomes elevated. This significance is readily seen in the application of conventional bearing capacity solutions to such soils. Due to the exponential dependence of the bearing capacity factors on the friction angle, small changes in the friction angle result in rather large changes in the computed bearing capacity when the average friction angle is large. These characteristics present particular problems when applying conventional theories, which rely on linear strength parameters, to predict bearing capacity. Previous studies (Perkins, 1995) have demonstrated the non-linear dependence of bearing capacity on foundation width and the inadequacy of conventional theories in predicting this behavior.

These characteristics and studies have prompted the search for advanced theories which directly account for the non-linear failure criterion. One such theory is the extension of limit analysis to include non-linear failure behavior. Elements of non-linear limit analysis were introduced by Baker and Frydan (1983) and further developed by Zhang and Chen (1987). The theory developed by Zhang and Chen (1987) is redefined in this paper within the specific context of the bearing capacity problem on level ground and is applied to predict results from centrifuge bearing capacity experiments. Details of the numerical algorithm are provided.

---

[1]Assistant Professor, Dept. of Civil Eng., Montana State University, Bozeman, MT, 59717

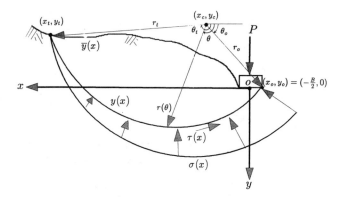

Figure 1: Slip Surface Geometry

Slip Mechanism Geometry

The mechanism for plastic collapse beneath a continuous shallow foundation is illustrated in Figure 1. The $x - y$ coordinate axes are chosen to have their origin at the center of the footing such that the resultant, externally applied foundation load, P, acts through the axes origin. The function $\bar{y}(x)$ describes the soil surface profile while the function $y(x)$ describes the geometry of the slip surface. The slip surface is taken to be a sliding and rotational discontinuity having a center of rotation denoted by the coordinates x_c and y_c. The geometry of the mechanism can also be described by the polar coordinates θ and $r(\theta)$. The mechanism initiates at the far right foundation edge, denoted by the cartesian coordinates $x = x_o = -B/2$ and $y(x) = y_o = 0$, where B is the foundation width, or by the polar coordinates $\theta = \theta_o$ and $r(\theta) = r_o$. The slip mechanism terminates along the soil surface profile at the point denoted by $x = x_t$, $y(x) = y_t$, $\theta = \theta_t$ and $r(\theta) = r_t$. Along the slip surface a normal stress distribution, $\sigma(x)$, is developed which controls the development of a tangential shearing stress $\tau(x)$. The non-linear relationship between the shear and normal stress is presented in a later section.

Solution Formulation

The principle of virtual work provides the basis for the theorems of limit analysis. For the problem geometry depicted in Figure 1 the total virtual work can be expressed as

$$W = \delta u H + \delta v V + \delta \omega M = 0 \qquad (1)$$

where δu and δv describe incremental virtual displacements in the x and y directions, respectively and $\delta \omega$ describes an incremental virtual rotation about the center of rotation of the slip mechanism. The forces H and V describe the resultant forces in the x and y directions, respectively, while the moment M defines the resultant moment about the center of rotation. These forces and moments embody the externally applied load, P and the internal stresses developed within the soil mass. Referring to Figure 1, H, V and M may be expressed as

$$H = \int_S (\sigma \cos\theta - \tau \sin\theta)dS \tag{2}$$

$$V = P - \int_S (\sigma \sin\theta + \tau \cos\theta)dS + \int_{x_o}^{x_t} (y - \overline{y})\gamma dx \tag{3}$$

$$M = \int_S [(\sigma \cos\theta - \tau \sin\theta)y + (\sigma \sin\theta + \tau \cos\theta)x]\,dS - \int_{x_o}^{x_t} (y - \overline{y})\gamma x dx \tag{4}$$

where S denotes the surface area of the slip mechanism. Recognizing that $dS = dx/\sin\theta$ and $\cot\theta = dy/dx$, equations 2, 3 and 4 may be rewritten as

$$H = \int_{x_o}^{x_t} (\sigma \frac{dy}{dx} - \tau)dx \tag{5}$$

$$V = P - \int_{x_o}^{x_t} \left[\sigma + \tau \frac{dy}{dx} - (y - \overline{y})\gamma\right] dx \tag{6}$$

$$M = \int_{x_o}^{x_t} \left[(\sigma \frac{dy}{dx} - \tau)y + (\sigma + \tau \frac{dy}{dx})x - (y - \overline{y})\gamma x\right] dx \tag{7}$$

Letting the virtual displacement $\delta v = 1$ in equation 1, while δu and $\delta \omega$ adjust accordingly, equations 5, 6 and 7 may be substituted into equation 1 to yield an equation for the foundation load P.

$$P = \int_{x_o}^{x_t} \{\left[\sigma + \tau \frac{dy}{dx} - (y - \overline{y})\gamma\right] - \delta u(\sigma \frac{dy}{dx} - \tau)$$
$$+ \delta\omega \left[-(\sigma \frac{dy}{dx} - \tau)y - (\sigma + \tau \frac{dy}{dx})x + (y - \overline{y})\gamma x\right]\}dx \tag{8}$$

Alternatively, equation 8 may be expressed as

$$P = \int_{x_o}^{x_t} gdx = G[y, \sigma, y', \delta u, \delta\omega] \tag{9}$$

where

$$g = \delta\omega(\tau g_1 + \sigma g_2 - \gamma g_3) \tag{10}$$

and g_1, g_2 and g_3 are given by

$$g_1 = \left(\frac{\delta u}{\delta\omega} + y\right) + \frac{dy}{dx}\left(\frac{1}{\delta\omega} - x\right) \tag{11}$$

$$g_2 = \left(\frac{1}{\delta\omega} - x\right) - \frac{dy}{dx}\left(\frac{\delta u}{\delta\omega} + y\right) \tag{12}$$

$$g_3 = \left(\frac{1}{\delta\omega} - x\right)(y - \overline{y}) \tag{13}$$

The integral g is stationary when its first variation vanishes. The necessary condition for g to be stationary is given by Euler's differential equations of variational calculus.

$$\frac{d}{dx}\left(\frac{\partial g}{\partial \sigma'}\right) - \frac{\partial g}{\partial \sigma} = 0 \tag{14}$$

$$\frac{d}{dx}\left(\frac{\partial g}{\partial y'}\right) - \frac{\partial g}{\partial y} = 0 \tag{15}$$

A polar coordinate transformation is accomplished by recognizing that

$$x = x_c - r\cos\theta \tag{16}$$

$$y = y_c + r\sin\theta \tag{17}$$

while

$$x_c = \frac{\delta v}{\delta\omega} = \frac{1}{\delta\omega} \tag{18}$$

$$y_c = -\frac{\delta u}{\delta\omega} \tag{19}$$

which yields the transformation equations

$$x = \frac{1}{\delta\omega} - r\cos\theta \tag{20}$$

$$y = -\frac{\delta u}{\delta\omega} + r\sin\theta \tag{21}$$

From equations 10- 14, 20 and 21 it can be shown

$$\frac{dr}{d\theta} = r\frac{d\tau}{d\sigma} \tag{22}$$

while from equations 10- 13, 15, 20, 21 and 22 the following result is obtained.

$$\frac{d\sigma}{d\theta} = \gamma r\cos\theta - 2\tau \tag{23}$$

Equations 22 and 23 constitute the two conditions necessary for a minimum footing load, from equation 9, necessary to cause slip. Chen and Liu (1990) demonstrated that the minimization of the footing load P implies that equilibrium has been satisfied. This conclusion allows for a particular solution to checked for equilibrium, indicating whether a minimum footing load has been obtained. The equilibrium condition may be specified as

$$E = \sqrt{H^2 + V^2} \tag{24}$$

Numerical Solution

Equations 22 and 23 represent two ordinary differential equations in terms of r and σ and may be solved by a Runge-Kutta numerical procedure. Equations 22 and 23 are rewritten in functional form as

$$\frac{dr}{d\theta} = f(\theta, r, \sigma) \tag{25}$$

$$\frac{d\sigma}{d\theta} = h(\theta, r, \sigma) \tag{26}$$

These equations are subject to the initial conditions

$$r(\theta_o) = r_o \tag{27}$$

$$\sigma(\theta_o) = \sigma_o \tag{28}$$

New values of r and θ are obtained from the equations

$$r_{n+1} = r_n + \frac{1}{6}j(k_1 + 2k_2 + 2k_3 + k_4) \tag{29}$$

$$\sigma_{n+1} = \sigma_n + \frac{1}{6}j(m_1 + 2m_2 + 2m_3 + m_4) \tag{30}$$

where

$$j = \theta_{n+1} - \theta_n \tag{31}$$

and the coefficients are determined by evaluating the functions f and h, given in equations 25 and 26 for the values given below.

$$k_i = f(\theta_i, r_i, \sigma_i) \quad m_i = h(\theta_i, r_i, \sigma_i) \tag{32}$$

$$
\begin{array}{lll}
\theta_1 = \theta_n & r_1 = r_n & \sigma_1 = \sigma_n \\
\theta_2 = \theta_n + \frac{1}{2}j & r_2 = r_n + \frac{1}{2}jk_1 & \sigma_2 = \sigma_n + \frac{1}{2}jm_1 \\
\theta_3 = \theta_n + \frac{1}{2}j & r_3 = r_n + \frac{1}{2}jk_2 & \sigma_3 = \sigma_n + \frac{1}{2}jm_2 \\
\theta_4 = \theta_n + j & r_4 = r_n + jk_3 & \sigma_4 = \sigma_n + jm_3
\end{array}
\tag{33}
$$

Calculation of the forces P, H and V in equations 9, 5 and 6, respectively, requires an expression for the failure criterion. A non-linear failure criterion specified in the Mohr diagram stress space and suggested by Zhang and Chen (1987) is used and is given as

$$\tau = c\left(1 - \frac{\sigma}{\sigma_t}\right)^M \tag{34}$$

where c is the intercept with the τ axis and σ_t represents the isotropic tensile strength, or the intersection with the σ axis, and is taken as negative in tension. The corresponding values of c and σ_t define the steepness of the line while M defines the degree of curvature. If $M = 1$ the criterion corresponds to the linear Mohr-Coulomb criterion while the criterion becomes increasingly non-linear as $M < 1$.

Numerical Implementation

A numerical program has been written to solve for the footing load necessary to cause soil collapse. In the main program, the material properties, c, σ_t, M and γ, the footing width, B, and the soil surface, given by the function $\overline{y}(x)$, are specified. An initial normal stress, σ_o, corresponding to the slip surface at the initiation point (i.e. at the point $x_o = -B/2$, $y_o = 0$) is assumed, while a range of values over which this initial normal stress may vary is specified. A center of rotation, given by the coordinates x_c

and y_c, is assumed, while a range over which these coordinates may vary is specified as appropriate to the problem geometry.

A four level nested looping process is initiated to converge on a final result giving the minimum footing load corresponding to some initial value of the normal stress on the slip surface at the initiation point and some center of rotation. The two most inner loops correspond to iterations of the assumed coordinates x_c and y_c over their specified ranges. The next loop level allows for different values of the assumed initial normal stress to be specified.

For each combination of x_c, y_c and σ_o the polar coordinates of the initiation point (θ_o, r_o) are calculated from the transformation equations 16 and 17. Setting the rotation angle increment, j, allows values of r and σ to be evaluated at each new point from equations 29 and 30 by evaluating the coefficients given in equations 32 and 33. The functions f and h are expressed by combining equations 22, 23, 25, 26 and 34 to yield

$$k_i = -\frac{r_i c M (1 - \sigma_i/\sigma_t)^{(M-1)}}{\sigma_t} \tag{35}$$

$$m_i = \gamma r_i \cos(\theta_i) - 2c(1 - \sigma_i/\sigma_t)^M \tag{36}$$

Corresponding x, y coordinates are determined at each new point from the transformation equations. The rotation angle is incremented until the slip surface reaches the ground surface. Correction approaches are taken to prevent the last incremental slip surface from extending above the ground surface. The forces H, P and V are calculated from equations 5, 9 and 6, where the virtual rotation and displacement, $\delta\omega$ and δu, respectively, are evaluated from equations 18 and 19. The resulting error in equilibrium is then calculated from equation 24.

During the course of looping for the two inner loops corresponding to different rotation centers, the rotation center resulting in the minimum equilibrium error is noted. Maintaining the same initial value for σ_o, the initial center of rotation is redefined as the point noted above and the two inner loops are repeated using a range for the variation in x_c and y_c that is $1/3$ the values in the previous looping process. This process is repeated until the equilibrium error is less than a specified tolerance.

The above described process is repeated for the next value of σ_o, corresponding to the 3rd level loop. For each new value of σ_o the resulting footing load, P, is noted. If the range for σ_o has been properly chosen it will be seen that P reaches some minimum value within the interior limits of this range. The 4th level loop repeats the above process by taking the first assumed value of σ_o to be equal to the value corresponding to the minimum footing load from the prior loop and reducing the range for assumed values of σ_o by $1/2$ the prior range. This final loop is repeated until the difference in calculated footing loads for values of σ_o within a particular range is less than a specified tolerance. The resulting footing load then meets the conditions of an upper bound while satisfying equilibrium.

Table 1: Bearing Capacity Predictions

Depth of Footing ($*B$)	Footing Width (cm)	Experimental q_{ult} (kPa)	Predicted q_{ult} (kPa)		
			NLLA	Vesić Tangent ϕ, c	Vesić Avg. ϕ, c
0	49.4	6700	12,200	7050	9870
0	34.8	5400	9780	5680	9170
0	10.9	1750	4810	2880	8010
0.5	49.2	6620	14,160	8790	10,900
0.5	34.9	3850	11,390	7110	9940
0.5	10.8	2740	5460	3430	8240
1	49.4	5810	16,130	10,900	12,000
1	35.2	4190	13,000	8480	10,700
1	10.7	1470	6030	4030	8470

Predictions of Experimental Results

The program described above has been used to predict the bearing capacity of a shallow foundation resting on a very dense, highly angular, dry, silty sand. Experimental results, to which the predictions can be compared, have been generated through centrifuge testing. The experimental program and results are described by Perkins (1995). The material failure envelope, descried in terms of peak measures obtained from conventional triaxial compression tests at confining pressures ranging from 1-750 kPa, have allowed the non-linear failure criterion to be defined according to equation 34, giving the parameters $c = 0.2$ kPa, $\sigma_t = 0.1$ kPa and $M = 0.93817$. While the failure envelope may not appear to have an appreciable curvature the secant measure of the friction angle varies from 60° to 47° for the range of confining pressures given above. Due to the elevated levels of the friction angle, small changes in the friction angle have a large impact on the resulting bearing capacity.

The results of the non-linear limit analysis (NLLA) and the experiments are given in Table 1. For purposes of comparison, the linear bearing capacity solution of Vesić (1973) has also been used to predict the experimental results. Two sets of linear strength parameters were used in this method. The first set relied upon choosing a cohesion and friction angle corresponding to a tangent line to the failure envelope at a level of normal stress representative of the footing problem being analyzed. In this way, the friction angle decreased while the cohesion increased for problems where the footing width increased or the depth of footing embedment increased. The second set relied on assessing the average friction angle and cohesion over a broad range of confining stress, resulting in the parameters $\phi = 47.5°$ and $c = 40.3$ kPa used for all the footing configurations. Further details of these methods are provided by Perkins (1995).

The results in Table 1 are expressed in Figure 2 by computing the amount by which each solution overpredicts the experimental values. In this figure, the data points labeled "ASP" represent results from Vesi{'c's approach using Average Strength

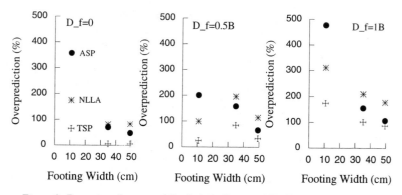

Figure 2: Percentage Increase of Predictions Compared To Experimental Results

Parameters, while "TSP" corresponds to Tangent Strength Parameters. From Figure 2 it is seen that using average strength parameters results in excessive overpredictions as the footing width decreases and as the depth of embedment increases. The non-linear limit analysis solution results in greater levels of overprediction in comparison to using tangent strength parameters yet the trend in the results is nearly identical between the two solutions.

Conclusion

The formulation of a solution for the bearing capacity of shallow foundations where the curvature of the material failure envelope is explicitly included has been performed using the theorems of limit analysis. While the solution is more theoretically realistic, comparison of the solution to centrifuge experimental results and to more conventional solution approaches where the material non-linearity is indirectly accounted for, suggests that the method works as well as conventional methods in terms of capturing experimental trends. The non-linear limit analysis yields prediction magnitudes, however, in excess of those from the conventional approach.

References

Baker, R. and Frydman, S. (1983), "Upper Bound Limit Analysis of Soil With Non-Linear Failure Criterion", *Soils and Foundations*, Jap. Soc. Soil Mech. and Fnd. Eng., V. 23, No.4, pp.34-42.

Chen, W.F. and Liu, X.L. (1990), "Limit Analysis in Soil Mechanics", Elsevier Sci. Pub. B.V., The Netherlands, 477p.

Perkins, S.W. (1995), "Bearing Capacity of Highly Frictional Material", Article Submitted to the *ASTM Geotechnical Testing J.*.

Vesić, A.S. (1973), "Analysis of Ultimate Loads of Shallow Foundations", *ASCE J. Soil Mech. and Fnd. Div*, V.99, No.SM1, pp.45-73.

Zhang, X.J. and Chen, W.F. (1987), "Stability Analysis of Slopeds With General Non-linear Failure Criterion", *Int. J. Num. Ana. Meth. Geomech.*, V.11, pp.33-50.

SLOPAS: A New Computer Program for Generalized Slope Stability Analysis

By R.D. Espinoza and J.D. Bray[1]

ABSTRACT

A new computer program, called SLOPAS, which provides a generalized framework for slope stability analysis, was developed. The algorithm contained in this program is based on a new unified formulation which satisfies both moment and force equilibrium. With this general formulation, most commonly used methods of slices can be derived as special cases. As a result, a flexible computer program that allows the analysis of slope stability problems considering different hypotheses for the interslice forces is available for the researcher and practitioner. Computed factors of safety, as well as other quantities of interest such as the internal forces, obtained using conventional methods that are based on different assumptions can be compared directly and objectively. This feature has important practical implications, because the choice of the stability procedures employed can sometimes result in significant differences in the computed factor of safety. The program has been implemented for personal computers with a user friendly, menu-driven interface that makes it an attractive tool for practitioners and teachers. In this paper, the structure of the program SLOPAS is described and an illustrative case example is analyzed to highlight the program's capabilities.

INTRODUCTION

Applications of limit equilibrium theory have proven to be useful tools in assessing the stability of slopes. Years of experience in evaluating the safety of natural and constructed slopes using analytical procedures based on limit equilibrium theory have improved the understanding of the parameters involved in the computation of safety factors. Slope stability analysis, using limit equilibrium methods, has evolved from the ordinary method of slices (Fellenius, 1936), which satisfy only global moment equilibrium, to methods that rigorously satisfy force and moment equilibrium conditions for each slice (e.g. Morgenstern and Price, 1965; Spencer, 1967; Sarma, 1979; Espinoza et al. 1994). As the evaluation of a slope's safety using rigorous methods involves the solution of simultaneous nonlinear equations, computer codes have been developed to implement the analytical procedures. With the advent of high speed personal computers with large memory capacity and powerful graphical capabilities at a relatively low cost, a number of slope stability analysis software packages have been implemented for microcomputers. The currently available computers programs basically differ in the procedure utilized for the solution of the factor of safety and the facilities available to handle complex geometric conditions.

It has been shown elsewhere (Espinoza et al., 1992 and 1994) that most current procedures can be encompassed within a single framework, and hence, a single program that handles the different procedures can be implemented. The reader are referred to the papers

---

[1]   Post-Doctoral Researcher and Assistant Professor, Department of civil Engineering, University of California, Berkeley, CA 94720.

mentioned above for a full description of the unified framework. There are significant advantages to having a single program which presents a unified framework for all previous slope stability procedures. Such a program allows for meaningful comparisons among different available methods, it greatly reduces the number of input-output operations that would be necessary if different programs were to be used, and using only one program allows the analyst to become familiar with its advantages and disadvantages. SLOPAS is such a program and a description of its features is presented in this paper. It was developed using a TURBO PASCAL programming language which takes advantage of its structured format and powerful graphical capabilities.

FACTORS OF SAFETY

Considering global horizontal force equilibrium of a potential sliding mass (Fig. 1a), an expression for the factor of safety $(F_{ff})^2$ is obtained (Espinoza et al., 1994):

$$F_{ff} = \frac{\sum [\{c\Delta l + (W\cos\alpha - u\Delta l)\tan\phi\} m_\alpha]}{\sum [W\sin\alpha\ m_\alpha] + E_a - E_d + \sum [\Delta Q] + I_{ff}} \tag{1a}$$

where:

$$I_{ff} = \sum [\Delta T \tan(\phi_m - \alpha)] \tag{1b}$$

$$m_\alpha = \frac{\sec\alpha}{(1 + \tan\alpha \tan\phi_m)} \tag{2}$$

and ϕ and c are the effective stress Mohr-Coulomb parameters; $\Delta T = T_{i+1} - T_i$; T_i denote the right internal tangential forces applied on the ith section (Fig. 1b); $\tan\phi_m = \tan\phi/F_{ff}$; $W = W_s + \Delta V$; W_s is the weight of the slice; W is the resultant vertical force acting on the slice; ΔQ and ΔV are the applied external horizontal and vertical forces, respectively; Δl denotes the length of arc **bc**; α the angle between the tangent at the center of the base and the horizontal; and E_a and E_d are the boundary forces on the ends of the potential sliding mass (Fig. 1a).

Similarly, establishing global moment equilibrium about an arbitrary point Ω (Fig. 1a) of the forces acting in each slice, an expression for factor of safety $(F_{mm})^2$ can be obtained (Espinoza et al., 1994) as:

$$F_{mm} = \frac{\sum [\{c\Delta l \bar{Y} + (W - u\Delta x)\tan\phi\ Y'\} m_\alpha]}{-\sum [W\bar{X}] - M_a - M_d + \sum [(W - u\Delta x)m_\alpha X'] + \sum [\Delta Q Y_Q] + \sum [u\Delta l\ X'] + I_{mm}} \tag{3a}$$

where

$$I_{mm} = \sum [\Delta T(Y'\tan\phi_m - X')m_\alpha] \tag{3b}$$

and $\tan\phi_m = \tan\phi/F_{mm}$; Δx is the width of the slice; $\bar{X}_i$ and $\bar{Y}_i$ are the horizontal and vertical distances, respectively, measured from the midpoint of the slice base to Ω (Fig. 1a); and X' and Y' are the distances parallel and perpendicular to the slice base inclination, respectively, measured from the midpoint of the slice base to Ω (Fig. 1a). Y_Q is the vertical distance between the point of application of the horizontal force, ΔQ and Ω. M_a and M_d are the moments due to forces, E_a, T_a and E_d, T_d, respectively, acting on the slope boundaries (Fig. 1a). A complete solution for the factor of safety is obtained when F_{ff} equals F_{mm} and full force and moment equilibrium conditions are satisfied. As the functions m_α, I_{ff} and I_{mm} depend on the safety factor, an iterative procedure is required to render a solution for the system of equations.

It is interesting to note in the two expressions for the factor of safety, Eqs. (1a) and (3a), that the internal shear force, $T(x)$, has been confined to a single term (I_{ff} and I_{mm}, respectively). These unknowns quantities have to be evaluated to compute the values of F_{ff}

---

[2] The subscripts *ff* and *mm* for the safety factor (F) indicate whether the derived expression is based on global force or moment equilibrium, respectively.

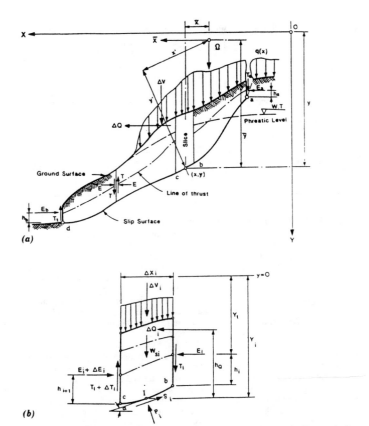

Fig. 1 (a) Typical Cross Section of Slope with Noncircular Failure Surface; (b) Forces Acting on Single Slice

and F_{mm}. If the distribution of the internal shear forces, $T(x)$, were known, both terms, I_{ff} and I_{mm}, could be determined. An expression for $T(x)$ can be obtained from moment equilibrium of a single slice (Janbu, 1954):

$$T(x) = E(x) \tan\alpha - \frac{dM(x)}{dx} + \frac{dQ}{dx} h_Q \tag{4}$$

where $M(x)=E(x)h(x)$ and $h(x)$ is the line of thrust. In general, the distribution of $T(x)$ is assumed and Eq. (4) is used to compute the location of the line of thrust. A number of different hypotheses have been postulated regarding the distribution of the internal shear forces which lead to the different expressions of safety factors available in the literature (e.g. Janbu, 1954; Bishop, 1955; Morgenstern and Price, 1965; Spencer, 1967). Most of these procedures can be evaluated objectively within the unified framework in which the primary hypotheses are classified in three main groups (Table 1). By choosing the appropriate hypothesis (I, II or III) and by assuming the appropriate form of the arbitrary function, $f(x)$, most of the current procedures can be emulated by SLOPAS.

Table 1 - Summary of the Methods Available in SLOPAS

Hypothesis Description	Group	Type	Internal Force Relationship	Reference
Direction of the internal forces, $T(x)/E(x)$, is assumed	I	(a)	$T(x) = \lambda f(x) E(x)$	Morgenstern & Price (1965)
		(b)	$T(x) = [f_o(x) + \lambda f(x)] E(x)$	Chen & Morgenstern (1983)
		(c)	$T(x) = \lambda f(x)\{c_{avg} H(x) + [E(x) - P_w(x)]\tan\phi_{avg}\}$	Sarma (1979)[†]
Moment equilibrium equation of a slice, Eq. (4), is used and the moment distribution, $M(x)$, is assumed	II	(a)	$M(x) = E(x)h(x);$ and $h(x) = \lambda f(x) H(x)$	Janbu (1954)[†]
		(b)	$M(x) = \lambda f(x)$	Espinoza et al. (1994)
Internal tangential force shape, $T(x)$, is assumed	III	(a)	$T(x) = \lambda f(x)$	Correia (1989)
		(b)	$T(x) = \lambda f(x)[c_{avg} H(x) + (k-r_u)\dfrac{\gamma H^2}{2}\tan\phi_{avg}]$	Sarma (1973)

[†] Generalized in SLOPAS

FEATURES OF THE SLOPAS PROGRAM

Taking advantage of the PASCAL language structure to facilitate future modifications, the computer code SLOPAS has been divided in a number of subroutines which are grouped according to their respective functions. The program is written in a user friendly environment with the different options presented in a menu-driven system which enable the analyst to easily interact with the program's features. Creating and editing input files is easily performed. Before a safety factor is computed, information regarding the geometry, strength parameters at the base of a slice, internal forces, weight of slice, applied external forces and several other important parameters are saved to a file for subsequent analysis.

SLOPAS has other features that differentiates this program from existing ones. For example, it can simulate most of the existing methods of slices. The special development for the expressions of the factor of safety allows the choice of most current methods of analysis within a single framework. This facilitates the comparison of different methods in terms of not only the value of the factor of safety but also of other parameters such as the variation of internal forces, the value of the scaling parameter and so forth (see Espinoza et al., 1992).

The program SLOPAS is a software package that is continually upgraded. Currently, SLOPAS has a number of important capabilities which are grouped as follows: (1) those related to generating the soil's stratigraphic and geometric representation; (2) those related to the calculation of the factor of safety and internal forces; and (3) those related to the input/output facilities. A description of these follows:

(1)   Describing Stratigraphic and Geometric Characteristics

· SLOPAS can handle complicated geometric and stratigraphic conditions. The X-axis is defined positive to the left and the Y-axis is defined positive pointing downwards (Fig. 2). The geometry can be defined in any of the four quadrants. A layer is described by providing the coordinates of the lower boundary. For example, Layer 1 in Fig. 2 is defined by points L1, L2 and L3, and Layer 2 is defined by points L2, L4 and L5. The last layer (e.g. Layer 4 in Fig. 2) is assumed to be infinitely deep and the lower boundary need not be specified. Coordinates describing the soil layers, ground surface, groundwater level and failure surface must be input from left to right (Fig. 2). For example, the points defining the ground surface must be input in the following order: G1, G2, G3 and G4. The soil parameters for each layer for both in situ and saturated conditions are then specified.

· The user specifies the minimum number of slices (Nslices) to be used in the analysis. The actual number of slices are computed taking into consideration a slope's critical points (Fig. 2). An interior point is defined as a point that lies within the failing mass. Critical points (Xcrit) are: (a) all points defining soil layers and the water table that are interior, (b) points of intersection of a layer with the slip surface, (c) points of intersection of the water table with the slip surface, soil layers and ground surface if they are interior, (d) all points defining the failure surface. The actual number of slices used in the calculation may be larger than that specified by the analyst. After all critical points have been found and sorted in ascending order, the number of slices between two consecutive critical points ($Xcrit_i$ and $Xcrit_{i+1}$ is computed as $N_s=(Xcrit_{i+1}-Xcrit_i)/b_{max}$, where $b_{max}=(X_d-X_a)/Nslices$.

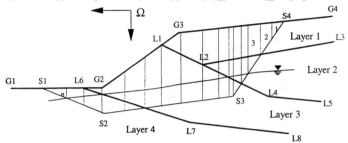

Fig. 2 Slope Layout

·   After all the slices have been determined, SLOPAS computes the weight of the slice taking into consideration the effective weights. The weights of the different layers composing the slice are added from top to bottom. Therefore, the numbering of the soil layers must be such that in any given slice, layer numbers are in ascending, not necessarily consecutive, order from top to bottom (Fig. 2).
·   Two failure modes (a circular and a general slip surface) can be examined. To define a circular slip surface, the center of the circle and the radius of the circle (or a point of the circle) is required. For general slip surfaces, the coordinates of the failure surface are provided.
·   The pore pressure can be represented by a phreatic level or by the values of the pore pressures parameter r_u (Bishop and Morgenstern, 1960). The r_u parameter can be described as constant within each layer or as variable along the failure surface.
·   Externally applied loads, as well as pseudo-static earthquake loads, can be considered in the analysis. The sign convention adopted for applied forces is positive in the direction of gravity and away from the slope (Fig. 1).
·   Values of cohesion linearly increasing with depth can be considered.

Once the geometry of the problem has been defined, the characteristics of each slice that depend on the geometry and the external applied forces are calculated. These are the weight of the slice, values of the strength parameter at the base of the slice, external vertical and horizontal loads and so forth. If no changes in the geometry or external forces are specified, these values are maintained as constant throughout the analysis.

(2) Performing an Analysis
Once the characteristic of the slices have been computed, the factor of safety for a given interslice force condition can be evaluated. The main features being:
·   The three hypotheses (I, II, and III) described in Table 1 have been implemented in the program. Each hypothesis has additional options (see Table 1). Hypothesis Ib or Ic requires supplementary data. In Hypothesis Ib, the analyst must specify the internal force ratio, $T(x)/E(x)$, at the toe and crest. This ratio can be related to the stress ratio at those points (Chen and Morgenstern, 1983). If Hypothesis Ic is chosen, the analyst must specify the average internal soil strength parameters.
·   The methods that can be emulated by SLOPAS are those proposed by Janbu (1954), Bishop (1955), Lowe & Karafiath (1960), Morgenstern & Price (1965), Spencer (1967), Sarma (1973 and 1979).

· Once the hypothesis that describes the internal force variation has been chosen, the analyst can choose up to six types of functions, $f_i(x)$. These are:

(a) $f_1(x) = 1$

(b) $f_2(x) = (\tan\alpha + \tan\beta)$, where $\tan\alpha$ and $\tan\beta$ are the failure and ground surface slopes

(c) $f_3(x) = \sin(\chi\pi)$, where $\chi = (x-x_a)/(x_a-x_d)$

(d) $f_4(x) = \tan\alpha$

(e) $f_5(x)$ = piece-wise distribution.

(f) $f_6(x,m,n) = A \chi^m (1-\chi)^n$, where $\chi = (x-x_a)/(x_a-x_d)$ and $A=(m+n)^{(m+n)} m^{-m} n^{-n}$

If $f_6(x,m,n)$ is chosen, the parameters m and n need to be specified by the user. This parameters represents the skewness of the function. For values of $m<n$, $f_6(x)$ will be skewed towards the crest of the slope.

· The value of λ_i for which F_{mm} equals F_{ff} can be computed automatically by the program. In addition, λ_i can be input manually and the corresponding values of F_{ff} and F_{mm} can be computed. This allows the simulation of procedures that assume a specific value of λ_i or is useful for cases in which convergence is not achieved with the automatic procedure.

· Simplified procedures to compute the factor of safety is also available.

· The global equilibrium of moment is taken with respect to a user's specified location. For circular failure surfaces, this point is usually taken as the center of the circle. Theoretically, the factor of safety is independent of the location of the center of moments. However, due to round-off errors in the numerical scheme, the particular location chosen may influence the convergency of the solution.

(3) Input and Output Features

As mentioned earlier, an input data processor has been developed within the framework of this program. This avoids the need of special programs for pre-processing the input data or post-processing the output data. The results obtained for each run can be sent to a user specified file for subsequent postprocessing. Likewise, the graphical capabilities of the PASCAL language is used to provide on-screen views of the geometry of the analyzed problem as well as the interslice force distribution, normal and shear stress along the failure surface, line of thrust ratio, and interslice force inclination. The menu system has been furnished with a concise descriptions of each of the available options of the program SLOPAS to assist the user while executing the program.

EXAMPLE PROBLEM

The slope shown in Fig. 3 was analyzed by Leschinsky (1990) and it is used in this paper to illustrate some of the capabilities of SLOPAS as well as to highlight some of the implications of using one type of slope stability analytical procedures versus another. Fig. 4 shows the variation of the internal forces for the different cases analyzed. The reported factor of safety using a variational calculus procedure was stated to be 1.263 (Leschinsky, 1990). Using Spencer's (1967) method with the center of moments located as shown in Fig. 3, the same author reported a factor of safety equal to 1.414. Factors of safety were computed using the same location of the center of moments shown in Fig. 3 for the different hypotheses considered and are summarized in Table 2. Using SLOPAS, a value of safety factor equal to 1.412 was obtained in Case 1 which represents Spencer's hypothesis, which is hypothesis Ia with $f(x)=1$. The values of safety factor for the different hypotheses range between 1.286 and 1.549 (a 20% variation). The maximum and minimum values were obtained using Hypothesis Ib with non-zero stress ratio at the toe and the crest, respectively (Table 2). The minimum value of factor of safety (1.286) and the internal force distribution compares well with that obtained using variational calculus (i.e. 1.263) by Leshchinsky (1990). This example shows that the factor of safety is not always insensitive to the assumptions regarding the internal forces.

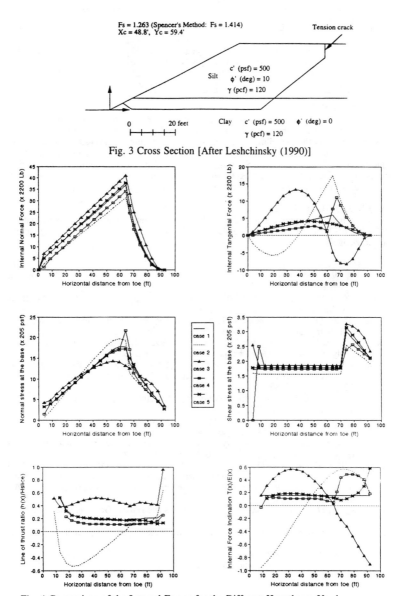

Fig. 3 Cross Section [After Leshchinsky (1990)]

Fig. 4 Comparison of the Internal Forces for the Different Hypotheses Used

Table 2 - Example

Case	Hypothesis	T/E_{toe}	T/E_{crest}	$f_i(x)$	F	λ	I_{ff}	I_{mm}
1	Ia	-	-	$f_1(x)$	1.412	0.161	-4.96	-3.06
2	Ib	-1.0	0.0	$f_6(x,1.0,2)$	1.549	0.863	-10.94	-100.88
3	Ib	0.0	-1.0	$f_6(x,1.5,1)$	1.286	0.920	1.59	105.70
4	IIa	-	-	$f_1(x)$	1.401	0.100	-4.67	5.72
5	IIIa	-	-	$f_3(x)$	1.366	4.166	-2.71	33.90

CONCLUSIONS

The availability of the present computer programs provides the analyst with a powerful tool to asses the factor of safety for analytical procedures based on different the assumptions. As the prevailing belief that the factor of safety is insensitive to assumption regarding the internal forces for methods that satisfy all conditions of equilibrium is at best questionable, the usefulness of SLOPAS is evident. The SLOPAS program for stability analysis of slopes is a versatile program with a friendly interface. Simulation of most current procedures can be done with ease, and fruitful comparisons can then be performed.

ACKNOWLEDGEMENTS

Financial support was provided by the national Science Foundation under Grant No. BCS-9157083 and the David and Lucile Packard Foundation, and this support is gratefully acknowledged.

REFERENCES

Bishop, A. W. (1955). The use of the slip circle in stability analysis of slopes. Geotechnique 5, No 1, 7-17.

Chen, Z.Y. & Morgenstern, N. (1983). Extensions to the generalized method of slices for stability analysis. Can. Geot. J.,20,104-119.

Correia, R. M. (1988). A limit equilibrium method for slope stability analysis. Proc. Fifth Intl. Symp. on landslides. Laussanne, pp. 595-598.

Espinoza, R.D., Repetto, P.C., and Muhunthan, B. (1992). General framework for stability analysis of slopes. Geotechnique, London, England, 42(4), 603-615.

Espinoza, R.D., Bourdeau, P.L. and Muhunthan, B. (1994). Unified formulation for stability analysis of slopes with general slip surfaces. J. of Geotech. Engineering, ASCE, Vol. 120, No 7, pp. 1185-1204.

Fellenius (1936). Calculation of the stability of earth dams. 2nd Cong. on Large Dams, 4, Washington, pp. 445-462.

Janbu, N. (1954). Application of composite slip surface for stability analysis. Proc. European conf. on Stability of Earth slopes, Sweden, 3, 43-49.

Leshchinsky, D. (1990). Slope stability analysis generalized approach. ASCE, Journal of Geotechnical Engineering, Vol. 116, No 5, pp. 851-867.

Lowe, J. III & Karafiath, L. (1960). Stability of earth dams upon drawdown. Proc. First Panamerican Conference on Soil Mechanics and Foundation Engineering, Mexico.

Morgenstern, N. R. and Price V.E. (1965). The analysis of the stability of general slip surfaces. Geotechnique, 15, 70-93.

Sarma, S.K. (1973). Stability analysis of embankments and slopes. Geotechnique, 23, 3, 423-433.

Sarma, S.K. (1979). Stability analysis of embankments and slopes. ASCE Geot. Div. Journal, 105, GT2, 1151-1524.

Spencer, E. (1967). A method of analysis of the stability of embankments assuming parallel inter-slice forces. Geotechnique 17, No 1, 11-26.

A GIS BASED TECHNIQUE FOR HYDROLOGIC MODELING

Jeffrey D. Jorgeson[1], Associate Member

Abstract

This paper details a Geographic Information System (GIS) based technique for hydrologic modeling using triangulated irregular networks (TINs) as the basis for watershed delineation, computation of watershed geometric attributes, and assignment of model parameters for use in the HEC-1 runoff model. A watershed modeling system called GeoShed, developed by the Engineering Computer Graphics Laboratory at Brigham Young University (BYU) in cooperation with the U.S. Army Engineer Waterways Experiment Station (WES), uses digital elevation data to generate a TIN, delineates watershed boundaries, and automatically computes the geometric attributes of basins. GeoShed provides a complete graphical interface for creating HEC-1 input files, running HEC-1, and viewing HEC-1 output. An application is presented where this technique was used to develop an HEC-1 model using GIS data to study the effects of land use changes for the 2,200 square kilometer watershed of the West Fork Cedar River in Iowa.

Introduction

Due to the recent widespread availability of spatially varying digital data for elevation, land use and soils, and the ease with which those data can be interpreted with GIS, new techniques in hydrologic modeling are emerging. Although spatially varying 2-dimensional hydrologic models may ultimately provide the best means of using these spatially varying data, such models may not be commonly available or accepted for several years or more. Meanwhile, well accepted and established lumped parameter models, such as HEC-1, are

---

[1] Research Hydraulic Engineer, U.S. Army Engineer Waterways Experiment Station, 3909 Halls Ferry Road, Vicksburg, MS 39180-6199

the most common type of model in use today. Many recent
advances in the use of lumped parameter models have been
in the development of new techniques through which
digital GIS data can be used to generate input
parameters for these models. One such technique using
TINs is detailed in this paper.

TIN Modeling

     An accurate digital terrain model can be one of the
most important aspects of computerized hydrologic
modeling. The topography of a watershed is not only
important in defining the movement of water, it also
influences many other aspects of the hydrologic system
(Wolock 1994). Thus, an accurate terrain model to
represent the topography is essential. From that
terrain model, many characteristics of the area being
modeled, such as the watershed boundary, surface area,
and average slope, can be computed. There are several
methods of constructing digital terrain models. Two of
the major techniques are the grid-based method and the
triangulated irregular network (TIN) method, and each of
these methods has certain advantages and disadvantages.

     For watershed modeling, TINs offer several
advantages over grid based methods. TINs are well
suited for watershed delineation and computation of
basin geometric parameters for use in hydrologic
modeling and can perform better than a grid (Defloriani
1986, Jones and Nelson 1992). With grids, however,
precision is often lacking in the definition of specific
watershed boundaries (Moore 1993). Thus, for hydrologic
modeling applications, the TIN structure can provide a
superior representation of the land surface and an
excellent framework for watershed analyses.

     To construct a TIN, a set of data points with x,y,z
coordinates is required. Many different algorithms have
been presented for triangulating a set of points to form
a TIN (Watson and Philip 1984, Jones 1990), most of
which use the Delauney criterion to guide the
triangulation process. The Delauney criterion is
satisfied when the circumcircle of the three vertices of
a triangle does not encompass any other vertices. The
Delauney criterion minimizes the existence of long, thin
triangles and the triangles that do make up the TIN are
as equiangular as possible (Nelson 1994). Figure 1
shows a sample TIN from a set of scattered data points.

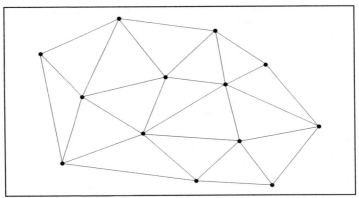

Figure 1. Sample TIN

Watershed Analysis with TINs

Once a TIN has been generated, it can be used for watershed analysis. Nelson (1994) set forth an algorithm that delineates watershed boundaries using a TIN model. The fundamental aspect of this method is tracing flow paths on the TIN. Jones et al. (1990) showed that flow paths can be constructed from any point on a TIN by following the path of maximum downward gradient, an example of which is shown in Figure 2.

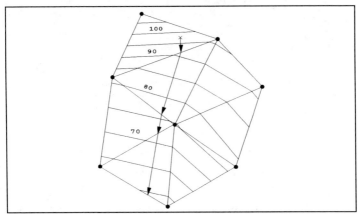

Figure 2.   Path of Maximum Downward Gradient

By following this path, the flowpath can be traced to a
terminus, which can be a watershed outlet or any
location in the TIN such as a subbasin outlet. The
drainage basin for any given terminus is comprised of
the set of triangles that contribute flow to that
terminus, and the perimeter of that set of triangles is
defined as the watershed boundary. After the desired
watershed and subbasins have been defined using this
technique, the area, average slope, maximum flow
distance, and other geometric parameters of the basins
are easily computed from the TIN. Those geometric
parameters can serve as input data for a hydrologic
model such as HEC-1.

Data Requirements

The basic input data required to generate a TIN is
digital elevation data with x,y,z coordinates for each
point. One of the most readily available sources of
elevation data for large areas within the United States
are digital elevation models (DEMs) from the United
States Geological Survey (USGS). USGS 1 Degree DEMs
were used for the work presented herein, and provided
data points on a three arcsecond grid with elevations in
meters relative to mean sea level (USGS 1987).

Additional data required for an HEC-1 input file
include rainfall, loss rate parameters, and unit
hydrograph parameters. GeoShed provides tools for
quickly assigning these input variables. For example,
the triangles in a TIN can be assigned land use and soil
classifications, and based upon those classifications a
weighted Soil Conservation Service (SCS) curve number is
automatically computed for each subbasin. Also, rain
gage data can be entered by geographic location, and
GeoShed automatically distributes the rainfall over each
subbasin based upon the Thiessen polygons. Similar
capabilities are provided to enter all data needed to
generate the HEC-1 input file (BYU 1994).

Sample Application

In support of the Scientific Assessment and
Strategy Team (SAST), which was formed to provide
technical support related to flooding in the Upper
Mississippi River Basin in 1993, a study was conducted
on the West Fork Cedar River in Iowa to evaluate the
effect of several proposed alternatives on reducing the
peak flows from the watershed. The GeoShed modeling
system was used to create a TIN, set up an HEC-1 model,
verify the model to several observed events, and model
the proposed alternatives.

Watershed Model

A TIN model of the watershed for the West Fork
Cedar River was modeled with GeoShed using 1 Degree DEM
data. To encompass the entire watershed, portions of
four DEMs were joined, and the resultant DEM grid was
filtered using a curvature based technique presented by
Southard (1990). This filtering technique selects
points within the DEM grid which represent peaks, pits,
valleys, ridges and breaks in slope. Those points
formed the TIN from which the watershed was delineated.

In addition to elevation data, GeoShed requires a
stream network to complete the watershed model. USGS
1:100,000 scale digital line graph (DLG) files were
imported into GeoShed for creation of the stream
network. The watershed was then divided into smaller
subbasins using the delineation technique outlined
earlier, and the geometric attributes of those subbasins
were computed. The West Fork Cedar River Watershed, its
subbasins, and the area, basin slope, and maximum flow
distance for each subbasin are shown in Figure 3.

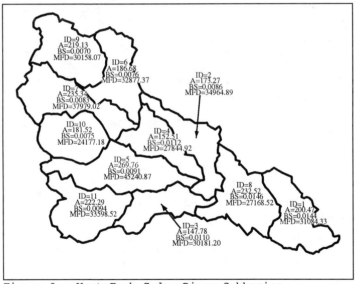

Figure 3.   West Fork Cedar River Subbasins

HEC-1 Model

The HEC-1 model simulates the runoff response of a watershed to a given precipitation event. The HEC-1 manual provides detailed information on HEC-1 (U.S. Army Corps of Engineers, 1990), and the specific requirements of HEC-1 will not be detailed here. However, the options used in modeling the West Fork Cedar River, with specific emphasis on the model parameters directly applicable to TINs, will be presented.

Basin Area. For each subbasin, the areas of the triangles in that subbasin were summed to determine the total subbasin area in square kilometers.

Loss Method. The SCS Curve Number (CN) method of computing losses was chosen. A land use and soil classification was assigned to each triangle in the TIN, and a CN value was assigned based upon those values. GeoShed then automatically computed a weighted CN value for each subbasin.

Unit Hydrograph Method. The SCS Dimensionless Unit Hydrograph method was used, and the hydrograph is described in HEC-1 by a single parameter called TLAG. The following equation (USDA 1985) relates TLAG to several of the geometric attributes computed by GeoShed for each subbasin,

$$TLAG = \frac{L^{0.8} * (R + 1)^{0.7}}{1900 * S^{0.5}}$$

where, L = maximum flow distance, R = (1000/CN)-10, and S = average watershed slope. Using the maximum flow distances and slopes computed by GeoShed, TLAG was computed for each subbasin.

Routing Method. The Muskingum-Cunge routing method was selected for this model. Eight-point cross sections were entered for each routing reach, and the stream lengths and slopes were computed by GeoShed from the TIN.

Rainfall. Rain gages in the area were entered into GeoShed, which computed Thiessen polygons and distributed the rainfall over each subbasin.

The HEC-1 model of the West Fork Cedar River watershed was verified to several observed events using precipitation data from rain gages located in and near the watershed and streamflow data from the gaging station at the outlet. After minor adjustments to the

CN values for antecedent moisture conditions, the model produced hydrographs at the watershed outlet that matched the peak and general shape of the observed hydrographs for each event.

The use of this TIN based method with GeoShed greatly facilitated development of the watershed model for this study. The TIN watershed model was created in a matter of a few days, which eliminated the need to manually calculate basin areas, slopes, stream lengths, and other geometric properties. The ability to automatically generate HEC-1 input files with GeoShed also proved extremely useful during the model verification process.

Conclusions

The use of GIS based methods for hydrologic modeling, as well as other types of environmental modeling, is a topic of great interest and relevance throughout the civil engineering profession. The work presented here details a tool currently available for use by practicing engineers. As sources of data for use in hydrologic modeling continue to increase and improve, techniques of interpreting and using those data must be tested and evaluated. The case study detailed in this paper shows one such technique for a given set of data, but the methodology presented can easily be adapted and refined for use in other hydrologic modeling applications.

Acknowledgements

The work presented herein, unless otherwise noted, was conducted at the U.S. Army Engineer Waterways Experiment Station. Permission was granted by the Chief of Engineers to publish this information. The contents of this paper are not to be used for advertisement, publication, or promotional purposes, and citations of trade names does not constitute an official endorsement or approval of the use of such commercial products.

References

Brigham Young University (BYU). (1994). GEOSHED Reference Manual, Engineering Computer Graphics Laboratory, Provo, UT.

DeFloriani, L., Falcidieno, B., Pienovi, C., Allen, D., and Nagy, G. (1986). "A Visibility Based Model for Terrain Features," Proceedings of the 2nd International Symposium on Spatial Data Handling, 235-250.

Jones, Norman L. (1990). "Solid modelling of earth
    masses for applications in geotechnical engineering,"
    PhD dissertation, University of Texas at Austin,
    Austin, TX.

Jones, Norman L., and Nelson, J. (1992). "Drainage
    Analysis Using Triangulated Irregular Networks", In
    Proceedings of the 8th Conference on Computing in
    Civil Engineering and GIS Symposium, 719-726, ASCE.

Jones, Norman L., Wright, Stephen G., and Maidment,
    David R. (1990). "Watershed Delineation with
    Triangle-Based Terrain Models," Journal of Hydraulic
    Engineering, 116(10), 1232-1251, ASCE.

Moore, I.D., Grayson, R.B., and Ladson, A.R. (1993).
    "Digital Terrain Modelling: A Review of Hydrological,
    Geomorphological, and Biological Applications," In
    Terrain Analysis and Distributed Modelling in
    Hydrology, 7-34, John Wiley & Sons, New York, NY.

Nelson, E. James, Jones, Norman L., and Miller A.W.
    (1994). "Algorithm for Precise Drainage-Basin
    Delineation," Journal of Hydraulic Engineering.
    120(3), 298-312, ASCE.

Southard, David A. (1990). "Piecewise Planar Surface
    Models from Sampled Data," In Scientific
    Visualization of Physical Phenomena, 667-680,
    Springer-Verlag, New York, NY.

U.S. Army Corps of Engineers. (1990). HEC-1 Flood
    Hydrograph Package User's Manual, Hydrologic
    Engineering Center, Davis, CA.

U.S. Department of Agriculture (USDA). (1985). National
    Engineering Handbook, Section 4, Hydrology (NEH-4),
    Soil Conservation Service.

U.S. Geological Survey (USGS). (1987). Digital
    Elevation Models - Data Users Guide. U.S. Department
    of Interior, Reston, VA.

Watson, D.F., and Philip, G.M. (1984). "Systematic
    triangulations," Computer Vision, Graphics, and Image
    Processing, 26, 217-223.

Wolock, David M., and Price, Curtis V. (1994). "Effects
    of Digital Elevation Model Map Scale and Data
    Resolution on a Topography-Based Watershed Model,"
    Water Resources Research, 30(11), 3041-3052.

GIS Applications in the Box Elder Watershed Study

James T. Wulliman[1]

Abstract

A comprehensive stormwater master planning project was undertaken for a 19,400-hectare (75-square-mile) study area adjacent to the new Denver International Airport. Watershed planning information was developed regarding hydrology, floodplains, stream stability, riparian resources, and water quality. Geographic Information System (GIS) techniques were used extensively in the project to enhance accuracy and efficiency. The GIS approach followed a "use the best tool for the job" strategy, employing an assortment of technologies and programs. These included workstation GIS, CAD/CAE software, a database, conventional hydrology/hydraulics models, and a variety of interface/processing programs developed specifically for the project. The GIS techniques utilized are described and both sponsor and consultant perspectives on their use are discussed.

Introduction

The Upper Box Elder Creek Outfall Systems Planning Study was undertaken to provide comprehensive stormwater master planning for a 19,400-hectare (75-square-mile) study area east of Denver, Colorado. Though currently undeveloped, the watershed areas adjacent to the new Denver International Airport are likely to be urbanized. The study was prepared for the Urban Drainage and Flood Control District (UDFCD), the City of Aurora, Arapahoe County, and Adams County. Stormwater planning information was developed covering hydrology, floodplains, stream stability, riparian resources, and water quality. Six alternative stormwater management plans were formulated and evaluated. Preliminary design of a recommended plan was prepared to help guide new development in the area.

---

[1]CH2M HILL, Inc., 6060 South Willow Drive, Englewood, CO 80111-5112.

For the hydrology analysis, the study area was divided into 615 subwatersheds averaging 32-hectares (78 acres) each to meet UDFCD criteria. Estimates of peak discharge were generated for six recurrence interval events (2-year through 500-year) for both existing and assumed future development conditions in the watershed. Each event was modeled with four different areal reduction factors to represent the effects of accumulated upstream watershed area. Thus, when combining the number of design permits with the number of required recurrence intervals and reduction factors, a total of about 30,000 discharge estimates needed to be managed to represent the hydrology of the study area. It was determined early in the project that the data management requirements of the job would be substantial; the printed output from the hydrology model alone created a stack of paper a meter high.

Because of the large number of subwatersheds and amount of data involved in the study, a decision was made to undertake as much of the project as possible using GIS techniques. The project agreement did not require using GIS techniques; they were used solely to enhance production efficiency. Other benefits of using GIS for this project became apparent; accuracy of results was significantly improved; and visualization of watershed characterization data and results was enhanced.

GIS Techniques

A six-block strategy of GIS modeling was employed to execute the project. The strategy is diagrammed in Figure 1. This strategy, based on using "the best tool for the job," included an assortment of technologies and programs rather than relying on just one program or tool to "do it all." The technologies used included workstation GIS, CAD/CAE software, a database, conventional hydrology/hydraulics models, and a variety of interface/processing programs developed specifically for the project.

The blocks shown in Figure 1 are explained in the following paragraphs.

Block 1 – GIS/CAD System

This block was used to characterize the watershed and floodplain corridors and prepare model input data. For the hydrology analysis, digital overlays were created to represent watershed topography, soils, and existing and assumed future land use. A triangulated irregular network (TIN) comprised of over one-half million facets and a terrain grid comprised of 15 million grid cells were computer generated from the topographic contours. Streamlines and subwatershed boundaries were derived from the terrain grid and were manually checked and adjusted as needed. Subwatershed area, flowline length, weighted

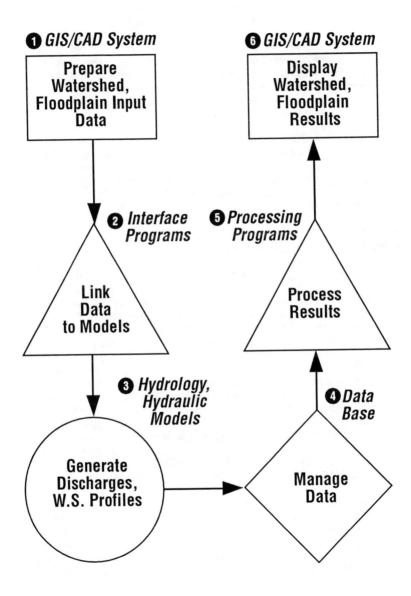

Figure 1—GIS Modeling Strategy

slope, centroidal length, imperviousness, and infiltration rate were measured, calculated, and written to a hydrologic input file using GIS routines.

For the hydraulic analysis, detailed terrain models were created along the floodplain corridors from 2-foot contour interval mapping. Representative cross sections were drawn on screen in plan view and limits between channel and overbank sections were indicated on cross-section views. From this information, complete X1 and GR records were encoded for use in the U.S. Army Corps of Engineers' HEC-2 program.

Workstation ARC/INFO, developed by ESRI, was used for the GIS work involved in characterizing the watershed and developing hydrologic input files. Microstation and INROADS, developed by Intergraph, was used for the creation of TINs and the HEC-2 cross-section coding. The HEC-2 routine was developed during the project as a team effort between Intergraph and CH2M HILL. Intergraph wrote all of the code and CH2M HILL created the application approach and tested beta versions of the program.

Block 2 – Interface Programs

Several interface programs were developed during the project to link data to the hydrology and hydraulic models. The programs reduced the manual effort involved in running the models. For the hydrology analysis, one interface program appended the appropriate rainfall information to the input file, executed the hydrology model, extracted all of the peak discharge results, and wrote the results to a file for later import into a database. The program repeated this cycle for as many rainstorm events as needed.

Another interface program was developed for an evaluation of alternative detention strategies. The program sized detention ponds to achieve a user specified outflow rate (typically the peak discharge for undeveloped watershed conditions). This program functioned by making an initial estimate of the storage-discharge relationship for any specified detention pond based on upstream drainage area and imperviousness. The interface program then executed the hydrology model, checked the results, adjusted the storage-discharge curve, and repeated the process until the target discharge was reached.

A third interface program linked hydrology results to HEC-2. Cross-section identifiers for the project were selected based on their location along a stream alignment. Hydrologic design points were also assigned a location along a stream. The interface program queried HEC-2 input files, located each cross section relative to the two adjacent design points, interpolated discharges, and inserted QT records with the appropriate discharges at each cross section.

Keystoke errors were eliminated and consistency between the hydrologic and hydraulic modeling was ensured.

Block 3—Hydrology, Hydraulic Models

Programs used to perform the hydrologic and hydraulic analyses for the project were the standard models specified by the sponsors. These included UDFCD's Colorado Urban Hydrograph Procedure (CUHP) and the Storm Water Management Model (UDSWM2) for hydrology. The District's CUHP program is a unit hydrograph approach based on local rainfall/runoff characteristics. UDSWM2 is a modified version of the runoff block of the Environmental Protection Agency's SWMM program. For the hydraulic analysis, HEC-2 was used. No modifications were made to any of these standard programs; the GIS techniques employed on the project involved only the input to and output from these standard models.

Block 4—Database

A database program was used to organize and manage the hydrology information generated during the project. After importing the raw results from the hydrology modeling, the database was used to sort through the discharges associated with the four areal reductions and assemble one set of discharges representing the appropriate reduction factors. The database was also used to organize and print report-quality tables of watershed and stream characterization data and hydrologic results. The meter-high stack of hydrology output never had to be printed out; it was replaced by a summary table that was 20 pages long. The database program used for the project was Paradox by Borland.

Block 5—Processing Programs

Several programs were developed during the project to process the results of the hydrologic and hydraulic modeling. These included a routine that produced scaled graphs of peak discharges versus stream station (distance along the stream) for the six return period events. These graphs provided a pictorial representation of the flows used in the HEC-2 modeling. Scaled hydrographs were also produced.

Another processing program was developed to facilitate the evaluation of alternative watershed and stream improvement plans. After generating the hydrology for each alternative, this program was used to calculate costs for channel improvements, detention facilities, and roadcrossings based on the appropriate peak discharges and sizes and lengths of improvements.

Processing programs were developed for the hydraulic analysis, also. One program was used to extract results from HEC-2 output files and print report-quality summary tables of the information. Another program used a portion of these data to create a base CAD file for profile drawings showing stream invert, water surface profiles, cross-section locations, and bridge information.

Block 6—GIS/CAD System

The GIS/CAD system was used to visualize the watershed and floodplain information produced during the project. This information included watershed topography, soils, existing and assumed future land use, roadways, stream locations, subwatershed boundaries, floodplains, and recommended improvement plans. Sponsors were offered digital files of these overlays for incorporation into their own GIS or CAD system.

Sponsor Perspectives

The project sponsors generally favored the use of the GIS techniques described above. The techniques enhanced the efficiency and accuracy of the project. It was important to the sponsors that the techniques did not alter the standard hydrology and hydraulic programs or the desired format of the project report. In this respect, the GIS techniques were somewhat "invisible."

The sponsors indicated that they are facing a general trend toward the expanding use of GIS. Two of the four sponsors already had GIS systems in place in their jurisdictions. There was a recognition that the trend toward GIS presented both concerns and opportunities. Using GIS requires upgraded hardware and software, staff training, new standards and quality control procedures, building a useful database, and overcoming the "learning curve." On the other hand, it eventually can enhance efficiency, accuracy, ease of updates, and visualization of information.

Consultant's Perspective

The consultant's perspective on using GIS techniques is favorable, though not without qualification. On the positive side, using GIS techniques can greatly reduce the manual effort involved in repetitive activities and calculations. Once calculation routines are developed and checked, the accuracy of results is significantly improved over manual methods. Visualization of results is enhanced as well.

On the other hand, GIS presents a whole new set of issues. Specialized hardware, software, and staff need to be added to the traditional project team of hydrology and hydraulic engineers. Developing these programs and routines requires a commitment of time and money. Quality control checking is critical to ensure that routines are technically correct and are functioning properly. Of great importance is learning what to automate and what not to automate. Certain tasks are more efficiently performed by the computer, while in other cases a person can outperform the machine.

Decision Support in M&R Planning of
U.S. Army Corps of Engineers Civil Works

by David T. McKay[1], Member ASCE

Abstract

The U.S. Army Corps of Engineers (USACE) is responsible for the upkeep and maintenance of many hydraulic structures. A variety of facilities in the USACE Civil Works program support inland waterways navigation, recreation, hydropower generation, flood control, and coastal protection. The Corps' mission within Civil Works is changing from that of constructing such facilities to that of maintaining them. The Repair, Evaluation, Maintenance and Rehabilitation (REMR) Research Program is seeking to discover and develop technologies that will extend the service life of USACE Civil Works structures. Within this area researchers are developing methodologies to provide consistent and objective condition assessment procedures for Civil Works structures. Such procedures coupled with microcomputer based database management, and Geographic Information Systems (GIS), provide decision support for cost effective planning of REMR type activities for Civil Works facilities. The blend of condition inspection procedures, condition rating systems, data analyses, GIS, database management, and automated reporting is called a REMR Management System. The philosophy underlying a REMR Management System is that the key to cost-effective maintenance is a good understanding of a facility's current condition, and an ability to predict future condition. This paper describes past and current efforts in the development of REMR Management Systems for USACE Civil Works, and describes the role played by computers in this effort.

USACE Civil Works

The Civil Works Program of the Army Corps of Engineers employs approximately 28,640 people in 48 Division and District offices in all 50 States, D.C., Puerto Rico, Virgin Islands, Guam, American Samoa, and the Northern Mariana Islands. Civil Works structures are seen along coastal and inland waterways. The structures are operated to maintain navigable waterways, manage recreation areas, control floods, store water, and generate hydropower. In the navigation area alone,

---

[1] Civil Engineer and Principal Investigator, U.S. Army Construction Engineering Research Laboratories; Engineering and Materials Division; P.O. Box 9005; Champaign, IL 61826-9005

the Corps operates roughly 660 locks, locks & dams, and dams; it maintains more than 13,000 riverine stone or timber training dikes and untold miles of riverine revetment or other stream bank stabilization structures; it is responsible for 669 inland and coastal ports and harbors (all figures above as of December 1992). Managing the construction, operation and the maintenance and repair (M&R) of this inventory is obviously a monumental task. Of note, 50% of the Corps' navigation locks, dams, and locks & dams will have reached their design life by the year 2000 [McDonald and Campbell, 1985].

The greater part of the USACE mission within Civil Works has been changing over recent years from that of new construction, to that of M&R of existing structures. To assist the Corps in this changing role, the Repair, Evaluation, Maintenance and Rehabilitation (REMR) research program, was initiated with funding of $35M and ran from 1984 to 1990. The objective of the program was to identify and develop new or existing technologies that will extend the service life of Civil Works. The successes of the REMR program (projected savings are $200M within 5 years) prompted a REMR II, which is again funded with $35M for 1991 to 1997. Seven Problem Areas within the program are addressed: Concrete & Steel Structures, Geotechnical, Hydraulics, Coastal, Environmental, Electrical & Mechanical Applications, and Operations Management. The REMR research and development (R&D) is spread throughout the Corps' R&D community, but the bulk of the work is performed at the Corps' Waterways Experiment Station (WES) and the Coastal Engineering Research Center (CERC) in Vicksburg, MS; the Cold Regions Research and Engineering Laboratory (CRREL) in Hanover, New Hampshire; and the Construction Engineering Research Laboratories (CERL) in Champaign, IL. The emerging REMR technologies are tested, fielded within the Corps, and tech transferred to the infrastructure community by a variety of T^2 avenues.

The work in the problem area of Operations Management is being done at CERL. Successes of Engineered Management Systems developed there, such as PAVER [Shahin and Kohn, 1981], and ROOFER [Shahin and Bailey, 1987], among others, motivated the development of similar systems for a variety of Civil Works Structures. The resulting product, a <u>REMR Management System</u>, provides quantification of engineering judgement during the condition assessment of Civil Works structures. The driving concept behind these systems is the idea that good management practice must be based, in a large part, on knowledge of the structure's condition, and its future condition. The current condition assessment practice within the Corps is known as a periodic inspection. Of course, the structures are monitored continuously during operation, but once every three to five years every civil works project (a lock and dam for e.g.) is completely inspected by a team of roughly a dozen USACE personnel, each with an expertise in structures, mechanical & electrical, hydraulic, or geotechnical areas; who make a subjective assessment of the project and write up a report with an ample supply of photographs. If conditions warrant, more detailed engineering analyses are conducted. These subjective reports work well within a District office for documenting condition, but at the Headquarters level, when the time comes to decide how to distribute scarce M&R dollars, they fall short of providing enough information for upper level managers to distinguish which projects are more deserving.

REMR Management Systems attempt to quantify engineering judgment during condition assessment, but also assist in quantifying the impact of M&R work, as well as helping managers analyze, or project the benefits of performing such work. This then is a decision support tool for M&R managers at all hierarchical levels within the Corps. A REMR Management System contains uniform, quantifiable condition assessment procedures; and an automated database program that manages a structures inventory, the condition inspection data, condition calculations and analyses, and automated reports. The primary goal of these systems is to remove subjectivity wherever possible, and make condition assessments based on objective, tangible, measurable observations. The Management System is intended to be used simply for decision support, and is not meant to be used to rank order or prioritize M&R projects based on condition alone.

The REMR Condition Index

At the heart of these systems is the Condition Index (CI). Figure 1 shows the REMR Condition Index Scale. This definition provides a consistent language, or definition with which to discuss condition. Using a number to describe condition has the obvious advantages of being easily stored in a computer; and can be manipulated in mathematical expressions. The CI is based on structural integrity, and on the ability of a structure to perform its function. This definition is designed to apply to all structures.

Zone	Condition Index	Condition Description	Recommended Action
1	85 to 100	*Excellent:* No noticeable defects. Some aging or wear may be visible.	Immediate action is not required.
1	70 to 84	*Good:* Only minor deterioration or defects are evident.	Immediate action is not required.
2	55 to 69	*Fair:* Some deterioration or defects are evident, but function is not significantly affected.	Economic analysis of repair alternatives is recommended to determine appropriate action.
2	40 to 54	*Marginal:* Moderate deterioration. Function is still adequate.	Economic analysis of repair alternatives is recommended to determine appropriate action.
3	25 to 39	*Poor:* Serious deterioration in at least some portions of the structure. Function is inadequate.	Detailed evaluation is required to determine the need for repair, rehabilitation, or reconstruction. Safety evaluation is recommended.
3	10 to 24	*Very Poor:* Extensive deterioration. Barely functional.	Detailed evaluation is required to determine the need for repair, rehabilitation, or reconstruction. Safety evaluation is recommended.
3	0 to 9	*Failed:* No longer functions. General failure or complete failure of a major structural component.	Detailed evaluation is required to determine the need for repair, rehabilitation, or reconstruction. Safety evaluation is recommended.

Figure 1. The REMR Condition Index Scale

The CI has 3 action zones and 7 condition levels. The purpose of the CI is to capture a "snapshot" of condition. The intent, when building the CI algorithms, is to capture as pure a picture of condition as possible. We strive to avoid incorporating anything predictive in the CI, such as age, or consequences of failure. (For e.g. - Two structures of identical condition might have entirely different consequences of failure, yet their CIs will be the same.) The CI should not be interpreted strictly as an indicator of a structure's need of repair; that is up to expert (and human) judgement. The CI should be likened to unbiased factual reporting: It should tell you only what the current condition is, but it should not endeavor to declare why, or predict what will follow; it should simply answer: what are the facts at this moment?

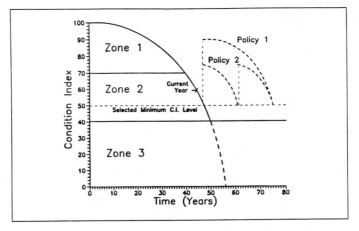

Figure 2. Theoretical Curve: Condition Index (CI) vs. Time

Gathering CIs thus, may enable the generation of curves such as the one shown in Figure 2, depicting the decay of a CI over time. The slopes of such curves would be characteristic of the structure whose condition is being tracked. If the CIs indeed are snapshots of condition, frozen in time, then the data might be used to *predict* condition. Managers could set minimum allowable condition levels, and be able to plan ahead with alternative M&R paths. But such curves are a long time in the making, and the CI has more immediate uses.

Corps District offices manage anywhere from 1 to approximately 20 projects (a lock and dam for e.g.). Doing REMR CI inspections allows them to establish engineering baselines with which they can track condition and perform comparisons to gauge deterioration. The data can be used to support budget requests, or can be used to argue against doing unnecessary preventative maintenance. The CI can assist in rank ordering projects. And the inspections themselves have discovered problems that were hidden or undetected up to that time.

At the Headquarters level, beginning in 1994, the CI is required input from all Districts requesting non-baseline M&R dollars. (Baseline M&R dollars are those that are provided for routine preventative maintenance.) The CI provides a Corps wide standard definition of condition. But the CI is not used as the sole deciding variable in comparing worthy projects for funding, it is one part of several parameters used in the decision making process. It cannot be overemphasized or repeated too often, that the CI does not remove professional judgment from the decision process at any level along the chain of command.

Building A CI Algorithm

Once a structure or one of its components has been selected for CI development, CERL investigators locate those people within the Corps who are responsible for the M&R of that type of structure. Experts from across the Corps are invited to participate, thus ensuring that all climates and operating environments are represented. On occasion expertise from outside the Corps is brought in as well. This group of people are then interviewed individually and collectively. They are asked to focus first upon each item that comprises the structure or component requiring M&R attention at some time. Then, for each item, the research teams asks what specific criteria, in the expert's engineering judgment, "raise the red flags," or first give warning that something may be very wrong with that item. This corresponds to a CI of 39, the cut off between Zones II and III of the REMR Condition Index Scale of Figure 1. This is the point that in the expert's opinion, further and more detailed engineering analysis is required on the structure; conditions warrant a much closer look. If, for example, a certain maximum allowable displacement is agreed upon by the expert panel, then all measurements of that particular displacement are compared to this; measurements approaching or exceeding the maximum allowable measurement are assigned CIs approaching or going below 39. The research team zeros in on other specific measurements that can be made, in order to quantify where the structure's condition is relative to the expert's judgement.

Expert rules in the form of equations and/or tables are constructed; and inspection procedures with emphasis on tangible, quantitative measurements are developed. Whenever possible, inspection procedures are designed to be performed on in-service facilities. Simplicity is strived for, objectivity is the goal; subjectivity cannot be completely avoided and subjective observations are used as seldom as possible. Inspections for nearly all systems developed to date can be completed in less that 2 person-days. To date, dial gauges, tape measures, rod and transit, a boat and a depth finder have been the most sophisticated equipment necessary. Equally important, the procedures must be repeatable. That is, different inspection teams must produce comparable results. All calculated and tabulated results are scaled to conform to the REMR CI definition.

Implementing REMR Management Systems

After the CI inspection procedures have been tested, and found to be meaningful and repeatable, software is developed and the completed REMR Management System is put into the field immediately. Week long training courses are conducted for 20 USACE personnel at a time. Three separate systems are covered during the week. The course focuses on performing the inspections and running the software. So far, CERL researchers have been doubling as course instructors, but steps are currently under way to institutionalize it within Corps' Proponent Sponsored Engineer Corp Training (PROSPECT) program as early as summer of 1996. As of this writing almost 200 people representing 30 Corps Districts have been trained on 8 different management systems.

Completed and Planned Systems

Since starting the development of REMR Management Systems in 1985, through many hard lessons and hands on experience, CERL researchers have learned how to produce a system in a little over a year. This includes forming an expert panel, developing the expert rules and inspection procedures, conducting field validation tests, writing the software, and teaching the training course. Systems that were completed as of September 1995 are

- REMR Management Systems for Navigation Structures -
Completed as of September 1995

Steel Sheet Pile Walls and Mooring Cells
Horizontally Framed Lock Miter Gates
Lock Sector Gates
Concrete Lockwall Monoliths
Concrete Dam & Spillway Monoliths and Piers
Lock Chamber Filling & Emptying Valves
Dam Tainter Gates
Timber Riverine Training Dikes
Stone Riverine Training Dikes & Revetment
Lock & Dam Gate and Valve Operating Equipment
Dam Roller Gates

1997 marks the final year of funding for the REMR program. More systems are on the drawing board, and are targeted for completion by then. Besides system development in the navigation arena, systems are also being developed for structures in coastal navigation/protection and hydropower and flood control generation arenas.

- REMR Management Systems for Navigation Structures -
To Be Completed by September 1997

Vertical Lift Gates
Earthen Embankment Dams
Dam Service Bridges

- REMR Management Systems for Coastal Structures -
To Be Completed by September 1997

Breakwaters
Jetties
Seawalls
Bulkheads
Revetment

- REMR Management Systems for Hydropower and Flood Control Structures -
To Be Completed by September 1997

Hydropower Stator Windings
Concrete Dams
Embankment Dams
Dam Gates
Power House

It should be noted that as full blown management systems, that is, systems that can indeed project future condition, can generate master work plans, can estimate dollar amounts or other economic impacts of M&R actions; these systems as fielded are far from complete. But it is significant that they are at least in place and ready, now, for further development. Ultimately, once all these systems have been completed, it is planned to integrate them into one of three single program packages: An Integrated System for Navigation Structures, Coastal Structures, and Hydropower Structures.

REMR Management System Software

The personal computer software is Disk Operating System (DOS) based, and is currently written in C, FORTRAN, and dBASE. The data input screens are usually built around the inspection form: where the inspection form features fill-in-the-blank and circle-appropriate-response actions, so does the software. Currently, a separate software application exists for each system.

Besides the CI inspection data storage, and CI calculations, the software will produce explanations as to why CI values were arrived at. Users can add their own experiences to this library. Consequence modelling can be performed, where one can specify repairs and see resulting CIs; and this in turn can be bundled with life cycle costs analysis to analyze the cost of achieving desired condition levels. Customized reports are available.

Object Oriented Tools Aid Software Development

Early in this program[2], the development of these systems was fairly linear; i.e. we worked on only one system at a time; starting with the CI algorithm and finishing with the system software. Little emphasis was given to the software at first, because our stronger focus was on developing the condition assessment procedures; the software simply served to support the condition assessment data. As more systems were completed, it became sorely evident that software development was going to be a significant hurdle. Software programming was a major drain on our limited manpower and money. With over a dozen navigation management systems planned for development, the software *had to be* user friendly and consistent in "look and feel" from one management system to the next. If field personnel were required to learn a new piece of software for each system, our entire program would face strong resistance and perhaps even fail altogether. With practice we learned to produce systems at a faster pace but we were still impeded by software development. Scarce funds forced us to do our own programming or rely on transient undergraduate and graduate students.

Recognizing the obvious importance of making each of these software programs professional in appearance, and the importance of uniform software "look and feel"; we sought to develop a set of Object Oriented software development tools, whose use would significantly reduce the time spent in software programming, and ensure that each management system we developed would look, operate, offer the same menus, and accept and report data in a like fashion. Internally, use of these tools ensured a conformity in database structure, and key function(s). This consistency across applications allows all of the them to be run off a single "data engine". I.e., all of the applications require the same set of executable files to manage their databases. This feature will facilitate the integration of all the separate programs into a single "master" program for navigation structures.

A contractor was hired to develop what became known as The REMR Tools. The REMR Tools are a collection of Objects that afford rapid construction of data entry user interfaces and databases. The goal of the contract was to develop the Objects such that they could 1) Build the User Interface and Databases (menu'd input screens for storing inspection and inventory data), 2) Perform CI Calculations, and 3) Allow the Rapid Construction of Reports. The Tools were to be designed so that

---

[2] The work described in this section pertains to navigation structures. Software for Coastal and Hydropower REMR Management Systems is not included in this discussion.

anybody with a basic understanding of programming techniques, could build a REMR Management System application, unaided and in a very reduced amount of time. The amount of written source code was to be minimal, with the command syntax chosen from menu-type pull-down lists. A contract was awarded for developmental work on parts 1 & 3. Limited success was achieved with Part 1, but the results of Part 3 were deemed too complicated and not flexible enough to be useful. It should be stated that during this time (1988-1992) similar packages, i.e. software that builds software, were just beginning to appear as commercial off-the-shelf products, but none were as suitable as our custom made Tools.

The REMR Tools, as developed, did allow us to build the front end of the management system applications with comparative speed. Once the learning curve for using the Tools is conquered, and the application is completely defined, a typical REMR Management system can now be produced in roughly 15 business days or less. The Tools are used for the front end of the application, that is, database creation and data input & storage. Fortran and C code is used for the CI calculations. A proprietary code that is embedded within the Tools is used for generating the hard copy reports. After completing several of these applications, the process is now almost mechanical.

Many lessons are learned every time a new system is developed. With each new system that is put into the field, needed improvements to the inspection procedures, the CI algorithm, and the software are discovered. Corrections to all of these modules are made immediately. Once all of the component systems have been developed and fielded, steps towards integrating all of the systems into a master program for navigation structures will be taken. In all likelihood, the modules will be rewritten using an off-the-shelf Object Oriented programming language, now that inexpensive and readily supportable packages are available.

GIS and REMR Management Systems

In the Spring of 1994 a concept demonstration incorporating Geographic Information Systems (GIS) into REMR Management Systems was put together. A 20 mile reach of the Arkansas River between 2 locks and dams was digitized. One of the coverages included all of the training dikes and revetment. (Training dikes are stone pile structures, constructed roughly perpendicular to the river bank. The purpose of the dike is to increase the water velocity at the riverward end, thus encouraging sediment transport and maintaining navigability.) Also included in this demonstration were scale drawings of concrete lockwalls. The point of this exercise was to use GIS to show the distribution of condition over a family of structures. The Environmental Systems Research Institute's (ESRI) PC Arc/Info was used.

Each dike and revetment and concrete lockwall monolith was color coded according to its CI. A relational database containing inventory and M&R data for each of the structures was linked to the polygon attribute tables; this allowed a "What is that?" point-and-click mouse interaction. All of the utilities in the ESRI GIS software and MicroSoft WINDOWS environment were available. (E.g. - Show all structures with a CI of 40 or below as purple.)   Various reports and analyses such as CI frequency distribution charts were presented.

The demo was included in presentations made at various USACE sites, including District Offices, a REMR workshop, and a REMR Research Program progress review meeting.   Everyone that saw the demo was impressed with the clarity and accuracy of the river image; and several people expressed excitement at the thought of having so much information on one's desktop.   But surprisingly, the river engineers for whom the demo was originally intended, felt the technology would not significantly facilitate their everyday work.

At this writing, no more work is budgeted for GIS development within REMR Management Systems, that is not to say that further work in this area is not anticipated.   However, the USACE Lower Mississippi Valley Division (LMVD) is currently automating channel improvement efforts on the Lower Mississippi and Atchafalaya Rivers using a combination of Computer Automated Design and Drafting (CADD), Global Positioning Systems (GPS) and GIS.   A master geospatial database for supporting channel stability and environmental activities has been initiated.   The resulting River Engineering and Environmental GIS (REEGIS) was established in 1993. REEGIS is a combination of CADD and GIS databases running on Intergraph systems software. LMVD expects to complete REEGIS by 1997.   It is hoped that REEGIS will be readily adaptable to other rivers and projects.

Growth Plans for REMR Management Systems

As this paper is being written, the funding picture for the support of these systems after the REMR research program ends is not clear.   But there are plans for growth in the making: A technical support center is being planned for software tech support, software upgrades, and handling of questions about field procedures. Ongoing training in usage of all systems will continue as long as there is demand.

Improvements to the system can come from outside the CERL research group as well as from within it.   One lesson was learned repeatedly during system development: that is, the system can always be improved.   New mechanical configurations that were not originally considered are discovered routinely in the field, and the CI algorithms must be adjusted accordingly.   Also, as time passes and once enough data is collected, there are several tried and proven probabilistic and statistical approaches to use for the purpose of predicting future condition, traffic loads, climate, and many other factors which may or may not have an economic impact on Civil Works operations.   These methodologies may be adapted to CI analyses as well.

New and emerging technologies will be incorporated into the systems.   Plans are on the drawing board for computerized "clip boards" making the transfer of inspection data to a desk top computer simply a matter of plugging the board into a socket.   Automated inspection devices, remote sensing techniques, and smart materials are also serious considerations for adoption.

Other Decision Support Tools

As emphasized by this paper, the CI is simply one of several parameters with which the allotment of M&R dollars is determined. Besides structure condition, other parameters must include traffic composition, lockage rates, climate, frequency of accidents, etc. Certainly, first hand familiarity with the field projects is a primary tool, but automated decision support is a welcome when working with such a large inventory. Two fairly new decision support programs used by USACE Headquarters are mentioned here: RELIABILITY [Department of the Army, 1992] and QUADRANT [Russell, Randolph, et al., 1993] analyses. Both systems attempt to estimate the economic benefits of performing the requested M&R, and the CI may optionally be employed in determining input to either system. Input to RELIABILITY requires calculating the probability of structural unsatisfactory performance. Input for QUADRANT requires estimates of initial structural condition in the form of a numerical index, not unlike the CI, and the change in overall condition based on a given dollar amount required for repairs. Because of limited space in this paper, they are referenced here for the interested reader to pursue.

Conclusion

The U.S. Army Corps of Engineers is working towards a standard for condition assessment of its Civil Works. Uniform and consistent condition assessment procedures produce numeric condition indicators called Condition Indices. As a part of standard operating procedure, Condition Indices are required from all Corps District offices requesting non-baseline M&R dollars. The indices serve the Districts as an engineering baseline to quantify condition and track trends in condition; they serve Headquarters level managers by ensuring that consistent definitions of condition are being used Corps wide; and the indices are helpful in developing input for other decision support schemes that measure the economic impact of performing particular M&R actions. Object Oriented programming techniques were employed to varying degrees of success in developing the supporting software. A GIS coverage of riverine dikes and revetments was produced; it contained condition indices as a coverage. The Lower Mississippi River Valley Division is developing REEGIS, a master CADD-GIS database that may serve as a template for other river basins.

REFERENCES

Department of The Army, Corps of Engineers Civil Works Engineering Division, (1992). "Engineering and Design - Reliability Assessment of Navigation Structures," Engineering Technical Letter ETL 1110-2-532, U.S. Army Corps of Engineers, 20 Massachusetts Ave. N.W. Washington, D.C. 20314-1000

McDonald, J.E., and Campbell, Sr., R.L. (1985). "The Condition of Corps of Engineers Civil Works Concrete Structures," Technical Report REMR CS-2, U.S. Army Waterways Experiment Station.

Russell, C.S., Randolph, M., et al., (1993) "QUADRANT: Incremental Analysis Methodology for Prioritizing O&M Projects (Locks & Dams) Report (R3)" Letter Report, Planning and Management Consultants, Ltd. P.O. Box 1316 Carbondale, IL

Shahin, M.Y., and Kohn, S.D., (1981). "Pavement Maintenance Management for Roads and Parking Lots," Technical Report M294/13, Vol II/ADA1120296, U.S. Army Construction Engineering Research Laboratories [USACERL].

Shahin, M.Y., Bailey, D.M., and Brotherson, D.E., (1987). "Membrane and Flashing Condition Indexes for Built-Up Roofs VOL II: Inspection and Distress Manual," Technical Report M87/13, Vol II/ADA190368, USACERL.

TOWARDS OBJECT DATABASES FOR PROJECT CONTROL
by
Tah, J. H. M.[1] and Howes, R.[2]

Abstract

This paper presents the analysis, and design of an object-oriented database to support a project planning and control knowledge-base system capable of monitoring and detecting the presence of time, cost, and quality performance deviations within construction activities and suggesting necessary corrective actions.

Project Data Management Problem Analysis

Previous research in the area of decision support for project management utilizing knowledge-based systems techniques have failed to make an impact in the construction industry. This can be attributed to the use of expert systems shells which are too restrictive and are suitable for handling problems in highly specialized domains. Construction management problems are characterized by a multiplicity of tasks or work packages which include such tasks as earthworks, temporary works, concreting, steelworks, drainage, tunneling, piling, and services. It is, therefore, difficult if not impossible to develop a convincing all embracing knowledge-based system for say construction planning or estimating using an expert system shell. The basis of our work is to acknowledge the existence of several different work packages each with its own knowledge requirements, and investigating the use of object-oriented knowledge representation and reasoning techniques together with the use of multi-agent and blackboard architectures to handle the disparate nature of construction problems. Furthermore, construction management problems are characterized by the presence of an abundance of quantitative or objective data. Solution strategies to these problems involve the use of these data together with

[1] Senior Lecturer, School of Construction Economics and Management, South Bank Univerity, Wandsworth Road, London SW8 2JZ, E-Mail: tahjh@uk.ac.sbu.vax.
[2] Professor and Head, School of Construction Economics and Management, South Bank Univerity, Wandsworth Road, London SW8 2JZ.

information of a qualitative nature, often derived from experience and previous cases. Therefore, new techniques should be capable of handling both quantitative and qualitative data characterized by vagueness and uncertainty.

Projects of whatever kind, particularly large and complex ones, can only be completed to any degree of satisfaction if they are purposefully directed and controlled. The capture and storage of accurate, up-to-date, and authentic information on the progressive realization of plans is vital to review progress and to decide management action. In the construction industry, the relationship between the construction processes, which involve the use of resources and the various parts of buildings produced, is not yet a part of a generally accepted structure.

The re-structuring of construction products and processes is a prerequisite in the development of project control systems. The information requirements for project planning and control can be categorized into general company database and decision-support knowledge. The general company database consists of corporate information which represents a model of the project managers environment and includes the following: the products that the company provides; the processes or actions that it performs; the resources that it uses; the sample plans or libraries that can be re-used; and the constraints and standards that must be enforced dictated by company policy. The decision support knowledge can be found within the following three fundamental knowledge-intensive stages in construction project planning and control: initial plan formulation; in-progress monitoring; and plan updating.

The Object-Oriented Data Model for Project Control

The object database model is an active, self-actuating database such that, data is acted upon as soon as data is input. The active nature of the ODBMS permits the system to constantly monitor the results of any analysis and signal the user when changes to the data create a situation that requires immediate attention, without having to make a decision to execute an application program against the database. This feature can be exploited for project performance monitoring and control purposes.

Fundamental to this work is the development of an integrated data or product model that represents the information requirements for total project design and construction (Iosifidis et. al., 1994). Figure 1. depicts a conceptualization of the top level classes of the integrated building product model currently being implemented, represented using the Object Modelling Technique (Rumbaugh et. al., 1991). Due to space restrictions only a selection of the classes have been presented and attributes and methods have been omitted for simplicity and clarity. The scope of the model has been limited to the architecture and structures disciplines as they are closely related to design activity and provide a substantial link with the project planning and control discipline. These disciplines are represented in the model in terms of the space, separation, structural, and the task systems. The space system represents all spaces in a building from a functional viewpoint. Spaces can be categorized as rooms or circulation spaces. A room can be a kitchen, office, lavatory, etc.. A circulation space can be a corridor, liftwell, etc.. Spaces are delimited by a number of separation (enclosures) elements and a separation element can enclose more than one space. Thus, the separation system is represents building components that are used as vertical (walls) or horizontal (floors, roofs) separators. The structural system represents members that provide support for the separation system and distributes both dead and live loading.

The task system represents the information required in the planning and controlling of the production process of the space, separation, and structural systems. The task system is centered around the construction project plan which consists of one or many construction planning activities or tasks. The diagram depicts a project as consisting of an aggregation of tasks. The recursive 'contains aggregation' relationship on task indicates that a task can contain further tasks, nested to an arbitrary depth, the number of potential levels being unlimited. This is very vital to provide the flexibility needed to allow the handling of projects with a deep hierarchical work breakdown structure. A task can be assigned to one or many resource groups. The assignment class associates tasks with resources and captures the number of units of each resource assigned. A resource group consists of one of more resources. A resource can be a plant, material, labour or sub-contractor resource.

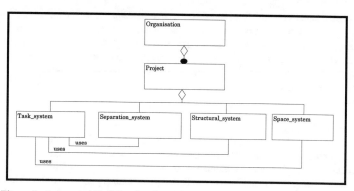

Figure 1. Integrated Building Product Model

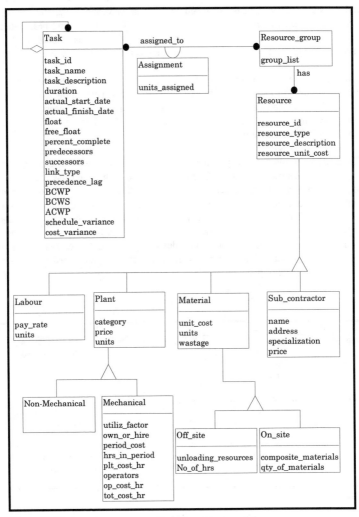

Figure 2. Task system resource allocation object model

The Progress Monitoring and Control Model

Progress monitoring includes schedule, cost, and quality performance monitoring and control of variances. Variances indicate the degree to which dates, resources, and costs differ from the baseline schedule. Current conventional systems are well advanced in terms of providing facilities for identifying critical and near-critical activities, actual and future progress, resource and cost overloads, cost and time variances using a variety of techniques including earned value analysis.

Although, these systems can identify variances they do not have facilities for logging events that may have given rise to the variances, performing diagnoses of the causes to deviations, and recommending corrective control measures. The events leading to performance deviations are to be found in the risks sources inherent within construction project activities.

There is an infinitely large potential number of risk sources that can have an impact on a construction project. The inter-relationships between risk sources and projects is depicted in the object model diagram in Figure 3. A large number of risk sources is being catalogued together with their counter remedial actions as indicated in Table 1. The intention is to develop matrices that map risks and impacts to remedial actions. A risk diagnoses knowledge-base will then use these matrices to relate identified performance deviations and the logged risk events or problems in project activities to risk sources and recommend remedial measures and early warnings of potential problems.

Table 1. A sample of risks, impacts, and remedial actions

Responsibility	Risk source	Potential impact	Remedial action
Employer	Variations, failure to provide site, failure to provide information.	Delays & cost overrun, claims.	Extension of time with recovery of overhead costs.
Contractor	Inadequate labour, inappropriate plant, remedial works,	Delays & cost overrun, low quality work, no compensation.	Increase labour force, acquire appropriate plant, schedule overtime.
Neither party	Strikes, riot, exceptional adverse weather, force majeure.	Delays & cost overrun, damage to works.	Extension of time to defray deduction of liquidated damages.

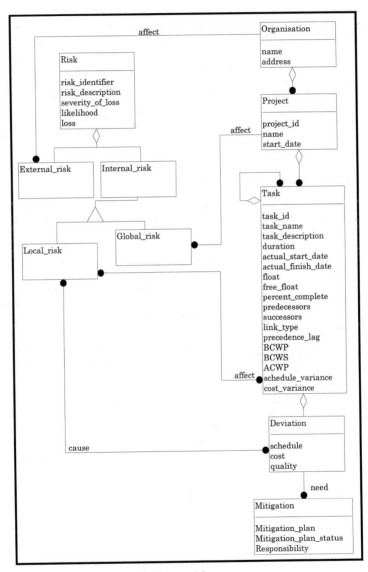

Figure 3. Project risk control object model

Probabilistic methods are currently being used for project risk analysis to represent data uncertainties. However, because of the individual nature of construction projects there is usually insufficient objective data to calculate the probability of occurrence of specific outcomes of risk events. A high degree of subjective judgment is usually required. More advanced systems utilize Monte-Carlo simulation methods (Howes et al., 1993) to attempt to address this problem. Experience in eliciting knowledge for risk analysis from practitioners (Tah et. al., 1993) has indicated that the parameters involved are best described qualitatively in linguistic terms using imprecise non-numeric quantification. The linguistic terms or variables can be translated into mathematical measures using fuzzy sets theory practitioners (Tah et. al., 1993). We have developed object-oriented fuzzy problem solving strategies for quantifying the linguistic matrices for control purposes (Wirba et. al., 1995).

Implementation and Further Work

The object models developed in this paper are being implemented in Visual C++ and the ObjectStore object database management system. ObjectStore provides persistent data storage in C++ in exactly the same way as transient data. It also provides amongst other facilities high productivity C and/or C++ development, complete distributed DBMS services, and support for cooperative group work.

An object-oriented CASE tool is used to develop the object diagrams and to generate C++ classes automatically. The C++ classes are then used by ObjectStore to generate the application schema. The process of generation of schema information is performed by the ObjectStore schema generator, and produces an ObjectStore database, known as the application schema database, as well as an assembly file and some source files. Several techniques are being experimented for populating the database schemas. In addition to direct data entry, project management information from windows-based packages are used to populate the schemas using dynamic data exchange mechanisms supported by visual C++. The inclusion of C++ based knowledge-based systems reasoning mechanisms is currently being investigated. The case for structuring a object-oriented knowledge-based system for project planning and control from a general purpose object-oriented database management system has been stated. This will allow the ODBMS to contain the rules for the subject domain and to use them to evaluate sets of data and reach a conclusion about them. Furthermore, the results of the analysis are retained in the object database for future use. The inheritance hierarchy and encapsulated methods features of the ODBMS concept can be used to construct a self-contained knowledge-base and to implement an object-based reasoning decision process from an object-oriented database.

Conclusions

This paper presents the analysis, and design of an object-oriented database to support a knowledge-based system for project planning and control capable of monitoring and detecting the presence of time, cost, and quality performance deviations within construction activities and suggesting necessary corrective actions. The object model developed captures information on construction resources, their assignment to construction tasks, and the risk sources that can have potentially disastrous impacts on activities. The model is being implemented within an object oriented database management environment with support for active logic as an extension being developed.   The research should produce an object-oriented knowledge-based building tool   which integrates logic-based and object-oriented programming paradigms incorporating fuzzy/approximate reasoning in a C++ environment.   This should establish the basis for the deployment of future knowledge-bases in different problem domains in the construction industry.

Acknowledgments

The support of the EPSRC in providing funding for this work under grant ref: GR/J42175 is gratefully acknowledged.

References

Howes, R., Fong, D., and Little, W. "Dynamic project control utilizing a new approach to project planning". CIB World Building Congress, May 92, Montreal, Canada.

Rumbaugh, J., Blaha, M., Premalani, W., Eddy, F., and Lorensen, W. "Object-oriented modelling and design". Prentice-Hall, Inc., Englewood Cliffs, New Jersey, 1991 .

Tah, J.H.M., Thorpe, A., and McCaffer, R. "Contractor project risks contingency allocation using linguistic approximation". Computing Systems in Engineering, Vol. 4, Nos. 2-3, pp. 281-293, 1993.

Iosifidis P., Tah, J.H.M. and Howes, R. "Utilizing product models for information sharing in an integrated CAD environment". CIB W78 Workshop on Computer Integrated Construction, Helsinki, Finland, August 22 -24, 1994.

Wirba E. N., Tah J. H. M., and Howes, R. "Towards an object-oriented approach to project control. ASCE second Congress on Computing in Civil Engineering, Atlanta, June 1995.

Towards an Object-Oriented Approach to Project Control.

Wirba E. N[1]., Tah J. H. M[2]., and Howes, R[3].

Abstract

The construction industry, in its quest to become more profitable, is increasingly turning to knowledge-based systems (KBSs) to exploit expertise. But present KBSs have largely failed to satisfy construction managers. This is due to a number of reasons: poor knowledge representation which limits cause-effect diagnosis; improper handling of uncertainty; large abstraction in the construction field that necessitates a special system architecture. Here, solutions to these problems are proposed, and an approach that employs object-orientation to represent construction knowledge on a blackboard architecture is explained. The uncertainty in construction knowledge is handled through fuzzy logic.

Introduction

Construction is one of the most important industries in Western economies, however, it is plagued by a multiplicity of problems. The construction industry thrives on investment, but investment is fraught with uncertainty, which leads to risk.. To minimise risk taking, strategies are needed to properly monitor and control construction projects. But this requires experts who are expensive to employ in sufficient numbers, therefore the trend is towards the exploitation of expertise in KBSs to help in project monitoring and control.

Here, shortcomings of KBSs in project control are reviewed, and the use of fuzzy logic to cater for uncertainty and vagueness in the knowledge is explained and demonstrated. A framework is discussed that combines knowledge

---

[1] Research Scholar, CEM, South Bank University, London.

[2] Senior Lecturer, CEM , South Bank University, London

[3] Professor and Head of School, School of Construction Economics and Management (CEM),, South Bank University, Wandsworth Road, London, SW8 2JZ. .

from planning and from site on a blackboard architecture to better manipulate the knowledge in construction management.

The Shortcomings of some Present KBSs in Construction Project Control

KBSs, unlike network-based tools, were envisaged to handle the detailed operations of project monitoring and control better, but present KBSs have not made a big impact in the field of construction project control for these reasons:

Most present KBSs concentrate on identifying the presence of deviations on project plans and not on explaining detailed planning and control operations. This is because most KBSs in construction project control typically manipulate data about projects and not the underlying knowledge (Ashley et al, 1987).

Most [construction project monitoring and control] KBSs are rule based in nature, expressing knowledge in the form: **if** <*condition*> **then** <*action*. Project management knowledge is non-modular in nature, and is severely limited when represented as if-then rules, a representation scheme which was originally designed to represent modular belief (Seiler, 1990).

Most KBSs in construction use probability theory to represent uncertainty, but subjective knowledge, as in construction project control, does not fit in properly with the traditional representational methods of Bayesian probability and certainty factors. Probabilistic methods purport a discreteness about the imprecision in the knowledge source that is not true of real systems. Fuzzy control strategies, unlike probabilistic ones, would best suit real systems since they attempt to remove linguistic barriers between humans, who think in fuzzy terms and machines, which take only exact instructions.

The knowledge and data in construction project control are diverse, and their representation on old expert system shells is limiting. This is due to the late arrival of KBS strategies in the field of construction project management.

The New Approach

The KBS that we are developing is designed so as to improve upon the shortcomings identified above. Our approach employs object-oriented paradigms to design and implement the knowledge system, and envisages easy tracing of causes and effects. The substructure construction has been taken as the area of interest in order to demonstrate the application of fuzzy logic to handle the uncertainty that abounds in this area. The use of object-orientation means that objects, which closely mirror real life entities are used, thus enhancing the performance of the KBS.

The Fuzzy Logic

Construction projects are generally affected by many risk factors resulting in time overruns, cost overruns and quality degradation. The likelihood of encountering such factors and the effect they would have on the project are usually expressed in linguistic form, using linguistic variables such as: *high*, *very low*, etc. Fuzzy set theory can be used to represent such linguistic variables and thus cater for uncertainty in the knowledge base of this system.

Fuzzy sets, unlike ordinary sets in mathematics, accept partial membership. Suppose X is our universe of discourse and comprises a set of elements x's ($\{x\}$) and, A is a subset of X, then $\mu_A(x)$ is a membership function of each element x associated with the subset A. Now, if

$$\mu_A(x) = \begin{cases} 0 & \text{if x does not belong to A} \\ 1 & \text{if x belongs to A} \end{cases} \tag{1}$$

then A is an ordinary set with sharp boundaries. However, if the membership function accepts values between 0 and 1, then A becomes a fuzzy set and no longer has sharp boundaries. A now is represented as a set of ordered pairs :

$$A = [x_1|\mu_A(x_1), x_2|\mu_A(x_2), \cdots, x_N|\mu_A(x_N)] \tag{2}$$

Like ordinary set theory, fuzzy set theory also has operations which can be performed on sets such as union, intersection , and difference, etc. The union $\cup$ of two fuzzy subsets A and B of a universe X has a membership function

$$\mu_{A \cup B}(x) = \max[\mu_A(x), \mu_B(x)] \tag{3}$$

The intersection $\cap$ of A and B has a membership function

$$\mu_{A \cap B}(x) = \min[\mu_A(x), \mu_B(x)] \tag{4}$$

The complement of a fuzzy set A, denoted $\tilde{A}$ will have a membership function

$$\mu_{\tilde{A}}(x) = 1 - \mu_A(x) \tag{5}$$

Application

Fuzzy algorithms have been devised, and a program written to handle fuzzy manipulations explained in this paper. With the present abundance in fuzzy logic-based programs, the success of a fuzzy system will not only depend on how flexible and easy it is to use, but also on how efficiently it uses computer resources, and on how well-suited it is to the problem area it is dealing with. In order to use computer memory economically, the program here has been designed to use linked lists. With linked lists, a fuzzy set can expand or shrink without the prior concern of the user, as of memory allocation. Another advantage brought about by the use of linked lists is that fuzzy sets can be defined to be of any shape, without having to be approximated to triangles or trapeziums, a process which reduces accuracy in many fuzzy systems. As indicated by equations (3) and (4), fuzzy logic is heavily influenced by minimum or maximum values, depending on the computation being carried out. This at first sight would appear to be a great constraint, but it ties in rather well with risk assessment, where the maximum risk value heavily influences the overall risk of a systems. Where the effects of maxima or minima are undesirable, averaging functions such as *meanAND*, *meanOR*, *productAND*, *productOR*, *yagerAND*, etc., as defined by Cox (1994) have been implemented and employed. Flexibility has been enhanced in our design by the use of object-oriented paradigms to design and write general fuzzy classes in C++ that are modular and are easily modified without making wholesale changes to the entire program. The program written handles a wide range of fuzzy logic operations used to perform fuzzy logic algebra on linguistic terms represented in the program. The fuzzy computations involve three major steps: the fuzzification process which converts linguistic variables into fuzzy sets; the calculation of a fuzzy weighted mean; and the defuzzification process which uses the Euclidean distance method (Tah et al 1994) to map the fuzzy weighted mean into a linguistic variable. Schmucker(1984) defines the fuzzy weighted

mean $\bar{W}$ of a sequence of integers R_i and a sequence of integer weights W_i as:

$$\bar{W} = \frac{\sum_{i=1}^{n} W_i \times R_i}{\sum_{i=1}^{n} W_i} \tag{6}$$

To enable the calculation of fuzzy weighted means if R_i and W_i are fuzzy sets, Zadeh's extension principle (Schmucker 1984, Tah et al 1994) is applied to equation (6). Equation (6) uses fuzzy logic addition, multiplication, and division. These are defined as follows:-

$$A + B = \max\{[x + y] | \min(\mu_A(x) + \mu_B(x) \quad 0 \le x, y \le 10)\} \tag{7}$$

$$A * B = \max\{[x * y] | \min(\mu_A(x) + \mu_B(x) \quad 0 \le x, y \le 10)\} \tag{8}$$

$$A \div B = \max\{[x \div y] | \min(\mu_A(x) + \mu_B(x) \quad 0 \le x, y \le 10)\} \tag{9}$$

where A and B are fuzzy sets defined as

$$A = \{x | \mu_A(x) \quad 0 \le x \le 10)\} \tag{10}$$

$$B = \{x | \mu_B(x) \quad 0 \le y \le 10)\} \tag{11}$$

The division operation is only meaningful if $(x \div y)$ is a whole number.

For the case where the project being carried out is a substructure construction, the uncertainty associated with the activities being carried out is high, and therefore contingencies have to be kept in place to cater for risk. The conditions that could affect the progress of a substructure construction include the following: encountering a high water table unexpectedly, therefore necessitating dewatering; a rock could be encountered aand would need to be excaavaated; the weight of the superstructure could make the rough collapse; bad weather could hinder progress of work. If a substructure project was being undertaken, and it was desired to find out the kind of effect the conditions summarised in the diagram below would have on the overall time taken to complete the project, the computation would proceed by converting the linguistic terms - low, medium, and high into fuzzy sets, thus:

$$Low = \left[0|1.0, 1|0.7, 2|0.5, 3|0.2, 4|0.0\right] \tag{12}$$

$$Medium = \left[0|0.2, 1|0.5, 2|1.0, 3|0.05, 4|0.2\right] \tag{13}$$

$$High = \left[0|0.0, 1|0.2, 2|0.5, 3|0.7, 4|1.0\right] \tag{14}$$

For the purpose of this presentation, fuzzy variable are represented over a given short range of numbers, here 0 to 4, to limit the computations required.

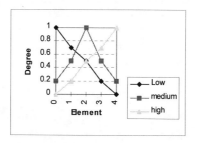

Figure1. Graphical representation of the fuzzy values low, medium, high.

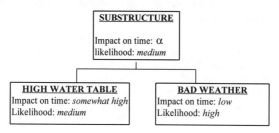

Figure 2. Conditions that could affect a substructure project

Somewhat is a fuzzy hedge function. Given a fuzzy set F;

$$somewhat \ F = F^{\frac{1}{2}}.$$ (15)

If we represent low by A, medium by B, and high by C, then, from equation (6),

$$\alpha = \frac{(C^{\frac{1}{2}} \times B) + (A \times C)}{(B + C)}$$ (16)

From computation,

$$C^{\frac{1}{2}} = \left[0|0.0,1|0.45,2|0.71,3|0.84,4|01.0\right.$$ (17)

$$C^{\frac{1}{2}} \times B = \left[\ 0|0.2,1|0.45,2|0.5,3|0.5,4|0.71,5|0.78,6|0.84,7|0.93,8|1.0,\right.$$
$$9|0.88,10|0.75,11|0.62,12|0.5,13|0.5,14|0.4,15|0.3,16|0.2\left.\right]$$ (18)

$$A \times C = \left[0|1.0,1|0.9,2|0.8,3|0.7,4|0.7,5|0.69,6|0.62,7|0.56,8|0.5,\right.$$
$$9|0.4,10|0.33,11|0.27,12|0.2,13|0.0,14|0.0,15|0.0,16|0.0\left.\right]$$ (19)

$$(C^{\frac{1}{2}} \times B) + (A \times C) = \left[\ 0|0.2,1|0.45,2|0.5,3|0.5,4|0.71,5|0.78,6|0.85,7|0.93,8|1.0,\right.$$
$$9|0.9,10|0.88,11|0.8,12|0.75,13|0.7,14|0.7,15|0.69,16|0.62,$$
$$17|0.62,18|0.56,19|0.5,20|0.5,21|0.5,22|0.4,23|0.4,24|0.33,$$
$$25|0.3,26|0.27,27|0.2,28|0.2,29|0.0,30|0.0,31|0.0,32|0.0\left.\right]$$ (20)

$$B + C = \left[\ 0|0.0,1|0.2,2|0.2,3|0.3,4|0.5,5|0.7,6|1.0,7|0.5,8|0.2\left.\right]\right.$$ (21)

$$\alpha = \left[\ 0|0.23,1|1.0,2|0.88,3|0.81,4|0.59,5|0.59,6|0.59,7|0.59,8|0.39,\right.$$
$$9|0.23,10|0.23,11|0.23,12|0.23,13|0.23,14|0.23,15|0.23,16|0.23,$$
$$17|0.23,18|0.23,19|0.23,20|0.23,21|0.23,22|0.23,23|0.23,24|0.23,$$
$$25|0.23,26|0.23,27|0.23,28|0.23,29|0.0,30|0.0,31|0.0,32|0.0\left.\right]$$ (22)

When defuzzified, α found to be closest to medium, since it gives the lowest Euclidean distance against medium, as shown in the table.

Linguistic term	Low	medium	high
Euclidean distance	1.20	0.71	1.00

The table shows that the impact on the overall time of a substructure construction is a moderate slowing down effect.

The Object-Oriented Design

The object-oriented design was embarked upon with the overriding aim of breaking the whole system down into smaller, less complex components which are resilient to change and are better able to evolve over time, thus lessening the amount of work to be done at a later stage when a change is being carried out on the system. A detailed object design for construction projects has been carried by Tah et al (1995); here only a logical framework of the system having the project and its components of work-package and plan-activities is discussed.

Every project is made up of several work packages, which in turn, are a collection of network activities. This gives an object-oriented design with a one-to-many relationship between project and work package, and between work package and network activity. This relationship helps in generating a work-breakdown structure, and brings a three-fold advantage: project needs are better defined, resources and time required to complete the project are improved, and establishing of a framework for constructing a network plan is made easier.

The Blackboard Architecture(BBA)

Construction is characterised by abstraction at many different levels resulting in an unusually large search space for the knowledge-base of construction project control KBS. The design of the BBA allows systems based on it to reduce large search spaces, and to be developed by several workers at one go. Each level of abstraction in construction project control can be implemented as a separate partition or knowledge source (KS) on a BBA, thus reducing the search space. If several individuals are working on the implementation, each individual would take on a knowledge source and implement it.

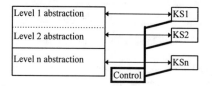

Figure 3. A blackboard architecture

The System Framework

A planning package such as POEM (Harris, 1993) will be used to plan the activities for the substructure. The activities will be scheduled and the plan information stored in databases according to work package, and used for comparison with the actual progress on site to determine deviations in performance or resource usage. The KBS will then suggest action to be taken to counter negative deviations. If the changes are positive, recommendations will be made to either enhance or maintain the positive deviation.

Conclusion

The limitations of KBSs in the area of construction project monitoring and control were reviewed, and solutions to these problems were proposed. A new approach to project monitoring and control that employs object-orientation, a blackboard architecture, and fuzzy logic was outlined. An example that shows how fuzzy logic could be used to cater for uncertainty and vagueness in the knowledge base was given. A framework that links our proposed system to a planning package and a database was explained. Such a systems would point out poor performances in projects and prescribe remedial action for them.

References

Ashley, D.B.; Levitt, R.E. Expert Systems in Construction: Work in Progress. *Journal of Computing in Civil Engineering*, **1**(4), October 1987, p303-11.

Cox, E. The Fuzzy Systems Handbook, Academic Press, New York, 1994.

Harris, E.C. POEM User Manual. Volumes I and II. E.C. Harris Computer Management, Essex, UK,1993.

Schmucker, K.J. Fuzzy set, natural language computations, and risk analysis. Computer Science Press, Rockville, Maryland, 1984.

Seiler, R.K. Reasoning about Uncertainty in Certain Expert Systems: Implications for Project Management Applications. *Project Management*, **8**(1), February 1990, p51-59

Tah J.H.M.; Thorpe A.; McCaffer R. Contractor Project Risks Contingency Allocation Using Linguistic Approximation. *Computing Systems in Civil Engineering,***4***(2-3),* 1993, p 281-293.

Tah, J.H.M.; Howes, R. Towards Object-oriented Database for Project Control. *1995,* (Also being presented at this conference).

OBJECT–ORIENTED MATERIALS MANAGEMENT SYSTEMS

By Hazem M. Elzarka,[1] Member, ASCE, and Lansford C. Bell,[2] Fellow, ASCE

ABSTRACT

The traditional capabilities of materials management computer systems (MMS) can be greatly enhanced through the use of object oriented methodology (OOM). Whereas the traditional MMS does link the various functions of materials management, there is also a need to integrate the MMS with design and scheduling systems. At present, most MMS are based on a relational database structure, and therefore are not easily integrated with external computer systems that are used for design and scheduling. The typical MMS could also be enhanced by incorporating knowledge–based systems to aid materials managers with decisions pertaining to purchase order consolidation, and commodity code generation. This paper briefly describes OOM, its application to MMS, and a prototype system that has been developed by the authors and reviewed by industry professionals.

INTRODUCTION

Properly designed materials management computer systems (MMS) can produce significant savings in construction project costs. Savings are associated with improved labor productivity, reductions in bulk materials surplus, cash flow savings, and reduction in required warehouse space (Bell and Stukhart 1987). Recognizing the benefits of materials management systems, many construction firms have invested in the development of such systems. Early development efforts focused on integrating the various functions of materials management (i.e. takeoff, requisition, purchasing, expediting, receiving, warehousing and distribution). Recently however, construction companies are realizing the importance of integrating their MMS with external computerized systems that perform tasks related to design and scheduling. Linking the MMS to design systems provides a mechanism for rapidly responding to design changes, which in turn has an impact on potential materials surplus. Linking the MMS to scheduling systems can effectively prioritize procurement activities and reduce

---

[1]Visit. Assist. Prof., Civil Eng. Dept., Clemson University, Clemson, SC 29634–0911.
[2]Prof., Civil Eng. Dept., Clemson University, Clemson, SC 29634–0911.

project durations. Also important is the integration of the MMS database with expert systems. Such integration allows the expert system to automatically retrieve and insert data from and into the database.

The relational database model has been utilized for many of the early MMS development efforts. The relational data model, although simple and easy to use, has some limitations. It does not meet all the data modeling and data manipulation requirements of engineering applications. Tables that are used in the relational model to structure data do not adequately represent real world entities. A concept that is frequently referred to as poor semantic expressiveness. The poor semantic expressiveness and the rigidity of the relational model makes upgrading and integrating the MMS with other engineering and construction computer systems a difficult task that requires a complete system redevelopment. The relational model has also complicated the integration of knowledge–based systems with the MMS because knowledge–based systems use modeling techniques that are different from the relational model. These techniques include semantic networks, frames, and production rules (Elzarka 1994).

From the above discussion it is clear that a more flexible modeling methodology is needed to allow for updates and modifications of the MMS in a modular and evolutionary fashion. The object–oriented methodology was selected because of its powerful data and knowledge representation and its well documented features of abstraction, encapsulation and inheritance. These features support system modularity and modifiability.

RESEARCH OBJECTIVES AND METHODOLOGY

The objective of the research described herein was to illustrate how the concepts of the object–oriented methodology (OOM) can be utilized to develop flexible MMS that can be integrated with design and scheduling systems and expanded to incorporate knowledge based systems.

To illustrate the advantages associated with the object oriented approach to MMS integration, a model for an object–oriented piping MMS was developed. The model was used to develop a prototype object–oriented MMS in a tutorial format using the C++ language. Design, scheduling, and knowledge–based integration applications are illustrated in the tutorial. Specific attributes of the tutorial described in this paper include automatic commodity code generation, automatic takeoff execution, intelligent purchase order consolidation, and schedule integration. The tutorial has been evaluated by a number of industry professionals who are familiar with construction materials management and data processing. Tutorial attributes judged to be of the most value by these industry professionals are also discussed.

OBJECT ORIENTED PROGRAMMING METHODOLOGY

A comprehensive description of the object oriented methodology may be found in several references (Cox 1986, Coad and Yourdon 1990). In the object–oriented methodology, all entities are modeled as *objects*. An *object* is a self–contained entity composed of data (*attributes*) and procedures (*methods*). The data describes the object's state while the methods describe its behavior. Methods and attributes are private to the object. Communication among *objects* is achieved by sending *messages* to one another. These *messages* invoke one or more *methods* inside the receiving *object*. *Methods* can respond to *messages* by updating local *attributes*, displaying *attributes'* values or sending *messages* to other *objects*. *Objects* that have the same definition of *attributes* and *methods* are classified under the same *class*.

The components of the object–oriented methodology (i.e. objects, classes, communication with messages) lead to some unique features that make this methodology one of the most powerful tools for software development. These features include abstraction, encapsulation, and inheritance. Abstraction is the process of deliberately omitting certain details about an object during its definition. These details are suppressed in order to focus attention on the essential characteristics of the object relative to the problem domain. Data abstraction is used in data modeling to reduce data complexity and to aid in understanding the essential properties of an object. Encapsulation is the concept of hiding the object's attributes and methods behind the message interface. Encapsulation supports system modifiability. An object design can be totally changed without affecting the remainder of the program. Encapsulation also supports modularity. New objects can be added without disturbing an existing system. Inheritance allows new classes to be defined from existing classes, thereby supporting code reusability. New classes can only be specified by their difference from an existing class, rather than having to be totally redefined.

OBJECT ORIENTED PIPING MMS MODEL

Benefits of the OOM can only be achieved if considerable effort is spent in developing a sound data model (Kim and Ibbs 1992). In modeling, elements of the system are represented as a group of objects, the attributes and methods of all objects are defined, and the relationships between objects are established. Figure 1 which utilizes the graphical notation introduced by Coad and Yourdon (1990) illustrates the proposed piping MMS model. In this notation, a class is graphically represented by a rectangle with rounded edges displaying the class name. As illustrated in Figure 1, several classes have been defined. A *component* class was needed to capture general information about piping components. As industrial projects include many types of piping components (i.e. pipes, fittings, flanges, gaskets, valves, etc.), these could directly translate into new classes. However, from a piping materials management point of view, only 2 classes are needed: a *pipe* class and a *fitting* class. The major difference between these two classes from the MMS point of view is the unit of measure. Pipe objects are measured in unit of length, whereas fittings are quantified by their total number.

Each piping component is part of a pipe line. The pipe line belongs to one system and is defined in one isometric. All components of the pipe line have the same material specification. A *Line* class was used to capture the pipe line information.

A piping system is composed of different pipe lines and serves a specific engineering purpose (i.e. heating, ventilation, etc.). A class *System* was introduced to capture each system's information. Each piping system is defined using several piping and instrumentation diagrams (p&ids). The p&ids show the different equipment needed in the process, piping to and from all units of engineered equipment, and the arrangement of the valves and instrumentation needed for operating the equipment. As the design of the piping system progresses, the p&id is used to develop isometrics with actual pipe dimensions from which the takeoff can be performed. Two classes *isometric* and *p&id* are used to capture information pertaining to p&ids, and isometrics. Since classes *isometric* and *p&id* have some commonalities (e.g. id, responsibility, due date, etc.) a class *drawing* is introduced to capture these commonalities.

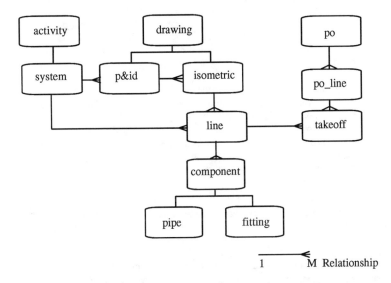

Figure 1. Object model for piping materials management system.

A class *takeoff* is introduced to record all important takeoff attributes and to contain the required takeoff methods. Once the takeoff information has been prepared, purchase orders (PO) can be generated. A *po* class is therefore defined. Another class, *po_line,* is introduced to capture all purchase order line item data. Finally, the class *activity* was introduced to enable the integration of the MMS with the scheduling. Some attributes of this class would be the start and finish date of the activity.

Figure 1 also shows class relationships. According to Coad and Yourdon (1990), class relationships can be of two types: classification relationships and assembly relationships. An example of a classification relationship is a pipe and a fitting as special piping components with some commonalities and some unique characteristics. In class *component* the common attributes and common methods of both *Pipe,* and *Fitting* are captured. In the object–oriented methodology, inheritance provides the mechanism for defining classification relationships.

The second type of relationships is assembly relationships. The assembly relationship can be one–to–one (1:1), one–to–many (1:M) or many–to–many (N:M). This relationship is graphically represented, as shown in Figure 1, by a straight line connecting two classes. The class that is connected to the straight line by several small lines represents the M side of the relationship whereas the class that is only connected by one line represents the 1 side of the relationship. In Figure 1, for example, the relationship between the *line* class and the *component* class is one to many. A pipe line contains many piping components, while a component belongs to only one line.

Another classification relationship exists between the *drawing, isometric* and *p&id* classes. The *activity* and *system* classes form a one–to–one assembly relationship where each activity is associated with only one system. *System* and *p&id* form a one–to–many (1 to M) assembly relationship where each system may be defined by many p&ids (at least one) while each p&id defines only one system. Similarly, the relationships between *p&id*, and *isometric*, and between *isometric* and *line* are one to many.

The quantity takeoff is executed at the pipeline level. Materials that have the same commodity codes and that belong to the same pipe line are consolidated into one takeoff object (takeoff line). Each pipe line can have many takeoff lines, as indicated by the assembly relationship between *line* and *takeoff* (Figure 1). Several takeoff lines are consolidated into one purchase order line during the purchase order generation process, which explains the one–to–many relationship between *takeoff* and *po_line*. Finally, many PO line items are grouped together under one PO.

OBJECT ORIENTED MMS TUTORIAL

The model described above was coded in a user–friendly tutorial format. Figure 2 illustrates the different modules and components of the tutorial and how they are related.

As is shown in Figure 2, the takeoff file is the central component of the system. All other modules are connected to the takeoff file.

Figure 2 also illustrates the two design interface features of the prototype. The first feature automatically executes the takeoff when new components are added, while the second feature updates the takeoff file when design changes are made.

When components are added, several database integrity checks are performed. Ensuring the integrity of the database is very important in engineering applications. Engineering databases contain data items that are highly interrelated and subject to changes and modifications throughout the project's history. For example, when asked about the pipe line number of a component, the user is not allowed to enter a line number that does not exist in the line database (referential integrity). After passing all the integrity checks, the commodity code of the component is calculated based on the values of its different attributes (i.e., size, line #, type, etc.).

The automatic generation of commodity codes is of extreme importance. A commodity code is a part number used to specify and purchase a commodity. Thousands of unique piping entities are incorporated into a typical industrial facility. Determining the correct commodity codes of these products requires expert knowledge. The ability of an MMS to automatically generate commodity codes will reduce the time utilized by management personnel in assigning these codes, allowing them to focus on other aspects of the materials management process.

After the commodity code generation, the prototype tutorial automatically adds the new component to the takeoff file. The automatic takeoff execution produces faster and more accurate takeoffs and can only be implemented if the MMS is integrated to the design system. Some of the companies contacted as part of this research effort have integrated their MMS and design computer systems. The main integration feature utilized by these companies systems is the automatic takeoff execution. The degree of the takeoff automation however is different from what is proposed in the prototype tutorial. Most existing MMS do not, for example, generate commodity codes. As a result, the commodity codes of all the piping components included in the design model have to be manually computed before executing the automatic takeoff.

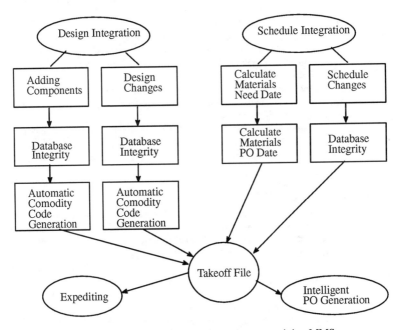

Figure 2 Modules and components of the prototype piping MMS.

As noted above, the prototype tutorial has another design integration feature that automatically updates the takeoff file in case a design change is made. This feature is not incorporated into most current integrated MMS. This feature is very important because it ensures that updates of design data are rapidly reflected in the takeoff file and eliminates problems related to material availability and surplus. Any design change (i.e., changing piping attributes, line material specification, etc.) modifies the commodity code of the component. The new commodity code is first generated as shown in Figure 2, and the takeoff file is then updated.

The scheduling interface of the tutorial also has two main features (Figure 2). The first feature uses schedule data (i.e., activities start dates and finish dates) to calculate the

COMPUTING IN CIVIL ENGINEERING

need dates of the materials. The materials need dates are calculated based on the start dates of their lines' activities. After calculating the materials need dates, their purchase order dates are calculated. The purchase order date, the date when the materials need to be purchased, is calculated by subtracting the lead time from the need date. Material lead time depends on several factors such as material specification and project location. Material lead times are calculated by the prototype using expert rules.

The second feature of the scheduling interface updates the takeoff file in case of a schedule change. This feature ensures that the need dates listed in the takeoff file always reflect the latest schedule. Materials affected by the schedule change will thus be available on site when needed.

The prototype performs other functions including intelligent purchase order generation and automatic generation of expediting reports. These functions are automatically executed by the prototype based on intelligent modules that constantly monitor the data stored in the takeoff file. For example, the "intelligent purchase order generation process module" periodically compares the purchase order dates of all takeoff lines to the current system date and consolidates those takeoff lines whose purchase order date is equal to the system date. This process ensures that purchase order dates are issued at optimum intervals.

Expediting reports are generated if a design or a schedule change modifies purchase order dates. These reports are automatically generated by the prototype and are saved in a text file. The user can browse or print this file at the end of the consultation.

TUTORIAL EVALUATION

The prototype tutorial was reviewed in detail by three industry professionals, all of whom are familiar with current MMS technology. The feedback from the industry reviewers indicated that the object–oriented methodology has potential when applied to the development of integrated materials management systems. The OOM not only provides powerful data and knowledge modeling capabilities that simplify the integration of MMS with design and scheduling, but it also allows for code reusability which can greatly reduce the cost of software development and permits the development of rapid prototypes that can be examined by potential users to identify necessary system modifications early in the development process. The ability of the OOM to encapsulate both data and knowledge in the "object" also simplifies software maintenance, and reduces maintenance costs which constitute a large portion of total data processing costs.

As to the values of the capabilities and linking interfaces of the tutorial, the reviewers agreed that the integration of MMS with design systems saves time and man–hours, and produces more accurate and reliable takeoff. Integration with design systems also allows design changes to be electronically communicated between project participants. Electronic communication becomes more important if the project is divided between offices that are thousands of miles away.

The reviewers indicated that the integration with scheduling systems is necessary to perform just–in–time (JIT) scheduling of procurement which dramatically improves cash flow. The reviewers stated that improving cash flow is especially important when purchasing major engineering equipment since the cost of major installed equipment is high compared to bulk materials. JIT scheduling of procurement also eliminates double handling and reduces the required warehouse space. Traditionally materials are handled twice, first as they are received in the field and stored in the warehouse, and second as they are issued to the crafts. In JIT procurement, materials are received when needed. The materials are checked for accuracy and forwarded directly to the work area.

The reviewers also agreed that expert systems technology has potential to improve the efficiency of the materials management process. They recommended that expert systems applications be developed to automate the most labor intensive and knowledge intensive processes. Good candidate expert systems can be developed to automatically generate commodity codes, compute material lead time, and select vendors. Overall the reviewers believe that the concepts illustrated in the tutorial can improve the efficiency of existing materials management computer systems. Some companies recently contacted by the authors are in the process of incorporating these concepts into their systems.

CONCLUSIONS

Current materials management systems can be enhanced by creating interface links to design and scheduling systems and by adding knowledge–based modules that can further automate the materials management process. This paper discussed several capabilities that can enhance the MMS including automatic commodity code generation, intelligent purchase order generation, expediting assistance, and integration to design and scheduling. To effectively integrate an MMS with design and scheduling, it is important to use an appropriate data modeling technique before beginning the programming effort. The object–oriented model is well suited to MMS development because of its well documented features of abstraction, encapsulation and inheritance. These features support modularity, modifiability, and reusability. The object–oriented model also provides many modeling capabilities needed to develop expert systems.

REFERENCES

Bell, L. C., and G. Stukhart (1987). "Costs and Benefits of Materials Management Systems," *Journal of Construction Engineering and Management*, ASCE, Vol. 113, No.2, June, pp. 222–234.

Coad, P., and E. Yourdon (1990). *Object–Oriented Analysis*. Prentice Hall, New Jersey.

Cox, B. J. (1986). *Object–Oriented Programming: an Evolutionary Approach*. Massachusetts: Addison–Wesley Publishing Company.

Elzarka, H. M., (1994). "Object–Oriented Methodology Applied to Materials Management Systems". Ph. D. Dissertation, Clemson University, Clemson, SC.

Kim, J. , and C. W. Ibbs (1992). "Comparing Object–Oriented and Relational Data Models for Project Control." *Journal of Computing in Civil Engineering*, ASCE, Vol.6, No. 3, July, pp. 348–369.

Proactive Object-Oriented Project Control System

Feniosky Peña-Mora,[1] Associate Member,
Robert Logcher,[2] Fellow, ASCE and Naoto Mine[3]

Abstract

This paper presents a broader, object-oriented, and knowledge based view of project data and its use in project planning and control. It is shown how data can be used more efficiently and in broader way than the typical single use for which it was collected. In particular, a system is provided which interprets a database of project plans and accomplishments to date, assesses project status, and presents feedback of problems by using object-oriented modeling and knowledge based interpretation of construction field data. By so doing, such a system improves forecasts of outcomes for the remainder of the project and improves planning for future projects.

1 Introduction

Efficiency and effectiveness of information processing on a construction project is very important to firms operating in the construction industry. While very large amounts of routine data are collected to document activities and characteristics of the project, to control financial transactions, and to measure the performance of people, equipment, and construction methods, much of this data is only used for a single task. For example, data collected for one reason, such as schedule evaluation, is usually not used for other purposes. This single use causes duplication of information. Opportunities for efficient use of formally collected information are often missed.

An initial effort to develop a new style of project control system has focused on a small area of project management. It has taken a well defined planning process, the Multi-Activity Chart (MAC), useful for optimizing resources in repetitive tasks [Matsumoto et al., 1986]. This method was developed by the Shimizu Corporation in its attempts to improve productivity of building construction. Shimizu researchers showed that the manufacturing industry realizes high productivity by repetitive work

[1] Assistant Professor, MIT Room 1-253, Intelligent Engineering Systems Laboratory, Department of Civil and Environmental Engineering, Massachusetts Institute of Technology, Cambridge, MA 02139

[2] Professor, MIT Room 1-253, Intelligent Engineering Systems Laboratory, Department of Civil and Environmental Engineering, Massachusetts Institute of Technology, Cambridge, MA 02139

[3] Senior Research Engineer, Construction Engineering Research Department, Institute of Technology, Shimizu Corporation, 4-17, Etchujima 3-Chome, Koto-Ku, Tokyo 135, JAPAN

using workers at fixed facilities with materials going through the facilities. To adapt this concept to building construction, where the construction work must take place at the final location of the materials because of their bulk and their permanence, one may reverse the flow and have the labor and equipment flowing by the final materials locations. The construction site is divided into locations which require similar work and the flow of work is planned through the locations so that the labor and equipment resources are used as efficiently as possible. In addition, "In order to bring construction near to industrial system, not only repetitive work but also a fixed members of workers and standardizing crew size is required" [Matsumoto et al., 1986]. An illustrative MAC, which takes in consideration all these factors,is shown in Figure 1 (adapted from [Matsumoto et al., 1986]).

Team	Crane	Carpenter			Steel Man	Helper
		Wall Form	Shoring Form	SD & ALW		
Num./Day	1	9	4	2	4	
1	Outside Big Wall Panel Setting / Inside Wall / SM Shoring	Outside Big Wall Panel Setting $n{\rightarrow}n{+}1$ / Inside Wall $n{\rightarrow}n{+}1$	SM Shoring	Flame Lumber Stripping n	Wall Re-bar Preassembly	Marking Carpenter (2)
2	Veranda / SD Deliver / BU Deliver		Veranda	Preparation / Delivering	$n{+}2$	$n{+}2$
3	PC Panel Setting / Wall Re-bar Erection	Inside Wall Panel Erection	Corridor & Veranda Form Erection $n{+}1$	SD, ALW Setting	PC Panel Setting n / Wall Re-bar Erection $n{+}2$	
4	(For C-part) $n{+}1$		n	Flame Lumber Assembling	Wall Re-bar Spacing	Slab Re-bar Steel Man (2)
5	Scaffolding / Wall Concrete Pouring $n{+}1$	Scaffolding / Wall Concrete Pouring $n{+}1$	Slab End Form & Misc. Form n		$n{+}2$ / Handrail $n{+}2$	n
6	Slab Concrete Pouring	Wall Form Stripping $n{+}1$			Wall Re-bar Preassembly $n{+}3$	Slab Concrete Pouring Mason (10) n

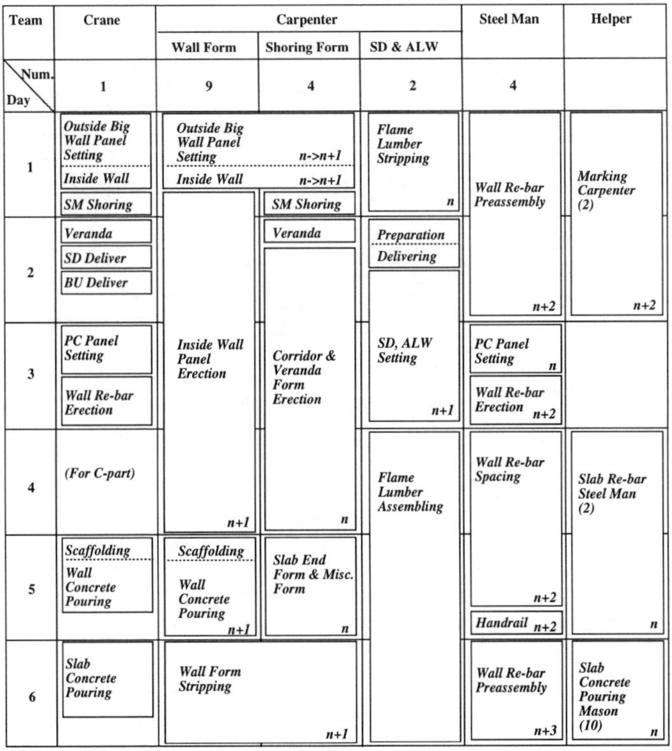

Figure 1: Multi-Activity Chart Illustration (adapted from [Matsumoto et. al., 1986]).

2   Multi-Activity Chart Object-Oriented Representation

This research developed an object structure for representing a MAC (see Figure 2). The planning part of this structure was then implemented using a knowledge representation system. The need for database interfacing became readily apparent. For each cycle, represented by cycle-characteristics, there are numerous cycle-activities to be carried out at each location. Similarly, there are numerous areas where the activities are performed. One can see that the number of location-activity objects expands rapidly and, if stored in the object space, would hinder performance of the system. Such information should be stored in a database until needed.

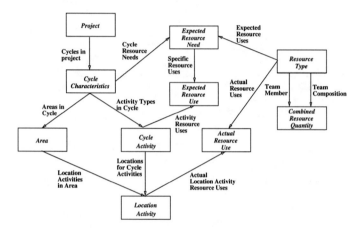

Figure 2: Class Structure for the Multi-Activity Chart.

In object-oriented programming, the principal elements are classes and instance objects. Classes are objects that can transmit all their properties to their progeny (instance objects created under the class). In a hierarchy, classes are higher than instance objects; they are like parents, transmitting to their progeny all their genetic information. Instance objects cannot pass their properties to other objects. The classes and the objects can be manipulated with message passing (methods). The methods are functions that make a specific class or object perform certain tasks.

In the program developed for the MAC, ten classes have been defined. They are the generic objects. Instance objects in unlimited number may be made from the generic objects and can be modified. The instance objects as well as generic objects have methods that allow them to create instance objects. However, an object created from an instance object cannot inherit the properties of its creator.

A method has been defined for the project class to make an instance object and input the project characteristics. The instance project can then execute a method to create a cycle-characteristics instance object. This object, in turn, can create area and cycle-activity objects. These will automatically create location-activity objects. The parent maintains three items: a list, the child, the name of the parent. These items extend throughout the network data structure.

The generic objects in this knowledge base receive the name of the object class itself. For example, generic project receives the name "project". The instance objects receive names selected by the user. For example, an instance of cycle-activity can be assigned the name of "frame_lumber_stripping". Each generic object has at least the number of slots that correspond to the record contents for the database developed by Logcher and Phelan ([Logcher and Phelan, 1989]).

In addition to the slots, the generic objects stores their relationship to other objects in a network structure. Three kinds of relationships are stored in each generic object: inheritance from a generic object or class, a link with a creator, and any other relationship link, called here "informational" (see Figure 3). These links are established when an instance object is created.

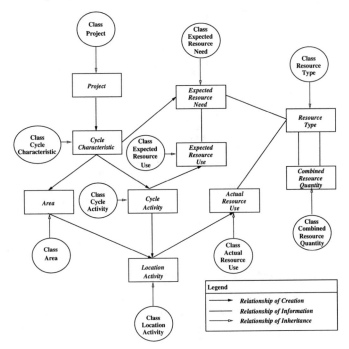

Figure 3: Relationships Among Generic and Instance Objects.

3 Object-Oriented Project Control System

The architecture of the object-oriented project control system is distinctive (see Figure 4). It consists of an Object-Oriented Database Management System (OODBMS), an Information Manager, a Context, Processes, Operators, a Knowledge Acquisition, and a User Interface modules. It incorporates the basic structure of an expert system [Firebaugh, 1989] and [Hendrikson and Au, 1989]. The database contains data both from an object model that is a result of the planning process, and from field reports or decisions that are results of the actual performance on the project. The Information Manager Module selectively stores project plans and status in and retrieves them from the OODBMS. This module limits the number of objects maintained at any time in the reasoning system. It stores in the OODBMS the large number of instance objects, and retrieves them only when they are required for analysis and/or update.

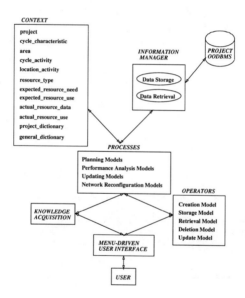

Figure 4: Architecture of an Object-Oriented Project Control System.

The Context Module in the knowledge representation system contains information on the particular project being considered, including the different balanced cycles (Line of Balance -LOB- [Mooder et al., 1983]) which compose the project, as well as each of the cycle components. These cycles have activities which are instantiated for each location in the project. These instantiations are "location activities" (a cycle activity at a particular location) with inherited information, such as resource needs, from their parent cycle. In the Context Module, information is stored in a series of hierarchically organized objects. Each object is linked to parent or child objects from which information can be retrieved. Objects are named and contain various data slots. These objects are organized to represent the current project plan, decisions made during planning, updating, and/or analysis processes. For more information about the information manager module refer to [Logcher and Phelan, 1989] and [Logcher and Peña, 1990].

The Process Module alters the system Context by creating objects or modifying their attributes by using facts and/or rules. The Module consists of the planning model, the performance analysis model, the updating model, and the network reconfiguration model. The planning model implements an object oriented representation of the network. It creates the planned schedule using activity on nodes. The performance analysis model operates on any new data added to or changed in the database since the last analysis, and works interactively with the user to determine the causes of any variance between planned and actual performances. The updating model takes recommendations of the performance analysis model and from the user to alter the planned schedule based on the actual performance. Finally, the network reconfiguration model converts the data supplied by the information manager from the OODBMS into the hierarchical structure needed in the knowledge representation system.

The Operator Module contains operators that create, delete, store, retrieve, and update the information stored or represented in the Context. The Menu Driven Interface Module controls with user commands the execution of the operators. The Knowledge Acquisition Module modifies the contents of the knowledge base requested by the user. This module provides a flexible and modifiable knowledge base, allowing the entry of new rules and procedures during run time and when all the rules in the knowledge base have failed to recognize the cause of the variance between planned and actual performance.

Focusing on the changes in a plan, the system makes use of the performance analysis model, the updating model, and the information manager module while working interactively with the user to update the project schedule and to determine the causes of any variance between planned and actual performances. In the update model the user sends a message to a cycle to update itself. The cycle asks the user for the name of the file which contains new field reports of actual use of the crew, equipment, and/or material. In this report, resources and the number of hours are specified for work during regular hours and over time hours. Then the cycle sends a request to the information manager to take the report file and select the resource use for the specific cycle being reviewed for a selected period. After the resources has been selected, the cycle sends a request to the information manager to load the location activities that are correlated with the actual resource data on the field report.

Once the location activities are in core, the cycle creates a record of actual resource use for each resource data being used. When an actual resource use is created, the actual start of the location activity is set. The system also sets the actual productivity of the resource to the planned productivity of the resource unless there is information that indicates that assumption is possibly incorrect.

If the actual resource use object for the resource given in the file report exists, it is updated. If the report supplies a percentage use of the resource for that location activity, the actual productivity of the resource is calculated, thus using a more realistic measure of performance. Thereafter, the planned and inferred percentage completed are calculated. First, the planned percent is calculated. This is the percentage of work that should have been finished by that time given the estimated conditions. Second, the inferred percentage completed is calculated based on the crew productivity and total hours worked (regular hours and over time hours) during this period.

After each actual resource use object has been created or updated for the period of revision, the location activities are updated. First, the location activity determines its duration. This is achieved by taking the expected resource uses for the cycle activity which is its parent, and analyzing the duration over which these resources are going to be used. The durations are calculated based on the resource used and the number of hours that this resource has been in use. If a percentage completed was reported, the program will update the duration as well as the start and finish times. If a percentage completion figure is not reported, the previous actual duration, actual start time and actual finish time are left as previously reported. This is done for the actual performance on the location activity since no new information from the field is reported to change the existent information. However, the information from the field reports always affects the inferred time. The inferred records give the user the schedule inferred based on reports. These reports take into consideration the actual field performance. If a percentage completion figure is given in a period report, the actual and inferred records will be equal since all the relevant data was given. On the other hand, if no percent complete is given, the computer speculates based on the known information about the state of the location activity. The inferred start of the location activity is calculated based on the inferred productivity of the crew and the amount of work left to be done.

Each resource that is needed for a location activity is studied and its contribution is analyzed. After all the input from the resources are analyzed by the system, the least favorable scenario is considered for setting the duration of the location activity. Then the location activity determines the actual and inferred start and finish times and assesses its own status.

This information will identify location activities that need further study. If further study is required, the system initiate the inference engine to analyze the causes of the difference between planned and actual performances as well as to suggest ways to recover any delay. The system will be able to delete all the location activities that do not need further analysis. Thus, memory is freed for performance analysis and schedule recovery. For more information on schedule recovery see [Peña Mora et al., 1994].

The updating model and performance analysis model in the processes module have not been fully implemented. They operate only in the location activity and the location network but do not consider the affect of each location on the rest of the project.

4   Conclusion

The next step in this research will take into consideration any variance from the planned schedule to improve re-allocation of resources and to change the schedule. Currently, the system is capable of automating the generation of the planning schedule. The system can produce a report of the schedule of any cycle activity. Additionally, the database communications problems have been solved and after their creation cycle activities can be stored in the database. To infer progress and problems with the construction work, these cycle activities can also be read as needed to apply knowledge to actual resource data.

References

[Firebaugh, 1989] Firebaugh, M. (1989). *Artificial Intelligence: A Knowledge-Based Approach*. PWS-Kent Publishing Co., Boston, MA.

[Hendrikson and Au, 1989] Hendrikson, C. and Au, T. (1989). *Project Management For Construction*. Prentice-Hall, Inc., Englewood Cliffs, NJ.

[Logcher and Peña, 1990] Logcher, R. and Peña, F. (1990). Phase I: Report on Cooperative Research on Construction Production, Monitoring, and Control. IESL Technical Report, MIT, Cambridge, MA.

[Logcher and Phelan, 1989] Logcher, R. and Phelan, S. (1989). Initial Report of Research on New Forms of Construction Productivity, Monitoring, and Control. IESL Technical Report, MIT, Cambridge, MA.

[Matsumoto et al., 1986] Matsumoto, S., Mine, N., and Uchiyama, Y. (1986). Work Scheduling and Management Method for Building Construction. Shimizu Technical Research Bulletin No. 5, Shimizu Corporation, Tokyo, Japan.

[Mooder et al., 1983] Mooder, J., Phillips, C., and Davis, E. (1983). *Project Management With CPM, PERT, and Precedence Diagramming*. Van Nostrand ReinHold, Co., Inc., New York, NY.

[Peña Mora et al., 1994] Peña Mora, F., Logcher, R., and McManus, T. (1994). SCHEREC: SCHedule RECovery system. In *Proceedings of the 1st ASCE Congress on Computing in Civil Engineering*. ASCE.

An Inductive Diagnostic Knowledge System for Rebar Corrosion

Hani G. Melhem[1] A.M., ASCE and Srinath Nagaraja[2]

Abstract

Corrosion of reinforcing steel in concrete structures exposed to chloride has become a major problem for bridge and structural engineers. In artificial intelligence, several inductive learning techniques supported by well formulated theories and algorithms have been developed over the years. An inductive knowledge base system for rebar corrosion is being developed in which the C4.5 programs have been chosen to perform the induction. The necessary attributes have been identified to support the knowledge base structuring. They are broadly classified as material factors, structural factors, and environmental factors. Data supporting these attributes form the basis for pattern recognition in the machine learning process. In this paper, the supporting inductive programs are reviewed and discussed, and the scope, purpose, architecture and development of the Rebar Corrosion system are presented.

Introduction

The Nation's civil infrastructure is seriously affected by the problem of concrete rebar corrosion. Bridges are a major part of the infrastructure and are particularly affected by this problem. The cost of repairing deteriorating bridges is one of the biggest challenges facing highway agencies, and is increasing very rapidly. The deicing salts penetrate the concrete and corrode the reinforcing steel resulting eventually in internal cracking and surface spalling of the concrete. This type of deterioration can occur on all concrete bridge components including decks, superstructure elements, and substructure elements.

One of the most widely used approaches of learning is deriving expertise knowledge from examples through the process of inductive learning. Inductive learning

---

[1]Assistant Professor, Dept. of Civil Engng., Kansas State University, Manhattan, KS 66506.

[2]Ph. D. Candidate, Dept. of Civil Engng., Kansas State University.

has been found to be very useful in several engineering domains. Two methods are commonly used in learning systems: *divide-and-conquer* or *covering*. In this investigation C4.5 programs (Quinlan, 1993) have been chosen to perform the induction. These programs have their ancestry tracing to ID3 (Quinlan, J. R.,1986) and AQ (Michalski, R.S., et al., 1983). There is a large amount of data pertaining to corrosion of steel in concrete.

This data is a result of experimental research in several countries (Tuutti, 1982). The data contains a number of variable dependencies and cannot be processed manually. After investigating the different machine learning approaches *induction from examples* was found to be the most appropriate method of obtaining rules from the available data. The inductive algorithm is the method which the computer program uses to induce rules from the training data. A summary of the various inductive learning techniques and their application to civil engineering has been done by the authors (Melhem and Nagaraja, 1995)

Purpose and Scope of System

The first immediate impact of this research is the validation and experimentation of the inductive learning techniques available, by means of building a machine learning expert system in the application domain of rebar corrosion. The application of the theory with actual engineering data from the field of rebar corrosion will strengthen many assumptions and solidify the applicability of inductive learning techniques. As a result of this research, new service life models of corrosion are explored and innovative machine learning techniques are implemented. The direct application of the inductive learning concepts readily available for a few years and awaiting for proper implementation will enhance the state of knowledge and the technical validity of artificial intelligence research.

The objectives of this research are to (1) develop new inductive learning based models for serviceability life of corroded structures, (2) validate existing models and their application to corrosion damage, and (3) incorporate the most reliable and most adequate models in the knowledge-based system. This knowledge-based system will be a comprehensive system with two main modules. It determines the degree of corrosion, and type of remedial action in the first, and serviceability life as well as life-cycle cost of the structure in the second.

From a general viewpoint, this research will lead to valuable long-term benefits to the state of knowledge and the engineering activity as well as to the different highway agencies and engineering firms. When adequately developed, expert systems such as the one presented here, can help identify, enhance, organize, and certainly save the knowledge of current professionals in the area of structural concrete engineering, and transfer their expertise to the less knowledgable person in this field. Special attention is given to the service life and strength considerations in concrete due to corrosion. Life-cycle and repair costs can be reduced by a judicious understanding of the extent of the problem and assessment of situation.

C4.5 Machine Learning Algorithm

The most important algorithm used in the C4.5 programs (Quinlan, 1993) consists of generating an initial decision tree from a set of training cases. The evaluating criteria for the ID3 algorithm (which is the forerunner of C4.5) is called *gain* and will be defined later. The information theory is used as the basis of this criterion and is given as follows: "The information conveyed by a message depends on its probability and can be measured in bits as minus the logarithm to the base 2 of the probability." Let S be a set of cases, and $freq(C_i,S)$ stands for the number of cases belonging to class C_i. The number of cases in the set S is denoted by $|S|$. If a case is selected at random from a set S of cases and is announced as belonging to some class C_j, this message will have the probability

$$\frac{freq(C_j,S)}{|S|} \tag{1}$$

and the information it conveys will be

$$-\log_2\left(\frac{freq(C_j,S)}{|S|}\right) \text{bits.} \tag{2}$$

To find the expected information from such a message pertaining to class membership, we sum over the classes in proportion to their frequencies in S, giving

$$info(S) = -\sum_{j=1}^{k} \frac{freq(C_j,S)}{|S|} \times \log_2\left(\frac{freq(C_j,S)}{|S|}\right) \text{bits.} \tag{3}$$

When applied to a set T of training cases, $info(T)$ measures the average amount of information needed to identify the class of a case in T. (This quantity is also known as the *entropy* of the set T). Considering a similar measurement after T has been partitioned in accordance with the n outcomes of a test X, the expected information requirement can be found as the weighted sum over the subsets as

$$info_X(T) = \sum_{i=1}^{n} \frac{|T_i|}{|T|} \times info(T_i). \tag{4}$$

The quantity

$$gain(X) = info(T) - info_X(T) \tag{5}$$

measures the information that is gained by partitioning T in accordance with the test X. Although the gain criterion gives quite good results, it has a serious deficiency: it has a strong bias in favor of tests with many outcomes. The bias inherent in the gain criterion can be rectified by a kind of normalization in which the apparent gain attributable to tests with many outcome is adjusted. Consider the information content of a message pertaining to a case that indicates the outcome of the test rather than the class to which the case belongs. By analogy with the definition of *info(S)*, we have

$$split\ info(X) = -\sum_{i-1}^{n} \frac{|T_i|}{|T|} \times \log_2(\frac{|T_i|}{|T|}) \tag{6}$$

This represents the potential information generated by dividing T into n subsets, whereas the information gain measures the information relevant to the classification that arises from the same division. Thus,

$$gain\ ratio(X) = \frac{gain(X)}{split\ info(X)} \tag{7}$$

expresses the proportion of information generated by the split that is useful, i.e., that appears helpful for classification. If the split is neartrivial, split information will be small and this ratio will be unstable. To avoid this, the gain ratio criterion selects a test to maximize the ratio above, subjected to the constraint that the information gain must be large --at least as great as the average gain over all the tests examined.

Concrete Rebar Corrosion Mechanism

Corrosion of steel in concrete structures exposed to chloride is a major concern to bridge and structural engineers. The problem is evidenced by large scale premature failures of reinforced concrete structures in a fraction of their design life. Many of these failures have been attributed to the corrosion of the reinforcing steel and are sufficiently severe to require refurbishing or replacement of the structure (Tonini, 1976).

When steel is exposed to an aerated alkaline solution corresponding to that found in the pores of well formulated concrete (whose pH is about 13), it corrodes to form a solid corrosion product. This product (an iron oxide) forms continuously, adherently and coherently on the metal surface and serves to stifle any further corrosion. The corrosion mechanism can be schematically represented as in Fig. 1.

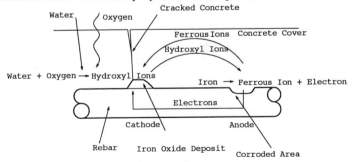

Fig. 1 Corrosion Mechanism

Service Life modeling using Inductive Learning

In general service life can be divided into two main parts: initiation period and corrosion periods (Schiessl, 1988; Tuutti, 1982). In the initiation period, the passivity of steel is destroyed and corrosion is initiated. The corrosion period begins from the depassivation stage onwards until the final corroded state is reached. The service life and life-cycle cost are the two most important outcomes that structural and bridge engineers would usually look for. These two factors for untreated and protected, repaired and rehabilitated bridges has been studied in detail in the Strategic Highway Research Program (Purvis et al., 1994; Weyers et al., 1994). The source of information for these studies are: historical data, informed opinion, and laboratory or field studies. The developed models in these studies use the statistical regression methods to come up with the predicted values. These models will serve as the generators of examples for the inductive process in the development of the corrosion system. Data that will be used are pertaining to the material factors such as cement type, presence of chloride ions, oxygen content, resistance of concrete, porosity, concrete cover, and crack occurrence. The structural factors encompass dimensions, type of structure, and functionality. The environmental factors include the moisture content, the temperature, and the snowfall attributes.

System Architecture and Development

A decision system for the assessment of corrosion of steel bars in reinforced concrete structures is under development (Nagaraja, S.). The system is composed of two main modules (Fig. 2). The first module determines the final decisions, namely, the degree of corrosion and type of remedial action. The second module predicts the serviceability life and life-cycle cost of the structure. The latter takes into consideration the important aspects of deterioration rate of rebar, the acceptable limit of deterioration, and the loss in load carrying capacity of the structure, in addition to several other important factors. The system is being developed and implemented with different types of data, both numeric and symbolic in nature.

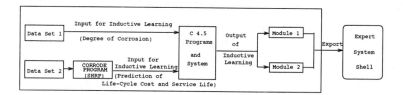

Fig. 2 Architecture of System

Several attributes or corrosion variables have been chosen to support the knowledge base structuring. These are broadly classified as material factors, structural factors, and environmental factors. These factors and their attributes are shown in Fig. 3. Data supporting these attributes form the basis for pattern recognition in the machine learning process. The data are pre-processed using the CORRODE (Version 1.0) in order to get the decision pertaining to the serviceability life and lifecycle cost module (Purvis et al., 1994).

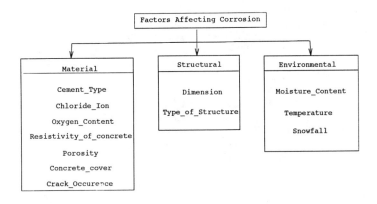

Fig. 3 Factors affecting rebar corrosion

Conclusion

Existing prediction models are based on experimental results and domain experts agree that it is very difficult to estimate the corrosion intensity in a corroding structure. In order to remove anomalies in corrosion intensity values, machine learning techniques can be applied to understand and model the serviceability life and life cycle costs. Also the developed system could be used for on-site monitoring of corrosion intensity of damaged structure. When completely developed, the machine learning expert system will constitute both an advisory system to be used by practitioners in real world evaluations, and a training aid to guide engineers and inspectors leaders in detecting and correcting deficiencies in concrete structures. It is expected that the expert system will be able to handle the most frequently encountered cases and routine situations, which will save the time and effort of the domain experts and higher-rank officials and permit them to concentrate on the more serious and uncommon situations.

References

Quinlan R. (1993). *C4.5: Programs for machine learning.* Morgan Kaufmann Publishers,CA.
Quinlan, J. R.(1986). "Induction of decision trees," *Machine Learning*, 1, 81-106.

Michalski, R.S., Carbonell, J. G., and Mitchell, T.M. (eds.) (1983). *Machine Learning: An Artificial Intelligence Approach*, Morgan Kaufmann pub. Inc., Los Altos, CA.

Tuutti, K. (ed.) (1982). *Corrosion of steel in concrete*, Swedish cement and concrete research institute

Melhem H.G. and Nagaraja S. (1995). "Inductive learning and its application to civil engineering systems", Submitted to *Civil Engineering Systems*.

Tonini, D.E. (ed.) (1976). *Chloride corrosion of steel in concrete. STP-629*, ASTM Publications.

Schiessl, P. (ed.)(1988). *Corrosion of steel in concrete.* Report of the Technical Committee 60-CSC RILEM, Chapman and Hall, London.

Purvis R., Babaei K., Clear K.C and Markow M.J. (1994). *Life-cycle Cost Analysis for Protection and Rehabilitation of Concrete Bridges Relative to Reinforcement Corrosion.* SHRP-S-377 Report, Strategic Highway Research Program, Washington D.C.

Weyers, R.E., Fitch, M.G., Larsen, E.P., Al-Qadi, I.L., Chamberlin, W.P., and Hoffman, P.C. (1994). *Concrete Bridge Protection and Rehabilitation: Chemical and Physical Techniques – Service Life Estimates.* SHRP-S-668 Report, Strategic Highway Research Program, Washington D.C.

Nagaraja, S. *A Rebar Corrosion Decision System using Machine Learning.* Ph.D. Dissertation (in preparation).

Machine Learning of Bridge Design Rules: A Case Study

Qin Chen[1], Tomasz Arciszewski[2], Associate Member, ASCE

Abstract

The paper describes an experimental application of a machine learning system to the automated knowledge acquisition in the area of preliminary design of cable-stayed bridges. Design examples were produced using ICADC, a knowledge based system for the preliminary design of cable-stayed bridges which was developed for the practical design purpose at Dalian University of Technology in the People's Republic of China. Learning experiments were conducted using INLEN, a learning system based on the AQ15 learning algorithm, both developed at George Mason University. The entire process of knowledge acquisition is described, including the preparation of examples, learning process, and knowledge verification which was conducted using various empirical error rates. Also, the final conclusions and an outline of future research are provided.

Introduction

The paper describes an experimental application of a machine learning system to the automated knowledge acquisition in the area of preliminary design of cable-stayed bridges. The application area for the study has been selected for significant technical reasons. Cable-stayed bridges are emerging in many cases as attractive structural systems, and the number of such bridges under consideration is rapidly growing. However, the current practice of the large bridge design, and of cable-stayed bridges in particular, is still not entirely formalized and it significantly relies on considerable expertise because bridge design is a formidable interdisciplinary task. Bridge designers, especially those without much expertise, can substantially benefit from a knowledge-based system that provides design advice using knowledge which has been learned from the past experience. Such a system can provide the most relevant information for the particular design concept at hand and it can partially automate the process of preliminary bridge design.

---

[1] Professor, Dalian University of Technology, Dalian, People's Republic of China, Visiting Associate Professor, Center for Machine Learning and Inference, George Mason University, Fairfax, VA 22030, USA.
[2] Associate Professor of Urban Systems Engineering, Systems Engineering Department, School of Information Technology and Engineering, Center for Machine Learning and Inference, George Mason University, Fairfax, VA 22030, USA.

There have been several attempts so far to construct knowledge-based systems for this domain. Chen (1988; 1989) reported on a knowledge based system, called "ICADC," for the preliminary design of cable-stayed bridges. Reich and Fenves (1992) implemented BRIDGER, an application of the clustering algorithm CLUSTER initially developed by Fisher (1987). BRIDGER was used to acquire cable-stayed bridge design knowledge in the form of decision trees. Also, Wu and Chen (1992) used a neural network to learning about cable-stayed bridge design.

The key to building knowledge-based systems for bridge design is the acquistion of design knowledge, and only machine learning has the potential for alleviating the complexity of manual knowledge acquisition (Reich and Fenves, 1989). There have been several previous studies on the application of machine learning to preliminary design knowledge acquisition: Mackenzie and Gero, 1987; McLaughlin and Gero, 1987; Arciszewski et al., 1987; Lu and Chen, 1987; Reich and Fenves, 1988; Gero et al., 1989; Hajela, 1989; Whitehall et al., 1990; Reich, 1991; 1992; Murlidharan et al., 1992; Reich and Fenves, 1992; Maher, 1992; Grierson and Pak, 1992; Mustafa and Arciszewski, 1992; Wu and Chen, 1992; Milzner and Harbecke, 1992; Arciszewski and Dybala, 1992; Maher and Li, 1992; 1993; Garrett and Ivezic, 1993; Ivezic and Garrett, 1993; Maher and Kundu, 1993; Arciszewski et al., 1994. However, these studies mostly demonstrated the initial feasibility of machine learning in design knowledge acquisition, and much more work is required to develop a methodological fundation for the practical applications of the automated bridge design knowledge acquistion.

In this paper, initial results of a research project are reported in which machine learning was used to produce decision rules from a collection of examples which were prepared utilizing ICADC. The learning system used, as well as the research methodology, results, and conclusions are presented.

INLEN System

INLEN (Inference and Learning), an experimental learning system developed at George Mason University Center for Machine Learning and Experience, has been used in the research reported in this paper. It is based on the learning algorithm AQ15, originally proposed by Michalski (1986). The AQ15 learns classification rules from training instances consisting of sample patterns and their correct classification. The algorithm seeks to find the most general rule in the rule space that discriminates training instances in class c_i from all training instances in all other classes c_j (i#j). The learned rules are called discrimination rules. The representation language used in AQ15 is VL_1, an extension of the propositional calculus. VL_1 is a fairly rich language that includes conjunction, disjunction, and set-membership operators. Consequently, the rule space of all possible VL_1 discrimination rules is quite large.

The algorithm utilizes STAR methodology (Michalski 1983). *A STAR of the event e against the event set E*, is defined as a set of all maximally general conjunctive expressions that cover event *e* and that do not cover any of the examples in set E, where: *e* (an event) is a positive example of a concept to be learned and *E* is a set of some counter examples of this concept. In practical problems a star of an event may contain a very large number of descriptions. Consequently, such a theoretical star is replaced by a *bounded star* that contains no more than a fixed number of descriptions. These descriptions are selected as the most preferable descriptions, according to the preference criterion defined in the problem background knowledge. A general algorithm utilizing STAR methodology can be described as follows (Michalski 1983): *Step1*. Randomly select a positive example. *Step2*. Generate a bounded star of that example against the set of negative examples. In the process of star generation, apply generalization rules, task specific rules, heuristic for generating new descriptors supplied by problem background knowledge, and definitions of previously learned concepts. *Step3*. In the obtained star, find a description with the highest preference according to the assumed preference criterion. *Step4*. If the description found completely covers set of positive examples, then go to Step6. *Step5*. Otherwise, reduce the set of positive examples to contain those events not covered by the learned description and repeat the whole process from Step1. The disjunction of all generated descriptions is complete and becomes a consistent concept description. *Step6*. As a final step apply various reformulation rules defined in background knowledge in order to obtain simpler expressions.

The central step in the above methodology is the generation of a bounded star. This can be done using a variety of methods. Thus, the above STAR methodology can be viewed as a general schema for implementing various learning methods and strategies. Details about the generation of a bounded star are described in (Michalski 1983). The AQ15 algorithm which directly employs the STAR methodology, can be tuned using many input parameters which control the execution of the program. For example, two possible modes of operation, generalization, and specialization can be used. The generalization mode induces rules as general as possible; i.e., they involve the minimum number of extended selectors, each with the maximum number of values. The specialization mode generates rules as simple as possible; i.e., with the maximum number of extended selectors and the minimum number of values. Redundant values are removed from extended selectors in the rule.

Knowledge Acquisition Process

The knowledge acquisition process used consisted of five major stages:

1. **Acquisition of Bridge Design Knowledge.** In this stage, the available general and specific knowledge on cable-stayed bridges was used by the first author in cooperation with a group of Chinese experts to develop formal knowledge for the bridge preliminary design. The knowledge was prepared in the form of decision rules. The acquisition process was initiated in the mid-1980s, and the majority of work was completed in the late 1980s.

2. **Building a Knowledge-Based System.** This stage involved the development of the ICADC computer program. The work was completed in the early 1990s by the first author. This stage is described in (Chen et al. 1989).

3. **Building the Knowledge Representation Space for Machine Learning.** The work in this stage included the identification of relevant attributes and their feasible values describing cable-stayed bridges and bridge design specifications. This was a continuation of the earlier work conducted in Stage 1, but the analysis of attributes and their hierarchies, in this case, was more systematic. The work was initiated in June of 1994 by the authors in cooperation with Ryszard Michalski of the Machine Learning and Inference Center at George Mason University and continued during the summer of 1994.

4. **Preparation of examples.** A collection of 170 examples of ICADC-generated preliminary design concepts under various design assumptions were prepared. It was completed by the first author in the Fall of 1994.

5. Induction of decision rules from examples. In this stage, INLEN was used by the first author to produce decision rules. The work was done in the winter of 1994 and the spring of 1995.

Stages 1 and 2 were conducted at Dalian University of Technology while stages 3 through 5 were completed at the Center for Machine Learning and Inference of George Mason University. In this process, ICADC is used to generate examples and INLEN is used to induce decision rules.

Building Knowledge Representation

A system of 82 attributes have been identified for the problem of preliminary design for cable-stayed bridges. These attributes are based on a general cable-stayed bridge description proposed and used by Chen in the development of ICADC (Chen 1988; 1989; 1992). The attributes are divided into two groups: 33 specification attributes, or independent attributes, and 49 design attributes, or dependent attributes. The specification attributes describe a given design problem (situation) while the design attributes describe its solution, in this case in the form of a bridge concept. Attributes are interrelated and therefore were presented in the form of hierarchical structures whose parts are shown as examples in Figs. 1 and 2. Also, as an example, feasible values of several attributes are given in Table 1.

The decision rules obtained are divided into several classes corresponding to the values of individual design attributes. For instance, for the decision attribute *Bridge System* , the following values were considered: *single-tower, twin-towers,* and *multi-towers*. Therefore, the decision rules that describe design solutions with single-tower are called "*single-tower rules.*" Similarly, "*twin-towers rules* " and "*multi-towers rules* " are defined.

NEEDS TO BE SATISFIED

number of traffic lanes
horizontal clearance
vertical clearance
sight-seeing hall
completion time
airline over
purpose

LOCATION

Social Aspects
    cultural background
    rural or urban area
Physical Aspects
    earthquake
        basic intensity
    hydrology
        high level
        low level
        duration of dry season
    geology
        bedrock depth
        depth of low quality rock
    terrain
        obstacle to cross
        width
        island
        width of left flood land
    climate
        basic wind pressure
        difference in temperature

FIG. 1. Hierarchy of Specification Attributes for Needs and Location

CONFIGURATION

span
girder elevation
bridge system
main span
side span
connection between
    girder, towers, and piers
cable layout
    main view
    side view
layout of towers
approach span
location of main pier
    foundation

SUPER-STRUCTURE

Girder
    shape of cross section
    width
    depth
    area of cross section
    dead load
    moment of inertia
    concrete grade
    reinforcing scheme
    damping device
    expansion joint
Towers
    height
    shape of cross section
    attached device
Cables
    interval length between cables
    cross section area *
        $\sin(\text{inclination angle } \text{ß})$
    anchorage
    damping block
    protection coat

FIG. 2. Hierarchy of Design Attributes for Configuration and Superstructure

TABLE 1. Values of Several Selected Attributes

Attributes	Values
Number of traffic lanes	2, 3, 4, 5, 6, 7, 8
Vertical clearance	0, 0-5, 5-10, 10-15, 15-20, 20-25, 25-
Rural or urban area	rural-area, urban-area
Basic intensity	6, 7, 8, 9
High level	0,0-5, 5-10, 10-15, 15-20, 20-25, ..., 30-
Bedrock depth	0, 0-5, 5-10, 10-20, 20-30, ..., 80
Obstacle to cross	rail-station, river, lake, gulf, valley
Island	yes, no
Basic wind pressure	10-15, ..., 25-30, 30-35, 35-40, 40-45
Span	100-125, 125-150, ..., 775-800, 800-
Bridge system	single-tower, twin-towers, multi-towers
Cable layout	harp-like, radial, fan-like
Layout of towers	1pole, ..., inverseY, A, H, gate-like
Girder shape	Inverse Trapizoid Boxes With Cantilevers, ...,Open Twin Triangular Boxes

Methodology of Experiments

    As discussed in the Introduction section several projects regarding the use of machine learning in design knowledge acquisition have been completed. In all these projects, however, learning systems were used as classification tools and in each case only a single decision (design) attribute was considered.

In the project reported here, 49 decision attributes were identified and the objective was to find decision rules which would relate groups of independent (specification) attributes and their values to the individual categories of all 49 decision attributes. In this way, we dealt with 49 separate design knowledge acquisition cases. For each case a specific single design attribute was considered, but all remaining design attributes were examined to identify the other related design attributes which should be included in the representation space. It was done in order to learn more meaningful decision rules which would include all relevant design attributes and which would better reflect the domain understanding the human experts have. For example, for the design attribute *Cable Layout*, six additional design attributes were included in the representation space, including *Total Bridge Span, Main Span, Side Span, Girder Elevation, Bridge System,* and *Connection between Towers, Girders and Piers.*

Knowledge Analysis

In the knowledge analysis conducted, two alternative and complementary types of analysis were used: partial knowledge validation and knowledge verification. In the first case, learned decision rules were compared with rules prepared by a group of Chinese bridge design experts as the part of the manual bridge design knowledge acquisition process which was conducted for the development of ICADC (Chen et al. 1989). In this way, learned knowledge was considered in the context of the state of the art, but this was done indirectly, through the use of human-developed decision rules. Therefore, this type of knowledge validation was called "Partial Knowledge Validation." In the second case of knowledge verification, learned knowledge was verified only in the context of examples which were used for learning. Verification was based on the predictive perfomance of INLEN which was measured using various empirical error rates (Arciszewski et al. 1992). Results of both parts of the knowledge analysis are briefly discussed below.

Partial Knowledge Validation

Individual automatically learned decision rules were compared with similar decision rules contained in the knowledge base of ICADC. Surprisingly, many rules were similar in terms of the attributes used and their values, but rules produced by INLEN were less complex. However, our comparison demonstrated that in most cases rules were equivalent. As an example, a single decision rule produced by INLEN is shown below:

If:

    RighFLandWid is 0_25 or 25_50 and
    EquakeRisk is <= 7 and
    GriderElev is 30_40

Then :

    Bridge System is Single Tower

Knowledge Verification

In the initial stage of our research knowledge was verified using the overall empirical error rate (Weiss and Kulikowski, 1991, Arciszewski et al., 1992). This error rate was determined for seven different design attributes and the range of variation between 0 and 7% was established. The results were good, considering the complexity of the problem and the relatively small number of examples.

Conclusions and Future Research

The research reported in this paper is a part of a long-term effort and the results are initial. However, these results have already demonstrated the feasibility of the application of machine learning in preliminary bridge design knowledge acquisition. The decision rules produced as the result of learning are simple, and their design interpretation is relatively clear.

Considering the significance of learning bridge design knowledge for the use in knowledge-based systems for the practical bridge design, more research effort is planned. In our case, a rigourous knowledge verification and the continuation of knowledge validation will be conducted.

In the case of knowledge verification, several empirical error rates, including overall, commission and omission error rates, will be used to construct learning curves for the majority of decision attributes. Also, various methods of sampling will be utilized, including leave-one-out, ten-fold and hold-out sampling. In all cases, learning will be conducted for several randomly generated sequences of examples to minimize the effect of randomness on learning and to determine the learning envelopes for the individual learning experiments.

The continuation of knowledge validation will be much more difficult than knowledge verification. The automatically produced decision rules will be compared with those proposed by human experts, and all major differencies will be analyzed. Also, learning with various learning parameters, which can be controlled in INLEN, is planned in order to produce rules close to those manually acquired.

The ultimate objective of the project is to acquire knowledge which could be used, at least partially, to improve the knowledge base of ICADC. In this way, the research reported here, and its planned continuation, will contribute to progress in the practical applications of knowledge-based systems in structural design.

Acknowledgements

The authors would like to thank Richard Michalski, Professor and Director of the Center for Machine Learning and Inference at George Mason University for his contribution to the project, and Ken Kaufman, a graduate student in the Center for his assisstance in the experiments reported here. Also, the authors are grateful to Kerry Sears, a student in Urban Systems Engineering Program at George Mason University for her assistance in the preparation of the manuscript of our paper.

This research was conducted in the Center for Machine Learning and Inference at George Mason University. The Center's research is supported in part by the Advanced Research Projects Agency under Grant No. N00014-91-J_1854, administered by the Office of Naval Research, and the Grant No. F49620-92-J-

0549, administered by the Air Force Office of Scientific Research, in part by the Office of Naval Research under Grant No. N00014-91-J-1351, and in part by the National Science Foundation under Grants No. IRI-9020266, CDA-9309725, and DMS 94-96192.

References

Arciszewski, T., Dybala, T., and Wnek, J. (1992). "A Method for Evaluation of Learning Systems." *J. Knowledge Engrg. "Heuristics,"* 5(4), 22-31.

Arciszewski, T., and Mustafa, M. (1989). "Inductive Learning Process: the User's Perspective." *Machine Learning*, R. Forsyth (ed.), Chapman and Hall, New York, N.Y., 39-61.

Arciszewski, T., Mustafa, M., and Ziarko, W. (1987). "A Methodology of Design Knowledge Acquisition for Use in Learning Expert Systems." *Int. J. Man-Machine Studies*, 27, 23-32.

Chen, Q., and Zhong, W. (1987). "An Initial Design System Based on Artificial Intelligence for Cable-Stayed Bridges." *Proc. of Intl. Conference on Cable-Stayed Bridges*, Bangkok, Thailand.

Chen, Q., Zhong, W., Cheng, G., and Qiu, C. (1989). "An Expert System for Design and Construction of Cable-Stayed Bridges." *Proc. of the 2nd East Asia-Pacific Conference on Structural Engineering and Construction*, Chiang Mai, Thailand.

Garrett, J. H., and Ivezic, N. (1992). "A Neural Network Approach for Acquiring and using synthesis knowledge." *Intelligent Engineering through Artificial Neural Networks*, Vol. 2, C.H. Dagli, L.I. Burke, and Y.C. Shin, eds., ASME Press, New York, N.Y., 767-772.

Maher, M.L., and Li, H. (1992). "Automatically Learning Preliminary Design Knowledge from Design Examples.", *Microcomp. in Civ. Engrg.*, 7(1), 73-80.

Michalski, R.S., Mozetic, I., Hong, J., and Lavrac, N. (1986). "The Multi-Purpose Incremental Learning system AQ15 and its Testing Application to Three Medical Domains", *Proc. of AAAI-86*, Philadelphia, PA.

Michalski, R.S. (1983). "Theory and Methodology of Inductive Learning," *Machine Learning: An Artiificial Intelligence Approach,* Michalski, R.S. Carbonell, J.G., Michell T.M., (Eds), Tioga Publishing.

Michalski, R.S. Mozetic, I., Hong, J., Lavrac, N. (1986). The AQ15 Inductive Learning System: An Overview and Experiments, George Mason University, *Reports of Machine Learning and Inference Center,* No. MLI-86-6.

Wu, W., and Chen, Q. (1992). "A Neural Network for the Pattern Choice of Cable-Stayed Bridges", *Proc. of 4th International Conference on Development on Education, Practice, and Promotion of Compuional Methods in Engineering Using Small Computers*, Dalian, China.

Fuzzy Navigation in an Unknown Environment

Seungho Lee [1] S.M., Boong-yeol Ryoo[1] S.M., and Teresa M. Adams [2], A.M.

Abstract

A fuzzy navigation method that combines the tangent algorithm for path planning with control rules based on fuzzy logic is presented. This fuzzy navigation method requires no physical contact between a car and the obstacles along its path. Consequently, the formulation assumes the use of ultrasonic sensor feedback. Anticipated applications are in construction.

The tangent algorithm for autonomous navigation potentially finds the shortest path but completely fails to find its target when obstacles are placed continuously in certain configurations. The tangent algorithm is improved by introducing a tracking mode for the autonomous navigation.

Introduction

Collision-free path planning of mobile robots remains an important problem in robotics [4]. Two basic models for the collision-free path planning on a two dimensional surface are path planning with incomplete information, and path planning with complete information. In path planning with complete information, perfect information about the mobile car and the obstacles is known a priori. In path planning with incomplete information, the mobile car obtains local information from sensor feedback. This paper deals with path planning with incomplete information, a constraint of construction automation.

Recently, fuzzy logic control (FLC), based on fuzzy logic, has been used to control mobile cars [1,7,8]. An FLC is a set of linguistic rules that describe the controlling strategy [3]. There are four methods to design fuzzy control rules [6]:

1. the operator's experience,

2. the controlling engineer's knowledge,

3. fuzzy modeling of the operator's controlling actions, and

---

1. Graduate Assistant, Dept. of Civil and Environ. Engrg. Univ. of Wisconsin-Madison
2. Assistant Professor, Dept. of Civil and Environ. Engrg. Univ. of Wisconsin-Madison.
   1415 Johnson Dr. Madison, WI 53706

4. fuzzy modeling of the process.

The fuzzy navigation formulation presented in this paper is based on three rule matrices for FLC: toward destination, obstacle avoidance, and tracking. The fuzzy control rules are derived from the authors' driving knowledge. The FLC matrices presented extend work done by Kong and Kosko [2].

This paper presents the formulation of the fuzzy navigation and illustrates its application through an example. The fuzzy navigation method comprises three navigation modes: Toward Destination, Searching and Tracking. Toward Destination Mode navigates the car toward its destination and is controlled by the toward destination fuzzy rule matrix. The Searching Mode navigates the car, while avoiding obstacles, to find a clear path toward its destination. The Searching Mode uses the toward destination and obstacle avoidance fuzzy rule matrices. The Tracking Mode navigates the perimeter of an obstacle and is controlled by the tracking and obstacle avoidance fuzzy rule matrices.

Navigation Parameters

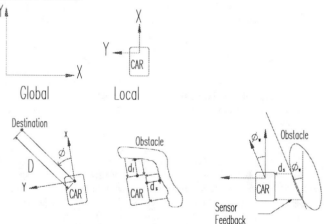

Figure 1. Navigation Parameters

Figure 1 shows the physical navigation parameters. The car has two front sensors and four side sensors, two on each side. The parameters are defined as follows.

D	The distance between the car and the destination
ϕ	The angle between the forward direction of the car and the destination in local coordinates
d_f	The minimum front sensor feedback distance.
d_s	The minimum side sensor feedback distance.

θ   The defuzzified steering angle measured from the forward direction
of the car

Fuzzy Rule Matrix - Rule Generation

Toward Destination Fuzzy Rule Matrix

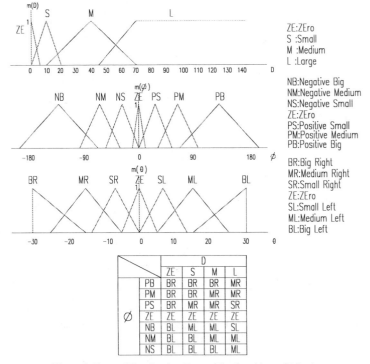

ZE:ZEro
S :Small
M :Medium
L :Large

NB:Negative Big
NM:Negative Medium
NS:Negative Small
ZE:ZEro
PS:Positive Small
PM:Positive Medium
PB:Positive Big

BR:Big Right
MR:Medium Right
SR:Small Right
ZE:ZEro
SL:Small Left
ML:Medium Left
BL:Big Left

		D			
		ZE	S	M	L
	PB	BR	BR	BR	MR
	PM	BR	BR	MR	MR
	PS	BR	MR	MR	SR
∅	ZE	ZE	ZE	ZE	ZE
	NB	BL	ML	ML	SL
	NM	BL	BL	ML	ML
	NS	BL	BL	BL	ML

Figure 2. Toward Destination Fuzzy Membership and Matrix

Figure 2 shows the fuzzy membership functions for D, ɸ and θ and the fuzzy
toward destination rule matrix. The reader should consult reference [2] for a
description of the fuzzy operations. The input variables are ɸ and D. The output
variable is θ. To control the steering angle, the fuzzy memberships corresponding to
near the destination, small |ɸ|, are narrower than the fuzzy memberships corresponding
to far from the destination. The closer to the destination the car is, the bigger the turn
required. If the distance between the car and the destination is small and ɸ is large,
then a small turn will cause the car to bypass its destination. The fuzzy memberships
of θ were adjusted in the fuzzy rule matrix to correct this problem.

For example, the fuzzy rule for the first matrix element is written,

Rule (1,1): If ϕ (PB) and D (ZE) then θ (BR).

If the value of ϕ has some Positive Big fuzzy membership and D has some ZEro fuzzy membership, then θ has some Big Right membership.

Obstacle Avoidance Fuzzy Rule Matrix

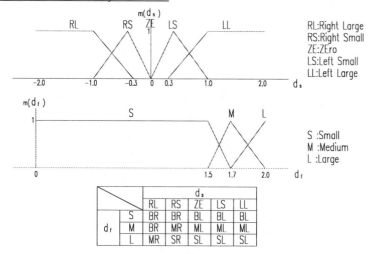

Figure 3. Obstacle Avoidance Fuzzy Membership and Matrix

Figure 3 shows the fuzzy membership functions for d_f and d_s and the obstacle avoidance rule matrix. The input variables are d_f and d_s. The output variable is θ. The front distance d_f, determines the turn size required to avoid the obstacle. The side distance d_s, determines the direction of the steering angle. The membership function for θ is shown in Figure 2. The car uses the two front sensor feedbacks to determine whether there is an obstacle. If there is an obstacle, the car uses the obstacle avoidance rule matrix to determine the steering angle. With the feedback from four side sensors, FLC determines the direction of the steering angle.

From the obstacle avoidance rule matrix, if the distance from the side sensor feedback has membership RL and RS, then the car turns right. If the side sensor feedback has membership LS and LL, the car turns left. If there is no feedback from the side sensors, but there is any feedback from the forward sensors, by default the car turns left.

<u>Tracking Fuzzy Rule Matrix</u>

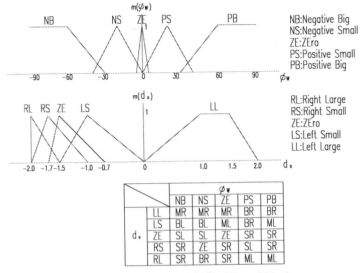

NB:Negative Big
NS:Negative Small
ZE:ZEro
PS:Positive Small
PB:Positive Big

RL:Right Large
RS:Right Small
ZE:ZEro
LS:Left Small
LL:Left Large

		ϕ_w				
		NB	NS	ZE	PS	PB
d_s	LL	MR	MR	MR	BR	BR
	LS	BL	BL	ML	BR	ML
	ZE	SL	SL	ZE	SR	SR
	RS	SR	ZE	SR	SL	SR
	RL	SR	BR	SR	ML	ML

Figure 4. Tracking Fuzzy Membership and Matrix

Figure 4 shows the fuzzy membership function for ϕ_w and d_s and the tracking fuzzy rule matrix. The input variables are ϕ_w and d_s. The output variable is θ. The tracking fuzzy rule matrix is based on a different membership function for d_s than the obstacle avoidance matrix. This is because the car tracks obstacles clockwise and thus must be more sensitive to variable distances on its right hand side. Thus, the large membership of LL is a simplification assumption. Tracking the surface of an obstacle is based on the tangent of the obstacle surface. The tangent line is derived from feedback of the two right side sensors. The car follows a tracking path that is parallel to the wall of the obstacle.

Tangent Algorithm

The FLC used a tangent algorithm for path planning. The following are parameters of the tangent algorithm.

Θ The angle between the forward direction of the car and the destination in global coordinates.

Θ_{start} The initial value of Θ.

$\Theta_{current}$ The value of Θ at any position on the navigation path.

Θ_{ref} The value of $\Theta_{current}$ at the start of the Tracking Mode (when an obstacle is encountered).

$\Theta_{compare}$ $|\Theta_{current} - \Theta_{ref}|$

T   An index for the tracking points while navigating around an
                obstacle.

Θ_T The value of Θ at the tracking point.

L   The point at which the car leaves Tracking Mode and resumes the
                Toward Destination mode.

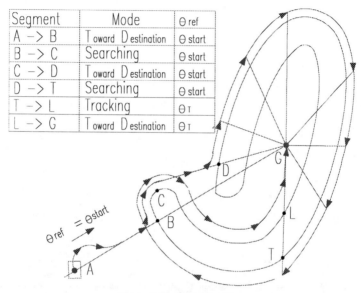

Segment	Mode	Θ ref
A -> B	Toward Destination	Θ start
B -> C	Searching	Θ start
C -> D	Toward Destination	Θ start
D -> T	Searching	Θ start
T -> L	Tracking	Θ T
L -> G	Toward Destination	Θ T

Figure 5. Tangent Algorithm

To find its destination, the car uses a Tangent Algorithm. Figure 5 shows how
the car finds the target. Points of interest are labeled on the diagram. The car starts in
Toward Destination Mode at A but, the obstacle prevents the car from going directly
toward its destination (point G). When it meets the obstacle at B, it changes its mode
to Searching Mode. At C, the car resumes Toward Destination Mode. At D, the car
meets the obstacle again. The car has no global information that the obstacle is
continuous. Consequently at D, the car changes again to Searching Mode. At each
position along the obstacle from D to T, the car computes $\Theta_{compare}$ to determine
whether its current direction is away from its destination. At T, $\Theta_{compare}$ is greater than
180°, thus the car defines T as a Tracking point and changes to Tracking Mode. The
car begins to track the perimeter of the obstacle until it reaches L, along the line
between T and the destination. L is defined as the leave point. At L, the car resumes
Toward Destination Mode.

Pseudo code for the Tangent Algorithm for path planning with FLC is as follows. The algorithm always begins in Toward destination Mode.

<u>Toward Destination Mode</u>

1) Set $\Theta_{ref} = \Theta_{start.}$
2) Move toward the destination until one of the following occurs:
   a) The destination is reached. The procedure stops.
   b) An obstacle is encountered. The car changes to Searching Mode.

<u>Searching Mode</u>

1) Calculate $\Theta_{current,}$ the current car orientation.
2) Follow the perimeter of the obstacle until one of the following occurs:
   a) The destination is reached. The procedure stops.
   b) The direction toward the destination clears. The car changes to Toward Destination Mode.
   c) $\Theta_{compare}$ is larger than 180°. The car changes to Tracking Mode.

<u>Tracking Mode</u>

1) Define the current point as a Tracking point, T, and calculate the line from the Tracking point to the destination. Follow the perimeter of the obstacle until one of the following occurs:
   a) The destination is reached. The procedure stops.
   b) The car meets the line from the Tracking point to the destination. Define the current point as a Leave point, L. Reset $\Theta_{ref} = \Theta_T$. The car changes to Toward Destination Mode.
   c) The car returns to T. The procedure stops. There is no solution because a closed curve along the obstacle is completed.

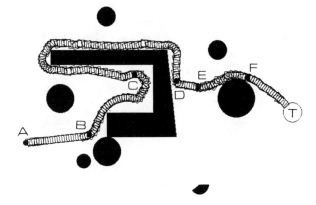

Figure 6. General Example of Fuzzy Navigation by Tangent Algorithm

Sample Application

The fuzzy navigation that combines the tangent algorithm with fuzzy control rules was programmed in the C language. Figure 6 shows a screen snapshot of a simulation example. In this example, the car navigates around a concave object. For convex obstacle, the Toward Destination and Searching Modes are required. For concave obstacles, the Tracking Mode is also required.

In Figure 6, the car starts at point A toward the target, T. From B to C, it uses Searching Mode. C is a Tracking point and the start of Tracking Mode. D, is a Leave point. and leaves the concave obstacle. From E to F, the car uses Searching Mode. At F, the car resumes Toward Destination Mode to reach the target, T.

Conclusion

In this paper, we presents a fuzzy navigation system for unknown environments. The method finds the destination based on local information from sensors without global maps. It stores very little local information and thus uses little memory. The method can be implemented in real time because the FLC can handle many possible unknown situations with rather simple computations. The fuzzy navigation method presented does not require search over the entire perimeter of obstacles. As a result, the method may produce shorter paths than other methods.

References

[1] Gasos, Garcia-Alegre, and Garcia. (1991). "Fuzzy strategies for the navigation of autonomous mobile Cars," In, Terano (ed.), *Fuzzy Engineering toward Human Friendly Systems*, ISO Press, pp. 1024-1034.

[2] Kong and Kosko. (1989). "Comparison of fuzzy and neural truck backer-upper control systems," In, Kosko (ed.), *Neural Networks and Fuzzy Systems*, Prentice-Hall, pp. 339-344.

[3] Lee. (1990). "Fuzzy logic in control systems: fuzzy logic controller-Part I," IEEE Transactions on Systems, Man, and Cybernetics, 20(2):404-418, Mar/Apr.

[4] Lumelsky. (1987). "Algorithmic and complexity issues of car in an uncertain environment," Journal of Complexity, 3:146-182.

[5] Mamdami.(1974). "Application of fuzzy algorithm for control of simple dynamic plant," Proc. IEE, 121(12):1585-1588.

[6] Sugeno and Murakami. (1985). "An experimental study on fuzzy parking control using a model car," In, Sugeno (ed.), *Industrial Applications of Fuzzy Control*, North-Holland, pp. 125-138.

[7] Zadeh. (1965). "Fuzzy sets," Information and Control, 8:338-353.

[8] Zadeh. (1968). "Fuzzy algorithm," Information and Control, 12:94-102.

Application of Hierarchical and Modular Simulation to a Bridge Planning Project

Anil Sawhney[1] and Simaan M. AbouRizk, A.M. ASCE[2]

Abstract

This paper describes the planning of a bridge project using hierarchical and modular simulation models. Hierarchical Simulation Modeling (HSM) method developed by the authors was used to plan the bridge project. The paper summarizes the planning steps that were performed using HSM. The objectives of this study were to test and validate the HSM modeling concepts, to demonstrate the feasibility of project level simulation and to illustrate the potential benefits of simulation-based planning for construction projects.

Introduction and Background

Traditional planning tools like CPM and PERT induce inherent deficiencies in the project plan due to their underlying assumptions and limitations. They are founded on the premise that planning a project requires identifying the tasks, assigning deterministic duration to these tasks, linking them in a predecessor-successor relationship and superficially superimposing resource allocation. On the contrary a construction project is characterized by the random nature of the conditions under which it is implemented and by the dynamic nature of the resource utilization. Therefore, to realistically plan construction projects and produce accurate schedules it is essential to model these important aspects. Computer simulation is a tool that can be used to model these random phenomena and to integrate variability into project planning techniques. Van Slyke (1963) and Pritsker (1979) discussed the initial developments in this area. Pritsker et. al. (1989) provide numerous illustrations of the application of simulation for project planning. PROMAX (Dabbas and Halpin (1982)) initiated attempts to utilize simulation for planning of construction projects. Using PROMAX, a project is scheduled by developing a CPM network in which some of the activities have individual CYCLONE models attached to them. Recently CIPROS, an object-oriented, interactive system for developing discrete-event simulation networks and simulating construction plans, was developed by Odeh (Odeh, Tommelein, and Carr, 1992). The basic drawback of these systems

[1] Assistant Professor, Construction Engineering and Management, College of Engineering and Applied Science, Western Michigan University, Kalamazoo, MI 49008-5064.
[2] Alberta Construction Industry Professor, Construction Engineering and Management, Department of Civil Engineering, University of Alberta, Edmonton, Alberta, Canada T6G 2G7.

is that they attempt to mimic CPM or PERT based techniques in one form or the other. This inadvertently produces a less efficient tool.

Motivated by the above described research effort, a simulation-based method for planning of construction projects that incorporates the characteristics of construction projects was developed. The method is called Hierarchical Simulation Modeling (HSM) (Sawhney and AbouRizk 1994). As part of this research effort the authors performed the planning of a bridge project that is described in this paper.

Overview of HSM

Planning of construction projects using the Hierarchical Simulation Modeling (HSM) method requires the accomplishment of the following steps:

1. Developing the project breakdown by identifying the operations and processes.
2. Defining the resources allocated to the project.
3. Defining the logical relations between the various project components.
4. Developing the process models using the CYCLONE simulation method.
5. Analyzing the simulation models and generation of the results.

The HSM framework provides a guideline for performing these tasks. HSM uses a symbolic graphical format for the development of a project plan. Modeling elements for the development of work breakdown structure, process models and links for operation sequencing are available to the modeler.

Description of the bridge project

The bridge across the Peace River 18 km northeast of Weberville in the province of Alberta (Canada), was contracted by the Alberta Transportation and Utilities (ATU), in the following manner:

1. Contract 1: The construction of the sub-structure of the bridge was the scope of this contract.
2. Contract 2: Off-site fabrication of the steel girders, delivery, erection and final positioning of the girders constituted the second contract.
3. Contract 3: Construction of the reinforced concrete deck, approaches and other miscellaneous work was part of the third contract.

The construction of the bridge started in 1989 and was completed in 1991. This study focused on the sub-structure part of the project. The bridge has seven spans, 5 spans of 112 m, 1 span of 82 m and 1 span of 92 m, with a total length of 734 m. The bridge has a grade of 0.7% sloping down from the west bank towards the east bank of the river. The specifications for the contract provided two options for the construction of the substructure. The first option involved for each pier a spread footing foundation while the second option used concrete caissons for the pier foundations. The bridge was constructed using the first option.

Figure 1 shows a schematic representation of a typical bridge pier. The dimension of the piers varies based on their location. Table 1 provides an approximate quantity estimate for the bridge (ATU, 1989).

Pier-1 of the bridge was constructed on the west shore and constituted of the installation of steel piles, blinding layer, pier footing, pier shaft and diaphragm, and pier cap. Since Pier-1 was constructed on the shore hence no pedestal was required. All of the remaining piers had pedestals of varying heights. The excavation process used a hydraulic excavator. Upon completion of the excavation, steel piles were driven and the blinding layer was placed. Footing formwork, rebar and concreting processes were then performed sequentially. The pier shaft was constructed in two lifts and then the cap was constructed.

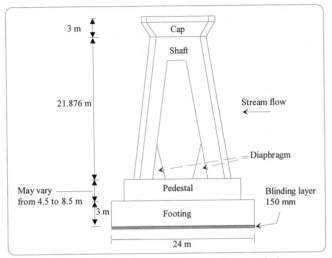

Figure 1: Schematic representation of a typical pier

Pier-2 was accessible from the west bank and as such did not require the construction of a berm. Similarly Pier-6 was accessible from the east bank and did not require the construction of a berm. The construction details for Pier-2 to 6 were exactly the same. For Pier-3 a berm was created from the west bank whereas for Pier-5 and Pier-4 the berm was constructed from the east bank. Piers 2 to 6 are typical piers and were constructed by excavating in the river to the top of the shale surface, placing the blinding layer, pouring the concrete footing, pedestal (with varying height depending on the depth of the river), pier shaft and diaphragm, and pier cap. For the typical piers, a steel frame was utilized for constructing a cofferdam. The construction of the cofferdam was started after the construction of the respective berm was completed. The steel frame was floated in place, positioned and anchored. This was followed by driving of spuds and steel sheet piles. During the cofferdam piling activity for the cofferdam process continuous dewatering was essential. Upon completion of the cofferdam, excavation was performed using a clamshell. This was followed by the placement of the blinding layer, and the remaining operations as described for Pier-1. The construction of Pier-2, 3, 4, 5 and

6 required casting of pedestals of varying heights. The two abutments and wing walls were cast on steel piles. The bridge deck is supported by four 4.545 m deep steel girders with diaphragms and lateral bracing. The girders were shop-fabricated, shipped to the site and erected by launching from the west bank. The bridge was launched using a launching nose and roller arrangement. The launched height of the girders was 800 mm higher than the final elevation. The girders were than jacked down to their final position. The reinforced concrete bridge deck had a polymer waterproofing membrane and an asphalt concrete wearing surface.

Table 1: Quantity estimate for Peace River Bridge (ATU, 1989)

Item	Unit	Sub-structure	Super-structure
Compacted fill and grading west end	Lump sum	-	-
Excavation and backfill east end	Lump sum	-	-
Excavation - structural	Lump sum	-	
Backfill - compacted granular	Lump sum	-	-
Steel H piling (HP 310 x 94)			
- Set-up	m	152	-
- Splice	splice	256	-
- Drive	pile	3,358	-
Pre-drill hole for H-piling - abutments	Lump sum	-	-
Concrete - Class A (Pier-1)	m^3	792	-
Concrete - Class C (Abutments)	m^3	751	-
Concrete - Class C (Pier shafts)	m^3	5,888	-
Concrete - Class C (Deck)	m^3	-	2,740
Reinforcing Steel - Plain	kg	700,509	159,955
Reinforcing Steel - Epoxy coated	kg	22,066	340,137
Structural steel	t	-	4,376
Bridge rail	m	-	1,537
Miscellaneous iron	Lump sum	-	-
Polymer waterproofing membrane	m^2	311	8,166
Asphaltic Concrete Wearing Surface	m^2	425	8,166

Development of project plan using HSM

Planning for the bridge project was performed using the step by step approach outlined in the HSM method. A brief description of these steps is discussed to provide the essence of the planning effort.

Project breakdown

In this step various operations and processes of the project were identified in a top-down fashion. The work breakdown structure for the sub-structure part is shown in Figure 2.

For the purpose of this study it was decided to identify the construction of the six piers and the two abutments as level-1 operations. Due to the construction detailing of the piers and abutments the modularity feature of HSM was extensively utilized. This feature allows the modeler to define part of the work breakdown structure as a modular model that can then be referenced by more than one operation. For example, at "level-1", construction of the two abutments and six piers were identified as operations. In the substructure phase of the bridge construction, the abutments were built to the lower level of the wing walls so as to allow launching of the girders. Therefore only the installation of steel piles and bridge seats was completed during the substructure-phase of the project. As such the Abutment-1 and Abutment-2 operations were broken down into the "piling" process. This is illustrated in the work breakdown structure by the piling processes attached to the two abutment operations. The piling process is a modular process that is referenced by both abutments. A generic piling process model was defined and was then used for Abutment-1 with 32 steel piles and for Abutment-2 with 30 steel piles. In order to simplify the work breakdown structure, two modular operations were defined for the project.

The construction details of Pier-1 were defined in a modular operation "PM1". This modular operation had "excavation", "piling", "blinding layer", "footing", "shaft" and "cap" operations were defined as level-1 operations. Corresponding processes were defined for the "excavation", "piling", "blinding layer", "footing" and "cap" operations. The "shaft" operation was further divided into "formwork", "rebar" and "concreting" operations which in turn had the corresponding processes attached to them.

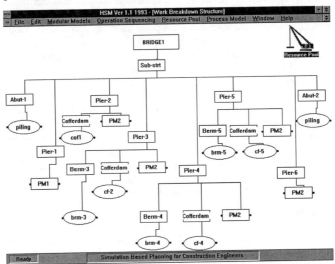

Figure 2: Work breakdown structure for substructure of the bridge project

As the same construction method was adopted for Piers 2, 3, 4, 5 and 6 a modular operation called "PM2" was designed and was referenced by these piers. The modular operation "PM2" consists of level-1 operations that include "blinding-layer", "footing", "pedestal", "shaft" and "cap" operations. These operations were then divided into the respective processes.

Resource definition

Resources required for the project were identified in the second step. A total of 31 resources were identified for the construction of the sub-structure. The information provided by the modeler included the quantity of each type of resource and the process numbers to which the resource is allocated in a decreasing order of priority. The priority for various processes is internally utilized by HSM during the simulation experiment in scheduling various work tasks that require common resources.

Operation sequencing

The next step in the plan development involved a description of the implementation strategy for the project by defining logical relationships between the various project components. HSM allows the use of four types of logical relationships that include serial, parallel, cyclic and hammock. These four links provide the modeler with the flexibility to model different scenarios normally experienced on construction projects. The governing rules that avoid conflicts and reduce redundancies are as follows:

1. Operations with common parents are sequenced as one group.
2. Once an operation has been defined as hammock, no other links can be provided for that operation.
3. For connecting more than one operations by a cyclic link the modeler has to first connect them serially.

Using the above rules operation sequencing for all the levels of the work breakdown structure were prepared. Figure 3 shows the operation sequencing for level 1. The operations are separated into two groups one being the piers constructed from the west bank and the other being the piers constructed from the east bank. On the west bank work was started at Abutment-1 and Pier-3 simultaneously. This is represented by a parallel link between these two operations. After the completion of Abutment-1, construction of Pier-1 was started which was then followed by Pier-2 and Abutment-2 in parallel. Pier-1 operation is therefore serially linked to Pier-2 and Abutment-2 operations. The construction on the east bank was started at Pier-5 which was followed by Pier-4 and 6. Therefore Pier-5 is serially linked to Pier-4 and Pier-6. The Abutment-2 operation began after the completion of Pier-1 operation.

Process models for the Bridge Project

The next step in the plan development involved defining all the process models identified for the bridge project. Some of these process models were modular and were referenced by the corresponding processes. The process models were developed using CYCLONE (Halpin (1977)). Three enhancements have been made to the CYCLONE method to allow project level simulation. The enhancements are as follows:

1. modeling of resources which are defined at the project level.

2. modeling of process interdependencies.
3. resource identification.

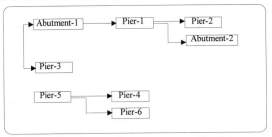

Figure 3: Operation sequencing at level-1

These enhancements were achieved by addition of four modeling elements that include:

1. Allocate Resource Node: captures the required resource from the project level resource library and allocates it to the process for completion of the work tasks.
2. Free Resource Node: frees the resource captured at the Allocate Resource Node back to the common library.
3. Successor Element: belongs to the QUE node class and replaces a simple QUE that initializes control units, consumable materials or construction components in the successor process.
4. Predecessor Element: belongs to the FUNCTION node class and corresponds to the Successor Element.

In Figure 4 the process model for the "berm" process is shown. Two resources are required to perform this process which include a dozer and five trucks. The trucks are allocated to the berm process at an Allocate Resource Node and are modeled to arrive at the site every 15 minutes. The use of five trucks and 15 minute arrival rate was based on the productivity factors the contractor had from previous similar projects. The trucks dump their dirt load directly into the river and exit the site. Four truck loads are allowed to accumulate before the dozer starts pushing the dirt and shaping the berm.

Simulation and Generation of results

The information provided in the first four steps constitutes the project level simulation models for the sub-structure of the Peace River Bridge project. These models were translated into equivalent SLAMSYSTEM models and the simulation was performed for these translated models. A project completion time of 36.91 weeks based on a seven work day week and one eight hour shift per day was obtained from this "pilot" simulation run. The results of the pilot run were validated using a CPM schedule for the project that was developed using Primavera Project Planner (Primavera, 1990). After the validation of the pilot run the simulation models were embellished to include stochastic duration estimates. Further experimentation was performed for different resource availability conditions and different operation sequencing options.

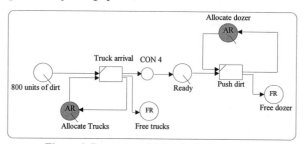

Figure 4: Process model for the berm process

Summary

The paper described the simulation based planning of the Peace River Bridge project. The five steps that form part of the HSM method were described to show the actual planning process. The exercise showed that the planning method allows the modeler to readily experiment with the project plan as compared to the existing tools. It allows modeling of external factors like weather, learning effects of management decision.

References

ATU (1989). "Alberta Transportation and Utilities (Bridge Engineering Branch) - Peace River bridge plans and specifications", Edmonton, Alberta.

Dabbas M. and Halpin D. W. (1982). "Integrated Project and Process Management" Journal of the Construction Division, ASCE Vol. 109.

Halpin, D. W. (1977). "CYCLONE: Method for Modeling of Job Site Processes" Journal of the Construction Division, ASCE, 103(3):489-499

Odeh A.M., Tommelein I.D., and Carr R. I. (1992). "Knowledge-Based Simulation of Construction Plans" Proceedings of the 8th Conf. on Computing in Civil Engineering, ASCE, Dallas, Texas. 1042-1049.

Primavera (1990). Primavera Project Planner - Reference Version 5.0, Primavera Systems Inc., BalaCynwyd, PA.

Pritsker A.A.B., Sigal, C.E. and Hammesfahr R.D.J. (1989) " SLAM II - Network Models for Decision Support", Prentice-Hall, Inc., NJ.

Pritsker, A.A.B. (1979), "Modeling and Analysis using Q-GERT Networks", Wiley and Pritsker Associates, New York and West Lafayette, IN.

Sawhney, A. and AbouRizk, S. (1994) "HSM - Simulation based planning method for construction projects", Journal of Construction Engineering and Management, ASCE (in press).

van Slyke, R.M. (1963) "Monte Carlo Methods and the PERT problem", Operations Research, Vol. 11, pp. 839-860.

SIMULATION IN THE ALBERTA CONSTRUCTION INDUSTRY

Brenda McCabe[1], Simaan M. AbouRizk[2] and Jingsheng Shi[3]

Abstract
The first section of this paper discusses a survey of a number of construction companies collaborating in a research and development project with University of Alberta. The focus of the survey was applications of simulation as a planning tool in construction. In the second section of the paper, three applications of simulation are reviewed. Each application used a different simulation method to meets its objectives.

Introduction
Simulation has been defined as "the development of a mathematical-logical model of a system and the experimental manipulation of the model on a computer" (Pritsker 1985). Although simulation has been accepted by the research community as a tool for construction, the construction industry has not been as receptive. In some cases, computerization of construction is strongly resisted.

At University of Alberta, a research and development project is being undertaken in collaboration with the construction industry. Forty companies contributed money toward this research activity. Some wanted full participation and, therefore, added in-kind contributions while others were interested in an enhancement of research activity in general and were not actively involved.

A detailed survey of the companies was conducted. Familiarization with their operations took place over a four month period, with special focus on potential applications of simulation. During these discussions with the industry, several themes recurred.

1. Approximately half of the collaborating companies are not sophisticated computer users. Computer use rarely went beyond document handling and basic spreadsheet use. This made simulation as a tool very difficult to promote.

2. The attitude of management toward computerization set the mood for the company

[1] Graduate Student, Construction Engineering & Management, Department of Civil Engineering, University of Alberta, Edmonton, Alberta, Canada, T6G 2G7

[2] Alberta Construction Industry Professor, Construction Engineering & Management, Department of Civil Engineering, University of Alberta, Edmonton, Alberta, Canada, T6G 2G7

[3] Ph.D. Candidate and Research Associate, Construction Engineering & Management, Department of Civil Engineering, University of Alberta, Edmonton, Alberta, Canada, T6G 2G7

despite the computer abilities of the employees or management. This applied to both cases where management wanted computerization and where it did not.

3. Most of their voiced needs centred around estimating and cost control, both of which have been researched and advanced to a respectable level for many years. It appears that this information was not getting to the construction industry. Interestingly, although this was perceived to be a problem, resistance to new tools remained strong.

4. There was pessimism aimed at the suggestion of using simulation. This attitude remained strong until they saw a demonstration of an application of simulation.

5. Several companies were interested in exploring the applications of simulation. Typically, these were major contracting companies who appreciated the importance of research for construction or had a research division within the company. They emphasized the need for the tool to be user-friendly and for input and output to be in a graphical format.

The remainder of this paper will explore specific simulation applications in construction. Three case studies will be discussed. Each takes place in a different sector of the construction industry and each uses a different simulation modeling technique to achieve its objectives.

Advanced Project Planning (AP2)

AP2 is a simulation-based project planning tool being developed at University of Alberta. It is based on the Hierarchical Simulation Modeling (HSM) method (Sawhney, 1994). HSM uses a multi-level breakdown structure of the construction project proposed by Halpin and Woodhead (1976). Each level becomes progressively more detailed as the project, operations, processes and finally work tasks are defined. From the work breakdown structure, a model of the system is developed using a simulation graphical network in AP2.

Once the model is outlined, CYCLONE models are created for the lowest level of the breakdown. In AP2, stochastic functions may be assigned to the task durations, providing a truer representation of the actual system. By conducting a number of runs of the model, risk analysis of the project cost and duration and sensitivity analysis of resource utilization can be performed.

AP2 currently uses a general purpose simulation system to carry the simulation experiment. A simulation environment of its own is being planned to facilitate transportability. The focus of the interface is to make the project model definitions easy to construct and lend itself to use by construction practitioners. Some of the features that will facilitate this include graphical interfaces for modeling, libraries of processes and operations that can interface with other parts of a model, and resource libraries to track historical data relevant to the company's own needs.

A Case Study: This planning tool was tested on a warehouse and office building construction project for Wadeco Oilfield Services Ltd. in Nisku, on the outskirts of Edmonton, Alberta (Thomas,1994). Forest Contract Management Ltd. of Edmonton, was contracted to erect the prefabricated steel structure according to the instruction provided by the manufacturer Robertson Building Systems - a Division of H. H. Robertson Inc. of Hamilton, Ontario.

The building covers a total area of 1860 m^2 and is supported on cast-in-place piles, perimeter grade beams and interior pile caps. A sandblasting bay required 1.2 m upstand walls to facilitate sand removal. Precast concrete sumps were installed in the wash bay, paint bay and truck shop. Only the building erection was modeled in this study i.e. the electrical and mechanical were not modeled. The work breakdown structure of the project was modeled as shown in Figure 1. A typical process model network is also shown in Figure 2. In the interest of brevity, the entire network will not be shown. The complete report is available on request.

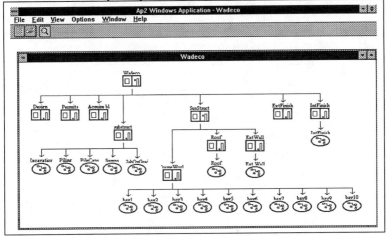

Figure 1: Wadeco Project Work Breakdown Structure

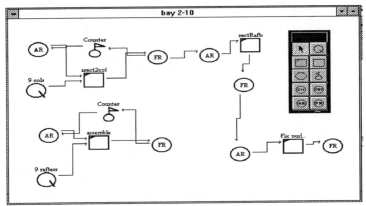

Figure 2: Sample of Simulation Network

The construction task durations were modeled using triangular distributions. The distribution parameters (low, mode, high) were provided by the project manager, who was experienced in this type of construction. A sample of the durations used in the foundation portion of the simulation are given in Table 1 for illustration.

The construction was broken into three main processes: substructure, superstructure and interior finishing. The model was constructed and fifty simulation runs were made. The mean and standard deviation of the duration times of select processes were then calculated from the output data as shown in Table 2.

AP2 was found to be very useful as a planning tool for this type of construction. It illustrated the preparation of a work schedule which recognized the stochastic nature of construction. As well, a sensitivity analysis of resource availability and costs could be performed. The contractor builds many of these structures, thereby making it cost-effective to build such a model, which could be used in subsequent projects of this type with minor adjustments.

Table 1: Process Duration Estimates

Process	Task	Units	Duration/unit (hrs)		
			Low	Mode	High
Piles, pilecaps & grade beams	Placing Rebar	cu. yd.	0.056	0.08	0.104
	Excavating	each	0.35	0.5	0.65
	Piling	kg	0.0042	0.006	0.0078
	Placing Formwork	sq. ft.	0.007	0.01	0.013
	Placing Concrete	m^3	0.231	0.33	0.429

Table 2: Simulation Durations of Processes

Activity	Duration in Hours			
	Mean	Std Deviation	Minimum	Maximum
Excavation	9.0	0.09	8.7	9.2
Piling	16.3	0.27	15.7	16.8
Grade Beams	50.7	0.64	49.1	52.3
Backfill	17.1	0.11	16.8	17.4
Place Rebar	30.0	0.43	28.9	30.9
Place Forms	0.8	0.022	0.77	0.85
Place Slab	20.0	0.6	18.8	21.6

Resource-Based Modeling For Earthmoving Applications

Resource-based modeling (RBM) methodology was developed for simplifying the modeling process of construction simulation by Shi and AbouRizk (1994) at the University of Alberta. The basic concepts come from hierarchical and modular modeling concepts which are developed by Zeiglar (1984 and 1987) for general purpose simulation.

Many real-world systems are characterized by dynamic resource interactions. Although each construction system is unique, the operating processes of its component resources are usually somewhat generic. The RBM methodology creates a resource library which consists of various resource models. These atomic models will be the basic blocks for constructing an entire simulation model through established linking structures. Project site conditions are incorporated into the simulation model during the modeling process. For earthmoving projects, all types of earthwork equipment including excavators, loaders, dozers, trucks, spreaders, and compactors are defined as the basic atomic models in the model library. Project site conditions can be incorporated into these atomic models to form project-specific models. Through intelligent linking structures that are based on the characteristics of earthmoving, the simulation model can be constructed with the user specifying required resources and project site conditions.

A Case Study: The RBM methodology is specially developed for construction practitioners who are familiar with construction operations, resources, and project site conditions, but not necessarily with simulation. A real example is used in this section to demonstrate the modeling process. This earthmoving project needs four types of equipment: a CAT-D11N tractor, a Hitachi EX-1800 backhoe, CAT-777 trucks, and a CAT-D9N tractor.

To construct a simulation model for this project using RBM, the user specifies the equipment as resources. Then he/she is required to provide information about the project. First, the user should define the makeup of the project by resource processes. Four resource processes are defined: soil pushing at the cut location, loading soil onto trucks, hauling and spreading. These processes have to be associating with specified resources: CAT-D11N for soil pushing, EX-1800 for loading, CAT-777s for hauling, and CAT-D9N for spreading. The project site conditions that affect the cyclic times of resource processes should then be specified. Finally the RBM environment will combine all information with atomic models and automatically identify appropriate linking structures to assemble the entire simulation model.

Upon experimenting with the model, the production per hour and unit cost can be obtained. By changing the truck numbers from two to eight and repeating the experiment, two curves, production and unit cost versus truck number, can be plotted as shown in Figure 3.

Optimization Of Aggregate Production Plants

An aggregate production system generally includes three operations: material transportation, size reduction and size separation. Material transportation is widely handled by conveyors. Conveyors are normally not key components because they are only constrained by capacities. Material size reduction involves crushing stones. Different types of crushers are available for different crushing purposes, including Jaw Crushers, Impact Crushers, Cone Crushers, and Roll Crushers. Screens are used to separate materials into different size categories. The general objectives of an aggregate plant are to maximize system production with high quality product. Four problems have to be solved in order to achieve these goals.

• Consistency. In order to produce a high quality product, an aggregate plant should operate continuously and consistently. All flow streams in the system should be consistent because variant input to a screen will affect the efficiency of the screen. Variant input to a crusher will affect crushing efficiency as well as result in variant gradation of crushed material. The consistent quality of product can be produced through consistent operations.

• System setting. Crushers and screens can have different settings. The closed side opening of a crusher will determine its capacity and the gradation of its output stream. The openings of the deck screens will determine the screen's capacity and the portions of output streams. As different combinations of these settings in a plant result in different production rates, which combination will maximize the system's production?

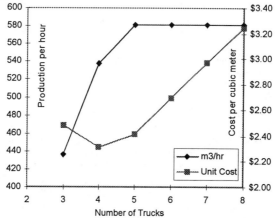

Figure 3: Production And Unit Cost Via Truck Number

• System design. Before setting up a plant, a study should be done to determine how many screens and crushers are necessary, and what types and sizes are appropriate according to the raw material and the required final product.

• Pit selection. Where a user has the option of selecting pits, which pit should be chosen before others?

Simulation and mathematical programming techniques have been attempted to solve this problem. A mathematical model was developed by Hancher and Havers (1972). It calculates the flow streams including weights and gradations from the source of the plant to the final product. With the user specifying the settings of system components and fed raw materials including weight and gradation, the production of the system and gradations of all flow streams can be obtained. One of the obvious limitations of this model is that it can not maximize the production of a system in one run.

The simulation model created for this application is based on the mathematical model given in Hancher and Havers (1972). The main focus was on providing an object-oriented, event-driven graphical interface and integrated structure that limits the user-end programming, enhances reporting and updated information about modeling components. The implementation of the model is in Visual C++ for the Windows environment. The model is also enhanced by providing an optimization component based on linear programming.

If settings of system components, weights of flow streams and gradations of flow streams are treated as decision variables, a nonlinear mathematical model can be constructed. By solving this model, the optimal settings for each component in the system as well as the maximal production of the system can be located. It is usually expensive to solve a non-linear mathematical model. By removing the component settings from decision variables (i.e. assuming each component is fixed at one setting), a linear model can be obtained. In solving this linear model, the maximal production of the system with the provided settings can be obtained. To achieve the optimal setting combination, sensitivity analysis can be done by modifying the linear models.

A Case Study: A loader is used to feed raw aggregate material into a hopper. Materials consistently flow out of it and are transported to the first screen by a conveyor. A two deck screen separates the raw materials into two streams: undersize stream with size less than 3/8 in. (10 mm) is waste; the other is routed to the second screen, which separate the materials into three streams: oversize to a crusher, undersize to final product pile, and medium size to the third screen which separates the materials into two streams: undersize to product pile, and large size to another crusher. The entire system is schematically illustrated in Figure 4. The gradation of the fed raw material is listed in Table 3. Table 4 gives the settings and capacities of three screens and two crushers. The required product has maximum size of 20 mm. There are two product streams in this system. They are X_8 and X_{12}. The objective is to maximize the production of the system i.e. Maximize $X8 + X12$.

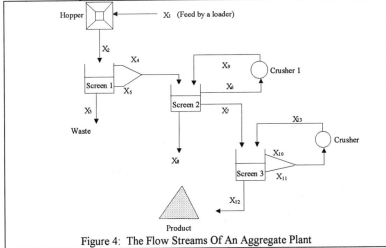

Figure 4: The Flow Streams Of An Aggregate Plant

Table 3: Gradation Of The Fed Raw Material

Size a_i (mm)	80	50	40	25	20	12.5	10	5	2.5	1.25	0.63	0.32	0.16	0.08
Percent passing (Z_{1i})	81.9	73.4	59.7	43.2	37.2	35.8	32.7	29.9	28.2	26.9	19.0	13.6	11.9	10.0

Table 4: Settings and capacities of screens and crushers

Component	Screen 1	Screen 2	Screen 3	Crusher 1	Crusher 2
Setting (mm) (Top/bottom)	50/10	40/20	25/20	50	20
Capacity (Top/bottom)	650/350	500/350	400/250	250	250

A linear model for this system can be constructed. Solving this model using LINDO, the following results can be obtained. The solution shows that the maximal production of the system is 266.2 tons/hr. The system can handle 396 tons raw material (X1). By analyzing the slack or surplus of capacity constraints, screen 3 is found to be the bottle neck of the system because its bottom deck has reached specified capacity (250 m³/hr.). If crusher 1's closed side opening is changed from 50 mm to 25 mm, the optimal solution shows that the maximal production of the system is 303.3 tons/hr, i.e. the same system's production rate can increase by 14%.

Summary
Simulation can be successfully used as a planning tool where stochastic variables best model reality, or where there are many variables and optimal solution is wanted. Three example cases, each using a different method of simulation to meet the needs of the specific project, were outlined. In general, the construction industry may be more receptive to the use of simulation in their operations if it is easy to use and graphically represented.

Acknowledgments
This project is funded by NSERC-CRD Grant # 166483 . We thank Forest Contract Management Ltd., North American Construction Group, and Lafarge Construction Materials Western Region for their participation.

References
Halpin, D.W., and Woodhead, R., (1976) "Planning and Analysis of Construction Operations", Wiley Interscience, New York, N.Y.

Hancher, D. E. And J. A. Havers (1972). "Mathematical Model of Aggregate Plant Production." A Construction Engineering and Management Research Report from the School of Civil Engineering, Purdue University, West Lafayette, Indiana.

Pritsker, A.A.B., (1985) "Introduction to Simulation and SLAM II, John Wiley and Sons, Inc., New York, N.Y.

Sawhney, A., (1994) "Simulation-based Planning for Construction", Ph.D. Dissertation, Construction Engineering & Management, University of Alberta, Edmonton, Alberta, Canada

Shi, J. and S. M. AbouRizk (1994). "A Resource-Based Simulation Approach with Application in Earthmoving/Strip Mining." *P. of the 1994 Winter Simulation Conference,* 1124-1129.

Thomas, T., (1994), "Project Planning Using the Hierarchical Simulation Modelling Method", M.Eng. Report, Construction Engineering & Management, University of Alberta, Edmonton, Alberta, Canada

Zeigler, B.P. (1976). "Theory of Modelling and Simulation." Wiley, N.Y. (Reissued by Krieger Pub. Co., Malabar, Fla. 1985).

Zeigler, B.P. (1987). "Hierarchical, modular discrete-event modeling in an object-oriented environment." *Simulation,* 49(5), 219-230.

Productivity Analysis Using Simulation

Mert Paksoylu, MSCE[1]
Daniel W. Halpin, Ph.D.[2], Member, ASCE
Dulcy M. Abraham, Ph.D.[3], Associate Member, ASCE

Abstract

Computer simulation is a valuable management tool which is well suited to the study of resource driven processes. It gives the analyst an insight into resource interaction and assists in identifying the non-value adding tasks in a process. Simulation allows the modeler to experiment with and evaluate different scenarios for better planning and analysis of construction operations.

Simulation also allows the analyst to see the effect of uncertainty or risk associated with a given process. This paper examines the impact on productivity of varying process durations in the context of a pile driving operation. The use of simulation to evaluate the time and cost implications of modifying the process is also presented. To evaluate the impact of risk and uncertainty on the production achieved, the SIDES program was used for developing the statistical parameters for process time durations.

SIDES (Subjective & Interactive Duration Estimation System)

The SIDES program models construction activity durations as probabilistic distributions in situations where only limited data are available. External factors effecting site productivity such as weather, management efficiency, etc., can be integrated into the duration modeling using this system.

[1]Yesilbahar Sok, Serim Apt. No: 18/12, Goztepe, 81060, Istanbul, Turkey; Tel:1-011-90-216-3596331

[2]Professor and Head, Division of Construction Engineering and Management, Purdue University, West Lafayette, Indiana 47907-1294; Tel:(317) 494-2244, Fax (317)494-0644, E-Mail:halpin@ecn.purdue.edu

[3]Assistant Professor, Division of Construction Engineering and Management, Purdue University, West Lafayette, Indiana 47907-1294;Tel:(317)494-2239, Fax(317)494-0644, E-Mail:dulcy@ecn.purdue.edu

Subjective factors are reflected in terms of linguistic descriptors and fuzzy set theory is applied to incorporate the linguistic descriptors into the fitted distribution.

SIDES allows the modeler to select the factors that appear most likely to impact the duration. The factors included in SIDES are weather, labor experience or skill, labor availability, management interference, and levels of site congestion. Site personnel can be asked to evaluate these factors in linguistic terms such as "very small," "quite small," "small," "medium," "large," etc. Once the subjective estimates have been provided the program quantifies their impact on the duration using a fuzzy set approach (AbouRizk et al., 1993).

Project Description

A study using simulation and SIDES input modeling was conducted on a building construction site in Izmir, Turkey. The project involved the construction of 16 mid-rise buildings. Due to soil conditions at the site, the foundation system for each building consisted of 593 concrete piles, each pile having a mean length of 32 meters and a diameter of 0.6 meters. The construction of the piles employed the Vibrex Pile system developed in Holland. Vibrex type foundation piles are driven cast in place piles that have the characteristics of precast piles. They are an improved version of the Delta pile widely used in the Dutch and Belgian markets.

The process of installing the piles as observed in the field had the following characteristics:
(1) The project used a 24 hour work cycle consisting of three 8-hour labor shifts.
(2) Delays in the supply of concrete by transit mix truck were observed leading to an task called "wait for concrete."
(3) The operator of the pile rig was involved in four activities:
   (a) Positioning of the rig.
   (b) Positioning of the pile tube.
   (c) Driving of the tube.
   (d) Installation of the reinforcement.
(4) The reinforcement, concrete, and pile caps were supplied by the prime contractor.
(5) One cycle was considered complete when the steel tube was extracted by the vibrator mechanism.
(6) The process was investigated for one building foundation (e.g. 593 piles).

The Pile Driving Process

The process of installing a concrete pile foundation (shown in Figure 1) is characterized by a number of steps as follows:

(1) A steel tube is driven into the ground with a pile hammer. The steel tube is made watertight by means of a steel plate closing the bottom of the tube. The tube is driven into the ground until the desired foundation level is reached. This can be determined by the blow count and the characteristics of the soil.

(2) After checking the tube to ensure that no water is present, the reinforcement cage is placed in the tube. The cage consists of steel bars attached by stirrups and can be provided with rollers to ensure the cage remains properly aligned.

(3) A vibrator is clamped around the tube and as soon as the tube has been vibrated loose, it is extracted while concrete is placed. Due to vibrations, the liquid concrete flows out of the tube filling the space vacated as the tube is withdrawn. Vibration assists in the placement of the concrete. The quantity of concrete is determined by the dimensions of the pile shaft and is controlled during extraction of the tube.

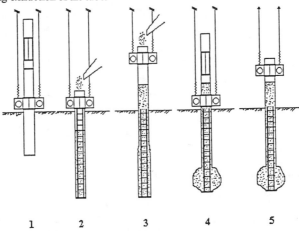

| 1 | 2 | 3 | 4 | 5 |

Figure 1:   The Vibrex Pile Installation System

A CYCLONE model for this process is shown in Figure 2. The process was simulated using the MicroCYCLONE program. For information regarding the CYCLONE system, the reader is referred to Halpin and Riggs (1992).

The main objectives of this simulation study were to determine:
(1) The time required to complete the pile installation for one building foundation.
(2) The per unit cost of driving one pile.
          In addition, the possibility of improving the process was considered.

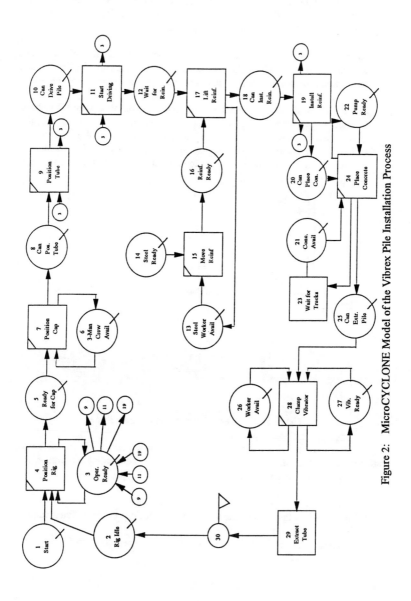

Figure 2: MicroCYCLONE Model of the Vibrex Pile Installation Process

Work Task Durations

Fifty observations for each activity were taken to establish probabilistic distributions for the work tasks involved in the pile installation process. Maximum, minimum values as well as the mean and the standard deviations for each distribution were obtained by the use of a spreadsheet.

Three duration sets were used for the analysis.

(1) The first simulation was performed using deterministic durations given in working minutes. The deterministic values used were the mean values of the 50 data observations obtained for each work activity.

(2) The next simulation used a triangular distribution for each work task. A maximum, minimum, and a mode value were required.

(3) The third simulation was performed using beta distribution. The beta distribution parameters were developed using the SIDES program.

These duration sets were used to study the sensitivity of productivity results as well as the per unit cost values.

Analysis of the simulation results of original process revealed that the productivity was about 0.87 piles/hour (using deterministic approach) which confirmed to the company representatives that there was room for improvement. The level of process detail captured in the CYCLONE model enabled the modeler to identify possible bottlenecks and sources of delays.

Revised Model

A revised model, incorporating the following modifications, was developed (Figure 3).

(1) The steel reinforcement that had to be repositioned from a staging area to each pile location was prepositioned to avoid double handling. This eliminated the activity "moving the reinforcement" in the model of Figure 2. The steel worker executing this task was also deleted.

(2) The delay in the arrival of concrete was eliminated by coordination with the general contractor's concrete plant. This eliminated the "waiting for concrete trucks" activity in the original model.

(3) The duration of the pile driving activity (COMBI 12) and the duration of the extracting of the steel tube activity (NORMAL 26) were decreased based on improved supervision.

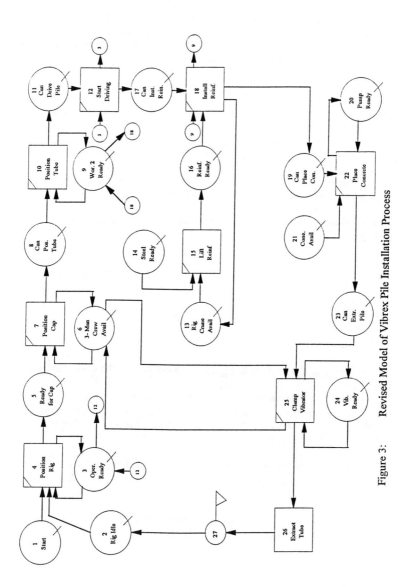

Figure 3:   Revised Model of Vibrex Pile Installation Process

In addition, simulation was conducted to see how sensitive the process would be to external factors as modeled by the SIDES program.

The SIDES program allowed modification of the parameters of the beta distribution based on verbal information provided by the pile foreman, and was able to reflect this "fuzzy" input in terms of modified beta distribution parameters (See AbouRizk et. al., 1993). The external factors effected the rate of driving and extracting the steel tube.

Results of the Simulation

Durations to complete each pile and a cost summary are given in Table 1. The original and the revised (improved) process were analyzed in terms of the three time duration sets (e.g. mean value, triangular distribution, and beta distribution). The deterministic simulation run gave an hourly productivity of 0.87 piles/hour and a per unit cost of $408.13 per pile. The beta distribution results in a productivity of 0.86 piles/hour at a $412 per pile.

The improved system (Figure 3) resulted in an hourly improvement of 0.15 piles per hour and a decrease in cost of $66.71 per pile (from data based on beta distributions). This yields a considerable saving in time and cost, since the site consists of 16 similar buildings (which corresponds to 9488 piles)

The time required for this work using the original model is 11032 hours or 92 working weeks (based on a five 24-hour work-week). The total cost of the original process is $3,909,056. The improved model decreased the time required for the pile installation to 9394 or 78 working weeks at a cost of $3,276,111. This results in a saving of $632,944.

Conclusion

Simulation with the MicroCYCLONE and SIDES programs proved to be an efficient way of analyzing productivity of a pile installation operation on a building project in Izmir, Turkey. The CYCLONE modeling environment provides an effective way of studying resource driven cyclic operations to determine where work flow improvements can be achieved. The SIDES provided a simple method by which site data regarding activity durations can be prepared for use with the MicroCYCLONE program. It also allowed the modeler to study the impact of changes in durations due to external factors as characterized in verbal formby site personnel.

TABLE 1: PRODUCTIVITY & COST SUMMARY

ORIGINAL PROCESS OBSERVED With;	PRODUCTIVITY (Piles/hr)	UNIT COST ($/pile)
Deterministic Durations :	0.87	408.13
Triangular Dist. Durations :	0.83	429.19
Beta Dist. Durations :	0.86	412.00
IMPROVED PROCESS OBSERVED With;		
Deterministic Durations :	1.02	342.34
Triangular Dist. Durations :	0.99	351.13
Beta Dist. Durations :	1.01	345.29
SENSITIVITY ANALYSIS OF THE IMPROVED CASE (Using Beta Distribution Durations)		
a) 2 Rigs and 2 Rig Operators :	1.95	300.15
b) 2 units for all Resources :	2.02	345.32
THE ESTIMATES OF PRODUCTIVITY AND COST USING SIDES ALTERED PAR. (Using Beta Distribution Durations)		
a) 1 unit for all Resources :	1.00	355.00

References

AbouRizk, S. M., Halpin, D. W. and Sawhney, A., (1993). "Construction Simulation and Distribution Fitting in a Data Deficient Environment," *Proceedings, ISUMA 93, The Second International Symposium on Uncertainty Modeling and Analysis*, University of Maryland, College Park, MD, April 25-28, pp. 493-499.

Bowles, Joseph A., (1988). *Foundation Analysis and Design*, 4th Edition, McGraw-Hill International Editions, Civil Engineering Series.

Halpin, Daniel W. and Riggs, Leland S., (1992). *Planning and Analysis of Construction Operations*, John Wiley & Sons, Inc.

Graphical User Interface for a Computerized PMS

V. Ravirala[1] (SM), D.A. Grivas[2] (M), D. Gaucas[3], and B.C. Schultz[4] (AM)

ABSTRACT

This paper describes the graphical user interface (GUI) for a computerized pavement management system (CPMS). The focus of the current CPMS development effort is the set of modules that support the pavement capital program development activity of the New York State Thruway Authority. Four major pavement management tasks for which a GUI has been implemented in CPMS are: (1) condition evaluation, (2) network sectioning, (3) treatment planning, and (4) goal-oriented network program optimization. The GUI has been designed to facilitate the problem-solving process using the following elements: (a) organization and representation of data, functions, and tasks; (b) flow among data, functions, and tasks; and (c) methods for conveying information graphically and supporting user interaction efficiently. The presented GUI design, with appealing graphical and consistent interaction features, provides two significant benefits: (1) it facilitates efficient implementation of methodological tasks using CPMS, and (2) it effectively communicates increasingly complex information of condition, modeling parameters, and optimization results.

INTRODUCTION

The graphical user interface (GUI) is a major component of a computerized pavement management system (CPMS) under development by the Rensselaer Polytechnic Institute and the New York State Thruway Authority (NYSTA). The development of the CPMS is pursued in a phased manner. The main objectives of this phased process are to: (a) allow individual methodologies and stand-alone software tools to mature prior to development of a prototype integrated system; (b) develop a prototype tool which will provide an integrated system with early products to assist agency operations; (c) generate early feedback on the methodological content and user interfaces of the prototype CPMS; and (d) develop an enhanced version of CPMS with refined methodologies and advanced GUI features [Grivas and Schutlz 1993a].

---

[1] Doctoral Candidate, [2] Professor, [3] Systems Developer, [4] Research Engineer
Department of Civil and Env. Engineering, Rensselaer Polytechnic Institute, Troy, NY 12180

The first phase of CPMS consisted of development of a relational database and a largely independent collection of programs running on personal computers with DOS as the operating system. Initially, PMS methodologies were implemented using spreadsheet and graphics tools and stand-alone programs written in C, Pascal, and FORTRAN. The second phase focused on development of a prototype tool, called WinPMS, that incorporated preliminary methodologies within a windows-based program manager [Grivas et al. 1993b]. WinPMS served as an interactive program manager with modules that performed pre- and post-processing of inputs/outputs. These modules were needed to establish linkages between methodologies and tools implemented in the first phase. The products of first two phases were used to conduct case studies and provide early assistance in pavement management activities. Subsequently, these products were evaluated by a team of researchers from RPI and NYSTA and feedback was provided on methodological content and user interfaces.

The current phase of CPMS development focuses on design and implementation of an advanced GUI, with additional enhancements to selected methodologies. The team identified integration of the goal-oriented optimization methodology for multi-year capital program development as a priority enhancement to CPMS. The optimization methodology was developed and implemented using stand-alone DOS software tools by Ravirala [1994]. The present paper describes the GUI design that supports the capital program methodology tasks implemented in CPMS.

GUI DESIGN

The CPMS implementation environment includes 486 PCs running Microsoft Windows 3.1 and the Visual Basic 3.0 programming language. The GUI design is organized around Visual Basic menus and forms. A form is a type of window that typically has an associated menu and is a container for graphical objects such as lists, grids, text boxes, charts, graphs, etc. The objects on a form support user actions and CPMS output. The form menu structure and the placement of command buttons on the form reflect both the hierarchy and ordering of tasks required to perform various pavement management activities. The graphical objects on each form are used to convey information and support user interaction in an efficient and appealing manner.

The focus of the current CPMS development effort is the set of sessions that support the pavement capital program development activity of the New York State Thruway Authority. Four major pavement management tasks for which a GUI has been implemented in CPMS are: (1) condition evaluation, (2) network sectioning, (3) treatment planning, and (4) goal-oriented network program optimization. The GUI has been designed to facilitate the problem-solving process using the following elements: (a) organization and representation of data, functions, and tasks; (b) flow among data, functions, and tasks; and (c) methods for conveying information graphically and supporting user interaction efficiently.

The GUI supports various types of *sessions* during which the CPMS is used to carry out different pavement management tasks. A *task* may be decomposed into *subtasks* that in turn may have an associated ordering. At the lowest decomposition level, a subtask is an *action* such as user input, a function that performs computation, or a graphical display of information. For each session, subtasks, relevant data, supporting computations and displays, and required user actions have been identified for the GUI design and implementation.

TASK DECOMPOSITION

The main tasks of the CPMS are organized around the top-level menus. Subtasks of each main task are represented by pull-down submenu items. An ordering, if relevant, is captured by the order of the menu items in the menu bar or submenu lists. In certain cases task ordering is enforced by enabling only those menu items that can be selected at a particular point in the task decomposition and execution. Each terminal menu item leads to a form which defines a new session. When a menu selection leads to a new session, a new menu of the corresponding form captures the subtasks of that session.

The top-level items in the main menu of CPMS and the pull-down submenu of the *Optimization* module are shown in Figure 1. The order of these items implies the general order in which the corresponding tasks would be accomplished to support capital program development. For example, the order of tasks for capital program development process is defined as: (1) assess the network condition, (2) select the sections to be analyzed, (3) identify treatment methods for each pavement condition state, (4) conduct an economic evaluation to estimate life-cycle costs and remaining life, and (5) develop optimal maintenance programs using the information generated from tasks 1 to 4.

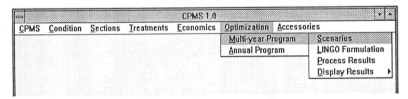

Figure 1: Main Menu of CPMS

Tasks 1 to 3 incorporate a state increment methodology developed by Ravirala [1994]. Economic analysis is performed using a state increment life-cycle cost analysis method described by Ravirala and Grivas [1994a]. The multi-year and annual program development processes use goal-oriented optimization methodologies [Ravirala 1994, Ravirala and Grivas 1994b]. Within the optimization process, annual program is developed after the multi-year program has been established. The sequence of tasks for the multi-year program development are: (1) define scenarios with different modeling parameters, goals, and constraints; (2) formulate the mathematical model and generate the input data files to be used by the LINGO optimization solver; (3) process the results obtained from LINGO; and (4) display the results using graphs and tables. As shown in Figure 1, the submenu items of *Multi-year Program* menu item reflect the precedence determined by the underlying methodology tasks.

Figure 2 shows the form layout and subtasks of the *Multi-year Program Scenarios* session. The purpose of this session is to enable the management to create multiple scenarios for the multi-year program by modifying the modeling parameters, strategies, goals, penalties, and constraints. In this particular session, the GUI design includes: (1) the menu of the parent form (also known as MDI form), (2) a command button bar on the parent form, and (3) command buttons on the child forms (also known as MDI children). In all three cases, placement of the items imply an ordering that can be made explicit by disabling and enabling the items accordingly. For example, the *Goals* and *Penalties* buttons are enabled only when the *Analysis Period* is specified in the *Parameter* form. This ensures that only for valid years within the analysis period the user enters the goals and associated penalties for deviating from the goals. The placement of graphical objects that support non-decomposable actions of a session such as list box selections or data grid entry can also imply a dependency ordering. For example, in Figure 2 the number of rows of inflation and interest rates to be filled depends on the analysis period selected from the list box. Apart from enforcing the order of action flow for the user to enter information, checks are made for inconsistencies in the user input when the OK button is pressed (OK and Cancel are placed at the bottom of each window and are not visible in Figure 2).

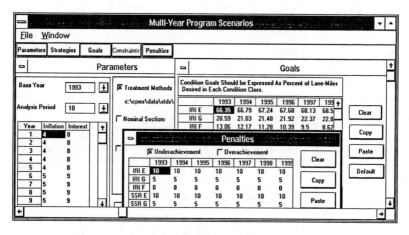

Figure 2: Illustration of Subtasks Within Scenarios Session

Multiple child forms corresponding to a session may be visible simultaneously as shown in Figure 2, although only one is considered to have *focus* and is brought to the front of the screen (*Penalties* screen). At runtime, a form can be given focus by clicking the mouse on that form. The child form can be instigated either by menu items of the parent form or by the command bar buttons. For example, Figure 3 shows a child form that is instigated by the *File-Load* menu command sequence.

If a form needs to accomplish its corresponding task completely before losing focus, it is made *modal* and any mouse clicks outside its window, such as menu selections, do not have an effect. This modal restriction is conveyed to the user by disabling other commands as shown in Figure 3. In this particular situation, the user will have to select a scenario to be "loaded" before the associated information can be modified using the command bar button sessions. The user can, however, enter a new scenario name in the drop-down list box (called combo box). Several different types of other graphical objects used to design CPMS GUI are presented next.

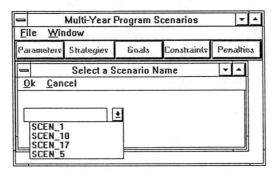

Figure 3: Modal Form that Requires Completion of a Task

GRAPHICAL OBJECTS

Within a session, actions such as user input, computation, or information display are supported by various graphical objects called *controls*. The GUI design attempts to employ controls in a way that helps the user to know intuitively what to do. Appropriate controls are used to represent data, functions, and tasks. Controls placement is designed to provide speedy access and facilitate comprehension of a session. In addition, the presentation style of each control is kept relatively consistent through out CPMS. Several general guidelines similar to those outlined in [Howlett 1994, Microsoft 1992] were followed in the design of session forms and in the application of controls. Some of the controls employed in CPMS that are shown in Figure 4 include drop-down list box (year selection), text box (milepost), label (example, "Division"), check box (various measures), simple list box (showing selected plots), and command buttons (example, "Delete"). Additional controls include a combo box which is a drop-down list box that allows user editing (as displayed in Figure 3), a grid with spreadsheet data entry (as shown for *Penalties* in Figure 2), and option buttons that allow one and only one choice to be selected ("All" button shown later in Figure 6). Another particularly useful control is a gauge, which provides feedback in terms of number of operations completed and/or percentage accomplished during time-intensive computation.

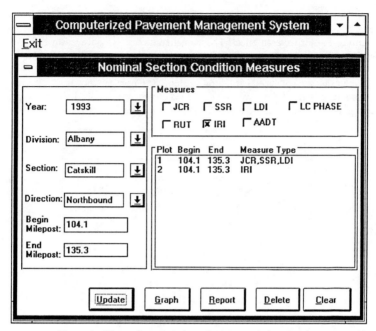

Figure 4: Illustration of Various GUI Controls in a Session

FORM LAYOUT

The form layouts of the CPMS attempt to organize controls and menu items (and therefore their corresponding data, functions, and tasks) in a manner that facilitates efficient accomplishment of the session's tasks. In Figure 4 for example, the data-input actions comprising the *Nominal Section Condition Measure* session flow from top to bottom and from user's left to right. The command buttons are placed in a sequence that represents the order of actions typically followed by the user after selecting measures to be plotted. Other CPMS sessions use a similar grouping of command buttons for consistency.

Sessions with parent forms that simultaneously display multiple child forms have window tiling and cascading features. This feature is demonstrated in Figure 5 which summarizes the investment levels, treatment distribution, and resulting condition (as measured by three different condition measures) of a specific multi-year program scenario. (Please note that the legend categories are currently distinguishable only in color). To create such a summary the user selects a scenario (as in Figure 3) and a set of plots to be generated (as in the list box of Figure 4). The style of each plotted graph can be changed by using the choices in Options menu item. Depending upon the type of data being plotted, CPMS allows line, 2D-pie, 3D-pie, 2D-bar, and 3D-bar styles.

These graphical features provide a powerful tool to compare and contrast the results from multiple scenario. For example, the user can select budget and IRI summary plots for competing scenarios and display them simultaneously. Figure 6 shows an example form that is used to select, preview, and print the graphs shown in Figure 5.

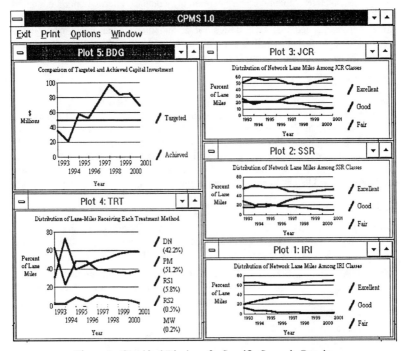

Figure 5: Graphical Display of a Specific Scenario Results

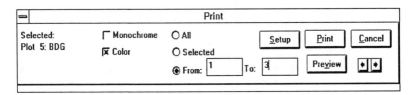

Figure 6: Form with GUI for Selecting, Previewing, and Printing Graphs

SUMMARY AND CONCLUSIONS

This paper presented the form-based graphical user interface (GUI) for a computerized pavement management system (CPMS). The focus of the current CPMS development effort is the set of modules that support the pavement capital program development activity of the New York State Thruway Authority. The presented GUI design, with appealing graphical and consistent interaction features, provides two significant benefits: (1) it facilitates efficient implementation of methodological tasks using CPMS, and (2) it effectively communicates increasingly complex information of condition, modeling parameters, and optimization results. On the basis of the findings of the study, it is concluded that GUI design for a computerized pavement management system can be facilitated by task decomposition in terms of sessions, menu items, user commands, and data entry controls.

ACKNOWLEDGMENT

The research presented in this paper was sponsored by the New York State Thruway Authority. It is part of a broader research effort to develop and implement a pavement management system for the Authority's pavements. The assistance and feedback provided by Thruway Maintenance personnel during this study is gratefully acknowledged. In particular, suggested enhancements to the CPMS GUI features identified by Mr. Richard Garrabrant have been greatly appreciated. Views and opinions expressed herein do not necessarily reflect those of the Authority.

REFERENCES

Grivas, D.A., and Schultz, B.C. (1993a). "Roadway Maintenance Programming and Monitoring within a Computerized Decision Framework." Proceedings of the XIIth IRF World Meeting, Madrid, Spain, Volume III, pp. 169-176.

Grivas, D.A., Schultz, B.C., and Tonias, D.E. (1993b). "An Integrated System for Pavement Management." Infrastructure Planning and Management, Proceedings of the ASCE Speciality Conference on Infrastructure Management: New Challenges, New Methods. J.L. Gifford, D.R., Uzarski, and McNeil, S., editors, Denver, Colorado, June 21-23, pp. 257-261.

Howlett, V. (1994). "Master the Principals of User Interface Design." Proceedings of the Visual Basic Insider's Technical Summit, Orlando, September 22 - 24, 1994, pp. 491 - 496.

Microsoft Corporation (1992). The Windows Interface: An Application Design Guide. Microsoft Windows Software Development Kit.

Ravirala, V. (1994). "Multi-Criteria Optimization Methodologies for Highway Program Development," Ph.D. Dissertation, Dept. of Civil and Environmental Engineering, Rensselaer Polytechnic Institute, Troy, New York.

Ravirala, V., and Grivas, D.A. (1994a). "State Increment Method of Life-cycle Cost Analysis for Highway Management." submitted to ASCE Journal of Infrastructure Systems.

Ravirala, V., and Grivas, D.A. (1994b). "A Non-preemptive Goal Programming Methodology for Developing an Annual Pavement Program." Preprint 940388, Transportation Research Board, Washington, D.C.

A Fuzzy Logic Model for Automated Decision Making
of Pavement Maintenance Management

Jian Chen[1] and Jiwan Gupta[2]

Abstract

The maintenance and rehabilitation of highway pavement need a large amount of manpower and resources. At present time, decision making of pavement rehabilitation is based upon the pavement performance data, such as roughness, distress, and friction. With integration of microcomputer and video camera technology, automated methods are developed to reduce the cost of manual visual inspection and to improve quality of road analysis techniques. This paper presents a computer model using fuzzy logic for automated decision making of pavement management system. The fuzzy logic model consists of three inputs and one output. The inputs are roughness, distress, and skid resistance respectively. The inputs are obtained from automated pavement data collection system. The corresponding output is maintenance level which is described by five membership functions, i.e, do nothing, low level maintenance, medium level maintenance, high level maintenance, and reconstruction. The fuzzy inputs and output are related by fuzzy rules which are key factors affecting precision of decision making.

Introduction

The maintenance and rehabilitation of highway pavements in the United State cost over $17 billion dollars a year, according to Guralnick *et al*, 1993. Automated methods have been developed to reduce the cost of manual visual inspection and to improve quality of road analysis

----

[1]Graduate Student, Department of Civil Engineering. The University of Toledo, 2801 W.Bancroft Street, Toledo, OH 43606.
[2]Professor, Department of Civil Engineering, The University of Toledo, 2801 W.Bancroft Street, Toledo, OH 43606.

techniques. As an important component of pavement management system, information of roughness, distress, and skid resistance of pavement is commonly utilized to evaluate pavement performance and to select strategies of maintenance and rehabilitation. Fuzzy logic model applies fuzzy reasoning through logic operations (such as AND, OR, and IF-THEN) to fuzzy sets which are represented by membership functions. The fuzzy logic model has the advantage of providing analytical description of the relationship between the system's input and output (Yager and Filev, 1994, p.231). Therefore, it is a potential tool of studying management strategy for pavement maintenance which requires expert experience to assess road conditions. Figure 1 shows a block of fuzzy logic system.

Figure 1. A Sketch of Fuzzy Logic System.

Fuzzy inputs are obtained by applying membership functions to crisp inputs and this process is called fuzzification. Rule evaluation, or inference engine, is a key unit in which expert knowledge is needed to obtain correct rules. After the rule evaluation, fuzzy inputs result in fuzzy outputs. Finally, these fuzzy outputs are converted to crisp outputs by a process called defuzzification in which weighted average approaches such as center of gravity method (COG) can be applied.

Inputs and Output of Fuzzy Logic System for Pavement Maintenance

The fuzzy logic system of pavement management presented in this paper includes three inputs and one output. The inputs are roughness, distress, and skid resistance. Maintenance level is the only output.

Roughness.

Roughness is the measure of ride quality of a road. Present serviceability rating (PSR, ranging from 0 to 5) can be used to describe roughness of a road section (Huang, 1993, p.427), as shown in Figure 2. For the scope of automated measurement of roughness, present serviceable index (PSI) which is a mathematical approximation of PSR within prescribed limits, is used to measure the roughness of a road instead of PSR. Based on the scale of

PSR, the fuzzy membership functions for roughness are built. Figure 3 presents five membership functions for the roughness.

Figure 2. Scale of Present Serviceability Rating (PSR).

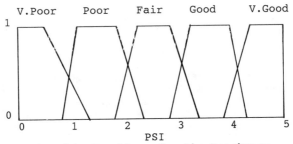

Figure 3. Membership Functions for the Roughness.

Distress

Although the types of pavement distress are also important, but due to difficulty of automated identification, this paper addresses only the levels of severity of distress. The levels of severity of distress are divided into three categories as listed below, and the unit of measure is "square meters" (Hawks etc., 1993, p.5):
a. low severity;
b. moderate severity;
c. high severity.

In this paper, the percent of distress area is selected as measure unit of severity. A sketch of three membership functions for low, moderate, and high severity is shown in Figure 4.

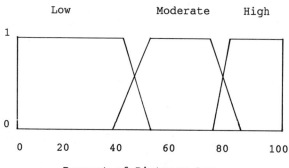

Figure 4. Membership Functions for the Levels of Severity
of Distress.

Skid Resistance

The skid resistance is characterized by skid number
which is defined as:

$$SN = 100 \ \mu = 100 \ (F/W)$$

where μ is the coefficient of friction, F is the tractive
force applied to the tire at the tire-pavement contact, and
W is the dynamic vertical load on the tire (Huang, 1993,
p.439). Because μ varies from 0 to 1, SN ranges from 0 to
100. Figure 5 shows three membership functions developed
for the skid resistance.

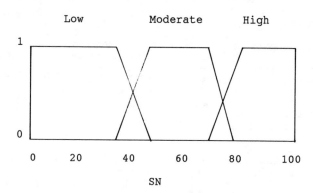

Figure 5. Membership Functions for Skid Resistance.

Maintenance Level

In the fuzzy logic model, the output generated is maintenance level. The maintenance level is divided into five categories according to the maintenance cost per unit area (such as dollars per square meters):
a. do nothing;
b. low level maintenance;
c. moderate level maintenance;
d. high level maintenance; and
e. reconstruction.

Figure 6 illustrates the sketch of five membership functions for the output.

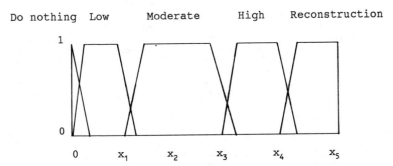

Figure 6. Membership Functions for the Maintenance Levels.

Fuzzy Rules and Defuzzification

The number of fuzzy rules is equal to the product of the numbers of all membership functions. So there are 5x3x3 = 45 rules available. Since some input combinations are unnecessary, it is possible to cut down the number of fuzzy rules by incorporating IF, AND, and THEN statements. The rule evaluation needs expert knowledge, For example, five rules are set up as following:

Rule 1. IF very good PSI AND low severity AND high skid number THEN do nothing;

Rule 2. IF very poor PSI AND high severity AND low skid number THEN reconstruction;

Rule 3. IF good PSI AND low severity AND high skid number THEN low level maintenance;

Rule 4. IF fair PSI AND moderate severity AND moderate skid
number THEN moderate level maintenance;

Rule 5. IF poor PSI AND high severity AND low skid number
THEN high level maintenance.

Defuzzification converts a fuzzy quantity represented by a
membership function to a crisp quantity (Vadiee, 1993,
p.105). In the fuzzy logic model, final crisp output is
cost per unit area.

Conclusions and Recommendations

The fuzzy logic model provides information for
decision making of pavement maintenance. Building
membership functions for inputs and outputs as well as
setting correct fuzzy rules are key points of applying
fuzzy logic to pavement management system. Although this
model considers only three inputs, roughness, distress, and
skid resistance, for the sake of high precision, other
pavement properties, such as deflection and distress types
of pavement should be taken into account in future work.
The pattern recognition and image processing are potential
tools for distinguishing different type of distress. In
future, the neural network method can be used to train
pavement data and to build optimal membership functions
(Yager and Filev, 1994, p.231-242). Furthermore, the
refining of fuzzy rules should consider field experience
and the human factors of decision making.

Appendix I.- References

Guralnick, S. A., Suen, E. S., and Smith, C. (1993).
"Automating   inspection of highway pavement surfaces."
Journal of Transportation Engineering, vol. 119, No.1, 1.

Hawks, N.F. et al, (1993). *Distress Identification Manual
for the Long-Term Pavement Performance Project,* SHRP-P-338,
Strategic Highway Research Program, National Research
Council, Washington, DC.

Huang, Y.H., (1993). *Pavement Analysis and Design,*
Prentice-Hall, Inc., New Jersey.

Vadiee, N.,(1993). "Fuzzy Rule-Based Expert Systems II." in
*Fuzzy logical and control-Software and hardware
applications,* Jamshidi, M., Vadiee,N., and Rose, T.J. eds.
PTR Prentice Hall, Englewood Cliffs, New Jersey.

Yager, R.R, and D.P. Filev, (1994). *Essentials of Fuzzy
Modeling and Control,* John Wiley & Sons, Inc. New York.

MICROCOMPUTER TOOL FOR MANAGEMENT OF UTILITY CUTS

Rajagopal S. Arudi[1], Vishwanath CVSA[2], Andrew Bodocsi[1]
Prahlad D. Pant[1] and Emin A. Aktan[1]

ABSTRACT

This paper describes the results of a study on the development of a Windows based microcomputer tool called 'UCMS' for the maintenance management of utility cuts in asphalt pavements. UCMS is a synthesis of field evaluation procedures, cost management and policy issues related to street pavement sections affected by utility cuts. Using the user input data, the software evaluates the condition of cuts and presents a report on budget requirement to upgrade their condition. In the event of budget limitations, the software provides a prioritized maintenance management schedule in the form of graphic and tabulated reports for various budget scenarios. At present, the software works as a stand-alone model. Attempts are being made to integrate UCMS with existing pavement management system.

INTRODUCTION

Utilities, like gas, electricity, telephone and waterworks, often make small openings in city street pavements to install or inspect utility services. Although guidelines have been developed by cities for opening and restoration of pavements, most cities have experienced additional maintenance costs due to poor restoration and pavement deterioration. Officials concerned with management of road networks with utility cuts are in need of practical methods that would address the impact of cuts on pavement performance, the economic evaluation of life cycle costs, and provide the basis for a realistic cost impact. A structured maintenance management program can answer these needs. As the existing pavement management systems do not consider the effect of utility

---

[1]Department of Civil and Env. Engineering, University of Cincinnati, PO Box 210071, Cincinnati OH 45221-0071
[2]Department of Operations Research, CWRU, Cleveland OH 44106

cuts, municipalities, like the City of Cincinnati, are currently seeking specific guidelines in the form of a Utility Cut Management System (UCMS).

GOALS OF UCMS

The goals of UCMS are to: 1) Identify the product most useful for evaluating the performance of utility cuts; 2) Differentiate between the quality of restoration by different utilities/sub-contractors; 3) Generate a comprehensive database; 4) Develop statistically calibrated models to predict future performance, life cycle cost and monetary impact; and 5) Address issues related to planning, investments and maintenance activities.

EVALUATING PERFORMANCE

The flow chart shown in Fig. 1 outlines the steps for: (a) evaluating the condition of cuts and surrounding pavements; and (b) development of a maintenance management program. In a regular pattern, deflection tests were first made to find the strength of the cut and surrounding pavement. Following this, periodic visual distress surveys were made to assess the present surface condition of the cuts.

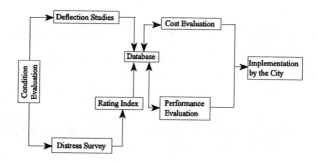

Fig. 1 Methodology

Deflection studies were made on 36 cuts in asphalt pavements using the standard Benkelman Beam along with a truck having a rear axle load of 80.1 kN and 482.7 kPa tire pressure in its rear dual wheels [Arudi et al., 1994]. Distress surveys were carried out on 75 cuts using the Delphi Method. The panel for the Delphi Study consisted of four engineers from the Highway Engineering Department and eleven inspectors from the Highway Maintenance Department of the City of Cincinnati. The survey was carried out in three rounds.

An important task of the study included the development of a rating index that can characterize the condition of a cut and the surrounding pavement by combining the individual distress data into a single number. Development of this index, designated as the Utility Cut Condition Index (UCCI), was made by using the Delphi Study data and a neural network methodology [Pant et al., 1993].

In the present system, deflection values at critical locations, at or near the cut, will be used as the criteria for estimating the extent of damage and the cost impact. The UCCI serves as a management tool to identify the time at which remedial action is to be implemented and the potential consequences of a decision.

PERFORMANCE MODELS AND LIFE PREDICTION

The performance prediction models require a variety of factors that affect the rate of deterioration. These factors include age of cut, restoration, traffic, backfill characteristics, pavement composition and construction quality. A cursory look at the City of Cincinnati's database indicated that information generally is limited to location, age, traffic, and name of the utility restoring the cut.

A list of 600 cuts in asphaltic concrete and macadam pavements was prepared by referring to permits. A team of trained personnel was assigned the task of conducting a Distress Survey. However, the team was able to locate only 94 cuts, many of the others had been resurfaced. The Distress Survey was performed using the Distress Manual, developed as a part of this study [University of Cincinnati, Cincinnati Infrastructure Institute, 1991]. The distress data were used in conjunction with the neural network model and the UCCI for each cut was computed. Statistical regression models, with UCCI as a function of age, were developed for cuts restored by various contractors. The general form of these models is:

$$UCCI = A + B * age + C * age^2 + D * age^3$$

IMPLEMENTATION ON A MICRO-COMPUTER

A comprehensive MS-Windows based software (UCMS Ver 1.0) has been developed in order to: 1) aid in the development of a comprehensive database, 2) select a prioritized listing of cuts to be maintained, 3) select the appropriate Maintenance and Rehabilitation (M&R) action, and 4) determine its cost impact. The basic context diagram is shown in Fig. 2.

Fig. 2 <u>Overview of Utility Cut Management System</u>

Input To The System: The inputs to this software are Cut Information, Distress Data (collected using the Distress Manual), Deflection Data, Traffic Data and Cost Data. UCMS contains one comprehensive screen, divided into three modules, for the entry of all the input data.

Cut Information: This module of the input to the system asks the user for the historical information on the cut. This information consists of the location of cut, date of survey, date of restoration, contractor's name, area of the cut, and type of pavement.

Distress Data: The distress data are primarily based on the visual inspection of the cuts. A user-friendly screen has been designed to enable the user to enter this data. Validation checks have been incorporated into the system to ensure that the user enters the appropriate data and error messages are displayed whenever an invalid entry is made into the system. Online help is available in the form of text as well as pictorial information to help the user to enter the data.

Deflection Data: As in the case of distress data, online help is available which explains in detail the information expected from the user. The deflection data is used to compute the overlay thickness required.

Traffic Data: The user is prompted for traffic information in terms of Average Daily Traffic (ADT), percent trucks and percent growth.

PROCESSING WITHIN THE SYSTEM

Computation of Overlay Thickness: Deflection information, if available, is used in the computation of the overlay thickness. If the deflection at any point within or near the cut is greater than the deflection at the control point, an overlay is recommended and the overlay thickness required for the excess deflection is computed using the Asphalt Institute Manual [The Asphalt Institute, 1983]. Since deflection is an objective measurement, the recommendation obtained through the use of deflection data overrides the recommendation based on the distress data.

Computation of UCCI: The distress data is used as an input to the neural model for the computation of UCCI. The UCCI values are added to the database for further processing.

Selection of M&R Action: When distress surveys were performed by four engineers and eleven inspectors from the City of Cincinnati, they were also asked to give their recommendation on the required maintenance and rehabilitation action for each cut. This information is used in the UCMS model to generate M&R actions based on the UCCI values.

Cost Computation: The model takes into account the labor, material and equipment costs involved in every M & R action for the cost computation. A facility has been provided to update the costs with the changing market prices.

OUTPUT FROM THE SYSTEM

The output from the system can be in three forms: 1) Individual Report; 2) Group Report; 3) Custom Report.

Individual Report: This report contains all of the information available in the database on any specific cut. This information includes general cut information, distress data, deflection data (if available), computed UCCI, computed overlay thickness, recommended M&R action and the cost implication. The software asks user to select a cut based on the cut location. This is done by presenting the user a list of all cut locations available in the database. After this selection the user can obtain this report either on the screen or on the printer. A sample copy of this report is shown in Fig. 3.

Group Report: This report is primarily aimed at assisting the engineer in policy decisions. It presents a histogram which consists of information on the number of cuts in various ranges of UCCI. It also gives the total amount of money required to rehabilitate the cuts. In addition, there is an option provided for analyzing various budget scenarios. In case of a budget limitation, the user can input the available budget and the software comes up with a revised histogram and an annual prioritized listing of the cuts to be rehabilitated based on their UCCI and the available budget. A sample copy of this report is shown in Fig. 4.

Custom Report: This is a customized report in which the user can select a list of cuts and obtain relevant information on this selected group of cuts like general cut information,

City of Cincinnati
Department of Public Works
INDIVIDUAL REPORT FOR LATERAL CUT

Cut Location: 1712 Bishop   **Between:** Jefferson   and M.L. King

Applicant: Water Works   **Cut Dimension:** 6 ft. x 5 ft.

Purpose: Inspection

Pavement: Asphalt   **Base Restored by:** _____

Surface Restored by: _____

Survey Date: 12/07/93   **Date of Restoration:** 04/15/83

DISTRESS INFORMATION

	Cut	Vicinity		Cut	Vicinity
Alligator Cr:	H	H	**Ravelling:**	M	
Edge Cr:			**Drop-Off:**	H	H
Transverse Cr:			**Edge Sep:**	M	
Potholes:			**Corner Br:**		
Rutting:			**UCCI**	37	

DEFLECTION INFORMATION

D1: 0.032"   **D2:** 0.064"   **D3:** 0.060"   **D4:** 0.054"

D5: 0.051"   **D6:** 0.057"   **D7:** 0.048"   **D8:** 0.045"

Pavement Surface Temperature: 77F

TRAFFIC INFORMATION

ADT: 4000   **Trucks:** 7%   **Growth Factor:** 3%

Recommended Action: Reconstruct   **Cost :** $1278.00

Fig. 3 Individual Report for Utility Cut

Utility Cut Histogram Report
UCCI vs. Number of Cuts

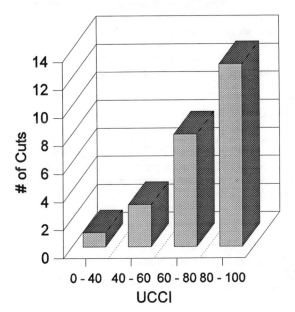

The total number of cuts is: 25

The total cost to rehabilitate the cuts is: $6220.44

Fig. 4 <u>Group Report on Condition and Budget Requirement</u>

UCCI, recommended action and cost. In addition to this, the total cost for rehabilitating this group of selected cuts is reported.

CONCLUDING REMARKS

The utility cut management system outlined in this paper is based on a detailed investigation of the strength and performance characteristics of utility cuts. The field evaluation procedure comprises of objective measurement of deflections and subjective measurement of visual distresses. The deflection measurements assisted in establishing the area of the pavement influenced by the cut and the cost impact. The subjective evaluation of condition lead to the development of a Utility Cut Condition Index. The UCCI is a valuable management tool for city managers to identify and prioritize candidate projects for maintenance. The UCMS considers all important facets of damage assessment, cost impact, maintenance programs, and is designed so that the technology can be easily transferred to other cities facing similar problems.

ACKNOWLEDGEMENTS

This study is a part of the study titled "Impact of Utility Cuts on Street Pavements" conducted by the Cincinnati Infrastructure Institute, University of Cincinnati, in cooperation with the City of Cincinnati and the American Public Works Association. The opinions and conclusions expressed are those of the authors. This report does not constitute a standard, specification or regulation.

REFERENCES

Arudi, R.S., P.D. Pant, A. Bodocsi and A.E. Aktan, "Planning and Implementing a Management System for Utility Cuts in City Streets", *Infrastructure: New Materials and Methods of Repair, Proceedings of the Third Materials Engineering Conference, ASCE,* San Diego November 1994.

Pant P.D., X. Zhou, R.S. Arudi, A. Bodocsi and A.E. Aktan, "Neural Network Bsed procedure for Condition Assessment of Utility Cuts in Flexible Pavements", *TRR 1399, TRB, National Research Council,* Washington D.C. 1993.

"Distress Identification Manual", *Cincinnati Infrastructure Institute, University of Cincinnati,* November 1991.

"Asphalt Overlays for Highways and Street Rehabilitation", *The Asphalt Institute, Manual Series No. 17,* June 1983.

A. Bodocsi, P.D. Pant, A.E. Aktan and R.S. Arudi, "Impact of Utility Cuts on Performance of Street Pavements", Draft Final Raport, Univ. of Cincinnati, January 1995.

FINITE ELEMENT MODELING TECHNIQUES
OF STEEL GIRDER BRIDGES

K. M. Tarhini[1], M. ASCE, M. Mabsout[2], M. Harajli[2], C. Tayar[3]

ABSTRACT

A typical one-span composite I-girder highway bridge was selected to compare the various finite element modeling techniques reported in the literature which could be adopted in evaluating a bridge superstructure. AASHTO design trucks (HS20) were positioned on the bridge in order to obtain maximum moments. The finite element program, SAP90, was used to perform the analysis along with its pre- and post-processing capabilities. The maximum wheel load distribution was obtained for each of the three-dimensional models using the maximum girder moment divided by maximum moment found in a simply supported beam due to a single line of wheel loads of the design truck. A summary of girder moments at critical sections and maximum load distribution factors for the four models are presented and compared with two published wheel load distribution formulas. This paper will assist structural engineers in selecting a finite element modeling technique in order to analyze new bridges or evaluate the load carrying capacity of existing bridges.

----

[1]   Department of Civil Engineering, Valparaiso University, Valparaiso, Indiana 46383

[2]   Department of Civil Engineering, American University of Beirut, P.O. Box 110236, Beirut, Lebanon

[3]   Graduate Student, AUB, and Structural Engineer, Dar Al-Handasah, Shair and Partners, Beirut, Lebanon

INTRODUCTION

A common type of highway bridge superstructure is the concrete slab placed on steel beams (I-girder). The analysis of these bridges is complicated by the general geometric boundaries and loading conditions. The current AASHTO bridge specifications (1992) contain a simple procedure used in the analysis and design of I-girder bridges. The AASHTO procedure is to isolate individual girders and calculate the maximum bending moments based on a single line of wheel loads from the design truck applied to the girder. The loads are positioned on the girder to produce maximum bending moment which is then multiplied by an empirical distribution factor (S/5.5 where S is girder spacing). Field testing has shown that many condemned bridges can carry loads in excess of the calculated values. Therefore, improving the bridge analysis methods will allow realistic calculations of the actual bridge strength. Numerous analytical techniques have been developed and applied to the analysis of concrete slab on steel girder bridges. Exact analytical solution is not possible without making simplifying assumptions. Realistic analysis of existing bridge superstructure may make it possible to better evaluate the load-carrying capacity and allow more bridges to continue in service or require minor repairs.

Recently AASHTO adopted new load distribution specifications on highway bridges (1994) as a result of research project NCHRP 12-26. These load distribution specifications are part of the new AASHTO LRFD Bridge Code (1994). It recommends the use of simplified formulae, graphical and simplified computer analysis, and/or detailed bridge deck finite element analysis in calculating the actual distribution of loads on highway bridges. Bridge engineers using the simplified formulae should understand the development process and limitations of these formulae. The formulae were developed for typical AASHTO design trucks, typical bridge types and dimensions. The new formulae are generally more complex than those previously recommended by AASHTO bridge specifications, but they also present a greater degree of accuracy. The adopted new formulae can be used to determine the wheel load distribution factors for moment and shear, in interior and exterior girders of straight or skewed, simply supported or continuous bridges. These formulae simplify the analysis process, do not require computer programs, and improve the accuracy of the results as long as the geometry of the bridge deck meets the applicable limitations imposed on these formulae.

Many graphical and simplified computer-based methods are available for calculating wheel load distribution. One popular technique is the use of design charts presented in the Ontario Highway Bridge Design Code (1983). Grillage analysis and simple computer programs such as SALOD (developed at the University of Florida by Hayes 1986) have become more acceptable than nomographs and design charts. The availability of powerful personal computers and finite element analysis software will assist bridge engineers in performing detailed bridge deck analysis and produce accurate results. However, the engineer needs to be familiar with the finite element method, mesh generation and its density, accurate modeling of the geometry and support conditions, selection of elements and their material properties, and loading positions that produce the maximum response in the bridge. Detailed finite element analysis of bridge decks could produce incorrect and inaccurate results if not carefully performed. The bending moment values may have to be calculated using the reported stresses by integrating over a cross-section. Bridge engineers are faced every day with unusual geometry and complex configuration of highway bridges which require detailed analysis and may not be allowed the use of simplified formulae or graphical procedures.

TYPICAL FEA MODELS

This paper presents a comparison of three-dimensional finite element modeling techniques employed by various researchers in predicting the actual behavior of I-girder bridges. The geometry of a bridge superstructure can be idealized for analysis in many different ways. It is in the idealization phase and the selection of finite element models that the greatest differences in approaches are encountered. Four published finite element modeling techniques considered for this study are:

a.   Hayes et al. (1986) idealized bridge superstructure using plate elements for the concrete slab and space frame elements for the steel girders. The centroid of each girder was coinciding with the centroid of concrete slab as shown in Figure 1.

b.   Imbsen and Nutt (1978) idealized bridge superstructure using ICES-STRUDL quadrilateral shell elements, with six degrees of freedom at each node, for the concrete slab. Girders were modeled using eccentrically connected space frame elements. The finite element model set up was similar to case "a" but rigid links were imposed to accommodate the eccentricity of the girders.

c.   Brockenbrough (1986) idealized bridge superstructure using MSC/NASTRAN quadrilateral shell elements for the concrete slab, flanges were modeled as space frame members, girder web was modeled using shell elements, and flange to deck eccentricity was modeled by a rigid link as shown in Figure 2.

d.   Tarhini and Frederick (1992) idealized bridge superstructure using ICES-STRUDL II finite element analysis. The concrete slab was modeled as an isotropic eight node brick element with

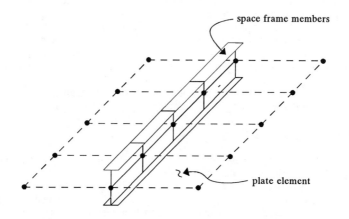

Figure 1.   Typical concrete deck and girder elements (case a)

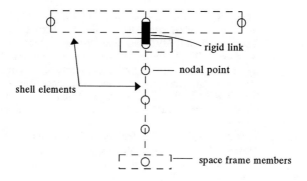

Figure 2.   Typical cross-section through part of finite element model (case c)

three degrees of freedom at each node. The steel girder flanges and web were modeled using quadrilateral shell elements with six degrees of freedom at each node.

SAP90 ANALYSIS

A typical one-span (56 ft) composite I-girder highway bridge was selected. The two-lane bridge deck was 30 ft (9.15 m) wide and consists of 7.5 inches (19 cm) reinforced concrete slab supported by four W36x160 steel girders spaced at 8 ft (2.45 m) as shown in Figure 3. The concrete compressive strength was selected to be 5 Ksi (34.5 MPa). Two AASHTO design trucks (HS20) were positioned simultaneously to produce the maximum moments at midspan as shown in Figures 3 and 4. This bridge was idealized using the modeling techniques described in the previous section. SAP90 was used in generating nodes, elements, and three-dimensional meshes of the bridge. A program was written to translate SAP90 stress output into plots and calculate the girder moments at critical sections.

The first model (Case a, Figure 1) idealized the concrete slab as a quadrilateral plate element with three degrees of freedom (d.o.f.) at each node. The girder was idealized as a beam or plane frame member. The second model (Case b) idealized the concrete slab as a quadrilateral shell element (SHELL with 5 d.o.f. at each node). The girder was idealized as a space frame (FRAME with 6 d.o.f. at each node). The nodes and elements for concrete slab were generated separately than the nodes and elements for girders. However, all girder nodes had the same coordinates as concrete slab nodes and SAP90 CONSTRAINTS option was used to link these common nodes. The third model (Case c, Figure 2) idealized the concrete slab as a quadrilateral shell element (SHELL with 5 d.o.f. at each node). The girder flanges were idealized as space frame elements (FRAME) and girder webs were idealized as a quadrilateral shell elements. Again, CONSTRAINTS option was used to link the common nodes at the interface between concrete slab and girder flanges. The fourth model (Case d) idealized the concrete slab as a brick element (SOLID with 3 d.o.f. at each node) and the girder flanges and web as quadrilateral shell elements (SHELL with 5 d.o.f. at each node). Generally, the discretization of all models were similar to the cases reported by Tarhini and Frederick (1992). The total moments at critical section for all cases showed about 3 or 5% error in the final results. The girder moments at critical sections for all four cases along with wheel load distribution factors are presented in Table 1. The maximum distribution factor was found in the second girder, as expected, due to the imposed loading conditions. The load distribution was calculated using maximum moment in the model divided by maximum moment found in a simply supported beam due to single line of a design truck (Figure 4).

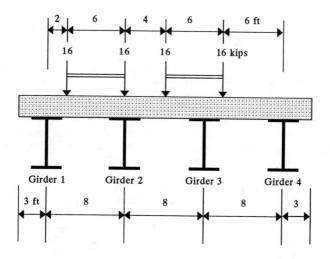

Figure 3.   Bridge Cross Section and Position of HS20 Trucks

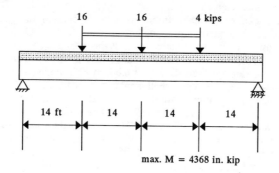

max. M = 4368 in. kip

Figure 4.   Longitudinal Direction

It appears from this preliminary study that a bridge engineer can model the concrete slab as shell elements and steel girders as space frame elements in order to obtain reasonable design moments. The wheel load distribution formula reported by Tarhini and Frederick (1992) gave a DF of 1.26. The new LRFD Bridge Design Specifications (1994) wheel load distribution factor was calculated to be 1.22. Both of these formulas agree with the three finite element analysis cases b, c, and d. Moreover, they were less than the previous AASHTO DF formula which gave 1.45 (DF = S/5.5).

SUMMARY

Bridge engineers can model highway bridge superstructures, similar to the ones presented in this paper, using quadrilateral shell elements for the concrete slab and space frame elements for the steel girders. Rigid links could be used to account for eccentricity. Cases c and d could be employed, for special bridge cross-sections, to represent the actual geometry, but they require longer time to prepare the input as well as more computer time. Stress distributions will be required at critical sections for Cases c and d to calculate the girder moments.

ACKNOWLEDGEMENT

This research project has been supported by Valparaiso University and the American University of Beirut. The authors appreciate the cooperation and support of the consulting firm Dar Al-Handasah, Shair and Partners for allowing the use of their SAP90 program.

REFERENCES

Brockenbrough, R. L. (1986). Distribution Factors for Curved I-girder Bridges. J. Struct. Engrg., ASCE, 112(10).

Guide Specification for Distribution of Loads for Highway Bridges (1994), American Association of State Highway and Transportation Officials, Washington, D.C.

Hays, C. O., Sessions, L. M., and Berry, A. J (1986). " Further Studies on Lateral Load Distribution Using Finite Element Methods." Transp. Res. Rec. (1072), Transportation Research Board, Washington, D.C.

Imbsen, R. A., and Nutt, R. V. (1978). Load Distribution Study on Highway Bridges Using STRUDL FEA Capabilities. Conf. on Computing in Civil Engineering, ASCE, New York, N.Y.

LRFD Bridge Design Specifications (1994). American Association of State Highway and Transportation Officials, Washington, D.C.

Ontario Highway Bridge Design Code (1983) 2nd Ed., Ontario Ministry of Transportation and Communications, Ontario, Canada.

Standard Specifications for Highway Bridges (1992), 15th Ed., American Association of State Highway and Transportation Officials, Washington, D.C.

Tarhini, K. M. and Frederick, G. R. (1992). Wheel Load Distribution on I-girder Highway Bridges. J. Struct. Engrg., ASCE, 118(5).

Table 1. SAP90 FEA Results

	GIRDER MOMENTS (INCH-KIP)				
	1	2	3	4	DF
Case a	4204	4433	4212	3764	1.015
Case b	4545	5095	4323	2570	1.166
Case c	4381	4968	4461	3211	1.137
Case d	4802	5206	4360	2630	1.192

Interactive Finite-Element Generation of Bridge Load Distribution Factors

Kyran Mish[1], Tim Osterkamp[2], Toorak Zokaie[3], and Roy Imbsen[4]

Abstract

This research is motivated by the desire to obtain more accurate wheel load distribution factors for bridge design than those obtained from the simple formulas in current use. The primary advantage of the simplified formulas is that the wheel load distribution factor can be obtained from a minimal set of input parameters (for example, girder spacing and bridge type). The main disadvantage is that they are inherently limited in their range of applicability. Through the utilization of more complex structural analyses, the range of applicability for wheel load distribution factors can be extended by a large margin, and more reliable and accurate results can be obtained. An application program (termed LDFAC, for Load Factor), has been implemented by the authors for automatic calculation of bridge wheel load distribution factors. The LDFAC program allows for more accurate analyses of several cases where simplified formulas are not applicable, and where a full three-dimensional analysis may not be warranted. These cases include bridges with skewed supports, continuous spans, and bridges where geometric parameters such as span length or girder spacing are outside the range of simplified formulas. The wheel load distribution factors obtained from LDFAC can be directly incorporated into the current design practice, providing added accuracy without complicating the existing design process.

[1] Adjunct Professor, Department of Civil Engineering, California State University, Chico, Chico, California 95929-0930

[2] Senior Bridge Engineer, Imbsen & Associates, Incorporated, 9912 Business Park Drive, Suite 130, Sacramento, California 95827

[3] Bridge Project Engineer, Imbsen & Associates, Incorporated

[4] President, Imbsen & Associates, Incorporated

Overview

Wheel load distribution factors are a useful means of simplifying the complexity of designing bridge decks (which form two-dimensional structures in plan) using only simple one-dimensional beam models. The load distribution factor allows the designer to consider the transverse distribution of loads in a bridge deck using factored loads applied to the longitudinal bridge members. For many bridges, sufficiently accurate estimates for wheel load factors can be obtained through the use of standard simplified formulas: for others (such as skewed or irregular bridges), the formulas may not yield accurate or conservative results. What is desired is a means to extend the utility of wheel load factors to these more complicated cases, without sacrificing the relative simplicity provided by the use of load factors for design (Zokaie, Osterkamp, and Imbsen, 1991).

By using the computer to construct and analyze an accurate model of bridge deck response, it is possible to synthesize wheel load distribution factors for a wide range of bridge deck geometries. The only serious limitation to using more complicated two- and three-dimensional computer models for analysis of the bridge deck is that many low-level analysis tasks (such as creating and interpreting a finite-element or matrix structural model for the bridge) must be handled by the bridge designer. In practice, it is possible to shield the designer from the details of the computational analysis by using a high-level interactive program interface (which requires only minimal input including simple geometric and material properties of the bridge deck), and performing most of the calculations in the background, so that the designer is presented only with simple input and output displays (Zokaie, Mish, and Imbsen, 1994).

The LDFAC software is based on a grillage and plate analysis, where a computer model of the bridge superstructure is used as the basis for structural response. Beam elements are used to model the behavior of the deck in the longitudinal direction, and Mindlin plate elements are used to model the transverse response. These plate elements are capable of capturing the shear deformations which become significant in common cellular sections (e.g., box girders or box beams), and they permit accurate results to be obtained even in the presence of highly skewed deck geometries. A textual menu-driven program interface is used to simplify preparation of input, and an integrated preprocessor is used to generate the model and the loadings from a simple and familiar set of input parameters. After the completion of the analysis, an integrated postprocessor is used to process the results and to compute the wheel load distribution factors. The postprocessor also permits the designer to compare the computed distribution factors with analogous formula-driven results. The schematic program architecture is shown in Figure 1 (note how this generic architecture resembles production finite-element analysis schemes).

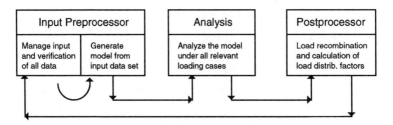

Figure 1: Architecture of the LDFAC Program

The overall effect of using the LDFAC package is that the designer can analyze a considerably broader range of bridge structures with only minor perturbations of the current design practice. Consideration of more accurate computational models for bridge analysis (such as fully three-dimensional plate, shell, and other continuum elements) for incorporation into the LDFAC program is an active area of research by the authors.

Program Description

The LDFAC program consists of several thousand lines of standard Fortran-77 source code, including a text-based menu system that permits interactive input of the data required to specify the geometry and material parameters appropriate for most classes of bridge cross section. The program provides sensible default values for many input parameters, allowing the designer to provide minimal geometric parameters (such as the number of spans, and a list of span lengths) in order to obtain a simplified bridge model. More accurate models can be obtained by providing more detailed input data sets, and these defining parameters can be archived for later re-use. A variety of loadings can be considered by the analysis component of the LDFAC program, including loads recovered from an extensible library of trucks. If the user prefers to place truck loads at a particular position on the bridge, the program analyzes that individual load case: otherwise, an automatic live-load generator is used to create sufficient load cases to estimate the desired wheel load distribution factors with reasonable accuracy.

Once all the input quantities have been specified, a two-dimensional finite-element analysis of the bridge deck is generated and solved. The LDFAC analysis is based on a grillage model: beam elements are used to model the behavior of the beams and the deck in the longitudinal direction, and beam or plate elements (depending on the presence of skew) are used to model the behavior of the deck in the transverse direction. Furthermore, the plate elements are capable of capturing the shear deformations which become significant in modeling of cellular sections (box girder and box beam). In the case of skew, continuum plate elements (based on Mindlin, or "shear-deformable" plate theory) are used in order to avoid inaccurate results commonly associated with skewed grillage meshes (Hambly, 1976). Details of the practical implementation of Mindlin plate theories can be found in most modern finite-element texts, e.g., the book by T.J. Hughes (Hughes, 1987).

One of the more novel schemes in the LDFAC program is an automatic live-load generator module that accurately estimates maximum factored response (with a user-specified set of multiple truck presence factors) for a variety of bridge geometries. The designer specifies the location of "Points-Of-Interest", (POI's), where maximal response is to be calculated, and the program's live-load generator determines truck numbers and placement in order to maximize shear and/or moment response. A simple flow chart for the automatic live-load generation scheme is shown in Figure 2:

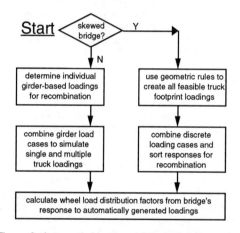

Figure 2: Automatic Live-Load Generation Procedure

The first step in the generation of live loads is to determine the longitudinal truck location so that a one-dimensional model of the bridge has an extremal response for the associated point-of-interest (i.e., maximum absolute value for shear response, maximum for positive moment response, and minimum for negative moment response). The truck selected by the user is moved incrementally forward and backwards over the entire bridge until those locations are found that extremize response for the associated point-of-interest. These optimal locations, and their corresponding extremal responses, are then saved for use in later calculations.

The next step in the live load generation process for straight bridges involves reduction of arbitrary truck loads into wheel line loads applied to the individual girders of the finite-element model. The transverse distribution of truck loads can be estimated by assuming that loads applied between girders will be redistributed to those girders according to a linear interpolation among the associated girders (i.e., simple beam reaction). The effect of this reduction is that exactly one load case per girder need be solved, with the overall response synthesized by appropriate recombination of the individual girder responses. Details of the entire decomposition process can be found in the NCHRP Phase III Report (Zokaie, Mish, and Imbsen, 1994).

After the various load cases have been solved, the postprocessor reconverts the girder wheel line loads into truck loads using transverse linear interpolation between adjacent girder load cases. This load recombination via transverse truck location is repeated for as many trucks as will fit on the bridge, and multiple truck loadings are combined according to code multiple-presence factors that may be input or modified by the designer. The resulting multi-lane and single-lane results are then divided by the corresponding one-dimensional responses in order to obtain the desired wheel load distribution factors.

Automatic load generation for skewed bridges is a more complex issue, involving a "brute-force" automated loading scheme in which an assortment of truck loads are applied to the bridge according to standard schemes of lane placement, and simplified schemes dictating where trucks should be located on the skewed deck in order to obtain maximal responses. Details of the skew load generator are found in the Phase II NCHRP Report (Zokaie, Osterkamp, and Imbsen, 1991), and Figure 3 shows a clarified version of the graphical methods for truck placement given in that reference.

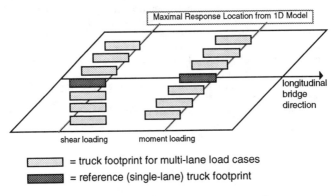

Figure 3: Truck Footprints for Skew Loading Case

For any bridge type, the load factors are calculated from the results of the automatic live-load generation process as ratios of girder response divided by analogous one-dimensional beam response. Once the wheel load distribution factors have been calculated, they are tabulated for each point-of-interest for interpretation and comparison to the analogous simplified formula result. A typical textual output screen is shown in Figure 4. More verbose text reports outlining load distribution calculations (as well as tabulating all relevant input data values) may be output for detailed analysis by the bridge designer.

```
LDFAC Version 1.0,  January 1994
     Out-of-Limits Results for Formula Are Marked With an Asterisk *

2 - Span  Straight  Box Girder  Bridge with          HS20TR Load

Multi-Lane Load Distribution Results

   Point-of-Interest Data          Analysis Results          Formula Results
   ----------------------          ----------------          ----------------
   No.  Type  Sp#  Sp%      Interior    Exterior      Interior    Exterior

     1   Shr   1    0        1.55738     1.54476       1.63696     1.57148
     2  +Mom   1    50       1.31429     0.93527       1.30712     1.14286
     3  -Mom   1    100      1.31990     1.07812       1.43783     1.25714

Press RETURN to continue :
```

Figure 4: Sample Output Screen Showing Distribution Factors

Although the LDFAC program substantially enlarges the class of bridges amenable to design and analysis using wheel load distribution factors, at present there are still several limitations that preclude its use in completely general circumstances. One important limitation is that only the bridge superstructure is modeled, so that substructural effects (such as the redistribution of moments from the superstructure into the supporting bent) cannot be considered without inclusion of some generalized support interaction effects. In addition, the present implementation of the program does not incorporate geometric effects due to bridge curvature (in plan) or superelevation. The authors are presently working to relax these limitations of the program, and to extend the analysis component to permit many types of bridge analysis applications without compromising the high-level interface presented to the designer. In addition, work is presently underway on the task of replacing the textual input and output routines (which were mandated by the desire to permit the program to be run on the widest variety of computer platforms) with a consistent graphical user interface (GUI).

Acknowledgments

The research presented in this document was performed under NCHRP Project 12-26/2 by Imbsen & Associates, Incorporated (IAI) of Sacramento, California. The main grillage/plate finite-element analysis portion of this computer program is based on a generic plane finite-element solver written by Mish and Haws (Mish and Haws, 1993) for the graduate finite-element course sequence at the University of California at Davis.

References

Hambly, E.C. (1976), "Bridge Deck Behavior", John Wiley and Sons, pp. 153-158

Hughes, T.J.R. (1987), "The Finite Element Method: Linear Static and Dynamic Finite Element Analysis", Prentice-Hall, pp. 310-382

Mish, K.D. and Haws, L. (1993), "A Generic Finite Element Model for Plane Problems", Proceedings of the Applied Computational Electromagnetics Society, March 1993

Zokaie, T., Mish, K.D., and Imbsen, R.A. (1994), "Distribution of Wheel Loads on Highway Bridges, Phase III", Final Report, NCHRP Project 12-26/3

Zokaie, T., Osterkamp, T.A., and Imbsen, R.A. (1991), "Distribution of Wheel Loads on Highway Bridges", Final Report, NCHRP Project 12-26/1

An Interactive Graphical Strut-and-Tie Application

Kyran Mish[1], Farid Nobari[2], and David Liu[3]

Abstract

Strut-and-tie models are useful idealizations for determining the shear capacity of reinforced concrete joints and structural members. These models consider the overall joint structural behavior to be localized into an equivalent truss composed of concrete struts, which carry compression forces, and reinforcing bar ties, which are loaded in tension. In theory, this simple truss analogy aids the designer in the placement of reinforcement suitable for all loading cases. In practice, the utility of the strut-and-tie model is limited by difficulties inherent in the construction of appropriate truss models for the joint. Such difficulties include determining optimal locations for truss members, determination of appropriate truss areas for concrete struts, and transformation of actual loads and displacements into equivalent conditions for the truss model. This research project utilizes an interactive computer visualization environment for creation and interpretation of strut-and-tie models for reinforced concrete joints.

Program Description

The fundamental premise of strut-and-tie modeling is that the complicated composite stress state of a reinforced concrete member can be reasonably modeled using a simplified truss analogy. Available steel reinforcement bars are modeled as tensile truss elements (ties), and the localized compression zones in the concrete are considered as compressive members (struts).

In order to apply the strut-tie analogy, the designer must perform several separate steps:

---

[1] Adjunct Professor, Department of Civil Engineering, California State University, Chico, Chico, California 95929-0930

[2] Senior Bridge Engineer, Imbsen & Associates, Incorporated, 9912 Business Park Drive, Suite 130, Sacramento, California 95827

[3] Bridge Project Engineer, Imbsen & Associates, Incorporated

(1) recognition of an underlying truss analogy within the concrete joint
(2) determination of member properties for each element of the truss
(3) determination of support or loading conditions for each vertex of the truss
(4) analysis of the resultant truss model to obtain strut/tie member forces
(5) comparison of the member stresses to limiting values for the members

Each of these steps involves considerable effort on the part of the designer. It is often particularly difficult to determine strut areas, since the effective cross-sectional area of available concrete can be difficult to estimate. Furthermore, determination of accurate loading conditions requires knowledge of the stress state adjacent to the joint (e.g., in the neighboring column or beam) in order to accumulate stresses into nodal truss loads. Some of the steps (such as the first) strongly depend upon the experience and creativity of the designer. Others (such as the fourth and fifth) are relatively straightforward to perform once an appropriate strut-tie model has been constructed.

Strut-and-tie models constitute a departure from more conventional design methods in that the designer performs a separate physical analysis of a complete structural subsystem in order to idealize the behavior of the member under consideration. The creation of the underlying truss model depends upon the designer's ability to estimate the response of the joint under consideration. Compression and tension regions must be isolated, and appropriate material and geometric properties synthesized. One way to aid the designer in synthesizing truss data is to incorporate a computational model of the joint into the strut-and-tie procedure. If an appropriate computer model for the joint can be constructed, the distribution of stress in the model can be used to estimate cross-sectional areas and applied loadings.

For the two-dimensional analysis and design of concrete beam-column joints, a plane stress idealization of the joint constitutes a good starting point for computational analysis of the joint model. The joint can be modeled as an isotropic linear-elastic material in order to determine an approximation for the flow of stresses in the actual joint (it is relatively straightforward to generalize this structural model to orthotropic materials). These calculated approximate stresses can be used as a backdrop in the construction of reasonable strut-and-tie models for the actual reinforced concrete joint. In order to isolate the designer from the tedious details of finite-element analysis (for example, constructing a finite-element mesh, synthesizing appropriate support conditions, and interpreting the results of the finite-element model), the program shields the designer from all unnecessary details of the underlying model. This isolation permits the joint to be modeled in very high-level terms (for example, resultant applied forces and overall joint dimensions), so that detailed understanding of finite-element modeling is not required for use.

An example of this high-level orientation is shown in Figure 1 below, where the type of joint to be modeled is selected from a modal dialog. Joint geometric and material properties are specified using similar menu commands. Support conditions appropriate for the joint are input using a more complex dialog, but the high-level orientation is followed throughout.

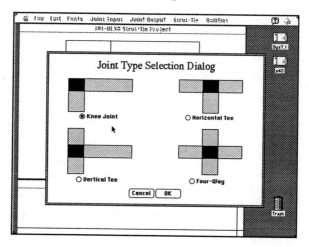

Figure 1: Selection of Type of Reinforced Concrete Joint

Once all geometric, material, support, and analysis data has been entered, the user initiates the actual construction and solution of the plane finite-element model by selecting the "Perform Plane Analysis" command from the "Joint Input" menu. Sensible default values for finite-element analysis (such as the relative element density in various parts of the joint) can be modified by designers comfortable with details of finite-element modeling principles. As the program progresses through the various steps of the analysis (namely constructing the model, generating the mesh, assembling and solving the finite-element equations, and constructing consistent nodal stresses and strains for postprocessing), appropriate progress indicators are written to the console window. Upon completion of the finite-element analysis phase, the program permits the designer to interpret the quality of the computer solution using standard visualization principles. Drawings of deformed mesh position, color contours of stress and strain, and various plots of principal stress trajectories can be used to visualize the computer solution, or to create a backdrop for construction of an appropriate strut-tie model for analysis.

Representative visualizations are shown in Figures 2 and 3. In the former, the trajectories of minimum principal stress are plotted over contours of that extremal stress component. In the latter, a close-up view of the joint region is used as a backdrop for construction of an equivalent strut-tie model

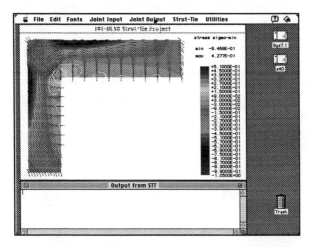

Figure 2: Minimum Principal Stress Trajectories in Knee Joint

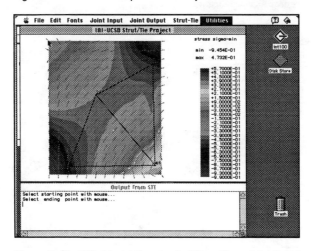

Figure 3: Strut-Tie Model Constructed Over Knee Joint Detail

Construction of an equivalent strut-tie system is facilitated by placing tension ties overlying a plot of maximal principal stress trajectories, and by locating compression struts over the display of minimal stress trajectories. Once the various truss members have been created, the more difficult job of calibrating the truss properties (both member section characteristics and nodal load resultants) must be addressed. In order to simplify these two steps, the program includes "probe" and "stress integration" facilities that allow sampling and/or integration of the stress field along a line between any two points in the model. The designer specifies end coordinates for a line using a pointing device (such as a mouse), and the program determines the normal and shear stress components along the given line segment. The resulting plot of stress distribution allows the designer to visualize the state of stress at various points along the line segment. This facility is useful for determining transitions between tension and compression (which is an important aid in estimating cross-sectional areas for the truss members), and for summing stresses to get resultant forces for application to truss nodes.

Figure 4 displays the result of integrating stresses over the compression region at the joint-beam connection of a knee joint during the joint opening loading. The upper curve in the main display shows the linear distribution of normal stress over the compression region (the top of the beam), while the lower curve graphs the distribution of shear stress. Note the display of the clipboard window at the lower right of Figure 4: all appropriate graphical interface elements (such as copying graphics to the clipboard) are supported by this interactive graphical application.

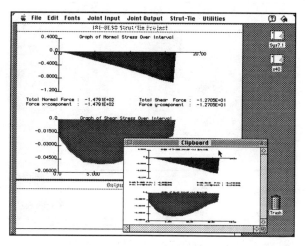

Figure 4: Resultant Stress Integration Over Selected Line Segment

The stress resultants obtained by graphically integrating along a line segment can be applied directly to the associated nodes of the strut-tie idealization as point loads: this process is illustrated by the dialog box shown in Figure 5. Note that the usual array of nodal loads or supports (pure force, rollers, and pinned connections) are selected, along with appropriate force boundary conditions obtained using the probe and integration tools.

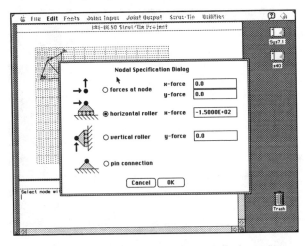

Figure 5: Selection of Truss Node Support/Loading Conditions

After the section properties are determined and the loads are applied, the truss model can be analyzed to determine the force state of the truss model. Figure 6 illustrates a graph of element overstress ratio (a color-coded display of the ratio of element force to element yield force). In this figure, the worst-case scenario for overstress ratio occurs in the diagonal strut connecting nodes two and four, and this overstress ratio (which is almost identical to one) indicates excellent agreement with an observed joint yield of 45 kips. The problem displayed in this figure (and in all the plots in this document) is taken from the calibration example given in the program user's manual (Mish, Nobari, and Liu, 1993), which is based upon large-scale computation simulations (Dameron, Kurkchubasche, and Rashid, 1992) of model tests conducted at U.C. San Diego (Priestley, 1993). For this calibration example, the simplified strut-tie analysis, when used by a reasonably experienced designer, predicted failure values for the joint within 10% of the those determined in laboratory simulations.

The authors are presently conducting further research into the practical utility of this interactive strut-tie modeling program. In addition to this planar example, full three-dimensional analyses are presently being explored with the use of more advanced graphical interaction and solution visualization schemes. It is expected that such two- and three-dimensional modeling applications will substantially simplify the task of detailing concrete designs for disturbed regions where modeling shear effects in reinforced concrete is of paramount importance.

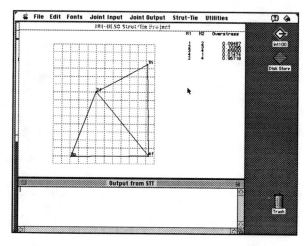

Figure 6: Display of Truss Member Overstress Ratios

Acknowledgments

This demonstration program has been assembled from a number of existing codes, each used successfully for some component of the overall solution process. The original concept for creating this program from these existing component sources was the pioneering paper of Alshegeir and Ramirez (Alshegeir and Ramirez, 1992). The main plane stress analysis portion of this computer program is based on a generic plane finite-element solver written by Mish and Haws (Mish and Haws, 1993) for the graduate finite-element course sequence at the University of California at Davis. The output modules (for plotting stresses and strains) were written by Mish for the same course sequence. The interactive truss analysis modules are adapted from a simple mouse-driven truss analysis program written by Mish for use in elementary statics and mechanics of materials courses. Funding for the development of this program (i.e., for the integration of the existing codes into a demonstration strut-and-tie analysis package) was provided by the California Department of Transportation and by the University of California at San Diego.

References

Alshegeir, A. and Ramirez, J. (1992), "Computer Graphics in Detailing Strut-Tie Models", ASCE Journal of Computing in Civil Engineering, Vol. 6, No. 2, April, 1992

Dameron, R.A., Kurkchubasche, and Rashid, Y.R. (1992), "Predictive Analysis of Outrigger Knee-Joint Hysteresis Tests: A Torsion/Flexure Test at U.C. Berkeley and Two Flexure Tests at U.C. San Diego", Work Performed Under Subcontract to U.C. Berkeley at to U.C. San Diego for the California Department of Transportation, 1992

Mish, K.D. and Haws, L. (1993), "A Generic Finite Element Model for Plane Problems", Proceedings of the Applied Computational Electromagnetics Society, March 1993

Mish, K.D., Nobari, F.S., and Liu, W.D. (1993), "User's Manual for MacStrut&Tie: An Interactive Graphical Demonstration Application for Strut-and-Tie Modeling of Reinforced Concrete Joints, Imbsen & Associates, Inc., Sacramento, California, October, 1993

Priestley, M.J.N. (1993), "Assessment and Design of Joints for Single-Level Bridges with Circular Columns", Structural Systems Research Project Report Number SSRP-93/02, Department of Applied Mechanics and Engineering Science, U.C. San Diego, La Jolla, California, February 1993

Bridge Design Software Development For The AASHTO Load And Resistance Factor Design Specifications

Richard Pickings[1], Karim Valimohamed[2], and Toorak Zokaie[3]

Abstract

The new AASHTO LRFD (Load and Resistance Factor Design) Specifications require a complex analysis of bridge structures. The complexity and practicality of using the specifications suggest that these analyses should be automated. Are programs written for previous AASHTO specifications amenable to programming the changes in the specifications? What are the issues that must be addressed? This paper describes our experiences with some of the issues associated with the implementation of LRFD superstructure loading into the AASHTO-BDS program. Analysis results for a typical noncomposite steel girder bridge are also presented.

Introduction

The new AASHTO LRFD Specifications [4] provide highway bridge design engineers with more rational and rigorous methods for the design and analysis of bridges than previous AASHTO Specifications. Many of these new methods are complex and need to be automated in order to promote widespread use of the LRFD Specifications in production environments.

We will discuss experiences encountered during the implementation of LRFD superstructure loading into an existing bridge design/analysis program called the AASHTO Bridge Design System; *BDS* henceforth. Many issues encountered during the BDS conversion may manifest themselves in other programs. We will attempt to share our insight for the development of new LRFD software.

---

[1] Bridge Research Engineer, Imbsen & Associates Inc., 9912 Business Park Drive # 130, Sacramento, CA 95827.
[2] Senior Engineer, M. ASCE, Imbsen & Associates Inc., 9912 Business Park Drive # 130, Sacramento, CA 95827.
[3] Bridge Project Engineer, M. ASCE, Imbsen & Associates Inc., 9912 Business Park Drive # 130, Sacramento, CA 95827.

The Load and Resistance Factor Design (LRFD) Specifications

The National Cooperative Highway Research Program (NCHRP) initiated a project in 1988 to develop a new set of specifications for bridge design. These specifications, which use a load and resistance factor design (LRFD) format, are complete and have been adopted by AASHTO as an alternative set of specifications that coexist with the 15th edition of the SLD/LFD specifications [3].

The LRFD specifications have improved consistency over previous versions. They have also resulted in more complex procedures. The differences between LRFD and SLD/LFD specifications are great enough that existing software programs cannot be utilized in most cases. Furthermore, the last decade has brought forth advanced software development technologies that can streamline and simplify the development process significantly. Many existing bridge design programs were developed in the 1960's using software development tools and techniques that may now be considered obsolete. It may not be feasible to upgrade these older programs both to new LRFD and software technology.

Software developers who implement LRFD programs face two types of challenges, technical and executive. Technical challenges focus on implementing an optimal engineering solution. Some of these include:

- Writing algorithms that correctly implement the LRFD Specifications.
- Interpretation of the Specifications. Parts of the Specifications do not contain adequate information for software implementation, other parts assume knowledge of 'standard practice'. In some cases, developers have to consult the authors or other experts to determine the original intent.
- Selection of optimal software tools and technologies.

Software development executives must consider short-term, long-term, and economic goals. Some issues that must be considered to achieve these goals are:

- The feasibility of recycling existing software.
- Customer needs such as cost, usability, hardware and software platforms, performance, and reliability.
- Internal requirements such as training and costs of development tools.

This paper will focus on technical challenges that were encountered during the implementation of LRFD superstructure loading into BDS.

Overview of AASHTO-BDS

The Bridge Design System is AASHTO's proprietary software product. It is being developed under the coordination of AASHTO with funding from approximately 30 states, the Federal Highway Administration (FHWA), and one Canadian Province.

The primary objective of BDS is to provide a comprehensive bridge design system that automates design by integrating the design of individual components. From the beginning, BDS was designed to be adaptable to changes in bridge design specifications.

Since its inception in 1986, the following BDS superstructure capabilities have been developed: an extensive geometrics subsystem; an analysis core; a model generator; a loading response (SLD/LFD/LRFD) processor; a specification checker; and, an output report generator.

Implementation of LRFD Superstructure Loading in AASHTO-BDS

Figure 1 shows the main procedural modules used in a BDS analysis. The legend indicates the level of influence that the specifications have on each module.

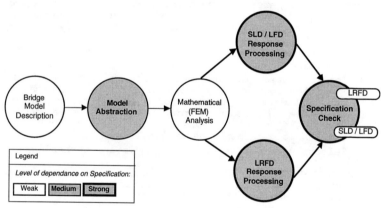

Figure 1
Modules Used in a BDS Analysis

BDS uses common bridge terminology, in the form of a Problem-Oriented-Language (POL), to describe the bridge model. Hence, the *bridge model description* has very little interaction with the specifications. Similarly, the finite element method (FEM) is simply a way of

performing a *mathematical analysis* on the model -- its use
is not defined by the specifications. The modules
performing *model abstraction, response processing, and
specification checking* are all dependent on the
specifications. Their implementation will be discussed in
detail below.

Model Abstraction
In BDS, model abstraction is the process of creating a
mathematical (finite element) model from the bridge model
description. This process involves two steps: 1) creation
of the analytical model from the bridge model, and
2) creation of the mathematical model from the analytical
model.

For superstructure analysis, the analytical model is a
single-girder (strip) model. BDS uses the same strip model
for both the LRFD and SLD/LFD specifications. However,
the live load distribution factors applied to live loading
are different.

Consequently, modifications are restricted to routines
that calculate LRFD distribution factors. These
modifications are straightforward.

Response Processing
LRFD's live loading and limit state response calculations
are considerably different from previous specifications
(SLD/LFD). The differences in algorithms and data
structures in BDS are so pervasive that the decision was
made to write a completely new response processor for
LRFD. The following sections outline some of these
differences.

LRFD Live Loading
The LRFD Specification uses the concept of a "notional"
truck. The term *notional* implies an entire range of
possible trucks, not a particular truck. The "notional"
truck is implemented as follows:
  1. Axles that do not contribute to the extreme effect
     under consideration are neglected.
  2. For the LRFD Design Vehicular Live Load, the extreme
     force effect is taken as the larger of:
      a. The effect of the design tandem and the lane
         load, or
      b. The effect of a variably spaced axle design
         truck and the lane load, or
      c. For negative moment and reactions, the effects
         caused by two fixed axle design trucks at a
         minimum spacing of 50 ft, and 90% of the lane
         load.

BDS uses an influence line live loading algorithm. Hence, item 1 is easily implemented by zeroing-out non-contributing portions of the influence prior to load application. However, the algorithm must keep track of the axles that contributed to the extreme effect when calculating corresponding force effects.

The calculation of maximum effects due to a truck with variably spaced axles was implemented using a Golden Section Search algorithm [5] that maximizes the effects due to two arbitrary fixed-axle trucks on the structure. Input to the algorithm includes a range in which the distance between the two trucks can vary. The general design of this algorithm also allows its use for maximizing effects of the "dual" truck configuration (item 2c).

LRFD Permit truck loading presented another challenge that deviated from previous specifications. The LRFD Specification states that: for lanes not loaded by the permit truck in multi-lane bridges, "*the other lanes should be assumed to be occupied by the vehicular live load as specified herein*". This implies that design trucks coexist on the bridge with the permit truck. The single-girder (strip method) analysis, as implemented in BDS, was developed to give accurate results for two live loading scenarios: a single truck passing over the bridge within a single lane (single-lane distribution factor) or; a single line of identical trucks side-by-side, one in each lane of the bridge (multi-lane distribution factor). The strip method is not applicable when different types of trucks are in each lane. To obtain accurate results for this case, a full three dimensional analysis is required. BDS uses a single-girder analysis and a single-lane distribution factor to analyze permit truck loading.

The LRFD Specification also states: "*the presence of the design lane load preceding and following the permit load in its lane should be considered*". The development of a reliable algorithm for maximizing truck and lane responses simultaneously proved to be very difficult. In BDS, the following approach was taken: first maximize the effects for the permit truck alone, then apply the lane load preceding and following the permit truck. Although not rigorous, this method will nearly always find the maximum results since local effects due to permit truck loading are larger than effects due to lane loading.

LRFD Limit States

The load combination algorithms used for LRFD limit states are very similar to those used for SLD/LFD load combinations. The major difference is that LRFD uses two load factors, maximum and minimum, for some permanent load cases. The maximum factor is to be used if the particular response maximizes the effect in question, and the minimum factor is to be used if it does not.

Implementation for a single-stage analysis is straightforward: maintain a maximum/minimum results envelope when combining all load cases into a limit state. However, the specification does not address issues concerned with multi-stage analyses. Should the factor selection be made on a stage-by-stage basis? This would not be rational. For example, it would be possible to use a factor of 1.25 for the component dead load case (DC) in Stage 1 and 0.90 in Stage 2. Using this interpretation, one could drastically change the results of an analysis by changing the definition of the events that occur in a given stage (consider the above example if both loads were applied in Stage 1). In BDS, unfactored results for each permanent load case are first summed for each stage, then load factors are applied.

Specification Checking

Currently, the specification checker has been programmed to evaluate the AASHTO SLD/LFD Specifications [3]. Rating capabilities have been programmed for the Maintenance Manual '83 [1] and the Guide Specifications '89 [2]. LRFD specification checking is forthcoming.

SICAD [6], a rule based system, is used to evaluate the specifications in BDS. SICAD facilitates the development of specification checking by providing a mechanism to separate the application program from the specification check. Consequently, changes to the specification can be made independently from the application program.

The application program establishes the context and classification variables that are used by the checker rule sets. This provides the software developer with the flexibility to separately define the scope of the application program and the rule set.

The specification checking infrastructure has been established in the BDS architecture. SICAD has proven itself to be a very powerful and flexible tool for this purpose. This will facilitate the implementation of LRFD specification checking.

Example LRFD Analysis -- Bridge Over Cherry Creek

This example demonstrates some of the LRFD analysis capabilities of BDS by examining an existing typical bridge in Niobrara County, Wyoming. The Bridge Over Cherry Creek is a three-span continuous, noncomposite, steel I-girder structure.

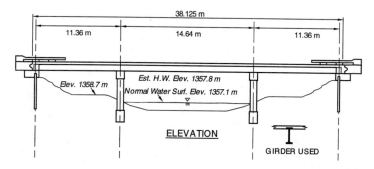

A continuous beam model was generated for a single-stage service analysis. Results compare an AASHTO 15th Edition [3] Service 1 combination using an HS25 Truck, and the LRFD [4] Service I limit state using the Design Vehicular Live Load. Moment response due to live loading and combinations are shown below:

- Live Loading

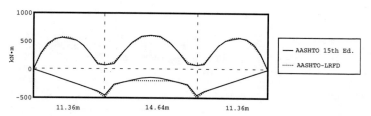

- Service Load Combination and Service-I Limit State

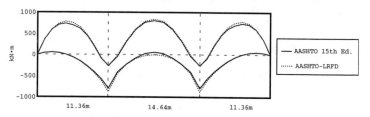

Examination of the above results reveals expected similarities between the two specifications when as HS25 Truck is used for the SLD analysis. The LRFD results are seen to be slightly more conservative, especially for negative moment due to live loading.

Summary and Conclusions
Superstructure analysis per the LRFD Specifications has been successfully incorporated into AASHTO-BDS and LRFD specification checking is forthcoming. The success of this project illustrates that existing bridge software can be upgraded for the LRFD specifications. Much of this success is due to the BDS architecture, which was designed to be adaptable to new specifications.

There are still some outstanding issues regarding interpretation of the specifications. However, preliminary results indicate that live load responses compare favorably to SLD/LFD.

Acknowledgments
The authors appreciate the effort of those who made this paper possible. In particular: FHWA, for funding the LRFD implementation; Dave Pope and the AASHTO-BDS Task Force, for technical guidance on the BDS Project; John Kulicki and NCHRP for assistance in interpreting the specifications; and Roy Imbsen, AASHTO-BDS Project Manager, for his support.

References
[1] AASHTO (1983). *Manual for Maintenance Inspection of Bridges.*
[2] AASHTO (1989). *Guide Specifications for Strength Evaluation of Existing Steel and Concrete Bridges.*
[3] AASHTO (1991). *Standard Specifications for Highway Bridges.* Fifteenth Edition.
[4] AASHTO (1994). *AASHTO LRFD Bridge Design Specifications.* First Edition.
[5] Press, W.H., Flannery, B.P., Teukolsky, S.A., and Vetterling, W.T. (1989). "Numerical Recipes -- The Art of Scientific Computing (FORTRAN Version)". Cambridge University Press.
[6] Lopez, L.A., Elam, S. and Reed, K. (1989). "Software Concept for Checking Engineering Designs for Conformance with Code and Standards". *Engineering with Computers* pp 63-78, Vol 5.

GeoFEAP for Geotechnical Engineering Analysis

R. David Espinoza, Jonathan D. Bray, Robert L. Taylor[1],
and Kenichi Soga[2]

ABSTRACT

GeoFEAP (Geotechnical Finite Element Analysis Program) is a general soil-structure interaction computer program for geotechnical applications. The program can model soil response due to different loading conditions including sequential construction, excavation, and consolidation. The material models currently available in GeoFEAP are briefly discussed. The results of an analysis can be represented in numerical and graphical format. A powerful feature of the program is the command language concept used in its development allowing the analyst/student to implement different solution strategies to simulate complex geotechnical problems.

INTRODUCTION

The computer software system, GeoFEAP, has been developed to make the finite element method (FEM) more accessible to geotechnical engineering practitioners and students. The geotechnical finite element analysis program GeoFEAP is based on the well-validated general purpose finite element program, FEAP, developed by R.L. Taylor, at the University of California at Berkeley (see Zienkiewicz and Taylor, 1989) with modifications undertaken to provide additional capabilities for solving problems of interest to geotechnical engineers. GeoFEAP is a general soil-structure interaction program for static analyses of two-dimensional nonlinear geotechnical problems. It can model sequential construction, excavation, and consolidation, and estimate the insitu stress state throughout the model. Currently, four types of elements are implemented in the program: (1) 3-node to 9-node soil elements, (2) 2-node bar elements, (3) 2-node beam elements, and (4) 4-node interface elements. With these type of element, the engineer can analyze earth structures, retaining walls, reinforced soil slopes, braced excavations and so forth.

Soil constitutive behavior can be approximated by an isotropic elastic model, a hyperbolic model (Duncan et al. 1980), an elastic perfectly plastic Drucker-Prager model or a modified Cam-clay model (Roscoe and Burland, 1968). Additionally, consolidation can be modelled using a linear elastic model or a Cam-clay model. The structural beam and bar elements are modelled using a linear isotropic elastic material. Interface elements allow modelling the

---

[1]   Post-Doctoral Researcher, Assistant Professor, and Professor, University of California, Berkeley, CA 94720

[2]   Lecturer, Cambridge University, UK

nonlinear behavior of soil-structure interfaces or shear planes within a soil mass. The GeoFEAP program structure was developed to provide flexibility to the analyst by providing a common framework for different material models. It readily allows modification of existing constitutive relationships or implementation of new material models within each element type.

An important feature of GeoFEAP is a command language concept used in the development of the program, allowing inclusion of a wide variety of solution algorithms, such as Newton-Raphson or modified Newton techniques. This flexibility is essential when solving problems using different stress-dependent material models. The solution scheme from pre-processing to post-processing is performed using a variety of commands, which allow performing the analysis in a batch mode, an interactive mode or a combination of the two. The program includes a powerful, integrated interactive graphic processor to allow the analyst to visualize and to interpret the FE results throughout the analysis process. The graphic processor includes plotting the deformed mesh, displacement vector, and sstress/ strain contours. A detailed description of the available commands and their implementation in GeoFEAP can be found elsewhere (Espinoza et al., 1995, Taylor, 1995). The program can be used in a variety of platforms such as workstations and personal computers.

In this paper, the GeoFEAP program structure, as well as some of its most important capabilities will be presented, an illustrative example problem will be used to highlight some of these capabilities. The use of the program in the classroom will be demonstrated.

SOIL MODELS

Soil elements are three to nine node, 2-D, isoparametric elements with compatible modes of displacement. Each node has two degrees of freedom (horizontal and vertical displacement). Soil element material properties during any solution increment are calculated based on (a) a linear elastic model, (b) a modified version of the nonlinear, stress-dependent hyperbolic model proposed by Duncan et al. (1980), (c) an elastic-perfectly plastic Drucker-Prager model or (d) the modified Cam-clay model (Roscoe and Burland, 1968). Soil element material properties during any increment are computed using the current stress state and stress history. A summary of the models available in GeoFEAP is presented in Bray et al. (1995). A brief description of the constitutive relationships implemented in GeoFEAP follows.

Linear Elastic Model

The simplest model to analyze the behavior of a soil structure under gravity and external loads is the linear elastic isotropic model. Although this type of model may not provide an accurate representation of the actual behavior of the soil across a wide range of loads and deformations, a number of geotechnical problems can be analyzed successfully with this soil model. The elastic soil model also allows the analyst to compare the computed results with closed form solutions available for some linear elastic problems. Additionally, as isotropic linear elastic problems are far less complicated than nonlinear ones (it requires only two parameters: the Young's modulus, E, and the Poisson's ratio, v), it allows the analyst to become familiar with the different features available in GeoFEAP prior to getting involved with more sophisticated soil constitutive models.

Incrementally Nonlinear Stress-Dependent Model

The usefulness of this model lies in its simplicity, with parameters (a total of 8) that can be obtained from standard triaxial compression tests, and its validity as it has been successfully applied in design of many geotechnical structures. The model is based on the assumption that the deviatoric stress ($q=\sigma_1-\sigma_3$) vs. strain (ε_1) curves for soils at different confining pressures (σ_3) can be approximated as hyperbolas (Fig. 1) where σ_1, σ_3 are the major and minor principal stresses, respectively, and ε_1 is the strain in the direction of σ_1 (Duncan et al., 1980). The hyperbola can be expressed in terms of the initial tangent modulus, E_i (which depends upon confining pressure, σ'_3), and the asymptotic value of the deviatoric stresses, $(\sigma_1-\sigma_3)_{ult}$. During primary loading, the tangent modulus increases with increasing confining stress (σ'_3) and decreases with increasing stress level SL [$=(\sigma_1-\sigma_3)/(\sigma_1-\sigma_3)_f$]. Primary loading is defined as all loading occurring at a stress state equal to or higher than all previous stress states SS [$= SL(\sigma_3/P_a)^{1/4}$ where P_a is the atmospheric pressure].

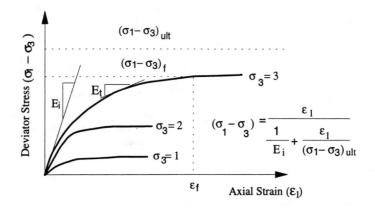

Fig. 1 Duncan hyperbolic model

The development of significant tension within soil elements is prevented by setting the element stiffness to very low values if the element is failing in tension. Tensile failure occurs when the minor principal stress is negative. Hence, this soil model achieves a near "no-tension" soil element, which is invaluable in realistic analysis of many geotechnical problems. In GeoFEAP, shear failure occurs when the stress level of an element exceeds 95% its shear strength. The soil modulus which fails in shear is modelled with a very low elastic modulus and a Poisson's ratio of 0.49. Lastly, the soil stiffness during unloading/reloading is assumed to be stress dependant and significantly stiffer than that during primary loading.

The nonlinear, stress-dependent soil behavior model used in GeoFEAP is a modified version of the hyperbolic stress-strain, and bulk modulus model proposed by Duncan et al. (1980). The original model was modified by Duncan et al. (1984) to: (a) provide improved modelling of bulk moduli at low stress levels and low confining stresses, (b) provide improved modelling of soil behavior during unloading/reloading, and (c) eliminate a source of potential computational instability for some types of incremental loading paths. These modifications improved the performance of the hyperbolic model with respect to analysis of the types of soil behavior associated with placement of soil, without affecting the model parameters.

The Drucker-Prager Model

The Drucker-Prager yield criterion has the form of a circular cone in principal stress space (Fig. 2) centered around the mean pressure p [$=(\sigma_1+\sigma_2+\sigma_3)/3$] and it is expressed in terms of mean pressure p and deviatoric stress q [$=(\frac{1}{2}((\sigma_1-\sigma_3)^2+(\sigma_1-\sigma_2)^2+(\sigma_2-\sigma_3)^2))^{\frac{1}{2}}$]. For elastic perfectly plastic materials, only four parameters (E, v, c and ϕ) are required to described the model. For the case of $\phi=0$, the Drucker-Prager model reduces to the von-Mises yield criterion.

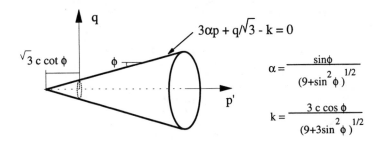

$$3\alpha p + q/\sqrt{3} - k = 0$$

$$\alpha = \frac{\sin\phi}{(9+\sin^2\phi)^{1/2}}$$

$$k = \frac{3\,c\cos\phi}{(9+3\sin^2\phi)^{1/2}}$$

Fig. 2 Drucker-Prager model

The Modified-Cam Clay model

The Cam-clay model was developed at Cambridge University to describe the elasto-plastic behavior of soils (Roscoe and Schofield, 1963). The significance of the model lies in the unified treatment of the volumetric and shearing response of saturated soils. The main feature of the model is the critical state concept which is based on the experimental observation that a soil undergoing shear eventually reaches a critical state in which the soil maintains constant volume under unlimited shear distortion while the effective stresses are stationary. The critical state is considered to be a function of the effective stresses and void ratio (e). The model also assumes the existence of an unique state boundary surface, which represents the

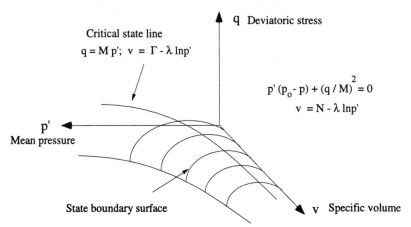

Critical state line

$$q = M\,p'; \quad v = \Gamma - \lambda \ln p'$$

$$p'\,(p_0 - p) + (q/M)^2 = 0$$

$$v = N - \lambda \ln p'$$

Fig. 3 Critical state line and state boundary surface

limit of possible states of a soil. The soil behaves elastically if a soil's state lies below the state boundary surface. When the soil's state is on the state boundary surface, the soil yields and responds as an elasto-plastic material. In this case, the plastic strains are computed using a hardening law and associated flow rule. To define a soil's state, the modified Cam-clay model uses the mean pressure, p', the deviatoric stress, q, and the specific volume, v [=1+e], which is the volume composed of an unit volume of soil material with surrounding voids. The modified Cam-clay model implemented in GeoFEAP requires six material parameters: M_c, λ, N, κ, v and f. In addition to these parameters, the overconsolidation ratio (OCR) and the coefficient of lateral stress at rest (K_o) are required to specify the initial in situ state within the soil mass.

In the modified Cam-clay model, the critical state is defined in three-dimensional p'-q-v space (Fig. 3) where M is the projection of the slope of the critical state line onto the p'-q plane, Γ is the specific volume at unit pressure at critical state, -λ is the slope of the projection of the critical state line onto the v-lnp' plane, p'_o is the isotropic preconsolidation pressure, and N is the specific volume at unit pressure on the isotropic virgin compression line. The modified Cam-clay soil model can also asses consolidation using a mix formulation to allow correct representation of the near incompressibility of an undrained loading of a saturated soil mixture. The dissipation of pore pressure due to consolidation and the resulting changes in the soil's mechanical response due to changes in effective stresses are captured by this mixed formulation when it is solved as a coupled problem. Capability of solving fully coupled problems is incorporated in GeoFEAP.

PERFORMING AN ANALYSIS AND POST-PROCESSING RESULTS

In many existing programs, only a limited number of fixed solution strategies are available to the analyst. Addressing all possible scenarios for simulating an actual construction sequence can be difficult with fixed solution algorithm programs. Moreover, there is not an unique algorithm that will render a solution for all cases when dealing with a system of nonlinear equations. For these reasons, a modular structure was implemented within GeoFEAP which provides a number of commands that allows the analyst to implement different solution strategies to simulate adequately a desired loading/construction sequence. GeoFEAP also allows for creating *procedures* which are a set of commands that are repeated several times during an analysis. This feature is important for first time users of GeoFEAP since the command language algorithm structure can be hidden within these procedures, thus allowing the students/analysts to concentrate initially on understanding the main concepts. A detailed description of GeoFEAP advanced features can be found in Taylor et al. (1995).

Output files can contain an excessive amount of data if all values are output for each load increment. If displacements and stresses at every node and Gauss point, respectively, are output by default, the interpretation of results may be cumbersome. A number of commands have been implemented within GeoFEAP to customize the output file, which allows the analyst to retain only essential information. GeoFEAP also includes more than 40 commands for graphic processing of the problem and for interactive visualization of the FE solution during the analysis process. The structure of GeoFEAP allows the analyst to go back and forth from the solution scheme to the graphic interface. The graphic post-processor includes plotting the deformed mesh, principal stress and strain contours, stress levels, deviatoric and volumetric strains, pore pressure variations (for consolidation problems) and displacement vectors. It also allows plotting stresses and displacements along a user specified line. Postscript files of the screen plots may also be obtained for inclusion in documents or other later uses.

USE IN THE CLASSROOM

The program is currently being used in the "Numerical Methods in Geomechanics" graduate course at the University of California at Berkeley. Well defined problems whose results can be compared with closed-form solutions are initially assigned to allow students to gain confidence in the program and become familiar with the different options available in

GeoFEAP. More sophisticated material models, as well as more complicated geometric and boundary conditions, are subsequently assigned. In the past, different computer codes were used for the different material models taught. Since learning a FE program is a time consuming process, using different computer codes in the same course did not allow for a broader coverage. As GeoFEAP provides a framework for different material models, more material can be covered in a unified fashion within the course.

APPLICATION EXAMPLE

This example shows how to build an embankment sequentially over a horizontal soil foundation. The geometry and stratigraphic characteristics are shown in Fig. 4. The corresponding mesh used in the analysis taking advantage of symmetry is shown in Fig. 5. Fig. 5 also shows the procedures used to perform the analysis. The soil response was modelled with the Duncan Hyperbolic

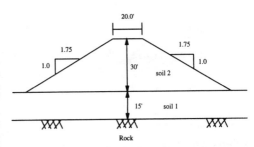

Fig. 4 Embankment Problem

parameters. The K_o coefficient and unit weight is used by GeoFEAP to compute initial stresses of the soil foundation assuming isotropic linear elasticity. Fig. 6 shows the variation of the horizontal stresses at different construction stages, and Fig. 7 shows the displacement vectors and deformed mesh (displacements have been magnified 20 fold).

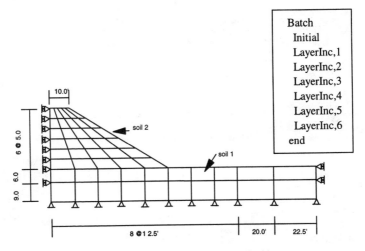

Fig. 5 Finite element mesh of Embankment Problem

CONCLUSIONS

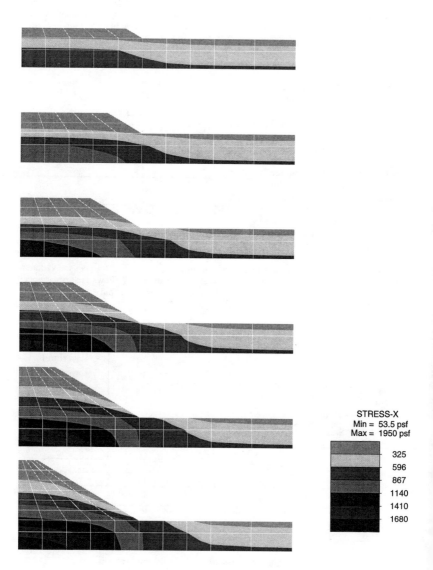

STRESS-X
Min = 53.5 psf
Max = 1950 psf

325
596
867
1140
1410
1680

Fig. 6 Horizonal Stress Distribution at different Stages of Construction

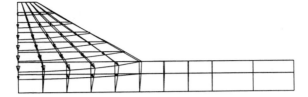

Fig. 7 Deformed mesh and displacement vectors

A powerful geotechnical finite element analysis program named GeoFEAP has been developed to make the FEM more accessible to geotechnical engineering students and practitioners. GeoFEAP allows for a discrete modeling of geotechnical structures including soil-structure interfaces. Several models are available in GeoFEAP to model the soil nonlinear response. The macro language structure implemented makes modelling of sequential construction and excavation relatively simple. The graphical capabilities, to which the analyst can access at any time during the problem execution, provides a powerful tool to the user.

ACKNOWLEDGEMENTS
Financial support was provided by the national Science Foundation under Grant No. BCS-9157083 and the David and Lucile Packard Foundation, and this support is gratefully acknowledged.

REFERENCES
Bray, J.D. Espinoza, R.D., Taylor, R.L., and Soga, K. (1995). GeoFEAP: Geotechnical Finite Element Program; Volume I-Theory, Draft Geotechnical Engineering Report No. UCB/GT/95-XX, Dept. of Civil Engineering, University of California, Berkeley, CA.

Duncan, J. M., Bryne, P., Wong, K. S. and Mabry, P. (1980). "Strength, Stress-Strain and Bulk Modulus Parameters for Finite Element Analyses of Stresses and Movements in Soil Masses," Geotechnical Engineering Research Report No. UCB/GT/80-01, University of California, Berkeley.

Duncan, J. M., Seed, R. B., Wong, K. S. and Ozawa, Y. (1984). "FEADAM84: A Computer Program for Finite Element Analysis of Dams," Department of Civil Engineering, Stanford University, Research Report No. SU/GT/84-03, November.

Espinoza, R.D., Taylor, R.L., Bray, J.D. and Soga, K. (1995). GeoFEAP: Geotechnical Finite Element Program; Volume II-User's Guide, Draft Geotechnical Engineering Report No. UCB/GT/95-XX, Dept. of Civil Engineering, University of California, Berkeley,CA.

Roscoe, K.H. and Burland, J.B. (1968). On the Generalized Stress-Strain Behaviour of Wet Clays. Engineering Plasticity. Cambridge University Press. pp. 535-609.

Taylor, R.L. (1995). GeoFEAP: Geotechnical Finite Element Program; Volume IV - Macro Commands, Draft Geotechnical Engineering Report No. UCB/GT/95-XX, Department of Civil Engineering, University of California, Berkeley, CA.

Taylor, R.L., Espinoza, R.D., and Bray, J.D. (1995). GeoFEAP: Geotechnical Finite Element Program; Volume III - Advance Capabilities, Draft Geotechnical Engineering Report No. UCB/GT/95-XX, Dept. of Civil Engineering, University of California, Berkeley, CA.

Zienkiewicz, O.C. and Taylor R. L. (1989). The Finite Element Method. 4th ed. Volume I. Mc Graw-Hill Book Co., London.

3-D FEM ANALYSIS OF EXCAVATION OF A SOIL-NAIL WALL

Khosrow S. Tabrizi[1], Nenad Gucunski[2] and M.H. Maher[3]

ABSTRACT

Soil nailing is an in-situ reinforcement technique to retain excavations constructed by successive increments. A 3-D FEM code is developed for simulation of this complex system by incorporating isoparametric elements to model soil, nail and their interface. Both material and geometrical nonlinearities are considered in the simulation. The developed code uses HISS plasticity model for soil and assumes elastic deformation of nails during soil-nail interaction. Soil to nail interface is modelled by using thin interface elements, which obey elastic-perfectly plastic behavior. The code is used to model actual field construction of soil-nailed walls for better understanding of their behavior during and after construction. In the analysis of typical soil-nailed walls, results presented consist of deformation, stress and nail force distributions.

INTRODUCTION

Soil-nailed walls have been in use for more than twenty years with first of such walls constructed in France in 1972. First documented use of permanent soil-nailed, however, was in Germany in 1979 (Mitchell and Villet, 1987). Soil-nail walls are considered to be a viable alternative for temporary and permanent support of

[1]Project Geotechnical Engineer, Bureau of Geotechnical Engineering, New Jersey Department of Transportation, 1035 Parkway Avenue, Trenton, NJ 08625.

[2]Assistant Professor, Department of Civil and Environmental Engineering, Rutgers University, Piscataway, NJ 08855-0909.

[3]Associate Professor, Department of Civil and Environmental Engineering, Rutgers University, Piscataway, NJ 08855-0909.

excavations. This is achieved through excavation in successive increments and introduction of passive inclusions known as nails. Commonly used limit equilibrium method for analysis of soil-nail structures provides information on the overall wall stability. Recent attempt on deformation analysis of soil-nail wall systems include finite element method (FEM) studies which simplify the true 3-dimensionality of the problem by treating it as a plain strain problem (Salama 1992).

Simulation of the excavation process for a soil-nail structure is presented by a 3-D FEM approach. This is a part of an on-going research by the authors in modelling of behavior and collapse mechanisms of soil-nail walls by finite elements. A typical situation is analyzed and results are presented in the form of face displacements, horizontal and vertical stresses distributions, and distributions of forces in the inclusions, i.e. nails.

DESCRIPTION OF THE FEM MODEL

Deformation, stress and force distributions of a soil-nailed structure are investigated for a model depicted in Fig. 1. A 0.9 m wide strip of a wall, 4.2 m long and 3.3 m high, is analyzed as a 3-D problem. Size of the soil elements are 0.3 m by 0.3 m by 0.6 m. Nail elements are defined as hexahedral elements having 0.03 m by 0.03 m cross section and 0.6 m long, surrounded by 0.01 m thick interface

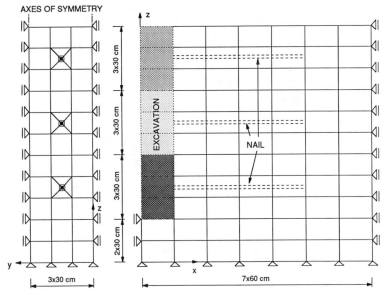

FIG. 1. A FEM Model of a Soil-Nailed Wall

elements. The total length of the nails is 2.4 m. Material and geometrical properties of interface elements and soil are described below.

Simulation Scenario

Actual field construction (excavation) of a soil-nailed wall is simulated by the four phases shown in Fig. 2. Three sets of excavations of 90 cm each are modelled. In phase I, in-situ stresses are calculated and utilized for the plasticity model described in the following section. To simplify model modifications as excavation progresses, nail and interface elements are introduced from the beginning, but with soil properties attributed to them. In phase II, the first set of excavation section, the removal of soil elements is modelled by assigning very low elastic properties to the elements in the excavation zone. In addition, forces equivalent and in the opposite direction to the weight and horizontal pressure from the removed elements are introduced. In phase III, nail and interface element properties are assigned as further excavation is simulated. The same procedure is followed in Phase IV for the third excavation section.

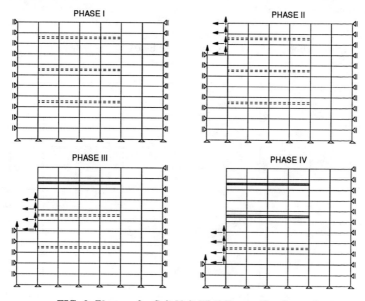

FIG. 2. Phases of a Soil-Nail Wall Excavation Scenario

Elasto-Plastic Model for Sand

A soil plasticity model identified as Hierarchial Single Surface (HISS) δ_1

model is used in the description of sand. This model is based on isotropic hardening with non-associative plasticity. The plastic potential function, defined by Desai *et al.* (1991), is

$$Q = \frac{J_{2D}}{p_a^2} - [\ \gamma(\frac{J_1}{p_a})^2 - \alpha_Q(\frac{J_1}{p_a})^n]\ (1 - \beta s_r)^m \qquad (1)$$

where: γ, β and m are ultimate behavior properties,
s_r is the stress ratio defined by $\sqrt{27}\ J_{3D}\ /\ 2J_{2D}^{1.5}$,
α_Q is the hardening parameter defined by $\alpha + \kappa(\alpha_0 - \alpha)(1 - r_v)$,
J_1 is the first stress tensor invariant,
J_{2D} is the second deviatoric stress tensor invariant, and
p_a is the atmospheric pressure.

Parameter α in the hardening parameter α_Q is $\alpha(\zeta, \zeta_v, \zeta_D, r_v, r_D)$ and it is defined by

$$\alpha = \frac{a_1}{\zeta^{n_1}} \qquad (2)$$

where: ζ is the trajectory of plastic strains,
ζ_v and ζ_D volumetric and deviatoric parts of ζ,
r_v and r_D ratios ζ_v / ζ and ζ_D / ζ, respectively,
a_1 and η_1 material constants, and
κ a parameter defined by $\kappa_1 + s_r \kappa_2$, where κ_1 and κ_2 are material constants.

Parameters used in the description of plastic behavior of sand are given in Table I. Elastic modulus and Poisson's ratio of sand are 50,000 kPa and 0.35.

TABLE I. Plastic Parameters for Sand and Clay

	SAND
γ	0.089
β	0.442
m	-0.5
n	3.0
a_1	0.00018
η_1	0.85
κ_1	0.2637
κ_2	-0.037

Soil-Nail Interface

Interaction between sand and nails is described by a set of thin layer interface elements (Desai et al. 1984). The behavior of these elements is modelled as elastic-perfectly plastic with respect to shearing in the plane parallel to the direction of the nails. Due to a relatively small scale of the model, the shear strength of the interface is assumed to be constant with height z of the model. Using the results of full scale pull-out tests on driven nails (Cartier and Gigan,1983), a shear strength of 44 kPa is assumed. Elastic properties of the interface elements are assumed to be equal to those of soil elements.

RESULTS

The behavior of the simulated model is described by horizontal and vertical stress distributions, stresses in the nails, and deformation patterns of the wall face. The following is the summary of the obtained results:

1) Stresses: It was observed that the existence of nails has no significant effect on vertical stress distributions as excavation progresses (Fig. 3a). On the other hand, a comparison of horizontal stress distributions for a nailed wall (Fig. 3b) and a pure soil excavation indicates that nails tend to retain a portion of the in-situ compressive horizontal stresses.

2) Nail stresses: A parabolic distribution of tensile stresses is observed in the nail elements. The stresses increase with every phase of excavation, and are higher at larger depths (Fig.4). This was also observed in FE simulations by Salama (1992).

3) Wall face deformations: Horizontal displacement patterns of the wall presented in Fig. 5 indicate that the displacements increase with depth in a linear fashion until the depth roughly equal to two thirds of the excavation depth. Maximum displacements rapidly increase with every excavation step. A comparison of the nailed and "no nailed" walls for Phase IV shows that nails slightly reduce face deformations during the excavation process. Based on comparisons for Phases II and III, it is anticipated that the difference would be more pronounced for a deeper excavation.

SUMMARY AND CONCLUSIONS

A 3-D FEM code is being developed to simulate construction and behavior of soil-nailed walls. The presented study deals with a FEM simulation of an excavation process in the construction of a soil-nail wall. Features of the developed FE model include: material nonlinearity for soil described by a HISS elastoplastic description of soil, description of soil-nail interaction by thin elastic-perfectly plastic interface elements, and geometrical nonlinearity. A comparison of the results for nailed and pure soil conditions is done to determine effects of nails on the behavior of the soil-nail system. Nails affect behavior in two ways: 1) they "conserve" a part of the in-situ horizontal compressive stresses after excavation and thus reduce a potential for tensile failure, and 2) they reduce the horizontal wall face deformations. Forces in nails increase with every stage of excavation, and are higher in nails at larger depths.

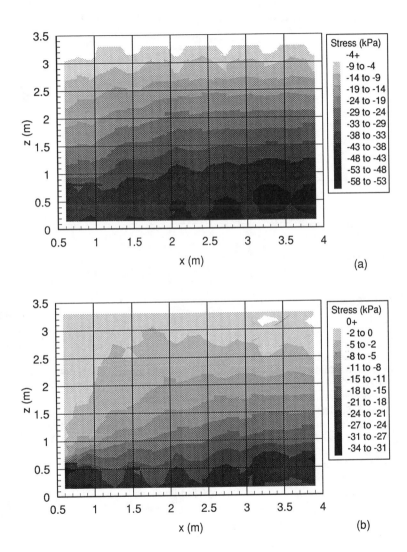

FIG 3. Vertical (a) and Horizontal (b) Stress Distributions After Phase IV

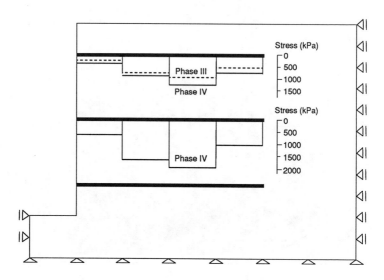

FIG. 4. Stresses in Nails

REFERENCES

Cartier, G. and Gigan, J.P. (1983), "Experiments and Observations on Soil Nailing Structures," *Proc. 8th European Conference on Soil Mechanics and Foundation Engineering*, Helsinki, Finland, 577-586.

Desai, C.S., Zaman, M.M., Lighterand, J.G., and Siriwardane, H.J. (1984), "Thin-Layer Elements for Interfaces and Joints," *Intl. Journal for Numerical and Analytical Methods in Geomechanics*, Vol. 8, pp. 19-43.

Desai, C.S., Sharma, K.G., Wathugala, G.W., and Rigby, D.B. (1989), "Implementation of Hierarchial Single Surface δ_0 and δ_1 Models in FE Procedure," *Intl. Journal for Numerical and Analytical Methods in Geomechanics*, Vol. 15, 649-680.

Mitchell, J.K. and Villet, W.C.B (1987), "Reinforcement of Earth Slopes and Embankments," *TRB-NCHRP Report No. 290*, Washington, D.C.

Salama, M.E. (1992), "Analysis of Soil Nailed Retaining Walls," Ph.D. Thesis, University of Illinois at Urbana-Champaign, Urbana, Illinois.

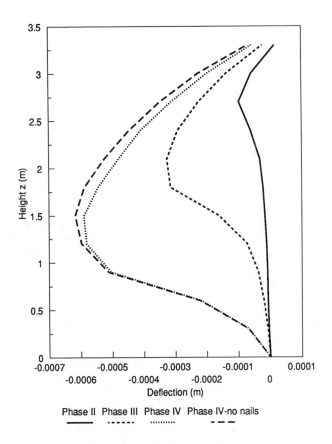

Phase II Phase III Phase IV Phase IV-no nails

FIG. 5. Horizontal Displacements of the Wall Face

Evaluation of Active Thrust on
Retaining Walls Using DDA

Max Y. Ma[1], Patrick Barbeau[1] and Dayakar Penumadu[2]

Introduction

Evaluation of active thrust on retaining wall is a common task in geotechnical practice and is often estimated using an approach based on limit equilibrium theory. This approach is difficult to use for a retaining wall with complicated loading and boundary conditions where analytical closed form solutions are difficult to obtain. The movement of wall due to the active thrust, distribution of lateral pressure by backfill and the interaction between wall and soil, which are important factors associated with the stability of a retaining system can not be fully considered in the limit equilibrium approach. Finite Element Method (FEM) has been used as an alternative in the past to analyze the retaining wall behavior to account for some of these factors. However, the shear failure of backfill soil features the non-linearity and discontinuity which violate some of the basic assumptions used in FEM. Moreover, FEM analysis requires information on the soil parameters, initial state of stress and the use of correct soil constitutive model and soil-wall interface elements. This increases the complexity of analysis and its interpretation.

The results of initial investigation on an alternative approach is presented in this paper to evaluate the active thrust on the retaining wall using Discontinuous Deformation Analysis (DDA). DDA was originally developed to provide a unique solution for large displacement computations and failure analysis of geologic systems consisting of discrete and discontinuous rock blocks (Shi, 1988). It can analyze the mechanical response of a rock block system under general loading and boundary conditions, with rigid body movement and deformation occurring simultaneously. Initial applications of DDA demonstrated that DDA is a powerful numerical tool for analyzing the boundary value problems in geotechnical engineering and has advantages over the limit equilibrium method (Chen and Pan, 1990).

----

[1] Former Student, Department of Civil and Environmental Engineering, Clarkson University, Potsdam, NY 13699-5710.
[2] Assistant Professor, Department of Civil and Environmental Engineering, Clarkson University, Potsdam, NY 13699-5710.

Like limit equilibrium method, a retaining system (which includes a wall, foundation of the wall and backfill soil) is represented as a system of discrete blocks. A solution of this system can be found without an assumption on the limit equilibrium status. The thrust on the wall is a function of wall movement and size, geometry and the shear strength of the backfill soil. In this paper, a retaining wall was analyzed by DDA and the results are compared with the solution obtained using limit equilibrium method.

Background

In contrast to other discontinuous methods (e.g., Discrete Element Method, Cundall and Strack, 1979), DDA is a displacement based method formulated on the basis of minimizing the total potential energy satisfying the equilibrium condition of moments, forces and stresses. Using displacement as a variable, it solves the equilibrium equation for a discontinuous system using dynamic formulation. It avoids the ill-conditioned stiffness matrix in static formulation due to the separation of blocks. Kinetic damping is applied in its static solution which effectively filters out the motions with higher modes to bring the system to a static condition (Chang and Acheampong, 1992). Since it is a dynamic formulation, time step is used and the equilibrium equation is updated and solved in each time step. Small displacements and deformations accumulated in each time step will lead to large displacements and deformations.

To obtain a complete solution, two conditions, equilibrium and compatibility are fulfilled. These conditions are automatically satisfied for every individual block by using a continuous displacement function. At contact points, the equilibrium condition is reached by minimizing of the potential energy due to contact forces. DDA implements compatibility condition by imposing the no-tension and no-penetration constraint between contacting blocks. It models the contact mechanism in a simple and effective way by using a contact spring. A contact spring is added when the blocks are in contact and is removed when blocks separate. However, whether a contact or separation exists is unknown when governing equations are assembled. Adding a spring at separation will impose a tension and removing a spring will result a penetration without contact force. Both cases violate the fundamental physical phenomena. It is thus necessary to impose the no-tension and no-penetration constraint to avoid such violation. Iteration named as open-close iteration has to be carried out by adding and/or removing springs in a trial and error manner until this constraint is satisfied.

Input parameters for DDA computation can be classified into five categories as listed in Table 1. For soil mechanics problems, mass is represented by DDA's element (block), which is linear elastic and the associated input are the values of Young's modulus, Poisson's ratio and initial stress state. The contact at the interface is represented by a contact spring in the normal and shear direction. The failure is governed by Mohr-Coulomb criteria. Tension strength (if desired) can also be included. A block consists of individual lines (segments) and the input parameters for geometry of a block are the coordination of these lines. The boundary condition in DDA can be displacement and/or force controlled and is imposed on the selected points in a block. The coordination of such points, magnitude and direction of displacements and forces are needed in the input. In addition, there are two other input parameters: assumed maximum displacement ratio and the time interval. Assumed maximum displacement ratio is a dimensionless quantity and is used for finding the possible contact at the current step based on the geometry of a block system. If this ratio is too large, it results in unnecessary contacts leading to large memory requirements and the time required for storing the contact information and computation. If it is too small some of the possible contacts may be undetected. To ensure the small displacements and the convergence of the open-close iteration, a small time interval is needed. If the solution within the time step does not converge within 6 iterations, the time interval will be automatically reduced to one third of its initial value by the DDA code. If the maximum displacement at one time step is three times larger than the distance for detecting the possible contact, the time interval can also be reduced. Complete details of these parameters are given by Shi (1993). Deformation at contact points are represented by the deformation of contact spring. If soil mass is divided by the block mesh, normal stiffness (k_n) of contact spring is chosen to be equal to the Young's modulus (E) of the soil. Shear stiffness (k_s) of spring is determined from Poisson's ratio (v) and k_n as $k_s = k_n(1+v)/2.0$ (Chang, 1991).

Table 1. Input Parameters in DDA.

Block (Soil Mass)	Young's Modulus, Poisson's Ratio, Density, Initial Stress
Interface	Normal/Shear Contact Stiffness, Friction Angle Cohesion and Tension
Geometry	Coordination of Blocks
Computation	Assumed Max Displacement Ratio, Time Step and Interval
Boundary Condition	Coordination of Points for Displacement Control or Force Control

Retaining Wall System using DDA

The stability of a soil retaining wall is often evaluated considering the potential for it to slide, overturn or based on the foundation bearing capacity. In evaluating the stability, it is important to predict the active lateral thrust on the wall due to soil accurately. Movement of several inches and even several feet are often of no concern as long as there is assurance that failure associated with larger motions will not suddenly occur (Lambe and Whitman, 1971). Thus the approach to the design of retaining wall in the past was to analyze the conditions that would exist at a collapse condition, and to apply suitable safety factors to prevent collapse regardless of the movement of the wall in most situations using limit equilibrium theory.

The active thrust, which develops as the backfill is placed with or without surcharge load, pushes the wall outwards. This outward motion is resisted by the sliding resistance along the base of the wall and by the passive resistance of the soil lying above the toe of the wall. The active thrust also tends to overturn the wall and is resisted by the weight of the wall and the vertical component of the active thrust. The weight of the wall is thus important in two ways: it resists overturning and it provides frictional sliding resistance at the base of the wall. A retaining wall together with the backfill and the soil that it supports is an indeterminate system. The magnitude of the forces that act upon a wall cannot be determined from static's alone. Thus the design of such a wall was traditionally on the analysis of the forces that would exist if the wall were to fail. In most cases, the failure is a result of fully mobilized shear strength on the failure surface within backfill, which means that a system of the wall and failed soil mass reaches the limit equilibrium condition. The failure surface is unknown and the active thrust depends on the assumed failure surface. The critical failure surface is the one that corresponds to maximum active thrust. Initial investigation of evaluating active thrust using DDA has been done for the following three retaining wall configurations (Table 2).

Case	φ	β	δ	θ
1	30	0	0	50.0,55.0,57.5,60.0,62.5,65.0,70.0
2	30	0	20	50.0,55.0,57.5,60.0,62.5,65.0,70.0
3	30	12	30	42.5,45.0,47.5,50.0,52.5,55.0,57.5,60.0

φ: friction angle of the backfill soil
β: inclination of the backfill surface
δ: friction angle between the wall and backfill soil
θ: inclination of the assumed failure plane

Table 2. Geometry and Friction Angle Used in the
Three Retaining Wall System Configurations

A typical retaining wall configuration used in the DDA code is shown in Figure 1. The continuous soil and wall system is represented by three discrete blocks. The corner where the three blocks meet was cut to allow the sliding without the 'corner effect' (Chen and Pan, 1990). A small amount of surface misalignment was introduced between the wall and soil blocks to avoid the vertex-vertex contact and the potential numerical errors associated with it.

Following three assumptions were used in the analysis of the retaining wall using DDA: 1) The rupture and the backfill surface are planar; 2) Failure wedge is rigid; and 3) Plane strain conditions are valid. The straight lines of block boundaries represent the plane rupture surface and the backfill surface. A large value of Young's Modulus (compared to the modulus of elasticity of dense sand) of 1 GPa was used for the failure wedge. Both plane strain and plane stress problems can be analyzed in DDA using appropriate elastic sub-matrix. A plane strain elastic sub-matrix is chosen in this study. In limit equilibrium theory, the shear stress is fully mobilized along the rupture surface (interface of soil-soil), and interface of the backfill soil and the wall. The fully mobilized shear stress in DDA means that sliding occur at the contact points. In a retaining wall system, the wall provide the resistance to prevent such sliding. The wall resistance depends on the self-weight and

The stiffness of the normal spring used is comparable with the modulus of elasticity of dense sand. The sensitivity study on the values of E and k_n used in the analysis (i.e. for E > 1 GPa and normal stiffness of contact spring (k_n) ranging from 20 MPa to 100 Mpa) on the active thrust calculated, indicate a small effect . As discussed before, the time interval is a crucial input parameter along with the assumed Maximum displacement ratio. A balance has to be made between less iteration steps and efficiency of computation, both of which depend on the value of time interval that is used. Figure 2 shows the active thrust computed for three different time intervals for the retaining wall system with soil friction angle of 30 degree, horizontal backfill surface, no wall friction and a 60 degree failure wedge. The active thrust did coverage to a constant value in all the three cases. However, the active thrust for 0.25 second interval is much higher than the converged solution at the initial stage because of the large time interval used which violated the small displacement assumption. If the time interval is too small (0.01 seconds), solution takes much longer to converge and may be due to the kinematic damping used in DDA's static solution. A value of 0.05 seconds was used as the time interval in this paper for all the cases.

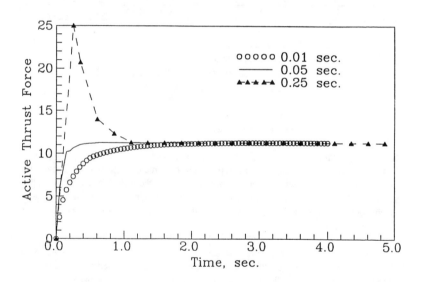

Figure 2. Effect of Time Interval on the Solution

Results and Conclusion

The active thrust (P) acting on the gravity retaining wall was calculated using the DDA code for a given trial wedge for the following three cases: 1) Horizontal backfill without wall friction; 2) Horizontal backfill with wall friction; and 3) Inclined backfill with wall friction. For the three cases of the gravity retaining wall, results obtained using DDA

friction angle between the wall and the foundation soil, which is assumed to be 30 degrees in this study. Reducing the wall density in DDA will reduce the self weight of the wall. DDA analysis involved the use of 100% of the wall density in the first iteration. This value was further reduced if the sliding along the rupture surface did not occur. During each iteration with different densities, the lateral thrust acting on the wall (F_w), the shear force at the bottom of the wall (F_S) and the status of the contact points (sliding or not sliding) are monitored. Based on this information, a retaining wall system can be identified to be stable, at limit equilibrium or unstable. The active thrust is equal to F_w corresponding to the limit equilibrium condition. It was found that when a system is near the limit equilibrium status, less than 1% of self weight can determine whether this system is either stable or unstable. The values of input parameters used in this study are listed in Table 3.

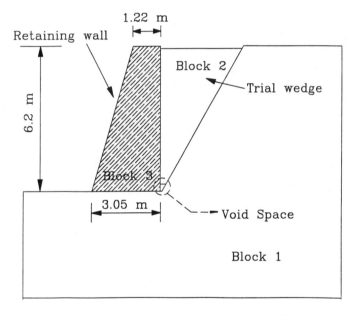

Figure 1. Configuration of a Simple Retaining Wall

Input Parameter	Values	Input Parameter	Values
Time Steps	40 to 50	Young's Modulus	1 GPa
Time Interval	0.05 sec.	Poisson Ratio	0.3
Assumed Max. Disp.	0.01	Normal Stiffness of	100 MPa
Ratio		Contact Spring	
Fixed Points	3	Soil Density	1.72 T/m^3

Table 3. Summary of Input Parameters

approach are compared with the limit equilibrium solution (Lambe and Whitman, 1979) in Figure 3, using the coefficient K $(= P / (1/2)\Upsilon H^2)$ for varying inclinations of the trial wedge.

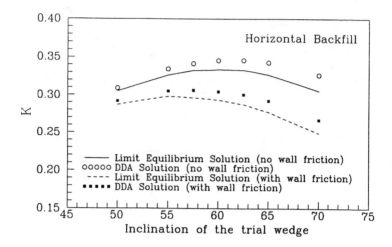

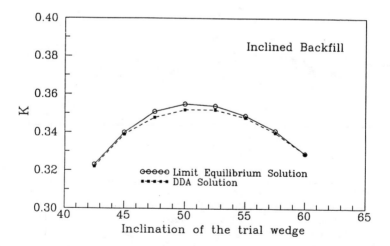

Figure 3: Comparison of Solution Obtained Using DDA
with Limit Equilibrium Approach for the Gravity Retaining Wall

The active thrust calculated for the three retaining wall configurations using the DDA approach compared well with the limit equilibrium solution. It is interesting to note that for a horizontal backfill, the active thrust predicted by DDA was consistently larger than the values obtained using limit equilibrium solution unlike the case with inclined backfill (Figure 3). The corners at which the two or more blocks meet was cut and this avoided 'locking effect' between the adjoining blocks at corners. The sensitivity analysis of input parameters associated with the simulation of the retaining wall block system (modulus of the block and normal spring stiffness) indicate a relatively small effect on the computed active thrust. This investigation revealed the potential for the use of DDA for solving retaining wall systems for simple geometry and loading conditions. This investigation provides incentive for using DDA for retaining walls with complex geometry and loading conditions (static and dynamic) and may prove to be a valuable tool for analyzing earth retaining systems.

Appendix I. References

Chang, C. S. (1991). "Discrete Element Method for Slope Stability Analysis," Journal of Geotechnical Engineering, ASCE, Vol. 118, No.12, 1889-1905.

Chang, C. S and Acheampong, K. B. (1993). "Accuracy and Stability for Static Analysis Using Dynamic Formulation in Discrete Element Methods," 2nd Int'l Conf. on Discrete Element Methods, MIT, Cambridge, 379-389.

Chen, W. F. and Pan, A. P. (1990). "Finite Element and Finite Block Methods in Geomechanics," Report. CE-STR-90-20, School of Civil Engineering, Purdue University, West Lafayette, IN, 1-14.

Cundall, P. A and Strack, O. D. (1979). "A Distinct Numerical Model for Granular Assemblies," Geotechnique, Vol. 29, 47-65.

Lambe, T. W. and Whitman, R. V. (1979). "Soil Mechanics, SI Version," John Wiley & Sons, 162-185.

Shi, G.-H. (1993) "User's Manual of Discontinuous Deformation Analysis Codes," WES-Geotechnical Lab Report, 1-30.

Shi, G.-H. (1988). "Discontinuous Deformation Analysis-A New Numerical Model for the Static and Dynamic Analysis of Block Systems," Ph.D. Dissertation, Department of Civil Engineering, University of California, Berkeley.

Temporary Earth Retaining Advisor (TERA);
A Selection-Type KBES

By Carlos Ferreira[1]

and

Ali Touran[2]

Abstract

This paper describes TERA, a selection-type knowledge based expert system that
assists the user in determining the preferred temporary earth retaining and bracing
system for a project excavation. TERA considers soldier piles and lagging, steel
sheeting, lime columns, slurry walls, soil nailing, unsupported excavation and
ground freezing as possible candidates for temporary earth retaining system. Based
on user responses regarding project conditions, TERA evaluates and rates each
temporary support system. The program provides textual and graphical explanation
for each temporary earth retaining system as they relate to the specific project
described by the user. It also evaluates the method of supporting the retaining
system by considering internal braces, tiebacks and cantilevered systems. Creating
a powerful graphical user interface was a major objective of this effort. The
program solicits user inputs by presenting a number of sketches, diagrams and tables
where the user can insert dimensions, depths and clearances, and choose or specify
soil and ground water conditions.

---

[1] Senior Engineer, Metcalf & Eddy, 30 Havard Mill Square, Wakefield MA
01880, (617) 224-6410.

[2] Associate Professor, Department of Civil Engineering, Northeastern
University,Boston, MA 02115. (617) 373-5508.

Introduction

Selecting a temporary earth retaining and bracing system for a project excavation is a problem faced by many contractors. There are many types of excavations and temporary earth retaining structures. There is not always a precise answer to the problem of requirement and selection of a cost effective temporary earth retaining structure. It depends on several variables and project constraints. Besides construction handbooks, technical literature, and OSHA regulations, the experience of experts plays an important role in the selection process. Selection of a temporary earth retaining structure doesn't have an absolute solution or algorithmic structure and is well suited to the application of a selection-type knowledge based expert system. This paper describes TERA Temporary Earth Retaining Advisor which is a selection-type knowledge based expert system that was developed to address this problem.

TERA considers the following temporary earth retaining systems; soldier piles and lagging, steel sheeting , lime columns, slurry walls, soil nailing, unsupported excavation and ground freezing. The following wall support systems are also considered in TERA; tiedback, internally braced, and unbraced excavations. Knowledge about these systems was acquired from several sources. A literature search was performed to document the knowledge related to each temporary earth retaining technology. The handbooks, design guides and periodicals listed in the list of references were used. Interviews were conducted with excavation and geotechnical engineering experts. All of this knowledge, experience, judgment and reasoning for selecting a temporary earth retaining system was grouped into the following five categories for analysis and comparison purposes: Description, Performance and Operational Considerations, Rules of Thumb, Advantages and Disadvantages.

Upon Review of all this knowledge it was determined that selection of a temporary earth retaining system depends on one or more of the following variables:
Existing soil conditions
Existing groundwater conditions
Geographic location
Topographical conditions
Existing Structures
Obstructions above ground
Obstructions below ground
Cost
Project Constraints:
How quickly excavation needs to be in place and for how long.
Size of the excavation; length, width, depth.
What types of equipment and labor are available.
Requirements for the work area inside the excavation.

Whether the temporary earth retaining structure is to be removed after construction or left in place.
Sensitivity of people and structure adjacent to the excavation.

Each one of these variables was considered for the rule base of each temporary earth retaining technology. Each variable's level of importance in determining a technology's feasibility varied from one technology to another. For example water conditions may be an important variable in determining the feasibility of unsupported excavation however it may be irrelevant or less important in determining the feasibility of slurry walls. Only variables that had an impact on a technology's feasibility were added to its rule base.

Rule Representation

Backward chaining was selected for representing the rule base of each technology. Each temporary earth retaining technology is considered to be completely feasible and is given a rating of 1 at the start of the analysis. Certainty Factors (weights) were assigned to each rule to incorporate the varying importance each variable has on particular technology's feasibility. A technology's rating is multiplied by a rule's certainty factor after each rule is fired. Responses to rules indicating a technology is feasible do not affect the technology's rating. Responses to rules indicating technology is not feasible reduce the technology's rating when multiplied by the rule's certainty factor. Certainty factors were assigned to rules with an understanding that a method is considered feasible at the end of the analysis if its final rating is greater than 0.64 and not feasible if its final rating is less than or equal to 0.64. The following is a list of rules and certainty factors utilized by TERA.

Rules and Weights for Selecting a Temporary Earth Retaining Technology

If horizontal obstruction < 50 ft. then 0.9 * SS; 0.9 * SPL.
If horizontal obstruction < = (depth /abs(tan(34))) then 0.64 * UE.
If (water depth < depth) and (dewater is "no" then 0.5 * UE; 0.5 * SPL; 0.5 * SN.
If (water velocity is > 3 ft/sec) and dewater is "no") and (water depth < depth) then 0.5 * GF.
If soil type is "Granular" then 0.9 * LC; 0.9 * SW.
If soil type is "Cohesive" then 0.9 * SN.
If (soil density is "loose" or "very loose" or "soft" or very soft") then 0.8 * SN.
If boulders exist then 0.8 * GF; 0.8 * SS.
If soil is uniform then 0.8 * SN.
If vertical obstruction < 12 then 0.5 * GF; 0.5 * SPL; 0.5 * SS; 0.5 * LC.
If vertical obstruction > 12 and < 25 then 0.95 * SPL; 0.95 * SS; 0.95 * LC.
If temporary structure cannot remain in place after construction completion the .50 * SW: 0.5 * LC.

If excavation depth > 5-ft. then 0.64 * UE.
If excavation depth > 80-ft. then 0.64 * LC.
If excavation depth > 40-ft. then 0.64 * SN.
If excavation depth > 200-ft then 0.80 * SW.
If excavation depth > 200-ft. then 0.80 * SS.
If excavation depth > 200-ft then 0.80 * SPL.
If excavation depth > 1000-ft then 0.80 * GF.

Note: SS - steel sheeting rating; SPL - soldier piles and lagging rating; UE - unsupported excavation rating; SN = soil nailing rating; GF - ground freezing rating; LC - lime columns rating; SW - slurry wall rating. Rules referenced from Fang, H. (1991), Pile Buck Inc. (1992), Pile Buck Inc. (1987), Ratay, R. T. (1984), and Carson, B. A. (1980).

Rules and Weights for Selecting a Wall Support Method

If depth of external underground horizontal obstruction > 10-ft. and distance to horizontal obstruction < 40-ft. and excavation depth > 10-ft. then 0.64 * tie.
If there are existing internal underground obstructing then 0.64 * brace.
If excavation must remain free of obstruction then 0.64 * brace.
If the elevation difference between opposite sides of the excavation is > 10-ft. then 0.8 * brace.
If excavation depth < 10-ft. then 0.64 * tie.
If excavation depth > 15-ft. then 0.64 * unbrace.
If excavation depth < 50-ft. then 0.80 * brace.

Note: Tie = tiedback rating; Unbrace = unbraced rating; Brace = Braced rating. Rules referenced from Fang, H. (1991), Pile Buck Inc. (1992), Pile Buck Inc. (1987), Ratay, R. T. (1984), Carson, B. A. (1980) and Schnabel, H. Jr. (1982)

Identification of Needs

The following was considered at the start of developing the expert system:
  1. Identification of users: Engineers, contractors and students.
  2. Identification of user requirements: Develop a prototype expert system that advises the user about which type of temporary earth retaining technology and wall support technology best meets their project constraints, while providing information on possible alternatives.
  3. Identification of hardware requirements: The prototype must be able to run on computer systems presently being used by the identified users, IBM PC or Macintosh compatible computers.
  4. Identification of software requirements: The prototype must be able to operate under the operating system being used by the identified users, Microsoft Windows. Also unlike other software developed so far in related fields, it should be capable of providing an efficient graphical user interface and be

compatible with recent deveopments in multimedia applications. This is very important given the fact that our objective was to develop a training tool for inexperienced engineers.

Given the above requirements, Toolbook by Asymetrix Corp (1991) was selected to develop TERA. Toolbook is an objected orientated software development tool which operates in the Microsoft Windows Environment. Software translation programs are available to translate application developed in Toolbook into Supercard version that runs on the Macintosh computer.

TERA Query Facilities

TERA has a graphical user interface which facilitates information solicitation from the user. TERA consists of two modules, one for determining the feasibility of the temporary earth retaining technologies and the other for determining the feasibility of wall support technologies. For both modules, user inputs are solicited by presenting a typical project excavation to the user, that has query buttons strategically located next to items requiring user responses. The query screen for the temporary earth retaining technologies is shown below.

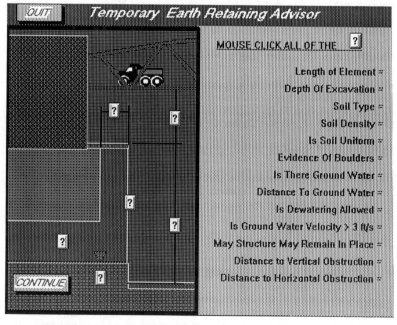

Figure 1. TERA Query Screen

The user invokes TERA to ask a question or present a new query screen by double clicking of the query buttons. Simple questions that can easily be answered by typing a response or choosing an answer from a list or table are asked using Windows dialogue boxes. Additional query screens were utilized when there were multiple questions relating to the same subject or when information was needed by the user to respond. In TERA, additional query screens were provided for soil conditions questions and ground water condition questions.

TERA Output Facilities

Once all of the information that is required by TERA is entered, a summary screen is presented listing all of the technologies considered and their resulting rating. Detail explanation screens are also provided for each technology. These screens give the there why the technology received it's rating, a general description and figure of the technology. The detailed explanation screen for the steel sheeting earth retaining technology is shown below.

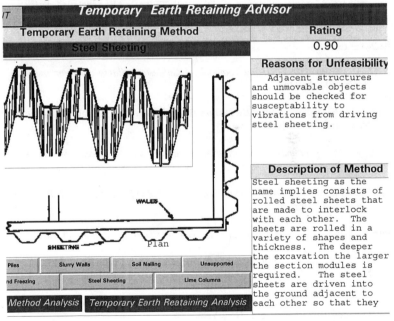

Figure 2. TERA Detailed Explanation Screen For Steel Sheeting

Each detailed explanation screen allows the user to navigate to any other technology's explanation screen, perform another temporary earth retaining wall interview session or start the wall support interview session. Similar query, summary and detailed explanation screens were provided for the wall support analysis module.

Conclusions

TERA demonstrates the potential role of a selection-type knowledge based expert system for analyzing potential temporary earth retaining methods. TERA can be used as an effective tool in the conceptual design phase and also as a training system for inexperienced engineers. Its graphical user interface facilitates soliciting and providing users information about their project and how the many available temporary earth retaining technologies apply to their project. TERA provides a platform on which the future work could be built upon. Possible future enhancements to the software are interfacing it with computer generated plan views of the project excavation and integrating design modules with TERA to allow detailed design of the selected temporary earth retaining system.

References

Asymetrix Corp. (1991), *Using Toolbook, A Guide to Building and Working With Books*, Washington.

Carson, B. A. (1980), *General Excavation Technologies*, R. E. Krieger Publications, New York, New York.

Fang, H. (1991), *Foundation Engineering Handbook*, Van Nostrand Reinhold, New York, New York

Pile Buck Inc. (1992), *Earth Support Systems and Retaining Structures, A Pile Buck Production*. Florida.

Pile Buck Inc. (1987), *Pile Buck Steel Sheet Piling Design Manual*, Florida.

Ratay, R. T. (1984), *Handbook of Temporary Structures in Construction*, McGraw-Hill, New York, New York.

Schnabel, H. Jr. (1982), *Tiebacks in Foundation Engineering and Construction*, McGraw-Hill, New York, New York.

Subject Index
Page number refers to first page of paper